Building Code Requirements for Structural Concrete (ACI 318-19)

An ACI Standard

Commentary on Building Code Requirements for Structural Concrete (ACI 318R-19)

Reported by ACI Committee 318

Jack P. Moehle, Chair

Gregory M. Zeisler, Secretary (Non-voting)

VOTING MEMBERS

Neal S. Anderson
Roger J. Becker
John F. Bonacci
Dean A. Browning
JoAnn P. Browning
James R. Cagley
Ned M. Cleland
Charles W. Dolan
Catherine E. French
Robert J. Frosch
Luis E. Garcia
Satyendra Ghosh
James R. Harris
Terence C. Holland
James O. Jirsa
Dominic J. Kelly
Gary J. Klein
Ronald Klemencic
William M. Klorman
Michael E. Kreger
Colin L. Lobo
Raymond Lui
Paul F. Mlakar
Michael C. Mota
Lawrence C. Novak
Carlos E. Ospina
Gustavo J. Parra-Montesinos
Randall W. Poston
Carin L. Roberts-Wollmann
Mario E. Rodriguez
David H. Sanders
Thomas C. Schaeffer
Stephen J. Seguirant
Andrew W. Taylor
John W. Wallace
James K. Wight
Sharon L. Wood
Loring A. Wyllie Jr.
Fernando Yanez

SUBCOMMITTEE MEMBERS

Theresa M. Ahlborn
F. Michael Bartlett
Asit N. Baxi
Abdeldjelil Belarbi
Allan P. Bommer
Sergio F. Brena
Jared E. Brewe
Nicholas J. Carino
Min Yuan Cheng
Ronald A. Cook
David Darwin
Curtis L. Decker
Jeffrey J. Dragovich
Jason L. Draper
Lisa R. Feldman
Damon R. Fick
David C. Fields
Anthony E. Fiorato
Rudolph P. Frizzi
Wassim M. Ghannoum
Harry A. Gleich
Zen Hoda
R. Brett Holland
R. Doug Hooton
Kenneth C. Hover
I-chi Huang
Matias Hube
Mary Beth D. Hueste
Jose M. Izquierdo-Encarnacion
Maria G. Juenger
Keith E. Kesner
Insung Kim
Donald P. Kline
Jason J. Krohn
Daniel A. Kuchma
James M. LaFave
Andres Lepage
Remy D. Lequesne
Ricardo R. Lopez
Laura N. Lowes
Frank Stephen Malits
Leonardo M. Massone
Steven L. McCabe
Ian S. McFarlane
Robert R. McGlohn
Donald F. Meinheit
Fred Meyer
Daniel T. Mullins
Clay J. Naito
William H. Oliver
Viral B. Patel
Conrad Paulson
Jose A. Pincheira
Mehran Pourzanjani
Santiago Pujol
Jose I. Restrepo
Nicolas Rodrigues
Andrea J. Schokker
Bahram M. Shahrooz
John F. Silva
Lesley H. Sneed
John F. Stanton
Bruce A. Suprenant
Miroslav Vejvoda
W. Jason Weiss
Christopher D. White

LIAISON MEMBERS

Raul D. Bertero*
Mario Alberto Chiorino
Juan Francisco Correal Daza*
Kenneth J. Elwood*
Luis B. Fargier-Gabaldon
Werner A. F. Fuchs*
Patricio Garcia*
Raymond Ian Gilbert
Wael Mohammed Hassan
Angel E. Herrera
Augusto H. Holmberg*
Hector Monzon-Despang
Ernesto Ng
Guney Ozcebe
Enrique Pasquel*
Guillermo Santana*
Ahmed B. Shuraim
Roberto Stark*
Julio Timerman
Roman Wan-Wendner

*Liaison members serving on various subcommittees.

CONSULTING MEMBERS

David P. Gustafson
Neil M. Hawkins
Robert F. Mast
Basile G. Rabbat
David M. Rogowsky

ACI 318-19 supersedes ACI 318-14, was adopted May 3, 2019, and published June 2019.

First printing: June 2019
ISBN: 978-1-64195-056-5
DOI: 10.14359/51716937

Building Code Requirements for Structural Concrete and Commentary

The technical committees responsible for ACI committee reports and standards strive to avoid ambiguities, omissions, and errors in these documents. In spite of these efforts, the users of ACI documents occasionally find information or requirements that may be subject to more than one interpretation or may be incomplete or incorrect. Users who have suggestions for the improvement of ACI documents are requested to contact ACI via the errata website at http://concrete.org/Publications/DocumentErrata.aspx. Proper use of this document includes periodically checking for errata for the most up-to-date revisions.

ACI committee documents are intended for the use of individuals who are competent to evaluate the significance and limitations of its content and recommendations and who will accept responsibility for the application of the material it contains. Individuals who use this publication in any way assume all risk and accept total responsibility for the application and use of this information.

All information in this publication is provided “as is” without warranty of any kind, either express or implied, including but not limited to, the implied warranties of merchantability, fitness for a particular purpose or non-infringement.

ACI and its members disclaim liability for damages of any kind, including any special, indirect, incidental, or consequential damages, including without limitation, lost revenues or lost profits, which may result from the use of this publication.

It is the responsibility of the user of this document to establish health and safety practices appropriate to the specific circumstances involved with its use. ACI does not make any representations with regard to health and safety issues and the use of this document. The user must determine the applicability of all regulatory limitations before applying the document and must comply with all applicable laws and regulations, including but not limited to, United States Occupational Safety and Health Administration (OSHA) health and safety standards.

Participation by governmental representatives in the work of the American Concrete Institute and in the development of Institute standards does not constitute governmental endorsement of ACI or the standards that it develops.

Order information: ACI documents are available in print, by download, through electronic subscription, or reprint, and may be obtained by contacting ACI.

ACI codes, specifications, and practices are made available in the ACI Collection of Concrete Codes, Specifications, and Practices. The online subscription to the ACI Collection is always updated, and includes current and historical versions of ACI’s codes and specifications (in both inch-pound and SI units) plus new titles as they are published. The ACI Collection is also available as an eight-volume set of books and a USB drive.

American Concrete Institute
38800 Country Club Drive
Farmington Hills, MI 48331
Phone: +1.248.848.3700
Fax: +1.248.848.3701

www.concrete.org

PREFACE TO ACI 318-19

The "Building Code Requirements for Structural Concrete" ("Code") provides minimum requirements for the materials, design, and detailing of structural concrete buildings and, where applicable, nonbuilding structures. This Code was developed by an ANSI-approved consensus process and addresses structural systems, members, and connections, including cast-in-place, precast, shotcrete, plain, nonprestressed, prestressed, and composite construction. Among the subjects covered are: design and construction for strength, serviceability, and durability; load combinations, load factors, and strength reduction factors; structural analysis methods; deflection limits; mechanical and adhesive anchoring to concrete; development and splicing of reinforcement; construction document information; field inspection and testing; and methods to evaluate the strength of existing structures.

The Code was substantially reorganized and reformatted in 2014, and this Code continues and expands that same organizational philosophy. The principal objectives of the reorganization were to present all design and detailing requirements for structural systems or for individual members in chapters devoted to those individual subjects, and to arrange the chapters in a manner that generally follows the process and chronology of design and construction. Information and procedures that are common to the design of multiple members are located in utility chapters. Additional enhancements implemented in this Code to provide greater clarity and ease of use include the first use of color illustrations and the use of color to help the user navigate the Code and quickly find the information they need. Special thanks to Bentley Systems, Incorporated, for use of their ProConcrete software to produce many of the figures found in the Commentary.

Uses of the Code include adoption by reference in a general building code, and earlier editions have been widely used in this manner. The Code is written in a format that allows such reference without change to its language. Therefore, background details or suggestions for carrying out the requirements or intent of the Code provisions cannot be included within the Code itself. The Commentary is provided for this purpose.

Some considerations of the committee in developing the Code are discussed in the Commentary, with emphasis given to the explanation of new or revised provisions. Much of the research data referenced in preparing the Code is cited for the user desiring to study individual questions in greater detail. Other documents that provide suggestions for carrying out the requirements of the Code are also cited.

Technical changes from ACI 318-14 to ACI 318-19 are outlined in the August 2019 issue of *Concrete International* and are marked in the text of this Code with change bars in the margins.

KEYWORDS

admixtures; aggregates; anchorage (structural); beam-column frame; beams (supports); caissons; cements; cold weather; columns (supports); combined stress; composite construction (concrete to concrete); compressive strength; concrete; construction documents; construction joints; continuity (structural); contraction joints; cover; curing; deep beams; deep foundations; deflections; drilled piers; earthquake-resistant structures; flexural strength; floors; footings; formwork (construction); hot weather; inspection; isolation joints; joints (junctions); joists; lightweight concretes; load tests (structural); loads (forces); mixture proportioning; modulus of elasticity; moments; piles; placing; plain concrete; precast concrete; prestressed concrete; prestressing steels; quality control; reinforced concrete; reinforcing steels; roofs; serviceability; shear strength; shotcrete; spans; splicing; strength analysis; stresses; structural analysis; structural design; structural integrity; structural walls; T-beams; torsion; walls; water; welded wire reinforcement.

INTRODUCTION

ACI 318-19, "Building Code Requirements for Structural Concrete," hereinafter called the Code or the 2019 Code, and ACI 318R-19, "Commentary," are presented in a side-by-side column format. These are two separate but coordinated documents, with Code text placed in the left column and the corresponding Commentary text aligned in the right column. Commentary section numbers are preceded by an "R" to further distinguish them from Code section numbers. The two documents are bound together solely for the user's convenience. Each document carries a separate enforceable and distinct copyright.

As the name implies, "Building Code Requirements for Structural Concrete" is meant to be used as part of a legally adopted building code and as such must differ in form and substance from documents that provide detailed specifications, recommended practice, complete design procedures, or design aids.

The Code is intended to cover all buildings of the usual types, both large and small. Requirements more stringent than the Code provisions may be desirable for unusual construction. The Code and Commentary cannot replace sound engineering knowledge, experience, and judgment.

A building code states only the minimum requirements necessary to provide for public health and safety. The Code is based on this principle. For any structure, the owner or the licensed design professional may require the quality of materials and construction to be higher than the minimum requirements necessary to protect the public as stated in the Code. However, lower standards are not permitted.

The Code has no legal status unless it is adopted by the government bodies having the police power to regulate building design and construction. Where the Code has not been adopted, it may serve as a reference to good practice even though it has no legal status.

The Code and Commentary are not intended for use in settling disputes between the owner, engineer, architect, contractor, or their agents, subcontractors, material suppliers, or testing agencies. Therefore, the Code cannot define the contract responsibility of each of the parties in usual construction. General references requiring compliance with the Code in the project specifications should be avoided because the contractor is rarely in a position to accept responsibility for design details or construction requirements that depend on a detailed knowledge of the design. Design-build construction contractors, however, typically combine the design and construction responsibility. Generally, the contract documents should contain all of the necessary requirements to ensure compliance with the Code. In part, this can be accomplished by reference to specific Code sections in the project specifications. Other ACI publications, such as "Specifications for Structural Concrete (ACI 301)" are written specifically for use as contract documents for construction.

The Commentary discusses some of the considerations of Committee 318 in developing the provisions contained in the Code. Emphasis is given to the explanation of new or revised provisions that may be unfamiliar to Code users. In addition, comments are included for some items contained in previous editions of the Code to make the present Commentary independent of the previous editions. Comments on specific provisions are made under the corresponding chapter and section numbers of the Code.

The Commentary is not intended to provide a complete historical background concerning the development of the Code, nor is it intended to provide a detailed résumé of the studies and research data reviewed by the committee in formulating the provisions of the Code. However, references to some of the research data are provided for those who wish to study the background material in depth.

The Commentary directs attention to other documents that provide suggestions for carrying out the requirements and intent of the Code. However, those documents and the Commentary are not a part of the Code.

The Commentary is intended for the use of individuals who are competent to evaluate the significance and limitations of its content and recommendations, and who will accept responsibility for the application of the information it contains. ACI disclaims any and all responsibility for the stated principles. The Institute shall not be liable for any loss or damage arising therefrom. Reference to the Commentary shall not be made in construction documents. If items found in the Commentary are desired by the licensed design professional to be a part of the contract documents, they shall be restated in mandatory language for incorporation by the licensed design professional.

It is recommended to have the materials, processes, quality control measures, and inspections described in this document tested, monitored, or performed by individuals holding the appropriate ACI Certification or equivalent, when available. The personnel certification programs of the American Concrete Institute and the Post-Tensioning Institute; the plant certification programs of the Precast/Prestressed Concrete Institute, the Post-Tensioning Institute, and the National Ready Mixed Concrete Association; and the Concrete Reinforcing Steel Institute's Voluntary Certification Program for Fusion-Bonded Epoxy Coating Applicator Plants are available for this purpose. In addition, "Standard Specification for Agencies Engaged in Construction Inspection, Testing, or Special Inspection" (ASTM E329-18) specifies performance requirements for inspection and testing agencies.

Design reference materials illustrating applications of the Code requirements are listed and described in the back of this document.

TABLE OF CONTENTS

PART 1: GENERAL

PART 2: LOADS & ANALYSIS

PART 3: MEMBERS

CHAPTER 9 BEAMS

CHAPTER 10 COLUMNS

CHAPTER 11 WALLS

CHAPTER 12 DIAPHRAGMS

CHAPTER 13 FOUNDATIONS

CHAPTER 14 PLAIN CONCRETE

PART 4: JOINTS/CONNECTIONS/ANCHORS

CHAPTER 15 BEAM-COLUMN AND SLAB-COLUMN JOINTS

CHAPTER 16 CONNECTIONS BETWEEN MEMBERS

CHAPTER 17 ANCHORING TO CONCRETE

PART 5: EARTHQUAKE RESISTANCE

CHAPTER 18
EARTHQUAKE-RESISTANT STRUCTURES

PART 6: MATERIALS & DURABILITY

CHAPTER 19
CONCRETE: DESIGN AND DURABILITY REQUIREMENTS

CHAPTER 20
STEEL REINFORCEMENT PROPERTIES, DURABILITY, AND EMBEDMENTS

PART 7: STRENGTH & SERVICEABILITY

CHAPTER 21
STRENGTH REDUCTION FACTORS

CHAPTER 22
SECTIONAL STRENGTH

CHAPTER 23
STRUT-AND-TIE METHOD

CHAPTER 24
SERVICEABILITY

PART 8: REINFORCEMENT

CHAPTER 25
REINFORCEMENT DETAILS

PART 9: CONSTRUCTION

CHAPTER 26
CONSTRUCTION DOCUMENTS AND INSPECTION

PART 10: EVALUATION

CHAPTER 27
STRENGTH EVALUATION OF EXISTING STRUCTURES

APPENDICES & REFERENCES

APPENDIX A
DESIGN VERIFICATION USING NONLINEAR RESPONSE HISTORY ANALYSIS

APPENDIX B
STEEL REINFORCEMENT INFORMATION

APPENDIX C
EQUIVALENCE BETWEEN SI-METRIC, MKS-METRIC, AND U.S. CUSTOMARY UNITS OF NONHOMOGENOUS EQUATIONS IN THE CODE

COMMENTARY REFERENCES

INDEX

Check 1-way shear

$b_1 = \frac{1}{2} FTG - \frac{1}{2} col - d = \frac{(13)(12)}{2} - \frac{12}{2} - 28 = 44''$

$b_2 = \frac{1}{2} FTG - \frac{1}{2} col - d = \frac{13(12)}{2} - \frac{30}{2} - 28 = 35$

$A_1 = \left(\frac{44}{12}\right)(13\ ft) = 47.75$ sf squares

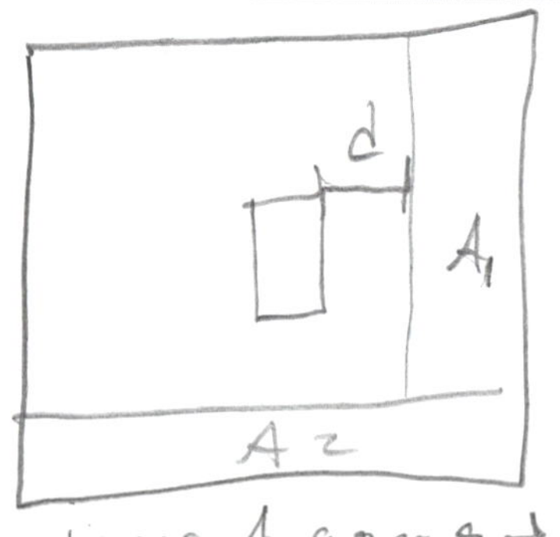

Larger A governs → A_1

Demand

$V_u = q_{nu} A = 5100\ psf(47.7) = 243^k$

Capacity

$V_c = 2\lambda\sqrt{f'_c}\ b_w d = 2(1)\sqrt{3000}\ (13' \times 12'')(28) = 478.8^k$

$\phi V_c = (0.75)(478.5^k = 359^k$

check $V_c < 5\lambda\sqrt{f'_c}\ b_w d$

$FS = \frac{\phi V_c}{V_u}$ Utilization $= \frac{1}{FS}$

CHAPTER 1—GENERAL

CODE

1.1—Scope of ACI 318

1.1.1 This chapter addresses (a) through (h):

(a) General requirements of this Code
(b) Purpose of this Code
(c) Applicability of this Code
(d) Interpretation of this Code
(e) Definition and role of the building official and the licensed design professional
(f) Construction documents
(g) Testing and inspection
(h) Approval of special systems of design, construction, or alternative construction materials

1.2—General

1.2.1 ACI 318, "Building Code Requirements for Structural Concrete," is hereafter referred to as "this Code."

1.2.2 In this Code, the general building code refers to the building code adopted in a jurisdiction. When adopted, this Code forms part of the general building code.

1.2.3 The official version of this Code is the English language version, using inch-pound units, published by the American Concrete Institute.

1.2.4 In case of conflict between the official version of this Code and other versions of this Code, the official version governs.

1.2.5 This Code provides minimum requirements for the materials, design, construction, and strength evaluation of structural concrete members and systems in any structure designed and constructed under the requirements of the general building code.

1.2.6 Modifications to this Code that are adopted by a particular jurisdiction are part of the laws of that jurisdiction, but are not a part of this Code.

1.2.7 If no general building code is adopted, this Code provides minimum requirements for the materials, design, construction, and strength evaluation of members and systems in any structure within the scope of this Code.

1.3—Purpose

1.3.1 The purpose of this Code is to provide for public health and safety by establishing minimum requirements for

COMMENTARY

R1.1—Scope of ACI 318

R1.1.1 This Code includes provisions for the design of concrete used for structural purposes, including plain concrete; concrete containing nonprestressed reinforcement, prestressed reinforcement, or both; and anchoring to concrete. This chapter includes a number of provisions that explain where this Code applies and how it is to be interpreted.

R1.2—General

R1.2.2 The American Concrete Institute recommends that this Code be adopted in its entirety.

R1.2.3 Committee 318 develops the Code in English, using inch-pound units. Based on that version, Committee 318 approved three other versions:

(a) In English using SI units (ACI 318M)
(b) In Spanish using SI units (ACI 318S)
(c) In Spanish using inch-pound units (ACI 318SUS).

Jurisdictions may adopt ACI 318, ACI 318M, ACI 318S, or ACI 318SUS.

R1.2.5 This Code provides minimum requirements and exceeding these minimum requirements is not a violation of the Code.

The licensed design professional may specify project requirements that exceed the minimum requirements of this Code.

R1.3—Purpose

R1.3.1 This Code provides a means of establishing minimum requirements for the design and construction of

CODE

strength, stability, serviceability, durability, and integrity of concrete structures.

1.3.2 This Code does not address all design considerations.

1.3.3 Construction means and methods are not addressed in this Code.

1.4—Applicability

1.4.1 This Code shall apply to concrete structures designed and constructed under the requirements of the general building code.

1.4.2 Provisions of this Code shall be permitted to be used for the assessment, repair, and rehabilitation of existing structures.

1.4.3 Applicable provisions of this Code shall be permitted to be used for structures not governed by the general building code.

1.4.4 The design of thin shells and folded plate concrete structures shall be in accordance with ACI 318.2, "Building Code Requirements for Concrete Thin Shells."

1.4.5 This Code shall apply to the design of slabs cast on stay-in-place, noncomposite steel decks.

COMMENTARY

structural concrete, as well as for acceptance of design and construction of concrete structures by the building officials or their designated representatives.

This Code does not provide a comprehensive statement of all duties of all parties to a contract or all requirements of a contract for a project constructed under this Code.

R1.3.2 The minimum requirements in this Code do not replace sound professional judgment or the licensed design professional's knowledge of the specific factors surrounding a project, its design, the project site, and other specific or unusual circumstances to the project.

R1.4—Applicability

R1.4.2 Specific provisions for assessment, repair, and rehabilitation of existing concrete structures are provided in ACI 562-19. Existing structures in ACI 562 are defined as structures that are complete and permitted for use.

R1.4.3 Structures such as arches, bins and silos, blast-resistant structures, chimneys, underground utility structures, gravity walls, and shielding walls involve design and construction requirements that are not specifically addressed by this Code. Many Code provisions, however, such as concrete quality and design principles, are applicable for these structures. Recommendations for design and construction of some of these structures are given in the following:

- "Code Requirements for Reinforced Concrete Chimneys and Commentary" (ACI 307-08)
- "Standard Practice for Design and Construction of Concrete Silos and Stacking Tubes for Storing Granular Materials" (ACI 313-97)
- "Code Requirements for Nuclear Safety-Related Concrete Structures and Commentary" (ACI 349)
- "Code for Concrete Containments" (ACI 359)

R1.4.5 In its most basic application, the noncomposite steel deck serves as a form, and the concrete slab is designed to resist all loads, while in other applications the concrete slab may be designed to resist only the superimposed loads. The design of a steel deck in a load-resisting application is given in "Standard for Non-Composite Steel Floor Deck"

CODE	COMMENTARY
	(SDI NC). The SDI standard refers to this Code for the design and construction of the structural concrete slab.
1.4.6 For one- and two-family dwellings, multiple single-family dwellings, townhouses, and accessory structures to these types of dwellings, the design and construction of cast-in-place footings, foundation walls, and slabs-on-ground in accordance with ACI 332 shall be permitted.	**R1.4.6** ACI 332 addresses only the design and construction of cast-in-place footings, foundation walls supported on continuous footings, and slabs-on-ground for limited residential construction applications. The 2015 IBC requires design and construction of residential post-tensioned slabs on expansive soils to be in accordance with PTI DC10.5-12, which provides requirements for slab-on-ground foundations, including soil investigation, design, and analysis. Guidance for the design and construction of post-tensioned slabs-on-ground that are not on expansive soils can be found in ACI 360R. Refer to R1.4.8.
1.4.7 This Code does not apply to the design and installation of concrete piles, drilled piers, and caissons embedded in ground, except as provided in (a) through (c): (a) For portions of deep foundation members in air or water, or in soil incapable of providing adequate lateral restraint to prevent buckling throughout their length (b) For precast concrete piles supporting structures assigned to Seismic Design Categories A and B (13.4) (c) For deep foundation elements supporting structures assigned to Seismic Design Categories C, D, E, and F (Ch. 13, 18.13)	**R1.4.7** The design and installation of concrete piles fully embedded in the ground is regulated by the general building code. The 2019 edition of the Code contains some provisions that previously were only available in the general building code. In addition to the provisions in this Code, recommendations for concrete piles are given in ACI 543R, recommendations for drilled piers are given in ACI 336.3R, and recommendations for precast prestressed concrete piles are given in "Recommended Practice for Design, Manufacture, and Installation of Prestressed Concrete Piling" (PCI 1993). Requirements for the design and construction of micropiles are not specifically addressed by this Code.
1.4.8 This Code does not apply to design and construction of slabs-on-ground, unless the slab transmits vertical loads or lateral forces from other portions of the structure to the soil.	**R1.4.8** Detailed recommendations for design and construction of slabs-on-ground and floors that do not transmit vertical loads or lateral forces from other portions of the structure to the soil are given in ACI 360R. This guide presents information on the design of slabs-on-ground, primarily industrial floors and the slabs adjacent to them. The guide addresses the planning, design, and detailing of the slabs. Background information on the design theories is followed by discussion of the soil support system, loadings, and types of slabs. Design methods are given for structural plain concrete, reinforced concrete, shrinkage-compensating concrete, and post-tensioned concrete slabs.
1.4.9 This Code does not apply to the design and construction of tanks and reservoirs.	**R1.4.9** Requirements and recommendations for the design and construction of tanks and reservoirs are given in ACI 350, ACI 334.1R, and ACI 372R.
1.4.10 This Code does not apply to composite design slabs cast on stay-in-place composite steel deck. Concrete used in the construction of such slabs shall be governed by this Code, where applicable. Portions of such slabs designed as reinforced concrete are governed by this Code.	**R1.4.10** In this type of construction, the steel deck serves as the positive moment reinforcement. The design and construction of concrete-steel deck slabs is described in "Standard for Composite Steel Floor Deck-Slabs" (SDI C). The standard refers to the appropriate portions of this Code for the design and construction of the concrete portion of the composite assembly. SDI C also provides guidance for design of composite-concrete-steel deck slabs. The design of negative moment reinforcement to create continuity at

CODE

COMMENTARY

supports is a common example where a portion of the slab is designed in conformance with this Code.

1.5—Interpretation

1.5.1 The principles of interpretation in this section shall apply to this Code as a whole unless otherwise stated.

1.5.2 This Code consists of chapters and appendixes, including text, headings, tables, figures, footnotes to tables and figures, and referenced standards.

1.5.3 The Commentary consists of a preface, introduction, commentary text, tables, figures, and cited publications. The Commentary is intended to provide contextual information, but is not part of this Code, does not provide binding requirements, and shall not be used to create a conflict with or ambiguity in this Code.

1.5.4 This Code shall be interpreted in a manner that avoids conflict between or among its provisions. Specific provisions shall govern over general provisions.

1.5.5 This Code shall be interpreted and applied in accordance with the plain meaning of the words and terms used. Specific definitions of words and terms in this Code shall be used where provided and applicable, regardless of whether other materials, standards, or resources outside of this Code provide a different definition.

1.5.6 The following words and terms in this Code shall be interpreted in accordance with (a) through (e):

(a) The word "shall" is always mandatory.
(b) Provisions of this Code are mandatory even if the word "shall" is not used.
(c) Words used in the present tense shall include the future.
(d) The word "and" indicates that all of the connected items, conditions, requirements, or events shall apply.
(e) The word "or" indicates that the connected items, conditions, requirements, or events are alternatives, at least one of which shall be satisfied.

1.5.7 In any case in which one or more provisions of this Code are declared by a court or tribunal to be invalid, that ruling shall not affect the validity of the remaining provisions of this Code, which are severable. The ruling of a court or tribunal shall be effective only in that court's jurisdiction, and shall not affect the content or interpretation of this Code in other jurisdictions.

1.5.8 If conflicts occur between provisions of this Code and those of standards and documents referenced in Chapter 3, this Code shall apply.

R1.5—Interpretation

R1.5.4 General provisions are broad statements, such as a building needs to be serviceable. Specific provisions, such as explicit reinforcement distribution requirements for crack control, govern over the general provisions.

R1.5.5 ACI Concrete Terminology (2018) is the primary resource to help determine the meaning of words or terms that are not defined in the Code. Dictionaries and other reference materials commonly used by licensed design professionals may be used as secondary resources.

R1.5.7 This Code addresses numerous requirements that can be implemented fully without modification if other requirements in this Code are determined to be invalid. This severability requirement is intended to preserve this Code and allow it to be implemented to the extent possible following legal decisions affecting one or more of its provisions.

CODE

1.6—Building official

1.6.1 All references in this Code to the building official shall be understood to mean persons who administer and enforce this Code.

1.6.2 Actions and decisions by the building official affect only the specific jurisdiction and do not change this Code.

1.6.3 The building official shall have the right to order testing of any materials used in concrete construction to determine if materials are of the quality specified.

1.7—Licensed design professional

1.7.1 All references in this Code to the licensed design professional shall be understood to mean the engineer in either 1.7.1.1 or 1.7.1.2.

1.7.1.1 The licensed design professional responsible for, and in charge of, the structural design work.

1.7.1.2 A specialty engineer to whom a specific portion of the structural design work has been delegated subject to the conditions of (a) and (b).

(a) The authority of the specialty engineer shall be explicitly limited to the delegated design work.
(b) The portion of design work delegated shall be well defined such that responsibilities and obligations of the parties are apparent.

1.8—Construction documents and design records

1.8.1 The licensed design professional shall provide in the construction documents the information required in Chapter 26 and that required by the jurisdiction.

1.8.2 Calculations pertinent to design shall be filed with the construction documents if required by the building official. Analyses and designs using computer programs shall be permitted provided design assumptions, user input, and computer-generated output are submitted. Model analysis shall be permitted to supplement calculations.

COMMENTARY

R1.6—Building official

R1.6.1 Building official is defined in 2.3.

R1.6.2 Only the American Concrete Institute has the authority to alter or amend this Code.

R1.7—Licensed design professional

R1.7.1 Licensed design professional is defined in 2.3.

R1.7.1.2(b) A portion of the design work may be delegated to a specialty engineer during the design phase or to the contractor in the construction documents. Examples of design work delegated to a specialty engineer or contractor include precast concrete and post-tensioned concrete design.

R1.8—Construction documents and design records

R1.8.1 The provisions of Chapter 26 for preparing project drawings and specifications are, in general, consistent with those of most general building codes. Additional information may be required by the building official.

R1.8.2 Documented computer output is acceptable instead of manual calculations. The extent of input and output information required will vary according to the specific requirements of individual building officials. However, if a computer program has been used, only skeleton data should normally be required. This should consist of sufficient input and output data and other information to allow the building official to perform a detailed review and make comparisons using another program or manual calculations. Input data should be identified as to member designation, applied loads, and span lengths. The related output data should include member designation and the shears, moments, and reactions at key points in the span. For column design, it is desirable to include moment magnification factors in the output where applicable.

The Code permits model analysis to be used to supplement structural analysis and design calculations. Documentation

CODE

COMMENTARY

of the model analysis should be provided with the related calculations. Model analysis should be performed by an individual having experience in this technique.

1.9—Testing and inspection

1.9.1 Concrete materials shall be tested in accordance with the requirements of Chapter 26.

1.9.2 Concrete construction shall be inspected in accordance with the general building code and in accordance with Chapter 26.

1.9.3 Inspection records shall include information in accordance with Chapter 26.

1.10—Approval of special systems of design, construction, or alternative construction materials

1.10.1 Sponsors of any system of design, construction, or alternative construction materials within the scope of this Code, the adequacy of which has been shown by successful use or by analysis or test, but which does not conform to or is not covered by this Code, shall have the right to present the data on which their design is based to the building official or to a board of examiners appointed by the building official. This board shall be composed of competent engineers and shall have authority to investigate the data so submitted, require tests, and formulate rules governing design and construction of such systems to meet the intent of this Code. These rules, when approved by the building official and promulgated, shall be of the same force and effect as the provisions of this Code.

R1.10—Approval of special systems of design, construction, or alternative construction materials

R1.10.1 New methods of design, new materials, and new uses of materials should undergo a period of development before being covered in a code. Hence, good systems or components might be excluded from use by implication if means were not available to obtain acceptance.

For special systems considered under this section, specific tests, load factors, deflection limits, and other pertinent requirements should be set by the board of examiners, and should be consistent with the intent of the Code.

The provisions of this section do not apply to model tests used to supplement calculations under 1.8.2 or to strength evaluation of existing structures under Chapter 27.

CHAPTER 2—NOTATION AND TERMINOLOGY

CODE

2.1—Scope

2.1.1 This chapter defines notation and terminology used in this Code.

2.2—Notation

a = depth of equivalent rectangular stress block, in.

a_v = shear span, equal to distance from center of concentrated load to either: (a) face of support for continuous or cantilevered members, or (b) center of support for simply supported members, in.

A_b = area of an individual bar or wire, in.^{2}

A_{bp} = area of the attachment base plate in contact with concrete or grout when loaded in compression, in.^{2}

A_{brg} = net bearing area of the head of stud, anchor bolt, or headed deformed bar, in.^{2}

A_c = area of concrete section resisting shear transfer, in.^{2}

A_{cf} = greater gross cross-sectional area of the two orthogonal slab-beam strips intersecting at a column of a two-way prestressed slab, in.^{2}

A_{ch} = cross-sectional area of a member measured to the outside edges of transverse reinforcement, in.^{2}

A_{cp} = area enclosed by outside perimeter of concrete cross section, in.^{2}

A_{cs} = cross-sectional area at one end of a strut in a strut-and-tie model, taken perpendicular to the axis of the strut, in.^{2}

A_{ct} = area of that part of cross section between the flexural tension face and centroid of gross section, in.^{2}

A_{cv} = gross area of concrete section bounded by web thickness and length of section in the direction of shear force considered in the case of walls, and gross area of concrete section in the case of diaphragms. Gross area is total area of the defined section minus area of any openings, in.^{2}

A_{cw} = area of concrete section of an individual pier, horizontal wall segment, or coupling beam resisting shear, in.^{2}

$A_{ef,sl}$ = effective bearing area of shear lug, in^2.

A_f = area of reinforcement in bracket or corbel resisting design moment, in.^{2}

A_g = gross area of concrete section, in.^{2} For a hollow section, A_g is the area of the concrete only and does not include the area of the void(s)

A_h = total area of shear reinforcement parallel to primary tension reinforcement in a corbel or bracket, in.^{2}

A_{hs} = total cross-sectional area of hooked or headed bars being developed at a critical section, in.^{2}

A_j = effective cross-sectional area within a joint in a plane parallel to plane of beam reinforcement generating shear in the joint, in.^{2}

A_ℓ = total area of longitudinal reinforcement to resist torsion, in.^{2}

$A_{\ell,min}$ = minimum area of longitudinal reinforcement to resist torsion, in.^{2}

COMMENTARY

R2.2—Notation

CODE

A_n = area of reinforcement in bracket or corbel resisting factored restraint force N_{uc}, in.^{2}

A_{nz} = area of a face of a nodal zone or a section through a nodal zone, in.^{2}

A_{Na} = projected influence area of a single adhesive anchor or group of adhesive anchors, for calculation of bond strength in tension, in.^{2}

A_{Nao} = projected influence area of a single adhesive anchor, for calculation of bond strength in tension if not limited by edge distance or spacing, in.^{2}

A_{Nc} = projected concrete failure area of a single anchor or group of anchors, for calculation of strength in tension, in.^{2}

A_{Nco} = projected concrete failure area of a single anchor, for calculation of strength in tension if not limited by edge distance or spacing, in.^{2}

A_o = gross area enclosed by torsional shear flow path, in.^{2}

A_{oh} = area enclosed by centerline of the outermost closed transverse torsional reinforcement, in.^{2}

A_{pd} = total area occupied by duct, sheathing, and prestressing reinforcement, in.^{2}

A_{ps} = area of prestressed longitudinal tension reinforcement, in.^{2}

A_{pt} = total area of prestressing reinforcement, in.^{2}

A_s = area of nonprestressed longitudinal tension reinforcement, in.^{2}

A_s' = area of compression reinforcement, in.^{2}

A_{sc} = area of primary tension reinforcement in a corbel or bracket, in.^{2}

$A_{se,N}$ = effective cross-sectional area of anchor in tension, in.^{2}

$A_{se,V}$ = effective cross-sectional area of anchor in shear, in.^{2}

A_{sh} = total cross-sectional area of transverse reinforcement, including crossties, within spacing s and perpendicular to dimension b_c, in.^{2}

A_{si} = total area of surface reinforcement at spacing s_i in the i-th layer crossing a strut, with reinforcement at an angle α_i to the axis of the strut, in.^{2}

$A_{s,min}$ = minimum area of flexural reinforcement, in.^{2}

A_{st} = total area of nonprestressed longitudinal reinforcement including bars or steel shapes, and excluding prestressing reinforcement, in.^{2}

A_t = area of one leg of a closed stirrup, hoop, or tie resisting torsion within spacing s, in.^{2}

A_{th} = total cross-sectional area of ties or stirrups confining hooked bars, in.^{2}

A_{tp} = area of prestressing reinforcement in a tie, in.^{2}

A_{tr} = total cross-sectional area of all transverse reinforcement within spacing s that crosses the potential plane of splitting through the reinforcement being developed, in.^{2}

A_{ts} = area of nonprestressed reinforcement in a tie, in.^{2}

COMMENTARY

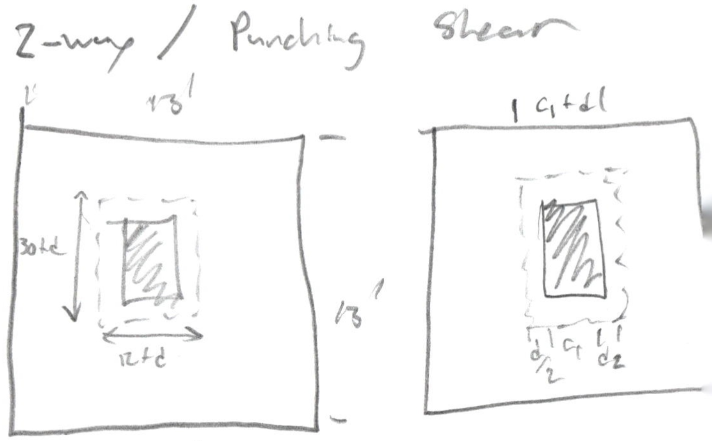

CODE

A_{tt} = total cross-sectional area of ties or stirrups acting as parallel tie reinforcement for headed bars, in.^{2}

A_v = area of shear reinforcement within spacing s, in.^{2}

A_{vd} = total area of reinforcement in each group of diagonal bars in a diagonally reinforced coupling beam, in.^{2}

A_{vf} = area of shear-friction reinforcement, in.^{2}

A_{vh} = area of shear reinforcement parallel to flexural tension reinforcement within spacing s_2, in.^{2}

$A_{v,min}$ = minimum area of shear reinforcement within spacing s, in.^{2}

A_{Vc} = projected concrete failure area of a single anchor or group of anchors, for calculation of strength in shear, in.^{2}

A_{Vco} = projected concrete failure area of a single anchor, for calculation of strength in shear, if not limited by corner influences, spacing, or member thickness, in.^{2}

A_1 = loaded area for consideration of bearing, strut, and node strength, in.^{2}

A_2 = area of the lower base of the largest frustum of a pyramid, cone, or tapered wedge contained wholly within the support and having its upper base equal to the loaded area. The sides of the pyramid, cone, or tapered wedge shall be sloped one vertical to two horizontal, in.^{2}

b = width of compression face of member, in.

b_c = cross-sectional dimension of member core measured to the outside edges of the transverse reinforcement composing area A_{sh}, in.

b_f = effective flange width, in.

b_o = perimeter of critical section for two-way shear in slabs and footings, in.

b_s = width of strut, in.

b_{sl} = width of shear lug, in.

b_{slab} = effective slab width, in.

b_t = width of that part of cross section containing the closed stirrups resisting torsion, in.

b_v = width of cross section at contact surface being investigated for horizontal shear, in.

b_w = web width or diameter of circular section, in.

b_1 = dimension of the critical section b_o measured in the direction of the span for which moments are determined, in.

b_2 = dimension of the critical section b_o measured in the direction perpendicular to b_1, in.

B_n = nominal bearing strength, lb

B_u = factored bearing load, lb

c = distance from extreme compression fiber to neutral axis, in.

c_{ac} = critical edge distance required to develop the basic strength as controlled by concrete breakout or bond of a post-installed anchor in tension in uncracked concrete without supplementary reinforcement to control splitting, in.

COMMENTARY

CODE

$c_{a,max}$ = maximum distance from center of an anchor shaft to the edge of concrete, in.

$c_{a,min}$ = minimum distance from center of an anchor shaft to the edge of concrete, in.

c_{a1} = distance from the center of an anchor shaft to the edge of concrete in one direction, in. If shear is applied to anchor, c_{a1} is taken in the direction of the applied shear. If tension is applied to the anchor, c_{a1} is the minimum edge distance. Where anchors subject to shear are located in narrow sections of limited thickness, see R17.7.2.1.2

c_{a2} = distance from center of an anchor shaft to the edge of concrete in the direction perpendicular to c_{a1}, in.

c_b = lesser of: (a) the distance from center of a bar or wire to nearest concrete surface, and (b) one-half the center-to-center spacing of bars or wires being developed, in.

c_c = clear cover of reinforcement, in.

c_{Na} = projected distance from center of an anchor shaft on one side of the anchor required to develop the full bond strength of a single adhesive anchor, in.

c_{sl} = distance from the centerline of the row of anchors in tension nearest the shear lug to the centerline of the shear lug measured in the direction of shear, in.

c_t = distance from the interior face of the column to the slab edge measured parallel to c_1, but not exceeding c_1, in.

c_1 = dimension of rectangular or equivalent rectangular column, capital, or bracket measured in the direction of the span for which moments are being determined, in.

c_2 = dimension of rectangular or equivalent rectangular column, capital, or bracket measured in the direction perpendicular to c_1, in.

C_m = factor relating actual moment diagram to an equivalent uniform moment diagram

d = distance from extreme compression fiber to centroid of longitudinal tension reinforcement, in.

d' = distance from extreme compression fiber to centroid of longitudinal compression reinforcement, in.

d_a = outside diameter of anchor or shaft diameter of headed stud, headed bolt, or hooked bolt, in.

d_a' = value substituted for d_a if an oversized anchor is used, in.

d_{agg} = nominal maximum size of coarse aggregate, in.

d_b = nominal diameter of bar, wire, or prestressing strand, in.

d_p = distance from extreme compression fiber to centroid of prestressed reinforcement, in.

COMMENTARY

c'_{a1} = limiting value of c_{a1} where anchors are located less than $1.5c_{a1}$ from three or more edges, in.; see Fig. R17.7.2.1.2

C = compressive force acting on a nodal zone, lb

d_{burst} = distance from the anchorage device to the centroid of the bursting force, T_{burst}, in.

CODE

d_{pile} = diameter of pile at footing base, in.

D = effect of service dead load

D_s = effect of superimposed dead load

D_w = effect of self-weight dead load of the concrete structural system

e_h = distance from the inner surface of the shaft of a J- or L-bolt to the outer tip of the J- or L-bolt, in.

e'_N = distance between resultant tension load on a group of anchors loaded in tension and the centroid of the group of anchors loaded in tension, in.; e'_N is always positive

e'_V = distance between resultant shear load on a group of anchors loaded in shear in the same direction, and the centroid of the group of anchors loaded in shear in the same direction, in.; e'_V is always positive

E = effect of horizontal and vertical earthquake-induced forces

E_c = modulus of elasticity of concrete, psi

E_{cb} = modulus of elasticity of beam concrete, psi

E_{cs} = modulus of elasticity of slab concrete, psi

EI = flexural stiffness of member, in.^{2}-lb

$(EI)_{eff}$ = effective flexural stiffness of member, in.^{2}-lb

E_p = modulus of elasticity of prestressing reinforcement, psi

E_s = modulus of elasticity of reinforcement and structural steel, excluding prestressing reinforcement, psi

f'_c = specified compressive strength of concrete, psi

$\sqrt{f'_c}$ = square root of specified compressive strength of concrete, psi

f'_{ci} = specified compressive strength of concrete at time of initial prestress, psi

$\sqrt{f'_{ci}}$ = square root of specified compressive strength of concrete at time of initial prestress, psi

f_{ce} = effective compressive strength of the concrete in a strut or a nodal zone, psi

f_d = stress due to unfactored dead load, at extreme fiber of section where tensile stress is caused by externally applied loads, psi

f_{dc} = decompression stress; stress in the prestressed reinforcement if stress is zero in the concrete at the same level as the centroid of the prestressed reinforcement, psi

f_{pc} = compressive stress in concrete, after allowance for all prestress losses, at centroid of cross section resisting externally applied loads or at junction of web and flange where the centroid lies within the flange, psi. In a composite member, f_{pc} is the resultant compressive stress at centroid of composite section, or at junction of web and flange where the centroid lies within the flange, due to both prestress

COMMENTARY

e_{anc} = eccentricity of the anchorage device or group of devices with respect to the centroid of the cross section, in.

Reinforce / Spacing limits

$As_{min} = 0.0018\ b(h) = (13' \times 12)(32'')$

$= 8.99$

Sec (24.4.3.2 & 7.6.1.1)

$As_{max} = 0.85 \left(\frac{f'_c}{f_y}\right) \beta_1 (d)(b)$

$S_{max} = \min(18 \text{ or } 3(h))$

$S_{min} = \begin{cases} 1 \\ d_b \\ \frac{4}{3} d_{agg} \end{cases}$

$\text{Spacing} = \frac{b_w - cc - \text{bars}}{\#\ \text{spaces} - 1}$

$= \frac{13(12'') - 2(3'') - 13(1)}{12}$

Bar development

$l_d = \frac{f_y \psi_t \psi_e}{20 \lambda \sqrt{f'_c}} d_b$

$l_{available} = S - \text{cover} = 72 - 3$

$l_{available} > l_d$ ✓

CODE

and moments resisted by precast member acting alone

f_{pe} = compressive stress in concrete due only to effective prestress forces, after allowance for all prestress losses, at extreme fiber of section if tensile stress is caused by externally applied loads, psi

f_{ps} = stress in prestressed reinforcement at nominal flexural strength, psi

f_{pu} = specified tensile strength of prestressing reinforcement, psi

f_{py} = specified yield strength of prestressing reinforcement, psi

f_r = modulus of rupture of concrete, psi

f_s = tensile stress in reinforcement at service loads, excluding prestressed reinforcement, psi

f_s' = compressive stress in reinforcement under factored loads, excluding prestressed reinforcement, psi

f_{se} = effective stress in prestressed reinforcement, after allowance for all prestress losses, psi

f_t = extreme fiber stress in the precompressed tension zone calculated at service loads using gross section properties after allowance of all prestress losses, psi

f_{uta} = specified tensile strength of anchor steel, psi

f_y = specified yield strength for nonprestressed reinforcement, psi

f_{ya} = specified yield strength of anchor steel, psi

f_{yt} = specified yield strength of transverse reinforcement, psi

F = effect of service load due to fluids with well-defined pressures and maximum heights

F_{nn} = nominal strength at face of a nodal zone, lb

F_{ns} = nominal strength of a strut, lb

F_{nt} = nominal strength of a tie, lb

F_{un} = factored force on the face of a node, lb

F_{us} = factored compressive force in a strut, lb

F_{ut} = factored tensile force in a tie, lb

h = overall thickness, height, or depth of member, in.

h_a = thickness of member in which an anchor is located, measured parallel to anchor axis, in.

h_{ef} = effective embedment depth of anchor, in.

$h_{ef,sl}$ = effective embedment depth of shear lug, in.

h_{sl} = embedment depth of shear lug, in.

h_{sx} = story height for story x, in.

h_u = laterally unsupported height at extreme compression fiber of wall or wall pier, in., equivalent to ℓ_u for compression members

COMMENTARY

f_{si} = stress in the i-th layer of surface reinforcement, psi

h_{anc} = dimension of anchorage device or single group of closely spaced devices in the direction of bursting being considered, in.

h'_{ef} = limiting value of h_{ef} where anchors are located less than $1.5h_{ef}$ from three or more edges, in.; refer to Fig. R17.6.2.1.2

CODE

h_w = height of entire wall from base to top, or clear height of wall segment or wall pier considered, in.

h_{wcs} = height of entire structural wall above the critical section for flexural and axial loads, in.

h_x = maximum center-to-center spacing of longitudinal bars laterally supported by corners of crossties or hoop legs around the perimeter of a column or wall boundary element, in.

H = effect of service load due to lateral earth pressure, ground water pressure, or pressure of bulk materials, lb

I = moment of inertia of section about centroidal axis, in.[4]

I_b = moment of inertia of gross section of beam about centroidal axis, in.[4]

I_{cr} = moment of inertia of cracked section transformed to concrete, in.[4]

I_e = effective moment of inertia for calculation of deflection, in.[4]

I_g = moment of inertia of gross concrete section about centroidal axis, neglecting reinforcement, in.[4]

I_s = moment of inertia of gross section of slab about centroidal axis, in.[4]

I_{se} = moment of inertia of reinforcement about centroidal axis of member cross section, in.[4]

k = effective length factor for compression members

k_c = coefficient for basic concrete breakout strength in tension

k_{cp} = coefficient for pryout strength

k_f = concrete strength factor

k_n = confinement effectiveness factor

K_{tr} = transverse reinforcement index, in.

ℓ = span length of beam or one-way slab; clear projection of cantilever, in.

ℓ_a = additional embedment length beyond centerline of support or point of inflection, in.

ℓ_{be} = length of boundary element from compression face of member, in.

ℓ_c = length of compression member, measured center-to-center of the joints, in.

ℓ_{cb} = arc length of bar bend along centerline of bar, in.

ℓ_d = development length in tension of deformed bar, deformed wire, plain and deformed welded wire reinforcement, or pretensioned strand, in.

ℓ_{dc} = development length in compression of deformed bars and deformed wire, in.

ℓ_{db} = debonded length of prestressed reinforcement at end of member, in.

COMMENTARY

Bearing

Capacity column

$\phi P_{nb} = \phi\, 0.85 (f'_c) A_1$

$0.65 (0.85)(5000)(12 \times 30)$

$P_u = 860$ k from earlier

Capacity Footing

$\phi P_{nb} = \phi (0.85) f'_c A_1 \sqrt{\frac{A_2}{A_1}}$

$A_2 = (142)(124)$

$A_1 = (12)(30)$

K_t = torsional stiffness of member; moment per unit rotation

K_{05} = coefficient associated with the 5 percent fractile

ℓ_{anc} = length along which anchorage of a tie must occur, in.

ℓ_b = width of bearing, in.

CODE

ℓ_{dh} = development length in tension of deformed bar or deformed wire with a standard hook, measured from outside end of hook, point of tangency, toward critical section, in.

ℓ_{dt} = development length in tension of headed deformed bar, measured from the bearing face of the head toward the critical section, in.

ℓ_e = load bearing length of anchor for shear, in.

ℓ_{ext} = straight extension at the end of a standard hook, in.

ℓ_n = length of clear span measured face-to-face of supports, in.

ℓ_o = length, measured from joint face along axis of member, over which special transverse reinforcement must be provided, in.

ℓ_{sc} = compression lap splice length, in.

ℓ_{st} = tension lap splice length, in.

ℓ_t = span of member under load test, taken as the shorter span for two-way slab systems, in. Span is the lesser of: (a) distance between centers of supports, and (b) clear distance between supports plus thickness h of member. Span for a cantilever shall be taken as twice the distance from face of support to cantilever end

ℓ_{tr} = transfer length of prestressed reinforcement, in.

ℓ_u = unsupported length of column or wall, in.

ℓ_w = length of entire wall, or length of wall segment or wall pier considered in direction of shear force, in.

ℓ_1 = length of span in direction that moments are being determined, measured center-to-center of supports, in.

ℓ_2 = length of span in direction perpendicular to ℓ_1, measured center-to-center of supports, in.

L = effect of service live load

L_r = effect of service roof live load

M_a = maximum moment in member due to service loads at stage deflection is calculated, in.-lb

M_c = factored moment amplified for the effects of member curvature used for design of compression member, in.-lb

M_{cr} = cracking moment, in.-lb

M_{cre} = moment causing flexural cracking at section due to externally applied loads, in.-lb

M_{max} = maximum factored moment at section due to externally applied loads, in.-lb

M_n = nominal flexural strength at section, in.-lb

M_{nb} = nominal flexural strength of beam including slab where in tension, framing into joint, in.-lb

M_{nc} = nominal flexural strength of column framing into joint, calculated for factored axial force, consistent with the direction of lateral forces considered, resulting in lowest flexural strength, in.-lb

M_{pr} = probable flexural strength of members, with or without axial load, determined using the properties of the member at joint faces assuming a tensile

COMMENTARY

M = moment acting on anchor or anchor group, in.-lb

CODE

stress in the longitudinal bars of at least $1.25f_y$ and a strength reduction factor ϕ of 1.0, in.-lb

M_{sa} = maximum moment in wall due to service loads, excluding $P\Delta$ effects, in.-lb

M_{sc} = factored slab moment that is resisted by the column at a joint, in.-lb

M_u = factored moment at section, in.-lb

M_{ua} = moment at midheight of wall due to factored lateral and eccentric vertical loads, not including $P\Delta$ effects, in.-lb

M_1 = lesser factored end moment on a compression member, in.-lb

M_{1ns} = factored end moment on a compression member at the end at which M_1 acts, due to loads that cause no appreciable sidesway, calculated using a first-order elastic frame analysis, in.-lb

M_{1s} = factored end moment on compression member at the end at which M_1 acts, due to loads that cause appreciable sidesway, calculated using a first-order elastic frame analysis, in.-lb

M_2 = greater factored end moment on a compression member. If transverse loading occurs between supports, M_2 is taken as the largest moment occurring in member. Value of M_2 is always positive, in.-lb

$M_{2,min}$ = minimum value of M_2, in.-lb

M_{2ns} = factored end moment on compression member at the end at which M_2 acts, due to loads that cause no appreciable sidesway, calculated using a first-order elastic frame analysis, in.-lb

M_{2s} = factored end moment on compression member at the end at which M_2 acts, due to loads that cause appreciable sidesway, calculated using a first-order elastic frame analysis, in.-lb

n = number of items, such as, bars, wires, monostrand anchorage devices, or anchors

n_ℓ = number of longitudinal bars around the perimeter of a column core with rectilinear hoops that are laterally supported by the corner of hoops or by seismic hooks. A bundle of bars is counted as a single bar

n_s = number of stories above the critical section

N_a = nominal bond strength in tension of a single adhesive anchor, lb

N_{ag} = nominal bond strength in tension of a group of adhesive anchors, lb

N_b = basic concrete breakout strength in tension of a single anchor in cracked concrete, lb

N_{ba} = basic bond strength in tension of a single adhesive anchor, lb

N_c = resultant tensile force acting on the portion of the concrete cross section that is subjected to tensile stresses due to the combined effects of service loads and effective prestress, lb

COMMENTARY

n_t = number of threads per inch

N = tension force acting on anchor or anchor group, lb

CODE

N_{cb} = nominal concrete breakout strength in tension of a single anchor, lb

N_{cbg} = nominal concrete breakout strength in tension of a group of anchors, lb

N_{cp} = basic concrete pryout strength of a single anchor, lb

N_{cpg} = basic concrete pryout strength of a group of anchors, lb

N_n = nominal strength in tension, lb

N_p = pullout strength in tension of a single anchor in cracked concrete, lb

N_{pn} = nominal pullout strength in tension of a single anchor, lb

N_{sa} = nominal strength of a single anchor or individual anchor in a group of anchors in tension as governed by the steel strength, lb

N_{sb} = side-face blowout strength of a single anchor, lb

N_{sbg} = side-face blowout strength of a group of anchors, lb

N_u = factored axial force normal to cross section occurring simultaneously with V_u or T_u; to be taken as positive for compression and negative for tension, lb

N_{ua} = factored tensile force applied to anchor or individual anchor in a group of anchors, lb

$N_{ua,g}$ = total factored tensile force applied to anchor group, lb

$N_{ua,i}$ = factored tensile force applied to most highly stressed anchor in a group of anchors, lb

$N_{ua,s}$ = factored sustained tension load, lb

N_{uc} = factored restraint force applied to a bearing connection acting perpendicular to and simultaneously with V_u, to be taken as positive for tension, lb

$N_{uc,max}$ = maximum restraint force that can be transmitted through the load path of a bearing connection multiplied by the load factor used for live loads in combinations with other factored load effects

p_{cp} = outside perimeter of concrete cross section, in.

p_h = perimeter of centerline of outermost closed transverse torsional reinforcement, in.

P_a = maximum allowable compressive strength of a deep foundation member, lb

P_c = critical buckling load, lb

P_n = nominal axial compressive strength of member, lb

$P_{n,max}$ = maximum nominal axial compressive strength of a member, lb

P_{nt} = nominal axial tensile strength of member, lb

$P_{nt,max}$ = maximum nominal axial tensile strength of member, lb

P_o = nominal axial strength at zero eccentricity, lb

P_{pu} = factored prestressing force at anchorage device, lb

P_s = unfactored axial load at the design, midheight section including effects of self-weight, lb

P_u = factored axial force; to be taken as positive for compression and negative for tension, lb

COMMENTARY

$P\delta$ = secondary moment due to individual member slenderness, in.-lb

CODE

$P\Delta$ = secondary moment due to lateral deflection, in.-lb
q_u = factored load per unit area, lb/ft^2
Q = stability index for a story
r = radius of gyration of cross section, in.
r_b = bend radius at the inside of a bar, in.
R = cumulative load effect of service rain load
s = center-to-center spacing of items, such as longitudinal reinforcement, transverse reinforcement, tendons, or anchors, in.
s_i = center-to-center spacing of reinforcement in the i-th direction adjacent to the surface of the member, in.
s_o = center-to-center spacing of transverse reinforcement within the length ℓ_o, in.
s_s = sample standard deviation, psi
s_w = clear distance between adjacent webs, in.
s_2 = center-to-center spacing of longitudinal shear or torsional reinforcement, in.
S = effect of service snow load
S_{DS} = 5 percent damped, spectral response acceleration parameter at short periods determined in accordance with the general building code
S_e = moment, shear, or axial force at connection corresponding to development of probable strength at intended yield locations, based on the governing mechanism of inelastic lateral deformation, considering both gravity and earthquake effects
S_m = elastic section modulus, in.^{3}
S_n = nominal moment, shear, axial, torsion, or bearing strength
S_y = yield strength of connection, based on f_y of the connected part, for moment, shear, torsion, or axial force, psi
t = wall thickness of hollow section, in.
t_f = thickness of flange, in.
t_{sl} = thickness of shear lug, in.
T = cumulative effects of service temperature, creep, shrinkage, differential settlement, and shrinkage-compensating concrete
T_{cr} = cracking torsional moment, in.-lb
T_t = total test load, lb
T_{th} = threshold torsional moment, in.-lb
T_n = nominal torsional moment strength, in.-lb
T_u = factored torsional moment at section, in.-lb
U = strength of a member or cross section required to resist factored loads or related internal moments and forces in such combinations as stipulated in this Code
v_c = stress corresponding to nominal two-way shear strength provided by concrete, psi

COMMENTARY

R = reaction, lb

T = tension force acting on a nodal zone in a strut-and-tie model, lb (T is also used to define the cumulative effects of service temperature, creep, shrinkage, differential settlement, and shrinkage-compensating concrete in the load combinations defined in 5.3.6.)
T_{burst} = tensile force in general zone acting ahead of the anchorage device caused by spreading of the anchorage force, lb

CODE

v_n = equivalent concrete stress corresponding to nominal two-way shear strength of slab or footing, psi

v_s = equivalent concrete stress corresponding to nominal two-way shear strength provided by reinforcement, psi

v_u = maximum factored two-way shear stress calculated around the perimeter of a given critical section, psi

v_{uv} = factored shear stress on the slab critical section for two-way action, from the controlling load combination, without moment transfer, psi

V_b = basic concrete breakout strength in shear of a single anchor in cracked concrete, lb

$V_{brg,sl}$ = nominal bearing strength of a shear lug in direction of shear, lb

V_c = nominal shear strength provided by concrete, lb

V_{cb} = nominal concrete breakout strength in shear of a single anchor, lb

V_{cbg} = nominal concrete breakout strength in shear of a group of anchors, lb

$V_{cb,sl}$ = nominal concrete breakout strength in shear of attachment with shear lugs, lb

V_{ci} = nominal shear strength provided by concrete where diagonal cracking results from combined shear and moment, lb

V_{cp} = nominal concrete pryout strength of a single anchor, lb

V_{cpg} = nominal concrete pryout strength of a group of anchors, lb

V_{cw} = nominal shear strength provided by concrete where diagonal cracking results from high principal tensile stress in web, lb

V_d = shear force at section due to unfactored dead load, lb

V_e = design shear force for load combinations including earthquake effects, lb

V_i = factored shear force at section due to externally applied loads occurring simultaneously with M_{max}, lb

V_n = nominal shear strength, lb

V_{nh} = nominal horizontal shear strength, lb

V_p = vertical component of effective prestress force at section, lb

V_s = nominal shear strength provided by shear reinforcement, lb

V_{sa} = nominal shear strength of a single anchor or individual anchor in a group of anchors as governed by the steel strength, lb

V_u = factored shear force at section, lb

V_{ua} = factored shear force applied to a single anchor or group of anchors, lb

COMMENTARY

V = shear force acting on anchor or anchor group, lb

$V_{\|}$ = maximum shear force that can be applied parallel to the edge, lb

$V_{\perp}$ = maximum shear force that can be applied perpendicular to the edge, lb

CODE

$V_{ua,g}$ = total factored shear force applied to anchor group, lb

$V_{ua,i}$ = factored shear force applied to most highly stressed anchor in a group of anchors, lb

V_{uh} = factored shear force along contact surface in composite concrete flexural member, lb

V_{us} = factored horizontal shear in a story, lb

$V_{u,x}$ = factored shear force at section in the x-direction, lb

$V_{u,y}$ = factored shear force at section in the y-direction, lb

$V_{n,x}$ = shear strength in the x-direction

$V_{n,y}$ = shear strength in the y-direction

w_c = density, unit weight, of normalweight concrete or equilibrium density of lightweight concrete, lb/ft^3

w_t = effective tie width in a strut-and-tie model, in.

w_u = factored load per unit length of beam or one-way slab, lb/in.

w/cm = water-cementitious materials ratio

W = effect of wind load

y_t = distance from centroidal axis of gross section, neglecting reinforcement, to tension face, in.

α = angle defining the orientation of reinforcement

α_c = coefficient defining the relative contribution of concrete strength to nominal wall shear strength

α_f = ratio of flexural stiffness of beam section to flexural stiffness of a width of slab bounded laterally by centerlines of adjacent panels, if any, on each side of the beam

α_{fm} = average value of α_f for all beams on edges of a panel

α_s = constant used to calculate V_c in slabs and footings

α_1 = minimum angle between unidirectional distributed reinforcement and a strut

β = ratio of long to short dimensions: clear spans for two-way slabs, sides of column, concentrated load or reaction area; or sides of a footing

β_b = ratio of area of reinforcement cut off to total area of tension reinforcement at section

β_c = confinement modification factor for struts and nodes in a strut-and-tie model

β_{dns} = ratio used to account for reduction of stiffness of columns due to sustained axial loads

β_{ds} = the ratio of maximum factored sustained shear within a story to the maximum factored shear in that story associated with the same load combination

β_n = factor used to account for the effect of the anchorage of ties on the effective compressive strength of a nodal zone

β_s = factor used to account for the effect of cracking and confining reinforcement on the effective compressive strength of the concrete in a strut

COMMENTARY

w_s = width of a strut perpendicular to the axis of the strut, in.

w_t = effective height of concrete concentric with a tie, used to dimension nodal zone, in.

$w_{t,max}$ = maximum effective height of concrete concentric with a tie, in.

W_a = service-level wind load, lb

α_f = $E_{cb}I_b/E_{cs}I_s$

CODE

β_1 = factor relating depth of equivalent rectangular compressive stress block to depth of neutral axis

γ_f = factor used to determine the fraction of M_{sc} transferred by slab flexure at slab-column connections

γ_p = factor used for type of prestressing reinforcement

γ_s = factor used to determine the portion of reinforcement located in center band of footing

γ_v = factor used to determine the fraction of M_{sc} transferred by eccentricity of shear at slab-column connections

δ = moment magnification factor used to reflect effects of member curvature between ends of a compression member

δ_c = wall displacement capacity at top of wall, in.

δ_s = moment magnification factor used for frames not braced against sidesway, to reflect lateral drift resulting from lateral and gravity loads

δ_u = design displacement, in.

Δ_{cr} = calculated out-of-plane deflection at midheight of wall corresponding to cracking moment M_{cr}, in.

Δ_n = calculated out-of-plane deflection at midheight of wall corresponding to nominal flexural strength M_n, in.

Δ_o = relative lateral deflection between the top and bottom of a story due to V_{us}, in.

Δf_p = increase in stress in prestressed reinforcement due to factored loads, psi

Δf_{ps} = stress in prestressed reinforcement at service loads less decompression stress, psi

Δ_r = residual deflection measured 24 hours after removal of the test load. For the first load test, residual deflection is measured relative to the position of the structure at the beginning of the first load test. For the second load test, residual deflection is measured relative to the position of the structure at the beginning of the second load test, in.

Δ_s = out-of-plane deflection due to service loads, in.

Δ_u = calculated out-of-plane deflection at midheight of wall due to factored loads, in.

Δ_x = design story drift of story x, in.

Δ_1 = maximum deflection, during first load test, measured 24 hours after application of the full test load, in.

Δ_2 = maximum deflection, during second load test, measured 24 hours after application of the full test load. Deflection is measured relative to the position of the structure at the beginning of the second load test, in.

COMMENTARY

Δf_{pt} = difference between the stress that can be developed in the prestressed reinforcement at the section under consideration and the stress required to resist factored bending moment at section, M_u/ϕ, psi

ε_{cu} = maximum usable strain at extreme concrete compression fiber

CODE

ε_t = net tensile strain in extreme layer of longitudinal tension reinforcement at nominal strength, excluding strains due to effective prestress, creep, shrinkage, and temperature

ε_{ty} = value of net tensile strain in the extreme layer of longitudinal tension reinforcement used to define a compression-controlled section

θ = angle between axis of strut, compression diagonal, or compression field and the tension chord of the members

λ = modification factor to reflect the reduced mechanical properties of lightweight concrete relative to normalweight concrete of the same compressive strength

λ_a = modification factor to reflect the reduced mechanical properties of lightweight concrete in certain concrete anchorage applications

λ_s = factor used to modify shear strength based on the effects of member depth, commonly referred to as the size effect factor.

λ_Δ = multiplier used for additional deflection due to long-term effects

μ = coefficient of friction

ξ = time-dependent factor for sustained load

ρ = ratio of A_s to bd

ρ' = ratio of A_s' to bd

ρ_ℓ = ratio of area of distributed longitudinal reinforcement to gross concrete area perpendicular to that reinforcement

ρ_p = ratio of A_{ps} to bd_p

ρ_s = ratio of volume of spiral reinforcement to total volume of core confined by the spiral, measured out-to-out of spirals

ρ_t = ratio of area of distributed transverse reinforcement to gross concrete area perpendicular to that reinforcement

ρ_v = ratio of tie reinforcement area to area of contact surface

ρ_w = ratio of A_s to $b_w d$

ϕ = strength reduction factor

ϕ_p = strength reduction factor for moment in pretensioned member at cross section closest to the end of the member where all strands are fully developed

τ_{cr} = characteristic bond stress of adhesive anchor in cracked concrete, psi

COMMENTARY

λ = in most cases, the reduction in mechanical properties is caused by the reduced ratio of tensile-to-compressive strength of lightweight concrete compared to normalweight concrete. There are instances in the Code where λ is used as a modifier to reduce expected performance of lightweight concrete where the reduction is not related directly to tensile strength.

ς = exponent symbol in tensile/shear force interaction equation

ϕ_K = stiffness reduction factor

σ = wall boundary extreme fiber concrete nominal compressive stress, psi

CODE

τ_{uncr} = characteristic bond stress of adhesive anchor in uncracked concrete, psi

$\psi_{brg,sl}$ = shear lug bearing factor used to modify bearing strength of shear lugs based on the influence of axial load

ψ_c = factor used to modify development length based on concrete strength

$\psi_{c,N}$ = breakout cracking factor used to modify tensile strength of anchors based on the influence of cracks in concrete

$\psi_{c,P}$ = pullout cracking factor used to modify pullout strength of anchors based on the influence of cracks in concrete

$\psi_{c,V}$ = breakout cracking factor used to modify shear strength of anchors based on the influence of cracks in concrete and presence or absence of supplementary reinforcement

$\psi_{cp,N}$ = breakout splitting factor used to modify tensile strength of post-installed anchors intended for use in uncracked concrete without supplementary reinforcement to account for the splitting tensile stresses

$\psi_{cp,Na}$ = bond splitting factor used to modify tensile strength of adhesive anchors intended for use in uncracked concrete without supplementary reinforcement to account for the splitting tensile stresses due to installation

ψ_e = factor used to modify development length based on reinforcement coating

$\psi_{ec,N}$ = breakout eccentricity factor used to modify tensile strength of anchors based on eccentricity of applied loads

$\psi_{ec,Na}$ = breakout eccentricity factor used to modify tensile strength of adhesive anchors based on eccentricity of applied loads

$\psi_{ec,V}$ = breakout eccentricity factor used to modify shear strength of anchors based on eccentricity of applied loads

$\psi_{ed,N}$ = breakout edge effect factor used to modify tensile strength of anchors based on proximity to edges of concrete member

$\psi_{ed,Na}$ = breakout edge effect factor used to modify tensile strength of adhesive anchors based on proximity to edges of concrete member

$\psi_{ed,V}$ = breakout edge effect factor used to modify shear strength of anchors based on proximity to edges of concrete member

ψ_g = factor used to modify development length based on grade of reinforcement

$\psi_{h,V}$ = breakout thickness factor used to modify shear strength of anchors located in concrete members with $h_a < 1.5c_{a1}$

ψ_o = factor used to modify development length of hooked and headed bars based on side cover and confinement

COMMENTARY

CODE

ψ_p = factor used to modify development length for headed reinforcement based on parallel tie reinforcement

ψ_r = factor used to modify development length based on confining reinforcement

ψ_s = factor used to modify development length based on reinforcement size

ψ_t = factor used to modify development length for casting location in tension

ψ_w = factor used to modify development length for welded deformed wire reinforcement in tension

Ω_o = amplification factor to account for overstrength of the seismic-force-resisting system determined in accordance with the general building code

Ω_v = overstrength factor equal to the ratio of M_{pr}/M_u at the wall critical section

ω_v = factor to account for dynamic shear amplification

2.3—Terminology

adhesive—chemical components formulated from organic polymers, or a combination of organic polymers and inorganic materials that cure if blended together.

admixture—material other than water, aggregate, cementitious materials, and fiber reinforcement used as an ingredient, which is added to grout, mortar, or concrete, either before or during its mixing, to modify the freshly mixed, setting, or hardened properties of the mixture.

aggregate—granular material, such as sand, gravel, crushed stone, iron blast-furnace slag, or recycled aggregates including crushed hydraulic cement concrete, used with a cementing medium to form concrete or mortar.

aggregate, lightweight—aggregate meeting the requirements of ASTM C330 and having a loose bulk density of 70 lb/ft^3 or less, determined in accordance with ASTM C29.

alternative cement—an inorganic cement that can be used as a complete replacement for portland cement or blended hydraulic cement, and that is not covered by applicable specifications for portland or blended hydraulic cements.

anchor—a steel element either cast into concrete or post-installed into a hardened concrete member and used to transmit applied loads to the concrete.

COMMENTARY

R2.3—Terminology

aggregate—The use of recycled aggregate is addressed in the Code in 2019. The definition of recycled materials in ASTM C33 is very broad and is likely to include materials that would not be expected to meet the intent of the provisions of this Code for use in structural concrete. Use of recycled aggregates including crushed hydraulic-cement concrete in structural concrete requires additional precautions. See 26.4.1.2.1(c).

aggregate, lightweight—In some standards, the term "lightweight aggregate" is being replaced by the term "low-density aggregate."

alternative cements—Alternative cements are described in the references listed in R26.4.1.1.1(b). Refer to 26.4.1.1.1(b) for precautions when using these materials in concrete covered by this Code.

anchor—Cast-in anchors include headed bolts, hooked bolts (J- or L-bolt), and headed studs. Post-installed anchors include expansion anchors, undercut anchors, screw anchors, and adhesive anchors; steel elements for adhesive anchors include threaded rods, deformed reinforcing bars, or internally threaded steel sleeves with external deformations. Anchor types are shown in Fig. R2.1.

CODE **COMMENTARY**

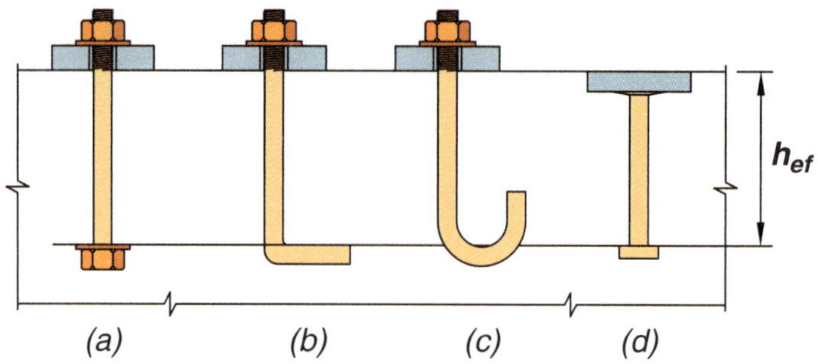

(A) Cast-in anchors: (a) hex head bolt with washer; (b) L-bolt; (c) J-bolt; and (d) welded headed stud.

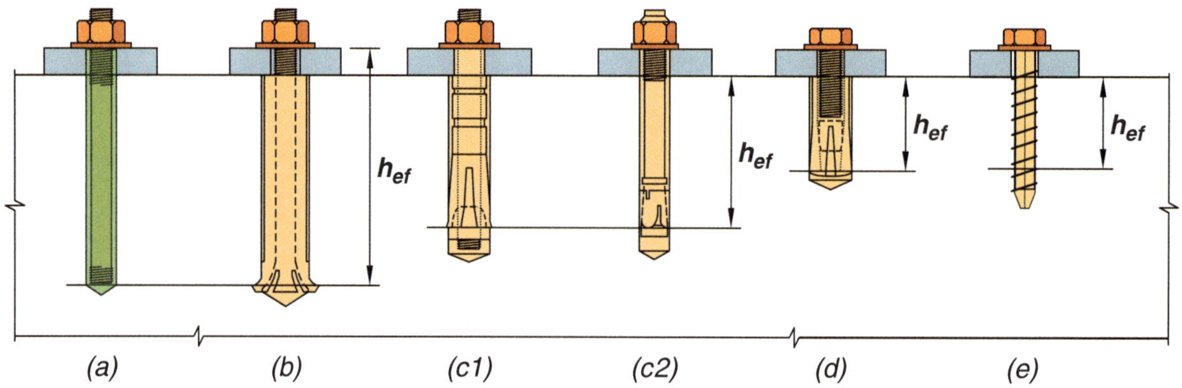

(B) Post-installed anchors: (a) adhesive anchor; (b) undercut anchor; (c) torque-controlled expansion anchors [(c1) sleeve-type and (c2) stud-type]; (d) drop-in type displacement-controlled expansion anchor; and (e) screw anchor.

***Fig. R2.1**—Types of anchors.*

anchor, adhesive—a post-installed anchor, inserted into hardened concrete with an anchor hole diameter not greater than 1.5 times the anchor diameter, that transfers loads to the concrete by bond between the anchor and the adhesive, and bond between the adhesive and the concrete.

anchor, adhesive—The design model included in Chapter 17 for adhesive anchors is based on the behavior of anchors with hole diameters not exceeding 1.5 times the anchor diameter. Anchors with hole diameters exceeding 1.5 times the anchor diameter behave differently and are therefore excluded from the scope of Chapter 17 and ACI 355.4. To limit shrinkage and reduce displacement under load, most adhesive anchor systems require the annular gap to be as narrow as practical while still maintaining sufficient clearance for insertion of the anchor element in the adhesive filled hole and ensuring complete coverage of the bonded area over the embedded length. The annular gap for reinforcing bars is generally greater than that for threaded rods. The required hole size is provided in the Manufacturer's Printed Installation Instructions (MPII).

anchor, cast-in—headed bolt, headed stud, or hooked bolt installed before placing concrete.

anchor, expansion—post-installed anchor, inserted into hardened concrete that transfers loads to or from the concrete by direct bearing or friction, or both.

anchor, expansion—Expansion anchors may be torque-controlled, where the expansion is achieved by a torque acting on the screw or bolt; or displacement controlled, where the expansion is achieved by impact forces acting on a sleeve or plug and the expansion is controlled by the length of travel of the sleeve or plug.

CODE

anchor, horizontal or upwardly inclined—Anchor installed in a hole drilled horizontally or in a hole drilled at any orientation above horizontal.

anchor, post-installed—anchor installed in hardened concrete; adhesive, expansion, screw, and undercut anchors are examples of post-installed anchors.

anchor, screw—a post-installed threaded, mechanical anchor inserted into hardened concrete that transfers loads to the concrete by engagement of the hardened threads of the screw with the grooves that the threads cut into the sidewall of a predrilled hole during anchor installation.

anchor, undercut—post-installed anchor that develops its tensile strength from the mechanical interlock provided by undercutting of the concrete at the embedded end of the anchor. Undercutting is achieved with a special drill before installing the anchor or alternatively by the anchor itself during its installation.

anchor group—a number of similar anchors having approximately equal effective embedment depths with spacing *s* between adjacent anchors such that the projected areas overlap.

anchor pullout strength—the strength corresponding to the anchoring device or a major component of the device sliding out from the concrete without breaking out a substantial portion of the surrounding concrete.

anchorage device—in post-tensioned members, the hardware used to transfer force from prestressed reinforcement to the concrete.

anchorage device, basic monostrand—anchorage device used with any single strand or a single 5/8 in. or smaller diameter bar that is in accordance with 25.8.1, 25.8.2, and 25.9.3.1(a).

anchorage device, basic multistrand—anchorage device used with multiple strands, bars, or wires, or with single bars larger than 5/8 in. diameter that satisfies 25.8.1, 25.8.2 and 25.9.3.1(b).

anchorage device, special—anchorage device that satisfies tests required in 25.9.3.1(c).

COMMENTARY

anchor, horizontal or upwardly inclined—Figure R2.2 illustrates the potential hole orientations for horizontal or upwardly inclined anchors.

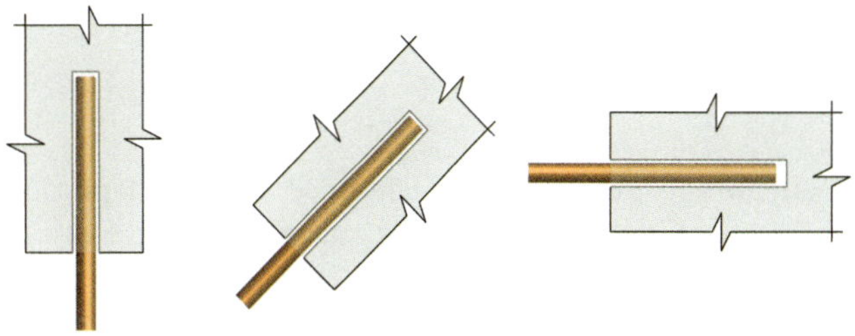

Fig. R2.2—Possible orientations of overhead, upwardly inclined, or horizontal anchors.

anchor, screw—The required predrilled hole size for a screw anchor is provided by the anchor manufacturer.

anchor group—For all potential failure modes (steel, concrete breakout, pullout, side-face blowout, and pryout), only those anchors susceptible to a particular failure mode should be considered when evaluating the strength associated with that failure mode.

anchorage device—Most anchorage devices for post-tensioning are standard manufactured devices available from commercial sources. In some cases, non-standard details or assemblages are developed that combine various wedges and wedge plates for anchoring prestressed reinforcement. Both standard and non-standard anchorage devices may be classified as basic anchorage devices or special anchorage devices as defined in this Code and AASHTO LRFDUS-US.

anchorage device, basic—Devices that are so proportioned that they can be checked analytically for compliance with bearing stress and stiffness requirements without having to undergo the acceptance-testing program required of special anchorage devices.

anchorage device, special—Special anchorage devices are any devices (monostrand or multistrand) that do not meet

CODE

anchorage zone—in post-tensioned members, portion of the member through which the concentrated prestressing force is transferred to concrete and distributed more uniformly across the section; its extent is equal to the largest dimension of the cross section; for anchorage devices located away from the end of a member, the anchorage zone includes the disturbed regions ahead of and behind the anchorage device.

attachment—structural assembly, external to the surface of the concrete, that transmits loads to or receives loads from the anchor.

B-region—portion of a member in which it is reasonable to assume that strains due to flexure vary linearly through section.

base of structure—level at which horizontal earthquake ground motions are assumed to be imparted to a building. This level does not necessarily coincide with the ground level.

beam—member subjected primarily to flexure and shear, with or without axial force or torsion; beams in a moment frame that forms part of the lateral-force-resisting system are predominantly horizontal members; a girder is a beam.

boundary element—portion along wall and diaphragm edge, including edges of openings, strengthened by longitudinal and transverse reinforcement.

breakout strength, concrete—strength corresponding to a volume of concrete surrounding the anchor or group of anchors separating from the member.

building official—term used to identify the Authority having jurisdiction or individual charged with administration and enforcement of provisions of the building code. Such terms as building commissioner or building inspector are variations of the title, and the term "building official" as used in this Code, is intended to include those variations, as well as others that are used in the same sense.

caisson—see drilled pier.

cementitious materials—materials that have cementing value if used in grout, mortar, or concrete, including portland cement, blended hydraulic cements, expansive cement, fly ash, raw or calcined natural pozzolan, slag cement, and silica fume, but excluding alternative cements.

collector—element that acts in axial tension or compression to transmit forces between a diaphragm and a vertical element of the lateral-force-resisting system.

column—member, usually vertical or predominantly vertical, used primarily to support axial compressive load, but that can also resist moment, shear, or torsion. Columns

COMMENTARY

the relevant PTI or AASHTO LFRDUS bearing stress and, where applicable, stiffness requirements. Most commercially marketed multi-bearing surface anchorage devices are special anchorage devices. As provided in 25.9.3, such devices can be used only if they have been shown experimentally to be in compliance with the AASHTO requirements. This demonstration of compliance will ordinarily be furnished by the device manufacturer.

anchorage zone—In post-tensioned members, the portion of the member through which the concentrated prestressing force is transferred to the concrete and distributed more uniformly across the section. Its extent is equal to the largest dimension of the cross section. For anchorage devices located away from the end of a member, the anchorage zone includes the disturbed regions ahead of and behind the anchorage devices. Refer to Fig. R25.9.1.1b.

cementitious materials—Cementitious materials permitted for use in this Code are addressed in 26.4.1.1. Fly ash, raw or calcined natural pozzolan, slag cement, and silica fume are considered supplementary cementitious materials.

CODE

used as part of a lateral-force-resisting system resist combined axial load, moment, and shear. See also **moment frame**.

column capital—enlargement of the top of a concrete column located directly below the slab or drop panel that is cast monolithically with the column.

compliance requirements—construction-related code requirements directed to the contractor to be incorporated into construction documents by the licensed design professional, as applicable.

composite concrete flexural members—concrete flexural members of precast or cast-in-place concrete elements, constructed in separate placements but connected so that all elements respond to loads as a unit.

compression-controlled section—cross section in which the net tensile strain in the extreme tension reinforcement at nominal strength is less than or equal to the compression-controlled strain limit.

compression-controlled strain limit—net tensile strain at balanced strain conditions.

concrete—mixture of portland cement or any other cementitious material, fine aggregate, coarse aggregate, and water, with or without admixtures.

concrete, all-lightweight—lightweight concrete containing only lightweight coarse and fine aggregates that conform to ASTM C330.

concrete, lightweight—concrete containing lightweight aggregate and having an equilibrium density, as determined by ASTM C567, between 90 and 135 lb/ft^3.

concrete, nonprestressed—reinforced concrete with at least the minimum amount of nonprestressed reinforcement and no prestressed reinforcement; or for two-way slabs, with less than the minimum amount of prestressed reinforcement.

concrete, normalweight—concrete containing only coarse and fine aggregates that conform to ASTM C33 and having a density greater than 135 lb/ft^3.

concrete, plain—structural concrete with no reinforcement or with less than the minimum amount of reinforcement specified for reinforced concrete.

concrete, precast—structural concrete element cast elsewhere than its final position in the structure.

concrete, prestressed—reinforced concrete in which internal stresses have been introduced by prestressed reinforcement to reduce potential tensile stresses in concrete resulting from loads, and for two-way slabs, with at least the minimum amount of prestressed reinforcement.

COMMENTARY

compliance requirements—Although primarily directed to the contractor, the compliance requirements are also commonly used by others involved with the project.

concrete, nonprestressed—Nonprestressed concrete usually contains no prestressed reinforcement. Prestressed two-way slabs require a minimum level of compressive stress in the concrete due to effective prestress in accordance with 8.6.2.1. Two-way slabs with less than this minimum level of precompression are required to be designed as nonprestressed concrete.

concrete, normalweight—Normalweight concrete typically has a density (unit weight) between 135 and 160 lb/ft^3, and is normally taken as 145 to 150 lb/ft^3.

concrete, plain—The presence of reinforcement, nonprestressed or prestressed, does not exclude the member from being classified as plain concrete, provided all requirements of Chapter 14 are satisfied.

concrete, prestressed—Classes of prestressed flexural members are defined in 24.5.2.1. Prestressed two-way slabs require a minimum level of compressive stress in the concrete due to effective prestress in accordance with 8.6.2.1. Although the behavior of a prestressed member with unbonded tendons may vary from that of members with continuously bonded prestressed reinforcement, bonded and unbonded prestressed concrete are combined with nonprestressed concrete under the generic term "reinforced concrete." Provisions common to both prestressed and

CODE

concrete, reinforced—structural concrete reinforced with at least the minimum amounts of nonprestressed reinforcement, prestressed reinforcement, or both, as specified in this Code.

concrete, sand-lightweight—lightweight concrete containing only normalweight fine aggregate that conforms to ASTM C33 and lightweight coarse aggregate that conforms to ASTM C330.

concrete, steel fiber-reinforced—concrete containing a prescribed amount of dispersed, randomly oriented, discontinuous deformed steel fibers.

concrete-filled pipe piles—steel pipe with a closed end that is driven for its full length in contact with the surrounding soil, or a steel pipe with an open end that is driven for its full length and the soil cleaned out; for both installation procedures, the pipe is subsequently filled with reinforcement and concrete.

concrete strength, specified compressive, (f_c')—compressive strength of concrete used in design and evaluated in accordance with provisions of this Code, psi; wherever the quantity f_c' is under a radical sign, the square root of numerical value only is intended, and the result has units of psi.

connection—region of a structure that joins two or more members; a connection also refers to a region that joins members of which one or more is precast.

connection, ductile—connection between one or more precast elements that experiences yielding as a result of the earthquake design displacements.

connection, strong—connection between one or more precast elements that remains elastic while adjoining members experience yielding as a result of earthquake design displacements.

construction documents—written and graphic documents and specifications prepared or assembled for describing the location, design, materials, and physical characteristics of the elements of a project necessary for obtaining a building permit and construction of the project.

contraction joint—formed, sawed, or tooled groove in a concrete structure to create a weakened plane and regulate the location of cracking resulting from the dimensional change of different parts of the structure.

cover, specified concrete—distance between the outermost surface of embedded reinforcement and the closest outer surface of the concrete.

crosstie—a continuous reinforcing bar having a seismic hook at one end and a hook not less than 90 degrees with at least a $6d_b$ extension at the other end. The hooks shall engage peripheral longitudinal bars. The 90-degree hooks

COMMENTARY

nonprestressed concrete are integrated to avoid overlapping and conflicting provisions.

concrete, reinforced—Includes members satisfying the requirements for nonprestressed and prestressed concrete.

concrete, sand-lightweight—By Code terminology, sand-lightweight concrete is lightweight concrete with all of the fine aggregate replaced by sand. This definition may not be in agreement with usage by some material suppliers or contractors where the majority, but not all, of the lightweight fines are replaced by sand. For proper application of the Code provisions, the replacement limits should be stated, with interpolation if partial sand replacement is used.

CODE

of two successive crossties engaging the same longitudinal bars shall be alternated end for end.

cutoff point—point where reinforcement is terminated.

D-region—portion of a member within a distance ***h*** of a force discontinuity or a geometric discontinuity.

design displacement—total calculated lateral displacement expected for the design-basis earthquake.

design information—project-specific information to be incorporated into construction documents by the licensed design professional, as applicable.

design load combination—combination of factored loads and forces.

design story drift ratio—relative difference of design displacement between the top and bottom of a story, divided by the story height.

development length—length of embedded reinforcement, including pretensioned strand, required to develop the design strength of reinforcement at a critical section.

discontinuity—abrupt change in geometry or loading.

distance sleeve—sleeve that encases the center part of an undercut anchor, a torque-controlled expansion anchor, or a displacement-controlled expansion anchor, but does not expand.

drilled piers or caissons—cast-in-place concrete foundation elements with or without an enlarged base (bell), constructed by excavating a hole in the ground and filling with reinforcement and concrete. Drilled piers or caissons are considered as uncased cast-in-place concrete drilled or augered piles, unless they have permanent steel casing, in which case they are considered as metal cased concrete piles.

drop panel—projection below the slab used to reduce the amount of negative reinforcement over a column or the minimum required slab thickness, and to increase the slab shear strength.

duct—conduit, plain or corrugated, to accommodate prestressing reinforcement for post-tensioning applications.

ductile coupled structural wall—see **structural wall, ductile coupled**.

durability—ability of a structure or member to resist deterioration that impairs performance or limits service life of the structure in the relevant environment considered in design.

edge distance—distance from the edge of the concrete surface to the center of the nearest anchor.

COMMENTARY

design displacement—The design displacement is an index of the maximum lateral displacement expected in design for the design-basis earthquake. In documents such as ASCE/SEI 7 and the International Building Code, the design displacement is calculated using static or dynamic linear elastic analysis under code-specified actions considering effects of cracked sections, effects of torsion, effects of vertical forces acting through lateral displacements, and modification factors to account for expected inelastic response. The design displacement generally is greater than the displacement calculated from design-level forces applied to a linear-elastic model of the building.

CODE

effective depth of section—distance measured from extreme compression fiber to centroid of longitudinal tension reinforcement.

effective embedment depth—overall depth through which the anchor transfers force to or from the surrounding concrete; effective embedment depth will normally be the depth of the concrete failure surface in tension applications; for cast-in headed anchor bolts and headed studs, the effective embedment depth is measured from the bearing contact surface of the head.

effective prestress—stress remaining in prestressed reinforcement after losses in 20.3.2.6 have occurred.

effective stiffness—stiffness of a structural member accounting for cracking, creep, and other nonlinear effects.

embedments—items embedded in concrete, excluding reinforcement as defined in Chapter 20 and anchors as defined in Chapter 17. Reinforcement or anchors welded, bolted or otherwise connected to the embedded item to develop the strength of the assembly, are considered to be part of the embedment.

embedments, pipe—embedded pipes, conduits, and sleeves.

embedment length—length of embedded reinforcement provided beyond a critical section.

equilibrium density—density of lightweight concrete determined in accordance with ASTM C567.

expansion sleeve—outer part of an expansion anchor that is forced outward by the center part, either by applied torque or impact, to bear against the sides of the predrilled hole. See also **anchor, expansion**.

extreme tension reinforcement—layer of prestressed or nonprestressed reinforcement that is the farthest from the extreme compression fiber.

finite element analysis—a numerical modeling technique in which a structure is divided into a number of discrete elements for analysis.

five percent fractile—statistical term meaning 90 percent confidence that there is 95 percent probability of the actual strength exceeding the nominal strength.

foundation seismic ties—elements used to sufficiently interconnect foundations to act as a unit. Elements may consist of grade beams, slabs-on-ground, or beams within a slab-on-ground.

headed deformed bars—deformed bars with heads attached at one or both ends.

COMMENTARY

effective embedment depth—Effective embedment depths for a variety of anchor types are shown in Fig. R2.1. For post-installed mechanical anchors, the value h_{ef} is obtained from the ACI 355.2 product evaluation report provided by the manufacturer.

five percent fractile—The determination of the coefficient $\boldsymbol{K_{05}}$ associated with the 5 percent fractile, $\bar{x} - K_{05}s_s$ depends on the number of tests, $\boldsymbol{n}$, used to calculate the sample mean, $\bar{x}$, and sample standard deviation, s_s. Values of $\boldsymbol{K_{05}}$ range, for example, from 1.645 for $n = \infty$, to 2.010 for $n = 40$, and 2.568 for $n = 10$. With this definition of the 5 percent fractile, the nominal strength in Chapter 17 is the same as the characteristic strength in ACI 355.2 and ACI 355.4.

headed deformed bars—The bearing area of a headed deformed bar is, for the most part, perpendicular to the bar axis. In contrast, the bearing area of the head of headed stud reinforcement is a nonplanar spatial surface of revolution, as shown in Fig. R20.4.1. The two types of reinforcement differ in other ways. The shanks of headed studs are smooth, not deformed as with headed deformed bars. The

CODE

headed bolt—cast-in steel anchor that develops its tensile strength from the mechanical interlock provided by either a head or nut at the embedded end of the anchor.

headed stud—a steel anchor conforming to the requirements of AWS D1.1 and affixed to a plate or similar steel attachment by the stud arc welding process before casting; also referred to as a **welded headed stud**.

headed shear stud reinforcement—reinforcement consisting of individual headed studs or groups of studs, with anchorage provided by a head at each end, or by a head at one end and a common base rail consisting of a steel plate or shape at the other end.

hooked bolt—cast-in anchor anchored mainly by bearing of the 90-degree bend (L-bolt) or 180-degree bend (J-bolt) against the concrete, at its embedded end, and having a minimum e_h equal to $3d_a$.

hoop—closed tie or continuously wound tie, made up of one or several reinforcement elements, each having seismic hooks at both ends. A closed tie shall not be made up of interlocking headed deformed bars. See 25.7.4.

inspection—observation, verification, and required documentation of the materials, installation, fabrication, erection, or placement of components and connections to determine compliance with construction documents and referenced standards.

inspection, continuous—the full-time observation, verification, and required documentation of work in the area where the work is being performed.

inspection, periodic—the part-time or intermittent observation, verification, and required documentation of work in the area where the work is being performed.

isolation joint—separation between adjoining parts of a concrete structure, usually a vertical plane at a designed location such as to interfere least with performance of the structure, yet such as to allow relative movement in three directions and avoid formation of cracks elsewhere in the concrete, and through which all or part of the bonded reinforcement is interrupted.

jacking force—in prestressed concrete, temporary force exerted by a device that introduces tension into prestressing reinforcement.

joint—portion of structure common to intersecting members.

licensed design professional—an individual who is licensed to practice structural design as defined by the statutory requirements of the professional licensing laws of the state or jurisdiction in which the project is to be constructed, and who is in responsible charge for all or part of the structural design.

COMMENTARY

minimum net bearing area of the head of a headed deformed bar is permitted to be as small as four times the bar area. In contrast, the minimum stud head area is not specified in terms of the bearing area, but by the total head area which must be at least 10 times the area of the shank.

joint—The effective cross-sectional area of a joint of a special moment frame, A_j, for shear strength calculations is given in 15.4.2.4.

licensed design professional—May also be referred to as "registered design professional" in other documents; a licensed design professional in responsible charge of the design work is often referred to as the "engineer of record" (EOR).

CODE

load—forces or other actions that result from the weight of all building materials, occupants, and their possessions, environmental effects, differential movement, and restrained dimensional changes; permanent loads are those loads in which variations over time are rare or of small magnitude; all other loads are variable loads.

load, dead—(a) the weights of the members, supported structure, and permanent attachments or accessories that are likely to be present on a structure in service; or (b) loads meeting specific criteria found in the general building code; without load factors.

load, factored—load, multiplied by appropriate load factors.

load, live—(a) load that is not permanently applied to a structure, but is likely to occur during the service life of the structure (excluding environmental loads); or (b) loads meeting specific criteria found in the general building code; without load factors.

load, roof live—a load on a roof produced: (a) during maintenance by workers, equipment, and materials, and (b) during the life of the structure by movable objects, such as planters or other similar small decorative appurtenances that are not occupancy related; or loads meeting specific criteria found in the general building code; without load factors.

load, self-weight dead—weight of the structural system, including the weight of any bonded concrete topping.

load, service—all loads, static or transitory, imposed on a structure or element thereof, during the operation of a facility, without load factors.

load, superimposed dead—dead loads other than the self-weight that are present or are considered in the design.

load effects—forces and deformations produced in structural members by applied loads or restrained volume changes.

load path—sequence of members and connections designed to transfer the factored loads and forces in such combinations as are stipulated in this Code, from the point of application or origination through the structure to the final support location or the foundation.

Manufacturer's Printed Installation Instructions (MPII)—published instructions for the correct installation of an adhesive anchor under all covered installation conditions as supplied in the product packaging.

metal cased concrete piles—thin-walled steel pipe, steel shell, or spiral-welded metal casing with a closed end that is driven for its full length in contact with the surrounding soil, left permanently in place, and subsequently filled with reinforcement and concrete.

modulus of elasticity—ratio of normal stress to corresponding strain for tensile or compressive stresses below proportional limit of material.

moment frame—frame in which beams, slabs, columns, and joints resist forces predominantly through flexure, shear, and axial force; beams or slabs are predominantly horizontal

COMMENTARY

loads—A number of definitions for loads are given as the Code contains requirements that are to be met at various load levels. The terms "dead load" and "live load" refer to the unfactored, sometimes called "service" loads specified or defined by the general building code. Service loads (loads without load factors) are to be used where specified in this Code to proportion or investigate members for adequate serviceability. Loads used to proportion a member for adequate strength are defined as factored loads. Factored loads are service loads multiplied by the appropriate load factors for required strength except wind and earthquake which are already specified as strength loads in ASCE/SEI 7. The factored load terminology clarifies where the load factors are applied to a particular load, moment, or shear value as used in the Code provisions.

load effects—Stresses and strains are directly related to forces and deformations and are considered as load effects.

CODE

or nearly horizontal; columns are predominantly vertical or nearly vertical.

moment frame, intermediate—cast-in-place beam-column frame or two-way slab-column frame without beams complying with 18.4.

moment frame, ordinary—cast-in-place or precast concrete beam-column or slab-column frame complying with 18.3.

moment frame, special—cast-in-place beam-column frame complying with 18.2.3 through 18.2.8; and 18.6 through 18.8. A precast beam-column frame complying with 18.2.3 through 18.2.8 and 18.9.

net tensile strain—the tensile strain at nominal strength exclusive of strains due to effective prestress, creep, shrinkage, and temperature.

nodal zone—volume of concrete around a node that is assumed to transfer strut-and-tie forces through the node.

node—point in a strut-and-tie model where the axes of the struts, ties, and concentrated forces acting on the joint intersect.

node, curved bar—the bend region of a continuous reinforcing bar (or bars) that defines a node in a strut-and-tie model.

one-way construction—members designed to be capable of supporting all loads through bending in a single direction; see also **two-way construction**.

panel, shotcrete mockup—a shotcrete specimen that simulates the size and detailing of reinforcement in a proposed structural member for preconstruction evaluation of the nozzle operator's ability to encase the reinforcement.

panel, shotcrete test—a shotcrete specimen prepared in accordance with ASTM C1140 for evaluation of shotcrete.

pedestal—member with a ratio of height-to-least lateral dimension less than or equal to 3 used primarily to support axial compressive load; for a tapered member, the least lateral dimension is the average of the top and bottom dimensions of the smaller side.

plastic hinge region—length of frame element over which flexural yielding is intended to occur due to earthquake design displacements, extending not less than a distance h from the critical section where flexural yielding initiates.

post-tensioning—method of prestressing in which prestressing reinforcement is tensioned after concrete has hardened.

precast concrete piles—driven piles that may be either prestressed concrete or conventionally reinforced concrete.

precompressed tension zone—portion of a prestressed member where flexural tension, calculated using gross section properties, would occur under service loads if the prestress force was not present.

pretensioning—method of prestressing in which prestressing reinforcement is tensioned before concrete is cast.

COMMENTARY

one-way construction—Joists, beams, girders, and some slabs and foundations are considered one-way construction.

panel, shotcrete mockup—Shotcrete mockup panels are used for preconstruction evaluation and are either sawed or cored, or both, to evaluate if the reinforcement has been adequately encased.

panel, shotcrete test—Shotcrete test panels are typically used to evaluate a shotcrete mixture, to qualify a nozzle operator, to verify surface finish, and to provide specimens for compressive or flexural strength testing.

CODE

projected area—area on the free surface of the concrete member that is used to represent the greater base of the assumed rectilinear failure surface.

projected influence area—rectilinear area on the free surface of the concrete member that is used to calculate the bond strength of adhesive anchors.

pryout strength, concrete—strength corresponding to formation of a concrete spall behind short, stiff anchors displaced in the direction opposite to the applied shear force.

reinforcement—steel element or elements embedded in concrete and conforming to 20.2 through 20.4. Prestressed reinforcement in external tendons is also considered reinforcement.

reinforcement, anchor—reinforcement used to transfer the design load from the anchors into the structural member.

reinforcement, bonded prestressed—pretensioned reinforcement or prestressed reinforcement in a bonded tendon.

reinforcement, deformed—deformed bars, welded bar mats, deformed wire, and welded wire reinforcement conforming to 20.2.1.3, 20.2.1.5, or 20.2.1.7, excluding plain wire.

reinforcement, nonprestressed—bonded reinforcement that is not prestressed.

reinforcement, plain—bars or wires conforming to 20.2.1.4 or 20.2.1.7 that do not conform to definition of deformed reinforcement.

reinforcement, prestressed—prestressing reinforcement that has been tensioned to impart forces to concrete.

reinforcement, prestressing—high-strength reinforcement such as strand, wire, or bar conforming to 20.3.1.

reinforcement, supplementary—reinforcement that acts to restrain the potential concrete breakout but is not designed to transfer the design load from the anchors into the structural member.

reinforcement, welded deformed steel bar mat—mat conforming to 20.2.1.5 consisting of two layers of deformed bars at right angles to each other welded at the intersections.

reinforcement, welded wire—plain or deformed wire fabricated into sheets or rolls conforming to 20.2.1.7.

Seismic Design Category—classification assigned to a structure based on its occupancy category and the severity of the design earthquake ground motion at the site, as defined by the general building code. Also denoted by the abbreviation SDC.

seismic-force-resisting system—portion of the structure designed to resist earthquake effects required by the general

COMMENTARY

reinforcement, anchor—Anchor reinforcement is designed and detailed specifically for the purpose of transferring anchor loads from the anchors into the member. Hairpins are generally used for this purpose (refer to 17.5.2.1(a) and 17.5.2.1(b)); however, other configurations that can be shown to effectively transfer the anchor load are acceptable.

reinforcement, deformed—Deformed reinforcement is defined as that meeting the reinforcement specifications in this Code. No other reinforcement qualifies. This definition permits accurate statement of development lengths. Bars or wire not meeting the deformation requirements or welded wire reinforcement not meeting the spacing requirements are "plain reinforcement," for code purposes, and may be used only for spirals.

reinforcement, supplementary—Supplementary reinforcement has a configuration and placement similar to anchor reinforcement but is not specifically designed to transfer loads from the anchors into the member. Stirrups, as used for shear reinforcement, may fall into this category.

CODE

building code using the applicable provisions and load combinations.

seismic hook—hook on a stirrup, hoop, or crosstie having a bend not less than 135 degrees, except that circular hoops shall have a bend not less than 90 degrees; hooks shall have an extension of at least $6d_b$, but not less than 3 in. The hooks shall engage the longitudinal reinforcement and the extension shall project into the interior of the stirrup or hoop.

shear cap—projection below the slab used to increase the slab shear strength.

shear lug—a steel element welded to an attachment base plate to transfer shear to concrete by bearing.

sheathing—material encasing prestressing reinforcement to prevent bonding of the prestressing reinforcement with the surrounding concrete, to provide corrosion protection, and to contain the corrosion-inhibiting coating.

shotcrete—mortar or concrete placed pneumatically by high velocity projection from a nozzle onto a surface.

shotcrete, dry-mix—shotcrete in which most of the mixing water is added to the concrete ingredients at the nozzle.

shotcrete, wet-mix—shotcrete in which the concrete ingredients, including water, are mixed before introduction into the delivery hose.

side-face blowout strength, concrete—strength of anchors with deep embedment and thin side-face cover such that spalling occurs on the side face around the embedded head without breakout occurring at the top concrete surface.

slab-beam strip—in two-way prestressed slabs, the width of the floor system, including both the slab and beam if applicable, bounded laterally by adjacent panel centerlines for an interior slab-beam strip, or by adjacent panel centerline and slab edge for an exterior slab-beam strip.

spacing, clear—least dimension between the outermost surfaces of adjacent items.

span length—distance between supports.

special seismic systems—structural systems that use special moment frames, special structural walls, or both.

specialty engineer—a licensed design professional to whom a specific portion of the design work has been delegated.

specialty insert—predesigned and prefabricated cast-in anchors specifically designed for attachment of bolted or slotted connections.

spiral reinforcement—continuously wound reinforcement in the form of a cylindrical helix.

steel element, brittle—element with a tensile test elongation of less than 14 percent, or reduction in area of less than 30 percent at failure.

steel element, ductile—element with a tensile test elongation of at least 14 percent and reduction in area of at least 30 percent; steel element meeting the requirements of ASTM A307 shall be considered ductile; except as modified by for earthquake effects, deformed reinforcing bars meeting

COMMENTARY

sheathing—Typically, sheathing is a continuous, seamless, high-density polyethylene material extruded directly on the coated prestressing reinforcement.

shotcrete—Terms such as gunite and sprayed concrete are sometimes used to refer to shotcrete.

specialty insert—Specialty inserts are devices often used for handling, transportation, erection, and anchoring elements; specialty inserts are not within the scope of this Code.

steel element, brittle—The 14 percent elongation should be measured over the gauge length specified in the appropriate ASTM standard for the steel.

steel element, ductile—The 14 percent elongation should be measured over the gauge length specified in the appropriate ASTM standard for steel. Due to concerns over fracture in cut threads, it should be verified that threaded deformed reinforcing bars satisfy the strength requirements of 25.5.7.1.

CODE

the requirements of ASTM A615, A706, or A955 shall be considered as ductile steel elements.

stirrup—reinforcement used to resist shear and torsion forces in a member; typically deformed bars, deformed wires, or welded wire reinforcement either single leg or bent into L, U, or rectangular shapes and located perpendicular to, or at an angle to, longitudinal reinforcement. See also **tie**.

strength, design—nominal strength multiplied by a strength reduction factor ϕ.

strength, nominal—strength of a member or cross section calculated in accordance with provisions and assumptions of the strength design method of this Code before application of any strength reduction factors.

strength, required—strength of a member or cross section required to resist factored loads or related internal moments and forces in such combinations as stipulated in this Code.

stretch length—length of anchor, extending beyond concrete in which it is anchored, subject to full tensile load applied to anchor, and for which cross-sectional area is minimum and constant.

structural concrete—concrete used for structural purposes, including plain and reinforced concrete.

structural diaphragm—member, such as a floor or roof slab, that transmits forces acting in the plane of the member to vertical elements of the lateral-force-resisting system. A structural diaphragm may include chords and collectors as part of the diaphragm.

structural integrity—ability of a structure through strength, redundancy, ductility, and detailing of reinforcement to redistribute stresses and maintain overall stability if localized damage or significant overstress occurs.

structural system—interconnected members designed to meet performance requirements.

structural truss—assemblage of reinforced concrete members subjected primarily to axial forces.

structural wall—wall proportioned to resist combinations of moments, shears, and axial forces in the plane of the wall; a shear wall is a structural wall.

structural wall, ductile coupled—a seismic-force-resisting-system complying with 18.10.9.

structural wall, ordinary reinforced concrete—a wall complying with Chapter 11.

structural wall, ordinary plain concrete—a wall complying with Chapter 14.

COMMENTARY

stirrup—The term "stirrup" is usually applied to transverse reinforcement in beams or slabs and the term "ties" or "hoops" to transverse reinforcement in compression members.

strength, nominal—Nominal or specified values of material strengths and dimensions are used in the calculation of nominal strength. The subscript n is used to denote the nominal strengths; for example, nominal axial load strength P_n, nominal moment strength M_n, and nominal shear strength V_n. For additional discussion on the concepts and nomenclature for strength design, refer to the Commentary of Chapter 22.

strength, required—The subscript u is used only to denote the required strengths; for example, required axial load strength P_u, required moment strength M_u, and required shear strength V_u, calculated from the applied factored loads and forces. The basic requirement for strength design may be expressed as follows: design strength $\geq$ required strength; for example, $\phi P_n \geq P_u$; $\phi M_n \geq M_u$; $\phi V_n \geq V_u$. For additional discussion on the concepts and nomenclature for strength design, refer to the Commentary of Chapter 22.

stretch length—Length of an anchor over which inelastic elongations are designed to occur under earthquake loadings. Examples illustrating stretch length are shown in Fig. R17.10.5.3.

2 Not. & Term.

CODE

structural wall, intermediate precast—a wall complying with 18.5.

structural wall, special—a cast-in-place structural wall in accordance with 18.2.3 through 18.2.8 and 18.10; or a precast structural wall in accordance with 18.2.3 through 18.2.8 and 18.11.

strut—compression member in a strut-and-tie model representing the resultant of a parallel or a fan-shaped compression field.

strut, boundary—strut located along the boundary of a member or discontinuity region.

strut, interior—strut not located along the boundary of a member or discontinuity region.

strut-and-tie model—truss model of a member or of a D-region in such a member, made up of struts and ties connected at nodes and capable of transferring the factored loads to the supports or to adjacent B-regions.

tendon—in post-tensioned members, a tendon is a complete assembly consisting of anchorages, prestressing reinforcement, and sheathing with coating for unbonded applications or ducts filled with grout for bonded applications.

tendon, bonded—tendon in which prestressed reinforcement is continuously bonded to the concrete through grouting of ducts embedded within the concrete cross section.

tendon, external—a tendon external to the member concrete cross section in post-tensioned applications.

tendon, unbonded—tendon in which prestressed reinforcement is prevented from bonding to the concrete. The prestressing force is permanently transferred to the concrete at the tendon ends by the anchorages only.

tension-controlled section—a cross section in which the net tensile strain in the extreme tension steel at nominal strength is greater than or equal to ε_{ty} + **0.003**.

tie—(a) reinforcing bar or wire enclosing longitudinal reinforcement; a continuously wound transverse bar or wire in the form of a circle, rectangle, or other polygonal shape without reentrant corners enclosing longitudinal reinforcement; see also **stirrup**, **hoop**; (b) tension element in a strut-and-tie model.

transfer—act of transferring stress in prestressed reinforcement from jacks or pretensioning bed to concrete member.

COMMENTARY

structural wall, intermediate precast—Requirements of 18.5 are intended to result in an intermediate precast structural wall having minimum strength and toughness equivalent to that for an ordinary reinforced concrete structural wall of cast-in-place concrete. A precast concrete wall not satisfying the requirements of 18.5 is considered to have ductility and structural integrity less than that for an intermediate precast structural wall.

structural wall, special—Requirements of 18.2.3 through 18.2.8 and 18.11 are intended to result in a special precast structural wall having minimum strength and toughness equivalent to that for a special reinforced concrete structural wall of cast-in-place concrete.

strut, boundary—A boundary strut is intended to apply to the flexural compression zone of a beam, wall, or other member. Boundary struts are not subject to transverse tension and are therefore stronger than interior struts (Fig. R23.2.1).

strut, interior—Interior struts are subject to tension, acting perpendicular to the strut in the plane of the model, from shear (Fig. R23.2.1).

tendon, external—In new or existing post-tensioned applications, a tendon totally or partially external to the member concrete cross section, or inside a box section, and attached at the anchor device and deviation points.

CODE

transfer length—length of embedded pretensioned reinforcement required to transfer the effective prestress to the concrete.

two-way construction—members designed to be capable of supporting loads through bending in two directions; some slabs and foundations are considered two-way construction. See also **one-way construction**.

uncased cast-in-place concrete drilled or augered piles—piles with or without an enlarged base (bell) that are constructed by either drilling a hole in the ground, or by installing a temporary casing in the ground and cleaning out the soil, and subsequently filling the hole with reinforcement and concrete.

wall—a vertical element designed to resist axial load, lateral load, or both, with a horizontal length-to-thickness ratio greater than 3, used to enclose or separate spaces.

wall segment—portion of wall bounded by vertical or horizontal openings or edges.

wall segment, horizontal—segment of a structural wall, bounded vertically by two openings or by an opening and an edge.

wall segment, vertical—segment of a structural wall, bounded horizontally by two openings or by an opening and an edge; wall piers are vertical wall segments.

wall pier—a vertical wall segment within a structural wall, bounded horizontally by two openings or by an opening and an edge, with ratio of horizontal length to wall thickness (ℓ_w/b_w) less than or equal to 6.0, and ratio of clear height to horizontal length (h_w/ℓ_w) greater than or equal to 2.0.

water-cementitious materials ratio—ratio of mass of water, excluding that absorbed by the aggregate, to the mass of cementitious materials in a mixture, stated as a decimal.

work—the entire construction or separately identifiable parts thereof that are required to be furnished under the construction documents.

yield strength—specified minimum yield strength or yield point of reinforcement; yield strength or yield point shall be determined in tension according to applicable ASTM standards as modified by this Code.

COMMENTARY

wall segment, horizontal—A horizontal wall segment is shown in Fig. R18.10.4.5.

wall pier—Wall piers are vertical wall segments with dimensions and reinforcement intended to result in shear demand being limited by flexural yielding of the vertical reinforcement in the pier.

CHAPTER 3—REFERENCED STANDARDS

CODE

3.1—Scope

3.1.1 Standards, or specific sections thereof, cited in this Code, including Annex, Appendixes, or Supplements where prescribed, are referenced without exception in this Code, unless specifically noted. Cited standards are listed in the following with their serial designations, including year of adoption or revision.

3.2—Referenced standards

3.2.1 *American Association of State Highway and Transportation Officials (AASHTO)*

LRFDUS-8—LRFD Bridge Design Specifications, 8th Edition, 2017, Articles 5.8.4.4.2, 5.8.4.4.3, and 5.8.4.5

LRFDCONS-4—LRFD Bridge Construction Specifications, Fourth Edition, 2017, Article 10.3.2.3

3.2.2 *American Concrete Institute (ACI)*

301-16—Specifications for Structural Concrete, Article 4.2.3

318.2-19—Building Code Requirements for Concrete Thin Shells and Commentary

332-14—Residential Code Requirements for Structural Concrete and Commentary

355.2-19—Qualification of Post-Installed Mechanical Anchors in Concrete and Commentary

355.4-11—Qualification of Post-Installed Adhesive Anchors in Concrete

369.1-17—Standard Requirements for Seismic Evaluation and Retrofit of Existing Concrete Buildings (369.1-17) and Commentary

374.1-05—Acceptance Criteria for Moment Frames Based on Structural Testing

423.7-14—Specification for Unbonded Single-Strand Tendon Materials

437.2-13—Code Requirements for Load Testing of Existing Concrete Structures and Commentary

550.3-13—Design Specification for Unbonded Post-Tensioned Precast Concrete Special Moment Frames Satisfying ACI 374.1 and Commentary

550.4-18—Qualification of Precast Concrete Diaphragm Connections and Reinforcement at Joints for Earthquake Loading and Commentary

550.5-18—Code Requirements for the Design of Precast Concrete Diaphragms for Earthquake Motions and Commentary

ITG-5.1-07—Acceptance Criteria for Special Unbonded Post-Tensioned Precast Structural Walls Based on Validation Testing

ITG-5.2-09—Requirements for Design of a Special Unbonded Post-Tensioned Precast Wall Satisfying ACI ITG-5.1 and Commentary

COMMENTARY

R3.1—Scope

R3.1.1 In this Code, references to standard specifications or other material are to a specific edition of the cited document. This is done by using the complete serial designation for the referenced standard including the title that indicates the subject and year of adoption. All standards referenced in this Code are listed in this chapter, with the title and complete serial designation. In other sections of the Code, referenced standards are abbreviated to include only the serial designation without a title or date. These abbreviated references correspond to specific standards listed in this chapter.

R3.2—Referenced standards

R3.2.1 *American Association of State Highway and Transportation Officials (AASHTO)*

Three articles of the AASHTO LRFD Specifications for Highway Bridge Design (AASHTO LRFDUS) and one article of the AASHTO LRFD Construction Specifications (AASHTO LRFDCONS) are cited in Chapters 2 and 25 of this Code.

R3.2.2 *American Concrete Institute (ACI)*

Article 4.2.3 of ACI 301 is referenced for the method of mixture proportioning cited in 26.4.3.1(b).

Prior to 2014, the provisions of ACI 318.2 were specified in Chapter 19 of the ACI 318 Building Code.

ACI 355.2 contains qualification requirements for testing and evaluating post-installed expansion, screw, and undercut anchors for use in both cracked and uncracked concrete.

ACI 355.4 contains qualification requirements for testing and evaluating adhesive anchors for use in both cracked and uncracked concrete.

ACI 423.7 requires the use of encapsulated tendon systems for applications subject to this Code.

CODE

3.2.3 *American Society of Civil Engineers (ASCE)*

ASCE/SEI 7-16—Minimum Design Loads for Buildings and Other Structures, Sections 2.3.2, Load Combinations Including Flood Loads; and 2.3.3, Load Combinations Including Atmospheric Ice Loads

3.2.4 *ASTM International*

A184/A184M-17—Standard Specification for Welded Deformed Steel Bar Mats for Concrete Reinforcement

A307-14$^{\varepsilon1}$—Standard Specification for Carbon Steel Bolts, Studs, and Threaded Rod 60000 PSI Tensile Strength

A370-18—Standard Test Methods and Definitions for Mechanical Testing of Steel Products

A416/A416M-18—Standard Specification for Low-Relaxation, Seven-Wire Steel Strand for Prestressed Concrete

A421/A421M-15—Standard Specification for Uncoated Stress-Relieved Steel Wire for Prestressed Concrete, including Supplementary Requirement SI, Low-Relaxation Wire and Relaxation Testing

A615/A615M-18$^{\varepsilon1}$—Standard Specification for Deformed and Plain Carbon-Steel Bars for Concrete Reinforcement

A706/A706M-16—Standard Specification for Deformed and Plain Low-Alloy Steel Bars for Concrete Reinforcement

A722/A722M-18—Standard Specification for Uncoated High-Strength Steel Bars for Prestressing Concrete

A767/A767M-16—Standard Specification for Zinc-Coated (Galvanized) Steel Bars for Concrete Reinforcement

A775/A775M-17—Standard Specification for Epoxy-Coated Steel Reinforcing Bars

A820/A820M-16—Standard Specification for Steel Fibers for Fiber-Reinforced Concrete

A884/A884M-14—Standard Specification for Epoxy-Coated Steel Wire and Welded Wire Reinforcement

A934/A934M-16—Standard Specification for Epoxy-Coated Prefabricated Steel Reinforcing Bars

A955/A955M-18b—Standard Specification for Deformed and Plain Stainless-Steel Bars for Concrete Reinforcement

A970/A970M-18—Standard Specification for Headed Steel Bars for Concrete Reinforcement, including Annex A1 Requirements for Class HA Head Dimensions

A996/A996M-16—Standard Specification for Rail-Steel and Axle-Steel Deformed Bars for Concrete Reinforcement

A1022/A1022M-16b—Standard Specification for Deformed and Plain Stainless Steel Wire and Welded Wire for Concrete Reinforcement

A1035/A1035M-16b—Standard Specification for Deformed and Plain, Low-Carbon, Chromium, Steel Bars for Concrete Reinforcement

A1044/A1044M-16a—Standard Specification for Steel Stud Assemblies for Shear Reinforcement of Concrete

A1055/A1055M-16—Standard Specification for Zinc and Epoxy Dual-Coated Steel Reinforcing Bars

A1060/A1060M-16b—Standard Specification for Zinc-Coated (Galvanized) Steel Welded Wire Reinforcement, Plain and Deformed, for Concrete

COMMENTARY

R3.2.3 *American Society of Civil Engineers (ASCE)*

The two specific sections of ASCE 7 are referenced for the purposes cited in 5.3.9 and 5.3.10.

R3.2.4 *ASTM International*

The ASTM standards listed are the latest editions at the time these code provisions were adopted. ASTM standards are revised frequently relative to the revision cycle for the Code. Current and historical editions of the referenced standards can be obtained from ASTM International. Use of an edition of a standard other than that referenced in the Code obligates the user to evaluate if any differences in the nonconforming edition are significant to use of the standard.

Many of the ASTM standards are combined standards as denoted by the dual designation, such as ASTM A36/A36M. For simplicity, these combined standards are referenced without the metric (M) designation within the text of the Code and Commentary. In this provision, however, the complete designation is given because that is the official designation for the standard.

CODE

A1064/A1064M-18a—Standard Specification for Carbon-Steel Wire and Welded Wire Reinforcement, Plain and Deformed, for Concrete

C29/C29M-17a—Standard Test Method for Bulk Density ("Unit Weight") and Voids in Aggregate

C31/C31M-19—Standard Practice for Making and Curing Concrete Test Specimens in the Field

C33/C33M-18—Standard Specification for Concrete Aggregates

C39/C39M-18—Standard Test Method for Compressive Strength of Cylindrical Concrete Specimens

C42/C42M-18a—Standard Test Method for Obtaining and Testing Drilled Cores and Sawed Beams of Concrete

C94/C94M-18—Standard Specification for Ready-Mixed Concrete

C138-17a—Standard Test Method for Density (Unit Weight), Yield, and Air Content (Gravimetric) of Concrete

C150/C150M-19a—Standard Specification for Portland Cement

C172/C172M-17—Standard Practice for Sampling Freshly Mixed Concrete

C173/C173M-16—Standard Test Method for Air Content of Freshly Mixed Concrete by the Volumetric Method

C192-18—Standard Practice for Making and Curing Concrete Test Specimens in the Laboratory

C231/C231M-17a—Standard Test Method for Air Content of Freshly Mixed Concrete by the Pressure Method

C260/C260M-10a(2016)—Standard Specification for Air-Entraining Admixtures for Concrete

C330/C330M-17a—Standard Specification for Lightweight Aggregates for Structural Concrete

C469/C469M-14—Standard Test Method for Static Modulus of Elasticity and Poisson's Ratio of Concrete in Compression

C494/C494M-17—Standard Specification for Chemical Admixtures for Concrete

C567/C567M-14—Standard Test Method for Determining Density of Structural Lightweight Concrete

C595/C595M-19—Standard Specification for Blended Hydraulic Cements

C618-19—Standard Specification for Coal Fly Ash and Raw or Calcined Natural Pozzolan for Use in Concrete

C685/C685M-17a—Standard Specification for Concrete Made by Volumetric Batching and Continuous Mixing

C845/C845M-18—Standard Specification for Expansive Hydraulic Cement

C989/C989M-18a—Standard Specification for Slag Cement for Use in Concrete and Mortars

C1012/C1012M-18b—Standard Test Method for Length Change of Hydraulic-Cement Mortars Exposed to a Sulfate Solution

C1017/C1017M-13$^{\varepsilon 1}$—Standard Specification for Chemical Admixtures for Use in Producing Flowing Concrete

C1077-17—Standard Practice for Agencies Testing Concrete and Concrete Aggregates for Use in Construction and Criteria for Testing Agency Evaluation

COMMENTARY

CODE

C1116/C1116M-10a(2015)—Standard Specification for Fiber-Reinforced Concrete

C1140-11—Standard Practice for Preparing and Testing Specimens from Shotcrete Test Panels

C1141-15—Standard Specification for Admixtures for Shotcrete

C1157/C1157M-17—Standard Performance Specification for Hydraulic Cement

C1218/C1218M-17—Standard Test Method for Water-Soluble Chloride in Mortar and Concrete

C1240-15—Standard Specification for Silica Fume Used in Cementitious Mixtures

C1436-13—Standard Specification for Materials for Shotcrete

C1480-07(2012)—Standard Specification for Packaged, Pre-Blended, Dry, Combined Materials for Use in Wet or Dry Shotcrete Application

C1580-15—Standard Test Method for Water-Soluble Sulfate in Soil

C1582/C1582M-11(2017)$^{\varepsilon 1}$—Standard Specification for Admixtures to Inhibit Chloride-Induced Corrosion of Reinforcing Steel in Concrete

C1602/C1602M-18—Standard Specification for Mixing Water Used in the Production of Hydraulic Cement Concrete

C1604-05(2012)—Standard Test Method for Obtaining and Testing Drilled Cores of Shotcrete

C1609/C1609M-12—Standard Test Method for Flexural Performance of Fiber-Reinforced Concrete (Using Beam with Third-Point Loading)

C1797-17—Standard Specification for Ground Calcium Carbonate and Aggregate Mineral Fillers for use in Hydraulic Cement Concrete

D516-16—Standard Test Method for Sulfate Ion in Water

D4130-15—Standard Test Method for Sulfate Ion in Brackish Water, Seawater, and Brines

3.2.5 *American Welding Society (AWS)*

D1.1/D1.1M: 2015—Structural Welding Code – Steel

D1.4/D1.4M: 2018—Structural Welding Code – Reinforcing Steel

COMMENTARY

CHAPTER 4—STRUCTURAL SYSTEM REQUIREMENTS

CODE / COMMENTARY

4.1—Scope

4.1.1 This chapter shall apply to design of structural concrete in structures or portions of structures defined in Chapter 1.

R4.1—Scope

This chapter was added to the 2014 Code to introduce structural system requirements. Requirements more stringent than the Code provisions may be desirable for unusual construction or construction where enhanced performance is appropriate. The Code and Commentary must be supplemented with sound engineering knowledge, experience, and judgment.

4.2—Materials

4.2.1 Design properties of concrete shall be selected to be in accordance with Chapter 19.

R4.2—Materials

Chapter 3 identifies the referenced standards permitted for design. Chapters 19 and 20 establish properties of concrete and steel reinforcement permitted for design. Chapter 26 presents construction requirements for concrete materials, proportioning, and acceptance of concrete.

4.2.1.1 Design properties of shotcrete shall conform to the requirements for concrete except as modified by provisions of the Code.

R4.2.1.1 Shotcrete is considered to behave and have properties similar to concrete unless otherwise noted. Sections where use of shotcrete is specifically addressed in this Code are shown in Table R4.2.1.1. Additional information on shotcrete can be found in ACI 506R and ACI 506.2.

Table R4.2.1.1—Sections in Code with shotcrete provisions

Topic covered	Section
Freezing and thawing	19.3.3.3 through 19.3.3.6
Reinforcement	25.2.7 through 25.2.10, 25.5.1.6, and 25.5.1.7
Where shotcrete is required or permitted	26.3.1, 26.3.2
Materials	26.4.1.2, 26.4.1.4, and 26.4.1.6
Proportioning mixtures	26.4.3
Documentation of mixtures	26.4.4.1
Placement and consolidation	26.5.2.1
Curing	26.5.3
Joints	26.5.6
Evaluation and acceptance	26.12

4.2.2 Design properties of reinforcement shall be selected to be in accordance with Chapter 20.

4.3—Design loads

4.3.1 Loads and load combinations considered in design shall be in accordance with Chapter 5.

R4.3—Design loads

R4.3.1 The provisions in Chapter 5 are based on ASCE/SEI 7. The design loads include, but are not limited to, dead loads, live loads, snow loads, wind loads, earthquake effects, prestressing effects, crane loads, vibration, impact, shrinkage, temperature changes, creep, expansion of shrinkage-compensating concrete, and predicted unequal settlement of supports. Other project-specific loads may be specified by the licensed design professional.

CODE

4.4—Structural system and load paths

4.4.1 The structural system shall include (a) through (g), as applicable:

(a) Floor construction and roof construction, including one-way and two-way slabs
(b) Beams and joists
(c) Columns
(d) Walls
(e) Diaphragms
(f) Foundations
(g) Joints, connections, and anchors as required to transmit forces from one component to another

4.4.2 Design of structural members including joints and connections given in 4.4.1 shall be in accordance with Chapters 7 through 18.

4.4.3 It shall be permitted to design a structural system comprising structural members not in accordance with 4.4.1 and 4.4.2, provided the structural system is approved in accordance with 1.10.1.

4.4.4 The structural system shall be designed to resist the factored loads in load combinations given in 4.3 without exceeding the appropriate member design strengths, considering one or more continuous load paths from the point of load application or origination to the final point of resistance.

4.4.5 Structural systems shall be designed to accommodate anticipated volume change and differential settlement.

COMMENTARY

R4.4—Structural system and load paths

R4.4.1 Structural concrete design has evolved from emphasizing the design of individual members to designing the structure as an entire system. A structural system consists of structural members, joints, and connections, each performing a specific role or function. A structural member may belong to one or more structural systems, serving different roles in each system and having to meet all the detailing requirements of the structural systems of which they are a part. Joints and connections are locations common to intersecting members or are items used to connect one member to another, but the distinction between members, joints, and connections can depend on how the structure is idealized. Throughout this chapter, the term "members" often refers to "structural members, joints, and connections."

Although the Code is written considering that a structural system comprises these members, many alternative arrangements are possible because not all structural member types are used in all building structural systems. The selection types of the members to use in a specific project and the role or roles these member types play is made by the licensed design professional complying with requirements of the Code.

R4.4.2 In the chapter for each type of structural member, requirements follow the same general sequence and scope, including general requirements, design limits, required strength, design strength, reinforcement limits, reinforcement detailing, and other requirements unique to the type of member.

R4.4.3 Some materials, structural members, or systems that may not be recognized in the prescriptive provisions of the Code may still be acceptable if they meet the intent of the Code. Section 1.10.1 outlines the procedures for obtaining approval of alternative materials and systems.

R4.4.4 The design should be based on members and connections that provide design strengths not less than the strengths required to transfer the loads along the load path. The licensed design professional may need to study one or more alternative paths to identify weak links along the sequence of elements that constitute each load path.

R4.4.5 The effects of column and wall creep and shrinkage, restraint of creep and shrinkage in long roof and floor systems, creep caused by prestress forces, volume changes caused by temperature variation, as well as potential damage to supporting members caused by these volume changes should be considered in design. Reinforcement, closure strips, or expansion joints are common ways of accommodating these effects. Minimum shrinkage and temperature reinforcement controls cracking to an acceptable level in many concrete structures of ordinary proportions and exposures.

CODE

COMMENTARY

Differential settlement or heave may be an important consideration in design. Geotechnical recommendations to allow for nominal values of differential settlement and heave are not normally included in design load combinations for ordinary building structures.

4.4.6 *Seismic-force-resisting system*

4.4.6.1 Every structure shall be assigned to a Seismic Design Category in accordance with the general building code or as determined by the building official in areas without a legally adopted building code.

4.4.6.2 Structural systems designated as part of the seismic-force-resisting system shall be restricted to those systems designated by the general building code or as determined by the building official in areas without a legally adopted building code.

4.4.6.3 Structural systems assigned to Seismic Design Category A shall satisfy the applicable requirements of this Code. Structures assigned to Seismic Design Category A are not required to be designed in accordance with Chapter 18.

4.4.6.4 Structural systems assigned to Seismic Design Category B, C, D, E, or F shall satisfy the requirements of Chapter 18 in addition to applicable requirements of other chapters of this Code.

4.4.6.5 Structural members assumed not to be part of the seismic-force-resisting system shall be permitted, subject to the requirements of 4.4.6.5.1 and 4.4.6.5.2.

4.4.6.5.1 In structures assigned to Seismic Design Category B, C, D, E, or F, the effects of those structural members on the response of the system shall be considered and accommodated in the structural design.

4.4.6.5.2 In structures assigned to Seismic Design Category B, C, D, E, or F, the consequences of damage to those structural members shall be considered.

4.4.6.5.3 In structures assigned to Seismic Design Category D, E, or F, structural members not considered part of

R4.4.6 *Seismic-force-resisting system*

R4.4.6.1 Design requirements in the Code are based on the seismic design category to which the structure is assigned. In general, the seismic design category relates to seismic risk level, soil type, occupancy, and building use. Assignment of a building to a seismic design category is under the jurisdiction of a general building code rather than this Code. In the absence of a general building code, ASCE/SEI 7 provides the assignment of a building to a seismic design category.

R4.4.6.2 The general building code prescribes, through ASCE/SEI 7, the types of structural systems permitted as part of the seismic-force-resisting system based on considerations such as seismic design category and building height. The seismic design requirements for systems assigned to Seismic Design Categories B through F are prescribed in Chapter 18. Other systems can be used if approved by the building official.

R4.4.6.3 Structures assigned to Seismic Design Category A are subject to the lowest seismic hazard. Chapter 18 does not apply.

R4.4.6.4 Chapter 18 contains provisions that are applicable depending on the seismic design category and on the seismic-force-resisting system used. Not all structural member types have specific requirements in all seismic design categories. For example, Chapter 18 does not include requirements for structural walls in Seismic Design Categories B and C, but does include special provisions for Seismic Design Categories D, E, and F.

R4.4.6.5 In Seismic Design Categories D, E, and F, structural members not considered part of the seismic-force-resisting system are required to be designed to accommodate drifts and forces that occur as the building responds to an earthquake.

CODE

the seismic-force-resisting system shall meet the applicable requirements in Chapter 18.

4.4.6.6 Effects of nonstructural members shall be accounted for as described in 18.2.2.1 and consequences of damage to nonstructural members shall be considered.

4.4.6.7 Design verification of earthquake-resistant concrete structures using nonlinear response history analysis shall be in accordance with Appendix A.

4.4.7 *Diaphragms*

4.4.7.1 Diaphragms, such as floor or roof slabs, shall be designed to resist simultaneously both out-of-plane gravity loads and in-plane lateral forces in load combinations given in 4.3.

4.4.7.2 Diaphragms and their connections to framing members shall be designed to transfer forces between the diaphragm and framing members.

4.4.7.3 Diaphragms and their connections shall be designed to provide lateral support to vertical, horizontal, and inclined elements.

4.4.7.4 Diaphragms shall be designed to resist applicable lateral loads from soil and hydrostatic pressure and other loads assigned to the diaphragm by structural analysis.

4.4.7.5 Collectors shall be provided where required to transmit forces between diaphragms and vertical elements.

4.4.7.6 Diaphragms that are part of the seismic-force-resisting system shall be designed for the applied forces. In structures assigned to Seismic Design Category D, E, and F, the diaphragm design shall be in accordance with Chapter 18.

4.5—Structural analysis

4.5.1 Analytical procedures shall satisfy compatibility of deformations and equilibrium of forces.

4.5.2 The methods of analysis given in Chapter 6 shall be permitted.

COMMENTARY

R4.4.6.6 Although the design of nonstructural elements for earthquake effects is not included in the scope of this Code, the potential negative effects of nonstructural elements on the structural behavior need to be considered in Seismic Design Categories B, C, D, E, and F. Interaction of nonstructural elements with the structural system—for example, the short-column effect—had led to failure of structural members and collapse of some structures during earthquakes in the past.

R4.4.7 *Diaphragms*

Floor and roof slabs play a dual role by simultaneously supporting gravity loads and transmitting lateral forces in their own plane as a diaphragm. General requirements for diaphragms are provided in Chapter 12, and roles of the diaphragm described in the Commentary to that chapter. Additional requirements for design of diaphragms in structures assigned to Seismic Design Categories D, E, and F are prescribed in Chapter 18.

R4.4.7.5 All structural systems must have a complete load path in accordance with 4.4.4. The load path includes collectors where required.

R4.5—Structural analysis

The role of analysis is to estimate the internal forces and deformations of the structural system and to establish compliance with the strength, serviceability, and stability requirements of the Code. The use of computers in structural engineering has made it feasible to perform analysis of complex structures. The Code requires that the analytical procedure used meets the fundamental principles of equilibrium and compatibility of deformations, permitting a number of analytical techniques, including the strut-and-tie method required for discontinuity regions, as provided in Chapter 6.

CODE

4.6—Strength

4.6.1 Design strength of a member and its joints and connections, in terms of moment, shear, torsional, axial, and bearing strength, shall be taken as the nominal strength S_n multiplied by the applicable strength reduction factor ϕ.

4.6.2 Structures and structural members shall have design strength at all sections, ϕS_n, greater than or equal to the required strength U calculated for the factored loads and forces in such combinations as required by this Code or the general building code.

COMMENTARY

R4.6—Strength

The basic requirement for strength design may be expressed as follows:

design strength ≥ required strength

$$\phi S_n \geq U$$

In the strength design procedure, the level of safety is provided by a combination of factors applied to the loads and strength reduction factors ϕ applied to the nominal strengths.

The strength of a member or cross section, calculated using standard assumptions and strength equations, along with nominal values of material strengths and dimensions, is referred to as nominal strength and is generally designated S_n. Design strength or usable strength of a member or cross section is the nominal strength reduced by the applicable strength reduction factor ϕ. The purpose of the strength reduction factor is to account for the probability of under-strength due to variations of in-place material strengths and dimensions, the effect of simplifying assumptions in the design equations, the degree of ductility, potential failure mode of the member, the required reliability, and significance of failure and existence of alternative load paths for the member in the structure.

This Code, or the general building code, prescribes design load combinations, also known as factored load combinations, which define the way different types of loads are multiplied (factored) by individual load factors and then combined to obtain a factored load U. The individual load factors and additive combination reflect the variability in magnitude of the individual loads, the probability of simultaneous occurrence of various loads, and the assumptions and approximations made in the structural analysis when determining required design strengths.

A typical design approach, where linear analysis is applicable, is to analyze the structure for individual unfactored load cases, and then combine the individual unfactored load cases in a factored load combination to determine the design load effects. Where effects of loads are nonlinear—for example, in foundation uplift—the factored loads are applied simultaneously to determine the nonlinear, factored load effect. The load effects relevant for strength design include moments, shears, torsions, axial forces, bearing forces, and punching shear stresses. Sometimes, design displacements are determined for factored loads. The load effects relevant for service design include stresses and deflections.

In the course of applying these principles, the licensed design professional should be aware that providing more strength than required does not necessarily lead to a safer structure because doing so may change the potential failure mode. For example, increasing longitudinal reinforcement area beyond that required for moment strength as derived from analysis without increasing transverse reinforcement could increase the probability of a shear failure occurring

CODE

COMMENTARY

prior to a flexural failure. Excess strength may be undesirable for structures expected to behave inelastically during earthquakes.

4.7—Serviceability

4.7.1 Evaluation of performance at service load conditions shall consider reactions, moments, shears, torsions, and axial forces induced by prestressing, creep, shrinkage, temperature change, axial deformation, restraint of attached structural members, and foundation settlement.

4.7.2 For structures, structural members, and their connections, the requirements of 4.7.1 shall be deemed to be satisfied if designed in accordance with the provisions of the applicable member chapters.

R4.7—Serviceability

Serviceability refers to the ability of the structural system or structural member to provide appropriate behavior and functionality under the actions affecting the system. Serviceability requirements address issues such as deflections and cracking, among others. Serviceability considerations for vibrations are discussed in R6.6.3.2.2 and R24.1.

Except as stated in Chapter 24, service-level load combinations are not defined in this Code, but are discussed in Appendix C of ASCE/SEI 7-16. Appendixes to ASCE/SEI 7 are not considered mandatory parts of the standard.

4.8—Durability

4.8.1 Concrete mixtures shall be designed in accordance with the requirements of 19.3.2 and 26.4, considering applicable environmental exposure to provide required durability.

4.8.2 Reinforcement shall be protected from corrosion in accordance with 20.5.

R4.8—Durability

The environment where the structure will be located will dictate the exposure category for materials selection, design details, and construction requirements to minimize potential for premature deterioration of the structure caused by environmental effects. Durability of a structure is also impacted by the level of preventative maintenance, which is not addressed in the Code.

Chapter 19 provides requirements for protecting concrete against major environmental causes of deterioration.

4.9—Sustainability

4.9.1 The licensed design professional shall be permitted to specify in the construction documents sustainability requirements in addition to strength, serviceability, and durability requirements of this Code.

4.9.2 The strength, serviceability, and durability requirements of this Code shall take precedence over sustainability considerations.

R4.9—Sustainability

The Code provisions for strength, serviceability, and durability are minimum requirements to achieve a safe and durable concrete structure. The Code permits the owner or the licensed design professional to specify requirements higher than the minimums mandated in the Code. Such optional requirements can include higher strengths, more restrictive deflection limits, enhanced durability, and sustainability provisions.

4.10—Structural integrity

4.10.1 *General*

4.10.1.1 Reinforcement and connections shall be detailed to tie the structure together effectively and to improve overall structural integrity.

R4.10—Structural integrity

R4.10.1 *General*

R4.10.1.1 It is the intent of the structural integrity requirements to improve redundancy and ductility through detailing of reinforcement and connections so that, in the event of damage to a major supporting element or an abnormal loading, the resulting damage will be localized and the structure will have a higher probability of maintaining overall stability.

Integrity requirements for selected structural member types are included in the corresponding member chapter in the sections noted.

4.10.2 *Minimum requirements for structural integrity*

4.10.2.1 Structural members and their connections shall be in accordance with structural integrity requirements in Table 4.10.2.1.

R4.10.2 *Minimum requirements for structural integrity*

Structural members and their connections referred to in this section include only member types that have specific requirements for structural integrity. Notwithstanding,

CODE

Table 4.10.2.1—Minimum requirements for structural integrity

Member type	Section
Nonprestressed one-way cast-in-place slabs	7.7.7
Nonprestressed two-way slabs	8.7.4.2
Prestressed two-way slabs	8.7.5.6
Nonprestressed two-way joist systems	8.8.1.6
Cast-in-place beam	9.7.7
Nonprestressed one-way joist system	9.8.1.6
Precast joints and connections	16.2.1.8

COMMENTARY

detailing requirements for other member types address structural integrity indirectly.

4.11—Fire resistance

4.11.1 Structural concrete members shall satisfy the fire protection requirements of the general building code.

4.11.2 Where the general building code requires a thickness of concrete cover for fire protection greater than the concrete cover specified in 20.5.1, such greater thickness shall govern.

R4.11—Fire resistance

Additional guidance on fire resistance of structural concrete is provided by ACI 216.1.

4.12—Requirements for specific types of construction

R4.12—Requirements for specific types of construction

This section contains requirements that are related to specific types of construction. Additional requirements that are specific to member types appear in the corresponding member chapters.

4.12.1 *Precast concrete systems*

4.12.1.1 Design of precast concrete members and connections shall include loading and restraint conditions from initial fabrication to end use in the structure, including form removal, storage, transportation, and erection.

4.12.1.2 Design, fabrication, and construction of precast members and their connections shall include the effects of tolerances.

R4.12.1 *Precast concrete systems*

All requirements in the Code apply to precast systems and members unless specifically excluded. In addition, some requirements apply specifically to precast concrete. This section contains specific requirements for precast systems. Other sections of this Code also provide specific requirements, such as required concrete cover, for precast systems.

Precast systems differ from monolithic systems in that the type of restraint at supports, the location of supports, and the induced stresses in the body of the member vary during fabrication, storage, transportation, erection, and the final interconnected configuration. Consequently, the member design forces to be considered may differ in magnitude and direction with varying critical sections at various stages of construction. For example, a precast flexural member may be simply supported for dead load effects before continuity at the supporting connections is established and may be a continuous member for live or environmental load effects due to the moment continuity created by the connections after erection.

R4.12.1.2 For guidance on including the effects of tolerances, refer to the *PCI Design Handbook* (PCI MNL 120).

CODE

4.12.1.3 When precast members are incorporated into a structural system, the forces and deformations occurring in and adjacent to connections shall be included in the design.

4.12.1.4 Where system behavior requires in-plane loads to be transferred between the members of a precast floor or wall system, (a) and (b) shall be satisfied:

(a) In-plane load paths shall be continuous through both connections and members.
(b) Where tension loads occur, a load path of steel or steel reinforcement, with or without splices, shall be provided.

4.12.1.5 Distribution of forces that act perpendicular to the plane of precast members shall be established by analysis or test.

4.12.2 *Prestressed concrete systems*

4.12.2.1 Design of prestressed members and systems shall be based on strength and on behavior at service conditions at all critical stages during the life of the structure from the time prestress is first applied.

4.12.2.2 Provisions shall be made for effects on adjoining construction of elastic and plastic deformations, deflections, changes in length, and rotations due to prestressing. Effects of temperature change, restraint of attached structural members, foundation settlement, creep, and shrinkage shall also be considered.

4.12.2.3 Stress concentrations due to prestressing shall be considered in design.

4.12.2.4 Effect of loss of area due to open ducts shall be considered in computing section properties before grout in post-tensioning ducts has attained design strength.

4.12.2.5 Post-tensioning tendons shall be permitted to be external to any concrete section of a member. Strength and serviceability design requirements of this Code shall be used to evaluate the effects of external tendon forces on the concrete structure.

COMMENTARY

R4.12.1.5 Concentrated and line loads can be distributed among members provided the members have sufficient torsional stiffness and shear can be transferred across joints. Torsionally stiff members such as hollow-core or solid slabs will provide better load distribution than torsionally flexible members such as double tees with thin flanges. The actual distribution of the load depends on many factors discussed in detail in LaGue (1971), Johnson and Ghadiali (1972), Pfeifer and Nelson (1983), Stanton (1987, 1992), *PCI Manual for the Design of Hollow Core Slabs and Walls* (PCI MNL 126), Aswad and Jacques (1992), and the *PCI Design Handbook* (PCI MNL 120). Large openings can cause significant changes in distribution of forces.

R4.12.2 *Prestressed concrete systems*

Prestressing, as used in the Code, may apply to pretensioning, bonded post-tensioning, or unbonded post-tensioning. All requirements in the Code apply to prestressed systems and members, unless specifically excluded. This section contains specific requirements for prestressed concrete systems. Other sections of this Code also provide specific requirements, such as required concrete cover for prestressed systems.

Creep and shrinkage effects may be greater in prestressed than in nonprestressed concrete structures because of the prestressing forces and because prestressed structures typically have less bonded reinforcement. Effects of movements due to creep and shrinkage may require more attention than is normally required for nonprestressed concrete. These movements may increase prestress losses.

Design of externally post-tensioned construction should consider aspects of corrosion protection and fire resistance that are applicable to this structural system.

CODE

4.12.3 *Composite concrete flexural members*

4.12.3.1 This Code shall apply to composite concrete flexural members as defined in Chapter 2.

4.12.3.2 Individual members shall be designed for all critical stages of loading.

4.12.3.3 Members shall be designed to support all loads introduced prior to full development of design strength of composite members.

4.12.3.4 Reinforcement shall be detailed to minimize cracking and to prevent separation of individual components of composite members.

4.12.4 *Structural plain concrete systems*

4.12.4.1 The design of structural plain concrete members, both cast-in-place and precast, shall be in accordance with Chapter 14.

4.13—Construction and inspection

4.13.1 Specifications for construction execution shall be in accordance with Chapter 26.

4.13.2 Inspection during construction shall be in accordance with Chapter 26 and the general building code.

4.14—Strength evaluation of existing structures

4.14.1 Strength evaluation of existing structures shall be in accordance with Chapter 27.

COMMENTARY

R4.12.3 *Composite concrete flexural members*

This section addresses structural concrete members, either precast or cast-in-place, prestressed or nonprestressed, consisting of concrete cast at different times intended to act as a composite member when loaded after concrete of the last stage of casting has set. All requirements in the Code apply to these members unless specifically excluded. In addition, some requirements apply specifically to composite concrete flexural members. This section contains requirements that are specific to these elements and are not covered in the applicable member chapters.

R4.13—Construction and inspection

Chapter 26 has been organized to collect into one location the design information, compliance requirements, and inspection provisions from the Code that should be included in construction documents There may be other information that should be included in construction documents that is not covered in Chapter 26.

R4.14—Strength evaluation of existing structures

Requirements in Chapter 27 for strength evaluation of existing structures by physical load test address the evaluation of structures subjected to gravity loads only. Chapter 27 also covers strength evaluation of existing structures by analytical evaluation, which may be used for gravity as well as other loadings such as earthquake or wind.

Notes

CHAPTER 5—LOADS

CODE

5.1—Scope

5.1.1 This chapter shall apply to selection of load factors and combinations used in design, except as permitted in Chapter 27.

5.2—General

5.2.1 Loads shall include self-weight; applied loads; and effects of prestressing, earthquakes, restraint of volume change, and differential settlement.

5.2.2 Loads and Seismic Design Categories (SDCs) shall be in accordance with the general building code, or determined by the building official.

COMMENTARY

R5.2—General

R5.2.1 Provisions in the Code are associated with dead, live, wind, and earthquake loads such as those recommended in ASCE/SEI 7. The commentary to Appendix C of ASCE/SEI 7 provides service-level wind loads W_a for serviceability checks; however, these loads are not appropriate for strength design.

If the service loads specified by the general building code differ from those of ASCE/SEI 7, the general building code governs. However, if the nature of the loads contained in a general building code differs considerably from ASCE/SEI 7 loads, some provisions of this Code may need modification to reflect the difference.

R5.2.2 Seismic Design Categories (SDCs) in this Code are adopted directly from ASCE/SEI 7. Similar designations are used by the International Building Code (2018 IBC) and the National Fire Protection Association (NFPA 5000 2012). The BOCA National Building Code (BOCA 1999) and "The Standard Building Code" (SBC 1999) used seismic performance categories. The "Uniform Building Code" (IBCO 1997) relates seismic design requirements to seismic zones, whereas editions of ACI 318 prior to 2008 related seismic design requirements to seismic risk levels. Table R5.2.2 correlates SDC to seismic risk terminology used in ACI 318 for several editions before the 2008 edition, and to the various methods of assigning design requirements used in the United States under the various model building codes, the ASCE/SEI 7 standard, and the National Earthquake Hazard Reduction Program (NEHRP 1994).

Design requirements for earthquake-resistant structures in this Code are determined by the SDC to which the structure is assigned. In general, the SDC relates to seismic hazard level, soil type, occupancy, and building use. Assignment of a building to an SDC is under the jurisdiction of the general building code rather than this Code.

In the absence of a general building code that prescribes earthquake effects and seismic zoning, it is the intent of Committee 318 that application of provisions for earthquake-resistant design be consistent with national standards or model building codes such as ASCE/SEI 7, 2012 IBC, and NFPA 5000. The model building codes also specify overstrength factors Ω_o that are related to the seismic-force-resisting system used for the structure and design of certain elements.

CODE

COMMENTARY

Table R5.2.2—Correlation between seismic-related terminology in model codes

Code, standard, or resource document and edition	Level of seismic risk or assigned seismic performance or design categories as defined in the Code		
ACI 318-08, ACI 318-11, ACI 318-14, ACI 318-19; IBC of 2000, 2003, 2006, 2009, 2012, 2015, 2018; NFPA 5000 of 2003, 2006, 2009, 2012, 2015, 2018; ASCE 7-98, 7-02, 7-05, 7-10, 7-16; NEHRP 1997, 2000, 2003, 2009, 2015	SDC[1] A, B	SDC C	SDC D, E, F
ACI 318-05 and previous editions	Low seismic risk	Moderate/intermediate seismic risk	High seismic risk
BOCA National Building Code 1993, 1996, 1999; Standard Building Code 1994, 1997, 1999; ASCE 7-93, 7-95; NEHRP 1991, 1994	SPC[2] A, B	SPC C	SPC D, E
Uniform Building Code 1991, 1994, 1997	Seismic Zone 0, 1	Seismic Zone 2	Seismic Zone 3, 4

[1]SDC = seismic design category as defined in code, standard, or resource document.

[2]SPC = seismic performance category as defined in code, standard, or resource document.

5.2.3 Live load reductions shall be permitted in accordance with the general building code or, in the absence of a general building code, in accordance with ASCE/SEI 7.

5.3—Load factors and combinations

5.3.1 Required strength U shall be at least equal to the effects of factored loads in Table 5.3.1, with exceptions and additions in 5.3.3 through 5.3.13.

Table 5.3.1—Load combinations

Load combination	Equation	Primary load
$U = 1.4D$	(5.3.1a)	D
$U = 1.2D + 1.6L + 0.5(L_r \text{ or } S \text{ or } R)$	(5.3.1b)	L
$U = 1.2D + 1.6(L_r \text{ or } S \text{ or } R) + (1.0L \text{ or } 0.5W)$	(5.3.1c)	L_r or S or R
$U = 1.2D + 1.0W + 1.0L + 0.5(L_r \text{ or } S \text{ or } R)$	(5.3.1d)	W
$U = 1.2D + 1.0E + 1.0L + 0.2S$	(5.3.1e)	E
$U = 0.9D + 1.0W$	(5.3.1f)	W
$U = 0.9D + 1.0E$	(5.3.1g)	E

R5.3—Load factors and combinations

R5.3.1 The required strength U is expressed in terms of factored loads. Factored loads are the loads specified in the general building code multiplied by appropriate load factors. If the load effects such as internal forces and moments are linearly related to the loads, the required strength U may be expressed in terms of load effects multiplied by the appropriate load factors with the identical result. If the load effects are nonlinearly related to the loads, such as frame P-delta effects (Rogowsky and Wight 2010), the loads are factored before determining the load effects. Typical practice for foundation design is discussed in R13.2.6.1. Nonlinear finite element analysis using factored load cases is discussed in R6.9.3.

The factor assigned to each load is influenced by the degree of accuracy to which the load effect usually can be calculated and the variation that might be expected in the load during the lifetime of the structure. Dead loads, because they are more accurately determined and less variable, are assigned a lower load factor than live loads. Load factors also account for variability in the structural analysis used to calculate moments and shears.

The Code gives load factors for specific combinations of loads. In assigning factors to combinations of loading, some consideration is given to the probability of simultaneous occurrence. While most of the usual combinations of loadings are included, it should not be assumed that all cases are covered.

Due regard is to be given to the sign (positive or negative) in determining U for combinations of loadings, as one type of loading may produce effects of opposite sense to that produced by another type. The load combinations with $\mathbf{0.9}\boldsymbol{D}$ are included for the case where a higher dead load reduces the effects of other loads. The loading case may also be critical for tension-controlled column sections. In such a case, a reduction in compressive axial load or development of tension with or without an increase in moment may result in a critical load combination.

CODE

COMMENTARY

Consideration should be given to various combinations of loading to determine the most critical design condition. This is particularly true when strength is dependent on more than one load effect, such as strength for combined flexure and axial load or shear strength in members with axial load.

If unusual circumstances require greater reliance on the strength of particular members than circumstances encountered in usual practice, some reduction in the stipulated strength reduction factors ϕ or increase in the stipulated load factors may be appropriate for such members.

Rain load $\boldsymbol{R}$ in Eq. (5.3.1b), (5.3.1c), and (5.3.1d) should account for all likely accumulations of water. Roofs should be designed with sufficient slope or camber to ensure adequate drainage accounting for any long-term deflection of the roof due to the dead loads. If deflection of roof members may result in ponding of water accompanied by increased deflection and additional ponding, the design should ensure that this process is self-limiting.

Model building codes and design load references refer to earthquake forces at the strength level, and the corresponding load factor is 1.0 (ASCE/SEI 7; BOCA 1999; SBC 1999; UBC (ICBO 1997); 2018 IBC). In the absence of a general building code that prescribes strength level earthquake effects, a higher load factor on $\boldsymbol{E}$ would be required. The load effect $\boldsymbol{E}$ in model building codes and design load reference standards includes the effect of both horizontal and vertical ground motions (as $\boldsymbol{E_h}$ and $\boldsymbol{E_v}$, respectively). The effect for vertical ground motions is applied as an addition to or subtraction from the dead load effect ($\boldsymbol{D}$), and it applies to all structural elements, whether part of the seismic force-resisting system or not, unless specifically excluded by the general building code.

5.3.2 The effect of one or more loads not acting simultaneously shall be investigated.

5.3.3 The load factor on live load $\boldsymbol{L}$ in Eq. (5.3.1c), (5.3.1d), and (5.3.1e) shall be permitted to be reduced to 0.5 except for (a), (b), or (c):

(a) Garages
(b) Areas occupied as places of public assembly
(c) Areas where $\boldsymbol{L}$ is greater than 100 lb/ft^2

R5.3.3 The load modification factor in this provision is different than the live load reductions based on the loaded area that may be allowed in the general building code. The live load reduction, based on loaded area, adjusts the nominal live load ($\boldsymbol{L_0}$ in ASCE/SEI 7) to $\boldsymbol{L}$. The live load reduction, as specified in the general building code, can be used in combination with the 0.5 load factor specified in this provision.

5.3.4 If applicable, $\boldsymbol{L}$ shall include (a) through (f):

(a) Concentrated live loads
(b) Vehicular loads
(c) Crane loads
(d) Loads on hand rails, guardrails, and vehicular barrier systems
(e) Impact effects
(f) Vibration effects

CODE

5.3.5 If wind load W is provided at service-level loads, **1.6*W*** shall be used in place of **1.0*W*** in Eq. (5.3.1d) and (5.3.1f), and **0.8*W*** shall be used in place of **0.5*W*** in Eq. (5.3.1c).

5.3.6 The structural effects of forces due to restraint of volume change and differential settlement T shall be considered in combination with other loads if the effects of T can adversely affect structural safety or performance. The load factor for T shall be established considering the uncertainty associated with the likely magnitude of T, the probability that the maximum effect of T will occur simultaneously with other applied loads, and the potential adverse consequences if the effect of T is greater than assumed. The load factor on T shall not have a value less than 1.0.

5.3.7 If fluid load F is present, it shall be included in the load combination equations of 5.3.1 in accordance with (a), (b), (c), or (d):

(a) If F acts alone or adds to the effects of D, it shall be included with a load factor of 1.4 in Eq. (5.3.1a).
(b) If F adds to the primary load, it shall be included with a load factor of 1.2 in Eq. (5.3.1b) through (5.3.1e).
(c) If the effect of F is permanent and counteracts the primary load, it shall be included with a load factor of 0.9 in Eq. (5.3.1g).

COMMENTARY

R5.3.5 In ASCE/SEI 7-05, wind loads are consistent with service-level design; a wind load factor of 1.6 is appropriate for use in Eq. (5.3.1d) and (5.3.1f) and a wind load factor of 0.8 is appropriate for use in Eq. (5.3.1c). ASCE/SEI 7-16 prescribes wind loads for strength-level design and the wind load factor is 1.0. Design wind speeds for strength-level design are based on storms with mean recurrence intervals of 300, 700, and 1700 years depending on the risk category of the structure. The higher load factors in 5.3.5 apply where service-level wind loads corresponding to a 50-year mean recurrence interval are used for design.

R5.3.6 Several strategies can be used to accommodate movements due to volume change and differential settlement. Restraint of such movements can cause significant member forces and moments, such as tension in slabs and shear forces and moments in vertical members. Forces due to T effects are not commonly calculated and combined with other load effects. Rather, designs rely on successful past practices using compliant structural members and ductile connections to accommodate differential settlement and volume change movement while providing the needed resistance to gravity and lateral loads. Expansion joints and construction closure strips are used to limit volume change movements based on the performance of similar structures. Shrinkage and temperature reinforcement, which may exceed the required flexural reinforcement, is commonly proportioned based on gross concrete area rather than calculated force.

Where structural movements can lead to damage of nonductile elements, calculation of the predicted force should consider the inherent variability of the expected movement and structural response.

A long-term study of the volume change behavior of precast concrete buildings (Klein and Lindenberg 2009) recommends procedures to account for connection stiffness, thermal exposure, member softening due to creep, and other factors that influence T forces.

Fintel et al. (1986) provides information on the magnitudes of volume change effects in tall structures and recommends procedures for including the forces resulting from these effects in design.

CODE

(d) If the effect of F is not permanent but, when present, counteracts the primary load, F shall not be included in Eq. (5.3.1a) through (5.3.1g).

5.3.8 If lateral earth pressure H is present, it shall be included in the load combination equations of 5.3.1 in accordance with (a), (b), or (c):

(a) If H acts alone or adds to the primary load effect, it shall be included with a load factor of 1.6.
(b) If the effect of H is permanent and counteracts the primary load effect, it shall be included with a load factor of 0.9.
(c) If the effect of H is not permanent but, when present, counteracts the primary load effect, H shall not be included.

5.3.9 If a structure is in a flood zone, the flood loads and the appropriate load factors and combinations of ASCE/SEI 7 shall be used.

5.3.10 If a structure is subjected to forces from atmospheric ice loads, the ice loads and the appropriate load factors and combinations of ASCE/SEI 7 shall be used.

5.3.11 Required strength U shall include internal load effects due to reactions induced by prestressing with a load factor of 1.0.

5.3.12 For post-tensioned anchorage zone design, a load factor of 1.2 shall be applied to the maximum prestressing reinforcement jacking force.

5.3.13 Load factors for the effects of prestressing used with the strut-and-tie method shall be included in the load combination equations of 5.3.1 in accordance with (a) or (b):

(a) A load factor of 1.2 shall be applied to the prestressing effects where the prestressing effects increase the net force in struts or ties.
(b) A load factor of 0.9 shall be applied to the prestressing effects where the prestressing effects reduce the net force in struts or ties.

COMMENTARY

R5.3.8 The required load factors for lateral pressures from soil, water in soil, and other materials, reflect their variability and the possibility that the materials may be removed. The commentary of ASCE/SEI 7 includes additional useful discussion pertaining to load factors for H.

R5.3.9 Areas subject to flooding are defined by flood hazard maps, usually maintained by local governmental jurisdictions.

R5.3.10 Ice buildup on a structural member increases the applied load and the projected area exposed to wind. ASCE/SEI 7 provides maps of probable ice thicknesses due to freezing rain, with concurrent 3-second gust speeds, for a 50-year return period.

R5.3.11 For statically indeterminate structures, the internal load effects due to reactions induced by prestressing forces, sometimes referred to as secondary moments, can be significant (Bondy 2003; Lin and Thornton 1972; Collins and Mitchell 1997).

R5.3.12 The load factor of 1.2 applied to the maximum tendon jacking force results in a design load of about 113 percent of the specified prestressing reinforcement yield strength, but not more than 96 percent of the nominal tensile strength of the prestressing reinforcement. This compares well with the maximum anchorage capacity, which is at least 95 percent of the nominal tensile strength of the prestressing reinforcement.

5 Loads

Notes

CHAPTER 6—STRUCTURAL ANALYSIS

CODE

6.1—Scope

6.1.1 This chapter shall apply to methods of analysis, modeling of members and structural systems, and calculation of load effects.

COMMENTARY

R6.1—Scope

The provisions of this chapter apply to analyses used to determine load effects for design.

Section 6.2 provides general requirements that are applicable for all analysis procedures.

Section 6.2.4 directs the licensed design professional to specific analysis provisions that are not contained in this chapter. Sections 6.2.4.1 and 6.2.4.2 identify analysis provisions that are specific to two-way slabs and walls.

Section 6.3 addresses modeling assumptions used in establishing the analysis model.

Section 6.4 prescribes the arrangements of live loads that are to be considered in the analysis.

Section 6.5 provides a simplified method of analysis for nonprestressed continuous beams and one-way slabs that can be used in place of a more rigorous analysis when the stipulated conditions are satisfied.

Section 6.6 includes provisions for a comprehensive linear elastic first-order analysis. The effects of cracked sections and creep are included in the analysis through the use of effective stiffnesses.

Section 6.7 includes provisions for linear elastic second-order analysis. Inclusion of the effects of cracking and creep is required.

Section 6.8 includes provisions for inelastic analysis.

Section 6.9 includes provisions for the use of the finite element method.

6.2—General

6.2.1 Members and structural systems shall be permitted to be modeled in accordance with 6.3.

6.2.2 All members and structural systems shall be analyzed to determine the maximum load effects including the arrangements of live load in accordance with 6.4.

6.2.3 Methods of analysis permitted by this chapter shall be (a) through (e):

(a) The simplified method for analysis of continuous beams and one-way slabs for gravity loads in 6.5
(b) Linear elastic first-order analysis in 6.6
(c) Linear elastic second-order analysis in 6.7
(d) Inelastic analysis in 6.8
(e) Finite element analysis in 6.9

R6.2—General

R6.2.3 A first-order analysis satisfies the equations of equilibrium using the original undeformed geometry of the structure. When only first-order results are considered, slenderness effects are not accounted for. Because these effects can be important, 6.6 provides procedures to calculate both individual member slenderness ($P\delta$) effects and sidesway ($P\Delta$) effects for the overall structure using the first-order results.

A second-order analysis satisfies the equations of equilibrium using the deformed geometry of the structure. If the second-order analysis uses nodes along compression members, the analysis accounts for slenderness effects due to lateral deformations along individual members, as well as sidesway of the overall structure. If the second-order analysis uses nodes at the member intersections only, the analysis captures the sidesway effects for the overall structure but neglects individual member slenderness effects. In this case, the moment magnifier method (6.6.4) is used to determine individual member slenderness effects.

CODE

COMMENTARY

An inelastic analysis i) represents the nonlinear stress-strain response of the materials composing the structure; ii) satisfies compatibility of deformations; and iii) satisfies equilibrium in the undeformed configuration for first-order analysis or in the deformed configuration for second-order analysis.

Finite element analysis was introduced in the 2014 Code to explicitly recognize a widely used analysis method.

6.2.4 Additional analysis methods that are permitted include 6.2.4.1 through 6.2.4.4.

6.2.4.1 Two-way slabs shall be permitted to be analyzed for gravity loads in accordance with (a) or (b):

(a) Direct design method for nonprestressed slabs
(b) Equivalent frame method for nonprestressed and prestressed slabs

R6.2.4.1 Code editions from 1971 to 2014 contained provisions for use of the direct design method and the equivalent frame method. These methods are well-established and are covered in available texts. These provisions for gravity load analysis of two-way slabs have been removed from the Code because they are considered to be only two of several analysis methods currently used for the design of two-way slabs. The direct design method and the equivalent frame method of the 2014 Code, however, may still be used for the analysis of two-way slabs for gravity loads.

6.2.4.2 Slender walls shall be permitted to be analyzed in accordance with 11.8 for out-of-plane effects.

6.2.4.3 Diaphragms shall be permitted to be analyzed in accordance with 12.4.2.

6.2.4.4 A member or region shall be permitted to be analyzed and designed using the strut-and-tie method in accordance with Chapter 23.

6.2.5 *Slenderness effects*

6.2.5.1 Slenderness effects shall be permitted to be neglected if (a) or (b) is satisfied:

(a) For columns not braced against sidesway

$$\frac{k\ell_u}{r} \leq 22 \qquad (6.2.5.1a)$$

(b) For columns braced against sidesway

$$\frac{k\ell_u}{r} \leq 34 + 12(M_1/M_2) \qquad (6.2.5.1b)$$

and

$$\frac{k\ell_u}{r} \leq 40 \qquad (6.2.5.1c)$$

where M_1/M_2 is negative if the column is bent in single curvature, and positive for double curvature.

R6.2.5 *Slenderness effects*

Second-order effects in many structures are negligible. In these cases, it is unnecessary to consider slenderness effects, and compression members, such as columns, walls, or braces, can be designed based on forces determined from first-order analyses. Slenderness effects can be neglected in both braced and unbraced systems, depending on the slenderness ratio ($k\ell_u/r$) of the member.

The sign convention for M_1/M_2 has been updated so that M_1/M_2 is negative if bent in single curvature and positive if bent in double curvature. This reflects a sign convention change from the 2011 Code.

The primary design aid to estimate the effective length factor k is the Jackson and Moreland Alignment Charts (Fig. R6.2.5.1), which provide a graphical determination of k for a column of constant cross section in a multi-bay frame (ACI SP-17(09); Column Research Council 1966).

Equations (6.2.5.1b) and (6.2.5.1c) are based on Eq. (6.6.4.5.1) assuming that a 5 percent increase in moments due to slenderness is acceptable (MacGregor et al. 1970).

CODE

If bracing elements resisting lateral movement of a story have a total stiffness of at least 12 times the gross lateral stiffness of the columns in the direction considered, it shall be permitted to consider columns within the story to be braced against sidesway.

COMMENTARY

As a first approximation, k may be taken equal to 1.0 in Eq. (6.2.5.1b) and (6.2.5.1c).

The stiffness of the lateral bracing is considered based on the principal directions of the framing system. Bracing elements in typical building structures consist of structural walls or lateral braces. Torsional response of the lateral-force-resisting system due to eccentricity of the structural system can increase second-order effects and should be considered.

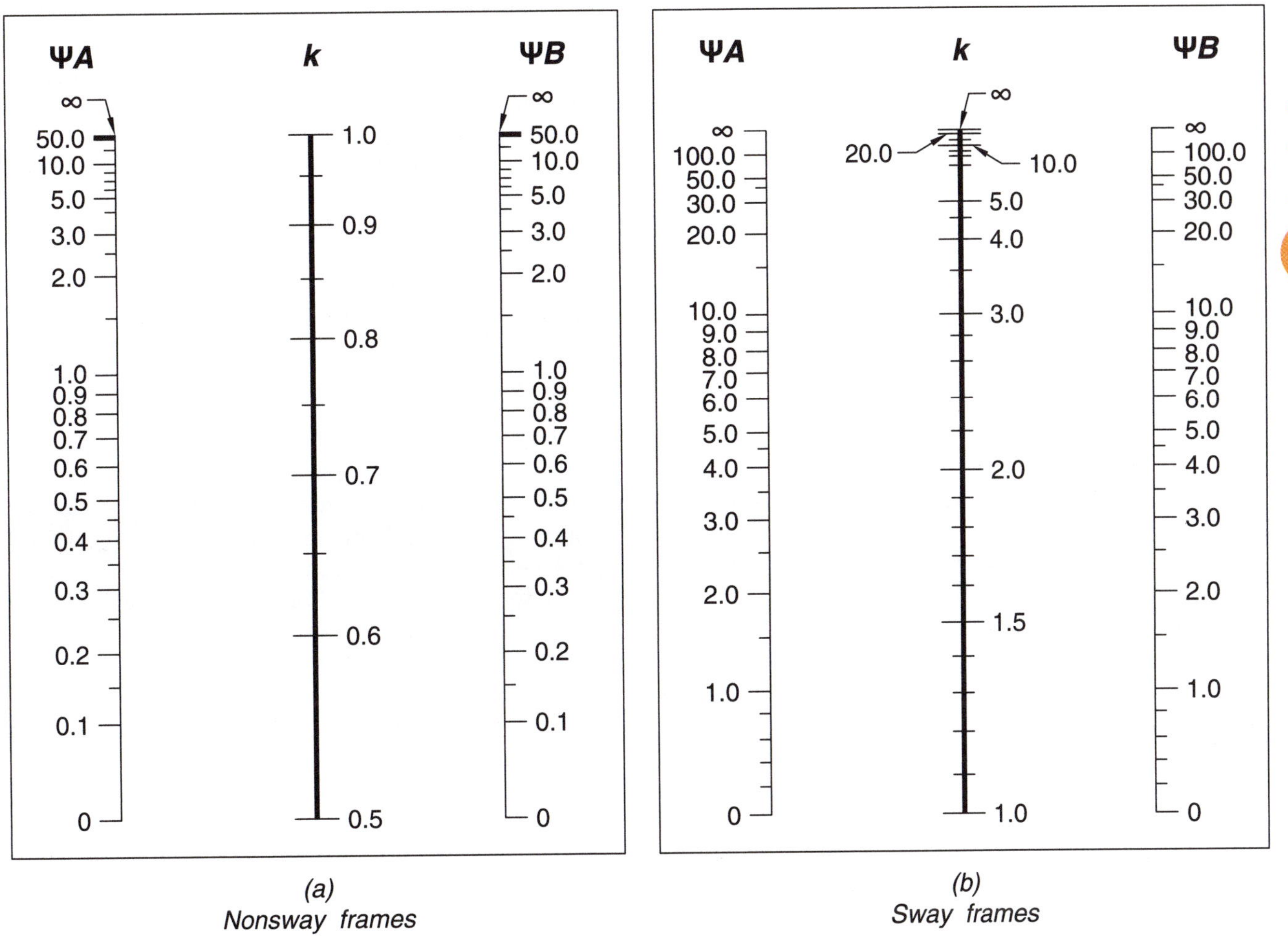

Ψ = ratio of $\sum(EI/\ell_c)$ of all columns to $\sum(EI/\ell)$ of beams in a plane at one end of a column

ℓ = span length of of beam measured center to center of joints

Fig. R6.2.5.1—Effective length factor **k**.

6.2.5.2 The radius of gyration, r, shall be permitted to be calculated by (a), (b), or (c):

(a) $$r = \sqrt{\frac{I_g}{A_g}} \quad (6.2.5.2)$$

(b) 0.30 times the dimension in the direction stability is being considered for rectangular columns

(c) 0.25 times the diameter of circular columns

CODE

6.2.5.3 Unless slenderness effects are neglected as permitted by 6.2.5.1, the design of columns, restraining beams, and other supporting members shall be based on the factored forces and moments considering second-order effects in accordance with 6.6.4, 6.7, or 6.8. M_u including second-order effects shall not exceed $1.4M_u$ due to first-order effects.

COMMENTARY

R6.2.5.3 Design considering second-order effects may be based on the moment magnifier approach (MacGregor et al. 1970; MacGregor 1993; Ford et al. 1981), an elastic second-order analysis, or a nonlinear second-order analysis. Figure R6.2.5.3 is intended to assist designers with application of the slenderness provisions of the Code.

End moments in compression members, such as columns, walls, or braces, should be considered in the design of adjacent flexural members. In nonsway frames, the effects of magnifying the end moments need not be considered in the design of adjacent beams. In sway frames, the magnified end moments should be considered in designing the adjoining flexural members.

Several methods have been developed to evaluate slenderness effects in compression members subject to biaxial bending. A review of some of these methods is presented in Furlong et al. (2004).

If the weight of a structure is high in proportion to its lateral stiffness, excessive $P\Delta$ effects, where secondary moments are more than 25 percent of the primary moments, may result. The $P\Delta$ effects will eventually introduce singularities into the solution to the equations of equilibrium, indicating physical structural instability (Wilson 1997). Analytical research (MacGregor and Hage 1977) on reinforced concrete frames showed that the probability of stability failure increases rapidly when the stability index Q, defined in 6.6.4.4.1, exceeds 0.2, which is equivalent to a secondary-to-primary moment ratio of 1.25. According to ASCE/SEI 7, the maximum value of the stability coefficient θ, which is close to the ACI stability coefficient Q, is 0.25. The value 0.25 is equivalent to a secondary-to-primary moment ratio of 1.33. Hence, the upper limit of 1.4 on the secondary-to-primary moment ratio was chosen.

CODE | COMMENTARY

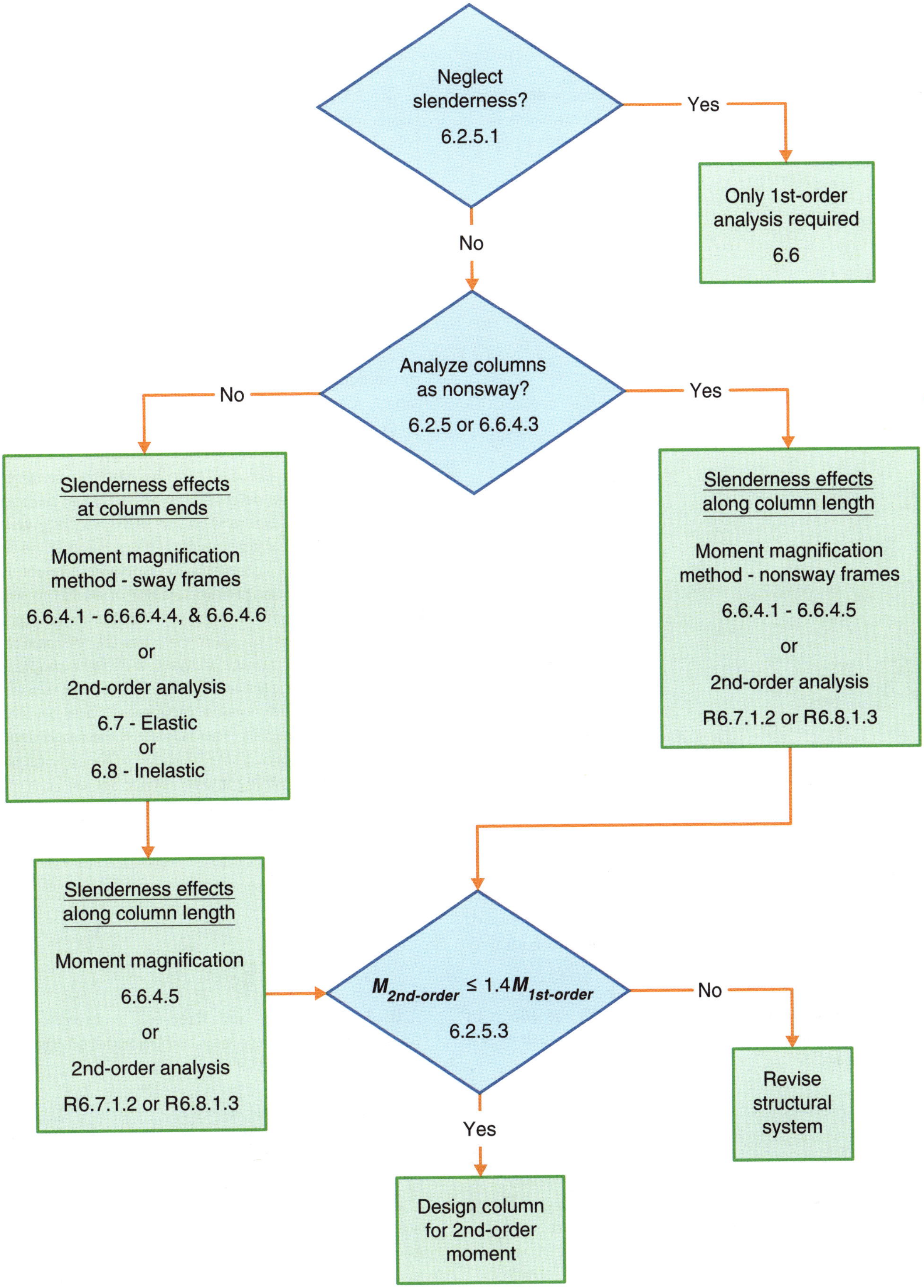

Fig. R6.2.5.3—Flowchart for determining column slenderness effects.

CODE

6.3—Modeling assumptions

6.3.1 *General*

6.3.1.1 Relative stiffnesses of members within structural systems shall be selected based on a reasonable set of assumptions. The assumptions shall be consistent throughout each analysis.

6.3.1.2 To calculate moments and shears caused by gravity loads in columns, beams, and slabs, it shall be permitted to use a model limited to the members in the level being considered and the columns above and below that level. It shall be permitted to assume far ends of columns built integrally with the structure to be fixed.

6.3.1.3 The analysis model shall consider the effects of variation of member cross-sectional properties, such as that due to haunches.

6.3.2 *T-beam geometry*

6.3.2.1 For nonprestressed T-beams supporting monolithic or composite slabs, the effective flange width b_f shall include the beam web width b_w plus an effective overhanging flange width in accordance with Table 6.3.2.1, where h is the slab thickness and s_w is the clear distance to the adjacent web.

COMMENTARY

R6.3—Modeling assumptions

R6.3.1 *General*

R6.3.1.1 Separate analyses with different stiffness assumptions may be performed for different objectives such as to check serviceability and strength criteria or to bound the demands on elements where stiffness assumptions are critical.

Ideally, the member stiffnesses E_cI and GJ should reflect the degree of cracking and inelastic action that has occurred along each member before yielding. However, the complexities involved in selecting different stiffnesses for all members of a frame would make frame analyses inefficient in the design process. Simpler assumptions are required to define flexural and torsional stiffnesses.

For braced frames, relative values of stiffness are important. A common assumption is to use $0.5I_g$ for beams and I_g for columns.

For sway frames, a realistic estimate of I is desirable and should be used if second-order analyses are performed. Guidance for the choice of I for this case is given in 6.6.3.1.

Two conditions determine whether it is necessary to consider torsional stiffness in the analysis of a given structure: 1) the relative magnitude of the torsional and flexural stiffnesses; and 2) whether torsion is required for equilibrium of the structure (equilibrium torsion) or is due to members twisting to maintain deformation compatibility (compatibility torsion). In the case of equilibrium torsion, torsional stiffness should be included in the analysis. It is, for example, necessary to consider the torsional stiffnesses of edge beams. In the case of compatibility torsion, torsional stiffness usually is not included in the analysis. This is because the cracked torsional stiffness of a beam is a small fraction of the flexural stiffness of the members framing into it. Torsion should be considered in design as required in Chapter 9.

R6.3.1.3 Stiffness and fixed-end moment coefficients for haunched members may be obtained from the Portland Cement Association (1972).

R6.3.2 *T-beam geometry*

R6.3.2.1 In ACI 318-11, the width of the slab effective as a T-beam flange was limited to one-fourth the span. The Code now allows one-eighth of the span on each side of the beam web. This was done to simplify Table 6.3.2.1 and has negligible impact on designs.

CODE

Table 6.3.2.1—Dimensional limits for effective overhanging flange width for T-beams

Flange location	Effective overhanging flange width, beyond face of web	
Each side of web	Least of:	$8h$
		$s_w/2$
		$\ell_n/8$
One side of web	Least of:	$6h$
		$s_w/2$
		$\ell_n/12$

6.3.2.2 Isolated nonprestressed T-beams in which the flange is used to provide additional compression area shall have a flange thickness greater than or equal to $\mathbf{0.5}b_w$ and an effective flange width less than or equal to $\mathbf{4}b_w$.

6.3.2.3 For prestressed T-beams, it shall be permitted to use the geometry provided by 6.3.2.1 and 6.3.2.2.

COMMENTARY

R6.3.2.3 The empirical provisions of 6.3.2.1 and 6.3.2.2 were developed for nonprestressed T-beams. The flange widths in 6.3.2.1 and 6.3.2.2 should be used unless experience has proven that variations are safe and satisfactory. Although many standard prestressed products in use today do not satisfy the effective flange width requirements of 6.3.2.1 and 6.3.2.2, they demonstrate satisfactory performance. Therefore, determination of an effective flange width for prestressed T-beams is left to the experience and judgment of the licensed design professional. It is not always considered conservative in elastic analysis and design considerations to use the maximum flange width as permitted in 6.3.2.1.

6.4—Arrangement of live load

6.4.1 For the design of floors or roofs to resist gravity loads, it shall be permitted to assume that live load is applied only to the level under consideration.

6.4.2 For one-way slabs and beams, it shall be permitted to assume (a) and (b):

(a) Maximum positive M_u near midspan occurs with factored L on the span and on alternate spans

(b) Maximum negative M_u at a support occurs with factored L on adjacent spans only

6.4.3 For two-way slab systems, factored moments shall be calculated in accordance with 6.4.3.1, 6.4.3.2, or 6.4.3.3, and shall be at least the moments resulting from factored L applied simultaneously to all panels.

6.4.3.1 If the arrangement of L is known, the slab system shall be analyzed for that arrangement.

6.4.3.2 If L is variable and does not exceed $\mathbf{0.75}D$, or the nature of L is such that all panels will be loaded simultaneously, it shall be permitted to assume that maximum M_u at

R6.4—Arrangement of live load

R6.4.2 The most demanding sets of design forces should be established by investigating the effects of live load placed in various critical patterns.

CODE

all sections occurs with factored L applied simultaneously to all panels.

6.4.3.3 For loading conditions other than those defined in 6.4.3.1 or 6.4.3.2, it shall be permitted to assume (a) and (b):

(a) Maximum positive M_u near midspan of panel occurs with 75 percent of factored L on the panel and alternate panels
(b) Maximum negative M_u at a support occurs with 75 percent of factored L on adjacent panels only

6.5—Simplified method of analysis for nonprestressed continuous beams and one-way slabs

6.5.1 It shall be permitted to calculate M_u and V_u due to gravity loads in accordance with this section for continuous beams and one-way slabs satisfying (a) through (e):

(a) Members are prismatic
(b) Loads are uniformly distributed
(c) $L \leq 3D$
(d) There are at least two spans
(e) The longer of two adjacent spans does not exceed the shorter by more than 20 percent

6.5.2 M_u due to gravity loads shall be calculated in accordance with Table 6.5.2.

COMMENTARY

R6.4.3.3 The use of only 75 percent of the full factored live load for maximum moment loading patterns is based on the fact that maximum negative and maximum positive live load moments cannot occur simultaneously and that redistribution of maximum moments is thus possible before failure occurs. This procedure, in effect, permits some local overstress under the full factored live load if it is distributed in the prescribed manner, but still ensures that the design strength of the slab system after redistribution of moment is not less than that required to resist the full factored dead and live loads on all panels.

R6.5—Simplified method of analysis for nonprestressed continuous beams and one-way slabs

R6.5.2 The approximate moments and shears give reasonable values for the stated conditions if the continuous beams and one-way slabs are part of a frame or continuous construction. Because the load patterns that produce critical values for moments in columns of frames differ from those for maximum negative moments in beams, column moments should be evaluated separately.

Table 6.5.2—Approximate moments for nonprestressed continuous beams and one-way slabs

Moment	Location	Condition	M_u
Positive	End span	Discontinuous end integral with support	$w_u\ell_n^2/14$
		Discontinuous end unrestrained	$w_u\ell_n^2/11$
	Interior spans	All	$w_u\ell_n^2/16$
Negative[1]	Interior face of exterior support	Member built integrally with supporting spandrel beam	$w_u\ell_n^2/24$
		Member built integrally with supporting column	$w_u\ell_n^2/16$
	Exterior face of first interior support	Two spans	$w_u\ell_n^2/9$
		More than two spans	$w_u\ell_n^2/10$
	Face of other supports	All	$w_u\ell_n^2/11$
	Face of all supports satisfying (a) or (b)	(a) slabs with spans not exceeding 10 ft (b) beams where ratio of sum of column stiffnesses to beam stiffness exceeds 8 at each end of span	$w_u\ell_n^2/12$

[1]To calculate negative moments, ℓ_n shall be the average of the adjacent clear span lengths.

CODE

6.5.3 Moments calculated in accordance with 6.5.2 shall not be redistributed.

6.5.4 V_u due to gravity loads shall be calculated in accordance with Table 6.5.4.

Table 6.5.4—Approximate shears for nonprestressed continuous beams and one-way slabs

Location	V_u
Exterior face of first interior support	$1.15w_u\ell_n/2$
Face of all other supports	$w_u\ell_n/2$

6.5.5 Floor or roof level moments shall be resisted by distributing the moment between columns immediately above and below the given floor in proportion to the relative column stiffnesses considering conditions of restraint.

COMMENTARY

R6.5.5 This section is provided to make certain that moments are included in column design. The moment refers to the difference between the end moments of the members framing into the column and exerted at the column centerline.

6.6—Linear elastic first-order analysis

6.6.1 *General*

6.6.1.1 Slenderness effects shall be considered in accordance with 6.6.4, unless they are allowed to be neglected by 6.2.5.1.

6.6.1.2 Redistribution of moments calculated by an elastic first-order analysis shall be permitted in accordance with 6.6.5.

R6.6—Linear elastic first-order analysis

R6.6.1 *General*

R6.6.1.1 When using linear elastic first-order analysis, slenderness effects are calculated using the moment magnifier approach (MacGregor et al. 1970; MacGregor 1993; Ford et al. 1981).

6.6.2 *Modeling of members and structural systems*

6.6.2.1 Floor or roof level moments shall be resisted by distributing the moment between columns immediately above and below the given floor in proportion to the relative column stiffnesses and considering conditions of restraint.

6.6.2.2 For frames or continuous construction, consideration shall be given to the effect of floor and roof load patterns on transfer of moment to exterior and interior columns, and of eccentric loading due to other causes.

6.6.2.3 It shall be permitted to simplify the analysis model by the assumptions of (a), (b), or both:

(a) Solid slabs or one-way joist systems built integrally with supports, with clear spans not more than 10 ft, shall be permitted to be analyzed as continuous members on knife-edge supports with spans equal to the clear spans of the member and width of support beams otherwise neglected.

R6.6.2 *Modeling of members and structural systems*

R6.6.2.1 This section is provided to make certain that moments are included in column design if members have been proportioned using 6.5.1 and 6.5.2. The moment refers to the difference between the end moments of the members framing into the column and exerted at the column centerline.

R6.6.2.3 A common feature of modern frame analysis software is the assumption of rigid connections. Section 6.6.2.3(b) is intended to apply to intersecting elements in frames, such as beam-column joints.

CODE

(b) For frames or continuous construction, it shall be permitted to assume the intersecting member regions are rigid.

6.6.3 *Section properties*

6.6.3.1 *Factored load analysis*

6.6.3.1.1 Moment of inertia and cross-sectional area of members shall be calculated in accordance with Tables 6.6.3.1.1(a) or 6.6.3.1.1(b), unless a more rigorous analysis is used. If sustained lateral loads are present, I for columns and walls shall be divided by $(1 + \beta_{ds})$, where β_{ds} is the ratio of maximum factored sustained shear within a story to the maximum factored shear in that story associated with the same load combination.

Table 6.6.3.1.1(a)—Moments of inertia and cross-sectional areas permitted for elastic analysis at factored load level

Member and condition		Moment of inertia	Cross-sectional area for axial deformations	Cross-sectional area for shear deformations
Columns		$0.70I_g$	$1.0A_g$	b_wh
Walls	Uncracked	$0.70I_g$		
	Cracked	$0.35I_g$		
Beams		$0.35I_g$		
Flat plates and flat slabs		$0.25I_g$		

COMMENTARY

R6.6.3 *Section properties*

R6.6.3.1 *Factored load analysis*

For lateral load analysis, either the stiffnesses presented in 6.6.3.1.1 or 6.6.3.1.2 can be used. These provisions both use values that approximate the stiffness for reinforced concrete building systems loaded to near or beyond the yield level, and have been shown to produce reasonable correlation with both experimental and detailed analytical results (Moehle 1992; Lepage 1998). For earthquake-induced loading, the use of 6.6.3.1.1 or 6.6.3.1.2 may require a deflection amplification factor to account for inelastic deformations. In general, for effective section properties, E_c may be calculated or specified in accordance with 19.2.2, the shear modulus may be taken as $0.4E_c$, and areas may be taken as given in Table 6.6.3.1.1(a).

R6.6.3.1.1 The values of I and A have been chosen from the results of frame tests and analyses, and include an allowance for the variability of the calculated deflections. The moments of inertia are taken from MacGregor and Hage (1977), which are multiplied by a stiffness reduction factor $\phi_K = 0.875$ (refer to R6.6.4.5.2). For example, the moment of inertia for columns is $0.875(0.80I_g) = 0.70I_g$.

The moment of inertia of T-beams should be based on the effective flange width defined in 6.3.2.1 or 6.3.2.2. It is generally sufficiently accurate to take I_g of a T-beam as $2I_g$ for the web, $2(b_wh^3/12)$.

If the factored moments and shears from an analysis based on the moment of inertia of a wall, taken equal to $0.70I_g$, indicate that the wall will crack in flexure, based on the modulus of rupture, the analysis should be repeated with $I = 0.35I_g$ in those stories where cracking is predicted using factored loads.

The values of the moments of inertia were derived for nonprestressed members. For prestressed members, the moments of inertia may differ depending on the amount, location, and type of reinforcement, and the degree of cracking prior to reaching ultimate load. The stiffness values for prestressed concrete members should include an allowance for the variability of the stiffnesses.

The equations in Table 6.6.3.1.1(b) provide more refined values of I considering axial load, eccentricity, reinforcement ratio, and concrete compressive strength as presented in Khuntia and Ghosh (2004a,b). The stiffnesses provided in these references are applicable for all levels of loading, including service and ultimate, and consider a stiffness reduction factor ϕ_K comparable to that for the moment of inertias included in Table 6.6.3.1.1(a). For use at load levels

CODE

Table 6.6.3.1.1(b)—Alternative moments of inertia for elastic analysis at factored load

Member	Alternative value of I for elastic analysis		
	Minimum	I	Maximum
Columns and walls	$0.35I_g$	$\left(0.80+25\frac{A_{st}}{A_g}\right)\left(1-\frac{M_u}{P_uh}-0.5\frac{P_u}{P_o}\right)I_g$	$0.875I_g$
Beams, flat plates, and flat slabs	$0.25I_g$	$(0.10+25\rho)\left(1.2-0.2\frac{b_w}{d}\right)I_g$	$0.5I_g$

Notes: For continuous flexural members, I shall be permitted to be taken as the average of values obtained for the critical positive and negative moment sections. P_u and M_u shall be calculated from the load combination under consideration, or the combination of P_u and M_u that produces the least value of I.

COMMENTARY

other than ultimate, P_u and M_u should be replaced with their appropriate values at the desired load level.

6.6.3.1.2 For factored lateral load analysis, it shall be permitted to assume $I = 0.5I_g$ for all members or to calculate I by a more detailed analysis, considering the effective stiffness of all members under the loading conditions.

R6.6.3.1.2 The lateral deflection of a structure under factored lateral loads can be substantially different from that calculated using linear analysis, in part because of the inelastic response of the members and the decrease in effective stiffness. Selection of the appropriate effective stiffness for reinforced concrete frame members has dual purposes: 1) to provide realistic estimates of lateral deflection; and 2) to determine deflection-imposed actions on the gravity system of the structure. A detailed nonlinear analysis of the structure would adequately capture these two effects. A simple way to estimate an equivalent nonlinear lateral deflection using linear analysis is to reduce the modeled stiffness of the concrete members in the structure. The type of lateral load analysis affects the selection of appropriate effective stiffness values. For analyses with wind loading, where it is desirable to prevent nonlinear action in the structure, effective stiffnesses representative of pre-yield behavior may be appropriate. For earthquake-induced loading, the level of nonlinear deformation depends on the intended structural performance and earthquake recurrence interval.

Varying degrees of confidence can be obtained from a simple linear analysis based on the computational rigor used to define the effective stiffness of each member. This stiffness can be based on the secant stiffness to a point at or beyond yield or, if yielding is not expected, to a point before yield occurs.

6.6.3.1.3 For factored lateral load analysis of two-way slab systems without beams, which are designated as part of the seismic-force-resisting system, I for slab members shall be defined by a model that is in substantial agreement with results of comprehensive tests and analysis and I of other frame members shall be in accordance with 6.6.3.1.1 and 6.6.3.1.2.

R6.6.3.1.3 Analysis of buildings with two-way slab systems without beams requires that the model represents the transfer of lateral loads between vertical members. The model should result in prediction of stiffness in substantial agreement with results of comprehensive tests and analysis. Several acceptable models have been proposed to accomplish this objective (Vanderbilt and Corley 1983; Hwang and Moehle 2000; Dovich and Wight 2005).

6.6.3.2 *Service load analysis*

R6.6.3.2 *Service load analysis*

6.6.3.2.1 Immediate and time-dependent deflections due to gravity loads shall be calculated in accordance with 24.2.

6 Analysis

CODE

6.6.3.2.2 It shall be permitted to calculate immediate lateral deflections using a moment of inertia of 1.4 times $\boldsymbol{I}$ defined in 6.6.3.1, or using a more detailed analysis, but the value shall not exceed $\boldsymbol{I_g}$.

6.6.4 *Slenderness effects, moment magnification method*

6.6.4.1 Unless 6.2.5.1 is satisfied, columns and stories in structures shall be designated as being nonsway or sway. Analysis of columns in nonsway frames or stories shall be in accordance with 6.6.4.5. Analysis of columns in sway frames or stories shall be in accordance with 6.6.4.6.

6.6.4.2 The cross-sectional dimensions of each member used in an analysis shall be within 10 percent of the specified member dimensions in construction documents or the analysis shall be repeated. If the stiffnesses of Table 6.6.3.1.1(b) are used in an analysis, the assumed member reinforcement ratio shall also be within 10 percent of the specified member reinforcement in construction documents.

6.6.4.3 It shall be permitted to analyze columns and stories in structures as nonsway frames if (a) or (b) is satisfied:

COMMENTARY

R6.6.3.2.2 Analyses of deflections, vibrations, and building periods are needed at various service (unfactored) load levels (Grossman 1987, 1990) to determine the performance of the structure in service. The moments of inertia of the structural members in the service load analyses should be representative of the degree of cracking at the various service load levels investigated. Unless a more accurate estimate of the degree of cracking at service load level is available, it is satisfactory to use 1.0/0.70 = 1.4 times the moments of inertia provided in 6.6.3.1, not to exceed $\boldsymbol{I_g}$, for service load analyses. Serviceability considerations for vibrations are discussed in R24.1.

R6.6.4 *Slenderness effects, moment magnification method*

R6.6.4.1 This section describes an approximate design procedure that uses the moment magnifier concept to account for slenderness effects. Moments calculated using a first-order frame analysis are multiplied by a moment magnifier that is a function of the factored axial load $\boldsymbol{P_u}$ and the critical buckling load $\boldsymbol{P_c}$ for the column. For the sway case, the moment magnifier is a function of the sum of $\boldsymbol{P_u}$ of the story and the sum of $\boldsymbol{P_c}$ of the sway-resisting columns in the story considered. Nonsway and sway frames are treated separately. A first-order frame analysis is an elastic analysis that excludes the internal force effects resulting from deflections.

The moment magnifier design method requires the designer to distinguish between nonsway frames, which are designed according to 6.6.4.5, and sway frames, which are designed according to 6.6.4.6. Frequently this can be done by comparing the total lateral stiffness of the columns in a story to that of the bracing elements. A compression member, such as a column, wall, or brace, may be assumed nonsway if it is located in a story in which the bracing elements (structural walls, shear trusses, or other types of lateral bracing) have such substantial lateral stiffness to resist the lateral deflections of the story that any resulting lateral deflection is not large enough to affect the column strength substantially. If not readily apparent without calculations, 6.6.4.3 provides two possible ways of determining if sway can be neglected.

R6.6.4.3 In 6.6.4.3(a), a story in a frame is classified as nonsway if the increase in the lateral load moments resulting from $\boldsymbol{P\Delta}$ effects does not exceed 5 percent of the first-order moments (MacGregor and Hage 1977). Section 6.6.4.3(b) provides an alternative method of determining if a frame is

CODE

(a) The increase in column end moments due to second-order effects does not exceed 5 percent of the first-order end moments

(b) $\boldsymbol{Q}$ in accordance with 6.6.4.4.1 does not exceed 0.05

6.6.4.4 *Stability properties*

6.6.4.4.1 The stability index for a story, $\boldsymbol{Q}$, shall be calculated by:

$$Q = \frac{\Sigma P_u \Delta_o}{V_{us} \ell_c} \qquad (6.6.4.4.1)$$

where $\sum \boldsymbol{P_u}$ and $\boldsymbol{V_{us}}$ are the total factored vertical load and horizontal story shear, respectively, in the story being evaluated, and Δ_o is the first-order relative lateral deflection between the top and the bottom of that story due to $\boldsymbol{V_{us}}$.

6.6.4.4.2 The critical buckling load $\boldsymbol{P_c}$ shall be calculated by:

$$P_c = \frac{\pi^2 (EI)_{eff}}{(k\ell_u)^2} \qquad (6.6.4.4.2)$$

6.6.4.4.3 The effective length factor $\boldsymbol{k}$ shall be calculated using $\boldsymbol{E_c}$ in accordance with 19.2.2 and $\boldsymbol{I}$ in accordance with 6.6.3.1.1. For nonsway members, $\boldsymbol{k}$ shall be permitted to be taken as 1.0, and for sway members, $\boldsymbol{k}$ shall be at least 1.0.

6.6.4.4.4 For columns, $\boldsymbol{(EI)_{eff}}$ shall be calculated in accordance with (a), (b), or (c):

(a) $$(EI)_{eff} = \frac{0.4E_c I_g}{1+\beta_{dns}} \qquad (6.6.4.4.4a)$$

(b) $$(EI)_{eff} = \frac{(0.2E_c I_g + E_s I_{se})}{1+\beta_{dns}} \qquad (6.6.4.4.4b)$$

(c) $$(EI)_{eff} = \frac{E_c I}{1+\beta_{dns}} \qquad (6.6.4.4.4c)$$

COMMENTARY

classified as nonsway based on the stability index for a story, $\boldsymbol{Q}$. In calculating $\boldsymbol{Q}$, $\sum \boldsymbol{P_u}$ should correspond to the lateral loading case for which $\sum \boldsymbol{P_u}$ is greatest. A frame may contain both nonsway and sway stories.

If the lateral load deflections of the frame are calculated using service loads and the service load moments of inertia given in 6.6.3.2.2, it is permissible to calculate $\boldsymbol{Q}$ in Eq. (6.6.4.4.1) using 1.2 times the sum of the service gravity loads, the service load story shear, and 1.4 times the first-order service load story deflections.

R6.6.4.4 *Stability properties*

R6.6.4.4.2 In calculating the critical axial buckling load, the primary concern is the choice of a stiffness $\boldsymbol{(EI)_{eff}}$ that reasonably approximates the variations in stiffness due to cracking, creep, and nonlinearity of the concrete stress-strain curve. Section 6.6.4.4.4 may be used to calculate $\boldsymbol{(EI)_{eff}}$.

R6.6.4.4.3 The effective length factor for a compression member, such as a column, wall, or brace, considering braced behavior, ranges from 0.5 to 1.0. It is recommended that a $\boldsymbol{k}$ value of 1.0 be used. If lower values are used, the calculation of $\boldsymbol{k}$ should be based on analysis of the frame using $\boldsymbol{I}$ values given in 6.6.3.1.1. The Jackson and Moreland Alignment Charts (Fig. R6.2.5.1) can be used to estimate appropriate values of $\boldsymbol{k}$ (ACI SP-17(09); Column Research Council 1966).

R6.6.4.4.4 The numerators of Eq. (6.6.4.4.4a) to (6.6.4.4.4c) represent the short-term column stiffness. Equation (6.6.4.4.4b) was derived for small eccentricity ratios and high levels of axial load. Equation (6.6.4.4.4a) is a simplified approximation to Eq. (6.6.4.4.4b) and is less accurate (Mirza 1990). For improved accuracy, $\boldsymbol{(EI)_{eff}}$ can be approximated using Eq. (6.6.4.4.4c).

Creep due to sustained loads will increase the lateral deflections of a column and, hence, the moment magnification. Creep effects are approximated in design by reducing the stiffness $\boldsymbol{(EI)_{eff}}$ used to calculate $\boldsymbol{P_c}$ and, hence, $\boldsymbol{\delta}$, by dividing the short-term $\boldsymbol{EI}$ provided by the numerator of Eq. (6.6.4.4.4a) through (6.6.4.4.4c) by $\boldsymbol{(1 + \beta_{dns})}$. For

CODE

where β_{dns} shall be the ratio of maximum factored sustained axial load to maximum factored axial load associated with the same load combination and I in Eq. (6.6.4.4.4c) is calculated according to Table 6.6.3.1.1(b) for columns and walls.

COMMENTARY

simplification, it can be assumed that β_{dns} = **0.6**. In this case, Eq. (6.6.4.4.4a) becomes $(EI)_{eff} = 0.25E_cI_g$.

In reinforced concrete columns subject to sustained loads, creep transfers some of the load from the concrete to the longitudinal reinforcement, increasing the reinforcement stresses. In the case of lightly reinforced columns, this load transfer may cause the compression reinforcement to yield prematurely, resulting in a loss in the effective EI. Accordingly, both the concrete and longitudinal reinforcement terms in Eq. (6.6.4.4.4b) are reduced to account for creep.

6.6.4.5 *Moment magnification method: Nonsway frames*

6.6.4.5.1 The factored moment used for design of columns and walls, M_c, shall be the first-order factored moment M_2 amplified for the effects of member curvature.

$$M_c = \delta M_2 \quad (6.6.4.5.1)$$

6.6.4.5.2 Magnification factor δ shall be calculated by:

$$\delta = \frac{C_m}{1 - \frac{P_u}{0.75P_c}} \geq 1.0 \quad (6.6.4.5.2)$$

6.6.4.5.3 C_m shall be in accordance with (a) or (b):

(a) For columns without transverse loads applied between supports

$$C_m = 0.6 - 0.4\frac{M_1}{M_2} \quad (6.6.4.5.3a)$$

where M_1/M_2 is negative if the column is bent in single curvature, and positive if bent in double curvature. M_1 corresponds to the end moment with the lesser absolute value.

(b) For columns with transverse loads applied between supports.

$$C_m = 1.0 \quad (6.6.4.5.3b)$$

R6.6.4.5 *Moment magnification method: Nonsway frames*

R6.6.4.5.2 The 0.75 factor in Eq. (6.6.4.5.2) is the stiffness reduction factor ϕ_K, which is based on the probability of understrength of a single isolated slender column. Studies reported in Mirza et al. (1987) indicate that the stiffness reduction factor ϕ_K and the cross-sectional strength reduction ϕ factors do not have the same values. These studies suggest the stiffness reduction factor ϕ_K for an isolated column should be 0.75 for both tied and spiral columns. In the case of a multistory frame, the column and frame deflections depend on the average concrete strength, which is higher than the strength of the concrete in the critical single understrength column. For this reason, the value of ϕ_K implicit in I values in 6.6.3.1.1 is 0.875.

R6.6.4.5.3 The factor C_m is a correction factor relating the actual moment diagram to an equivalent uniform moment diagram. The derivation of the moment magnifier assumes that the maximum moment is at or near midheight of the column. If the maximum moment occurs at one end of the column, design should be based on an equivalent uniform moment C_mM_2 that leads to the same maximum moment at or near midheight of the column when magnified (MacGregor et al. 1970).

The sign convention for M_1/M_2 has been updated to follow the right hand rule convention; hence, M_1/M_2 is negative if bent in single curvature and positive if bent in double curvature. This reflects a sign convention change from the 2011 Code.

In the case of columns that are subjected to transverse loading between supports, it is possible that the maximum moment will occur at a section away from the end of the member. If this occurs, the value of the largest calculated moment occurring anywhere along the member should be used for the value of M_2 in Eq. (6.6.4.5.1). C_m is to be taken as 1.0 for this case.

CODE

6.6.4.5.4 M_2 in Eq. (6.6.4.5.1) shall be at least $M_{2,min}$ calculated according to Eq. (6.6.4.5.4) about each axis separately.

$$M_{2,min} = P_u(0.6 + 0.03h) \quad (6.6.4.5.4)$$

If $M_{2,min}$ exceeds M_2, C_m shall be taken equal to 1.0 or calculated based on the ratio of the calculated end moments M_1/M_2, using Eq. (6.6.4.5.3a).

COMMENTARY

R6.6.4.5.4 In the Code, slenderness is accounted for by magnifying the column end moments. If the factored column moments are small or zero, the design of slender columns should be based on the minimum eccentricity provided in Eq. (6.6.4.5.4). It is not intended that the minimum eccentricity be applied about both axes simultaneously.

The factored column end moments from the structural analysis are used in Eq. (6.6.4.5.3a) in determining the ratio M_1/M_2 for the column when the design is based on the minimum eccentricity. This eliminates what would otherwise be a discontinuity between columns with calculated eccentricities less than the minimum eccentricity and columns with calculated eccentricities equal to or greater than the minimum eccentricity.

CODE

6.6.4.6 *Moment magnification method: Sway frames*

6.6.4.6.1 Moments M_1 and M_2 at the ends of an individual column shall be calculated by (a) and (b).

(a) $M_1 = M_{1ns} + \delta_s M_{1s}$ (6.6.4.6.1a)

(b) $M_2 = M_{2ns} + \delta_s M_{2s}$ (6.6.4.6.1b)

6.6.4.6.2 The moment magnifier δ_s shall be calculated by (a), (b), or (c). If δ_s exceeds 1.5, only (b) or (c) shall be permitted:

(a) $\delta_s = \dfrac{1}{1-Q} \geq 1$ (6.6.4.6.2a)

(b) $\delta_s = \dfrac{1}{1-\dfrac{\Sigma P_u}{0.75\Sigma P_c}} \geq 1$ (6.6.4.6.2b)

(c) Second-order elastic analysis

where $\sum P_u$ is the summation of all the factored vertical loads in a story and $\sum P_c$ is the summation for all sway-resisting columns in a story. P_c is calculated using Eq. (6.6.4.4.2) with k determined for sway members from 6.6.4.4.3 and $(EI)_{eff}$ from 6.6.4.4.4 with β_{ds} substituted for β_{dns}.

COMMENTARY

R6.6.4.6 *Moment magnification method: Sway frames*

R6.6.4.6.1 The analysis described in this section deals only with plane frames subjected to loads causing deflections in that plane. If the lateral load deflections involve significant torsional displacement, the moment magnification in the columns farthest from the center of twist may be underestimated by the moment magnifier procedure. In such cases, a three-dimensional second-order analysis should be used.

R6.6.4.6.2 Three different methods are allowed for calculating the moment magnifier. These approaches include the Q method, the sum of P concept, and second-order elastic analysis.

(a) Q method:

The iterative $P\Delta$ analysis for second-order moments can be represented by an infinite series. The solution of this series is given by Eq. (6.6.4.6.2a) (MacGregor and Hage 1977). Lai and MacGregor (1983) show that Eq. (6.6.4.6.2a) closely predicts the second-order moments in a sway frame until δ_s exceeds 1.5.

The $P\Delta$ moment diagrams for deflected columns are curved, with Δ related to the deflected shape of the columns. Equation (6.6.4.6.2a) and most commercially available second-order frame analyses have been derived assuming that the $P\Delta$ moments result from equal and opposite forces of $P\Delta/\ell_c$ applied at the bottom and top of the story. These forces give a straight-line $P\Delta$ moment diagram. The curved $P\Delta$ moment diagrams lead to lateral displacements on the order of 15 percent larger than those from the straight-line $P\Delta$ moment diagrams. This effect can be included in Eq. (6.6.4.6.2a) by writing the denominator as $(1 - 1.15Q)$ rather than $(1 - Q)$. The 1.15 factor has been omitted from Eq. (6.6.4.6.2a) for simplicity.

If deflections have been calculated using service loads, Q in Eq. (6.6.4.6.2a) should be calculated in the manner explained in R6.6.4.3.

The Q factor analysis is based on deflections calculated using the I values from 6.6.3.1.1, which include the equivalent

6 Analysis

CODE | COMMENTARY

of a stiffness reduction factor ϕ_K. These I values lead to a 20 to 25 percent over-estimation of the lateral deflections that corresponds to a stiffness reduction factor ϕ_K between 0.80 and 0.85 on the $P\Delta$ moments. As a result, no additional ϕ factor is needed. Once the moments are established using Eq. (6.6.4.6.2a), selection of the cross sections of the columns involves the strength reduction factors ϕ from 21.2.2.

(b) Sum of P concept:

To check the effects of story stability, δ_s is calculated as an averaged value for the entire story based on use of $\sum P_u/\sum P_c$. This reflects the interaction of all sway-resisting columns in the story on the $P\Delta$ effects because the lateral deflection of all columns in the story should be equal in the absence of torsional displacements about a vertical axis. In addition, it is possible that a particularly slender individual column in a sway frame could have substantial midheight deflections, even if adequately braced against lateral end deflections by other columns in the story. Such a column is checked using 6.6.4.6.4.

The 0.75 in the denominator of Eq. (6.6.4.6.2b) is a stiffness reduction factor ϕ_K, as explained in R6.6.4.5.2.

In the calculation of $(EI)_{eff}$, β_{ds} will normally be zero for a sway frame because the lateral loads are generally of short duration. Sway deflections due to short-term loads, such as wind or earthquake, are a function of the short-term stiffness of the columns following a period of sustained gravity load.

For this case, the definition of β_{ds} in 6.6.3.1.1 gives $\beta_{ds} = 0$. In the unusual case of a sway frame where the lateral loads are sustained, β_{ds} will not be zero. This might occur if a building on a sloping site is subjected to earth pressure on one side but not on the other.

6.6.4.6.3 Flexural members shall be designed for the total magnified end moments of the columns at the joint.

R6.6.4.6.3 The strength of a sway frame is governed by stability of the columns and the degree of end restraint provided by the beams in the frame. If plastic hinges form in the restraining beam, as the structure approaches a failure mechanism, its axial strength is drastically reduced. This section requires the restraining flexural members to have enough strength to resist the total magnified column end moments at the joint.

6.6.4.6.4 Second-order effects shall be considered along the length of columns in sway frames. It shall be permitted to account for these effects using 6.6.4.5, where C_m is calculated using M_1 and M_2 from 6.6.4.6.1.

R6.6.4.6.4 The maximum moment in a compression member, such as a column, wall, or brace, may occur between its ends. While second-order computer analysis programs may be used to evaluate magnification of the end moments, magnification between the ends may not be accounted for unless the member is subdivided along its length. The magnification may be evaluated using the procedure outlined in 6.6.4.5.

6.6.5 *Redistribution of moments in continuous flexural members*

6.6.5.1 Except where approximate values for moments are used in accordance with 6.5, where moments have been

R6.6.5 *Redistribution of moments in continuous flexural members*

Redistribution of moments is dependent on adequate ductility in plastic hinge regions. These plastic hinge regions

CODE

calculated in accordance with 6.8, or where moments in two-way slabs are determined using pattern loading specified in 6.4.3.3, reduction of moments at sections of maximum negative or maximum positive moment calculated by elastic theory shall be permitted for any assumed loading arrangement if (a) and (b) are satisfied:

(a) Flexural members are continuous
(b) $\varepsilon_t \geq \mathbf{0.0075}$ at the section at which moment is reduced

6.6.5.2 For prestressed members, moments include those due to factored loads and those due to reactions induced by prestressing.

6.6.5.3 At the section where the moment is reduced, redistribution shall not exceed the lesser of $\mathbf{1000}\varepsilon_t$ percent and 20 percent.

6.6.5.4 The reduced moment shall be used to calculate redistributed moments at all other sections within the spans such that static equilibrium is maintained after redistribution of moments for each loading arrangement.

6.6.5.5 Shears and support reactions shall be calculated in accordance with static equilibrium considering the redistributed moments for each loading arrangement.

COMMENTARY

develop at sections of maximum positive or negative moment and cause a shift in the elastic moment diagram. The usual result is a reduction in the values of maximum negative moments in the support regions and an increase in the values of positive moments between supports from those calculated by elastic analysis. However, because negative moments are typically determined for one loading arrangement and positive moments for another (6.4.3 provides an exception for certain loading conditions), economies in reinforcement can sometimes be realized by reducing maximum elastic positive moments and increasing negative moments, thus narrowing the envelope of maximum negative and positive moments at any section in the span (Bondy 2003). Plastic hinges permit utilization of the full capacity of more cross sections of a flexural member at ultimate loads.

The Code permissible redistribution is shown in Fig. R6.6.5. Using conservative values of limiting concrete strains and lengths of plastic hinges derived from extensive tests, flexural members with small rotation capacities were analyzed for redistribution of moments up to 20 percent, depending on the reinforcement ratio. As shown, the permissible redistribution percentages are conservative relative to the calculated percentages available for both f_y **= 60 ksi and 80 ksi**. Studies by Cohn (1965) and Mattock (1959) support this conclusion and indicate that cracking and deflection of beams designed for redistribution of moments are not significantly greater at service loads than for beams designed by the distribution of moments according to elastic theory. Also, these studies indicate that adequate rotational capacity for the redistribution of moments allowed by the Code is available if the members satisfy 6.6.5.1.

The provisions for redistribution of moments apply equally to prestressed members (Mast 1992).

The elastic deformations caused by a nonconcordant tendon change the amount of inelastic rotation required to obtain a given amount of redistribution of moments. Conversely, for a beam with a given inelastic rotational capacity, the amount by which the moment at the support may be varied is changed by an amount equal to the secondary moment at the support due to prestressing. Thus, the Code requires that secondary moments caused by reactions generated by prestressing forces be included in determining design moments.

Redistribution of moments as permitted by 6.6.5 is not appropriate where approximate values of bending moments are used, such as provided by the simplified method of 6.5.

Redistribution of moments is also not appropriate for two-way slab systems that are analyzed using the pattern loadings given in 6.4.3.3. These loadings use only 75 percent of the full factored live load, which is based on considerations of moment redistribution.

CODE

COMMENTARY

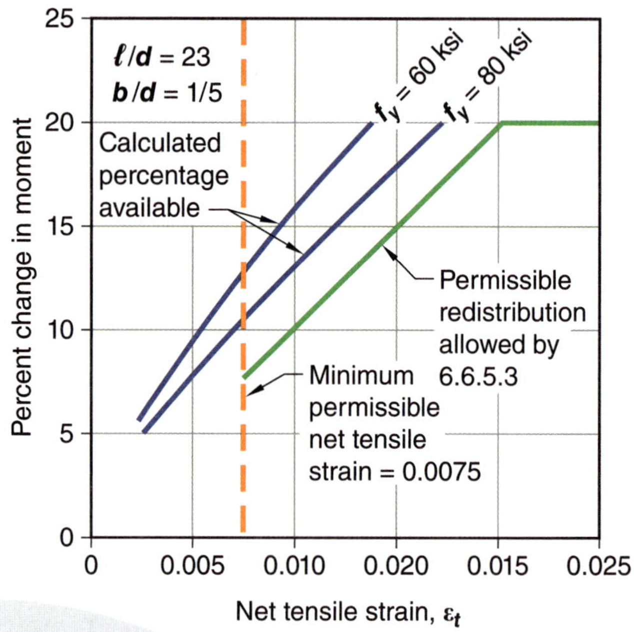

Fig. R6.6.5—Permissible redistribution of moments for minimum rotation capacity.

6.7—Linear elastic second-order analysis

6.7.1 *General*

6.7.1.1 A linear elastic second-order analysis shall consider the influence of axial loads, presence of cracked regions along the length of the member, and effects of load duration. These considerations are satisfied using the cross-sectional properties defined in 6.7.2.

R6.7—Linear elastic second-order analysis

R6.7.1 *General*

In linear elastic second-order analyses, the deformed geometry of the structure is included in the equations of equilibrium so that $P\Delta$ effects are determined. The structure is assumed to remain elastic, but the effects of cracking and creep are considered by using an effective stiffness EI. In contrast, linear elastic first-order analysis satisfies the equations of equilibrium using the original undeformed geometry of the structure and estimates $P\Delta$ effects by magnifying the column-end sway moments using Eq. (6.6.4.6.2a) or (6.6.4.6.2b).

R6.7.1.1 The stiffnesses EI used in an analysis for strength design should represent the stiffnesses of the members immediately prior to failure. This is particularly true for a second-order analysis that should predict the lateral deflections at loads approaching ultimate. The EI values should not be based solely on the moment-curvature relationship for the most highly loaded section along the length of each member. Instead, they should correspond to the moment-end rotation relationship for a complete member.

To allow for variability in the actual member properties in the analysis, the member properties used in analysis should be multiplied by a stiffness reduction factor ϕ_K less than 1. The cross-sectional properties defined in 6.7.2 already include this stiffness reduction factor. The stiffness reduction factor ϕ_K may be taken as 0.875. Note that the overall stiffness is further reduced considering that the modulus of elasticity of the concrete, E_c, is based on the specified concrete compressive strength, while the sway deflections

CODE | COMMENTARY

are a function of the average concrete strength, which is typically higher.

6.7.1.2 Slenderness effects along the length of a column shall be considered. It shall be permitted to calculate these effects using 6.6.4.5.

R6.7.1.2 The maximum moment in a compression member may occur between its ends. In computer analysis programs, columns may be subdivided using nodes along their length to evaluate slenderness effects between the ends. If the column is not subdivided along its length, slenderness effects may be evaluated using the nonsway moment magnifier method specified in 6.6.4.5 with member-end moments from the second-order elastic analysis as input. Second-order analysis already accounts for the relative displacement of member ends.

6.7.1.3 The cross-sectional dimensions of each member used in an analysis to calculate slenderness effects shall be within 10 percent of the specified member dimensions in construction documents or the analysis shall be repeated.

6.7.1.4 Redistribution of moments calculated by an elastic second-order analysis shall be permitted in accordance with 6.6.5.

6.7.2 *Section properties*

R6.7.2 *Section properties*

6.7.2.1 *Factored load analysis*

6.7.2.1.1 It shall be permitted to use section properties calculated in accordance with 6.6.3.1.

6.7.2.2 *Service load analysis*

R6.7.2.2 *Service load analysis*

6.7.2.2.1 Immediate and time-dependent deflections due to gravity loads shall be calculated in accordance with 24.2.

6.7.2.2.2 Alternatively, it shall be permitted to calculate immediate deflections using a moment of inertia of 1.4 times I given in 6.6.3.1, or calculated using a more detailed analysis, but the value shall not exceed I_g.

R6.7.2.2.2 Refer to R6.6.3.2.2.

6.8—Inelastic analysis

6.8.1 *General*

R6.8—Inelastic analysis

R6.8.1 *General*

6.8.1.1 An inelastic analysis shall consider material nonlinearity. An inelastic first-order analysis shall satisfy equilibrium in the undeformed configuration. An inelastic second-order analysis shall satisfy equilibrium in the deformed configuration.

R6.8.1.1 Material nonlinearity may be affected by multiple factors including duration of loads, shrinkage, and creep.

6.8.1.2 An inelastic analysis procedure shall have been shown to result in calculation of strength and deformations that are in substantial agreement with results of physical tests of reinforced concrete components, subassemblages, or structural systems exhibiting response mechanisms consistent with those expected in the structure being designed.

R6.8.1.2 Substantial agreement should be demonstrated at characteristic points on the reported response. The characteristic points selected should depend on the purpose of the analysis, the applied loads, and the response phenomena exhibited by the component, subassemblage, or structural system. For nonlinear analysis to support design under

CODE

COMMENTARY

service-level loading, characteristic points should represent loads and deformations less than those corresponding to yielding of reinforcement. For nonlinear analysis to support design or assess response under design-level loading, characteristic points should represent loads and deformations less than those corresponding to yielding of reinforcement as well as points corresponding to yielding of reinforcement and onset of strength loss. Strength loss need not be represented if design loading does not extend the response into the strength-loss range. Typically, inelastic analysis to support design should employ specified material strengths and mean values of other material properties and component stiffnesses. Nonlinear response history analysis to verify the design of earthquake-resistant concrete structures should employ expected material strengths, expected material properties, and expected component stiffnesses, as specified in A.6.2.

6.8.1.3 Unless slenderness effects are permitted to be neglected in accordance with 6.2.5.1, an inelastic analysis shall satisfy equilibrium in the deformed configuration. It shall be permitted to calculate slenderness effects along the length of a column using 6.6.4.5.

R6.8.1.3 Refer to R6.7.1.2.

6.8.1.4 The cross-sectional dimensions of each member used in an analysis to calculate slenderness effects shall be within 10 percent of the specified member dimensions in construction documents or the analysis shall be repeated.

6.8.1.5 Redistribution of moments calculated by an inelastic analysis shall not be permitted.

R6.8.1.5 Section 6.6.5 allows for redistribution of moments calculated using elastic analysis to account for inelastic response of the system. Moments calculated by inelastic analysis explicitly account for inelastic response; therefore, further redistribution of moments is not appropriate.

6.9—Acceptability of finite element analysis

6.9.1 Finite element analysis to determine load effects shall be permitted.

6.9.2 The finite element model shall be appropriate for its intended purpose.

R6.9—Acceptability of finite element analysis

R6.9.1 This section was introduced in the 2014 Code to explicitly recognize a widely used analysis method.

R6.9.2 The licensed design professional should ensure that an appropriate analysis model is used for the particular problem of interest. This includes selection of computer software program, element type, model mesh, and other modeling assumptions.

A large variety of finite element analysis computer software programs are available, including those that perform static, dynamic, elastic, and inelastic analyses.

The element types used should be capable of determining the response required. Finite element models may have beam-column elements that model structural framing members, such as beams and columns, along with plane stress elements; plate elements; and shell elements, brick elements, or both, that are used to model the floor slabs, mat foundations, diaphragms, walls, and connections. The model mesh size selected should be capable of determining

CODE

6.9.3 For inelastic analysis, a separate analysis shall be performed for each factored load combination.

6.9.4 The licensed design professional shall confirm that the results are appropriate for the purposes of the analysis.

6.9.5 The cross-sectional dimensions of each member used in an analysis shall be within 10 percent of the specified member dimensions in construction documents or the analysis shall be repeated.

6.9.6 Redistribution of moments calculated by an inelastic analysis shall not be permitted.

COMMENTARY

the structural response in sufficient detail. The use of any set of reasonable assumptions for member stiffness is allowed.

R6.9.3 For an inelastic finite element analysis, the rules of linear superposition do not apply. To determine the ultimate member inelastic response, for example, it is not correct to analyze for service loads and subsequently combine the results linearly using load factors. A separate inelastic analysis should be performed for each factored load combination.

Notes

Strip Footing Examples

$d = h - cc - b_d = 12 - 2 - 0.5$ (# bars)

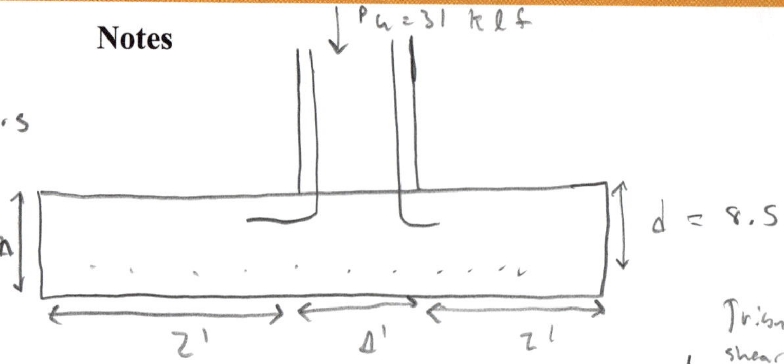

Determine Bars for Flexure

$q_u = \frac{P_u}{A} = \frac{31 \text{ klf}}{(5)(1)} = 6.2 \frac{\text{ksf}}{\text{ft}}$

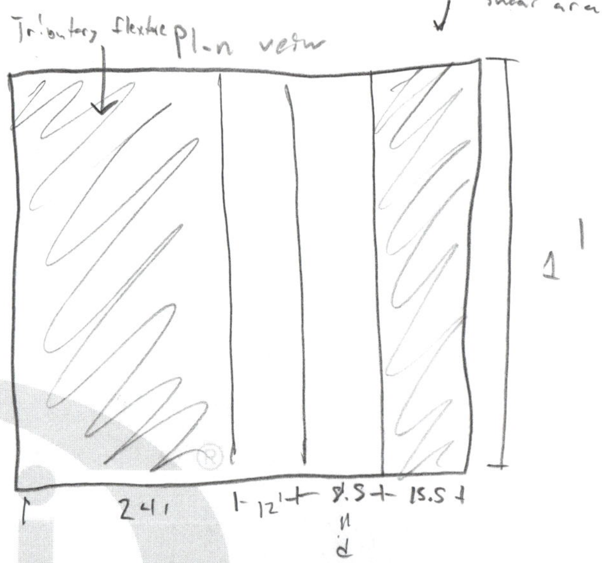

$M_u = q_{nu}(b)(f)\left(\frac{f}{2}\right)$

$M_u = 6.2(1)(2')\left(\frac{2}{2}\right) = 12.4 \frac{\text{k-ft}}{\text{ft}}$

$\phi = 0.9$ (tension controlled)

$$\left[\frac{\phi f_y^2}{2(0.85)f'_c b}\right] A_s^2 + \left[-\phi f_y d\right] A_s + M_u = 0$$

(b = 12")

$$S = \frac{A_b}{A_s}(12'')$$

Look at Spread foot example to check $A_{s,min}$, $A_{s,max}$, S_{min}, S_{max}

Design Example:

Roof $D_L = 28$ psf

Roof $S_L = 25$ psf

Wall $D_L = 65$ psf (12" CMU)

2'x1' structural plain strip footing

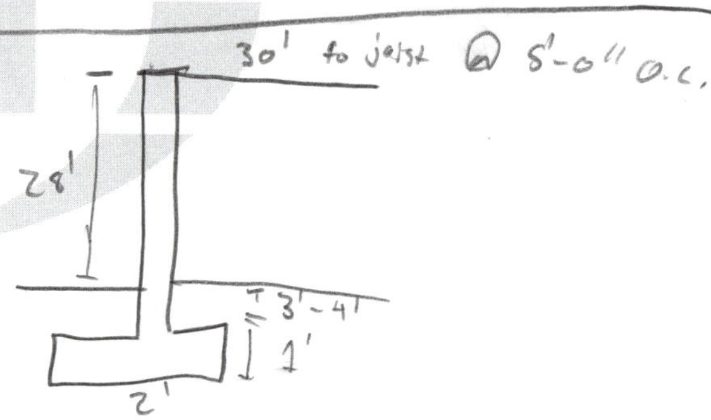

1.) Find W_u

$d_{L,wall} = (28 \text{ psf})(15') + (65)(28) = 2240$ plf

$W_{SL} = 25(15) = 375$ psf

$V_1 = 1.2(2240) + (1.6)(375) = 3288$ plf $= 3288 \frac{\text{psf}}{\text{ft}}$

Find $q_{nu} = \frac{W_u}{A} = \frac{3288}{(2')(1')} = 1644$ psf/ft

Flexural Demand

$M_u = [q_{nu}(f)\left(\frac{f}{2}\right)] = [1644\left(\frac{9}{12}\right)]\left(\frac{1}{2}\cdot\frac{9}{12}\right)$ = 462 $\frac{\text{ft-lb}}{\text{ft}}$

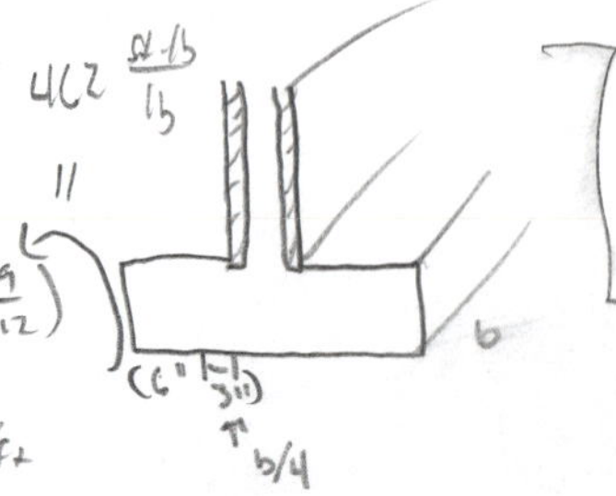

$\text{Capacity} = \frac{(bh^2)}{6} = \frac{1}{6}(12'')(12''-2'')^2 = 200 \text{ in}^3/\text{ft}$

$\phi M_n = 0.65\sqrt{3000}(200) = 32863 \frac{\text{in-lb}}{\text{ft}} = 2739 \frac{\text{ft-lb}}{\text{ft}}$

For plain Concrete! See example of strip footing

Longitudinal Reinforcing

$A_{s,min} = 0.0018\,bh = 0.0018(24)(12) = 0.52 \text{ in}^2$ (3) #4 0.6 in/ft

CHAPTER 7—ONE-WAY SLABS

7.1—Scope

7.1.1 This chapter shall apply to the design of nonprestressed and prestressed slabs reinforced for flexure in one direction, including:

(a) Solid slabs
(b) Slabs cast on stay-in-place, noncomposite steel deck
(c) Composite slabs of concrete elements constructed in separate placements but connected so that all elements resist loads as a unit
(d) Precast, prestressed hollow-core slabs

R7.1—Scope

R7.1.1 The design and construction of composite slabs on steel deck is described in "Standard for Composite Steel Floor Deck – Slabs" (SDI C).

Provisions for one-way joist systems are provided in Chapter 9.

7.2—General

7.2.1 The effects of concentrated loads, slab openings, and voids within the slab shall be considered in design.

7.2.2 *Materials*

7.2.2.1 Design properties for concrete shall be selected to be in accordance with Chapter 19.

7.2.2.2 Design properties for steel reinforcement shall be selected to be in accordance with Chapter 20.

7.2.2.3 Materials, design, and detailing requirements for embedments in concrete shall be in accordance with 20.6.

7.2.3 *Connection to other members*

7.2.3.1 For cast-in-place construction, beam-column and slab-column joints shall satisfy Chapter 15.

7.2.3.2 For precast construction, connections shall satisfy the force transfer requirements of 16.2.

R7.2—General

R7.2.1 Concentrated loads and slab openings create local moments and shears and may cause regions of one-way slabs to have two-way behavior. The influence of openings through the slab and voids within the slab (for example ducts) on flexural and shear strength as well as deflections is to be considered, including evaluating the potential for critical sections created by the openings and voids.

7.3—Design limits

7.3.1 *Minimum slab thickness*

7.3.1.1 For solid nonprestressed slabs not supporting or attached to partitions or other construction likely to be damaged by large deflections, overall slab thickness $\boldsymbol{h}$ shall not be less than the limits in Table 7.3.1.1, unless the calculated deflection limits of 7.3.2 are satisfied.

R7.3—Design limits

R7.3.1 *Minimum slab thickness*

The basis for minimum thickness for one-way slabs is the same as that for beams. Refer to R9.3.1 for additional information.

CODE

COMMENTARY

Table 7.3.1.1—Minimum thickness of solid nonprestressed one-way slabs

Support condition	Minimum h[1]
Simply supported	$\ell/20$
One end continuous	$\ell/24$
Both ends continuous	$\ell/28$
Cantilever	$\ell/10$

[1]Expression applicable for normalweight concrete and f_y = 60,000 psi. For other cases, minimum h shall be modified in accordance with 7.3.1.1.1 through 7.3.1.1.3, as appropriate.

7.3.1.1.1 For f_y other than 60,000 psi, the expressions in Table 7.3.1.1 shall be multiplied by $(\mathbf{0.4} + f_y/\mathbf{100{,}000})$.

7.3.1.1.2 For nonprestressed slabs made of lightweight concrete having w_c in the range of 90 to 115 lb/ft^3, the expressions in Table 7.3.1.1 shall be multiplied by the greater of (a) and (b):

(a) $\mathbf{1.65 - 0.005}w_c$
(b) **1.09**

7.3.1.1.3 For nonprestressed composite slabs made of a combination of lightweight and normalweight concrete, shored during construction, and where the lightweight concrete is in compression, the modifier of 7.3.1.1.2 shall apply.

7.3.1.2 The thickness of a concrete floor finish shall be permitted to be included in h if it is placed monolithically with the floor slab or if the floor finish is designed to be composite with the floor slab in accordance with 16.4.

7.3.2 *Calculated deflection limits*

7.3.2.1 For nonprestressed slabs not satisfying 7.3.1 and for prestressed slabs, immediate and time-dependent deflections shall be calculated in accordance with 24.2 and shall not exceed the limits in 24.2.2.

7.3.2.2 For nonprestressed composite concrete slabs satisfying 7.3.1, deflections occurring after the member becomes composite need not be calculated. Deflections occurring before the member becomes composite shall be investigated, unless the precomposite thickness also satisfies 7.3.1.

R7.3.2 *Calculated deflection limits*

The basis for calculated deflections for one-way slabs is the same as that for beams. Refer to R9.3.2 for additional information.

7.3.3 *Reinforcement strain limit in nonprestressed slabs*

7.3.3.1 Nonprestressed slabs shall be tension-controlled in accordance with Table 21.2.2.

R7.3.3 *Reinforcement strain limit in nonprestressed slabs*

R7.3.3.1 The basis for a reinforcement strain limit for one-way slabs is the same as that for beams. Refer to R9.3.3 for additional information.

7.3.4 *Stress limits in prestressed slabs*

7.3.4.1 Prestressed slabs shall be classified as Class U, T, or C in accordance with 24.5.2.

CODE

7.3.4.2 Stresses in prestressed slabs immediately after transfer and at service loads shall not exceed the permissible stresses in 24.5.3 and 24.5.4.

7.4—Required strength

7.4.1 *General*

7.4.1.1 Required strength shall be calculated in accordance with the factored load combinations in Chapter 5.

7.4.1.2 Required strength shall be calculated in accordance with the analysis procedures in Chapter 6.

7.4.1.3 For prestressed slabs, effects of reactions induced by prestressing shall be considered in accordance with 5.3.11.

7.4.2 *Factored moment*

7.4.2.1 For slabs built integrally with supports, M_u at the support shall be permitted to be calculated at the face of support.

7.4.3 *Factored shear*

7.4.3.1 For slabs built integrally with supports, V_u at the support shall be permitted to be calculated at the face of support.

7.4.3.2 Sections between the face of support and a critical section located d from the face of support for nonprestressed slabs or $h/2$ from the face of support for prestressed slabs shall be permitted to be designed for V_u at that critical section if (a) through (c) are satisfied:

(a) Support reaction, in direction of applied shear, introduces compression into the end region of the slab
(b) Loads are applied at or near the top surface of the slab
(c) No concentrated load occurs between the face of support and critical section

7.5—Design strength

7.5.1 *General*

7.5.1.1 For each applicable factored load combination, design strength at all sections shall satisfy $\phi S_n \geq U$ including (a) and (b). Interaction between load effects shall be considered.

(a) $\phi M_n \geq M_u$
(b) $\phi V_n \geq V_u$

7.5.1.2 ϕ shall be determined in accordance with 21.2.

7.5.2 *Moment*

7.5.2.1 M_n shall be calculated in accordance with 22.3.

COMMENTARY

R7.4—Required strength

R7.4.3 *Factored shear*

R7.4.3.2 The requirements for the selection of the critical section for shear in one-way slabs are the same as those for beams. Refer to R9.4.3.2 for additional information.

R7.5—Design strength

R7.5.1 *General*

R7.5.1.1 Refer to R9.5.1.1.

R7.5.2 *Moment*

CODE

7.5.2.2 For prestressed slabs, external tendons shall be considered as unbonded tendons in calculating flexural strength, unless the external tendons are effectively bonded to the concrete section along the entire length.

7.5.2.3 If primary flexural reinforcement in a slab that is considered to be a T-beam flange is parallel to the longitudinal axis of the beam, reinforcement perpendicular to the longitudinal axis of the beam shall be provided in the top of the slab in accordance with (a) and (b). This provision does not apply to joist construction.

(a) Slab reinforcement perpendicular to the beam shall be designed to resist the factored load on the overhanging slab width assumed to act as a cantilever.
(b) Only the effective overhanging slab width in accordance with 6.3.2 need be considered.

7.5.3 *Shear*

7.5.3.1 V_n shall be calculated in accordance with 22.5.

7.5.3.2 For composite concrete slabs, horizontal shear strength V_{nh} shall be calculated in accordance with 16.4.

7.6—Reinforcement limits

7.6.1 *Minimum flexural reinforcement in nonprestressed slabs*

7.6.1.1 A minimum area of flexural reinforcement, $A_{s,min}$, of $\mathbf{0.0018}A_g$ shall be provided.

7.6.2 *Minimum flexural reinforcement in prestressed slabs*

7.6.2.1 For slabs with bonded prestressed reinforcement, total quantity of A_s and A_{ps} shall be adequate to develop a factored load at least 1.2 times the cracking load calculated on the basis of f_r as given in 19.2.3.

7.6.2.2 For slabs with both flexural and shear design strength at least twice the required strength, 7.6.2.1 need not be satisfied.

7.6.2.3 For slabs with unbonded tendons, the minimum area of bonded deformed longitudinal reinforcement, $A_{s,min}$, shall be:

$$A_{s,min} \geq 0.004A_{ct} \qquad (7.6.2.3)$$

COMMENTARY

R7.5.2.3 This provision applies only where a T-beam is parallel to the span of a one-way slab. For example, this beam might be used to support a wall or concentrated load that the slab alone cannot support. In that case, the primary slab reinforcement is parallel to the beam and the perpendicular reinforcement is usually sized for temperature and shrinkage. The reinforcement required by this provision is intended to consider "unintended" negative moments that may develop over the beam that exceed the requirements for temperature and shrinkage reinforcement alone.

R7.6—Reinforcement limits

R7.6.1 *Minimum flexural reinforcement in nonprestressed slabs*

R7.6.1.1 The required area of deformed or welded wire reinforcement used as minimum flexural reinforcement is the same as provided for shrinkage and temperature in 24.4.3.2. However, whereas shrinkage and temperature reinforcement is permitted to be distributed between the two faces of the slab as deemed appropriate for specific conditions, minimum flexural reinforcement should be placed as close as practicable to the face of the concrete in tension due to applied loads.

R7.6.2 *Minimum flexural reinforcement in prestressed slabs*

The requirements for minimum flexural reinforcement for prestressed one-way slabs are the same as those for prestressed beams. Refer to R9.6.2 for additional information.

CODE

where A_{ct} is the area of that part of the cross section between the flexural tension face and the centroid of the gross section.

7.6.3 *Minimum shear reinforcement*

7.6.3.1 A minimum area of shear reinforcement, $A_{v,min}$, shall be provided in all regions where $V_u > \phi V_c$. For precast prestressed hollow-core slabs with untopped h **> 12.5 in.**, $A_{v,min}$ shall be provided in all regions where $V_u > 0.5\phi V_{cw}$.

7.6.3.2 If shown by testing that the required M_n and V_n can be developed, 7.6.3.1 need not be satisfied. Such tests shall simulate effects of differential settlement, creep, shrinkage, and temperature change, based on a realistic assessment of these effects occurring in service.

7.6.3.3 If shear reinforcement is required, $A_{v,min}$ shall be in accordance with 9.6.3.4.

7.6.4 *Minimum shrinkage and temperature reinforcement*

7.6.4.1 Reinforcement shall be provided to resist shrinkage and temperature stresses in accordance with 24.4.

7.6.4.2 If prestressed shrinkage and temperature reinforcement in accordance with 24.4.4 is used, 7.6.4.2.1 through 7.6.4.2.3 shall apply.

7.6.4.2.1 For monolithic, cast-in-place, post-tensioned beam-and-slab construction, gross concrete area shall consist of the total beam area including the slab thickness and the slab area within half the clear distance to adjacent beam webs. It shall be permitted to include the effective force in beam tendons in the calculation of total prestress force acting on gross concrete area.

COMMENTARY

R7.6.3 *Minimum shear reinforcement*

The basis for minimum shear reinforcement for one-way slabs is the same as that for beams. Refer to R9.6.3 for additional information.

R7.6.3.1 Solid slabs and footings have less stringent minimum shear reinforcement requirements than beams because there is a possibility of load sharing between weak and strong areas. However, research (Angelakos et al. 2001; Lubell et al. 2004; Brown et al. 2006) has shown that deep, lightly reinforced one-way slabs, particularly if constructed with high-strength concrete or concrete having a small coarse aggregate size, may fail at shears less than V_c calculated from Eq. (22.5.5.1). One-way slabs subjected to concentrated loads are more likely to exhibit this vulnerability.

Results of tests on precast, prestressed hollow-core units (Becker and Buettner 1985; Anderson 1978) with $h \leq$ **12.5 in.** have shown shear strengths greater than those calculated by Eq. (22.5.6.3.1a) and Eq. (22.5.6.3.2). Results of tests on hollow-core units with h **> 12.5 in.** have shown that web-shear strengths in end regions can be less than strengths calculated by Eq. (22.5.6.3.2). In contrast, flexure-shear strengths in the deeper hollow-core units equaled or exceeded strengths calculated by Eq. (22.5.6.3.1a).

R7.6.3.2 The basis for the testing-based strength evaluation for one-way slabs is the same as that for beams. Refer to R9.6.3.3 for additional information.

R7.6.4 *Minimum shrinkage and temperature reinforcement*

R7.6.4.2 In prestressed monolithic beam-and-slab construction, at least one shrinkage and temperature tendon is required between beams, even if the beam tendons alone provide at least 100 psi average compressive stress as required by 24.4.4.1 on the gross concrete area as defined in 7.6.4.2.1. A tendon of any size is permissible as long as all other requirements of 7.6.4.2 and 7.7.6.3 are satisfied. Application of the provisions of 7.6.4.2 and 7.7.6.3 to monolithic, cast-in-place, post-tensioned, beam-and-slab construction is illustrated in Fig. R7.6.4.2.

Tendons used for shrinkage and temperature reinforcement should be positioned as close as practicable to the mid-depth

CODE

7.6.4.2.2 If slabs are supported on walls or not cast monolithically with beams, gross concrete area is the slab section tributary to the tendon or tendon group.

7.6.4.2.3 At least one tendon is required in the slab between faces of adjacent beams or walls.

COMMENTARY

of the slab. In cases where the shrinkage and temperature tendons are used for supporting the principal tendons, variations from the slab centroid are permissible; however, the resultant of the shrinkage and temperature tendons should not fall outside the middle third of the slab thickness.

The effects of slab shortening should be evaluated to ensure the effectiveness of the prestressing. In most cases, the low level of prestressing recommended should not cause difficulties in a properly detailed structure. Additional attention may be required where thermal effects become significant.

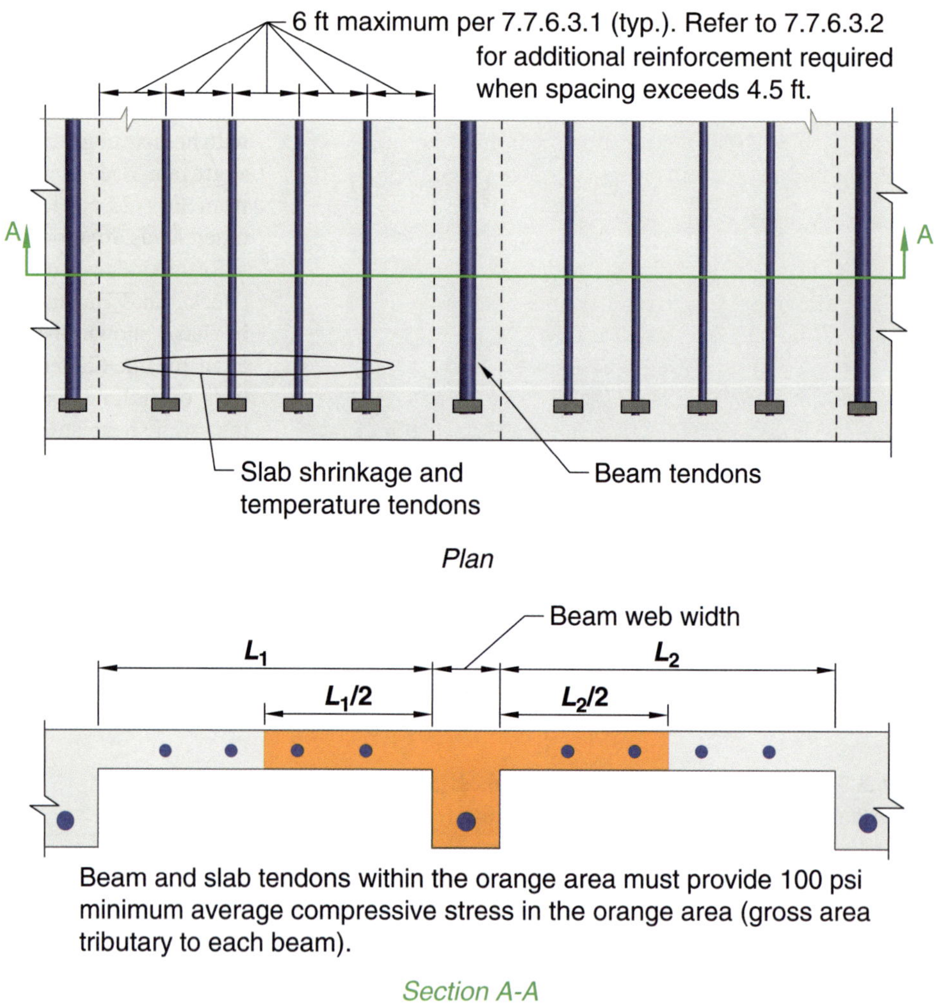

***Fig. R7.6.4.2**—Section through beams cast monolithically with slab.*

7.7—Reinforcement detailing

7.7.1 *General*

7.7.1.1 Concrete cover for reinforcement shall be in accordance with 20.5.1.

7.7.1.2 Development lengths of deformed and prestressed reinforcement shall be in accordance with 25.4.

R7.7—Reinforcement detailing

CODE

7.7.1.3 Splices of deformed reinforcement shall be in accordance with 25.5.

7.7.1.4 Bundled bars shall be in accordance with 25.6.

7.7.2 *Reinforcement spacing*

7.7.2.1 Minimum spacing s shall be in accordance with 25.2.

7.7.2.2 For nonprestressed and Class C prestressed slabs, spacing of bonded longitudinal reinforcement closest to the tension face shall not exceed s given in 24.3.

7.7.2.3 For nonprestressed and Class T and C prestressed slabs with unbonded tendons, maximum spacing s of deformed longitudinal reinforcement shall be the lesser of $3h$ and 18 in.

7.7.2.4 Maximum spacing, s, of reinforcement required by 7.5.2.3 shall be the lesser of $5h$ and 18 in.

7.7.3 *Flexural reinforcement in nonprestressed slabs*

7.7.3.1 Calculated tensile or compressive force in reinforcement at each section of the slab shall be developed on each side of that section.

7.7.3.2 Critical locations for development of reinforcement are points of maximum stress and points along the span where bent or terminated tension reinforcement is no longer required to resist flexure.

7.7.3.3 Reinforcement shall extend beyond the point at which it is no longer required to resist flexure for a distance at least the greater of d and $12d_b$, except at supports of simply-supported spans and at free ends of cantilevers.

7.7.3.4 Continuing flexural tension reinforcement shall have an embedment length at least ℓ_d beyond the point where bent or terminated tension reinforcement is no longer required to resist flexure.

7.7.3.5 Flexural tension reinforcement shall not be terminated in a tension zone unless (a), (b), or (c) is satisfied:

(a) $V_u \leq (2/3)\phi V_n$ at the cutoff point.
(b) For No. 11 bars and smaller, continuing reinforcement provides double the area required for flexure at the cutoff point and $V_u \leq (3/4)\phi V_n$.
(c) Stirrup area in excess of that required for shear is provided along each terminated bar or wire over a distance

COMMENTARY

R7.7.2 *Reinforcement spacing*

R7.7.2.3 Editions of ACI 318 prior to 2019 excluded the provisions of 7.7.2.3 for prestressed concrete. However, Class T and C slabs prestressed with unbonded tendons rely solely on deformed reinforcement for crack control. Consequently, the requirements of 7.7.2.3 have been extended to apply to Class T and C slabs prestressed with unbonded tendons.

R7.7.2.4 The spacing limitations for slab reinforcement are based on flange thickness, which for tapered flanges can be taken as the average thickness.

R7.7.3 *Flexural reinforcement in nonprestressed slabs*

Requirements for development of reinforcement in one-way slabs are similar to those for beams. Refer to R9.7.3 for additional information.

CODE

3/4*d* from the cutoff point. Excess stirrup area shall be not less than $60b_w s/f_{yt}$. Spacing s shall not exceed $d/(8\beta_b)$.

7.7.3.6 Adequate anchorage shall be provided for tension reinforcement where reinforcement stress is not directly proportional to moment, such as in sloped, stepped, or tapered slabs, or where tension reinforcement is not parallel to the compression face.

7.7.3.7 In slabs with spans not exceeding 10 ft, welded wire reinforcement, with wire size not exceeding W5 or D5, shall be permitted to be curved from a point near the top of slab over the support to a point near the bottom of slab at midspan, provided such reinforcement is continuous over, or developed at, the support.

7.7.3.8 *Termination of reinforcement*

7.7.3.8.1 At simple supports, at least one-third of the maximum positive moment reinforcement shall extend along the slab bottom into the support, except for precast slabs where such reinforcement shall extend at least to the center of the bearing length.

7.7.3.8.2 At other supports, at least one-fourth of the maximum positive moment reinforcement shall extend along the slab bottom into the support at least 6 in.

7.7.3.8.3 At simple supports and points of inflection, d_b for positive moment tension reinforcement shall be limited such that ℓ_d for that reinforcement satisfies (a) or (b). If reinforcement terminates beyond the centerline of supports by a standard hook or a mechanical anchorage at least equivalent to a standard hook, (a) or (b) need not be satisfied.

(a) $\ell_d \leq (1.3M_n/V_u + \ell_a)$ if end of reinforcement is confined by a compressive reaction
(b) $\ell_d \leq (M_n/V_u + \ell_a)$ if end of reinforcement is not confined by a compressive reaction

M_n is calculated assuming all reinforcement at the section is stressed to f_y and V_u is calculated at the section. At a support, ℓ_a is the embedment length beyond the center of the support. At a point of inflection, ℓ_a is the embedment length beyond the point of inflection, limited to the greater of d and $12d_b$.

7.7.3.8.4 At least one-third of the negative moment reinforcement at a support shall have an embedment length beyond the point of inflection at least the greatest of d, $12d_b$, and $\ell_n/16$.

7.7.4 *Flexural reinforcement in prestressed slabs*

COMMENTARY

R7.7.3.8 *Termination of reinforcement*

Requirements for termination of reinforcement in one-way slabs are similar to those for beams. Refer to R9.7.3.8 for additional information.

R7.7.4 *Flexural reinforcement in prestressed slabs*

CODE

7.7.4.1 External tendons shall be attached to the member in a manner that maintains the specified eccentricity between the tendons and the concrete centroid through the full range of anticipated member deflections.

7.7.4.2 If nonprestressed reinforcement is required to satisfy flexural strength, the detailing requirements of 7.7.3 shall be satisfied.

7.7.4.3 *Termination of prestressed reinforcement*

7.7.4.3.1 Post-tensioned anchorage zones shall be designed and detailed in accordance with 25.9.

7.7.4.3.2 Post-tensioning anchorages and couplers shall be designed and detailed in accordance with 25.8.

7.7.4.4 *Termination of deformed reinforcement in slabs with unbonded tendons*

7.7.4.4.1 Length of deformed reinforcement required by 7.6.2.3 shall be in accordance with (a) and (b):

(a) At least $\ell_n/3$ in positive moment areas and be centered in those areas
(b) At least $\ell_n/6$ on each side of the face of support

7.7.5 *Shear reinforcement*

7.7.5.1 If shear reinforcement is required, transverse reinforcement shall be detailed according to 9.7.6.2.

7.7.6 *Shrinkage and temperature reinforcement*

7.7.6.1 Shrinkage and temperature reinforcement in accordance with 7.6.4 shall be placed perpendicular to flexural reinforcement.

7.7.6.2 *Nonprestressed reinforcement*

7.7.6.2.1 Spacing of deformed shrinkage and temperature reinforcement shall not exceed the lesser of $5h$ and 18 in.

7.7.6.3 *Prestressed reinforcement*

7.7.6.3.1 Spacing of slab tendons required by 7.6.4.2 and the distance between face of beam or wall to the nearest slab tendon shall not exceed 6 ft.

7.7.6.3.2 If spacing of slab tendons exceeds 4.5 ft, additional deformed shrinkage and temperature reinforcement conforming to 24.4.3 shall be provided parallel to the tendons, except 24.4.3.4 need not be satisfied. In calculating the area of additional reinforcement, it shall be permitted to take the gross concrete area in 24.4.3.2 as the slab area

COMMENTARY

R7.7.4.4 *Termination of deformed reinforcement in slabs with unbonded tendons*

Requirements for termination of deformed reinforcement in one-way slabs with unbonded tendons are the same as those for beams. Refer to R9.7.4.4 for additional information.

R7.7.6 *Shrinkage and temperature reinforcement*

R7.7.6.3 *Prestressed reinforcement*

R7.7.6.3.2 Widely spaced tendons result in non-uniform compressive stresses near the slab edges. The additional reinforcement is to reinforce regions near the slab edge that may be inadequately compressed. Placement of this reinforcement is illustrated in Fig. R7.7.6.3.2.

CODE

between faces of beams. This shrinkage and temperature reinforcement shall extend from the slab edge for a distance not less than the slab tendon spacing.

COMMENTARY

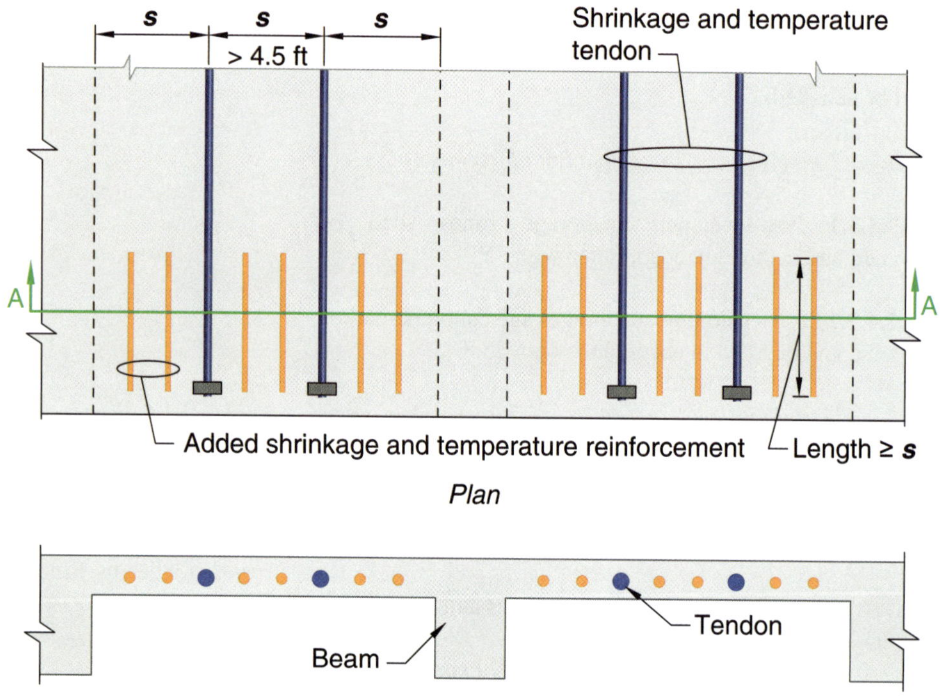

***Fig. R7.7.6.3.2**—Plan view at slab edge showing added shrinkage and temperature reinforcement.*

7.7.7 *Structural integrity reinforcement in cast-in-place one-way slabs*

7.7.7.1 Longitudinal structural integrity reinforcement consisting of at least one-quarter of the maximum positive moment reinforcement shall be continuous.

7.7.7.2 Longitudinal structural integrity reinforcement at noncontinuous supports shall be anchored to develop f_y at the face of the support.

7.7.7.3 If splices are necessary in continuous structural integrity reinforcement, the reinforcement shall be spliced near supports. Splices shall be mechanical or welded in accordance with 25.5.7 or Class B tension lap splices in accordance with 25.5.2.

R7.7.7 *Structural integrity reinforcement in cast-in-place one-way slabs*

Positive moment structural integrity reinforcement for one-way slabs is intended to be similar to that for beams. Refer to R9.7.7 for a discussion of structural integrity reinforcement for beams.

CHAPTER 8—TWO-WAY SLABS

CODE

8.1—Scope

8.1.1 This chapter shall apply to the design of nonprestressed and prestressed slabs reinforced for flexure in two directions, with or without beams between supports, including (a) through (d):

(a) Solid slabs
(b) Slabs cast on stay-in-place, noncomposite steel deck
(c) Composite slabs of concrete elements constructed in separate placements but connected so that all elements resist loads as a unit
(d) Two-way joist systems in accordance with 8.8

8.2—General

8.2.1 A slab system shall be permitted to be designed by any procedure satisfying equilibrium and geometric compatibility, provided that design strength at every section is at least equal to required strength, and all serviceability requirements are satisfied. The direct design method or the equivalent frame method is permitted.

COMMENTARY

R8.1—Scope

The design methods given in this chapter are based on analysis of the results of an extensive series of tests (Burns and Hemakom 1977; Gamble et al. 1969; Gerber and Burns 1971; Guralnick and LaFraugh 1963; Hatcher et al. 1965, 1969; Hawkins 1981; Jirsa et al. 1966; PTI DC20.8; Smith and Burns 1974; Scordelis et al. 1959; Vanderbilt et al. 1969; Xanthakis and Sozen 1963) and the well-established performance records of various slab systems. The fundamental design principles are applicable to all planar structural systems subjected to transverse loads. Several specific design rules, as well as historical precedents, limit the types of structures to which this chapter applies. General slab systems that may be designed according to this chapter include flat slabs, flat plates, two-way slabs, and waffle slabs. Slabs with paneled ceilings are two-way, wide-band, beam systems.

Slabs-on-ground that do not transmit vertical loads from other parts of the structure to the soil are excluded.

For slabs with beams, the explicit design procedures of this chapter apply only when the beams are located at the edges of the panel and when the beams are supported by columns or other essentially nondeflecting supports at the corners of the panel. Two-way slabs with beams in one direction, with both slab and beams supported by girders in the other direction, may be designed under the general requirements of this chapter. Such designs should be based upon analysis compatible with the deflected position of the supporting beams and girders.

For slabs supported on walls, the explicit design procedures in this chapter treat the wall as a beam of infinite stiffness; therefore, each wall should support the entire length of an edge of the panel (refer to 8.4.1.7). Walls of width less than a full panel length can be treated as columns.

R8.2—General

R8.2.1 This section permits a design to be based directly on fundamental principles of structural mechanics, provided it can be demonstrated explicitly that all strength and serviceability criteria are satisfied. The design of the slab may be achieved through the combined use of classic solutions based on a linearly elastic continuum, numerical solutions based on discrete elements, or yield-line analyses, including, in all cases, evaluation of the stress conditions around the supports in relation to shear, torsion, and flexure, as well as the effects of reduced stiffness of elements due to cracking and support geometry. The design of a slab system involves more than its analysis; any deviations in physical dimensions of the slab from common practice should be justified on the basis of knowledge of the expected loads and the reliability of the calculated stresses and deformations of the structure.

The direct design method and the equivalent frame method are limited in application to orthogonal frames subject to gravity loads only.

CODE

8.2.2 The effects of concentrated loads, slab openings, and slab voids shall be considered in design.

8.2.3 Slabs prestressed with an average effective compressive stress less than 125 psi shall be designed as nonprestressed slabs.

8.2.4 A drop panel in a nonprestressed slab, where used to reduce the minimum required thickness in accordance with 8.3.1.1 or the quantity of deformed negative moment reinforcement at a support in accordance with 8.5.2.2, shall satisfy (a) and (b):

(a) The drop panel shall project below the slab at least one-fourth of the adjacent slab thickness.
(b) The drop panel shall extend in each direction from the centerline of support a distance not less than one-sixth the span length measured from center-to-center of supports in that direction.

8.2.5 A shear cap, where used to increase the critical section for shear at a slab-column joint, shall project below the slab soffit and extend horizontally from the face of the column a distance at least equal to the thickness of the projection below the slab soffit.

8.2.6 *Materials*

8.2.6.1 Design properties for concrete shall be selected to be in accordance with Chapter 19.

8.2.6.2 Design properties for steel reinforcement shall be selected to be in accordance with Chapter 20.

8.2.6.3 Materials, design, and detailing requirements for embedments in concrete shall be in accordance with 20.6.

8.2.7 *Connections to other members*

8.2.7.1 Beam-column and slab-column joints shall satisfy Chapter 15.

8.3—Design limits

8.3.1 *Minimum slab thickness*

COMMENTARY

R8.2.2 Refer to R7.2.1.

R8.2.4 and R8.2.5 Drop panel dimensions specified in 8.2.4 are necessary when reducing the amount of negative moment reinforcement following 8.5.2.2 or to satisfy minimum slab thicknesses permitted in 8.3.1.1. If the dimensions are less than specified in 8.2.4, the projection may be used as a shear cap to increase the shear strength of the slab. For slabs with changes in thickness, it is necessary to check the shear strength at several sections (Refer to 22.6.4.1(b)).

R8.2.7 *Connections to other members*

Safety of a slab system requires consideration of the transmission of load from the slab to the columns by flexure, torsion, and shear.

R8.3—Design limits

R8.3.1 *Minimum slab thickness*

The minimum slab thicknesses in 8.3.1.1 and 8.3.1.2 are independent of loading and concrete modulus of elasticity, both of which have significant effects on deflections. These minimum thicknesses are not applicable to slabs with unusually heavy superimposed sustained loads or for concrete with modulus of elasticity significantly lower than that of ordinary normalweight concrete. Deflections should be calculated for such situations.

CODE

8.3.1.1 For nonprestressed slabs without interior beams spanning between supports on all sides, having a maximum ratio of long-to-short span of 2, overall slab thickness h shall not be less than the limits in Table 8.3.1.1, and shall be at least the value in (a) or (b), unless the calculated deflection limits of 8.3.2 are satisfied:

(a) Slabs without drop panels as given in 8.2.4.......... 5 in.
(b) Slabs with drop panels as given in 8.2.4.............. 4 in.

For f_y exceeding 80,000 psi, the calculated deflection limits in 8.3.2 shall be satisfied assuming a reduced modulus of rupture $f_r = 5\sqrt{f_c'}$.

COMMENTARY

R8.3.1.1 The minimum thicknesses in Table 8.3.1.1 are those that have been developed through the years. Use of longitudinal reinforcement with $f_y > 80{,}000$ psi may result in larger long-term deflections than in the case of $f_y < 80{,}000$ psi unless associated service stresses calculated for cracked sections are smaller than 40,000 psi. Careful calculation of deflections should be performed.

Table 8.3.1.1—Minimum thickness of nonprestressed two-way slabs without interior beams (in.)[1]

f_y, psi[2]	Without drop panels[3]			With drop panels[3]		
	Exterior panels		Interior panels	Exterior panels		Interior panels
	Without edge beams	With edge beams[4]		Without edge beams	With edge beams[4]	
40,000	$\ell_n/33$	$\ell_n/36$	$\ell_n/36$	$\ell_n/36$	$\ell_n/40$	$\ell_n/40$
60,000	$\ell_n/30$	$\ell_n/33$	$\ell_n/33$	$\ell_n/33$	$\ell_n/36$	$\ell_n/36$
80,000	$\ell_n/27$	$\ell_n/30$	$\ell_n/30$	$\ell_n/30$	$\ell_n/33$	$\ell_n/33$

[1]ℓ_n is the clear span in the long direction, measured face-to-face of supports (in.).
[2]For f_y between the values given in the table, minimum thickness shall be calculated by linear interpolation.
[3]Drop panels as given in 8.2.4.
[4]Slabs with beams between columns along exterior edges. Exterior panels shall be considered to be without edge beams if α_f is less than 0.8.

8.3.1.2 For nonprestressed slabs with beams spanning between supports on all sides, overall slab thickness h shall satisfy the limits in Table 8.3.1.2, unless the calculated deflection limits of 8.3.2 are satisfied.

Table 8.3.1.2—Minimum thickness of nonprestressed two-way slabs with beams spanning between supports on all sides

α_{fm}[1]	Minimum h, in.		
$\alpha_{fm} \le 0.2$	8.3.1.1 applies		(a)
$0.2 < \alpha_{fm} \le 2.0$	Greater of:	$\dfrac{\ell_n\left(0.8+\dfrac{f_y}{200{,}000}\right)}{36+5\beta(\alpha_{fm}-0.2)}$	(b)[1],[2]
		5.0	(c)
$\alpha_{fm} > 2.0$	Greater of:	$\dfrac{\ell_n\left(0.8+\dfrac{f_y}{200{,}000}\right)}{36+9\beta}$	(d)
		3.5	(e)

[1]α_{fm} is the average value of α_f for all beams on edges of a panel.
[2]ℓ_n is the clear span in the long direction, measured face-to-face of beams (in.).
[3]β is the ratio of clear spans in long to short directions of slab.

R8.3.1.2 For panels having a ratio of long-to-short span greater than 2, the use of expressions (b) and (d) of Table 8.3.1.2, which give the minimum thickness as a fraction of the long span, may give unreasonable results. For such panels, the rules applying to one-way construction in 7.3.1 should be used.

CODE

8.3.1.2.1 At discontinuous edges of slabs conforming to 8.3.1.2, an edge beam with $\alpha_f \geq$ **0.80** shall be provided, or the minimum thickness required by (b) or (d) of Table 8.3.1.2 shall be increased by at least 10 percent in the panel with a discontinuous edge.

8.3.1.3 The thickness of a concrete floor finish shall be permitted to be included in h if it is placed monolithically with the floor slab, or if the floor finish is designed to be composite with the floor slab in accordance with 16.4.

8.3.1.4 If single- or multiple-leg stirrups are used as shear reinforcement, the slab thickness shall be sufficient to satisfy the requirements for d in 22.6.7.1.

8.3.2 *Calculated deflection limits*

8.3.2.1 Immediate and time-dependent deflections shall be calculated in accordance with 24.2 and shall not exceed the limits in 24.2.2 for two-way slabs given in (a) through (c):

(a) Nonprestressed slabs not satisfying 8.3.1
(b) Nonprestressed slabs without interior beams spanning between the supports on all sides and having a ratio of long-to-short span exceeding 2.0
(c) Prestressed slabs

8.3.2.2 For nonprestressed composite concrete slabs satisfying 8.3.1.1 or 8.3.1.2, deflections occurring after the member becomes composite need not be calculated. Deflections occurring before the member becomes composite shall be investigated, unless the precomposite thickness also satisfies 8.3.1.1 or 8.3.1.2.

8.3.3 *Reinforcement strain limit in nonprestressed slabs*

8.3.3.1 Nonprestressed slabs shall be tension-controlled in accordance with Table 21.2.2.

8.3.4 *Stress limits in prestressed slabs*

8.3.4.1 Prestressed slabs shall be designed as Class U with $f_t \leq 6\sqrt{f'_c}$. Other stresses in prestressed slabs immedi-

COMMENTARY

R8.3.1.3 The Code does not specify an additional thickness for wearing surfaces subjected to unusual conditions of wear. The need for added thickness for unusual wear is left to the discretion of the licensed design professional.

A concrete floor finish may be considered for strength purposes only if it is cast monolithically with the slab. A separate concrete finish is permitted to be included in the structural thickness if composite action is provided in accordance with 16.4.

R8.3.2 *Calculated deflection limits*

R8.3.2.1 For prestressed flat slabs continuous over two or more spans in each direction, the span-thickness ratio generally should not exceed 42 for floors and 48 for roofs; these limits may be increased to 48 and 52, respectively, if calculations verify that both short- and long-term deflection, camber, and vibration frequency and amplitude are not objectionable.

Short- and long-term deflection and camber should be calculated and checked against serviceability requirements of the structure.

R8.3.2.2 If any portion of a composite member is prestressed, or if the member is prestressed after the components have been cast, the provisions of 8.3.2.1 apply and deflections are to be calculated. For nonprestressed composite members, deflections need to be calculated and compared with the limiting values in Table 24.2.2, only when the thickness of the member or the precast part of the member is less than the minimum thickness given in Table 8.3.1.1. In unshored construction, the thickness of concern depends on whether the deflection before or after the attainment of effective composite action is being considered.

R8.3.3 *Reinforcement strain limit in nonprestressed slabs*

R8.3.3.1 The basis for a reinforcement strain limit for two-way slabs is the same as that for beams. Refer to R9.3.3 for additional information.

CODE

ately after transfer and at service loads shall not exceed the permissible stresses in 24.5.3 and 24.5.4.

8.4—Required strength

8.4.1 *General*

8.4.1.1 Required strength shall be calculated in accordance with the factored load combinations in Chapter 5.

8.4.1.2 Required strength shall be calculated in accordance with the analysis procedures given in Chapter 6.

8.4.1.3 For prestressed slabs, effects of reactions induced by prestressing shall be considered in accordance with 5.3.11.

8.4.1.4 For a slab system supported by columns or walls, dimensions c_1, c_2, and ℓ_n shall be based on an effective support area. The effective support area is the intersection of the bottom surface of the slab, or drop panel or shear cap if present, with the largest right circular cone, right pyramid, or tapered wedge whose surfaces are located within the column and the capital or bracket and are oriented no greater than 45 degrees to the axis of the column.

8.4.1.5 A column strip is a design strip with a width on each side of a column centerline equal to the lesser of $\mathbf{0.25\ell_2}$ and $\mathbf{0.25\ell_1}$. A column strip shall include beams within the strip, if present.

8.4.1.6 A middle strip is a design strip bounded by two column strips.

8.4.1.7 A panel is bounded by column, beam, or wall centerlines on all sides.

8.4.1.8 For monolithic or fully composite construction supporting two-way slabs, a beam includes that portion of slab, on each side of the beam extending a distance equal to the projection of the beam above or below the slab, whichever is greater, but not greater than four times the slab thickness.

COMMENTARY

R8.4—Required strength

R8.4.1 *General*

R8.4.1.2 To determine service and factored moments as well as shears in prestressed slab systems, numerical analysis is required rather than simplified approaches such as the direct design method. The equivalent frame method of analysis as contained in the 2014 edition of the Code is a numerical method that has been shown by tests of large structural models to satisfactorily predict factored moments and shears in prestressed slab systems (Smith and Burns 1974; Burns and Hemakom 1977; Hawkins 1981; PTI DC20.8; Gerber and Burns 1971; Scordelis et al. 1959). The referenced research also shows that analysis using prismatic sections or other approximations of stiffness may provide erroneous and unsafe results. Moment redistribution for prestressed slabs is permitted in accordance with 6.6.5. PTI DC20.8 provides guidance for prestressed concrete slab systems.

R8.4.1.7 A panel includes all flexural elements between column centerlines. Thus, the column strip includes the beam, if any.

R8.4.1.8 For monolithic or fully composite construction, the beams include portions of the slab as flanges. Two examples of the rule are provided in Fig. R8.4.1.8.

CODE

8.4.1.9 Combining the results of a gravity load analysis with the results of a lateral load analysis shall be permitted.

COMMENTARY

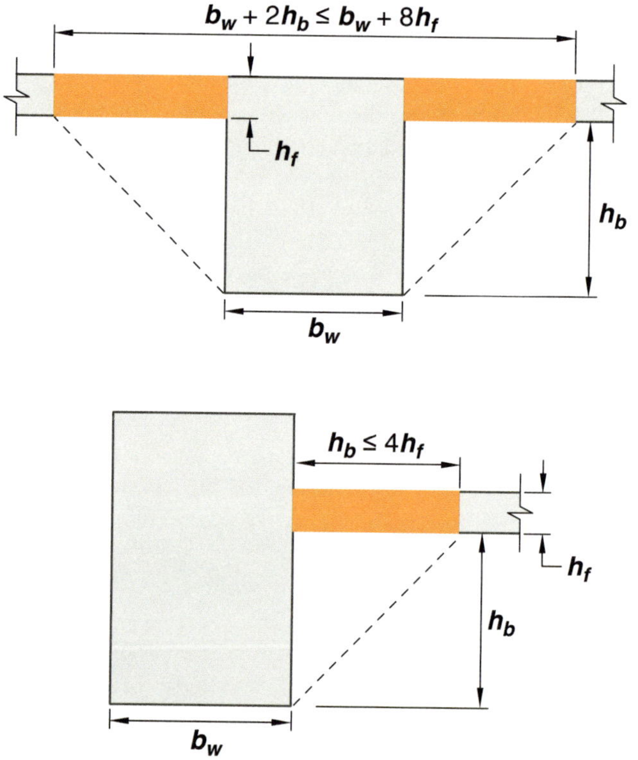

***Fig. R8.4.1.8**—Examples of the portion of slab to be included with the beam under 8.4.1.8.*

8.4.2 *Factored moment*

8.4.2.1 For slabs built integrally with supports, M_u at the support shall be permitted to be calculated at the face of support.

R8.4.2 *Factored moment*

8.4.2.2 *Factored slab moment resisted by the column*

8.4.2.2.1 If gravity, wind, earthquake, or other loads cause a transfer of moment between the slab and column, a fraction of M_{sc}, the factored slab moment resisted by the column at a joint, shall be transferred by flexure in accordance with 8.4.2.2.2 through 8.4.2.2.5.

8.4.2.2.2 The fraction of factored slab moment resisted by the column, $\gamma_f M_{sc}$, shall be assumed to be transferred by flexure, where γ_f shall be calculated by:

$$\gamma_f = \frac{1}{1+\left(\frac{2}{3}\right)\sqrt{\frac{b_1}{b_2}}} \qquad (8.4.2.2.2)$$

8.4.2.2.3 The effective slab width b_{slab} for resisting $\gamma_f M_{sc}$ shall be the width of column or capital plus a distance on each side in accordance with Table 8.4.2.2.3.

R8.4.2.2 *Factored slab moment resisted by the column*

R8.4.2.2.1 This section is concerned primarily with slab systems without beams.

R8.4.2.2.3 Unless measures are taken to resist the torsional and shear stresses, all reinforcement resisting that part of the moment to be transferred to the column by flexure should

CODE

COMMENTARY

be placed between lines that are one and one-half the slab or drop panel thickness, **1.5*h***, on each side of the column.

Table 8.4.2.2.3—Dimensional limits for effective slab width

	Distance on each side of column or capital	
Without drop panel or shear cap	Lesser	1.5*h* of slab
		Distance to edge of slab
With drop panel or shear cap	Lesser	1.5*h* of drop or cap
		Distance to edge of the drop or cap plus 1.5*h* of slab

8.4.2.2.4 For nonprestressed slabs, where the limitations on v_{uv} and ε_t in Table 8.4.2.2.4 are satisfied, γ_f shall be permitted to be increased to the maximum modified values provided in Table 8.4.2.2.4, where v_c is calculated in accordance with 22.6.5.

R8.4.2.2.4 Some flexibility in distribution of M_{sc} transferred by shear and flexure at both exterior and interior columns is possible. Interior, exterior, and corner columns refer to slab-column connections for which the critical perimeter for rectangular columns has four, three, and two sides, respectively.

At exterior columns, for M_{sc} resisted about an axis parallel to the edge, the portion of moment transferred by eccentricity of shear $\gamma_v M_{sc}$ may be reduced, provided that the factored shear at the column (excluding the shear produced by moment transfer) does not exceed 75 percent of the shear strength ϕv_c as defined in 22.6.5.1 for edge columns, or 50 percent for corner columns. Tests (Moehle 1988; ACI 352.1R) indicate that there is no significant interaction between shear and M_{sc} at the exterior column in such cases. Note that as $\gamma_v M_{sc}$ is decreased, $\gamma_f M_{sc}$ is increased.

At interior columns, some flexibility in distributing M_{sc} transferred by shear and flexure is possible, but with more severe limitations than for exterior columns. For interior columns, M_{sc} transferred by flexure is permitted to be increased up to 25 percent, provided that the factored shear (excluding the shear caused by the moment transfer) at the interior columns does not exceed 40 percent of the shear strength ϕv_c as defined in 22.6.5.1.

If the factored shear for a slab-column connection is large, the slab-column joint cannot always develop all of the reinforcement provided in the effective width. The modifications for interior slab-column connections in this provision are permitted only where the reinforcement required to develop $\gamma_f M_{sc}$ within the effective width has a net tensile strain ε_t not less than ε_{ty} + **0.008**, where the value of ε_{ty} is determined in 21.2.2. The use of Eq. (8.4.2.2.2) without the modification permitted in this provision will generally indicate overstress conditions on the joint. This provision is intended to improve ductile behavior of the slab-column joint. If reversal of moments occurs at opposite faces of an interior column, both top and bottom reinforcement should be concentrated within the effective width. A ratio of top-to-bottom reinforcement of approximately 2 has been observed to be appropriate.

Before the 2019 Code, the strain limits on ε_t in Table 8.4.2.2.4 were constants of 0.004 and 0.010. Beginning with the 2019 Code, to accommodate nonprestressed reinforcement

8 Two-way Slabs

CODE

COMMENTARY

of higher grades, these limits are replaced by the expressions $\boldsymbol{\varepsilon_{ty}}$ **+ 0.003** and $\boldsymbol{\varepsilon_{ty}}$ **+ 0.008**, respectively. The first expression is the same expression as used for the limit on ε_t for classification of tension-controlled members in Table 21.2.2; this expression is further described in Commentary R21.2.2. The second expression provides a limit on ε_t with Grade 60 reinforcement that is approximately the same value as the former constant of 0.010.

Table 8.4.2.2.4—Maximum modified values of γ_f for nonprestressed two-way slabs

Column location	Span direction	v_{uv}	ε_t (within b_{slab})	Maximum modified γ_f
Corner column	Either direction	$\leq 0.5\phi v_c$	$\geq \varepsilon_{ty} + 0.003$	1.0
Edge column	Perpendicular to the edge	$\leq 0.75\phi v_c$	$\geq \varepsilon_{ty} + 0.003$	1.0
	Parallel to the edge	$\leq 0.4\phi v_c$	$\geq \varepsilon_{ty} + 0.008$	$\dfrac{1.25}{1+\left(\dfrac{2}{3}\right)\sqrt{\dfrac{b_1}{b_2}}} \leq 1.0$
Interior column	Either direction	$\leq 0.4\phi v_c$	$\geq \varepsilon_{ty} + 0.008$	$\dfrac{1.25}{1+\left(\dfrac{2}{3}\right)\sqrt{\dfrac{b_1}{b_2}}} \leq 1.0$

8.4.2.2.5 Concentration of reinforcement over the column by closer spacing or additional reinforcement shall be used to resist moment on the effective slab width defined in 8.4.2.2.2 and 8.4.2.2.3.

8.4.2.2.6 The fraction of M_{sc} not calculated to be resisted by flexure shall be assumed to be resisted by eccentricity of shear in accordance with 8.4.4.2.

8.4.3 *Factored one-way shear*

8.4.3.1 For slabs built integrally with supports, V_u at the support shall be permitted to be calculated at the face of support.

8.4.3.2 Sections between the face of support and a critical section located d from the face of support for nonprestressed slabs and $h/2$ from the face of support for prestressed slabs shall be permitted to be designed for V_u at that critical section if (a) through (c) are satisfied:

(a) Support reaction, in direction of applied shear, introduces compression into the end regions of the slab.
(b) Loads are applied at or near the top surface of the slab.
(c) No concentrated load occurs between the face of support and critical section.

8.4.4 *Factored two-way shear*

R8.4.4 *Factored two-way shear*

The calculated shear stresses in the slab around the column are required to conform to the requirements of 22.6.

CODE

8.4.4.1 *Critical section*

8.4.4.1.1 Slabs shall be evaluated for two-way shear in the vicinity of columns, concentrated loads, and reaction areas at critical sections in accordance with 22.6.4.

8.4.4.1.2 Slabs reinforced with stirrups or headed shear stud reinforcement shall be evaluated for two-way shear at critical sections in accordance with 22.6.4.2.

8.4.4.2 *Factored two-way shear stress due to shear and factored slab moment resisted by the column*

8.4.4.2.1 For two-way shear with factored slab moment resisted by the column, factored shear stress v_u shall be calculated at critical sections in accordance with 8.4.4.1. Factored shear stress v_u corresponds to a combination of v_{uv} and the shear stress produced by $\gamma_v M_{sc}$, where γ_v is given in 8.4.4.2.2 and M_{sc} is given in 8.4.2.2.1.

8.4.4.2.2 The fraction of M_{sc} transferred by eccentricity of shear, $\gamma_v M_{sc}$, shall be applied at the centroid of the critical section in accordance with 8.4.4.1, where:

$$\gamma_v = 1 - \gamma_f \qquad (8.4.4.2.2)$$

8.4.4.2.3 The factored shear stress resulting from $\gamma_v M_{sc}$ shall be assumed to vary linearly about the centroid of the critical section in accordance with 8.4.4.1.

COMMENTARY

R8.4.4.2 *Factored two-way shear stress due to shear and factored slab moment resisted by the column*

R8.4.4.2.2 Hanson and Hanson (1968) found that where moment is transferred between a column and a slab, 60 percent of the moment should be considered transferred by flexure across the perimeter of the critical section defined in 22.6.4.1, and 40 percent by eccentricity of the shear about the centroid of the critical section. For rectangular columns, the portion of the moment transferred by flexure increases as the width of the face of the critical section resisting the moment increases, as given by Eq. (8.4.2.2.2).

Most of the data in Hanson and Hanson (1968) were obtained from tests of square columns. Limited information is available for round columns; however, these can be approximated as square columns having the same cross-sectional area.

R8.4.4.2.3 The stress distribution is assumed as illustrated in Fig. R8.4.4.2.3 for an interior or exterior column. The perimeter of the critical section, ***ABCD***, is determined in accordance with 22.6.4.1. The factored shear stress v_{uv} and factored slab moment resisted by the column M_{sc} are determined at the centroidal axis ***c-c*** of the critical section. The maximum factored shear stress may be calculated from:

$$v_{u,AB} = v_{uv} + \frac{\gamma_v M_{sc} c_{AB}}{J_c}$$

or

$$v_{u,CD} = v_{uv} - \frac{\gamma_v M_{sc} cCD}{J_c}$$

where γ_v is given by Eq. (8.4.4.2.2).

For an interior column, J_c may be calculated by:

J_c = property of assumed critical section analogous to polar moment of inertia

CODE

COMMENTARY

$$= \frac{d(c_1+d)^3}{6} + \frac{(c_1+d)d^3}{6} + \frac{d(c_2+d)(c_1+d)^2}{2}$$

Similar equations may be developed for J_c for columns located at the edge or corner of a slab.

The fraction of M_{sc} not transferred by eccentricity of the shear should be transferred by flexure in accordance with 8.4.2.2. A conservative method assigns the fraction transferred by flexure over an effective slab width defined in 8.4.2.2.3. Often, column strip reinforcement is concentrated near the column to accommodate M_{sc}. Available test data (Hanson and Hanson 1968) seem to indicate that this practice does not increase shear strength but may be desirable to increase the stiffness of the slab-column junction.

Test data (Hawkins 1981) indicate that the moment transfer strength of a prestressed slab-to-column connection can be calculated using the procedures of 8.4.2.2 and 8.4.4.2.

Where shear reinforcement has been used, the critical section beyond the shear reinforcement generally has a polygonal shape (Fig. R8.7.6(d) and (e)). Equations for calculating shear stresses on such sections are given in ACI 421.1R.

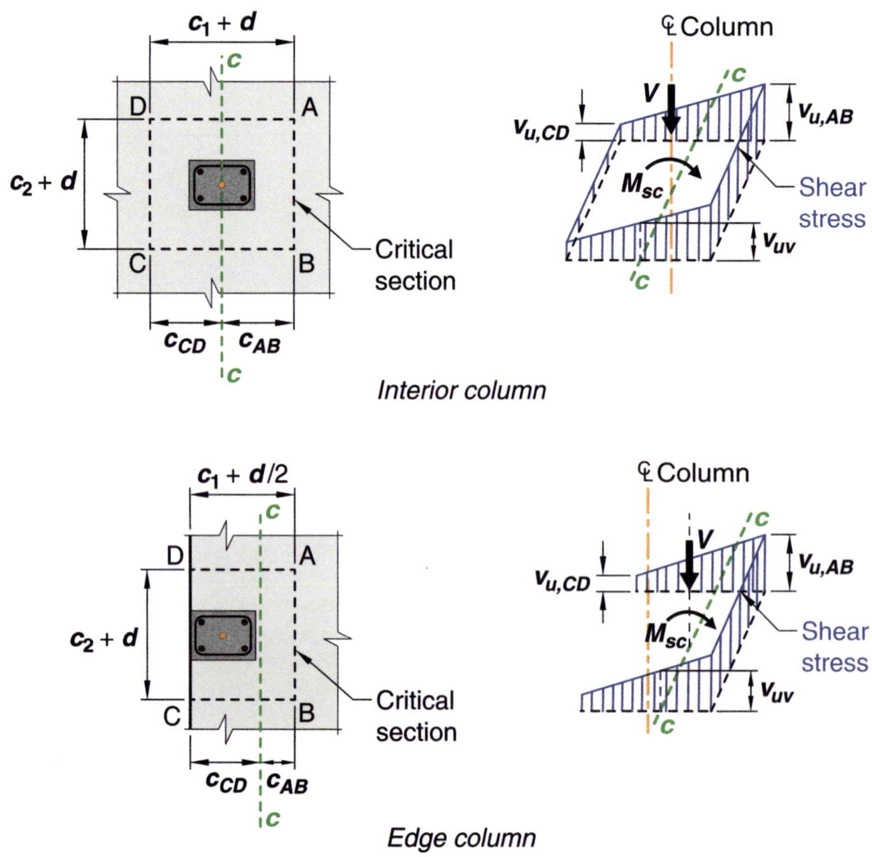

Fig. R8.4.4.2.3—Assumed distribution of shear stress.

CODE

8.5—Design strength

8.5.1 *General*

8.5.1.1 For each applicable factored load combination, design strength shall satisfy $\phi S_n \geq U$, including (a) through (d). Interaction between load effects shall be considered.

(a) $\phi M_n \geq M_u$ at all sections along the span in each direction
(b) $\phi M_n \geq \gamma_f M_{sc}$ within b_{slab} as defined in 8.4.2.2.3
(c) $\phi V_n \geq V_u$ at all sections along the span in each direction for one-way shear
(d) $\phi v_n \geq v_u$ at the critical sections defined in 8.4.4.1 for two-way shear

8.5.1.2 ϕ shall be in accordance with 21.2.

8.5.2 *Moment*

8.5.2.1 M_n shall be calculated in accordance with 22.3.

8.5.2.2 In calculating M_n for nonprestressed slabs with a drop panel, the thickness of the drop panel below the slab shall not be assumed to be greater than one-fourth the distance from the edge of drop panel to the face of column or column capital.

8.5.2.3 In calculating M_n for prestressed slabs, external tendons shall be considered as unbonded unless the external tendons are effectively bonded to the slab along its entire length.

8.5.3 *Shear*

8.5.3.1 Design shear strength of slabs in the vicinity of columns, concentrated loads, or reaction areas shall be the more severe of 8.5.3.1.1 and 8.5.3.1.2.

8.5.3.1.1 For one-way shear, where each critical section to be investigated extends in a plane across the entire slab width, V_n shall be calculated in accordance with 22.5.

8.5.3.1.2 For two-way shear, v_n shall be calculated in accordance with 22.6.

8.5.3.2 For composite concrete slabs, horizontal shear strength V_{nh} shall be calculated in accordance with 16.4.

8.5.4 *Openings in slab systems*

8.5.4.1 Openings of any size shall be permitted in slab systems if shown by analysis that all strength and serviceability requirements, including the limits on deflections, are satisfied.

COMMENTARY

R8.5—Design strength

R8.5.1 *General*

R8.5.1.1 Refer to R9.5.1.1.

R8.5.3 *Shear*

R8.5.3.1 Differentiation should be made between a long and narrow slab acting as a beam, and a slab subject to two-way action where failure may occur by punching along a truncated cone or pyramid around a concentrated load or reaction area.

CODE

8.5.4.2 As an alternative to 8.5.4.1, openings shall be permitted in slab systems without beams in accordance with (a) through (d).

(a) Openings of any size shall be permitted in the area common to intersecting middle strips, but the total quantity of reinforcement in the panel shall be at least that required for the panel without the opening.
(b) At two intersecting column strips, not more than one-eighth the width of column strip in either span shall be interrupted by openings. A quantity of reinforcement at least equal to that interrupted by an opening shall be added on the sides of the opening.
(c) At the intersection of one column strip and one middle strip, not more than one-fourth of the reinforcement in either strip shall be interrupted by openings. A quantity of reinforcement at least equal to that interrupted by an opening shall be added on the sides of the opening.
(d) If an opening is located closer than $\mathbf{4h}$ from the periphery of a column, concentrated load or reaction area, 22.6.4.3 shall be satisfied.

8.6—Reinforcement limits

8.6.1 *Minimum flexural reinforcement in nonprestressed slabs*

8.6.1.1 A minimum area of flexural reinforcement, $A_{s,min}$ of $\mathbf{0.0018}A_g$, or as defined in 8.6.1.2, shall be provided near the tension face of the slab in the direction of the span under consideration.

COMMENTARY

R8.6—Reinforcement limits

R8.6.1 *Minimum flexural reinforcement in nonprestressed slabs*

R8.6.1.1 The required area of deformed or welded wire reinforcement used as minimum flexural reinforcement is the same as that required for shrinkage and temperature in 24.4.3.2. However, whereas shrinkage and temperature reinforcement is permitted to be distributed between the two faces of the slab as deemed appropriate for specific conditions, minimum flexural reinforcement should be placed as close as practicable to the face of the concrete in tension due to applied loads.

Figure R8.6.1.1 illustrates the arrangement of minimum reinforcement required near the top of a two-way slab supporting uniform gravity load. The bar cutoff points are based on the requirements shown in Fig. 8.7.4.1.3.

To improve crack control and to intercept potential punching shear cracks with tension reinforcement, the licensed design professional should consider specifying continuous reinforcement in each direction near both faces of thick two-way slabs, such as transfer slabs, podium slabs, and mat foundations. Also refer to R8.7.4.1.3.

CODE

COMMENTARY

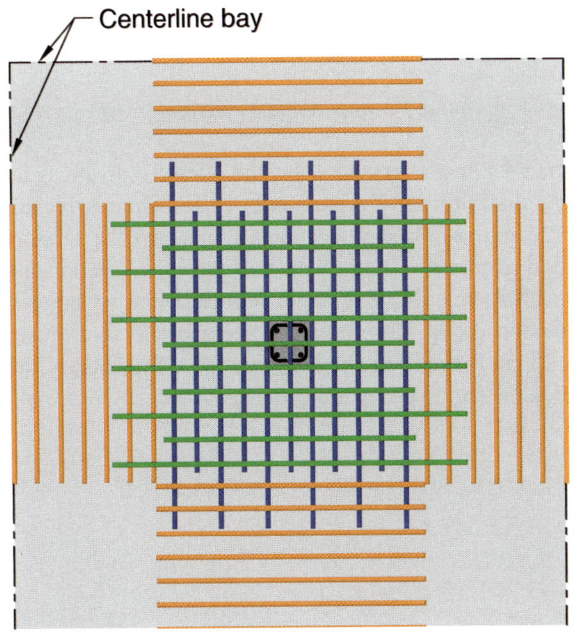

Fig. R8.6.1.1—Arrangement of minimum reinforcement near the top of a two-way slab.

8.6.1.2 If $v_{uv} > \phi 2\lambda_s\lambda\sqrt{f_c'}$ on the critical section for two-way shear surrounding a column, concentrated load, or reaction area, $A_{s,min}$, provided over the width b_{slab}, shall satisfy Eq. (8.6.1.2)

$$A_{s,min} = \frac{5v_{uv}b_{slab}b_o}{\phi\alpha_s f_y} \quad (8.6.1.2)$$

where b_{slab} is the width specified in 8.4.2.2.3, α_s is given in 22.6.5.3, ϕ is the value for shear, and λ_s is given in 22.5.5.1.3.

R8.6.1.2 Tests on interior column-to-slab connections with lightly reinforced slabs with and without shear reinforcement (Peiris and Ghali 2012; Hawkins and Ospina 2017; Widianto et al. 2009; Muttoni 2008; Dam et al. 2017) have shown that yielding of the slab flexural tension reinforcement in the vicinity of the column or loaded area leads to increased local rotations and opening of any inclined crack existing within the slab. In such cases, sliding along the inclined crack can cause a flexure-driven punching failure at a shear force less than the strength calculated by the two-way shear equations of Table 22.6.5.2 for slabs without shear reinforcement and less than the strength calculated in accordance with 22.6.6.3 for slabs with shear reinforcement.

Tests of slabs with flexural reinforcement less than $A_{s,min}$ have shown that shear reinforcement does not increase the punching shear strength. However, shear reinforcement may increase plastic rotations prior to the flexure-driven punching failure (Peiris and Ghali 2012).

Inclined cracking develops within the depth of the slab at a shear stress of approximately $2\lambda_s\lambda\sqrt{f_c'}$. At higher shear stresses, the possibility of a flexure-driven punching failure increases if $A_{s,min}$ is not satisfied. $A_{s,min}$ was developed for an interior column, such that the factored shear force on the critical section for shear equals the shear force associated with local yielding at the column faces.

To derive Eq. (8.6.1.2) the shear force associated with local yielding was taken as $8A_{s,min}f_yd/b_{slab}$ for an interior column connection (Hawkins and Ospina 2017) and generalized as $(\alpha_s/5)A_{s,min}f_yd/b_{slab}$ to account for edge and corner conditions. $A_{s,min}$ also needs to be provided at the periphery of drop panels and shear caps.

CODE

8.6.2 *Minimum flexural reinforcement in prestressed slabs*

8.6.2.1 For prestressed slabs, the effective prestress force $A_{ps}f_{se}$ shall provide a minimum average compressive stress of 125 psi on the slab section tributary to the tendon or tendon group. For slabs with varying cross section along the slab span, either parallel or perpendicular to the tendon or tendon group, the minimum average effective prestress of 125 psi is required at every cross section tributary to the tendon or tendon group along the span.

8.6.2.2 For slabs with bonded prestressed reinforcement, total quantity of A_s and A_{ps} shall be adequate to develop a factored load at least 1.2 times the cracking load calculated on the basis of f_r defined in 19.2.3.

8.6.2.2.1 For slabs with both flexural and shear design strength at least twice the required strength, 8.6.2.2 need not be satisfied.

8.6.2.3 For prestressed slabs, a minimum area of bonded deformed longitudinal reinforcement, $A_{s,min}$, shall be provided in the precompressed tension zone in the direction of the span under consideration in accordance with Table 8.6.2.3.

Table 8.6.2.3—Minimum bonded deformed longitudinal reinforcement $A_{s,min}$ in two-way slabs with bonded or unbonded tendons

Region	Calculated f_t after all losses, psi	$A_{s,min}$, in.[2]	
Positive moment	$f_t \leq 2\sqrt{f'_c}$	Not required	(a)
	$2\sqrt{f'_c} < f_t \leq 6\sqrt{f'_c}$	$\frac{N_c}{0.5f_y}$	(b)[1],[2]
Negative moment at columns	$f_t \leq 6\sqrt{f'_c}$	$0.00075A_{cf}$	(c)[2]

[1]The value of f_y shall not exceed 60,000 psi.

[2]For slabs with bonded tendons, it shall be permitted to reduce $A_{s,min}$ by the area of the bonded prestressed reinforcement located within the area used to determine N_c for positive moment, or within the width of slab defined in 8.7.5.3(a) for negative moment.

COMMENTARY

Commentary on size effect factor is provided in R22.5.5.1 and R22.6.5.2.

R8.6.2 *Minimum flexural reinforcement in prestressed slabs*

R8.6.2.1 The minimum average effective prestress of 125 psi was used in two-way test panels in the early 1970s to address punching shear concerns of lightly reinforced slabs. For this reason, the minimum effective prestress is required to be provided at every cross section.

If the slab thickness varies along the span of a slab or perpendicular to the span of a slab, resulting in a varying slab cross section, the 125 psi minimum effective prestress and the maximum tendon spacing is required at every cross section tributary to the tendon or group of tendons along the span, considering both the thinner and the thicker slab sections. This may result in higher than the minimum f_{pc} in thinner cross sections, and tendons spaced at less than the maximum in thicker cross sections along a span with varying thickness, due to the practical aspects of tendon placement in the field.

R8.6.2.2 This provision is a precaution against abrupt flexural failure developing immediately after cracking. A flexural member designed according to Code provisions requires considerable additional load beyond cracking to reach its flexural strength. Thus, considerable deflection would warn that the member strength is approaching. If the flexural strength were reached shortly after cracking, the warning deflection would not occur. Transfer of force between the concrete and the prestressed reinforcement, and abrupt flexural failure immediately after cracking, does not occur when the prestressed reinforcement is unbonded (ACI 423.3R); therefore, this requirement does not apply to members with unbonded tendons.

R8.6.2.3 Some bonded reinforcement is required by the Code in prestressed slabs to limit crack width and spacing at service load when concrete tensile stresses exceed the modulus of rupture and, for slabs with unbonded tendons, to ensure flexural performance at nominal strength, rather than performance as a tied arch. Providing the minimum bonded reinforcement as stipulated in this provision helps to ensure adequate performance.

The minimum amount of bonded reinforcement in two-way flat slab systems is based on reports by Joint ACI-ASCE Committee 423 (1958) and ACI 423.3R. Limited research available for two-way flat slabs with drop panels (Odello and Mehta 1967) indicates that behavior of these particular systems is similar to the behavior of flat plates.

For usual loads and span lengths, flat plate tests summarized in Joint ACI-ASCE Committee 423 (1958) and experience since the 1963 Code was adopted indicate satisfactory performance without bonded reinforcement in positive moment regions where $f_t \leq 2\sqrt{f'_c}$. In positive moment regions where $2\sqrt{f'_c} \leq f_t \leq 6\sqrt{f'_c}$, a minimum bonded reinforcement area

CODE | COMMENTARY

proportioned to resist N_c according to Eq. (8.6.2.3(b)) is required. The tensile force N_c is calculated at service load on the basis of an uncracked, homogeneous section.

Research on unbonded post-tensioned two-way flat slab systems (Joint ACI-ASCE Committee 423 1958, 1974; ACI 423.3R; Odello and Mehta 1967) shows that bonded reinforcement in negative moment regions, proportioned on the basis of 0.075 percent of the cross-sectional area of the slab-beam strip, provides sufficient ductility and reduces crack width and spacing. The same area of bonded reinforcement is required in slabs with either bonded or unbonded tendons. The minimum bonded reinforcement area required by Eq. (8.6.2.3(c)) is a minimum area independent of grade of reinforcement or design yield strength. To account for different adjacent tributary spans, this equation is given on the basis of slab-beam strips as defined in 2.3. For rectangular slab panels, this equation is conservatively based on the greater of the cross-sectional areas of the two intersecting slab-beam strips at the column. This ensures that the minimum percentage of reinforcement recommended by research is provided in both directions. Concentration of this reinforcement in the top of the slab directly over and immediately adjacent to the column is important. Research also shows that where low tensile stresses occur at service loads, satisfactory behavior has been achieved at factored loads without bonded reinforcement. However, the Code requires minimum bonded reinforcement regardless of service load stress levels to help ensure flexural continuity and ductility, and to limit crack widths and spacing due to overload, temperature, or shrinkage. Research on post-tensioned flat plate-to-column connections is reported in Smith and Burns (1974), Burns and Hemakom (1977), Hawkins (1981), PTI TAB.1, and Foutch et al. (1990).

Unbonded post-tensioned members do not inherently provide large capacity for energy dissipation under severe earthquake loadings because the member response is primarily elastic. For this reason, unbonded post-tensioned structural members reinforced in accordance with the provisions of this section should be assumed to resist only vertical loads and to act as horizontal diaphragms between energy-dissipating elements under earthquake loadings of the magnitude defined in 18.2.1.

8.7—Reinforcement detailing

8.7.1 *General*

8.7.1.1 Concrete cover for reinforcement shall be in accordance with 20.5.1.

8.7.1.2 Development lengths of deformed and prestressed reinforcement shall be in accordance with 25.4.

8.7.1.3 Splice lengths of deformed reinforcement shall be in accordance with 25.5.

R8.7—Reinforcement detailing

CODE

8.7.1.4 Bundled bars shall be detailed in accordance with 25.6.

8.7.2 *Flexural reinforcement spacing*

8.7.2.1 Minimum spacing s shall be in accordance with 25.2.

8.7.2.2 For nonprestressed solid slabs, maximum spacing s of deformed longitudinal reinforcement shall be the lesser of $2h$ and 18 in. at critical sections, and the lesser of $3h$ and 18 in. at other sections.

8.7.2.3 For prestressed slabs with uniformly distributed loads, maximum spacing s of tendons or groups of tendons in at least one direction shall be the lesser of $8h$ and 5 ft.

8.7.2.4 Concentrated loads and openings shall be considered in determining tendon spacing.

8.7.3 *Corner restraint in slabs*

8.7.3.1 At exterior corners of slabs supported by edge walls or where one or more edge beams have a value of α_f greater than 1.0, reinforcement at top and bottom of slab shall be designed to resist M_u per unit width due to corner effects equal to the maximum positive M_u per unit width in the slab panel.

8.7.3.1.1 Factored moment due to corner effects, M_u, shall be assumed to be about an axis perpendicular to the diagonal from the corner in the top of the slab and about an axis parallel to the diagonal from the corner in the bottom of the slab.

8.7.3.1.2 Reinforcement shall be provided for a distance in each direction from the corner equal to one-fifth the longer span.

8.7.3.1.3 Reinforcement shall be placed parallel to the diagonal in the top of the slab and perpendicular to the diagonal in the bottom of the slab. Alternatively, reinforcement shall be placed in two layers parallel to the sides of the slab in both the top and bottom of the slab.

COMMENTARY

R8.7.2 *Flexural reinforcement spacing*

R8.7.2.2 The requirement that the center-to-center spacing of the reinforcement be not more than two times the slab thickness applies only to the reinforcement in solid slabs, and not to reinforcement in joists or waffle slabs. This limitation is to ensure slab action, control cracking, and provide for the possibility of loads concentrated on small areas of the slab. Refer also to R24.3.

R8.7.2.3 This section provides specific guidance concerning tendon distribution that will permit the use of banded tendon distributions in one direction. This method of tendon distribution has been shown to provide satisfactory performance by structural research (Burns and Hemakom 1977).

R8.7.3 *Corner restraint in slabs*

R8.7.3.1 Unrestrained corners of two-way slabs tend to lift when loaded. If this lifting tendency is restrained by edge walls or beams, bending moments result in the slab. This section requires reinforcement to resist these moments and control cracking. Reinforcement provided for flexure in the primary directions may be used to satisfy this requirement. Refer to Fig. R8.7.3.1.

CODE | COMMENTARY

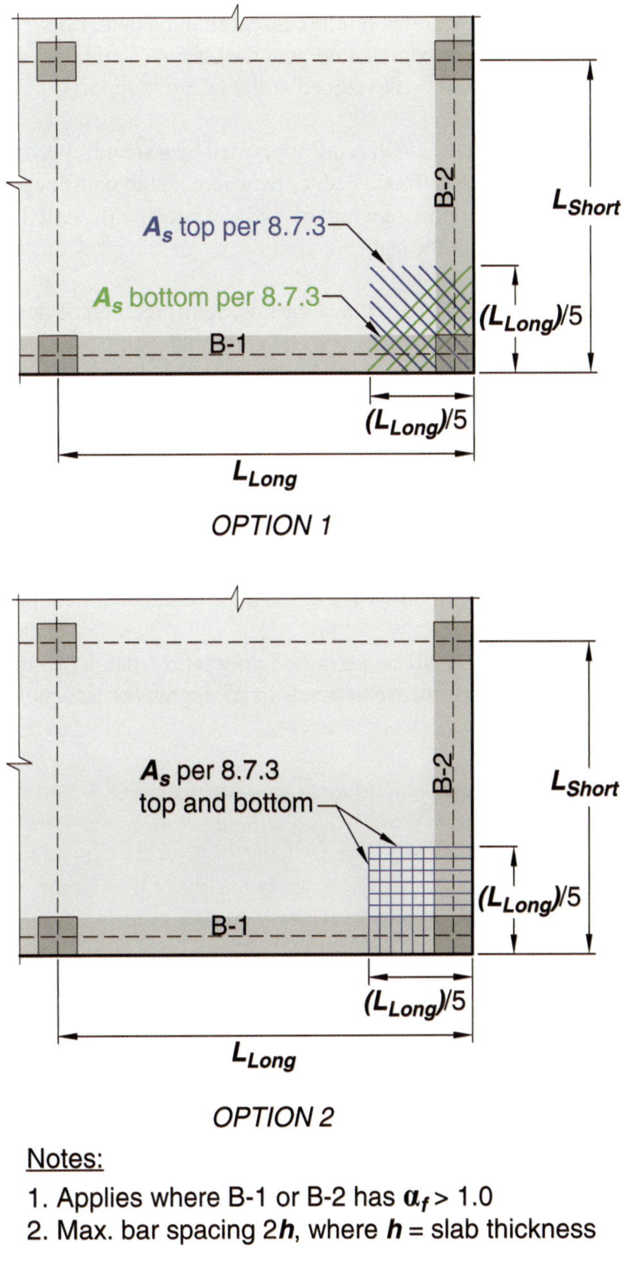

***Fig. R8.7.3.1**—Slab corner reinforcement.*

8.7.4 *Flexural reinforcement in nonprestressed slabs*

8.7.4.1 *Termination of reinforcement*

8.7.4.1.1 Where a slab is supported on spandrel beams, columns, or walls, anchorage of reinforcement perpendicular to a discontinuous edge shall satisfy (a) and (b):

(a) Positive moment reinforcement shall extend to the edge of slab and have embedment, straight or hooked, at least 6 in. into spandrel beams, columns, or walls

R8.7.4 *Flexural reinforcement in nonprestressed slabs*

R8.7.4.1 *Termination of reinforcement*

R8.7.4.1.1 and R8.7.4.1.2 Bending moments in slabs at spandrel beams may vary significantly. If spandrel beams are built solidly into walls, the slab approaches complete fixity. Without an integral wall, the slab could approach being simply supported, depending on the torsional rigidity of the spandrel beam or slab edge. These requirements provide for unknown conditions that might normally occur in a structure.

CODE

(b) Negative moment reinforcement shall be bent, hooked, or otherwise anchored into spandrel beams, columns, or walls, and shall be developed at the face of support

8.7.4.1.2 Where a slab is not supported by a spandrel beam or wall at a discontinuous edge, or where a slab cantilevers beyond the support, anchorage of reinforcement shall be permitted within the slab.

8.7.4.1.3 For slabs without beams, reinforcement extensions shall be in accordance with (a) through (c):

(a) Reinforcement lengths shall be at least in accordance with Fig. 8.7.4.1.3, and if slabs act as primary members resisting lateral loads, reinforcement lengths shall be at least those required by analysis.
(b) If adjacent spans are unequal, extensions of negative moment reinforcement beyond the face of support in accordance with Fig. 8.7.4.1.3 shall be based on the longer span.
(c) Bent bars shall be permitted only where the depth-to-span ratio permits use of bends of 45 degrees or less.

COMMENTARY

R8.7.4.1.3 The minimum lengths and extensions of reinforcement expressed as a fraction of the clear span in Fig. 8.7.4.1.3 were developed for slabs of ordinary proportions supporting gravity loads. These minimum lengths and extensions of bars may not be sufficient to intercept potential punching shear cracks in thick two-way slabs such as transfer slabs, podium slabs, and mat foundations. Therefore, the Code requires extensions for at least half of the column strip top bars to be at least $5d$. For slabs with drop panels, d is the effective depth within the drop panel. In these thick two-way slabs, continuous reinforcement in each direction near both faces is desirable to improve structural integrity, control cracking, and reduce creep deflections. As illustrated in Fig. R8.7.4.1.3, punching shear cracks, which can develop at angles as low as approximately 20 degrees, may not be intercepted by the tension reinforcement in thick slabs if this reinforcement does not extend to at least $5d$ beyond the face of the support. The $5d$ bar extension requirement governs where ℓ_n/h is less than approximately 15. For moments resulting from combined lateral and gravity loadings, these minimum lengths and extensions may not be sufficient.

Bent bars are seldom used and are difficult to place properly. Bent bars, however, are permitted provided they comply with 8.7.4.1.3(c). Further guidance on the use of bent bar systems can be found in 13.4.8 of the 1983 Code.

CODE **COMMENTARY**

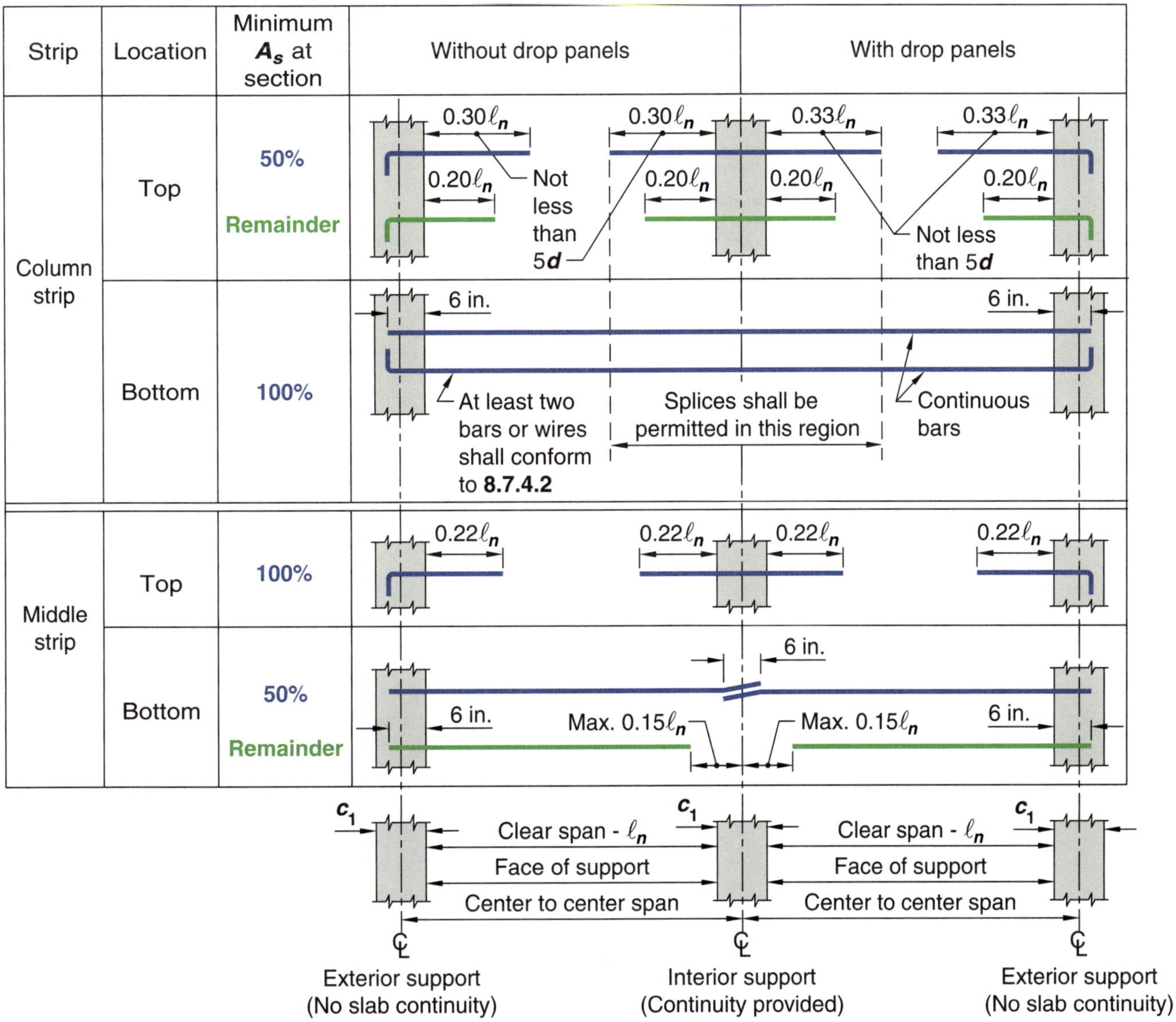

Fig. 8.7.4.1.3—Minimum extensions for deformed reinforcement in two-way slabs without beams.

CODE

COMMENTARY

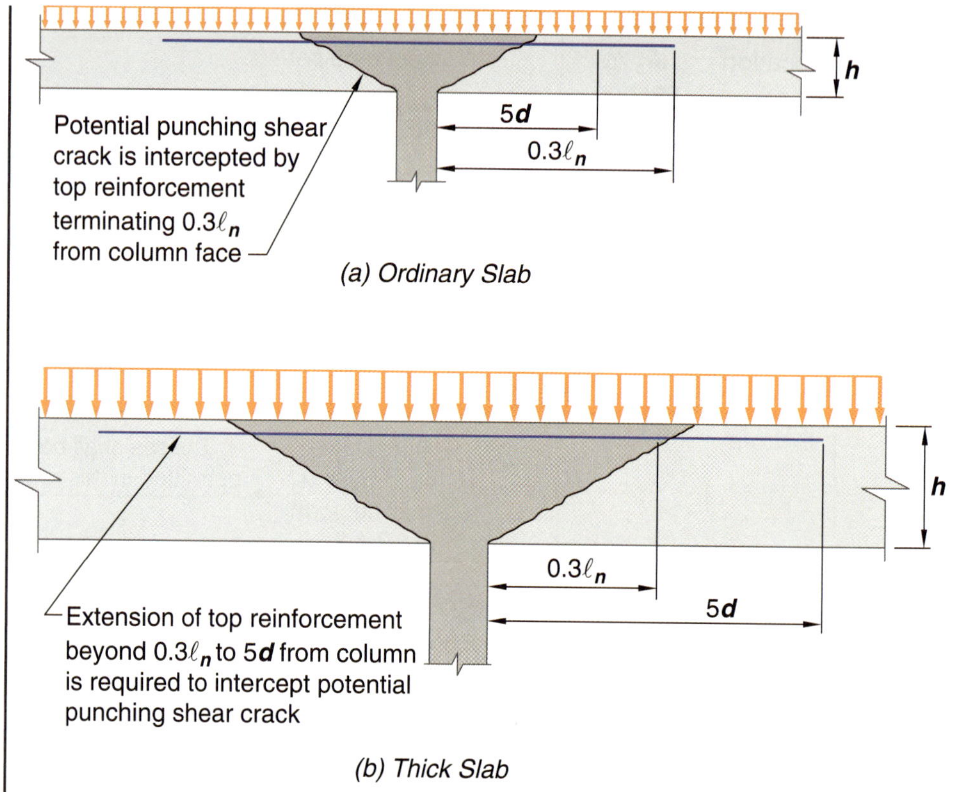

***Fig. R8.7.4.1.3**—Punching shear cracks in ordinary and thick slabs.*

8.7.4.2 *Structural integrity*

8.7.4.2.1 All bottom deformed bars or deformed wires within the column strip, in each direction, shall be continuous or spliced using mechanical or welded splices in accordance with 25.5.7 or Class B tension lap splices in accordance with 25.5.2. Splices shall be located in accordance with Fig. 8.7.4.1.3.

8.7.4.2.2 At least two of the column strip bottom bars or wires in each direction shall pass within the region bounded by the longitudinal reinforcement of the column and shall be anchored at exterior supports.

R8.7.4.2 *Structural integrity*

R8.7.4.2.1 and R8.7.4.2.2 The continuous column strip bottom reinforcement provides the slab some residual ability to span to the adjacent supports should a single support be damaged. The two continuous column strip bottom bars or wires through the column may be termed "integrity reinforcement," and are provided to give the slab some residual strength following a single punching shear failure at a single support (Mitchell and Cook 1984). Joint ACI-ASCE Committee 352 (ACI 352.1R) provides further guidance on the design of integrity reinforcement in slab-column connections. Similar provisions for slabs with unbonded tendons are provided in 8.7.5.6.

8.7.5 *Flexural reinforcement in prestressed slabs*

8.7.5.1 External tendons shall be attached to the slab in a manner that maintains the specified eccentricity between the tendons and the concrete centroid through the full range of anticipated member deflections.

8.7.5.2 If bonded deformed longitudinal reinforcement is required to satisfy flexural strength or for tensile stress conditions in accordance with Eq. (8.6.2.3(b)), the detailing requirements of 7.7.3 shall be satisfied.

R8.7.5 *Flexural reinforcement in prestressed slabs*

R8.7.5.2 Bonded reinforcement should be adequately anchored to develop the required strength to resist factored loads. The requirements of 7.7.3 are intended to provide adequate anchorage for tensile or compressive forces developed in bonded reinforcement by flexure under factored

CODE

8.7.5.3 Bonded longitudinal reinforcement required by Eq. (8.6.2.3(c)) shall be placed in the top of the slab, and shall be in accordance with (a) through (c):

(a) Reinforcement shall be distributed between lines that are **1.5*h*** outside opposite faces of the column support.
(b) At least four deformed bars, deformed wires, or bonded strands shall be provided in each direction.
(c) Maximum spacing ***s*** between bonded longitudinal reinforcement shall not exceed 12 in.

8.7.5.4 *Termination of prestressed reinforcement*

8.7.5.4.1 Post-tensioned anchorage zones shall be designed and detailed in accordance with 25.9.

8.7.5.4.2 Post-tensioning anchorages and couplers shall be designed and detailed in accordance with 25.8.

8.7.5.5 *Termination of deformed reinforcement in slabs with unbonded tendons*

8.7.5.5.1 Length of deformed reinforcement required by 8.6.2.3 shall be in accordance with (a) and (b):

(a) In positive moment areas, length of reinforcement shall be at least $\ell_n/3$ and be centered in those areas
(b) In negative moment areas, reinforcement shall extend at least $\ell_n/6$ on each side of the face of support

8.7.5.6 *Structural integrity*

8.7.5.6.1 Except as permitted in 8.7.5.6.3, at least two tendons with 1/2 in. diameter or larger strand shall be placed in each direction at columns in accordance with (a) or (b):

(a) Tendons shall pass through the region bounded by the longitudinal reinforcement of the column.
(b) Tendons shall be anchored within the region bounded by the longitudinal reinforcement of the column, and the anchorage shall be located beyond the column centroid and away from the anchored span.

8.7.5.6.2 Outside of the column and shear cap faces, the two structural integrity tendons required by 8.7.5.6.1 shall pass under any orthogonal tendons in adjacent spans.

COMMENTARY

loads in accordance with 22.3.2, or by tensile stresses at service load in accordance with Eq. (8.6.2.3(b)).

R8.7.5.5 *Termination of deformed reinforcement in slabs with unbonded tendons*

R8.7.5.5.1 The minimum lengths apply for bonded reinforcement required by 8.6.2.3, but not required for flexural strength in accordance with 22.3.2. Research (Odello and Mehta 1967) on continuous spans shows that these minimum lengths provide adequate behavior under service load and factored load conditions.

R8.7.5.6 *Structural integrity*

R8.7.5.6.1 Prestressing tendons that pass through the slab-column joint at any location over the depth of the slab suspend the slab following a punching shear failure, provided the tendons are continuous through or anchored within the region bounded by the longitudinal reinforcement of the column and are prevented from bursting through the top surface of the slab (ACI 352.1R).

R8.7.5.6.2 Between column or shear cap faces, structural integrity tendons should pass below the orthogonal tendons from adjacent spans so that vertical movements of the integrity tendons are restrained by the orthogonal tendons. Where tendons are distributed in one direction and banded in the orthogonal direction, this requirement can be satisfied by first placing the integrity tendons for the distributed tendon direction and then placing the banded tendons. Where tendons are

CODE

8.7.5.6.3 Slabs with tendons not satisfying 8.7.5.6.1 shall be permitted if bonded bottom deformed reinforcement is provided in each direction in accordance with 8.7.5.6.3.1 through 8.7.5.6.3.3.

8.7.5.6.3.1 Minimum bottom deformed reinforcement A_s in each direction shall be the larger of (a) and (b). The value of f_y shall be limited to a maximum of 80,000 psi:

(a) $$A_s = \frac{4.5\sqrt{f_c'}c_2d}{f_y} \quad (8.7.5.6.3.1a)$$

(b) $$A_s = \frac{300c_2d}{f_y} \quad (8.7.5.6.3.1b)$$

where c_2 is measured at the column faces through which the reinforcement passes.

8.7.5.6.3.2 Bottom deformed reinforcement calculated in 8.7.5.6.3.1 shall pass within the region bounded by the longitudinal reinforcement of the column and shall be anchored at exterior supports.

8.7.5.6.3.3 Bottom deformed reinforcement shall be anchored to develop f_y beyond the column or shear cap face.

8.7.6 *Shear reinforcement – stirrups*

8.7.6.1 Single-leg, simple-U, multiple-U, and closed stirrups shall be permitted as shear reinforcement.

8.7.6.2 Stirrup anchorage and geometry shall be in accordance with 25.7.1.

8.7.6.3 If stirrups are provided, location and spacing shall be in accordance with Table 8.7.6.3.

Table 8.7.6.3—First stirrup location and spacing limits

Direction of measurement	Description of measurement	Maximum distance or spacing, in.
Perpendicular to column face	Distance from column face to first stirrup	$d/2$
	Spacing between stirrups	$d/2$
Parallel to column face	Spacing between vertical legs of stirrups	$2d$

COMMENTARY

distributed in both directions, weaving of tendons is necessary and use of 8.7.5.6.3 may be an easier approach.

R8.7.5.6.3 In some prestressed slabs, tendon layout constraints make it difficult to provide the structural integrity tendons required by 8.7.5.6.1. In such situations, the structural integrity tendons can be replaced by deformed bar bottom reinforcement (ACI 352.1R).

R8.7.6 *Shear reinforcement – stirrups*

Research (Hawkins 1974; Broms 1990; Yamada et al. 1991; Hawkins et al. 1975; ACI 421.1R) has shown that shear reinforcement consisting of properly anchored bars or wires and single- or multiple-leg stirrups, or closed stirrups, can increase the punching shear resistance of slabs. The spacing limits given in 8.7.6.3 correspond to slab shear reinforcement details that have been shown to be effective. Section 25.7.1 gives anchorage requirements for stirrup-type shear reinforcement that should also be applied for bars or wires used as slab shear reinforcement. It is essential that this shear reinforcement engage longitudinal reinforcement at both the top and bottom of the slab, as shown for typical details in Fig. R8.7.6(a) to (c). Anchorage of shear reinforcement according to the requirements of 25.7.1 is difficult in slabs thinner than 10 in. Shear reinforcement consisting of vertical bars mechanically anchored at each end by a plate or head capable of developing the yield strength of the bars has been used successfully (ACI 421.1R).

In a slab-column connection for which moment transfer is negligible, the shear reinforcement should be symmetrical about the centroid of the critical section (Fig. R8.7.6(d)).

CODE

COMMENTARY

Spacing limits defined in 8.7.6.3 are also shown in Fig. R8.7.6(d) and (e).

At edge columns or for interior connections where moment transfer is significant, closed stirrups are recommended in a pattern as symmetrical as possible. Although the average shear stresses on faces ***AD*** and ***BC*** of the exterior column in Fig. R8.7.6(e) are lower than on face ***AB***, the closed stirrups extending from faces ***AD*** and ***BC*** provide some torsional strength along the edge of the slab.

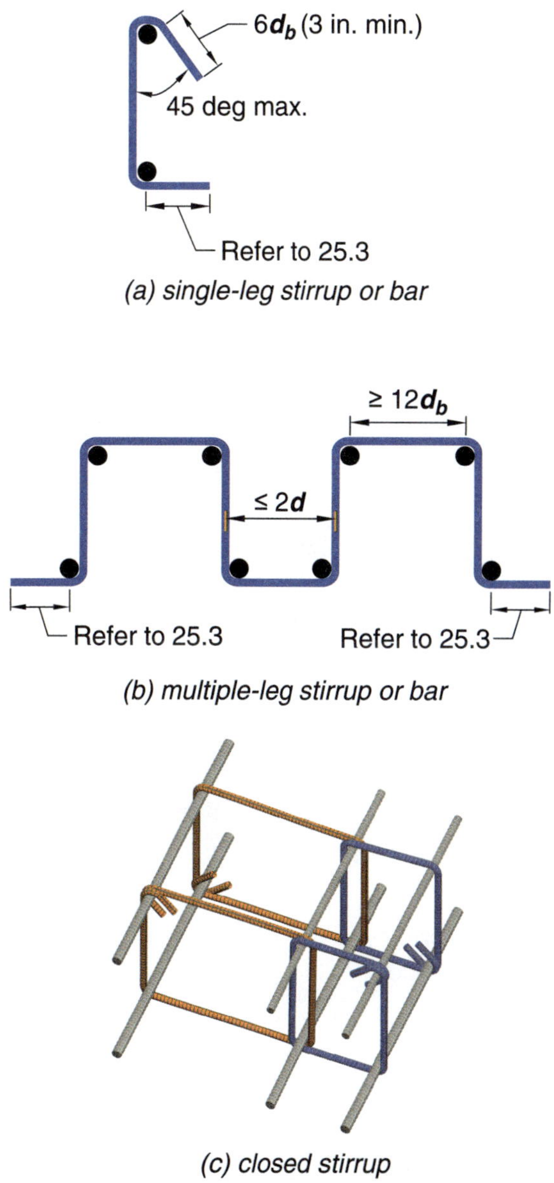

Fig. R8.7.6(a)-(c)—*Single- or multiple-leg stirrup-type slab shear reinforcement.*

CODE

COMMENTARY

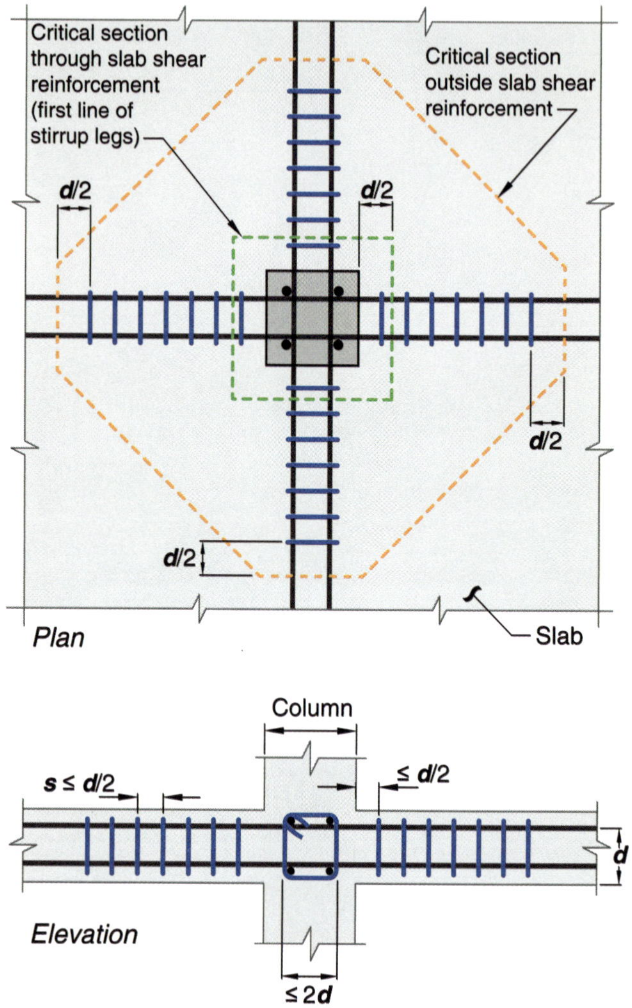

Fig. R8.7.6(d)—Arrangement of stirrup shear reinforcement, interior column.

CODE | COMMENTARY

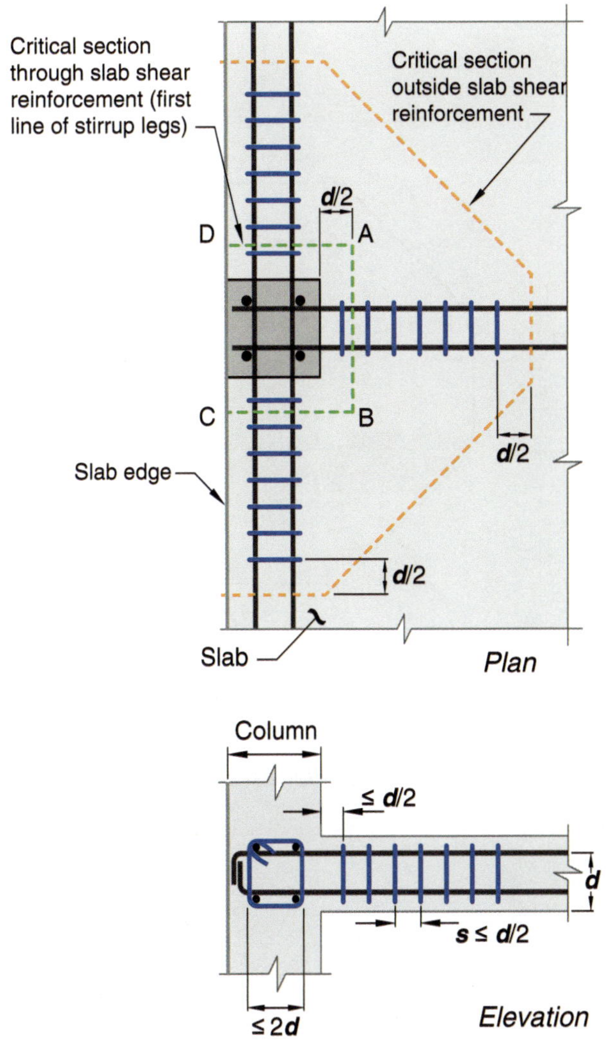

Fig. R8.7.6(e)—Arrangement of stirrup shear reinforcement, edge column.

8.7.7 *Shear reinforcement – headed studs*

8.7.7.1 Headed shear stud reinforcement shall be permitted if placed perpendicular to the plane of the slab.

8.7.7.1.1 The overall height of the shear stud assembly shall be at least the thickness of the slab minus the sum of (a) through (c):

(a) Concrete cover on the top flexural reinforcement
(b) Concrete cover on the base rail
(c) One-half the bar diameter of the flexural tension reinforcement

R8.7.7 *Shear reinforcement – headed studs*

Using headed stud assemblies as shear reinforcement in slabs requires specifying the stud shank diameter, the spacing of the studs, and the height of the assemblies for the particular applications.

Tests (ACI 421.1R) show that vertical studs mechanically anchored as close as possible to the top and bottom of slabs are effective in resisting punching shear. The bounds of the overall specified height achieve this objective while providing a reasonable tolerance in specifying that height, as shown in Fig. R20.5.1.3.5.

Compared with a leg of a stirrup having bends at the ends, a stud head exhibits smaller slip and, thus, results in smaller shear crack widths. The improved performance results in increased limits for shear strength and spacing between peripheral lines of headed shear stud reinforcement. Typical arrangements of headed shear stud reinforcement are shown in Fig. R8.7.7. The critical section beyond the shear reinforcement

CODE | COMMENTARY

generally has a polygonal shape. Equations for calculating shear stresses on such sections are given in ACI 421.1R.

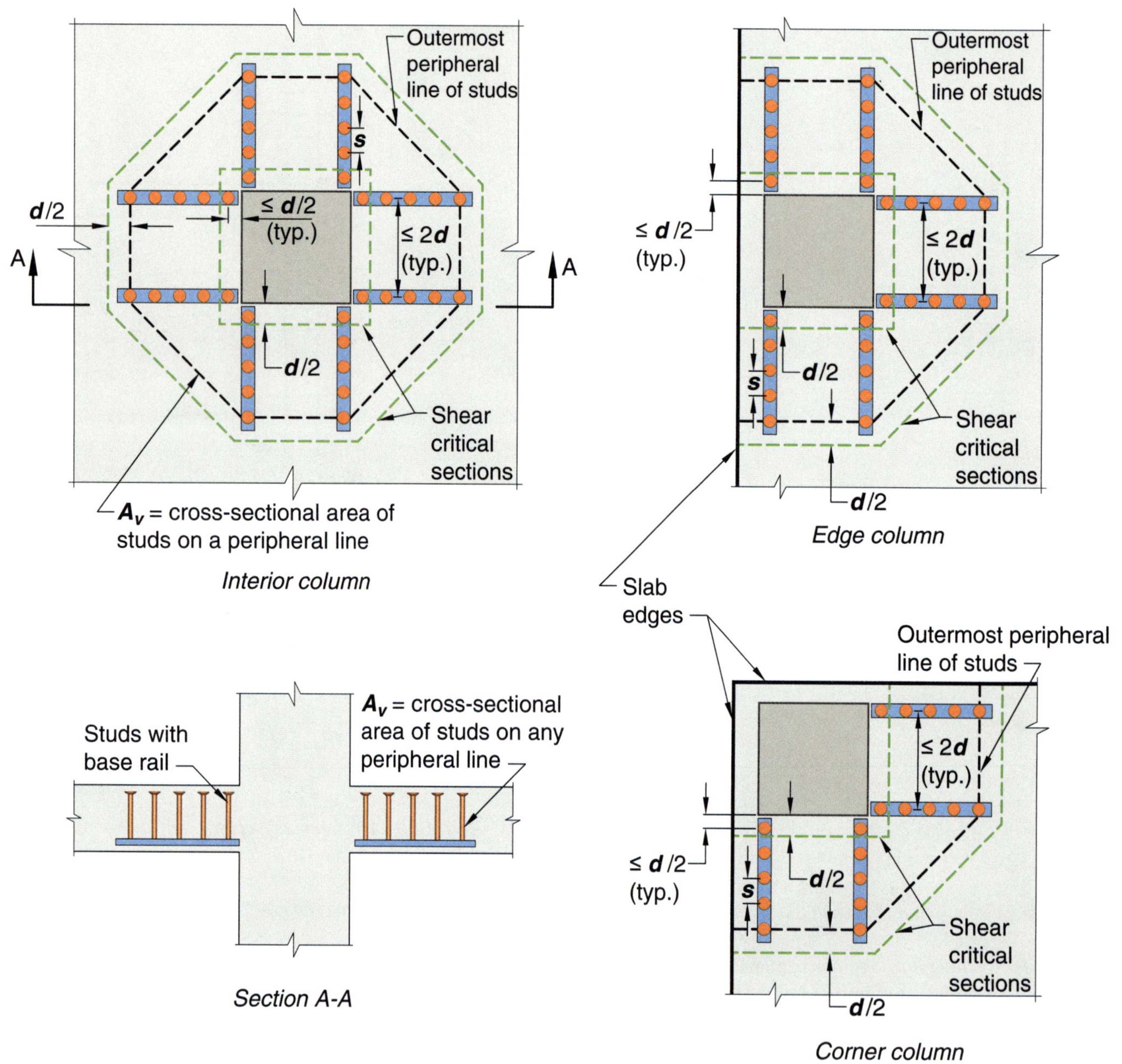

***Fig. R8.7.7**—Typical arrangements of headed shear stud reinforcement and critical sections.*

8.7.7.1.2 Headed shear stud reinforcement location and spacing shall be in accordance with Table 8.7.7.1.2.

R8.7.7.1.2 The specified spacings between peripheral lines of shear reinforcement are justified by experiments (ACI 421.1R). The clear spacing between the heads of the studs should be adequate to permit placing of the flexural reinforcement.

CODE

COMMENTARY

Table 8.7.7.1.2—Shear stud location and spacing limits

Direction of measurement	Description of measurement	Condition		Maximum distance or spacing, in.
Perpendicular to column face	Distance from column face to first peripheral line of shear studs	All		$d/2$
	Constant spacing between peripheral lines of shear studs	Nonprestressed slab with	$v_u \le \phi 6\sqrt{f_c'}$	$3d/4$
		Nonprestressed slab with	$v_u > \phi 6\sqrt{f_c'}$	$d/2$
		Prestressed slabs conforming to 22.6.5.4		$3d/4$
Parallel to column face	Spacing between adjacent shear studs on peripheral line nearest to column face	All		$2d$

8.8—Nonprestressed two-way joist systems

8.8.1 *General*

R8.8—Nonprestressed two-way joist systems

R8.8.1 *General*

The empirical limits established for nonprestressed reinforced concrete joist floors are based on successful past performance of joist construction using standard joist forming systems. For prestressed joist construction, this section may be used as a guide.

8.8.1.1 Nonprestressed two-way joist construction consists of a monolithic combination of regularly spaced ribs and a top slab designed to span in two orthogonal directions.

8.8.1.2 Width of ribs shall be at least 4 in. at any location along the depth.

8.8.1.3 Overall depth of ribs shall not exceed 3.5 times the minimum width.

8.8.1.4 Clear spacing between ribs shall not exceed 30 in.

R8.8.1.4 A limit on the maximum spacing of ribs is required because of the provisions permitting higher shear strengths and less concrete cover for the reinforcement for these relatively small, repetitive members.

8.8.1.5 V_c shall be permitted to be taken as 1.1 times the values calculated in 22.5.

R8.8.1.5 The increase in shear strength is justified on the basis of: 1) satisfactory performance of joist construction designed with higher calculated shear strength specified in previous Codes, which allowed comparable shear stresses; and 2) potential for redistribution of local overloads to adjacent joists.

8.8.1.6 For structural integrity, at least one bottom bar in each joist shall be continuous and shall be anchored to develop f_y at the face of supports.

8.8.1.7 Reinforcement area perpendicular to the ribs shall satisfy slab moment strength requirements, considering load concentrations, and shall be at least the shrinkage and temperature reinforcement area in accordance with 24.4.

CODE

COMMENTARY

8.8.1.8 Two-way joist construction not satisfying the limitations of 8.8.1.1 through 8.8.1.4 shall be designed as slabs and beams.

8.8.2 *Joist systems with structural fillers*

8.8.2.1 If permanent burned clay or concrete tile fillers of material having a unit compressive strength at least equal to f_c' in the joists are used, 8.8.2.1.1 and 8.8.2.1.2 shall apply.

8.8.2.1.1 Slab thickness over fillers shall be at least the greater of one-twelfth the clear distance between ribs and 1.5 in.

8.8.2.1.2 For calculation of shear and negative moment strength, it shall be permitted to include the vertical shells of fillers in contact with the ribs. Other portions of fillers shall not be included in strength calculations.

8.8.3 *Joist systems with other fillers*

8.8.3.1 If fillers not complying with 8.8.2.1 or removable forms are used, slab thickness shall be at least the greater of one-twelfth the clear distance between ribs and 2 in.

8.9—Lift-slab construction

8.9.1 In slabs constructed with lift-slab methods where it is impractical to pass the tendons required by 8.7.5.6.1 or the bottom bars required by 8.7.4.2 or 8.7.5.6.3 through the column, at least two post-tensioned tendons or two bonded bottom bars or wires in each direction shall pass through the lifting collar as close to the column as practicable, and be continuous or spliced using mechanical or welded splices in accordance with 25.5.7 or Class B tension lap splices in accordance with 25.5.2. At exterior columns, the reinforcement shall be anchored at the lifting collar.

CHAPTER 9—BEAMS

CODE

9.1—Scope

9.1.1 This chapter shall apply to the design of nonprestressed and prestressed beams, including:

(a) Composite beams of concrete elements constructed in separate placements but connected so that all elements resist loads as a unit
(b) One-way joist systems in accordance with 9.8
(c) Deep beams in accordance with 9.9

9.2—General

9.2.1 *Materials*

9.2.1.1 Design properties for concrete shall be selected to be in accordance with Chapter 19.

9.2.1.2 Design properties for steel reinforcement shall be selected to be in accordance with Chapter 20.

9.2.1.3 Materials, design, and detailing requirements for embedments in concrete shall be in accordance with 20.6.

9.2.2 *Connection to other members*

9.2.2.1 For cast-in-place construction, beam-column and slab-column joints shall satisfy Chapter 15.

9.2.2.2 For precast construction, connections shall satisfy the force transfer requirements of 16.2.

9.2.3 *Stability*

9.2.3.1 If a beam is not continuously laterally braced, (a) and (b) shall be satisfied:

(a) Spacing of lateral bracing shall not exceed 50 times the least width of compression flange or face.
(b) Spacing of lateral bracing shall take into account effects of eccentric loads.

9.2.3.2 In prestressed beams, buckling of thin webs and flanges shall be considered. If there is intermittent contact between prestressed reinforcement and an oversize duct, member buckling between contact points shall be considered.

9.2.4 *T-beam construction*

COMMENTARY

R9.1—Scope

R9.1.1 Composite structural steel-concrete beams are not covered in this chapter. Design provisions for such composite beams are covered in AISC 360.

R9.2—General

R9.2.3 *Stability*

R9.2.3.1 Tests (Hansell and Winter 1959; Sant and Bletzacker 1961) have shown that laterally unbraced reinforced concrete beams, even when very deep and narrow, will not fail prematurely by lateral buckling, provided the beams are loaded without lateral eccentricity that causes torsion.

Laterally unbraced beams are frequently loaded eccentrically or with slight inclination. Stresses and deformations by such loading become detrimental for narrow, deep beams with long unsupported lengths. Lateral supports spaced closer than $50b$ may be required for such loading conditions.

R9.2.3.2 In post-tensioned members where the prestressed reinforcement has intermittent contact with an oversize duct, the member can buckle due to the axial prestressing force, as the member can deflect laterally while the prestressed reinforcement does not. If the prestressed reinforcement is in continuous contact with the member being prestressed or is part of an unbonded tendon with the sheathing not excessively larger than the prestressed reinforcement, the prestressing force cannot buckle the member.

R9.2.4 *T-beam construction*

CODE

9.2.4.1 In T-beam construction, flange and web concrete shall be placed monolithically or made composite in accordance with 16.4.

9.2.4.2 Effective flange width shall be in accordance with 6.3.2.

9.2.4.3 For T-beam flanges where the primary flexural slab reinforcement is parallel to the longitudinal axis of the beam, reinforcement in the flange perpendicular to the longitudinal axis of the beam shall be in accordance with 7.5.2.3.

9.2.4.4 For torsional design according to 22.7, the overhanging flange width used to calculate A_{cp}, A_g, and p_{cp} shall be in accordance with (a) and (b):

(a) The overhanging flange width shall include that portion of slab on each side of the beam extending a distance equal to the projection of the beam above or below the slab, whichever is greater, but not greater than four times the slab thickness.

(b) The overhanging flanges shall be neglected in cases where the parameter A_{cp}^2/p_{cp} for solid sections or A_g^2/p_{cp} for hollow sections calculated for a beam with flanges is less than that calculated for the same beam ignoring the flanges.

COMMENTARY

R9.2.4.1 For monolithic or fully composite construction, the beam includes a portion of the slab as flanges.

R9.2.4.3 Refer to R7.5.2.3.

R9.2.4.4 Two examples of the section to be considered in torsional design are provided in Fig. R9.2.4.4.

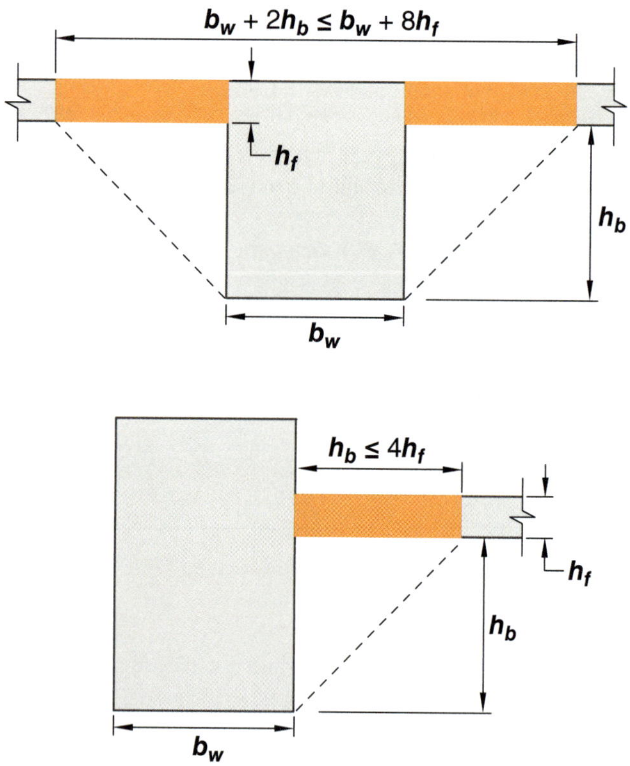

Fig. R9.2.4.4—Examples of the portion of slab to be included with the beam for torsional design.

9.3—Design limits

9.3.1 *Minimum beam depth*

9.3.1.1 For nonprestressed beams not supporting or attached to partitions or other construction likely to be damaged by large deflections, overall beam depth h shall satisfy the limits in Table 9.3.1.1, unless the calculated deflection limits of 9.3.2 are satisfied.

R9.3—Design limits

R9.3.1 *Minimum beam depth*

R9.3.1.1 For application of this provision to composite concrete beams, refer to R9.3.2.2.

CODE

Table 9.3.1.1—Minimum depth of nonprestressed beams

Support condition	Minimum h[1]
Simply supported	$\ell/16$
One end continuous	$\ell/18.5$
Both ends continuous	$\ell/21$
Cantilever	$\ell/8$

[1]Expressions applicable for normalweight concrete and f_y = 60,000 psi. For other cases, minimum h shall be modified in accordance with 9.3.1.1.1 through 9.3.1.1.3, as appropriate.

9.3.1.1.1 For f_y other than 60,000 psi, the expressions in Table 9.3.1.1 shall be multiplied by $(0.4 + f_y/100{,}000)$.

9.3.1.1.2 For nonprestressed beams made of lightweight concrete having w_c in the range of 90 to 115 lb/ft^3, the expressions in Table 9.3.1.1 shall be multiplied by the greater of (a) and (b):

(a) $1.65 - 0.005w_c$
(b) **1.09**

9.3.1.1.3 For nonprestressed composite beams made of a combination of lightweight and normalweight concrete, shored during construction, and where the lightweight concrete is in compression, the modifier of 9.3.1.1.2 shall apply.

9.3.1.2 The thickness of a concrete floor finish shall be permitted to be included in h if it is placed monolithically with the beam or if the floor finish is designed to be composite with the beam in accordance with 16.4.

9.3.2 *Calculated deflection limits*

9.3.2.1 For nonprestressed beams not satisfying 9.3.1 and for prestressed beams, immediate and time-dependent deflections shall be calculated in accordance with 24.2 and shall not exceed the limits in 24.2.2.

9.3.2.2 For nonprestressed composite concrete beams satisfying 9.3.1, deflections occurring after the member becomes composite need not be calculated. Deflections occurring before the member becomes composite shall be investigated unless the precomposite depth also satisfies 9.3.1.

COMMENTARY

R9.3.1.1.1 The modification for f_y is approximate, but should provide conservative results for typical reinforcement ratios and for values of f_y between 40,000 and 100,000 psi.

R9.3.1.1.2 The modification for lightweight concrete is based on the results and discussions in ACI 213R. No correction is given for concretes with w_c greater than 115 lb/ft^3 because the correction term would be close to unity in this range.

R9.3.2 *Calculated deflection limits*

R9.3.2.2 The limits in Table 9.3.1.1 apply to the entire depth of nonprestressed composite beams shored during construction so that, after removal of temporary supports, the dead load is resisted by the full composite section. In unshored construction, the beam depth of concern depends on if the deflection being considered occurs before or after the attainment of effective composite action.

Additional deflections due to excessive creep and shrinkage caused by premature loading should be considered. This is especially important at early ages when the moisture content is high and the strength is low.

The transfer of horizontal shear by direct bond is important if excessive deflection from slippage is to be prevented.

CODE

9.3.3 *Reinforcement strain limit in nonprestressed beams*

9.3.3.1 Nonprestressed beams with $P_u < 0.10 f_c' A_g$ shall be tension controlled in accordance with Table 21.2.2.

9.3.4 *Stress limits in prestressed beams*

9.3.4.1 Prestressed beams shall be classified as Class U, T, or C in accordance with 24.5.2.

9.3.4.2 Stresses in prestressed beams immediately after transfer and at service loads shall not exceed permissible stresses in 24.5.3 and 24.5.4.

9.4—Required strength

9.4.1 *General*

9.4.1.1 Required strength shall be calculated in accordance with the factored load combinations in Chapter 5.

9.4.1.2 Required strength shall be calculated in accordance with the analysis procedures in Chapter 6.

9.4.1.3 For prestressed beams, effects of reactions induced by prestressing shall be considered in accordance with 5.3.11.

9.4.2 *Factored moment*

9.4.2.1 For beams built integrally with supports, M_u at the support shall be permitted to be calculated at the face of support.

9.4.3 *Factored shear*

9.4.3.1 For beams built integrally with supports, V_u at the support shall be permitted to be calculated at the face of support.

9.4.3.2 Sections between the face of support and a critical section located d from the face of support for nonprestressed beams and $h/2$ from the face of support for prestressed beams shall be permitted to be designed for V_u at that critical section if (a) through (c) are satisfied:

(a) Support reaction, in direction of applied shear, introduces compression into the end region of the beam

COMMENTARY

Shear keys provide a means of transferring shear but will not be engaged until slippage occurs.

R9.3.3 *Reinforcement strain limit in nonprestressed beams*

R9.3.3.1 The effect of this limitation is to restrict the reinforcement ratio in nonprestressed beams to mitigate brittle flexural behavior in case of an overload. This limitation does not apply to prestressed beams. Before the 2019 Code, a minimum strain limit of 0.004 was specified for nonprestressed flexural members. Beginning with the 2019 Code, this limit is revised to require that the section be tension-controlled.

R9.4—Required strength

R9.4.3 *Factored shear*

R9.4.3.2 The closest inclined crack to the support of the beam in Fig. R9.4.3.2a will extend upward from the face of the support reaching the compression zone approximately d from the face of the support. If loads are applied to the top of the beam, the stirrups across this crack need only resist the shear force due to loads acting beyond d (right free body in Fig. R9.4.3.2a). The loads applied to the beam between the face of the support and the point d away from the face

CODE

(b) Loads are applied at or near the top surface of the beam

(c) No concentrated load occurs between the face of support and critical section

COMMENTARY

are transferred directly to the support by compression in the web above the crack. Accordingly, the Code permits design for a maximum factored shear V_u at a distance d from the support for nonprestressed beams and at a distance $h/2$ for prestressed beams.

In Fig. R9.4.3.2b, loads are shown acting near the bottom of a beam. In this case, the critical section is taken at the face of the support. Loads acting near the support should be transferred across the inclined crack extending upward from the support face. The shear force acting on the critical section should include all loads applied below the potential inclined crack.

Typical support conditions where the shear force at a distance d from the support may be used include:

(a) Beams supported by bearing at the bottom of the beam, such as shown in Fig. R9.4.3.2(c)

(b) Beams framing monolithically into a column, as illustrated in Fig. R9.4.3.2(d)

Typical support conditions where the critical section is taken at the face of support include:

(a) Beams framing into a supporting member in tension, such as shown in Fig. R9.4.3.2(e). Shear within the connection should also be investigated and special corner reinforcement should be provided.

(b) Beams for which loads are not applied at or near the top, as previously discussed and as shown in Fig. R9.4.3.2b.

(c) Beams loaded such that the shear at sections between the support and a distance d from the support differs radically from the shear at distance d. This commonly occurs in brackets and in beams where a concentrated load is located close to the support, as shown in Fig. R9.4.3.2(f).

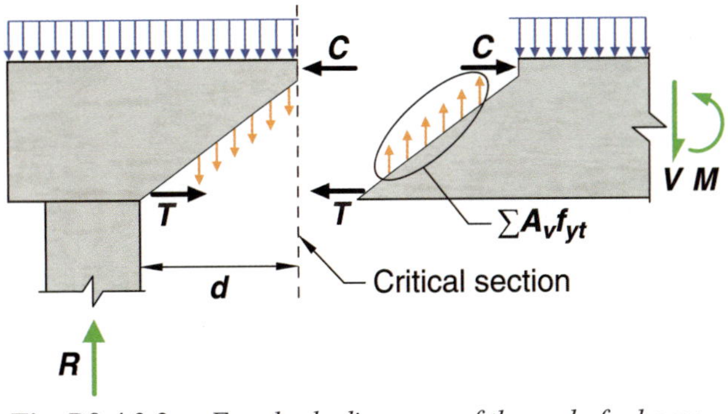

Fig. R9.4.3.2a—Free body diagrams of the end of a beam.

CODE

COMMENTARY

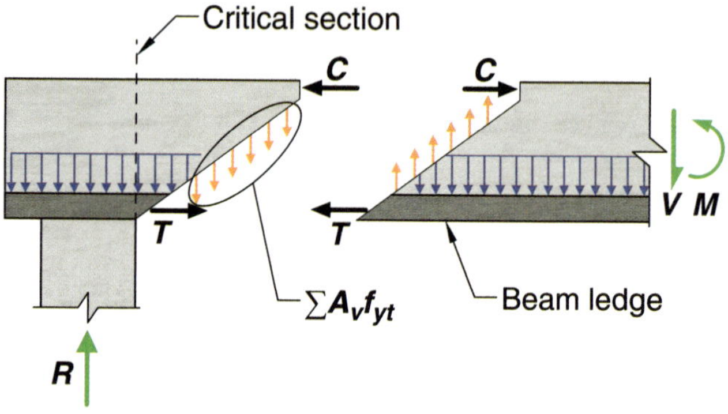

Fig. R9.4.3.2b—Location of critical section for shear in a beam loaded near bottom.

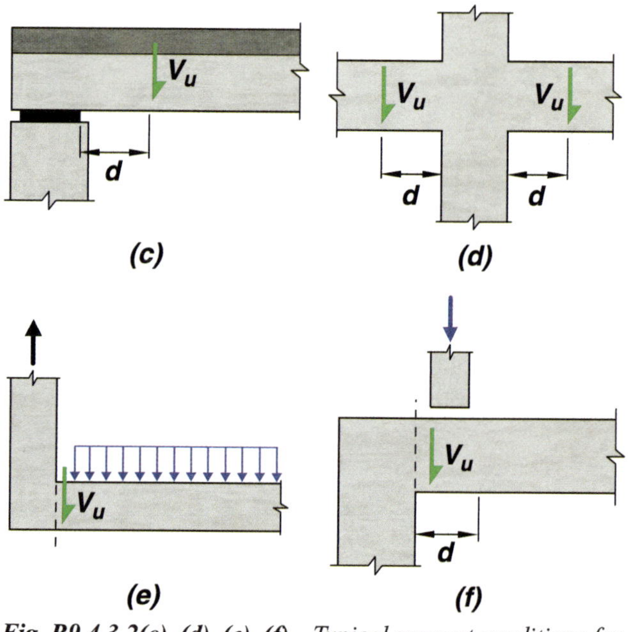

Fig. R9.4.3.2(c), (d), (e), (f)—Typical support conditions for locating factored shear force V_u.

9.4.4 *Factored torsion*

9.4.4.1 Unless determined by a more detailed analysis, it shall be permitted to take the torsional loading from a slab as uniformly distributed along the beam.

9.4.4.2 For beams built integrally with supports, T_u at the support shall be permitted to be calculated at the face of support.

9.4.4.3 Sections between the face of support and a critical section located d from the face of support for nonprestressed beams or $h/2$ from the face of support for prestressed beams shall be permitted to be designed for T_u at that critical section unless a concentrated torsional moment occurs within this distance. In that case, the critical section shall be taken at the face of the support.

R9.4.4 *Factored torsion*

R9.4.4.3 It is not uncommon for a beam to frame into one side of a girder near the support of the girder. In such a case, a concentrated shear and torsional moment are applied to the girder.

CODE

9.4.4.4 It shall be permitted to reduce T_u in accordance with 22.7.3.

9.5—Design strength

9.5.1 *General*

9.5.1.1 For each applicable factored load combination, design strength at all sections shall satisfy $\phi S_n \geq U$ including (a) through (d). Interaction between load effects shall be considered.

(a) $\phi M_n \geq M_u$
(b) $\phi V_n \geq V_u$
(c) $\phi T_n \geq T_u$
(d) $\phi P_n \geq P_u$

9.5.1.2 ϕ shall be determined in accordance with 21.2.

9.5.2 *Moment*

9.5.2.1 If $P_u < 0.10f_c'A_g$, M_n shall be calculated in accordance with 22.3.

9.5.2.2 If $P_u \geq 0.10f_c'A_g$, M_n shall be calculated in accordance with 22.4.

9.5.2.3 For prestressed beams, external tendons shall be considered as unbonded tendons in calculating flexural strength, unless the external tendons are effectively bonded to the concrete along the entire length.

9.5.3 *Shear*

9.5.3.1 V_n shall be calculated in accordance with 22.5.

9.5.3.2 For composite concrete beams, horizontal shear strength V_{nh} shall be calculated in accordance with 16.4.

9.5.4 *Torsion*

9.5.4.1 If $T_u < \phi T_{th}$, where T_{th} is given in 22.7, it shall be permitted to neglect torsional effects. The minimum reinforcement requirements of 9.6.4 and the detailing requirements of 9.7.5 and 9.7.6.3 need not be satisfied.

9.5.4.2 T_n shall be calculated in accordance with 22.7.

COMMENTARY

R9.5—Design strength

R9.5.1 *General*

R9.5.1.1 The design conditions 9.5.1.1(a) through (d) list the typical forces and moments that need to be considered. However, the general condition $\phi S_n \geq U$ indicates that all forces and moments that are relevant for a given structure need to be considered.

R9.5.2 *Moment*

R9.5.2.2 Beams resisting significant axial forces require consideration of the combined effects of axial forces and moments. These beams are not required to satisfy the provisions of Chapter 10, but are required to satisfy the additional requirements for ties or spirals defined in Table 22.4.2.1. For slender beams with significant axial loads, consideration should be given to slenderness effects as required for columns in 6.2.5.

R9.5.4 *Torsion*

CODE

9.5.4.3 Longitudinal and transverse reinforcement required for torsion shall be added to that required for the V_u, M_u, and P_u that act in combination with the torsion.

9.5.4.4 For prestressed beams, the total area of longitudinal reinforcement, A_s and A_{ps}, at each section shall be designed to resist M_u at that section, plus an additional concentric longitudinal tensile force equal to $A_\ell f_y$, based on T_u at that section.

9.5.4.5 It shall be permitted to reduce the area of longitudinal torsional reinforcement in the flexural compression

COMMENTARY

R9.5.4.3 The requirements for torsional reinforcement and shear reinforcement are added and stirrups are provided to supply at least the total amount required. Because the reinforcement area A_v for shear is defined in terms of all the legs of a given stirrup while the reinforcement area A_t for torsion is defined in terms of one leg only, the addition of transverse reinforcement area is calculated as follows:

$$\text{Total}\left(\frac{A_{v+t}}{s}\right) = \frac{A_v}{s} + 2\frac{A_t}{s} \qquad \text{(R9.5.4.3)}$$

If a stirrup group has more than two legs for shear, only the legs adjacent to the sides of the beam are included in this summation because the inner legs would be ineffective for resisting torsion.

The longitudinal reinforcement required for torsion is added at each section to the longitudinal reinforcement required for bending moment that acts concurrently with the torsion. The longitudinal reinforcement is then chosen for this sum, but should not be less than the amount required for the maximum bending moment at that section if this exceeds the moment acting concurrently with the torsion. If the maximum bending moment occurs at one section, such as midspan, while the maximum torsional moment occurs at another, such as the face of the support, the total longitudinal reinforcement required may be less than that obtained by adding the maximum flexural reinforcement, plus the maximum torsional reinforcement. In such a case, the required longitudinal reinforcement is evaluated at several locations.

R9.5.4.4 Torsion causes an axial tensile force in the longitudinal reinforcement balanced by the force in the diagonal concrete compression struts. In a nonprestressed beam, the tensile force must be resisted by longitudinal reinforcement having an axial tensile strength of $A_\ell f_y$. This reinforcement is in addition to the required flexural reinforcement and is distributed uniformly inside and around the perimeter of the closed transverse reinforcement so that the resultant of $A_\ell f_y$ acts along the axis of the member.

In a prestressed beam, the same approach (providing additional reinforcing bars with strength $A_\ell f_y$) may be followed, or overstrength of the prestressed reinforcement can be used to resist some of the axial force $A_\ell f_y$. The stress in the prestressed reinforcement at nominal strength will be between f_{se} and f_{ps}. A portion of the $A_\ell f_y$ force can be resisted by a force of $A_{ps}\Delta f_{pt}$ in the prestressed reinforcement. The stress required to resist the bending moment can be calculated as $M_u/(\phi 0.9 d_p A_{ps})$. For pretensioned strands, the stress that can be developed near the free end of the strand can be calculated using the procedure illustrated in Fig. R25.4.8.3.

R9.5.4.5 The longitudinal tension due to torsion is offset in part by the compression in the flexural compression zone,

CODE

zone by an amount equal to $M_u/(0.9df_y)$, where M_u occurs simultaneously with T_u at that section, except that the longitudinal reinforcement area shall not be less than the minimum required in 9.6.4.

9.5.4.6 For solid sections with an aspect ratio $h/b_t \geq 3$, it shall be permitted to use an alternative design procedure, provided the adequacy of the procedure has been shown by analysis and substantial agreement with results of comprehensive tests. The minimum reinforcement requirements of 9.6.4 need not be satisfied, but the detailing requirements of 9.7.5 and 9.7.6.3 apply.

9.5.4.7 For solid precast sections with an aspect ratio $h/b_t \geq 4.5$, it shall be permitted to use an alternative design procedure and open web reinforcement, provided the adequacy of the procedure and reinforcement have been shown by analysis and substantial agreement with results of comprehensive tests. The minimum reinforcement requirements of 9.6.4 and detailing requirements of 9.7.5 and 9.7.6.3 need not be satisfied.

9.6—Reinforcement limits

9.6.1 *Minimum flexural reinforcement in nonprestressed beams*

9.6.1.1 A minimum area of flexural reinforcement, $A_{s,min}$, shall be provided at every section where tension reinforcement is required by analysis.

9.6.1.2 $A_{s,min}$ shall be the larger of (a) and (b), except as provided in 9.6.1.3. For a statically determinate beam with a flange in tension, the value of b_w shall be the smaller of b_f and $2b_w$. The value of f_y shall be limited to a maximum of 80,000 psi.

(a) $\dfrac{3\sqrt{f_c'}}{f_y}b_w d$

COMMENTARY

allowing a reduction in the longitudinal torsional reinforcement required in the compression zone.

R9.5.4.6 An example of an alternative design that satisfies this provision can be found in Zia and Hsu (2004), which has been extensively and successfully used for design of precast, prestressed concrete spandrel beams with $h/b_t \geq 3$ and closed stirrups. The seventh edition of the *PCI Design Handbook* (PCI MNL-120) describes the procedure of Zia and Hsu (2004). This procedure was experimentally verified by the tests described in Klein (1986).

R9.5.4.7 The experimental results described in Lucier et al. (2011a) demonstrate that properly designed open web reinforcement is a safe and effective alternative to traditional closed stirrups for precast spandrels with $h/b_t \geq 4.5$. Lucier et al. (2011b) presents a design procedure that satisfies this provision for slender spandrels and describes the limited conditions to which the procedure applies.

R9.6—Reinforcement limits

R9.6.1 *Minimum flexural reinforcement in nonprestressed beams*

R9.6.1.1 This provision is intended to result in flexural strength exceeding the cracking strength by a margin. The objective is to produce a beam that will be able to sustain loading after the onset of flexural cracking, with visible cracking and deflection, thereby warning of possible overload. Beams with less reinforcement may sustain sudden failure with the onset of flexural cracking.

In practice, this provision only controls reinforcement design for beams which, for architectural or other reasons, are larger in cross section than required for strength. With a small amount of tension reinforcement required for strength, the calculated moment strength of a reinforced concrete section using cracked section analysis becomes less than that of the corresponding unreinforced concrete section calculated from its modulus of rupture. Failure in such a case could occur at first cracking and without warning. To prevent such a failure, a minimum amount of tension reinforcement is required in both positive and negative moment regions.

R9.6.1.2 If the flange of a section is in tension, the amount of tension reinforcement needed to make the strength of the reinforced section equal that of the unreinforced section is approximately twice that for a rectangular section or that of a flanged section with the flange in compression. A larger amount of minimum tension reinforcement is particularly necessary in cantilevers and other statically determinate beams where there is no possibility for redistribution of moments.

CODE

(b) $\frac{200}{f_y}b_w d$

9.6.1.3 If A_s provided at every section is at least one-third greater than A_s required by analysis, 9.6.1.1 and 9.6.1.2 need not be satisfied.

9.6.2 *Minimum flexural reinforcement in prestressed beams*

9.6.2.1 For beams with bonded prestressed reinforcement, total quantity of A_s and A_{ps} shall be adequate to develop a factored load at least 1.2 times the cracking load calculated on the basis of f_r defined in 19.2.3.

9.6.2.2 For beams with both flexural and shear design strength at least twice the required strength, 9.6.2.1 need not be satisfied.

9.6.2.3 For beams with unbonded tendons, the minimum area of bonded deformed longitudinal reinforcement $A_{s,min}$ shall be:

$$A_{s,min} = 0.004A_{ct} \quad (9.6.2.3)$$

where A_{ct} is the area of that part of the cross section between the flexural tension face and the centroid of the gross section.

9.6.3 *Minimum shear reinforcement*

9.6.3.1 For nonprestressed beams, minimum area of shear reinforcement, $A_{v,min}$, shall be provided in all regions where $V_u > \phi\lambda\sqrt{f_c'}b_w d$ except for the cases in Table 9.6.3.1. For these cases, at least $A_{v,min}$ shall be provided where $V_u > \phi V_c$.

Table 9.6.3.1—Cases where $A_{v,min}$ is not required if $V_u \le \phi V_c$

Beam type	Conditions
Shallow depth	$h \le 10$ in.
Integral with slab	$h \le$ greater of $2.5t_f$ or $0.5b_w$ and $h \le 24$ in.
Constructed with steel fiber-reinforced normalweight concrete conforming to 26.4.1.5.1(a), 26.4.2.2(i), and 26.12.7.1(a) and with $f_c' \le 6000$ psi	$h \le 24$ in. and $V_u \le \phi 2\sqrt{f_c'}b_w d$
One-way joist system	In accordance with 9.8

COMMENTARY

R9.6.2 *Minimum flexural reinforcement in prestressed beams*

R9.6.2.1 Minimum flexural reinforcement is required for reasons similar to nonprestressed beams as discussed in R9.6.1.1.

Abrupt flexural failure immediately after cracking does not occur when the prestressed reinforcement is unbonded (ACI 423.3R); therefore, this requirement does not apply to members with unbonded tendons.

R9.6.2.3 Minimum bonded reinforcement is required by the Code in beams prestressed with unbonded tendons to ensure flexural behavior at ultimate beam strength, rather than tied arch behavior, and to limit crack width and spacing at service load when concrete tensile stresses exceed the modulus of rupture. Providing minimum bonded reinforcement helps to ensure acceptable behavior at all loading stages. The minimum amount of bonded reinforcement is based on research comparing the behavior of bonded and unbonded post-tensioned beams (Mattock et al. 1971). The minimum bonded reinforcement area required by Eq. (9.6.2.3) is independent of reinforcement f_y.

R9.6.3 *Minimum shear reinforcement*

R9.6.3.1 Shear reinforcement restrains the growth of inclined cracking so that ductility of the beam is improved and a warning of failure is provided. In an unreinforced web, the formation of inclined cracking might lead directly to failure without warning. Such reinforcement is of great value if a beam is subjected to an unexpected tensile force or an overload.

The exception for beams constructed using steel fiber-reinforced concrete is intended to provide a design alternative to the use of shear reinforcement, as defined in 22.5.8.5, for beams with longitudinal flexural reinforcement in which V_u does not exceed $\phi 2\sqrt{f_c'}b_w d$. Chapter 26 specifies design information and compliance requirements that need to be incorporated into the construction documents when steel fiber-reinforced concrete is used for this purpose. Fiber-reinforced concrete beams with hooked or crimped steel fibers, in dosages as required by 26.4.2.2(i), have been shown through laboratory tests to exhibit shear strengths

CODE

COMMENTARY

greater than $3.5\sqrt{f_c'}\,b_wd$ (Parra-Montesinos 2006). There are no data for the use of steel fibers as shear reinforcement in concrete beams exposed to chlorides from deicing chemicals, salt, salt water, brackish water, seawater, or spray from these sources. Where steel fibers are used as shear reinforcement in corrosive environments, corrosion protection should be considered.

Joists are excluded from the minimum shear reinforcement requirement as indicated because there is a possibility of load sharing between weak and strong areas.

Even when V_u is less than $\phi\lambda\sqrt{f_c'}\,b_wd$, the use of some web reinforcement is recommended in all thin-web, post-tensioned members such as joists, waffle slabs, beams, and T-beams, to reinforce against tensile forces in webs resulting from local deviations from the design tendon profile and to provide a means of supporting the tendons in the design profile during construction. If sufficient support is not provided, lateral wobble and local deviations from the smooth parabolic tendon profile assumed in design may result during placement of the concrete. In such cases, the deviations in the tendons tend to straighten out when the tendons are stressed. This process may impose large tensile stresses in webs, and severe cracking may develop if no web reinforcement is provided. Unintended curvature of the tendons, and the resulting tensile stresses in webs, may be minimized by securely tying tendons to stirrups that are rigidly held in place by other elements of the reinforcement cage. The recommended maximum spacing of stirrups used for this purpose is the smaller of $1.5h$ or 4 ft. If applicable, the shear reinforcement provisions of 9.6.3 and 9.7.6.2.2 will require closer stirrup spacings.

For repeated loading of beams, the possibility of inclined diagonal tension cracks forming at stresses appreciably smaller than under static loading should be taken into account in design. In these instances, use of at least the minimum shear reinforcement expressed by 9.6.3.4 is recommended even though tests or calculations based on static loads show that shear reinforcement is not required.

9.6.3.2 For prestressed beams, a minimum area of shear reinforcement, $A_{v,min}$, shall be provided in all regions where $V_u > 0.5\phi V_c$ except for the cases in Table 9.6.3.1. For these cases, at least $A_{v,min}$ shall be provided where $V_u > \phi V_c$.

9.6.3.3 If shown by testing that the required M_n and V_n can be developed, 9.6.3.1 and 9.6.3.2 need not be satisfied. Such tests shall simulate effects of differential settlement, creep, shrinkage, and temperature change, based on a realistic assessment of these effects occurring in service.

R9.6.3.3 When a beam is tested to demonstrate that its shear and flexural strengths are adequate, the actual beam dimensions and material strengths are known. Therefore, the test strengths are considered the nominal strengths V_n and M_n. Considering these strengths as nominal values ensures that if the actual material strengths in the field were less than specified, or the member dimensions were in error such as to result in a reduced member strength, a satisfactory margin of safety will be retained due to the strength reduction factor ϕ.

CODE

9.6.3.4 If shear reinforcement is required and torsional effects can be neglected according to 9.5.4.1, $A_{v,min}$ shall be in accordance with Table 9.6.3.4.

Table 9.6.3.4—Required $A_{v,min}$

Beam type	$A_{v,min}/s$			
Nonprestressed and prestressed with $A_{ps}f_{se} < 0.4(A_{ps}f_{pu} + A_sf_y)$	Greater of:	$0.75\sqrt{f_c'}\frac{b_w}{f_{yt}}$		(a)
		$50\frac{b_w}{f_{yt}}$		(b)
Prestressed with $A_{ps}f_{se} \geq 0.4(A_{ps}f_{pu} + A_sf_y)$	Lesser of:	Greater of:	$0.75\sqrt{f_c'}\frac{b_w}{f_{yt}}$	(c)
			$50\frac{b_w}{f_{yt}}$	(d)
		$\frac{A_{ps}f_{pu}}{80f_{yt}d}\sqrt{\frac{d}{b_w}}$		(e)

COMMENTARY

R9.6.3.4 Tests (Roller and Russell 1990) have indicated the need to increase the minimum area of shear reinforcement as the concrete strength increases to prevent sudden shear failures when inclined cracking occurs. Therefore, expressions (a) and (c) in Table 9.6.3.4 provide for a gradual increase in the minimum area of transverse reinforcement with increasing concrete strength. Expressions (b) and (d) in Table 9.6.3.4 provide for a minimum area of transverse reinforcement independent of concrete strength and govern for concrete strengths less than 4400 psi.

Tests (Olesen et al. 1967) of prestressed beams with minimum web reinforcement based on 9.6.3.4 indicate that the lesser of $A_{v,min}$ from expressions (c) and (e) is sufficient to develop ductile behavior. Expression (e) is discussed in Olesen et al. (1967).

9.6.4 *Minimum torsional reinforcement*

9.6.4.1 A minimum area of torsional reinforcement shall be provided in all regions where $T_u \geq \phi T_{th}$ in accordance with 22.7.

9.6.4.2 If torsional reinforcement is required, minimum transverse reinforcement $(A_v + 2A_t)_{min}/s$ shall be the greater of (a) and (b):

(a) $0.75\sqrt{f_c'}\frac{b_w}{f_{yt}}$

(b) $50\frac{b_w}{f_{yt}}$

9.6.4.3 If torsional reinforcement is required, minimum area of longitudinal reinforcement $A_{\ell,min}$ shall be the lesser of (a) and (b):

(a) $\frac{5\sqrt{f_c'}A_{cp}}{f_y} - \left(\frac{A_t}{s}\right)p_h\frac{f_{yt}}{f_y}$

(b) $\frac{5\sqrt{f_c'}A_{cp}}{f_y} - \left(\frac{25b_w}{f_{yt}}\right)p_h\frac{f_{yt}}{f_y}$

R9.6.4 *Minimum torsional reinforcement*

R9.6.4.2 The differences in the definitions of A_v and A_t should be noted: A_v is the area of two legs of a closed stirrup, whereas A_t is the area of only one leg of a closed stirrup. If a stirrup group has more than two legs, only the legs adjacent to the sides of the beam are considered, as discussed in R9.5.4.3.

Tests (Roller and Russell 1990) of high-strength reinforced concrete beams have indicated the need to increase the minimum area of shear reinforcement to prevent shear failures when inclined cracking occurs. Although there are a limited number of tests of high-strength concrete beams in torsion, the equation for the minimum area of transverse closed stirrups has been made consistent with calculations required for minimum shear reinforcement.

R9.6.4.3 Under combined torsion and shear, the torsional cracking moment decreases with applied shear, which leads to a reduction in torsional reinforcement required to prevent brittle failure immediately after cracking. When subjected to pure torsion, reinforced concrete beam specimens with less than 1 percent torsional reinforcement by volume have failed at first torsional cracking (MacGregor and Ghoneim 1995). Equation 9.6.4.3(a) is based on a 2:1 ratio of torsion stress to shear stress and results in a torsional reinforcement volumetric ratio of approximately 0.5 percent (Hsu 1968). Tests of prestressed concrete beams have shown that a similar amount of longitudinal reinforcement is required.

CODE

9.7—Reinforcement detailing

9.7.1 *General*

9.7.1.1 Concrete cover for reinforcement shall be in accordance with 20.5.1.

9.7.1.2 Development lengths of deformed and prestressed reinforcement shall be in accordance with 25.4.

9.7.1.3 Splices of deformed reinforcement shall be in accordance with 25.5.

9.7.1.4 Along development and lap splice lengths of longitudinal bars with $f_y \geq$ **80,000 psi**, transverse reinforcement shall be provided such that K_{tr} shall not be smaller than $0.5d_b$.

9.7.1.5 Bundled bars shall be in accordance with 25.6.

9.7.2 *Reinforcement spacing*

9.7.2.1 Minimum spacing s shall be in accordance with 25.2.

9.7.2.2 For nonprestressed and Class C prestressed beams, spacing of bonded longitudinal reinforcement closest to the tension face shall not exceed s given in 24.3.

9.7.2.3 For nonprestressed and Class C prestressed beams with h exceeding 36 in., longitudinal skin reinforcement shall be uniformly distributed on both side faces of the beam for a distance $h/2$ from the tension face. Spacing of skin reinforcement shall not exceed s given in 24.3.2, where c_c is the clear cover from the skin reinforcement to the side face. It shall be permitted to include skin reinforcement in strength calculations if a strain compatibility analysis is made.

COMMENTARY

R9.7—Reinforcement detailing

R9.7.2 *Reinforcement spacing*

R9.7.2.3 For relatively deep beams, some reinforcement should be placed near the vertical faces of the tension zone to control cracking in the web (Frantz and Breen 1980; Frosch 2002), as shown in Fig. R9.7.2.3. Without such auxiliary reinforcement, the width of the cracks in the web may exceed the crack widths at the level of the flexural tension reinforcement.

The size of the skin reinforcement is not specified; research has indicated that the spacing rather than bar size is of primary importance (Frosch 2002). Bar sizes No. 3 to No. 5, or welded wire reinforcement with a minimum area of 0.1 in.^{2} per foot of depth, are typically provided.

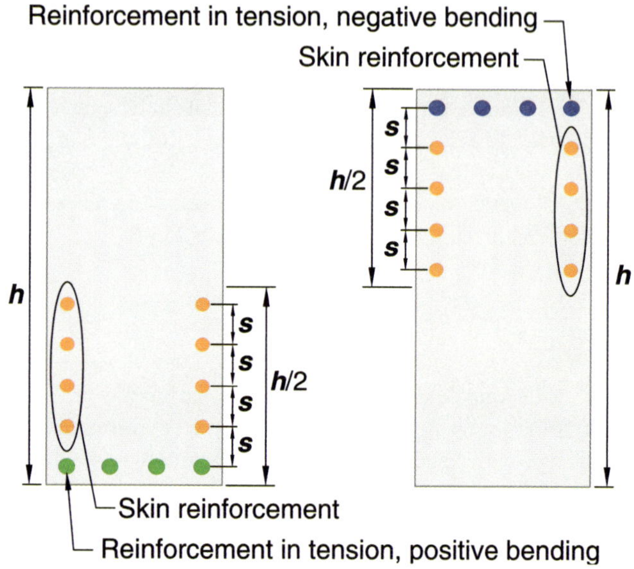

Fig. R9.7.2.3—Skin reinforcement for beams and joists with **h** > *36 in.*

9.7.3 *Flexural reinforcement in nonprestressed beams*

R9.7.3 *Flexural reinforcement in nonprestressed beams*

9.7.3.1 Calculated tensile or compressive force in reinforcement at each section of the beam shall be developed on each side of that section.

9.7.3.2 Critical locations for development of reinforcement are points of maximum stress and points along the span where bent or terminated tension reinforcement is no longer required to resist flexure.

R9.7.3.2 In Codes before 2014, one of the critical sections was defined as the location where adjacent reinforcement terminates or is bent. In the 2014 Code, this critical section is redefined as the location, "where bent or terminated tension reinforcement is no longer required to resist flexure."

Critical sections for a typical continuous beam are indicated with a "c" for points of maximum stress or an "x" for points where bent or terminated tension reinforcement is no longer required to resist flexure (Fig. R9.7.3.2). For uniform loading, the positive reinforcement extending into the support is more likely governed by the requirements of 9.7.3.8.1 or 9.7.3.8.3 than by development length measured from a point of maximum moment or the bar cutoff point.

CODE | **COMMENTARY**

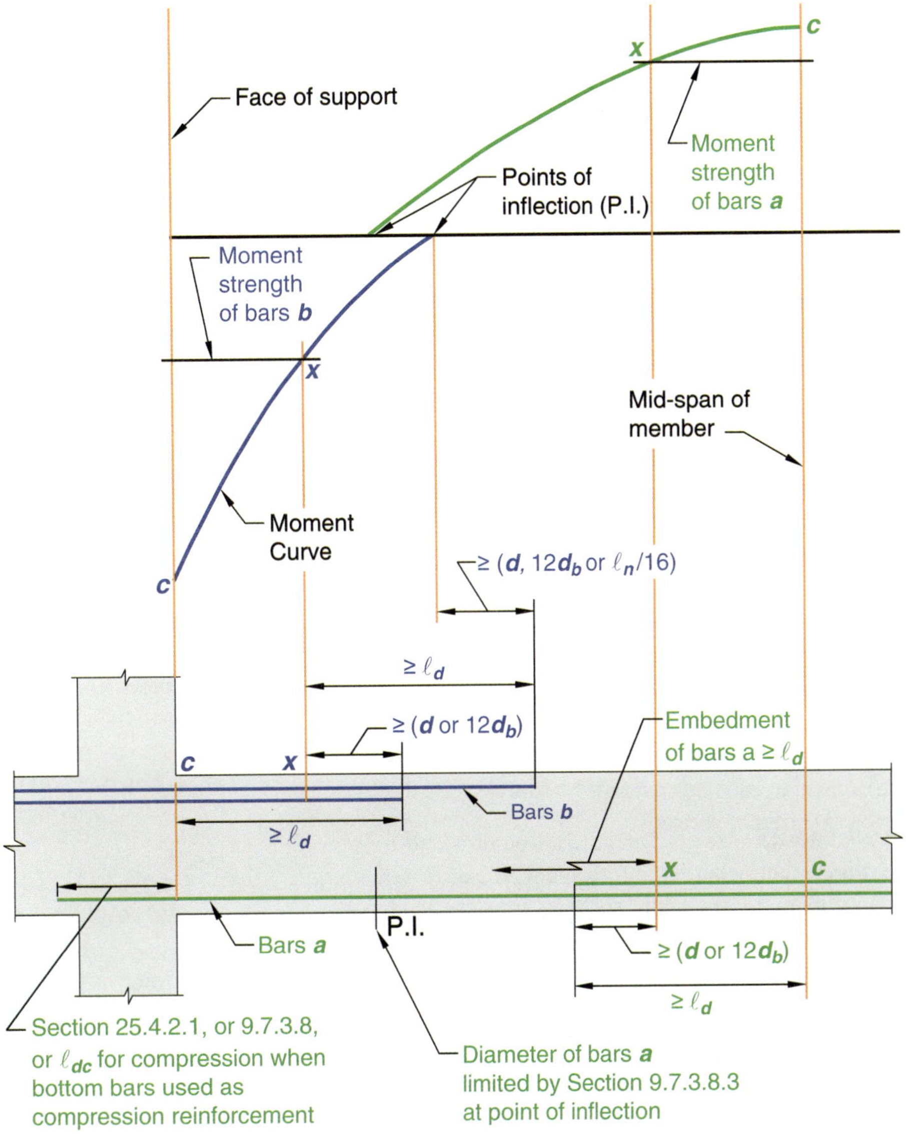

***Fig. R9.7.3.2**—Development of flexural reinforcement in a typical continuous beam.*

9.7.3.3 Reinforcement shall extend beyond the point at which it is no longer required to resist flexure for a distance equal to the greater of ***d*** and $\mathbf{12}\boldsymbol{d_b}$, except at supports of simply-supported spans and at free ends of cantilevers.

R9.7.3.3 The moment diagrams customarily used in design are approximate; some shifting of the location of maximum moments may occur due to changes in loading, settlement of supports, lateral loads, or other causes. A diagonal tension crack in a flexural member without stirrups may shift the location of the calculated tensile stress approximately a distance ***d*** toward a point of zero moment. If stirrups are provided, this effect is less severe, although still present to some extent.

To provide for shifts in the location of maximum moments, the Code requires the extension of reinforcement a distance ***d*** or $\mathbf{12}\boldsymbol{d_b}$ beyond the point at which it is calculated to be no longer required to resist flexure, except as noted. Cutoff points of bars to meet this requirement are illustrated in Fig. R9.7.3.2. If different bar sizes are used, the extension should be in accordance with the diameter of the bar being terminated.

CODE

9.7.3.4 Continuing flexural tension reinforcement shall have an embedment length at least ℓ_d beyond the point where bent or terminated tension reinforcement is no longer required to resist flexure.

9.7.3.5 Flexural tension reinforcement shall not be terminated in a tension zone unless (a), (b), or (c) is satisfied:

(a) $V_u \leq (2/3)\phi V_n$ at the cutoff point
(b) For No. 11 bars and smaller, continuing reinforcement provides double the area required for flexure at the cutoff point and $V_u \leq (3/4)\phi V_n$
(c) Stirrup or hoop area in excess of that required for shear and torsion is provided along each terminated bar or wire over a distance $3/4d$ from the cutoff point. Excess stirrup or hoop area shall be at least $60b_w s/f_{yt}$. Spacing s shall not exceed $d/(8\beta_b)$

9.7.3.6 Adequate anchorage shall be provided for tension reinforcement where reinforcement stress is not directly proportional to moment, such as in sloped, stepped, or tapered beams, or where tension reinforcement is not parallel to the compression face.

9.7.3.7 Development of tension reinforcement by bending across the web to be anchored or made continuous with reinforcement on the opposite face of beam shall be permitted.

9.7.3.8 *Termination of reinforcement*

9.7.3.8.1 At simple supports, at least one-third of the maximum positive moment reinforcement shall extend along the beam bottom into the support at least 6 in., except for precast beams where such reinforcement shall extend at least to the center of the bearing length.

9.7.3.8.2 At other supports, at least one-fourth of the maximum positive moment reinforcement shall extend along the beam bottom into the support at least 6 in. and, if the beam is part of the primary lateral-load-resisting system, shall be anchored to develop f_y at the face of the support.

9.7.3.8.3 At simple supports and points of inflection, d_b for positive moment tension reinforcement shall be limited such that ℓ_d for that reinforcement satisfies (a) or (b). If rein-

COMMENTARY

R9.7.3.4 Local peak stresses exist in the remaining bars wherever adjacent bars are cut off in tension regions. In Fig. R9.7.3.2, an "x" is used to indicate the point where terminated tension reinforcement is no longer required to resist flexure. If bars were cut off at this location (the required cutoff point is beyond location "x" in accordance with 9.7.3.3), peak stresses in the continuing bars would reach f_y at "x". Therefore, the continuing reinforcement is required to have a full ℓ_d extension as indicated.

R9.7.3.5 Reduced shear strength and loss of ductility when bars are cut off in a tension zone, as in Fig. R9.7.3.2, have been reported. The Code does not permit flexural reinforcement to be terminated in a tension zone unless additional conditions are satisfied. Flexural cracks tend to open at low load levels wherever any reinforcement is terminated in a tension zone. If the stress in the continuing reinforcement and the shear strength are each near their limiting values, diagonal tension cracking tends to develop prematurely from these flexural cracks. Diagonal cracks are less likely to form where shear stress is low (9.7.3.5(a)) or flexural reinforcement stress is low (9.7.3.5(b)). Diagonal cracks can be restrained by closely spaced stirrups (9.7.3.5(c)). These requirements are not intended to apply to tension splices that are covered by 25.5.

R9.7.3.7 A bar bent to the far face of a beam and continued there may be considered effective in satisfying 9.7.3.3 to the point where the bar crosses the mid-depth of the member.

R9.7.3.8 *Termination of reinforcement*

R9.7.3.8.1 Positive moment reinforcement is extended into the support to provide for some shifting of the moments due to changes in loading, settlement of supports, and lateral loads. It also enhances structural integrity.

For precast beams, tolerances and reinforcement cover should be considered to avoid bearing on plain concrete where reinforcement has been discontinued.

R9.7.3.8.2 Development of the positive moment reinforcement at the support is required for beams that are part of the primary lateral-load-resisting system to provide ductility in the event of moment reversal.

R9.7.3.8.3 The diameter of the positive moment tension reinforcement is limited to ensure that the bars are developed in a length short enough such that the moment capacity

CODE

forcement terminates beyond the centerline of supports by a standard hook or a mechanical anchorage at least equivalent to a standard hook, (a) or (b) need not be satisfied.

(a) $\ell_d \leq (1.3M_n/V_u + \ell_a)$ if end of reinforcement is confined by a compressive reaction
(b) $\ell_d \leq (M_n/V_u + \ell_a)$ if end of reinforcement is not confined by a compressive reaction

M_n is calculated assuming all reinforcement at the section is stressed to f_y, and V_u is calculated at the section. At a support, ℓ_a is the embedment length beyond the center of the support. At a point of inflection, ℓ_a is the embedment length beyond the point of inflection limited to the greater of d and $12d_b$.

COMMENTARY

is greater than the applied moment over the entire length of the beam. As illustrated in the moment diagram of Fig. R9.7.3.8.3(a), the slope of the moment diagram is V_u, while the slope of moment development is M_n/ℓ_d, where M_n is the nominal flexural strength of the cross section. By sizing the reinforcement such that the capacity slope M_n/ℓ_d equals or exceeds the demand slope V_u, proper development is provided. Therefore, M_n/V_u represents the available development length. Under favorable support conditions, a 30 percent increase for M_n/V_u is permitted when the ends of the reinforcement are confined by a compressive reaction.

The application of this provision is illustrated in Fig. R9.7.3.8.3(b) for simple supports and in Fig. R9.7.3.8.3(c) for points of inflection. For example, the bar size provided at a simple support is satisfactory only if the corresponding bar, ℓ_d, calculated in accordance with 25.4.2, does not exceed $1.3M_n/V_u + \ell_a$.

The ℓ_a to be used at points of inflection is limited to the effective depth of the member d or 12 bar diameters $(12d_b)$, whichever is greater. The ℓ_a limitation is provided because test data are not available to show that a long end anchorage length will be fully effective in developing a bar that has only a short length between a point of inflection and a point of maximum stress.

CODE

9.7.3.8.4 At least one-third of the negative moment reinforcement at a support shall have an embedment length beyond the point of inflection at least the greatest of d, $12d_b$, and $\ell_n/16$.

COMMENTARY

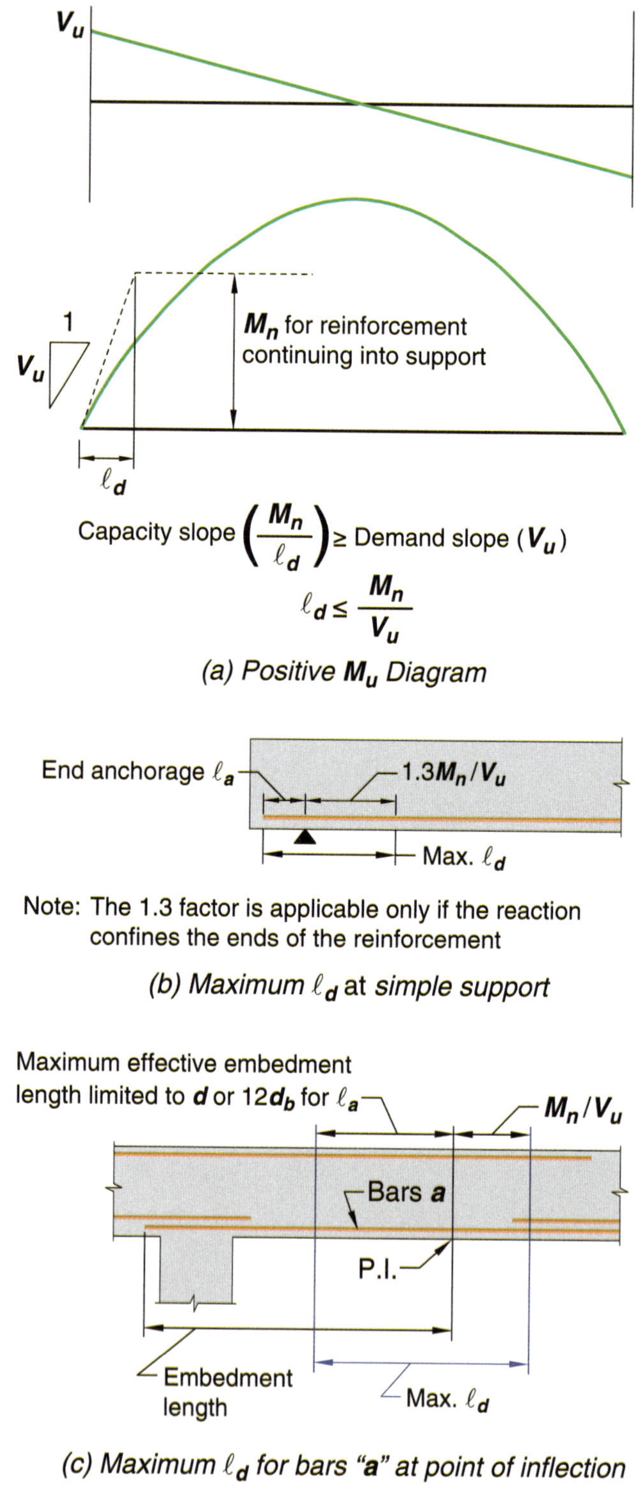

Fig. R9.7.3.8.3—Determination of maximum bar size according to 9.7.3.8.3.

CODE

9.7.4 *Flexural reinforcement in prestressed beams*

9.7.4.1 External tendons shall be attached to the member in a manner that maintains the specified eccentricity between the tendons and the concrete centroid through the full range of anticipated member deflections.

9.7.4.2 If nonprestressed reinforcement is required to satisfy flexural strength, the detailing requirements of 9.7.3 shall be satisfied.

9.7.4.3 *Termination of prestressed reinforcement*

9.7.4.3.1 Post-tensioned anchorage zones shall be designed and detailed in accordance with 25.9.

9.7.4.3.2 Post-tensioning anchorages and couplers shall be designed and detailed in accordance with 25.8.

9.7.4.4 *Termination of deformed reinforcement in beams with unbonded tendons*

9.7.4.4.1 Length of deformed reinforcement required by 9.6.2.3 shall be in accordance with (a) and (b):

(a) At least $\ell_n/3$ in positive moment areas and be centered in those areas

(b) At least $\ell_n/6$ on each side of the face of support in negative moment areas

9.7.5 *Longitudinal torsional reinforcement*

9.7.5.1 If torsional reinforcement is required, longitudinal torsional reinforcement shall be distributed around the perimeter of closed stirrups that satisfy 25.7.1.6 or hoops with a spacing not greater than 12 in. The longitudinal reinforcement shall be inside the stirrup or hoop, and at least one longitudinal bar or tendon shall be placed in each corner.

9.7.5.2 Longitudinal torsional reinforcement shall have a diameter at least 0.042 times the transverse reinforcement spacing, but not less than 3/8 in.

COMMENTARY

R9.7.4 *Flexural reinforcement in prestressed beams*

R9.7.4.1 External tendons are often attached to the concrete beam at various locations between anchorages, such as midspan, quarter points, or third points, for desired load balancing effects, for tendon alignment, or to address tendon vibration concerns. Consideration should be given to the effects caused by the tendon profile shifting in relationship to the concrete centroid as the member deforms under effects of post-tensioning and applied load.

R9.7.4.2 Nonprestressed reinforcement should be developed to achieve factored load forces. The requirements of 9.7.3 provide that bonded reinforcement required for flexural strength under factored loads is developed to achieve tensile or compressive forces.

R9.7.4.4 *Termination of deformed reinforcement in beams with unbonded tendons*

R9.7.4.4.1 The minimum lengths apply for bonded reinforcement required by 9.6.2.3. Research (Odello and Mehta 1967) on continuous spans shows that these minimum lengths provide satisfactory behavior under service load and factored load conditions.

R9.7.5 *Longitudinal torsional reinforcement*

R9.7.5.1 Longitudinal reinforcement is needed to resist the sum of the longitudinal tensile forces due to torsion. Because the force acts along the centroidal axis of the section, the centroid of the additional longitudinal reinforcement for torsion should approximately coincide with the centroid of the section. The Code accomplishes this by requiring the longitudinal torsional reinforcement be distributed around the perimeter of the closed stirrups. Longitudinal bars or tendons are required in each corner of the stirrups to provide anchorage for the stirrup legs. Corner bars have also been found to be effective in developing torsional strength and controlling cracks.

CODE

9.7.5.3 Longitudinal torsional reinforcement shall extend for a distance of at least $(b_t + d)$ beyond the point required by analysis.

9.7.5.4 Longitudinal torsional reinforcement shall be developed at the face of the support at both ends of the beam.

9.7.6 *Transverse reinforcement*

9.7.6.1 *General*

9.7.6.1.1 Transverse reinforcement shall be in accordance with this section. The most restrictive requirements shall apply.

9.7.6.1.2 Details of transverse reinforcement shall be in accordance with 25.7.

9.7.6.2 *Shear*

9.7.6.2.1 If required, shear reinforcement shall be provided using stirrups, hoops, or longitudinal bent bars.

COMMENTARY

R9.7.5.3 The distance $(b_t + d)$ beyond the point at which longitudinal torsional reinforcement is calculated to be no longer required is greater than that used for shear and flexural reinforcement because torsional diagonal tension cracks develop in a helical form. The same distance is required by 9.7.6.3.2 for transverse torsional reinforcement.

R9.7.5.4 Longitudinal torsional reinforcement required at a support should be adequately anchored into the support. Sufficient embedment length should be provided outside the inner face of the support to develop the needed tensile force in the bars or tendons. For bars, this may require hooks or horizontal U-shaped bars lapped with the longitudinal torsional reinforcement.

R9.7.6 *Transverse reinforcement*

R9.7.6.2 *Shear*

R9.7.6.2.1 If a reinforced concrete beam is cast monolithically with a supporting beam and intersects one or both side faces of a supporting beam, the soffit of the supporting beam may be subject to premature failure unless additional transverse reinforcement, commonly referred to as hanger reinforcement, is provided (Mattock and Shen 1992). The hanger reinforcement (Fig. R9.7.6.2.1), placed in addition to other transverse reinforcement, is provided to transfer shear from the end of the supported beam. Research indicates that if the bottom of the supported beam is at or above middepth of the supporting beam or if the factored shear transferred from the supported beam is less than $3\sqrt{f_c'}b_w d$, hanger reinforcement is not required.

CODE | COMMENTARY

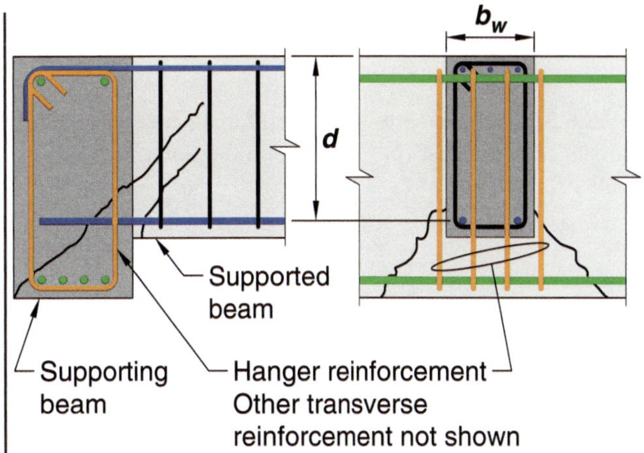

Fig. R9.7.6.2.1*—Hanger reinforcement for shear transfer.*

9.7.6.2.2 Maximum spacing of legs of shear reinforcement along the length of the member and across the width of the member shall be in accordance with Table 9.7.6.2.2.

Table 9.7.6.2.2—Maximum spacing of legs of shear reinforcement

Required V_s		Maximum *s*, in.			
		Nonprestressed beam		Prestressed beam	
		Along length	Across width	Along length	Across width
$\leq 4\sqrt{f'_c}b_wd$	Lesser of:	$d/2$	d	$3h/4$	$3h/2$
		24 in.			
$> 4\sqrt{f'_c}b_wd$	Lesser of:	$d/4$	$d/2$	$3h/8$	$3h/4$
		12 in.			

R9.7.6.2.2 Reduced stirrup spacing across the beam width provides a more uniform transfer of diagonal compression across the beam web, enhancing shear capacity. Laboratory tests (Leonhardt and Walther 1964; Anderson and Ramirez 1989; Lubell et al. 2009) of wide members with large spacing of legs of shear reinforcement across the member width indicate that the nominal shear capacity is not always achieved. The intent of this provision is to provide multiple stirrup legs across wide beams and one-way slabs that require stirrups.

9.7.6.2.3 Inclined stirrups and longitudinal bars bent to act as shear reinforcement shall be spaced so that every 45-degree line, extending ***d*/2** toward the reaction from mid-depth of member to longitudinal tension reinforcement, shall be crossed by at least one line of shear reinforcement.

9.7.6.2.4 Longitudinal bars bent to act as shear reinforcement, if extended into a region of tension, shall be continuous with longitudinal reinforcement and, if extended into a region of compression, shall be anchored ***d*/2** beyond mid-depth of member.

9.7.6.3 *Torsion*

9.7.6.3.1 If required, transverse torsional reinforcement shall be closed stirrups satisfying 25.7.1.6 or hoops.

R9.7.6.3 *Torsion*

R9.7.6.3.1 The stirrups are required to be closed because inclined cracking due to torsion may occur on all faces of a member.

In the case of sections subjected primarily to torsion, the concrete side cover over the stirrups spalls off at high torsional moments (Mitchell and Collins 1976). This renders lap-spliced stirrups ineffective, leading to a premature torsional failure (Behera and Rajagopalan 1969). Therefore,

9 Beams

closed stirrups should not be made up of pairs of U-stirrups lapping one another.

9.7.6.3.2 Transverse torsional reinforcement shall extend a distance of at least $(b_t + d)$ beyond the point required by analysis.

R9.7.6.3.2 The distance $(b_t + d)$ beyond the point at which transverse torsional reinforcement is calculated to be no longer required is greater than that used for shear and flexural reinforcement because torsional diagonal tension cracks develop in a helical form. The same distance is required by 9.7.5.3 for longitudinal torsional reinforcement.

9.7.6.3.3 Spacing of transverse torsional reinforcement shall not exceed the lesser of $p_h/8$ and 12 in.

R9.7.6.3.3 Spacing of the transverse torsional reinforcement is limited to ensure development of the torsional strength of the beam, prevent excessive loss of torsional stiffness after cracking, and control crack widths. For a square cross section, the $p_h/8$ limitation requires stirrups at approximately $d/2$, which corresponds to 9.7.6.2.

9.7.6.3.4 For hollow sections, the distance from the centerline of the transverse torsional reinforcement to the inside face of the wall of the hollow section shall be at least $0.5A_{oh}/p_h$.

R9.7.6.3.4 The transverse torsional reinforcement in a hollow section should be located in the outer half of the wall thickness effective for torsion where the wall thickness can be taken as A_{oh}/p_h.

9.7.6.4 *Lateral support of compression reinforcement*

R9.7.6.4 *Lateral support of compression reinforcement*

9.7.6.4.1 Transverse reinforcement shall be provided throughout the distance where longitudinal compression reinforcement is required. Lateral support of longitudinal compression reinforcement shall be provided by closed stirrups or hoops in accordance with 9.7.6.4.2 through 9.7.6.4.4.

R9.7.6.4.1 Compression reinforcement in beams should be enclosed by transverse reinforcement to prevent buckling.

9.7.6.4.2 Size of transverse reinforcement shall be at least (a) or (b). Deformed wire or welded wire reinforcement of equivalent area shall be permitted.

(a) No. 3 for longitudinal bars No. 10 and smaller
(b) No. 4 for longitudinal bars No. 11 and larger and for longitudinal bundled bars

9.7.6.4.3 Spacing of transverse reinforcement shall not exceed the least of (a) through (c):

(a) $16d_b$ of longitudinal reinforcement
(b) $48d_b$ of transverse reinforcement
(c) Least dimension of beam

9.7.6.4.4 Longitudinal compression reinforcement shall be arranged such that every corner and alternate compression bar shall be enclosed by the corner of the transverse reinforcement with an included angle of not more than 135 degrees, and no bar shall be farther than 6 in. clear on each side along the transverse reinforcement from such an enclosed bar.

9.7.7 *Structural integrity reinforcement in cast-in-place beams*

R9.7.7 *Structural integrity reinforcement in cast-in-place beams*

CODE

COMMENTARY

Experience has shown that the overall integrity of a structure can be substantially enhanced by minor changes in detailing of reinforcement and connections. It is the intent of this section of the Code to improve the redundancy and ductility in structures so that in the event of damage to a major supporting element or an abnormal loading event, the resulting damage may be localized and the structure will have a higher probability of maintaining overall stability.

With damage to a support, top reinforcement that is continuous over the support, but not confined by stirrups, will tend to tear out of the concrete and will not provide the catenary action required to bridge the damaged support. By making a portion of the bottom reinforcement continuous, catenary action can be provided.

If the depth of a continuous beam changes at a support, the bottom reinforcement in the deeper member should be terminated into the support with a standard hook or headed bar and the bottom reinforcement in the shallower member should be extended into and fully developed in the deeper member.

9.7.7.1 For beams along the perimeter of the structure, structural integrity reinforcement shall be in accordance with (a) through (c):

(a) At least one-quarter of the maximum positive moment reinforcement, but not less than two bars or strands, shall be continuous
(b) At least one-sixth of the negative moment reinforcement at the support, but not less than two bars or strands, shall be continuous
(c) Longitudinal structural integrity reinforcement shall be enclosed by closed stirrups in accordance with 25.7.1.6 or hoops along the clear span of the beam

R9.7.7.1 Requiring continuous top and bottom reinforcement in perimeter or spandrel beams provides a continuous tie around the structure. It is not the intent to require a tension tie of continuous reinforcement of constant size around the entire perimeter of a structure, but rather to require that one-half of the top flexural reinforcement required to extend past the point of inflection by 9.7.3.8.4 be further extended and spliced at or near midspan as required by 9.7.7.5. Similarly, the bottom reinforcement required to extend into the support in 9.7.3.8.2 should be made continuous or spliced with bottom reinforcement from the adjacent span. At noncontinuous supports, the longitudinal reinforcement is anchored as required by 9.7.7.4.

Figure R9.7.7.1 shows an example of a two-piece stirrup that satisfies the requirement of Sections 9.7.7.1(c) and 9.7.7.2(b). The 90-degree hook of the cap tie is located on the slab side so that it is better confined. Pairs of U-stirrups lapping one another as defined in 25.7.1.7 are not permitted in perimeter or spandrel beams. In the event of damage to the side concrete cover, the top longitudinal reinforcement may tend to tear out of the concrete and will not be adequately restrained by the exposed lap splice of the stirrup. Thus, the top longitudinal reinforcement will not provide the catenary action needed to bridge over a damaged region. Further, lapped U-stirrups will not be effective at high torsional moments as discussed in R9.7.6.3.1.

CODE

9.7.7.2 For other than perimeter beams, structural integrity reinforcement shall be in accordance with (a) or (b):

(a) At least one-quarter of the maximum positive moment reinforcement, but not less than two bars or strands, shall be continuous.

(b) Longitudinal reinforcement shall be enclosed by closed stirrups in accordance with 25.7.1.6 or hoops along the clear span of the beam.

9.7.7.3 Longitudinal structural integrity reinforcement shall pass through the region bounded by the longitudinal reinforcement of the column.

9.7.7.4 Longitudinal structural integrity reinforcement at noncontinuous supports shall be anchored to develop f_y at the face of the support.

9.7.7.5 If splices are necessary in continuous structural integrity reinforcement, the reinforcement shall be spliced in accordance with (a) and (b):

(a) Positive moment reinforcement shall be spliced at or near the support

(b) Negative moment reinforcement shall be spliced at or near midspan

9.7.7.6 Splices shall be mechanical or welded in accordance with 25.5.7 or Class B tension lap splices in accordance with 25.5.2.

COMMENTARY

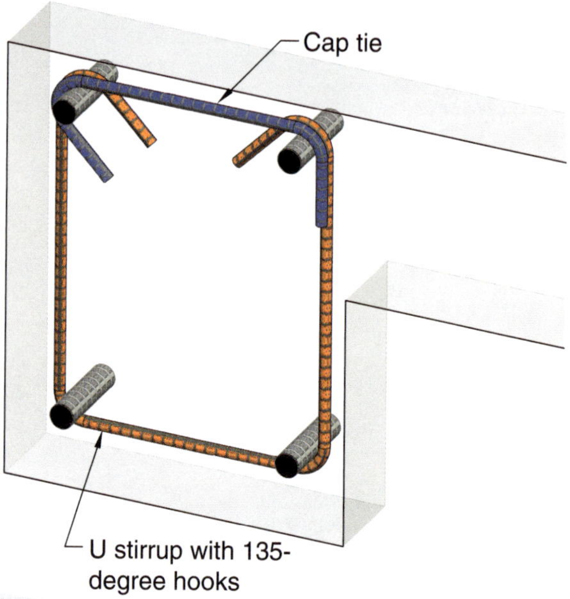

Fig. R9.7.7.1*—Example of a two-piece stirrup that complies with the requirements of 9.7.7.1(c) and 9.7.7.2(b).*

R9.7.7.2 At noncontinuous supports, the longitudinal reinforcement is anchored as required by 9.7.7.4.

R9.7.7.1 provides an example of a two-piece stirrup that satisfies 9.7.7.2(b).

R9.7.7.3 In the case of walls providing vertical support, the longitudinal reinforcement should pass through or be anchored in the wall.

CODE

9.8—Nonprestressed one-way joist systems

9.8.1 *General*

9.8.1.1 Nonprestressed one-way joist construction consists of a monolithic combination of regularly spaced ribs and a top slab designed to span in one direction.

9.8.1.2 Width of ribs shall be at least 4 in. at any location along the depth.

9.8.1.3 Overall depth of ribs shall not exceed 3.5 times the minimum width.

9.8.1.4 Clear spacing between ribs shall not exceed 30 in.

9.8.1.5 V_c shall be permitted to be taken as 1.1 times the value calculated in 22.5.

9.8.1.6 For structural integrity, at least one bottom bar in each joist shall be continuous and shall be anchored to develop f_y at the face of supports.

9.8.1.7 Reinforcement perpendicular to the ribs shall be provided in the slab as required for flexure, considering load concentrations, and shall be at least that required for shrinkage and temperature in accordance with 24.4.

9.8.1.8 One-way joist construction not satisfying the limitations of 9.8.1.1 through 9.8.1.4 shall be designed as slabs and beams.

9.8.2 *Joist systems with structural fillers*

9.8.2.1 If permanent burned clay or concrete tile fillers of material having a unit compressive strength at least equal to f_c' in the joists are used, 9.8.2.1.1 and 9.8.2.1.2 shall apply.

9.8.2.1.1 Slab thickness over fillers shall be at least the greater of one-twelfth the clear distance between ribs and 1.5 in.

9.8.2.1.2 For calculation of shear and negative moment strength, it shall be permitted to include the vertical shells of fillers in contact with the ribs. Other portions of fillers shall not be included in strength calculations.

9.8.3 *Joist systems with other fillers*

COMMENTARY

R9.8—Nonprestressed one-way joist systems

R9.8.1 *General*

The empirical limits established for nonprestressed reinforced concrete joist floors are based on successful past performance of joist construction using standard joist forming systems. For prestressed joist construction, this section may be used as guide.

R9.8.1.4 A limit on the maximum spacing of ribs is required because of the provisions permitting higher shear strengths and less concrete cover for the reinforcement for these relatively small, repetitive members.

R9.8.1.5 This increase in shear strength is justified on the basis of: 1) satisfactory performance of joist construction designed with higher calculated shear strengths specified in previous Codes which allowed comparable shear stresses; and 2) potential for redistribution of local overloads to adjacent joists.

CODE

9.8.3.1 If fillers not complying with 9.8.2.1 or removable forms are used, slab thickness shall be at least the greater of one-twelfth the clear distance between ribs and 2 in.

9.9—Deep beams

9.9.1 *General*

9.9.1.1 Deep beams are members that are loaded on one face and supported on the opposite face such that strut-like compression elements can develop between the loads and supports and that satisfy (a) or (b):

(a) Clear span does not exceed four times the overall member depth h

(b) Concentrated loads exist within a distance $2h$ from the face of the support

9.9.1.2 Deep beams shall be designed taking into account nonlinear distribution of longitudinal strain over the depth of the beam.

9.9.1.3 The strut-and-tie method in accordance with Chapter 23 is deemed to satisfy 9.9.1.2.

9.9.2 *Dimensional limits*

9.9.2.1 Except as permitted by 23.4.4, deep beam dimensions shall be selected such that:

$$V_u \le \phi 10\sqrt{f_c'}b_w d \tag{9.9.2.1}$$

9.9.3 *Reinforcement limits*

9.9.3.1 Distributed reinforcement along the side faces of deep beams shall be at least that required in (a) and (b):

(a) The area of distributed reinforcement perpendicular to the longitudinal axis of the beam, A_v, shall be at least $0.0025b_w s$, where s is the spacing of the distributed transverse reinforcement.

(b) The area of distributed reinforcement parallel to the longitudinal axis of the beam, A_{vh}, shall be at least $0.0025b_w s_2$, where s_2 is the spacing of the distributed longitudinal reinforcement.

9.9.3.2 The minimum area of flexural tension reinforcement, $A_{s,min}$, shall be determined in accordance with 9.6.1.

9.9.4 *Reinforcement detailing*

9.9.4.1 Concrete cover shall be in accordance with 20.5.1.

COMMENTARY

R9.9—Deep beams

R9.9.1 *General*

R9.9.1.1 The behavior of deep beams is discussed in Schlaich et al. (1987), Rogowsky and MacGregor (1986), Marti (1985), and Crist (1966). For a deep beam supporting gravity loads, this provision applies if the loads are applied on the top of the beam and the beam is supported on its bottom face. If the loads are applied through the sides or bottom of such a member, the strut-and-tie method, as defined in Chapter 23 should be used to design reinforcement to internally transfer the loads to the top of the beam and distribute them to adjacent supports.

R9.9.1.2 The Code does not contain detailed requirements for designing deep beams for moment, except that a nonlinear strain distribution should be considered. Guidance for the design of deep beams for flexure is given in Chow et al. (1953), Portland Cement Association (1946), and Park and Paulay (1975).

R9.9.2 *Dimensional limits*

R9.9.2.1 This limit imposes a dimensional restriction to control cracking under service loads and to guard against diagonal compression failures in deep beams.

R9.9.3 *Reinforcement limits*

R9.9.3.1 The minimum reinforcement requirements of this section are to be used irrespective of the method used for design and are intended to control the width and propagation of inclined cracks. Tests (Rogowsky and MacGregor 1986; Marti 1985; Crist 1966) have shown that vertical shear reinforcement, perpendicular to the longitudinal axis of the member, is more effective for member shear strength than horizontal shear reinforcement, parallel to the longitudinal axis of the member, in a deep beam; however, the specified minimum reinforcement is the same in both directions to control the growth and width of diagonal cracks.

R9.9.4 *Reinforcement detailing*

9.9.4.2 Minimum spacing for longitudinal reinforcement shall be in accordance with 25.2.

9.9.4.3 Spacing of distributed reinforcement required in 9.9.3.1 shall not exceed the lesser of $d/5$ and 12 in.

9.9.4.4 Development of tension reinforcement shall account for distribution of stress in reinforcement that is not directly proportional to the bending moment.

R9.9.4.4 In deep beams, the stress in the longitudinal reinforcement is more uniform along the length than that of a beam or region that is not deep. High reinforcement stresses normally limited to the center region of a typical beam can extend to the supports in deep beams. Thus, the ends of longitudinal reinforcement may require positive anchorage in the form of standard hooks, bar heads, or other mechanical anchorage at supports.

9.9.4.5 At simple supports, positive moment tension reinforcement shall be anchored to develop f_y at the face of the support. If a deep beam is designed using Chapter 23, the positive moment tension reinforcement shall be anchored in accordance with 23.8.2 and 23.8.3.

R9.9.4.5 The use of the strut-and-tie method for the design of deep beams illustrates that tensile forces in the bottom tie reinforcement need to be anchored at the face of the support. From this consideration, tie reinforcement should be continuous or developed at the face of the support (Rogowsky and MacGregor 1986).

9.9.4.6 At interior supports, (a) and (b) shall be satisfied:

(a) Negative moment tension reinforcement shall be continuous with that of the adjacent spans.

(b) Positive moment tension reinforcement shall be continuous or spliced with that of the adjacent spans.

Notes

Combined footings

Sizing example:

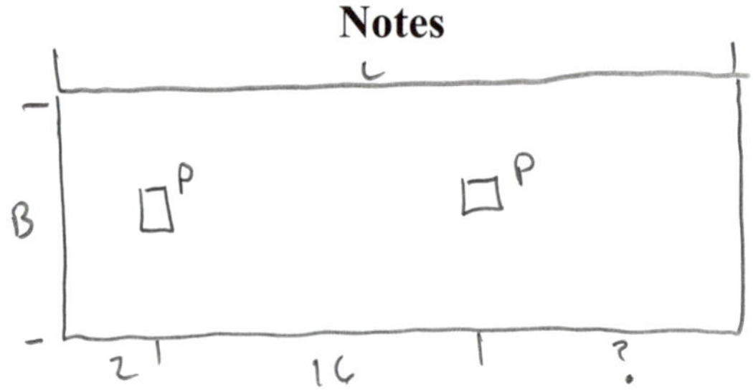

Soil $q_a = 4800$ psf

Service loads:

Column 1:

$D_L = 180^k$

$LL = 120^k$

Column 2:

$D_L = 265^k$

$LL = 120^k$

- Size footing to have uniform bearing pressure under SL

$P_1 = P_{DL} + P_{LL} = 180 + 120 = 300^k$

$P_2 = P_{DL} + P_{LL} = 265^k + 120^k = 385^k$

Locate Centroid (allowable! Not factored)

$$x_r = \frac{\Sigma P_i x_i}{\Sigma P_i} = \frac{300(2) + 385(18)}{300 + 385} = 11'$$

$$L = 2(x_r) = 2(11) = 22 \qquad ? = 4'$$

$$B = \frac{\Sigma P_i}{q_a(L)} = \frac{\Sigma P_i}{q_a L} = \frac{300_k + 385_k}{(4.8\ ksf)(22')} = 6.5'$$

Design example: Same foot consider DL+LL, LL on 1, LL on 2, DL only

(↑ $1.2 D_{L1} + 1.6 LL_1$; $1.2 D_{L1} + 0$)

$P_{u1} = 408^k$ $(1.2_{DL} + 1.6_{LL})$ $\quad x_2 = 10.9'$

$P_{u2} = 510^k$

$$\text{eccentricity} = e = \frac{L}{2} - x_r = \frac{22}{2} - 10.9 = 0.104$$

$$M_{total} = P_{total}(e) = (408 + 510)(0.104) = 95.5\ (ft\text{-}k)$$

Flexure formula's:

$$q = \frac{P}{A} \pm \frac{My}{I} = \frac{(918)}{(L \times B)} \pm \frac{(95.5)\ (x_r)}{(\frac{1}{2} b L^3)} = \begin{matrix} +6.60\ ksf \\ -6.24\ ksf \end{matrix}$$

Distributed load under footing

$$\omega = q B$$

$$\omega_{max} = (6.60)(6.5) = 42.9$$

$$\omega_{min} = (6.24)(6.5) = 40.6\ k/ft$$

$$\text{Slop} = \frac{\omega_{max} - \omega_{min}}{\ell}$$

Creates a moment envelope

CHAPTER 10—COLUMNS

CODE

10.1—Scope

10.1.1 This chapter shall apply to the design of nonprestressed and prestressed columns, including reinforced concrete pedestals.

10.1.2 Design of plain concrete pedestals shall be in accordance with Chapter 14.

10.2—General

10.2.1 *Materials*

10.2.1.1 Design properties for concrete shall be selected to be in accordance with Chapter 19.

10.2.1.2 Design properties for steel reinforcement shall be selected to be in accordance with Chapter 20.

10.2.1.3 Materials, design, and detailing requirements for embedments in concrete shall be in accordance with 20.6.

10.2.2 *Connection to other members*

10.2.2.1 For cast-in-place construction, beam-column and slab-column joints shall satisfy Chapter 15.

10.2.2.2 For precast construction, connections shall satisfy the force transfer requirements of 16.2.

10.2.2.3 Connections of columns to foundations shall satisfy 16.3.

10.3—Design limits

10.3.1 *Dimensional limits*

10.3.1.1 For columns with a square, octagonal, or other shaped cross section, it shall be permitted to base gross area considered, required reinforcement, and design strength on a circular section with a diameter equal to the least lateral dimension of the actual shape.

10.3.1.2 For columns with cross sections larger than required by considerations of loading, it shall be permitted to base gross area considered, required reinforcement, and design strength on a reduced effective area, not less than one-half the total area. This provision shall not apply to columns in special moment frames or columns not part of the seismic-force-resisting system required to be designed in accordance with Chapter 18.

10.3.1.3 For columns built monolithically with a concrete wall, the outer limits of the effective cross section of the

COMMENTARY

R10.1—Scope

R10.1.1 Composite structural steel-concrete columns are not covered in this chapter. Composite columns include both structural steel sections encased in reinforced concrete and hollow structural steel sections filled with concrete. Design provisions for such composite columns are covered in AISC 360.

R10.3—Design limits

R10.3.1 *Dimensional limits*

Explicit minimum sizes for columns are not specified to permit the use of reinforced concrete columns with small cross sections in lightly loaded structures, such as low-rise residential and light office buildings. If small cross sections are used, there is a greater need for careful workmanship, and shrinkage stresses have increased significance.

R10.3.1.2 In some cases, the gross area of a column is larger than necessary to resist the factored load. In those cases, the minimum reinforcement percentage may be calculated on the basis of the required area rather than the provided area, but the area of reinforcement cannot be less than 0.5 percent of the actual cross-sectional area.

CODE

column shall not be taken greater than 1.5 in. outside the transverse reinforcement.

10.3.1.4 For columns with two or more interlocking spirals, outer limits of the effective cross section shall be taken at a distance outside the spirals equal to the minimum required concrete cover.

10.3.1.5 If a reduced effective area is considered according to 10.3.1.1 through 10.3.1.4, structural analysis and design of other parts of the structure that interact with the column shall be based on the actual cross section.

10.4—Required strength

10.4.1 *General*

10.4.1.1 Required strength shall be calculated in accordance with the factored load combinations in Chapter 5.

10.4.1.2 Required strength shall be calculated in accordance with the analysis procedures in Chapter 6.

10.4.2 *Factored axial force and moment*

10.4.2.1 P_u and M_u occurring simultaneously for each applicable factored load combination shall be considered.

COMMENTARY

R10.4—Required strength

R10.4.2 *Factored axial force and moment*

R10.4.2.1 The critical load combinations may be difficult to discern without methodically checking each combination. As illustrated in Fig. R10.4.2.1, considering only the factored load combinations associated with maximum axial force (LC1) and with maximum bending moment (LC2) does not necessarily provide a code-compliant design for other load combinations such as LC3.

CODE

COMMENTARY

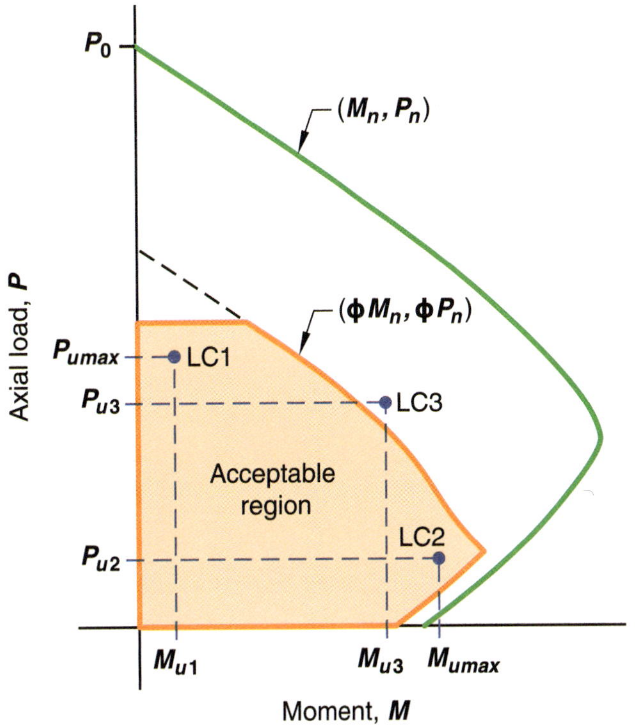

Fig. R10.4.2.1*—Critical column load combination.*

10.5—Design strength

10.5.1 *General*

10.5.1.1 For each applicable factored load combination, design strength at all sections shall satisfy $\phi S_n \geq U$, including (a) through (d). Interaction between load effects shall be considered:

(a) $\phi P_n \geq P_u$
(b) $\phi M_n \geq M_u$
(c) $\phi V_n \geq V_u$
(d) $\phi T_n \geq T_u$

10.5.1.2 ϕ shall be determined in accordance with 21.2.

10.5.2 *Axial force and moment*

10.5.2.1 P_n and M_n shall be calculated in accordance with 22.4.

10.5.3 *Shear*

10.5.3.1 V_n shall be calculated in accordance with 22.5.

10.5.4 *Torsion*

10.5.4.1 If $T_u \geq \phi T_{th}$, where T_{th} is given in 22.7, torsion shall be considered in accordance with Chapter 9.

R10.5—Design strength

R10.5.1 *General*

R10.5.1.1 Refer to R9.5.1.1.

R10.5.4 *Torsion*

Torsion acting on columns in buildings is typically negligible and is rarely a governing factor in the design of columns.

CODE

10.6—Reinforcement limits

10.6.1 *Minimum and maximum longitudinal reinforcement*

10.6.1.1 For nonprestressed columns and for prestressed columns with average f_{pe} < **225 psi**, area of longitudinal reinforcement shall be at least $\mathbf{0.01}A_g$ but shall not exceed $\mathbf{0.08}A_g$.

10.6.2 *Minimum shear reinforcement*

10.6.2.1 A minimum area of shear reinforcement, $A_{v,min}$, shall be provided in all regions where $V_u > \mathbf{0.5}\phi V_c$.

10.6.2.2 If shear reinforcement is required, $A_{v,min}$ shall be the greater of (a) and (b):

(a) $0.75\sqrt{f_c'}\dfrac{b_w s}{f_{yt}}$

(b) $50\dfrac{b_w s}{f_{yt}}$

10.7—Reinforcement detailing

10.7.1 *General*

10.7.1.1 Concrete cover for reinforcement shall be in accordance with 20.5.1.

10.7.1.2 Development lengths of deformed and prestressed reinforcement shall be in accordance with 25.4.

COMMENTARY

R10.6—Reinforcement limits

R10.6.1 *Minimum and maximum longitudinal reinforcement*

R10.6.1.1 Limits are provided for both the minimum and maximum longitudinal reinforcement ratios.

Minimum reinforcement—Reinforcement is necessary to provide resistance to bending, which may exist regardless of analytical results, and to reduce the effects of creep and shrinkage of the concrete under sustained compressive stresses. Creep and shrinkage tend to transfer load from the concrete to the reinforcement, and the resultant increase in reinforcement stress becomes greater as the reinforcement ratio decreases. Therefore, a minimum limit is placed on the reinforcement ratio to prevent reinforcement from yielding under sustained service loads (Richart 1933).

Maximum reinforcement—The amount of longitudinal reinforcement is limited to ensure that concrete can be effectively consolidated around the bars and to ensure that columns designed according to the Code are similar to the test specimens by which the Code was calibrated. The 0.08 limit applies at all sections, including splice regions, and can also be considered a practical maximum for longitudinal reinforcement in terms of economy and requirements for placing. Longitudinal reinforcement in columns should usually not exceed 4 percent if the column bars are required to be lap spliced, as the lap splice zone will have twice as much reinforcement if all lap splices occur at the same location.

R10.6.2 *Minimum shear reinforcement*

R10.6.2.1 The basis for the minimum shear reinforcement is the same for columns and beams. Refer to R9.6.3 for more information.

R10.7—Reinforcement detailing

CODE

10.7.1.3 Along development and lap splice lengths of longitudinal bars with $f_y \geq$ **80,000 psi**, transverse reinforcement shall be provided such that K_{tr} shall not be smaller than $\mathbf{0.5}d_b$.

10.7.1.4 Bundled bars shall be in accordance with 25.6.

10.7.2 *Reinforcement spacing*

10.7.2.1 Minimum spacing s shall be in accordance with 25.2.

10.7.3 *Longitudinal reinforcement*

10.7.3.1 For nonprestressed columns and for prestressed columns with average f_{pe} < **225 psi**, the minimum number of longitudinal bars shall be (a), (b), or (c):

(a) Three within triangular ties

(b) Four within rectangular or circular ties

(c) Six enclosed by spirals or for columns of special moment frames enclosed by circular hoops

10.7.4 *Offset bent longitudinal reinforcement*

10.7.4.1 The slope of the inclined portion of an offset bent longitudinal bar relative to the longitudinal axis of the column shall not exceed 1 in 6. Portions of bar above and below an offset shall be parallel to axis of column.

10.7.4.2 If the column face is offset 3 in. or more, longitudinal bars shall not be offset bent and separate dowels, lap spliced with the longitudinal bars adjacent to the offset column faces, shall be provided.

10.7.5 *Splices of longitudinal reinforcement*

10.7.5.1 *General*

10.7.5.1.1 Lap splices, mechanical splices, butt-welded splices, and end-bearing splices shall be permitted.

10.7.5.1.2 Splices shall satisfy requirements for all factored load combinations.

10.7.5.1.3 Splices of deformed reinforcement shall be in accordance with 25.5 and, if applicable, shall satisfy the requirements of 10.7.5.2 for lap splices or 10.7.5.3 for end-bearing splices.

COMMENTARY

R10.7.3 *Longitudinal reinforcement*

R10.7.3.1 At least four longitudinal bars are required when bars are enclosed by rectangular or circular ties. For other tie shapes, one bar should be provided at each apex or corner and proper transverse reinforcement provided. For example, tied triangular columns require at least three longitudinal bars, with one at each apex of the triangular ties. For bars enclosed by spirals, at least six bars are required.

If the number of bars in a circular arrangement is less than eight, the orientation of the bars may significantly affect the moment strength of eccentrically loaded columns and should be considered in design.

R10.7.5 *Splices of longitudinal reinforcement*

R10.7.5.1 *General*

R10.7.5.1.2 Frequently, the basic gravity load combination will govern the design of the column itself, but a load combination including wind or earthquake effects may induce greater tension in some column bars. Each bar splice should be designed for the maximum calculated bar tensile force.

R10.7.5.1.3 For the purpose of calculating ℓ_d for tension lap splices in columns with offset bars, Fig. R10.7.5.1.3 illustrates the clear spacing to be used.

CODE

COMMENTARY

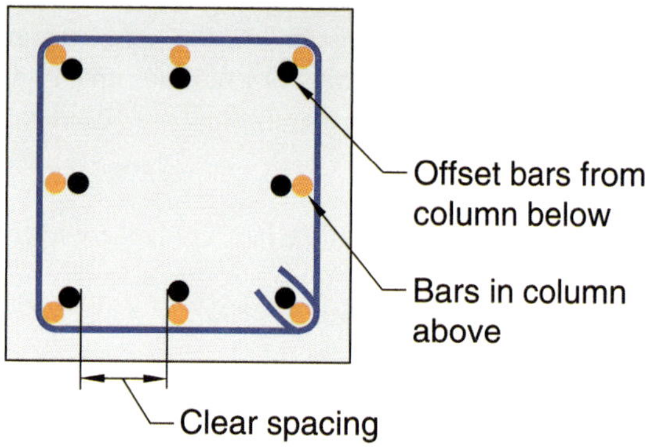

Fig. R10.7.5.1.3—Offset column bars.

10.7.5.2 *Lap splices*

R10.7.5.2 *Lap splices*

In columns subject to moment and axial force, tensile stresses may occur on one face of the column for moderate and large eccentricities as shown in Fig. R10.7.5.2. If such stresses occur, 10.7.5.2.2 requires tension splices to be used.

The splice requirements have been formulated on the basis that a compression lap splice has a tensile strength of at least $0.25f_y$. Therefore, even if columns bars are designed for compression according to 10.7.5.2.1, some tensile strength is inherently provided.

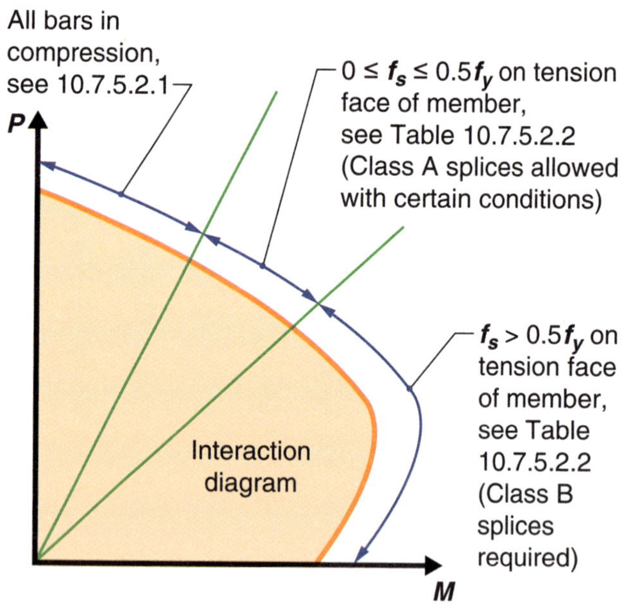

Fig. R10.7.5.2—Lap splice requirements for columns.

10.7.5.2.1 If the bar force due to factored loads is compressive, compression lap splices shall be permitted. It shall be permitted to decrease the compression lap splice length in accordance with (a) or (b), but the lap splice length shall be at least 12 in.

(a) For tied columns, where ties throughout the lap splice length have an effective area not less than $0.0015hs$ in both directions, lap splice length shall be permitted to be

R10.7.5.2.1 Reduced lap lengths are permitted if the splice is enclosed throughout its length by sufficient ties. The tie leg areas perpendicular to each direction are calculated separately. An example is provided in Fig. R10.7.5.2.1, where four legs are effective in one direction and two legs in the other direction.

Compression lap lengths may also be reduced if the lap splice is enclosed throughout its length by spirals due to increased splitting resistance.

CODE

multiplied by 0.83. Tie legs perpendicular to dimension h shall be considered in calculating effective area.

(b) For spiral columns, where spirals throughout the lap splice length satisfy 25.7.3, lap splice length shall be permitted to be multiplied by 0.75.

10.7.5.2.2 If the bar force due to factored loads is tensile, tension lap splices shall be in accordance with Table 10.7.5.2.2.

Table 10.7.5.2.2—Tension lap splice class

Tensile bar stress	Splice details	Splice type
$\leq 0.5f_y$	$\leq$50% bars spliced at any section and lap splices on adjacent bars staggered by at least ℓ_d	Class A
	Other	Class B
$>0.5f_y$	All cases	Class B

10.7.5.3 *End-bearing splices*

10.7.5.3.1 If the bar force due to factored loads is compressive, end-bearing splices shall be permitted provided the splices are staggered or additional bars are provided at splice locations. The continuing bars in each face of the column shall have a tensile strength at least $0.25f_y$ times the area of the vertical reinforcement along that face.

10.7.6 *Transverse reinforcement*

10.7.6.1 *General*

10.7.6.1.1 Transverse reinforcement shall satisfy the most restrictive requirements for reinforcement spacing.

COMMENTARY

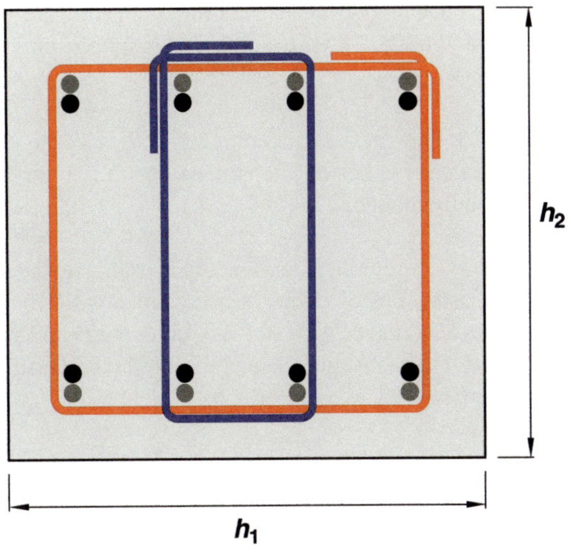

Direction 1: $4A_b \geq 0.0015h_1s$

Direction 2: $2A_b \geq 0.0015h_2s$

where A_b is the area of the tie

Fig. R10.7.5.2.1*—Example of application of 10.7.5.2.1(a).*

R10.7.5.3 *End-bearing splices*

R10.7.5.3.1 Details for end-bearing splices are provided in 25.5.6.

R10.7.6 *Transverse reinforcement*

R10.7.6.1 *General*

CODE

10.7.6.1.2 Details of transverse reinforcement shall be in accordance with 25.7.2 for ties, 25.7.3 for spirals, or 25.7.4 for hoops.

10.7.6.1.3 For prestressed columns with average $f_{pe} \geq$ **225 psi**, transverse ties or hoops need not satisfy the $\mathbf{16}d_b$ spacing requirement of 25.7.2.1.

10.7.6.1.4 Longitudinal reinforcement shall be laterally supported using ties or hoops in accordance with 10.7.6.2 or spirals in accordance with 10.7.6.3, unless tests and structural analyses demonstrate adequate strength and feasibility of construction.

10.7.6.1.5 If anchor bolts are placed in the top of a column or pedestal, the bolts shall be enclosed by transverse reinforcement that also surrounds at least four longitudinal bars within the column or pedestal. The transverse reinforcement shall be distributed within 5 in. of the top of the column or pedestal and shall consist of at least two No. 4 or three No. 3 ties or hoops.

10.7.6.1.6 If mechanical couplers or extended bars for connection to a precast element are placed in the ends of columns or pedestals, the mechanical couplers or extended bars shall be enclosed by transverse reinforcement. The transverse reinforcement shall be distributed within 5 in. of the ends of the column or pedestal and shall consist of at least two No. 4 or three No. 3 ties or hoops.

10.7.6.2 *Lateral support of longitudinal bars using ties or hoops*

10.7.6.2.1 In any story, the bottom tie or hoop shall be located not more than one-half the tie or hoop spacing above the top of footing or slab.

10.7.6.2.2 In any story, the top tie or hoop shall be located not more than one-half the tie or hoop spacing below the lowest horizontal reinforcement in the slab, drop panel, or shear cap. If beams or brackets frame into all sides of the column, the top tie or hoop shall be located not more than 3 in. below the lowest horizontal reinforcement in the shallowest beam or bracket.

COMMENTARY

R10.7.6.1.4 All longitudinal bars in compression should be enclosed within transverse reinforcement. Where longitudinal bars are arranged in a circular pattern, only one circular tie per specified spacing is required. This requirement can be satisfied by a continuous circular tie (helix), with the maximum pitch being equal to the required tie spacing.

It is prudent to provide a set of ties at each end of lap spliced bars, above and below end-bearing splices, and at minimum spacings immediately below sloping regions of offset bent bars.

Precast columns with cover less than 1-1/2 in., prestressed columns without longitudinal bars, columns of concrete with small size coarse aggregate, wall-like columns, and other unusual columns may require special designs for transverse reinforcement.

R10.7.6.1.5 and **R10.7.6.1.6** Confinement improves load transfer from the anchor bolts and mechanical couplers to the column or pedestal where concrete may crack in the vicinity of the bolts and mechanical couplers. Such cracking can occur due to unanticipated forces caused by temperature, restrained shrinkage, accidental impact during construction, and similar effects.

R10.7.6.2 *Lateral support of longitudinal bars using ties or hoops*

R10.7.6.2.2 For rectangular columns, beams or brackets framing into all four sides at the same elevation are considered to provide restraint over a joint depth equal to that of the shallowest beam or bracket. For columns with other shapes, four beams framing into the column from two orthogonal directions are considered to provide equivalent restraint.

CODE

10.7.6.3 *Lateral support of longitudinal bars using spirals*

10.7.6.3.1 In any story, the bottom of the spiral shall be located at the top of footing or slab.

10.7.6.3.2 In any story, the top of the spiral shall be located in accordance with Table 10.7.6.3.2.

Table 10.7.6.3.2 —Spiral extension requirements at top of column

Framing at column end	Extension requirements
Beams or brackets frame into all sides of the column	Extend to the level of the lowest horizontal reinforcement in members supported above.
Beams or brackets do not frame into all sides of the column	Extend to the level of the lowest horizontal reinforcement in members supported above. Additional column ties shall extend above termination of spiral to bottom of slab, drop panel, or shear cap.
Columns with capitals	Extend to the level at which the diameter or width of capital is twice that of the column.

10.7.6.4 *Lateral support of offset bent longitudinal bars*

10.7.6.4.1 Where longitudinal bars are offset, horizontal support shall be provided by ties, hoops, spirals, or parts of the floor construction and shall be designed to resist 1.5 times the horizontal component of the calculated force in the inclined portion of the offset bar.

10.7.6.4.2 If transverse reinforcement is provided to resist forces that result from offset bends, ties, hoops, or spirals shall be placed not more than 6 in. from points of bend.

10.7.6.5 *Shear*

10.7.6.5.1 If required, shear reinforcement shall be provided using ties, hoops, or spirals.

10.7.6.5.2 Maximum spacing of shear reinforcement shall be in accordance with Table 10.7.6.5.2.

Table 10.7.6.5.2—Maximum spacing of shear reinforcement

V_s	Maximum s, in.		
		Nonprestressed column	Prestressed column
$\leq 4\sqrt{f'_c}b_w d$	Lesser of:	$d/2$	$3h/4$
		24	
$> 4\sqrt{f'_c}b_w d$	Lesser of:	$d/4$	$3h/8$
		12	

COMMENTARY

R10.7.6.3 *Lateral support of longitudinal bars using spirals*

R10.7.6.3.2 Refer to R10.7.6.2.2.

10 Columns

Notes

CHAPTER 11—WALLS

CODE

11.1—Scope

11.1.1 This chapter shall apply to the design of nonprestressed and prestressed walls including (a) through (c):

(a) Cast-in-place
(b) Precast in-plant
(c) Precast on-site including tilt-up

11.1.2 Design of special structural walls shall be in accordance with Chapter 18.

11.1.3 Design of plain concrete walls shall be in accordance with Chapter 14.

11.1.4 Design of cantilever retaining walls shall be in accordance with Chapter 13.

11.1.5 Design of walls as grade beams shall be in accordance with 13.3.5.

11.1.6 Cast-in-place walls with insulating forms shall be permitted by this Code for use in one- or two-story buildings.

11.2—General

11.2.1 *Materials*

11.2.1.1 Design properties for concrete shall be selected to be in accordance with Chapter 19.

11.2.1.2 Design properties for steel reinforcement shall be selected to be in accordance with Chapter 20.

11.2.1.3 Materials, design, and detailing requirements for embedments in concrete shall be in accordance with 20.6.

11.2.2 *Connection to other members*

11.2.2.1 For precast walls, connections shall be designed in accordance with 16.2.

11.2.2.2 Connections of walls to foundations shall satisfy 16.3.

COMMENTARY

R11.1—Scope

R11.1.1 This chapter applies generally to walls as vertical and lateral force-resisting members. Provisions for in-plane shear in ordinary structural walls, as opposed to special structural walls conforming to 18.10, are included in this chapter.

R11.1.2 Special structural walls are detailed according to the provisions of 18.10. This Code uses the term "structural wall" as being synonymous with "shear wall." While the term "shear wall" is not defined in this Code, the definition of a structural wall in Chapter 2 states "a shear wall is a structural wall."

ASCE/SEI 7 defines a structural wall as a wall that meets the definition for a bearing wall or a shear wall. A bearing wall is defined as a wall that supports vertical load beyond a certain threshold value. A shear wall is defined as a wall, bearing or nonbearing, designed to resist lateral forces acting in the plane of the wall. ASCE/SEI 7 definitions are widely accepted.

R11.1.6 Specific design recommendations for cast-in-place walls constructed with insulating concrete forms are not provided in this Code. Guidance can be found in ACI 506R and PCA 100.

R11.2—General

CODE

11.2.3 *Load distribution*

11.2.3.1 Unless otherwise demonstrated by an analysis, the horizontal length of wall considered as effective for resisting each concentrated load shall not exceed the lesser of the center-to-center distance between loads, and the bearing width plus four times the wall thickness. Effective horizontal length for bearing shall not extend beyond vertical wall joints unless design provides for transfer of forces across the joints.

11.2.4 *Intersecting elements*

11.2.4.1 Walls shall be anchored to intersecting elements, such as floors and roofs; columns, pilasters, buttresses, or intersecting walls; and to footings.

11.2.4.2 For cast-in-place walls having $P_u > 0.2f_c'A_g$, the portion of the wall within the thickness of the floor system shall have specified compressive strength at least $0.8f_c'$ of the wall.

11.3—Design limits

11.3.1 *Minimum wall thickness*

11.3.1.1 Minimum wall thicknesses shall be in accordance with Table 11.3.1.1. Thinner walls are permitted if adequate strength and stability can be demonstrated by structural analysis.

Table 11.3.1.1—Minimum wall thickness *h*

Wall type	Minimum thickness *h*		
Bearing[1]	Greater of:	4 in.	(a)
		1/25 the lesser of unsupported length and unsupported height	(b)
Nonbearing	Greater of:	4 in.	(c)
		1/30 the lesser of unsupported length and unsupported height	(d)
Exterior basement and foundation[1]	7.5 in.		(e)

[1]Only applies to walls designed in accordance with the simplified design method of 11.5.3.

11.4—Required strength

11.4.1 *General*

11.4.1.1 Required strength shall be calculated in accordance with the factored load combinations in Chapter 5.

11.4.1.2 Required strength shall be calculated in accordance with the analysis procedures in Chapter 6.

COMMENTARY

R11.2.4 *Intersecting elements*

R11.2.4.1 Walls that do not depend on intersecting elements for support, do not have to be connected to those elements. It is not uncommon to separate massive retaining walls from intersecting walls to accommodate differences in deformations.

R11.2.4.2 The 0.8 factor reflects reduced confinement in floor-wall joints compared with floor-column joints under gravity loads.

R11.3—Design limits

R11.3.1 *Minimum wall thickness*

R11.3.1.1 The minimum thickness requirements need not be applied to bearing walls and exterior basement and foundation walls designed by 11.5.2 or analyzed by 11.8.

R11.4—Required strength

R11.4.1 *General*

CODE

11.4.1.3 Slenderness effects shall be calculated in accordance with 6.6.4, 6.7, or 6.8. Alternatively, out-of-plane slenderness analysis shall be permitted using 11.8 for walls meeting the requirements of that section.

11.4.1.4 Walls shall be designed for eccentric axial loads and any lateral or other loads to which they are subjected.

11.4.2 *Factored axial force and moment*

11.4.2.1 Walls shall be designed for the maximum factored moment M_u that can accompany the factored axial force for each applicable load combination. The factored axial force P_u at given eccentricity shall not exceed $\phi P_{n,max}$, where $P_{n,max}$ shall be as given in 22.4.2.1 and strength reduction factor ϕ shall be that for compression-controlled sections in 21.2.2. The maximum factored moment M_u shall be magnified for slenderness effects in accordance with 6.6.4, 6.7, or 6.8.

11.4.3 *Factored shear*

11.4.3.1 Walls shall be designed for the maximum in-plane V_u and out-of-plane V_u.

11.5—Design strength

11.5.1 *General*

11.5.1.1 For each applicable factored load combination, design strength at all sections shall satisfy $\phi S_n \geq U$, including (a) through (c). Interaction between axial load and moment shall be considered.

COMMENTARY

R11.4.1.3 The forces typically acting on a wall are illustrated in Fig. R11.4.1.3.

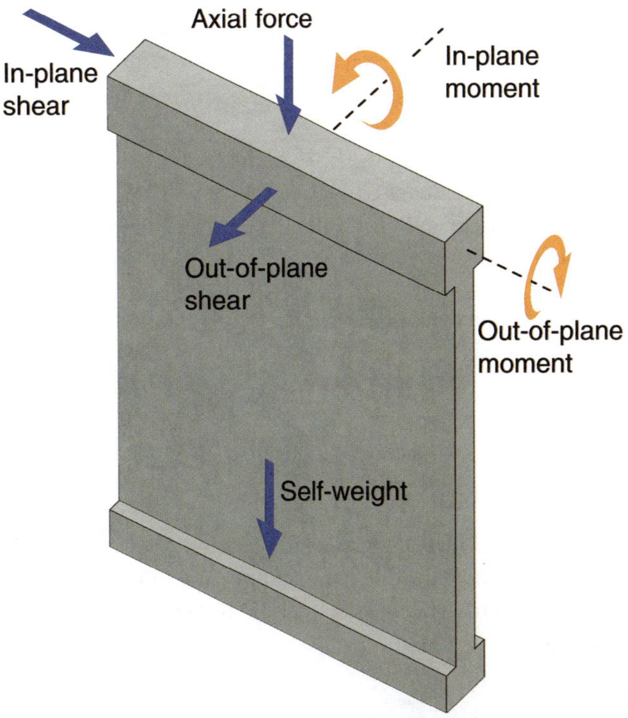

***Fig. R11.4.1.3**—In-plane and out-of-plane forces.*

R11.5—Design strength

Walls

11

CODE

(a) $\phi \boldsymbol{P_n} \geq \boldsymbol{P_u}$
(b) $\phi \boldsymbol{M_n} \geq \boldsymbol{M_u}$
(c) $\phi \boldsymbol{V_n} \geq \boldsymbol{V_u}$

11.5.1.2 ϕ shall be determined in accordance with 21.2.

11.5.2 *Axial load and in-plane or out-of-plane flexure*

11.5.2.1 For bearing walls, $\boldsymbol{P_n}$ and $\boldsymbol{M_n}$ (in-plane or out-of-plane) shall be calculated in accordance with 22.4. Alternatively, axial load and out-of-plane flexure shall be permitted to be considered in accordance with 11.5.3.

11.5.2.2 For nonbearing walls, $\boldsymbol{M_n}$ shall be calculated in accordance with 22.3.

11.5.3 *Axial load and out-of-plane flexure – simplified design method*

11.5.3.1 If the resultant of all factored loads is located within the middle third of the thickness of a solid wall with a rectangular cross section, $\boldsymbol{P_n}$ shall be permitted to be calculated by:

$$P_n = 0.55 f_c' A_g \left[1 - \left(\frac{k\ell_c}{32h}\right)^2\right] \quad (11.5.3.1)$$

COMMENTARY

R11.5.2 *Axial load and in-plane or out-of-plane flexure*

R11.5.2.2 Nonbearing walls, by definition, are not subject to any significant axial force; therefore, flexural strength is not a function of axial force.

R11.5.3 *Axial load and out-of-plane flexure – simplified design method*

R11.5.3.1 The simplified design method applies only to solid rectangular cross sections; all other shapes should be designed in accordance with 11.5.2.

Eccentric axial loads and moments due to out-of-plane forces are used to determine the maximum total eccentricity of the factored axial force $\boldsymbol{P_u}$. When the resultant axial force for all applicable load combinations falls within the middle third of the wall thickness (eccentricity not greater than $\boldsymbol{h/6}$) at all sections along the length of the undeformed wall, no tension is induced in the wall and the simplified design method may be used. The design is then carried out considering $\boldsymbol{P_u}$ as a concentric axial force. The factored axial force $\boldsymbol{P_u}$ should be less than or equal to the design axial strength $\phi \boldsymbol{P_n}$ calculated using Eq. (11.5.3.1).

Equation (11.5.3.1) results in strengths comparable to those determined in accordance with 11.5.2 for members loaded at the middle third of the thickness with different braced and restrained end conditions. Refer to Fig. R11.5.3.1.

CODE

COMMENTARY

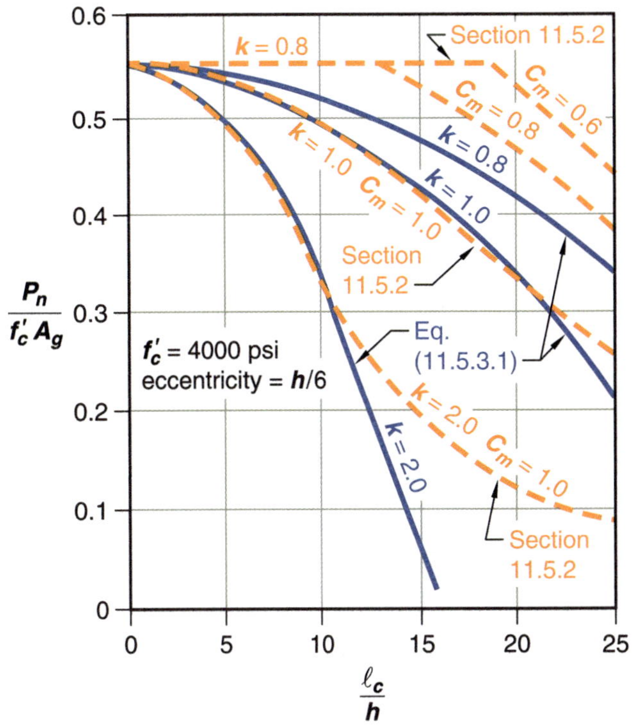

Fig. R11.5.3.1—Simplified design of walls, Eq. (11.5.3.1) versus 11.5.2.

11.5.3.2 Effective length factor k for use with Eq. (11.5.3.1) shall be in accordance with Table 11.5.3.2.

Table 11.5.3.2—Effective length factor *k* for walls

Boundary conditions	k
Walls braced top and bottom against lateral translation and:	
(a) Restrained against rotation at one or both ends (top, bottom, or both)	0.8
(b) Unrestrained against rotation at both ends	1.0
Walls not braced against lateral translation	2.0

11.5.3.3 P_n from Eq. (11.5.3.1) shall be reduced by ϕ for compression-controlled sections in 21.2.2.

11.5.3.4 Wall reinforcement shall be at least that required by 11.6.

11.5.4 *In-plane shear*

11.5.4.1 V_n shall be calculated in accordance with 11.5.4.2 through 11.5.4.4. Alternatively, for walls with $h_w/\ell_w < 2$, it shall be permitted to design for in-plane shear in accordance with the strut-and-tie method of Chapter 23. In all cases, reinforcement shall satisfy the limits of 11.6, 11.7.2, and 11.7.3.

R11.5.4 *In-plane shear*

R11.5.4.1 Shear in the plane of the wall is primarily of importance for structural walls with a small height-to-length ratio. The design of taller walls, particularly walls with uniformly distributed reinforcement, will likely be controlled by flexural considerations. Possible exceptions may occur in tall structural walls subject to strong earthquake excitation.

11.5.4.2 V_n at any horizontal section shall not exceed $8\sqrt{f_c'}A_{cv}$.

R11.5.4.2 This limit is imposed to guard against diagonal compression failure in structural walls. The coefficient used in this equation has been reduced from a value of 10 in ACI 318-14 to a value of 8 in ACI 318-19 because the effective shear area has been increased to $h\ell_w$, from hd used in prior editions of the Code.

11.5.4.3 V_n shall be calculated by:

$$V_n = (\alpha_c\lambda\sqrt{f_c'} + \rho_t f_{yt})A_{cv} \qquad (11.5.4.3)$$

where:
$\alpha_c = 3$ for $h_w/\ell_w \leq 1.5$
$\alpha_c = 2$ for $h_w/\ell_w \geq 2.0$
α_c varies linearly between 3 and 2 for $1.5 < h_w/\ell_w < 2.0$

R11.5.4.3 To improve consistency in the Code, the nominal in-plane shear strength equation in 11.5.4.3 now has the same form as the shear strength equation used in 18.10.4.1 for structural walls resisting seismic loads. Research results reported by Orakcal et al. (2009) indicate that nominal strengths calculated using Eq. (11.5.4.3) are similar to values obtained using equations from prior editions of the Code, and thus, provide a comparable level of safety.

11.5.4.4 For walls subject to a net axial tension, α_c in Eq. (11.5.4.3) shall be taken as:

$$\alpha_c = 2\left(1 + \frac{N_u}{500A_g}\right) \geq 0.0 \qquad (11.5.4.4)$$

where N_u is negative for tension.

R11.5.4.4 For structural walls where a net axial tension force is calculated for the entire wall section, the shear strength contribution attributed to the concrete is reduced and may be negligible. For these members, wall transverse reinforcement must be designed to resist most, if not all, of the factored shear force.

11.5.5 *Out-of-plane shear*

11.5.5.1 V_n shall be calculated in accordance with 22.5.

11.6—Reinforcement limits

11.6.1 If in-plane $V_u \leq 0.5\phi\alpha_c\lambda\sqrt{f_c'}A_{cv}$, minimum ρ_ℓ and minimum ρ_t shall be in accordance with Table 11.6.1. These limits need not be satisfied if adequate strength and stability can be demonstrated by structural analysis.

R11.6—Reinforcement limits

R11.6.1 Both horizontal and vertical shear reinforcement are required for all walls. The distributed reinforcement is identified as being oriented parallel to either the longitudinal or transverse axis of the wall. Therefore, for vertical wall segments, the notation used to describe the horizontal distributed reinforcement ratio is ρ_t, and the notation used to describe the vertical distributed reinforcement ratio is ρ_ℓ.

Transverse reinforcement is not required in precast, prestressed walls equal to or less than 12 ft in width because this width is less than that in which shrinkage and temperature stresses can build up to a magnitude requiring transverse reinforcement. In addition, much of the shrinkage occurs before the members are connected into the structure. Once in the final structure, the members are usually not as rigidly connected transversely as monolithic concrete; thus, the transverse restraint stresses due to both shrinkage and temperature change are significantly reduced.

The minimum area of wall reinforcement for precast walls has been used for many years and is recommended by the Precast/Prestressed Concrete Institute (PCI MNL-120) and the Canadian Precast Concrete Design Standard (2016). Reduced minimum reinforcement and greater spacings in 11.7.2.2 are allowed recognizing that precast wall panels have very little restraint at their edges during early stages

CODE

COMMENTARY

of curing and develop less shrinkage stress than comparable cast-in-place walls.

Table 11.6.1—Minimum reinforcement for walls with in-plane $V_u \le 0.5\phi\alpha_c\lambda\sqrt{f_c'}A_{cv}$

Wall type	Type of nonprestressed reinforcement	Bar/wire size	f_y, psi	Minimum longitudinal[1], ρ_ℓ	Minimum transverse, ρ_t
Cast-in-place	Deformed bars	≤ No. 5	≥60,000	0.0012	0.0020
			<60,000	0.0015	0.0025
		> No. 5	Any	0.0015	0.0025
	Welded-wire reinforcement	≤ W31 or D31	Any	0.0012	0.0020
Precast[2]	Deformed bars or welded-wire reinforcement	Any	Any	0.0010	0.0010

[1]Prestressed walls with an average effective compressive stress of at least 225 psi need not meet the requirement for minimum longitudinal reinforcement ρ_ℓ.

[2]In one-way precast, prestressed walls not wider than 12 ft and not mechanically connected to cause restraint in the transverse direction, the minimum reinforcement requirement in the direction normal to the flexural reinforcement need not be satisfied.

11.6.2 If in-plane $V_u \le 0.5\phi\alpha_c\lambda\sqrt{f_c'}A_{cv}$, (a) and (b) shall be satisfied:

(a) ρ_ℓ shall be at least the greater of the value calculated by Eq. (11.6.2) and 0.0025, but need not exceed ρ_t required for strength by 11.5.4.3.

$$\rho_\ell \ge 0.0025 + 0.5(2.5 - h_w/\ell_w)(\rho_t - 0.0025) \qquad (11.6.2)$$

(b) ρ_t shall be at least 0.0025

R11.6.2 For monotonically loaded walls with low height-to-length ratios, test data (Barda et al. 1977) indicate that horizontal shear reinforcement becomes less effective for shear resistance than vertical reinforcement. This change in effectiveness of the horizontal versus vertical reinforcement is recognized in Eq. (11.6.2); if h_w/ℓ_w is less than 0.5, the amount of vertical reinforcement is equal to the amount of horizontal reinforcement. If h_w/ρ_w is greater than 2.5, only a minimum amount of vertical reinforcement is required ($0.0025sh$).

11.7—Reinforcement detailing

11.7.1 *General*

11.7.1.1 Concrete cover for reinforcement shall be in accordance with 20.5.1.

11.7.1.2 Development lengths of deformed and prestressed reinforcement shall be in accordance with 25.4.

11.7.1.3 Splice lengths of deformed reinforcement shall be in accordance with 25.5.

11.7.2 *Spacing of longitudinal reinforcement*

11.7.2.1 Spacing s of longitudinal bars in cast-in-place walls shall not exceed the lesser of $3h$ and 18 in. If shear reinforcement is required for in-plane strength, spacing of longitudinal reinforcement shall not exceed $\ell_w/3$.

11.7.2.2 Spacing s of longitudinal bars in precast walls shall not exceed the lesser of (a) and (b):

(a) $5h$
(b) 18 in. for exterior walls or 30 in. for interior walls

If shear reinforcement is required for in-plane strength, s shall not exceed the smallest of $3h$, 18 in., and $\ell_w/3$.

CODE

11.7.2.3 For walls with thickness greater than 10 in., except single story basement walls and cantilever retaining walls, distributed reinforcement in each direction shall be placed in at least two layers, one near each face.

11.7.2.4 Flexural tension reinforcement shall be well distributed and placed as close as practicable to the tension face.

11.7.3 *Spacing of transverse reinforcement*

11.7.3.1 Spacing s of transverse reinforcement in cast-in-place walls shall not exceed the lesser of $3h$ and 18 in. If shear reinforcement is required for in-plane strength, s shall not exceed $\ell_w/5$.

11.7.3.2 Spacing s of transverse bars in precast walls shall not exceed the lesser of (a) and (b):

(a) $5h$
(b) 18 in. for exterior walls or 30 in. for interior walls

If shear reinforcement is required for in-plane strength, s shall not exceed the least of $3h$, 18 in., and $\ell_w/5$.

11.7.4 *Lateral support of longitudinal reinforcement*

11.7.4.1 If longitudinal reinforcement is required for compression and if A_{st} exceeds $0.01A_g$, longitudinal reinforcement shall be laterally supported by transverse ties.

11.7.5 *Reinforcement around openings*

11.7.5.1 In addition to the minimum reinforcement required by 11.6, at least two No. 5 bars in walls having two layers of reinforcement in both directions and one No. 5 bar in walls having a single layer of reinforcement in both directions shall be provided around window, door, and similarly sized openings. Such bars shall be anchored to develop f_y in tension at the corners of the openings.

11.8—Alternative method for out-of-plane slender wall analysis

11.8.1 *General*

11.8.1.1 It shall be permitted to analyze out-of-plane slenderness effects in accordance with this section for walls satisfying (a) through (e):

(a) Cross section is constant over the height of the wall
(b) Wall is tension-controlled for out-of-plane moment effect
(c) ϕM_n is at least M_{cr}, where M_{cr} is calculated using f_r as provided in 19.2.3
(d) P_u at the midheight section does not exceed $0.06f_c'A_g$

COMMENTARY

R11.8—Alternative method for out-of-plane slender wall analysis

R11.8.1 *General*

R11.8.1.1 This procedure is presented as an alternative to the requirements of 11.5.2.1 for the out-of-plane design of slender wall panels, where the panels are restrained against rotation at the top.

Panels that have windows or other large openings are not considered to have constant cross section over the height of the panel. Such walls are to be designed taking into account the effects of openings.

Many aspects of the design of tilt-up walls and buildings are discussed in ACI 551.2R and Carter et al. (1993).

CODE

(e) Calculated out-of-plane deflection due to service loads, Δ_s, including $P\Delta$ effects, does not exceed $\ell_c/150$

11.8.2 *Modeling*

11.8.2.1 The wall shall be analyzed as a simply supported, axially loaded member subject to an out-of-plane uniformly distributed lateral load, with maximum moments and deflections occurring at midheight.

11.8.2.2 Concentrated gravity loads applied to the wall above any section shall be assumed to be distributed over a width equal to the bearing width, plus a width on each side that increases at a slope of 2 vertical to 1 horizontal, but not extending beyond (a) or (b):

(a) The spacing of the concentrated loads
(b) The edges of the wall panel

11.8.3 *Factored moment*

11.8.3.1 M_u at midheight of wall due to combined flexure and axial loads shall include the effects of wall deflection in accordance with (a) or (b):

(a) By iterative calculation using

$$M_u = M_{ua} + P_u\Delta_u \qquad (11.8.3.1a)$$

where M_{ua} is the maximum factored moment at midheight of wall due to lateral and eccentric vertical loads, not including $P\Delta$ effects.
Δ_u shall be calculated by:

$$\Delta_u = \frac{5M_u\ell_c^2}{(0.75)48E_cI_{cr}} \qquad (11.8.3.1b)$$

where I_{cr} shall be calculated by:

$$I_{cr} = \frac{E_s}{E_c}\left(A_s + \frac{P_u}{f_y}\frac{h}{2d}\right)(d-c)^2 + \frac{\ell_w c^3}{3} \qquad (11.8.3.1c)$$

and the value of E_s/E_c shall be at least 6.
(b) By direct calculation using:

$$M_u = \frac{M_{ua}}{\left(1 - \frac{5P_u\ell_c^2}{(0.75)48E_cI_{cr}}\right)} \qquad (11.8.3.1d)$$

11.8.4 *Out-of-plane deflection – service loads*

COMMENTARY

R11.8.3 *Factored moment*

R11.8.3.1 The neutral axis depth c in Eq. (11.8.3.1c) corresponds to the following effective area of longitudinal reinforcement.

$$A_{se,w} = A_s + \frac{P_u}{f_y}\left(\frac{h/2}{d}\right)$$

R11.8.4 *Out-of-plane deflection – service loads*

Walls 11

CODE

11.8.4.1 Out-of-plane deflection due to service loads, Δ_s, shall be calculated in accordance with Table 11.8.4.1, where M_a is calculated by 11.8.4.2.

Table 11.8.4.1—Calculation of Δ_s

M_a	Δ_s	
$\leq(2/3)M_{cr}$	$\Delta_s = \left(\frac{M_a}{M_{cr}}\right)\Delta_{cr}$	(a)
$>(2/3)M_{cr}$	$\Delta_s = (2/3)\Delta_{cr} + \frac{(M_a - (2/3)M_{cr})}{(M_n - (2/3)M_{cr})}(\Delta_n - (2/3)\Delta_{cr})$	(b)

11.8.4.2 The maximum moment M_a at midheight of wall due to service lateral and eccentric vertical loads, including $P_s\Delta_s$ effects, shall be calculated by Eq. (11.8.4.2) with iteration of deflections.

$$M_a = M_{sa} + P_s\Delta_s \quad (11.8.4.2)$$

11.8.4.3 Δ_{cr} and Δ_n shall be calculated by (a) and (b):

(a) $$\Delta_{cr} = \frac{5M_{cr}\ell_c^2}{48E_cI_g} \quad (11.8.4.3a)$$

(b) $$\Delta_n = \frac{5M_n\ell_c^2}{48E_cI_{cr}} \quad (11.8.4.3b)$$

COMMENTARY

11.8.4.1 Test data (Athey 1982) demonstrate that out-of-plane deflections increase rapidly when the service-level moment exceeds **$2/3M_{cr}$**. A linear interpolation between Δ_{cr} and Δ_n is used to determine Δ_s to simplify the design of slender walls if **$M_a > 2/3M_{cr}$**.

Service-level load combinations are not defined in Chapter 5 of this Code, but they are discussed in Appendix C of ASCE/SEI 7. Appendixes to ASCE/SEI 7 are not considered mandatory parts of that standard. For calculating service-level lateral deflections of structures, Appendix C of ASCE/SEI 7 recommends using the following load combination:

$$D + 0.5L + W_a$$

in which W_a is wind load based on serviceability wind speeds provided in the commentary to Appendix C of ASCE/SEI 7. If the slender wall is designed to resist earthquake effects E, and E is based on strength-level earthquake effects, the following load combination is considered to be appropriate for evaluating the service-level lateral deflections

$$D + 0.5L + 0.7E$$

CHAPTER 12—DIAPHRAGMS

CODE

12.1—Scope

12.1.1 This chapter shall apply to the design of nonprestressed and prestressed diaphragms, including (a) through (d):

(a) Diaphragms that are cast-in-place slabs
(b) Diaphragms that comprise a cast-in-place topping slab on precast elements
(c) Diaphragms that comprise precast elements with end strips formed by either a cast-in-place concrete topping slab or edge beams
(d) Diaphragms of interconnected precast elements without cast-in-place concrete topping

COMMENTARY

R12.1—Scope

R12.1.1 Diaphragms typically are horizontal or nearly horizontal planar elements that serve to transfer lateral forces to vertical elements of the lateral-force-resisting system (Fig. R12.1.1). Diaphragms also tie the building elements together into a complete three-dimensional system and provide lateral support to those elements by connecting them to the lateral-force-resisting system. Typically, diaphragms also serve as floor and roof slabs, or as parking structure ramps and, therefore, support gravity loads. A diaphragm may include chords and collectors.

When subjected to lateral loads, such as the in-plane inertial loads acting on the roof diaphragm of Fig. R12.1.1, a diaphragm acts essentially as a beam spanning horizontally between vertical elements of the lateral-force-resisting system. The diaphragm thus develops in-plane bending moments, shears, and possibly other actions. Where vertical elements of the lateral-force-resisting system do not extend along the full depth of the diaphragm, collectors may be required to collect the diaphragm shear and transfer it to the vertical elements. The term "distributor" is sometimes used to describe a collector that transfers force from a vertical element of the lateral-force-resisting system into the diaphragm. This chapter describes minimum requirements for diaphragm and collector design and detailing, including configuration, analysis models, materials, and strength.

This chapter covers only the types of diaphragms listed in this provision. Other diaphragm types, such as horizontal trusses, are used successfully in buildings, but this chapter does not include prescriptive provisions for those other types.

CODE

COMMENTARY

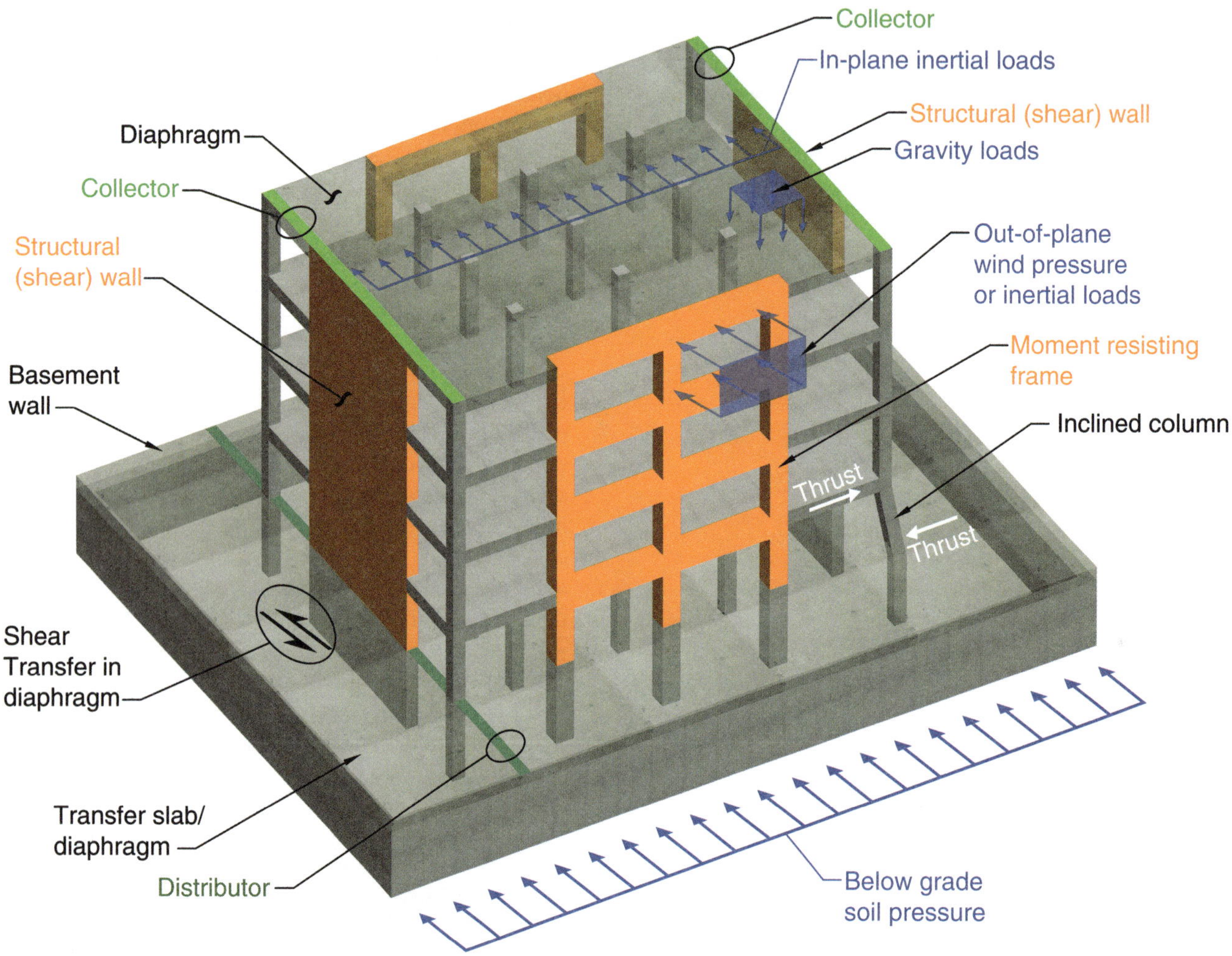

Fig. R12.1.1—Typical diaphragm actions.

12.1.2 Diaphragms in structures assigned to Seismic Design Category D, E, or F shall also satisfy requirements of 18.12.

12.2—General

12.2.1 Design shall consider forces (a) through (e):

(a) Diaphragm in-plane forces due to lateral loads acting on the building
(b) Diaphragm transfer forces
(c) Connection forces between the diaphragm and vertical framing or nonstructural elements
(d) Forces resulting from bracing vertical or sloped building elements
(e) Diaphragm out-of-plane forces due to gravity and other loads applied to the diaphragm surface

R12.2—General

R12.2.1 As partially illustrated in Fig. R12.1.1, diaphragms resist forces from several types of actions (Moehle et al. 2010):

(a) **Diaphragm in-plane forces**—Lateral forces from load combinations including wind, earthquake, and horizontal fluid or soil pressure generate in-plane shear, axial, and bending actions in diaphragms as they span between, and transfer forces to, vertical elements of the lateral-force-resisting system. For wind loading, lateral force is generated by wind pressure acting on building cladding that is transferred by diaphragms to the vertical elements. For earthquake loading, inertial forces are generated within the diaphragm and tributary portions of walls, columns, and other elements, and then transferred by diaphragms to the vertical elements. For buildings with subterranean levels, lateral forces are generated by soil pressure bearing against the basement walls; in a typical system, the basement walls span vertically between floors also serving as diaphragms,

CODE

COMMENTARY

which in turn distribute the lateral soil forces to other force-resisting elements.

(b) **Diaphragm transfer forces**—Vertical elements of the lateral-force-resisting system may have different properties over their height, or their planes of resistance may change from one story to another, creating force transfers between vertical elements. A common location where planes of resistance change is at grade level of a building with an enlarged subterranean plan; at this location, forces may transfer from the narrower tower into the basement walls through a podium diaphragm (refer to Fig. R12.1.1).

(c) **Connection forces**—Wind pressure acting on exposed building surfaces generates out-of-plane forces on those surfaces. Similarly, earthquake shaking can produce inertial forces in vertical framing and nonstructural elements such as cladding. These forces are transferred from the elements where the forces are developed to the diaphragm through connections.

(d) **Column bracing forces**—Architectural configurations sometimes require inclined columns, which can result in large horizontal thrusts acting within the plane of the diaphragms due to gravity and overturning actions. The thrusts can act in different directions depending on orientation of the column and whether it is in compression or tension. Where these thrusts are not balanced locally by other elements, the forces have to be transferred into the diaphragm so they can be transmitted to other suitable elements of the lateral-force-resisting system. Such forces are common and may be significant with eccentrically loaded precast concrete columns that are not monolithic with adjacent framing. The diaphragm also provides lateral support to columns not designed as part of the lateral-force-resisting system by connecting them to other elements that provide lateral stability for the structure.

(e) **Diaphragm out-of-plane forces**—Most diaphragms are part of floor and roof framing and, therefore, support gravity loads. The general building code may also require consideration of out-of-plane forces due to wind uplift pressure on a roof slab and vertical acceleration due to earthquake effects.

12.2.2 The effects of slab openings and slab voids shall be considered in design.

R12.2.2 Refer to R7.2.1.

12.2.3 *Materials*

12.2.3.1 Design properties for concrete shall be selected to be in accordance with Chapter 19.

12.2.3.2 Design properties for steel reinforcement shall be selected to be in accordance with Chapter 20.

12.3—Design limits

12.3.1 *Minimum diaphragm thickness*

R12.3—Design limits

R12.3.1 *Minimum diaphragm thickness*

CODE

12.3.1.1 Diaphragms shall have thickness as required for stability, strength, and stiffness under factored load combinations.

12.3.1.2 Floor and roof diaphragms shall have a thickness not less than that required for floor and roof elements in other parts of this Code.

12.4—Required strength

12.4.1 *General*

12.4.1.1 Required strength of diaphragms, collectors, and their connections shall be calculated in accordance with the factored load combinations in Chapter 5.

12.4.1.2 Required strength of diaphragms that are part of floor or roof construction shall include effects of out-of-plane loads simultaneous with other applicable loads.

12.4.2 *Diaphragm modeling and analysis*

12.4.2.1 Diaphragm modeling and analysis requirements of the general building code shall govern where applicable. Otherwise, diaphragm modeling and analysis shall be in accordance with 12.4.2.2 through 12.4.2.4.

12.4.2.2 Modeling and analysis procedures shall satisfy requirements of Chapter 6.

COMMENTARY

Diaphragms may be required to resist in-plane moment, shear, and axial force. For diaphragms that are entirely cast-in-place or comprise topping slabs composite with precast members, thickness of the entire diaphragm must be sufficient to resist these actions. For noncomposite topping slabs, thickness of the cast-in-place topping alone must be sufficient to resist these actions. Section 18.12 contains specific requirements for diaphragms in buildings assigned to Seismic Design Categories D, E, and F.

In addition to requirements for in-plane force resistance, diaphragms that are part of floor or roof construction must satisfy applicable requirements for slab or flange thickness.

R12.4—Required strength

Factored load combinations generally require consideration of out-of-plane loads that act simultaneously with diaphragm in-plane forces. For example, this is required where a floor beam also serves as a collector, in which case the beam is to be designed to resist axial forces acting as a collector and bending moments acting as a floor beam supporting gravity loads.

R12.4.2 *Diaphragm modeling and analysis*

R12.4.2.1 ASCE/SEI 7 includes diaphragm modeling requirements for some design conditions, such as design to resist wind and earthquake loads. Where ASCE/SEI 7 is adopted as part of the general building code, those requirements govern over provisions of this Code.

R12.4.2.2 Chapter 6 contains general requirements for analysis that are applicable to diaphragms. Diaphragms are usually designed to remain elastic or nearly elastic for forces acting within their plane under factored load combinations. Therefore, analysis methods satisfying theory of elastic analysis are generally acceptable. The provisions for elastic analysis in 6.6.1 through 6.6.3 can be applied.

Diaphragm in-plane stiffness affects not only the distribution of forces within the diaphragm, but also the distribution of displacements and forces among the vertical elements. Thus, the diaphragm stiffness model should be consistent with characteristics of the building. Where the diaphragm is very stiff compared to the vertical elements, as in a low aspect ratio, cast-in-place diaphragm supported by moment frames, it is acceptable to model the diaphragm as a completely rigid element. Where the diaphragm is flexible compared with the vertical elements, as in some jointed precast systems supported by structural walls, it may be acceptable to model the diaphragm as a flexible beam spanning between rigid supports. In other cases, it may be advisable to adopt a more detailed analytical model to account for the effects of diaphragm flexibility on the distribution of displacements and forces. Examples include buildings

CODE

COMMENTARY

in which diaphragm and vertical element stiffnesses have approximately the same value, buildings with large force transfers, and parking structures in which ramps connect between floors and act essentially as bracing elements within the building.

For diaphragms constructed of concrete slabs, ASCE/SEI 7 permits the assumption of a rigid diaphragm if the diaphragm aspect ratio falls within a prescribed limit, which is different for wind and earthquake loads, and if the structure has no horizontal irregularities. ASCE/SEI 7 provisions do not prohibit the rigid diaphragm assumption for other conditions, provided the rigid diaphragm assumption is reasonably consistent with anticipated behavior. Cast-in-place concrete diaphragms designed with the rigid-diaphragm assumption have a long history of satisfactory performance even though they may fall outside the ASCE/SEI 7 index values.

12.4.2.3 Any set of reasonable and consistent assumptions for diaphragm stiffness shall be permitted.

R12.4.2.3 For low-aspect-ratio diaphragms that are entirely cast-in-place or comprise a cast-in-place topping slab on precast elements, the diaphragm is often modeled as a rigid element supported by flexible vertical elements. However, effects of diaphragm flexibility should be considered where such effects will materially affect calculated design actions. Such effects should be considered for diaphragms that use precast elements, with or without a cast-in-place topping. Where large transfer forces occur, as outlined in R12.2.1(b), more realistic design forces can be obtained by modeling diaphragm in-plane stiffness. Diaphragms with long spans, large cutout areas, or other irregularities may develop in-plane deformations that should be considered in design (refer to Fig. R12.4.2.3a).

For a diaphragm considered rigid in its own plane, and for semi-rigid diaphragms, the diaphragm internal force distribution can be obtained by modeling it as a horizontal rigid beam supported on springs representing lateral stiffnesses of the vertical elements (refer to Fig. R12.4.2.3b). Effects of in-plane eccentricity between applied forces and vertical element resistances, resulting in overall building torsion, should be included in the analysis. Elements of the lateral-force-resisting system aligned in the orthogonal direction can participate in resisting diaphragm plan rotation (Moehle et al. 2010).

CODE

COMMENTARY

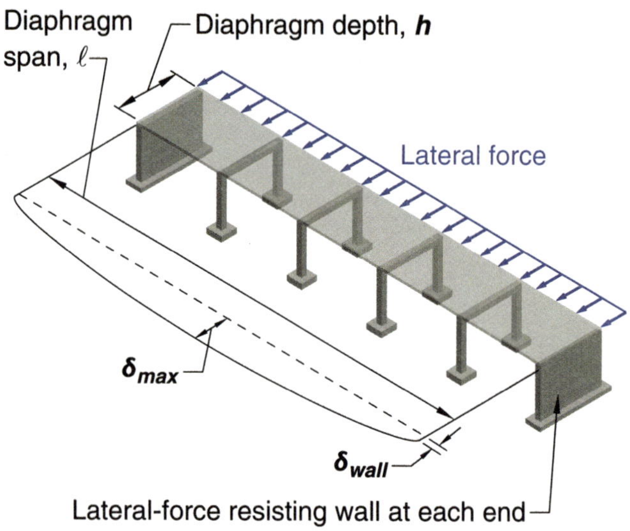

Fig. R12.4.2.3a—Example of diaphragm that might not be considered rigid in its plane.

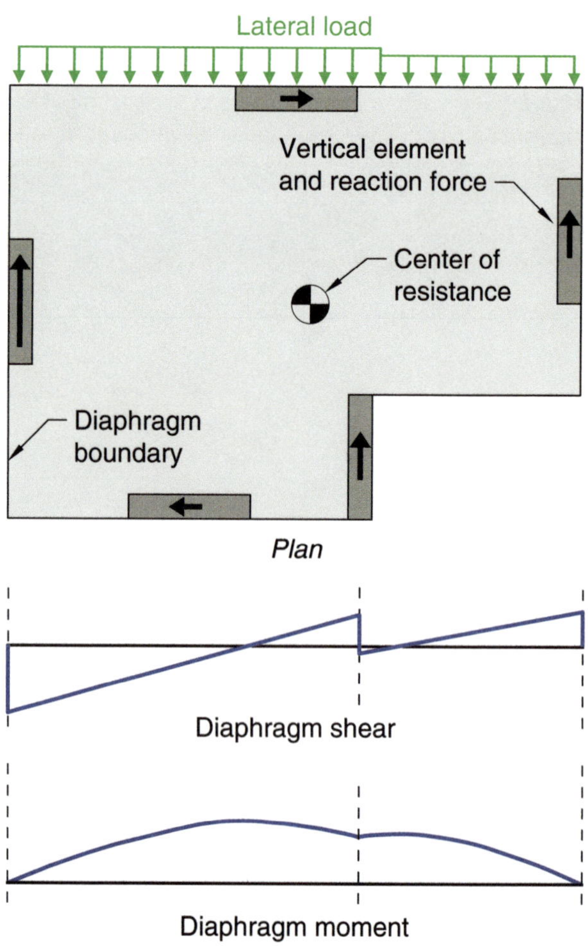

Fig. R12.4.2.3b—Diaphragm in-plane actions obtained by modeling the diaphragm as a horizontal rigid beam on flexible supports.

CODE

12.4.2.4 Calculation of diaphragm in-plane design moments, shears, and axial forces shall be consistent with requirements of equilibrium and with design boundary conditions. It shall be permitted to calculate design moments, shears, and axial forces in accordance with one of (a) through (e):

(a) A rigid diaphragm model if the diaphragm can be idealized as rigid
(b) A flexible diaphragm model if the diaphragm can be idealized as flexible
(c) A bounding analysis in which the design values are the envelope of values obtained by assuming upper bound and lower bound in-plane stiffnesses for the diaphragm in two or more separate analyses
(d) A finite element model considering diaphragm flexibility
(e) A strut-and-tie model in accordance with 23.2

COMMENTARY

R12.4.2.4 The rigid diaphragm model is widely used for diaphragms that are entirely cast-in-place and for diaphragms that comprise a cast-in-place topping slab on precast elements, provided flexible conditions are not created by a long span, by a large aspect ratio, or by diaphragm irregularity. For more flexible diaphragms, a bounding analysis is sometimes done in which the diaphragm is analyzed as a stiff or rigid element on flexible supports and as a flexible diaphragm on rigid supports, with the design values taken as the envelope of values from the two analyses. Finite element models can be suitable for any diaphragm, but are especially useful for irregularly shaped diaphragms and diaphragms resisting large transfer forces. Stiffness should be adjusted to account for expected concrete cracking under design loads. For jointed precast concrete diaphragms that rely on mechanical connectors, it may be necessary to include the joints and connectors in the finite element model. Strut-and-tie models may be used for diaphragm design. The strut-and-tie models should include considerations of force reversals that may occur under design load combinations.

12.5—Design strength

12.5.1 *General*

12.5.1.1 For each applicable factored load combination, design strengths of diaphragms and connections shall satisfy $\phi S_n \geq U$. Interaction between load effects shall be considered.

12.5.1.2 ϕ shall be determined in accordance with 21.2.

R12.5—Design strength

R12.5.1 *General*

R12.5.1.1 Design actions commonly include in-plane moment, with or without axial force; in-plane shear; and axial compression and tension in collectors and other elements acting as struts or ties. Some diaphragm configurations may result in additional types of design actions. For example, a diaphragm vertical step can result in out-of-plane bending, torsion, or both. The diaphragm is required to be designed for such actions where they occur in elements that are part of the load path.

Nominal strengths are prescribed in Chapter 22 for a diaphragm idealized as a beam or solid element resisting in-plane moment, axial force, and shear; and in Chapter 23 for a diaphragm or diaphragm segment idealized as a strut-and-tie system. Collectors and struts around openings can be designed as compression members subjected to axial force using provisions of 10.5.2 with the strength reduction factor for compression-controlled members in 21.2.2. For axial tension in such members, nominal tensile strength is $A_s f_y$, and the strength reduction factor is 0.90 as required for tension-controlled members in 21.2.2.

Diaphragms are designed under load combinations of 5.3. Where a diaphragm or part of a diaphragm is subjected to multiple load effects, the interaction of the load effects is to be considered. A common example is where a collector is built within a beam or slab that also resists gravity loads, in which case the element is designed for combined moment and axial force. Another example is where a connection is subjected to simultaneous tension and shear.

CODE

12.5.1.3 Design strengths shall be in accordance with (a), (b), (c), or (d):

(a) For a diaphragm idealized as a beam whose depth is equal to the full diaphragm depth, with moment resisted by boundary reinforcement concentrated at the diaphragm edges, design strengths shall be in accordance with 12.5.2 through 12.5.4.
(b) For a diaphragm or a diaphragm segment modeled as a strut-and-tie system, design strengths shall be in accordance with 23.3.
(c) For a diaphragm idealized with a finite-element model, design strengths shall be in accordance with Chapter 22. Nonuniform shear distributions shall be considered in design for shear. Collectors in such designs shall be provided to transfer diaphragm shears to the vertical elements of the lateral-force-resisting system.
(d) For a diaphragm designed by alternative methods, such methods shall satisfy the requirements of equilibrium and shall provide design strengths at least equal to required strengths for all elements in the load path.

12.5.1.4 It shall be permitted to use precompression from prestressed reinforcement to resist diaphragm forces.

12.5.1.5 If nonprestressed, bonded prestressing reinforcement is designed to resist collector forces, diaphragm shear, or tension due to in-plane moment, the value of steel stress used to calculate resistance shall not exceed the lesser of the specified yield strength and 60,000 psi.

12.5.2 *Moment and axial force*

12.5.2.1 It shall be permitted to design a diaphragm to resist in-plane moment and axial force in accordance with 22.3 and 22.4.

COMMENTARY

R12.5.1.3 Different design strength requirements apply depending on how the diaphragm load-path is idealized.

Section 12.5.1.3(a) addresses requirements for the common case where a diaphragm is idealized as a beam spanning between supports and resisting forces within its plane, with chord reinforcement at the boundaries to resist in-plane moment and axial force. If diaphragms are designed according to this model, then it is appropriate to assume that shear flow is uniform through the diaphragm depth. Diaphragm depth refers to the dimension measured in the direction of lateral forces within the plane of the diaphragm (refer to Fig. R12.4.2.3a). If vertical elements of the lateral-force-resisting system do not extend the full depth of the diaphragm, then collectors are required to transfer shear acting along the remaining portions of the diaphragm depth to the vertical elements. Sections 12.5.2 through 12.5.4 are based on this model. This design approach is acceptable even if some of the moment is resisted by precompression as provided by 12.5.1.4.

Sections 12.5.1.3(b) through (d) permit alternative methods for design of diaphragms. If diaphragms are designed to resist moment through distributed chords, or if diaphragms are designed according to stress fields determined by finite-element analysis, then non-uniform shear flow should be taken into account.

R12.5.1.4 In the typical case of a prestressed floor slab, prestressing is required, at a minimum, to resist the factored load combination $1.2D + 1.6L$, where L may have been reduced as permitted by the general building code. For wind or earthquake design, however, the gravity load to be resisted by prestressing is reduced because the governing load combination is $1.2D + f_1L + (W \text{ or } E)$, where f_1 is either 1.0 or 0.5 depending on the nature of L. Thus, only a portion of the effective prestress is required to resist the reduced gravity loads. The remainder of the effective prestress can be used to resist in-plane diaphragm moments. Additional moment, if any, is resisted by added reinforcement.

R12.5.1.5 Nonprestressed bonded prestressing reinforcement, either strand or bars, is sometimes used to resist diaphragm design forces. The imposed limit on assumed yield strength is to control crack width and joint opening. The Code does not include provisions for developing nonprestressed, bonded prestressing reinforcement. Stress limits for other provided reinforcement are prescribed in Chapter 20.

R12.5.2 *Moment and axial force*

R12.5.2.1 This section permits design for moment and axial force in accordance with the usual assumptions of 22.3 and 22.4, including the assumption that strains vary linearly through the depth of the diaphragm. In most cases, design for moment and axial force can be accomplished satisfacto-

CODE

12.5.2.2 It shall be permitted to resist tension due to moment by (a), (b), (c), or (d), or those methods in combination:

(a) Deformed bars conforming to 20.2.1
(b) Strands or bars conforming to 20.3.1, either prestressed or nonprestressed
(c) Mechanical connectors crossing joints between precast elements
(d) Precompression from prestressed reinforcement

12.5.2.3 Nonprestressed reinforcement and mechanical connectors resisting tension due to moment shall be located within ***h*/4** of the tension edge of the diaphragm, where ***h*** is diaphragm depth measured in the plane of the diaphragm at that location. Where diaphragm depth changes along the span, it shall be permitted to develop reinforcement into adjacent diaphragm segments that are not within the ***h*/4** limit.

COMMENTARY

rily using an approximate tension-compression couple with the strength reduction factor equal to 0.90.

R12.5.2.2 Bonded prestressing reinforcement used to resist in-plane moment and axial force can be either prestressed or nonprestressed. Mechanical connectors crossing joints between precast concrete elements are provided to complete a continuous load path for reinforcement embedded in those elements. The use of precompression from prestressed reinforcement is discussed in R12.5.1.4.

R12.5.2.3 Figure R12.5.2.3 illustrates permitted locations of nonprestressed reinforcement resisting tension due to moment and axial force. Where diaphragm depth changes along the span, it is permitted to develop tension reinforcement in adjacent sections even if the reinforcement falls outside the ***h*/4** limit of the adjacent section. In such cases, the strut-and-tie method or elastic plane stress analysis can be used to determine bar extensions and other reinforcement requirements to provide continuity across the step. The restriction on location of nonprestressed reinforcement and mechanical connectors is intended to control cracking and excessive joint opening that might occur near the edges if reinforcement or mechanical connectors were distributed throughout the diaphragm depth. The concentration of flexural tension reinforcement near the edge of the diaphragm also results in more uniform shear flow through the depth of the diaphragm.

There are no restrictions on placement of prestressed reinforcement provided to resist moment through precompression. In effect, the precompression determines a moment that the prestressed reinforcement can resist, with the remainder of the moment resisted by reinforcement or mechanical connectors placed in accordance with 12.5.2.3.

The Code does not require that diaphragm boundary elements resisting design flexural compression forces be detailed as columns. However, where a boundary element resists a large compressive force compared with axial strength, or is designed as a strut adjacent to an edge or opening, detailing with transverse reinforcement similar to column hoops should be considered.

CODE

COMMENTARY

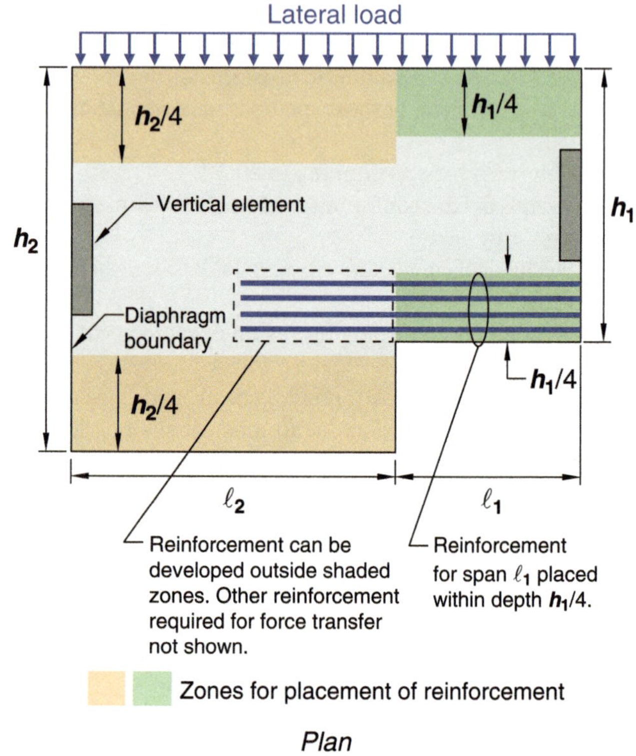

***Fig. R12.5.2.3**—Locations of nonprestressed reinforcement resisting tension due to moment and axial force according to 12.5.2.3.*

12.5.2.4 Mechanical connectors crossing joints between precast elements shall be designed to resist required tension under the anticipated joint opening.

R12.5.2.4 In an untopped precast diaphragm resisting in-plane forces and responding in the linear range, some joint opening (on the order of 0.1 in. or less) should be anticipated. A larger joint opening may occur under earthquake motions exceeding the design level. Mechanical connectors should be capable of maintaining design strength under the anticipated joint opening.

12.5.3 *Shear*

R12.5.3 *Shear*

12.5.3.1 This section shall apply to diaphragm in-plane shear strength.

R12.5.3.1 These provisions assume that diaphragm shear flow is approximately uniform over the diaphragm depth, as is the case where design is in accordance with 12.5.1.3(a). Where alternative approaches are used, local variations of in-plane shear through the diaphragm depth should be considered.

12.5.3.2 ϕ shall be 0.75, unless a lesser value is required by 21.2.4.

R12.5.3.2 A lower strength reduction factor may be required in Seismic Design Categories D, E, or F, or where special systems for earthquake resistance are used.

12.5.3.3 For a diaphragm that is entirely cast-in-place, V_n shall be calculated by Eq. (12.5.3.3).

$$V_n = A_{cv}(2\lambda\sqrt{f_c'} + \rho_t f_y) \qquad (12.5.3.3)$$

R12.5.3.3 This provision was adapted from the earthquake-resistant design provisions of 18.12.9. The term A_{cv} refers to the cross-sectional area of the effective deep beam that forms the diaphragm.

CODE

where A_{cv} is the gross area of concrete bounded by diaphragm web thickness and depth, reduced by void areas if present; the value of $\sqrt{f_c'}$ used to calculate V_n shall not exceed 100 psi; and ρ_t refers to the distributed reinforcement oriented parallel to the in-plane shear.

12.5.3.4 For a diaphragm that is entirely cast-in-place, cross-sectional dimensions shall be selected to satisfy Eq. (12.5.3.4).

$$V_u \leq \phi 8 A_{cv} \sqrt{f_c'} \qquad (12.5.3.4)$$

where the value of $\sqrt{f_c'}$ used to calculate V_n shall not exceed 100 psi.

12.5.3.5 For diaphragms that are cast-in-place concrete topping slabs on precast elements, (a) and (b) shall be satisfied:

(a) V_n shall be calculated in accordance with Eq. (12.5.3.3), and cross-sectional dimensions shall be selected to satisfy Eq. (12.5.3.4). A_{cv} shall be calculated using the thickness of the topping slab for noncomposite topping slab diaphragms and the combined thickness of cast-in-place and precast elements for composite topping slab diaphragms. For composite topping slab diaphragms, the value of f_c' in Eq. (12.5.3.3) and (12.5.3.4) shall not exceed the lesser of f_c' for the precast members and f_c' for the topping slab.

(b) V_n shall not exceed the value calculated in accordance with the shear-friction provisions of 22.9 considering the thickness of the topping slab above joints between precast elements in noncomposite and composite topping slab diaphragms and the reinforcement crossing the joints between the precast members.

12.5.3.6 For diaphragms that are interconnected precast elements without a concrete topping, and for diaphragms that are precast elements with end strips formed by either a cast-in-place concrete topping slab or edge beams, it shall be permitted to design for shear in accordance with (a), (b), or both.

(a) The nominal strength of grouted joints shall not exceed 80 psi. Reinforcement shall be designed to resist shear through shear-friction in accordance with 22.9. Shear-friction reinforcement shall be in addition to reinforcement designed to resist tension due to moment and axial force.

(b) Mechanical connectors crossing joints between precast elements shall be designed to resist required shear under anticipated joint opening.

12.5.3.7 For any diaphragm, where shear is transferred from the diaphragm to a collector, or from the diaphragm or collector to a vertical element of the lateral-force-resisting system, (a) or (b) shall apply:

COMMENTARY

R12.5.3.5 For diaphragms with cast-in-place topping slab on precast elements, the effective thickness in 12.5.3.5(a) is reduced to the topping slab thickness if the topping slab is not composite with the precast elements. Topping slabs tend to develop cracks above and along the joints between precast elements. Thus, 12.5.3.5(b) limits the shear strength to the shear-friction strength of the topping slab above the joints between the precast elements.

R12.5.3.6 This Code does not contain provisions for untopped diaphragms in buildings assigned to Seismic Design Categories D, E, and F. Diaphragm shear in untopped diaphragms can be resisted by using shear-friction reinforcement in grouted joints (FEMA P751). Required shear-friction reinforcement is in addition to reinforcement required by design to resist other tensile forces in the diaphragm, such as those due to diaphragm moment and axial force, or due to collector tension. The intent is to reduce joint opening while simultaneously resisting shear through shear-friction. Alternatively, or additionally, mechanical connectors can be used to transfer shear across joints of precast elements. In this case, some joint opening should be anticipated. The mechanical connectors should be capable of maintaining design strength under anticipated joint opening.

R12.5.3.7 In addition to having adequate shear strength within its plane, a diaphragm should be reinforced to transfer shear through shear-friction or mechanical connectors to collectors and to vertical elements of the lateral-force-resisting

CODE

(a) Where shear is transferred through concrete, the shear-friction provisions of 22.9 shall be satisfied.

(b) Where shear is transferred through mechanical connectors or dowels, effects of uplift and rotation of the vertical element of the lateral-force-resisting system shall be considered.

COMMENTARY

system. In diaphragms that are entirely cast-in-place, reinforcement provided for other purposes usually is adequate to transfer force from the diaphragm into the collectors through shear-friction. However, additional reinforcement may be required to transfer diaphragm or collector shear into vertical elements of the lateral-force-resisting system through shear-friction. Figure R12.5.3.7 illustrates a common detail of dowels provided for this purpose.

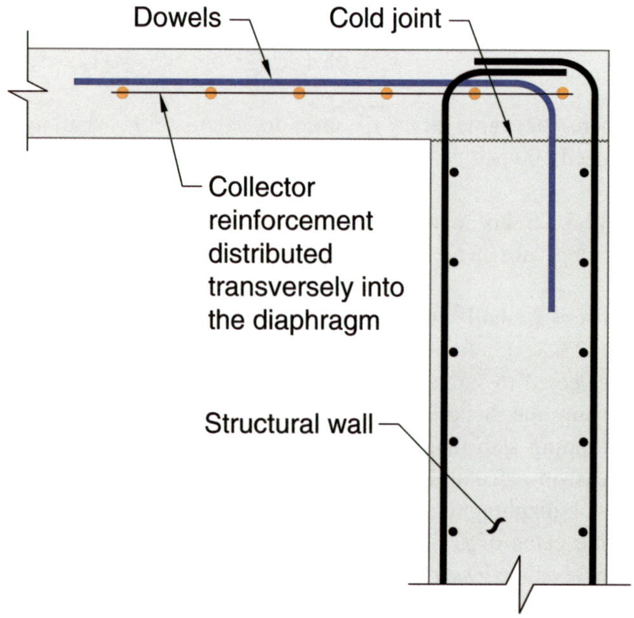

***Fig. R12.5.3.7**—Typical detail showing dowels provided for shear transfer to a structural wall through shear-friction.*

12.5.4 *Collectors*

12.5.4.1 Collectors shall extend from the vertical elements of the lateral-force-resisting system across all or part of the diaphragm depth as required to transfer shear from the diaphragm to the vertical element. It shall be permitted to discontinue a collector along lengths of vertical elements of the lateral-force-resisting system where transfer of design collector forces is not required.

R12.5.4 *Collectors*

A collector is a region of a diaphragm that transfers forces between the diaphragm and a vertical element of the lateral-force-resisting system. A collector can extend transversely into the diaphragm to reduce nominal stresses and reinforcement congestion, as shown in Fig. R12.5.3.7. Where a collector width extends into the slab, the collector width on each side of the vertical element should not exceed approximately one-half the contact length between the collector and the vertical element.

R12.5.4.1 The design procedure in 12.5.1.3(a) models the diaphragm as a full-depth beam with uniform shear flow. If vertical elements of the lateral-force-resisting system do not extend the full depth of the diaphragm, then collectors are required to transfer shear acting along the remaining portions of the diaphragm depth to the vertical element, as shown in Fig. R12.5.4.1. Partial-depth collectors can also be considered, but a complete force path should be designed that is capable of transmitting all forces from the diaphragm to the collector and into the vertical elements (Moehle et al. 2010).

CODE | COMMENTARY

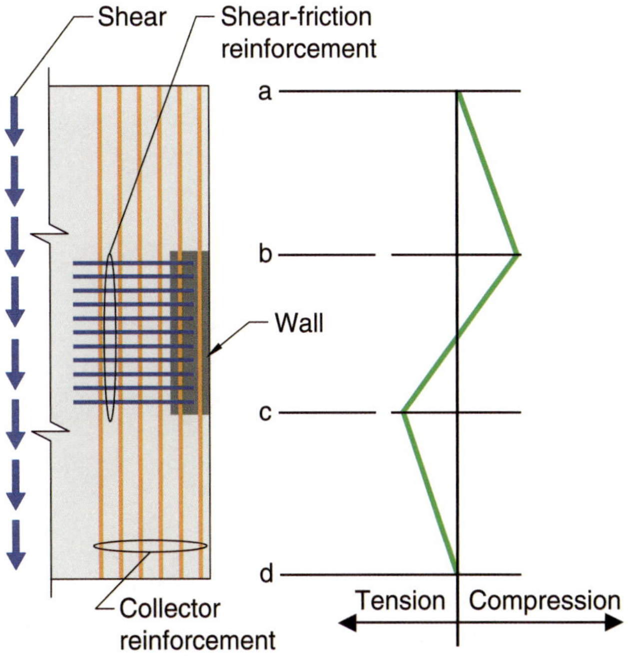

(a) Collector and shear-friction reinforcement

(b) Collector tension and compression forces

Fig. R12.5.4.1—Full-depth collector and shear-friction reinforcement required to transfer collector force into wall.

12.5.4.2 Collectors shall be designed as tension members, compression members, or both, in accordance with 22.4.

R12.5.4.2 Tension and compression forces in a collector are determined by the diaphragm shear forces they transmit to the vertical elements of the lateral-force-resisting system (refer to Fig. R12.5.4.1). Except as required by 18.12.7.6, the Code does not require that collectors resisting design compressive forces be detailed as columns. However, in structures where collectors resist large compressive forces compared with axial strength, or are designed as struts passing adjacent to edges or openings, detailing with transverse reinforcement similar to column hoops should be considered. Such detailing is required by 18.12.7.6 for some diaphragms in buildings assigned to Seismic Design Categories D, E, and F.

12.5.4.3 Where a collector is designed to transfer forces to a vertical element, collector reinforcement shall extend along the vertical element at least the greater of (a) and (b):

(a) The length required to develop the reinforcement in tension
(b) The length required to transmit the design forces to the vertical element through shear-friction in accordance with 22.9, through mechanical connectors, or through other force transfer mechanisms

R12.5.4.3 In addition to having sufficient development length, the collector reinforcement should be extended as needed to fully transfer its forces into the vertical elements of the lateral-force-resisting system. A common practice is to extend some of the collector reinforcement the full length of the vertical element, such that collector forces can be transmitted uniformly through shear-friction (refer to Fig. R12.5.4.1). Figure R12.5.4.3 shows an example of collector reinforcement extended as required to transfer forces into three frame columns.

12 Diaphragms

Note: Collector reinforcement should extend as required to transfer forces into the vertical element and should be developed at critical sections.

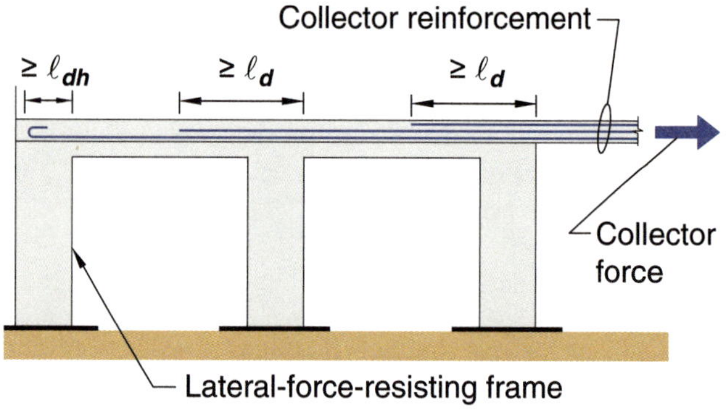

Fig. R12.5.4.3—Schematic force transfer from collector into vertical element of the lateral-force-resisting system.

12.6—Reinforcement limits

12.6.1 Reinforcement to resist shrinkage and temperature stresses shall be in accordance with 24.4.

12.6.2 Except for slabs-on-ground, diaphragms that are part of floor or roof construction shall satisfy reinforcement limits for one-way slabs in accordance with 7.6 or two-way slabs in accordance with 8.6, as applicable.

12.6.3 Reinforcement designed to resist diaphragm in-plane forces shall be in addition to reinforcement designed to resist other load effects, except reinforcement designed to resist shrinkage and temperature load effects shall be permitted to also resist diaphragm in-plane forces

12.7—Reinforcement detailing

12.7.1 *General*

12.7.1.1 Concrete cover for reinforcement shall be in accordance with 20.5.1.

12.7.1.2 Development lengths of deformed and prestressed reinforcement shall be in accordance with 25.4, unless longer lengths are required by Chapter 18.

12.7.1.3 Splices of deformed reinforcement shall be in accordance with 25.5.

12.7.1.4 Bundled bars shall be in accordance with 25.6.

12.7.2 *Reinforcement spacing*

12.7.2.1 Minimum spacing s of reinforcement shall be in accordance with 25.2.

R12.7—Reinforcement detailing

R12.7.1 *General*

R12.7.1.1 For a structure assigned to Seismic Design Category D, E, or F, concrete cover may be governed by the requirements of 18.12.7.7.

R12.7.2 *Reinforcement spacing*

R12.7.2.1 For a structure assigned to Seismic Design Category D, E, or F, spacing of confining reinforcement in collectors may be governed by the requirements of 18.12.7.6.

CODE

12.7.2.2 Maximum spacing ***s*** of deformed reinforcement shall be the lesser of five times the diaphragm thickness and 18 in.

12.7.3 *Diaphragm and collector reinforcement*

12.7.3.1 Except for slabs-on-ground, diaphragms that are part of floor or roof construction shall satisfy reinforcement detailing of one-way slabs in accordance with 7.7 or two-way slabs in accordance with 8.7, as applicable.

12.7.3.2 Calculated tensile or compressive force in reinforcement at each section of the diaphragm or collector shall be developed on each side of that section.

12.7.3.3 Reinforcement provided to resist tension shall extend beyond the point at which it is no longer required to resist tension at least ℓ_d, except at diaphragm edges and at expansion joints.

COMMENTARY

R12.7.3 *Diaphragm and collector reinforcement*

R12.7.3.2 Critical sections for development of reinforcement generally are at points of maximum stress, at points where adjacent terminated reinforcement is no longer required to resist design forces, and at other points of discontinuity in the diaphragm.

R12.7.3.3 For a beam, the Code requires flexural reinforcement to extend the greater of d and $12d_b$ past points where it is no longer required for flexure. These extensions are important for a beam to protect against development or shear failure that could result from inaccuracies in calculated locations of tensile stress. Similar failures in diaphragms have not been reported. To simplify design and avoid excessively long bar extensions that could result if the beam provisions were applied to diaphragms, this provision only requires that tension reinforcement extend ℓ_d beyond points where it is no longer required to resist tension.

Notes

CHAPTER 13—FOUNDATIONS

13.1—Scope

13.1.1 This chapter shall apply to the design of nonprestressed and prestressed foundations, including shallow foundations (a) through (f), deep foundations (g) through (i), and retaining walls (j) and (k):

(a) Strip footings
(b) Isolated footings
(c) Combined footings
(d) Mat foundations
(e) Grade beams
(f) Pile caps
(g) Piles
(h) Drilled piers
(i) Caissons
(j) Cantilever retaining walls
(k) Counterfort and buttressed cantilever retaining walls

COMMENTARY

R13.1—Scope

While requirements applicable to foundations are provided in this chapter, the majority of requirements used for foundation design are found in other chapters of the Code. These other chapters are referenced in Chapter 13. However, the applicability of the specific provisions within these other chapters may not be explicitly defined for foundations.

R13.1.1 Examples of foundation types covered by this chapter are illustrated in Fig. R13.1.1. Stepped and sloped footings are considered to be subsets of other footing types.

The 2019 edition of the Code contains provisions for the design of deep foundations. These provisions are based in part on similar provisions that were previously included in ASCE/SEI 7 and the IBC.

CODE

COMMENTARY

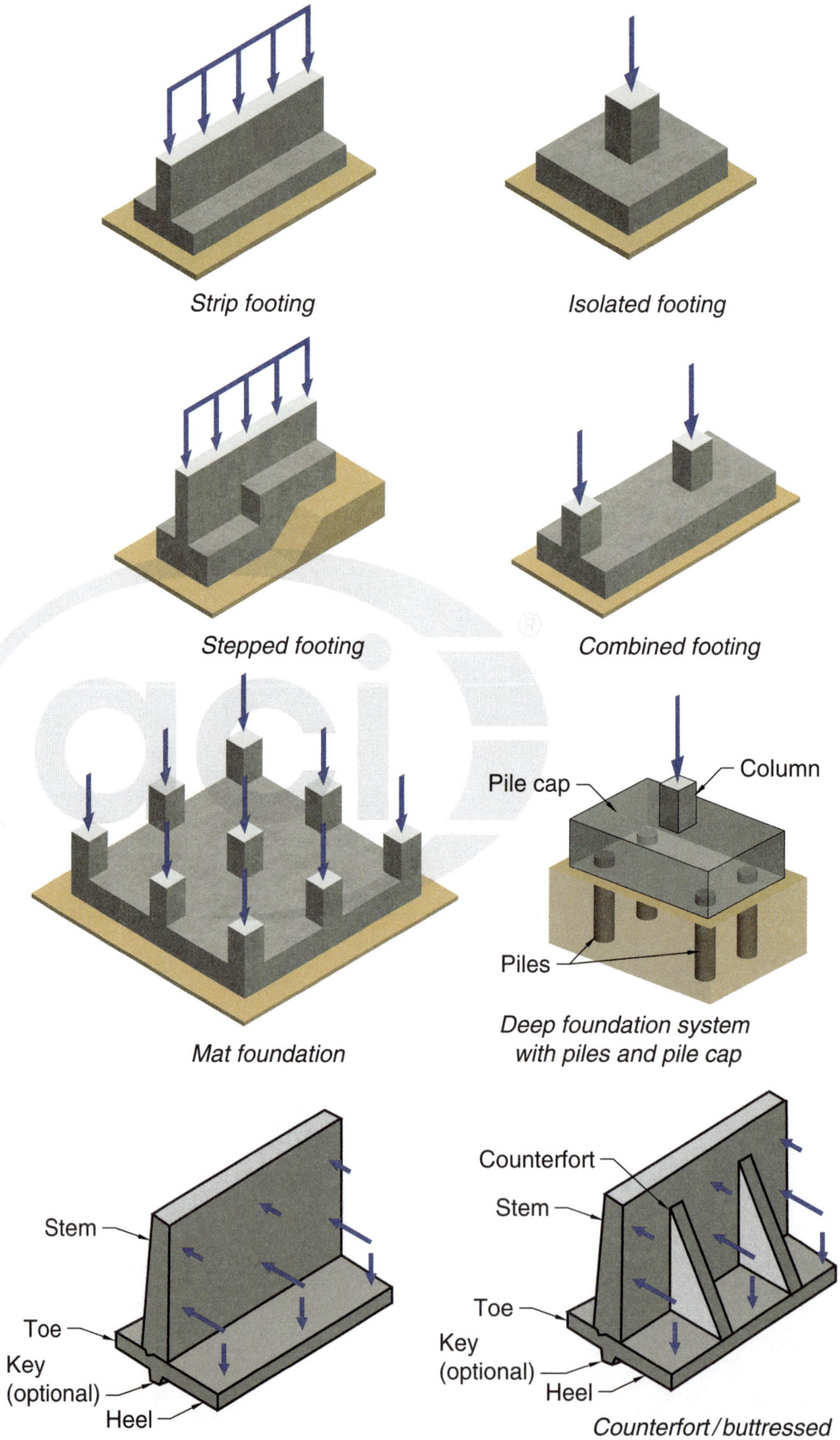

***Fig. R13.1.1**—Types of foundations.*

13.1.2 Foundations excluded by 1.4.7 are excluded from this chapter.

CODE

13.2—General

13.2.1 *Materials*

13.2.1.1 Design properties for concrete shall be selected to be in accordance with Chapter 19.

13.2.1.2 Design properties for steel reinforcement shall be selected to be in accordance with Chapter 20.

13.2.1.3 Materials, design, and detailing requirements for embedments in concrete shall be in accordance with 20.6.

13.2.2 *Connection to other members*

13.2.2.1 Design and detailing of cast-in-place and precast column, pedestal, and wall connections to foundations shall be in accordance with 16.3.

13.2.3 *Earthquake effects*

13.2.3.1 Structural members extending below the base of the structure that are required to transmit forces resulting from earthquake effects to the foundation shall be designed in accordance with 18.2.2.3.

13.2.3.2 For structures assigned to Seismic Design Category (SDC) C, D, E, or F, foundations resisting earthquake-induced forces or transferring earthquake-induced forces between structure and ground shall be designed in accordance with 18.13.

13.2.4 *Slabs-on-ground*

13.2.4.1 Slabs-on-ground that transmit vertical loads or lateral forces from other parts of the structure to the ground shall be designed and detailed in accordance with applicable provisions of this Code.

13.2.4.2 Slabs-on-ground that transmit lateral forces as part of the seismic-force-resisting system shall be designed in accordance with 18.13.

13.2.5 *Plain concrete*

13.2.5.1 Plain concrete foundations shall be designed in accordance with Chapter 14.

13.2.6 *Design criteria*

COMMENTARY

R13.2—General

R13.2.3 *Earthquake effects*

R13.2.3.1 The base of a structure, as defined in analysis, does not necessarily correspond to the foundation or ground level, or to the base of a building as defined in the general building code for planning (for example, for height limits or fire protection requirements). Details of columns and walls extending below the base of a structure to the foundation are required to be consistent with those above the base of the structure. For additional discussion of the design of foundations for earthquake effects, see R18.13.1.

R13.2.4 *Slabs-on-ground*

Slabs-on-ground often act as a diaphragm to hold the building together at the ground level and minimize the effects of out-of-phase ground motion that may occur over the footprint of the building. In these cases, the slab-on-ground should be adequately reinforced and detailed. As required in Chapter 26, construction documents should clearly state that these slabs-on-ground are structural members so as to prohibit sawcutting of such slabs.

R13.2.6 *Design criteria*

CODE

13.2.6.1 Foundations shall be proportioned for bearing effects, stability against overturning and sliding at the soil-foundation interface in accordance with the general building code.

COMMENTARY

R13.2.6.1 Permissible soil pressures or permissible deep foundation strengths are determined by principles of soil mechanics and in accordance with the general building code. The size of the base area of a footing on soil or the number and arrangement of deep foundation members are established by using allowable geotechnical strength and service-level load combinations or by using nominal geotechnical strength with resistance factor and factored load combinations.

Only the calculated end moments at the base of a column or pedestal require transfer to the footing. The minimum moment requirement for slenderness considerations given in 6.6.4.5 need not be considered for transfer of forces and moments to footings.

13.2.6.2 For one-way shallow foundations, two-way isolated footings, or two-way combined footings and mat foundations, it is permissible to neglect the size effect factor specified in 22.5 for one-way shear strength and 22.6 for two-way shear strength.

13.2.6.3 Foundation members shall be designed to resist factored loads and corresponding induced reactions except as permitted by 13.4.2.

R13.2.6.3 To design a footing or pile cap for strength, the induced reactions due to factored loads applied to the foundation should be determined. For a single concentrically-loaded spread footing, the soil pressure due to factored loading is calculated as the factored load divided by the base area of the footing. For the case of footings or mats with eccentric loading, applied factored loads may be used to determine soil pressures. For pile caps or mats supported by deep foundations, applied factored loads may be used to determine member reactions. However, the resulting pressures or reactions may be incompatible with the geotechnical design resulting in unacceptable subgrade reactions or instability (Rogowsky and Wight 2010). In such cases, the design should be adjusted in coordination with the geotechnical engineer.

Only the calculated end moments at the base of a column or pedestal require transfer to the footing. The minimum moment requirements for slenderness considerations given in 6.6.4.5 need not be considered for transfer of forces and moments to footings.

13.2.6.4 Foundation systems shall be permitted to be designed by any procedure satisfying equilibrium and geometric compatibility.

R13.2.6.4 Foundation design is permitted to be based directly on fundamental principles of structural mechanics, provided it can be demonstrated that all strength and serviceability criteria are satisfied. Design of the foundation may be achieved through the use of classic solutions based on a linearly elastic continuum, numerical solutions based on discrete elements, or yield-line analyses. In all cases, analyses and evaluation of the stress conditions at points of load application or pile reactions in relation to shear and torsion, as well as flexure, should be included.

13.2.6.5 Foundation design in accordance with the strut-and-tie method, Chapter 23, shall be permitted.

R13.2.6.5 An example of the application of this provision is a pile cap similar to that shown in Fig. R13.1.1. Pile caps may be designed using a three-dimensional strut-and-tie

CODE

COMMENTARY

model satisfying Chapter 23 (Adebar et al. 1990) provided the shear force limits of 23.4.4 are also satisfied.

Figure R13.2.6.5 illustrates the application of the shear force limits of 23.4.4 and the provisions of 13.2.7.2 for one-way shear design of a spread footing using the strut-and-tie method. Soil pressure within ***d*** from the face of the column or wall does not contribute to shear across the critical crack (Uzel et al. 2011), but the soil pressure within ***d*** contributes to the bending moment at the face of the column or wall.

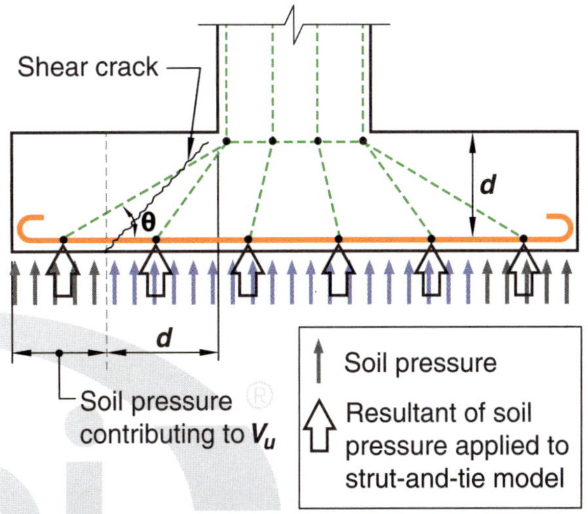

Fig. R13.2.6.5*—One-way shear design of a spread footing using the strut-and-tie method.*

13.2.6.6 External moment on any section of a strip footing, isolated footing, or pile cap shall be calculated by passing a vertical plane through the member and calculating the moment of the forces acting over the entire area of member on one side of that vertical plane.

13.2.7 *Critical sections for shallow foundations and pile caps*

R13.2.7 *Critical sections for shallow foundations and pile caps*

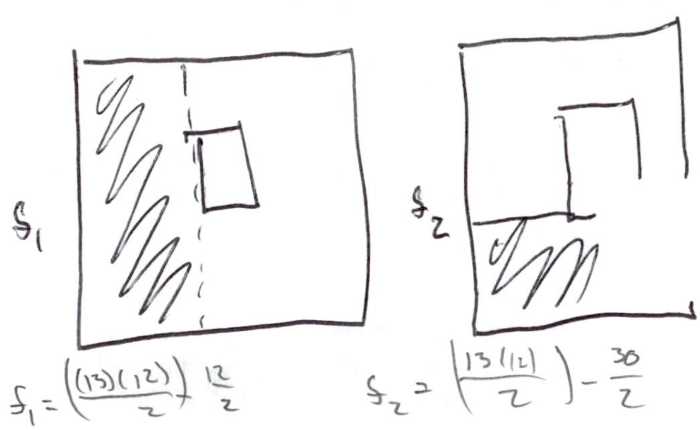

13.2.7.1 M_u at the supported member shall be permitted to be calculated at the critical section defined in accordance with Table 13.2.7.1.

Table 13.2.7.1—Location of critical section for M_u

Supported member	Location of critical section
Column or pedestal	Face of column or pedestal
Column with steel base plate	Halfway between face of column and edge of steel base plate
Concrete wall	Face of wall
Masonry wall	Halfway between center and face of masonry wall

13.2.7.2 The location of critical section for factored shear in accordance with 7.4.3 and 8.4.3 for one-way shear or 8.4.4.1 for two-way shear shall be measured from the location of the critical section for M_u in 13.2.7.1.

R13.2.7.2 The shear strength of a footing is determined for the more severe condition of 8.5.3.1.1 and 8.5.3.1.2. The critical section for shear is measured from the face of the supported member (column, pedestal, or wall), except for masonry walls and members supported on steel base plates.

COMMENTARY

Calculation of shear requires that the soil reaction be obtained from factored loads, and the design strength be in accordance with Chapter 22.

Where necessary, shear around individual piles may be investigated in accordance with 8.5.3.1.2. If shear perimeters overlap, the modified critical perimeter b_o should be taken as that portion of the smallest envelope of individual shear perimeters that will actually resist the critical shear for the group under consideration. One such situation is illustrated in Fig. R13.2.7.2.

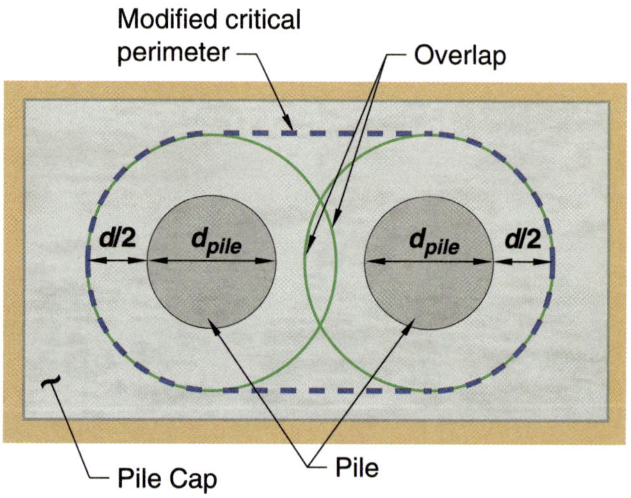

Fig. R13.2.7.2*—Modified critical perimeter for shear with overlapping critical perimeters.*

CODE

13.2.7.3 Circular or regular polygon-shaped concrete columns or pedestals shall be permitted to be treated as square members of equivalent area when locating critical sections for moment, shear, and development of reinforcement.

13.2.8 *Development of reinforcement in shallow foundations and pile caps*

13.2.8.1 Development of reinforcement shall be in accordance with Chapter 25.

13.2.8.2 Calculated tensile or compressive force in reinforcement at each section shall be developed on each side of that section.

13.2.8.3 Critical sections for development of reinforcement shall be assumed at the same locations as given in 13.2.7.1 for maximum factored moment and at all other vertical planes where changes of section or reinforcement occur.

13.2.8.4 Adequate anchorage shall be provided for tension reinforcement where reinforcement stress is not directly proportional to moment, such as in sloped, stepped, or tapered foundations; or where tension reinforcement is not parallel to the compression face.

CODE

13.3—Shallow foundations

13.3.1 *General*

13.3.1.1 Minimum base area of foundation shall be proportioned to not exceed the permissible bearing pressure when subjected to forces and moments applied to the foundation. Permissible bearing pressures shall be determined through principles of soil or rock mechanics in accordance with the general building code, or other requirements as determined by the building official.

13.3.1.2 Overall depth of foundation shall be selected such that the effective depth of bottom reinforcement is at least 6 in.

13.3.1.3 In sloped, stepped, or tapered foundations, depth and location of steps or angle of slope shall be such that design requirements are satisfied at every section.

13.3.2 *One-way shallow foundations*

13.3.2.1 The design and detailing of one-way shallow foundations, including strip footings, combined footings, and grade beams, shall be in accordance with this section and the applicable provisions of Chapter 7 and Chapter 9.

13.3.2.2 Reinforcement shall be distributed uniformly across entire width of one-way footings.

13.3.3 *Two-way isolated footings*

13.3.3.1 The design and detailing of two-way isolated footings shall be in accordance with this section and the applicable provisions of Chapter 7 and Chapter 8.

13.3.3.2 In square two-way footings, reinforcement shall be distributed uniformly across entire width of footing in both directions.

13.3.3.3 In rectangular footings, reinforcement shall be distributed in accordance with (a) and (b):

(a) Reinforcement in the long direction shall be distributed uniformly across entire width of footing.
(b) For reinforcement in the short direction, a portion of the total reinforcement, $\gamma_s A_s$, shall be distributed uniformly over a band width equal to the length of short side of footing, centered on centerline of column or pedestal. Remainder of reinforcement required in the short direction, $(1 - \gamma_s)A_s$, shall be distributed uniformly outside the center band width of footing, where γ_s is calculated by:

$$\gamma_s = \frac{2}{(\beta+1)} \qquad (13.3.3.3)$$

where β is the ratio of long to short side of footing.

COMMENTARY

R13.3—Shallow foundations

R13.3.1 *General*

R13.3.1.1 General discussion on the sizing of shallow foundations is provided in R13.2.6.1.

R13.3.1.3 Anchorage of reinforcement in sloped, stepped, or tapered foundations is addressed in 13.2.8.4.

R13.3.3 *Two-way isolated footings*

R13.3.3.3 To minimize potential construction errors in placing bars, a common practice is to increase the amount of reinforcement in the short direction by $2\beta/(\beta + 1)$ and space it uniformly along the long dimension of the footing (CRSI Handbook 1984; Fling 1987).

13 Foundations

CODE

13.3.4 *Two-way combined footings and mat foundations*

13.3.4.1 The design and detailing of combined footings and mat foundations shall be in accordance with this section and the applicable provisions of Chapter 8.

13.3.4.2 The direct design method shall not be used to design combined footings and mat foundations.

13.3.4.3 Distribution of bearing pressure under combined footings and mat foundations shall be consistent with properties of the soil or rock and the structure, and with established principles of soil or rock mechanics.

13.3.4.4 Minimum reinforcement in nonprestressed mat foundations shall be in accordance with 8.6.1.1.

13.3.5 *Walls as grade beams*

13.3.5.1 The design of walls as grade beams shall be in accordance with the applicable provisions of Chapter 9.

13.3.5.2 If a grade beam wall is considered a deep beam in accordance with 9.9.1.1, design shall satisfy the requirements of 9.9.

13.3.5.3 Grade beam walls shall satisfy the minimum reinforcement requirements of 11.6.

13.3.6 *Wall components of cantilever retaining walls*

13.3.6.1 The stem of a cantilever retaining wall shall be designed as a one-way slab in accordance with the applicable provisions of Chapter 7.

13.3.6.2 The stem of a counterfort or buttressed cantilever retaining wall shall be designed as a two-way slab in accordance with the applicable provisions of Chapter 8.

13.3.6.3 For walls of uniform thickness, the critical section for shear and flexure shall be at the interface between the stem and the footing. For walls with a tapered or varied thickness, shear and moment shall be investigated throughout the height of the wall.

13.4—Deep foundations

13.4.1 *General*

COMMENTARY

R13.3.4 *Two-way combined footings and mat foundations*

R13.3.4.1 Detailed recommendations for design of combined footings and mat foundations are reported by ACI 336.2R. Also refer to Kramrisch and Rogers (1961).

R13.3.4.2 The direct design method is a method used for the design of two-way slabs. Refer to R6.2.4.1.

R13.3.4.3 Design methods using factored loads and strength reduction factors ϕ can be applied to combined footings or mat foundations, regardless of the bearing pressure distribution.

R13.3.4.4 To improve crack control due to thermal gradients and to intercept potential punching shear cracks with tension reinforcement, the licensed design professional should consider specifying continuous reinforcement in each direction near both faces of mat foundations.

R13.3.6 *Wall components of cantilever retaining walls*

R13.3.6.2 Counterfort or buttressed cantilever retaining walls tend to behave more in two-way action than in one-way action; therefore, additional care should be given to crack control in both directions.

R13.3.6.3 In general, the joint between the wall stem and the footing will be opening under lateral loads; therefore, the critical section should be at the face of the joint. If hooks are required to develop the wall flexural reinforcement, hooks should be located near the bottom of the footing with the free end of the bars oriented toward the opposite face of the wall (Nilsson and Losberg 1976).

R13.4—Deep foundations

R13.4.1 *General*

CODE

13.4.1.1 Number and arrangement of deep foundation members shall be determined such that forces and moments applied to the foundation do not exceed the permissible deep foundation strength. Permissible deep foundation strength shall be determined through principles of soil or rock mechanics in accordance with the general building code, or other requirements as determined by the building official.

13.4.1.2 Design of deep foundation members shall be in accordance with 13.4.2 or 13.4.3.

13.4.2 *Allowable axial strength*

13.4.2.1 It shall be permitted to design a deep foundation member using load combinations for allowable stress design in ASCE/SEI 7, Section 2.4, and the allowable strength specified in Table 13.4.2.1 if (a) and (b) are satisfied:

(a) The deep foundation member is laterally supported for its entire height

(b) The applied forces cause bending moments in the deep foundation member less than the moment due to an accidental eccentricity of 5 percent of the member diameter or width

Table 13.4.2.1—Maximum allowable compressive strength for deep foundation members

Deep foundation member type	Maximum allowable compressive strength [1]	
Uncased cast-in-place concrete drilled or augered pile	$P_a = 0.3f_c'A_g + 0.4f_yA_s$	(a)
Cast-in-place concrete pile in rock or within a pipe, tube, or other permanent metal casing that does not satisfy 13.4.2.3	$P_a = 0.33f_c'A_g + 0.4f_yA_s$ [2]	(b)
Metal cased concrete pile confined in accordance with 13.4.2.3	$P_a = 0.4f_c'A_g$	(c)
Precast nonprestressed concrete pile	$P_a = 0.33f_c'A_g + 0.4f_yA_s$	(d)
Precast prestressed concrete pile	$P_a = (0.33f_c' - 0.27f_{pc})A_g$	(e)

[1] A_g applies to the gross cross-sectional area. If a temporary or permanent casing is used, the inside face of the casing shall be considered the concrete surface.

[2] A_s does not include the steel casing, pipe, or tube.

13.4.2.2 If 13.4.2.1(a) or 13.4.2.1(b) is not satisfied, a deep foundation member shall be designed using strength design in accordance with 13.4.3.

13.4.2.3 Metal cased cast-in-place concrete deep foundation members shall be considered to be confined if (a) through (f) are satisfied:

(a) Design shall not use the casing to resist any portion of the axial load imposed.

(b) Casing shall have a sealed tip and shall be mandrel-driven.

COMMENTARY

R13.4.1.1 General discussion on selecting the number and arrangement of piles, drilled piers, and caissons is provided in R13.2.6.1.

R13.4.2 *Allowable axial strength*

R13.4.2.1 Potential changes to lateral support of the deep foundation member due to liquefaction, excavation, or other causes, should be considered.

The values in the Table 13.4.2.1 represent an upper bound for well understood soil conditions with quality workmanship. A lower value for the maximum allowable compressive strength may be appropriate, depending on soil conditions and the construction and quality control procedures used. For auger-grout piles, where grout is placed through the stem of a hollow-stem auger as it is withdrawn from the soil, the strength coefficient of 0.3 is based on a strength reduction factor of 0.6. The designer should carefully consider the reliable grout strength, grout strength testing methods, and the minimum cross-sectional area of the pile, accounting for soil conditions and construction procedures. Additional information is provided in ACI 543R.

R13.4.2.3 The basis for this allowable strength is the added strength provided to the concrete by the confining action of the steel casing. This strength applies only to non-axial load-bearing steel where the stress in the steel is taken in hoop tension instead of axial compression. In this Code, steel pile casing is not to be considered in the design of the pile to resist a portion of the pile axial load. Provisions for

13 Foundations

CODE

(c) Thickness of the casing shall not be less than manufacturer's standard gauge No. 14 (0.068 in.).
(d) Casing shall be seamless, or provided with seams of strength equal to the basic material, and be of a configuration that will provide confinement to the cast-in-place concrete.
(e) Ratio of yield strength of the steel casing to f_c' shall be at least 6, and yield strength shall be at least 30,000 psi.
(f) Nominal diameter of the member shall be less than or equal to 16 in.

13.4.2.4 The use of allowable strengths greater than those specified in Table 13.4.2.1 shall be permitted if accepted by the building official in accordance with 1.10 and justified by load tests.

13.4.3 *Strength design*

13.4.3.1 Strength design in accordance with this section is permitted for all deep foundation members.

13.4.3.2 The strength design of deep foundation members shall be in accordance with 10.5 using the compressive strength reduction factors of Table 13.4.3.2 for axial load without moment, and the strength reduction factors of Table 21.2.1 for tension, shear, and combined axial force and moment. The provisions of 22.4.2.4 and 22.4.2.5 shall not apply to deep foundations.

Table 13.4.3.2—Compressive strength reduction factors ϕ for deep foundation members

Deep foundation member type	Compressive strength reduction factors ϕ	
Uncased cast-in-place concrete drilled or augered pile[1]	0.55	(a)
Cast-in-place concrete pile in rock or within a pipe, tube,[2] or other permanent casing that does not satisfy 13.4.2.3	0.60	(b)
Cast-in-place concrete-filled steel pipe pile[3]	0.70	(c)
Metal cased concrete pile confined in accordance with 13.4.2.3	0.65	(d)
Precast-nonprestressed concrete pile	0.65	(e)
Precast-prestressed concrete pile	0.65	(f)

[1]The factor of 0.55 represents an upper bound for well understood soil conditions with quality workmanship. A lower value for the strength reduction factor may be appropriate, depending on soil conditions and the construction and quality control procedures used.
[2]For wall thickness of the steel pipe or tube less than 0.25 in.
[3]Wall thickness of the steel pipe shall be at least 0.25 in.

13.4.4 *Cast-in-place deep foundations*

13.4.4.1 Cast-in-place deep foundations that are subject to uplift or where M_u is greater than $0.4M_{cr}$ shall be reinforced, unless enclosed by a structural steel pipe or tube.

COMMENTARY

members designed to be composite with steel pipe or casing are covered in AISC 360.

Potential corrosion of the metal casing should be considered; provision is based on a non-corrosive environment.

R13.4.2.4 Geotechnical and load test requirements for deep foundation members can be found in the IBC.

R13.4.3 *Strength design*

R13.4.3.2 The strength design of deep foundation members is discussed in detail in ACI 543R.

If cast-in-place concrete drilled or augered piles are subject to flexure, shear, or tension loads, the strength reduction factors should be adjusted accordingly, considering the soil conditions, quality-control procedures that will be implemented, likely workmanship quality, and local experience. Guidance for adjustment factors is provided in ACI 543R.

R13.4.4 *Cast-in-place deep foundations*

CODE

13.4.4.2 Portions of deep foundation members in air, water, or soils not capable of providing adequate restraint throughout the member length to prevent lateral buckling shall be designed as columns in accordance with the applicable provisions of Chapter 10.

13.4.5 *Precast concrete piles*

13.4.5.1 Precast concrete piles supporting buildings assigned to SDC A or B shall satisfy the requirements of 13.4.5.2 through 13.4.5.6.

13.4.5.2 Longitudinal reinforcement shall be arranged in a symmetrical pattern.

13.4.5.3 For precast nonprestressed piles, longitudinal reinforcement shall be provided according to (a) and (b):

(a) Minimum of 4 bars
(b) Minimum area of $\mathbf{0.008}A_g$

13.4.5.4 For precast prestressed piles, the effective prestress in the pile shall provide a minimum average compressive stress in the concrete in accordance with Table 13.4.5.4.

Table 13.4.5.4—Minimum compressive stress in precast prestressed piles

Pile length, ft	Minimum compressive stress, psi
Pile length ≤ 30	400
30 < Pile length ≤ 50	550
Pile length > 50	700

13.4.5.5 For precast prestressed piles, the effective prestress in the pile shall be calculated based on an assumed total loss of 30,000 psi in the prestressed reinforcement.

13.4.5.6 The longitudinal reinforcement shall be enclosed by transverse reinforcement according to Table 13.4.5.6(a) and shall be spaced according to Table 13.4.5.6(b):

Table 13.4.5.6(a)—Minimum transverse reinforcement size

Least horizontal pile dimension h, in.	Minimum wire size transverse reinforcement[1]
$h \leq 16$	W4, D4
$16 < h < 20$	W4.5, D5
$h \geq 20$	W5.5, D6

[1]If bars are used, minimum of No. 3 bar applies to all values of h.

COMMENTARY

R13.4.5 *Precast concrete piles*

R13.4.5.6 The minimum transverse reinforcement required in this section is typically sufficient for driving and handling stresses. These provisions for precast concrete piles in SDC A and B are based on information from PCI *Recommended Practice for the Design, Manufacture, and Installation of Prestressed Concrete Piling* (1993) and the *PCI Bridge Design Manual*, Chapter 20 (2004). Minimum reinforcement requirements for precast concrete piles supporting buildings assigned to SDC C, D, E, and F are defined in 18.13.5.10.

13 Foundations

CODE

Table 13.4.5.6(b)—Maximum transverse reinforcement spacing

Reinforcement location in the pile	Maximum center-to-center spacing, in.
First five ties or spirals at each end of pile	1
24 in. from each end of pile	4
Remainder of pile	6

13.4.6 *Pile caps*

13.4.6.1 Overall depth of pile cap shall be selected such that the effective depth of bottom reinforcement is at least 12 in.

13.4.6.2 Factored moments and shears shall be permitted to be calculated with the reaction from any pile assumed to be concentrated at the centroid of the pile section.

13.4.6.3 Except for pile caps designed in accordance with 13.2.6.5, the pile cap shall be designed such that (a) is satisfied for one-way foundations and (a) and (b) are satisfied for two-way foundations.

(a) $\phi V_n \geq V_u$, where V_n shall be calculated in accordance with 22.5 for one-way shear, V_u shall be calculated in accordance with 13.4.2.7, and ϕ shall be in accordance with 21.2
(b) $\phi v_n \geq v_u$, where v_n shall be calculated in accordance with 22.6 for two-way shear, v_u shall be calculated in accordance with 13.4.2.7, and ϕ shall be in accordance with 21.2

13.4.6.4 If the pile cap is designed in accordance with the strut-and-tie method as permitted in 13.2.6.5, the effective concrete compressive strength of the struts, f_{ce}, shall be calculated in accordance with 23.4.3, where $\boldsymbol{\beta_s = 0.60\lambda}$, and λ is in accordance with 19.2.4.

13.4.6.5 Calculation of factored shear on any section through a pile cap shall be in accordance with (a) through (c):

(a) Entire reaction from any pile with its center located $d_{pile}/2$ or more outside the section shall be considered as producing shear on that section.
(b) Reaction from any pile with its center located $d_{pile}/2$ or more inside the section shall be considered as producing no shear on that section.
(c) For intermediate positions of pile center, the portion of the pile reaction to be considered as producing shear on the section shall be based on a linear interpolation between full value at $d_{pile}/2$ outside the section and zero value at $d_{pile}/2$ inside the section.

COMMENTARY

R13.4.6 *Pile caps*

R13.4.6.4 It is typically necessary to take the effective concrete compressive strength from expression (d) or (f) in Table 23.4.3(a) because it is generally not practical to provide confining reinforcement satisfying 23.5 in a pile cap.

R13.4.6.5 If piles are located inside the critical sections d or $d/2$ from face of column, for one-way or two-way shear, respectively, an upper limit on the shear strength at a section adjacent to the face of the column should be considered. The *CRSI Handbook* (1984) offers guidance for this situation.

CHAPTER 14—PLAIN CONCRETE

CODE

14.1—Scope

14.1.1 This chapter shall apply to the design of plain concrete members, including (a) and (b):

(a) Members in building structures
(b) Members in non-building structures such as arches, underground utility structures, gravity walls, and shielding walls

14.1.2 This chapter shall not govern the design of cast-in-place concrete piles and piers embedded in ground.

14.1.3 Plain concrete shall be permitted only in cases (a) through (d):

(a) Members that are continuously supported by soil or supported by other structural members capable of providing continuous vertical support
(b) Members for which arch action provides compression under all conditions of loading
(c) Walls
(d) Pedestals

14.1.4 Plain concrete shall be permitted for a structure assigned to Seismic Design Category (SDC) D, E, or F, only in cases (a) and (b):

(a) Footings supporting cast-in-place reinforced concrete or reinforced masonry walls, provided the footings are reinforced longitudinally with at least two continuous reinforcing bars. Bars shall be at least No. 4 and have a total area of not less than 0.002 times the gross cross-sectional area of the footing. Continuity of reinforcement shall be provided at corners and intersections.
(b) Foundation elements (i) through (iii) for detached one- and two-family dwellings not exceeding three stories and constructed with stud bearing walls:
(i) Footings supporting walls
(ii) Isolated footings supporting columns or pedestals
(iii) Foundation or basement walls not less than 7-1/2 in. thick and retaining no more than 4 ft of unbalanced fill.

14.1.5 Plain concrete shall not be permitted for columns and pile caps.

COMMENTARY

R14.1—Scope

R14.1.2 Structural elements, such as cast-in-place plain concrete piles and piers in ground or other material sufficiently stiff to provide adequate lateral support to prevent buckling, are not covered by the Code. Such elements are covered by the general building code.

R14.1.3 Because the strength and structural integrity of structural plain concrete members is based solely on the member size, concrete strength, and other concrete properties, use of structural plain concrete should be limited to members:

(a) That are primarily in a state of compression
(b) That can tolerate random cracks without detriment to their structural integrity
(c) For which ductility is not an essential feature of design

The tensile strength of concrete can be used in design of structural plain concrete members. Tensile stresses due to restraint from creep, shrinkage, or temperature effects are to be considered to avoid uncontrolled cracks or structural failure. For residential construction within the scope of ACI 332, refer to 1.4.6.

R14.1.5 Because plain concrete lacks the necessary ductility that columns should possess, and because a random crack in an unreinforced column will most likely endanger its structural integrity, the Code does not permit use of plain concrete for columns. It does allow its use for pedestals limited to a ratio of unsupported height to least lateral dimension of 3 or less (refer to 14.1.3(d) and 14.3.3).

14.2—General

14.2.1 *Materials*

14.2.1.1 Design properties for concrete shall be selected to be in accordance with Chapter 19.

14.2.1.2 Steel reinforcement, if required, shall be selected to be in accordance with Chapter 20.

14.2.1.3 Materials, design, and detailing requirements for embedments in concrete shall be in accordance with 20.6.

14.2.2 *Connection to other members*

14.2.2.1 Tension shall not be transmitted through outside edges, construction joints, contraction joints, or isolation joints of an individual plain concrete element.

14.2.2.2 Walls shall be braced against lateral translation.

14.2.3 *Precast*

14.2.3.1 Design of precast members shall consider all loading conditions from initial fabrication to completion of the structure, including form removal, storage, transportation, and erection.

14.2.3.2 Precast members shall be connected to transfer lateral forces into a structural system capable of resisting such forces.

14.3—Design limits

14.3.1 *Bearing walls*

14.3.1.1 Minimum bearing wall thickness shall be in accordance with Table 14.3.1.1.

R14.2—General

R14.2.2 *Connection to other members*

R14.2.2.2 Provisions for plain concrete walls are applicable only for walls laterally supported in such a manner as to prohibit relative lateral displacement at top and bottom of individual wall elements. The Code does not cover walls without horizontal support to prohibit relative displacement at top and bottom of wall elements. Such laterally unsupported walls are to be designed as reinforced concrete members in accordance with the Code.

R14.2.3 *Precast*

Precast structural plain concrete members are considered subject to all limitations and provisions for cast-in-place concrete contained in this chapter.

The approach to contraction or isolation joints is expected to be somewhat different than for cast-in-place concrete because the major portion of shrinkage in precast members occurs prior to erection. To ensure stability, precast members should be connected to other members. The connection should transfer no tension.

R14.3—Design limits

R14.3.1 *Bearing walls*

Plain concrete walls are commonly used for basement wall construction for residential and light commercial buildings located in areas of low seismic risk. Although the Code imposes no absolute maximum height limitation on the use of plain concrete walls, experience with use of plain concrete in relatively minor structures should not be extrapolated to using plain concrete walls in multistory construction and other major structures where differential settlement, wind,

CODE

Table 14.3.1.1—Minimum thickness of bearing walls

Wall type	Minimum thickness	
General	Greater of:	5.5 in.
		1/24 the lesser of unsupported length and unsupported height
Exterior basement	7.5 in.	
Foundation	7.5 in.	

COMMENTARY

earthquake, or other unforeseen loading conditions require the walls to possess some ductility and ability to maintain integrity when cracked. For such conditions, ACI Committee 318 strongly encourages the use of walls designed in accordance with Chapter 11.

14.3.2 *Footings*

14.3.2.1 Footing thickness shall be at least 8 in.

14.3.2.2 Base area of footing shall be determined from unfactored forces and moments transmitted by footing to soil and permissible soil pressure selected through principles of soil mechanics.

R14.3.2 *Footings*

R14.3.2.1 Thickness of plain concrete footings of usual proportions will typically be controlled by flexural strength (extreme fiber stress in tension not greater than ($\phi 5\lambda\sqrt{f_c'}$) rather than shear strength (refer to R14.5.5.1). For footings cast against soil, overall thickness ***h*** used for strength calculations is specified in 14.5.1.7.

14.3.3 *Pedestals*

14.3.3.1 Ratio of unsupported height to average least lateral dimension shall not exceed 3.

R14.3.3 *Pedestals*

R14.3.3.1 The height-thickness limitation for plain concrete pedestals does not apply for portions of pedestals embedded in soil capable of providing lateral restraint.

14.3.4 *Contraction and isolation joints*

14.3.4.1 Contraction or isolation joints shall be provided to divide structural plain concrete members into flexurally discontinuous elements. The size of each element shall be selected to limit stress caused by restraint to movements from creep, shrinkage, and temperature effects.

14.3.4.2 The number and location of contraction or isolation joints shall be determined considering (a) through (f):

R14.3.4 *Contraction and isolation joints*

R14.3.4.1 Joints in plain concrete construction are an important design consideration. In reinforced concrete, reinforcement is provided to resist the stresses due to restraint of creep, shrinkage, and temperature effects. In plain concrete, joints are the only means of controlling, and thereby relieving, the buildup of such tensile stresses. A plain concrete member should therefore be small enough, or divided into smaller elements by joints, to control the buildup of internal stresses. The joint may be a contraction joint or isolation joint. A minimum 25 percent reduction of member thickness is typically sufficient for contraction joints to be effective. The jointing should be such that no axial tension or flexural tension can be developed across a joint after cracking, if applicable—a condition referred to as flexural discontinuity. Where random cracking due to creep, shrinkage, and temperature effects will not affect structural integrity and is otherwise acceptable (such as transverse cracks in a continuous wall footing), transverse contraction or isolation joints should not be necessary.

CODE

(a) Influence of climatic conditions
(b) Selection and proportioning of materials
(c) Mixing, placing, and curing of concrete
(d) Degree of restraint to movement
(e) Stresses due to loads to which an element is subjected
(f) Construction techniques

14.4—Required strength

14.4.1 *General*

14.4.1.1 Required strength shall be calculated in accordance with the factored load combinations defined in Chapter 5.

14.4.1.2 Required strength shall be calculated in accordance with the analysis procedures in Chapter 6.

14.4.1.3 No flexural continuity due to tension shall be assumed between adjacent structural plain concrete elements.

14.4.2 *Walls*

14.4.2.1 Walls shall be designed for an eccentricity corresponding to the maximum moment that can accompany the axial load but not less than $\mathbf{0.10}\boldsymbol{h}$, where $\boldsymbol{h}$ is the wall thickness.

14.4.3 *Footings*

14.4.3.1 *General*

14.4.3.1.1 For footings supporting circular or regular polygon-shaped concrete columns or pedestals, it shall be permitted to assume a square section of equivalent area for determining critical sections.

14.4.3.2 *Factored moment*

14.4.3.2.1 The critical section for $\boldsymbol{M_u}$ shall be located in accordance with Table 14.4.3.2.1.

Table 14.4.3.2.1—Location of critical section for M_u

Supported member	Location of critical section
Column or pedestal	Face of column or pedestal
Column with steel base plate	Halfway between face of column and edge of steel base plate
Concrete wall	Face of wall
Masonry wall	Halfway between center and face of masonry wall

COMMENTARY

R14.4—Required strength

R14.4.1 *General*

R14.4.1.1 Plain concrete members are proportioned for adequate strength using factored loads and forces. When the design strength is exceeded, the cross section should be increased or the specified strength of concrete increased, or both, or the member designed as a reinforced concrete member in accordance with the Code. An increase in concrete section may have a detrimental effect; stress due to load will decrease but stresses due to creep, shrinkage, and temperature effects may increase.

CODE

14.4.3.3 *Factored one-way shear*

14.4.3.3.1 For one-way shear, critical sections shall be located $\boldsymbol{h}$ from (a) and (b), where $\boldsymbol{h}$ is the footing thickness.

(a) Location defined in Table 14.4.3.2.1
(b) Face of concentrated loads or reaction areas

14.4.3.3.2 Sections between (a) or (b) of 14.4.3.3.1 and the critical section for shear shall be permitted to be designed for V_u at the critical section for shear.

14.4.3.4 *Factored two-way shear*

14.4.3.4.1 For two-way shear, critical sections shall be located so that the perimeter $\boldsymbol{b_o}$ is a minimum but need not be closer than $\boldsymbol{h/2}$ to (a) through (c):

(a) Location defined in Table 14.4.3.2.1
(b) Face of concentrated loads or reaction areas
(c) Changes in footing thickness

14.4.3.4.2 For square or rectangular columns, concentrated loads, or reaction areas, the critical section for two-way shear shall be permitted to be calculated assuming straight sides.

14.5—Design strength

14.5.1 *General*

14.5.1.1 For each applicable factored load combination, design strength at all sections shall satisfy $\phi S_n \geq U$, including (a) through (d). Interaction between load effects shall be considered.

(a) $\phi M_n \geq M_u$
(b) $\phi P_n \geq P_u$
(c) $\phi V_n \geq V_u$
(d) $\phi B_n \geq B_u$

14.5.1.2 ϕ shall be determined in accordance with 21.2.

14.5.1.3 Tensile strength of concrete shall be permitted to be considered in design.

COMMENTARY

R14.4.3.4 *Factored two-way shear*

R14.4.3.4.1 The critical section defined in this provision is similar to that defined for reinforced concrete elements in 22.6.4.1, except that for plain concrete, the critical section is based on $\boldsymbol{h}$ rather than $\boldsymbol{d}$.

R14.5—Design strength

R14.5.1 *General*

R14.5.1.1 Refer to R9.5.1.1.

R14.5.1.2 The strength reduction factor ϕ for plain concrete design is the same for all strength conditions. Because both flexural tensile strength and shear strength for plain concrete depend on the tensile strength characteristics of the concrete, with no reserve strength or ductility possible due to the absence of reinforcement, equal strength reduction factors for both bending and shear are considered appropriate.

R14.5.1.3 Flexural tension may be considered in design of plain concrete members to resist loads, provided the calculated stress does not exceed the permissible stress, and construction, contraction, or isolation joints are provided to relieve the resulting tensile stresses due to restraint of creep, shrinkage, and temperature effects.

CODE

14.5.1.4 Flexure and axial strength calculations shall be based on a linear stress-strain relationship in both tension and compression.

14.5.1.5 λ for lightweight concrete shall be in accordance with 19.2.4.

14.5.1.6 No strength shall be assigned to steel reinforcement.

14.5.1.7 When calculating member strength in flexure, combined flexure and axial load, or shear, the entire cross section shall be considered in design, except for concrete cast against soil where overall thickness h shall be taken as 2 in. less than the specified thickness.

14.5.1.8 Unless demonstrated by analysis, horizontal length of wall to be considered effective for resisting each vertical concentrated load shall not exceed center-to-center distance between loads, or bearing width plus four times the wall thickness.

14.5.2 *Flexure*

14.5.2.1 M_n shall be the lesser of Eq. (14.5.2.1a) calculated at the tension face and Eq. (14.5.2.1b) calculated at the compression face:

$$M_n = 5\lambda\sqrt{f_c'}S_m \quad (14.5.2.1a)$$

$$M_n = 0.85f_c'S_m \quad (14.5.2.1b)$$

where S_m is the corresponding elastic section modulus.

14.5.3 *Axial compression*

14.5.3.1 P_n shall be calculated by:

$$P_n = 0.60f_c'A_g\left[1-\left(\frac{\ell_c}{32h}\right)^2\right] \quad (14.5.3.1)$$

14.5.4 *Flexure and axial compression*

14.5.4.1 Unless permitted by 14.5.4.2, member dimensions shall be proportioned to be in accordance with Table 14.5.4.1, where M_n is calculated in accordance with Eq. (14.5.2.1b) and P_n is calculated in accordance with Eq. (14.5.3.1).

COMMENTARY

R14.5.1.7 The reduced overall thickness h for concrete cast against earth is to allow for unevenness of excavation and for some contamination of the concrete adjacent to the soil.

R14.5.2 *Flexure*

R14.5.2.1 Equation (14.5.2.1b) may control for nonsymmetrical cross sections.

R14.5.3 *Axial compression*

R14.5.3.1 Equation (14.5.3.1) is presented to reflect the general range of braced and restrained end conditions encountered in plain concrete elements. The effective length factor was omitted as a modifier of ℓ_c, the vertical distance between supports, because this is conservative for walls with assumed pin supports that are required to be braced against lateral translation as in 14.2.2.2.

R14.5.4 *Flexure and axial compression*

CODE

Table 14.5.4.1—Combined flexure and axial compression

Location	Interaction equation	
Tension face	$\frac{M_u}{S_m} - \frac{P_u}{A_g} \leq \phi 5\lambda\sqrt{f'_c}$	(a)
Compression face	$\frac{M_u}{\phi M_n} + \frac{P_u}{\phi P_n} \leq 1.0$	(b)

14.5.4.2 For walls of solid rectangular cross section where $M_u \leq P_u(h/6)$, M_u need not be considered in design and P_n is calculated by:

$$P_n = 0.45 f'_c A_g \left[1 - \left(\frac{\ell_c}{32h}\right)^2\right] \quad (14.5.4.2)$$

COMMENTARY

R14.5.4.2 If the resultant load falls within the middle third of the wall thickness, plain concrete walls may be designed using the simplified Eq. (14.5.4.2). Eccentric loads and lateral forces are used to determine the total eccentricity of the factored axial force P_u. Equation (14.5.4.2) reflects the range of braced and restrained end conditions encountered in wall design. The limitations of 14.2.2.2, 14.3.1.1, and 14.5.1.8 apply whether the wall is proportioned by 14.5.4.1 or by 14.5.4.2.

14.5.5 *Shear*

14.5.5.1 V_n shall be calculated in accordance with Table 14.5.5.1.

Table 14.5.5.1—Nominal shear strength

Shear action	Nominal shear strength V_n		
One-way	$\frac{4}{3}\lambda\sqrt{f'_c}b_w h$		(a)
Two-way	Lesser of:	$\left(1+\frac{2}{\beta}\right)\left(\frac{4}{3}\lambda\sqrt{f'_c}b_o h\right)$ [1]	(b)
		$2\left(\frac{4}{3}\lambda\sqrt{f'_c}b_o h\right)$	(c)

[1]β is the ratio of long side to short side of concentrated load or reaction area.

R14.5.5 *Shear*

R14.5.5.1 Proportions of plain concrete members usually are controlled by tensile strength rather than shear strength. Shear stress (as a substitute for principal tensile stress) rarely will control. However, because it is difficult to foresee all possible conditions where shear may have to be investigated, such as shear keys, Committee 318 maintains the investigation of this basic stress condition.

The shear requirements for plain concrete assume an uncracked section. Shear failure in plain concrete will be a diagonal tension failure, occurring when the principal tensile stress near the centroidal axis becomes equal to the tensile strength of the concrete. Because the major portion of the principal tensile stress results from shear, the Code safeguards against tension failure by limiting the permissible shear at the centroidal axis as calculated from the equation for a section of homogeneous material:

$$v = VQ/Ib$$

where v and V are the shear stress and shear force, respectively, at the section considered; Q is the statical moment of the area above or below the centroid of the gross section calculated about the centroidal axis; I is the moment of inertia of the gross section; and b is the section width where shear stress is being calculated.

14.5.6 *Bearing*

14.5.6.1 B_n shall be calculated in accordance with Table 14.5.6.1.

14 Plain Conc.

CODE

COMMENTARY

Table 14.5.6.1—Nominal bearing strength

Relative geometric conditions	B_n		
Supporting surface is wider on all sides than the loaded area	Lesser of:	$\sqrt{A_2/A_1}(0.85f_c'A_1)$	(a)
		$2(0.85f_c'A_1)$	(b)
Other	$0.85f_c'A_1$		(c)

14.6—Reinforcement detailing

14.6.1 At least two No. 5 bars shall be provided around window, door, and similarly sized openings. Such bars shall extend at least 24 in. beyond the corners of openings or shall be anchored to develop f_y in tension at the corners of the openings.

CHAPTER 15—BEAM-COLUMN AND SLAB-COLUMN JOINTS

CODE

15.1—Scope

15.1.1 This chapter shall apply to the design and detailing of cast-in-place beam-column and slab-column joints.

15.2—General

15.2.1 Beam-column joints shall satisfy the detailing provisions of 15.3 and strength requirements of 15.4.

15.2.2 Beam-column and slab-column joints shall satisfy 15.5 for transfer of column axial force through the floor system.

15.2.3 If gravity load, wind, earthquake, or other lateral forces cause transfer of moment at beam-column joints, the shear resulting from moment transfer shall be considered in the design of the joint.

15.2.4 At corner joints between two members, the effects of closing and opening moments within the joint shall be considered.

15.2.5 If a beam framing into the joint and generating joint shear has depth exceeding twice the column depth, analysis and design of the joint shall be based on the strut-and-tie method in accordance with Chapter 23 and (a) and (b) shall be satisfied:

(a) Design joint shear strength determined in accordance with Chapter 23 shall not exceed ϕV_n calculated in accordance with 15.4.2.
(b) Detailing provisions of 15.3 shall be satisfied.

15.2.6 A column extension assumed to provide continuity through a beam-column joint in the direction of joint shear considered shall satisfy (a) and (b):

(a) The column extends above the joint at least one column depth, h, measured in the direction of joint shear considered.
(b) Longitudinal and transverse reinforcement from the column below the joint is continued through the extension.

15.2.7 A beam extension assumed to provide continuity through a beam-column joint in the direction of joint shear considered shall satisfy (a) and (b):

COMMENTARY

R15.1—Scope

A joint is the portion of a structure common to intersecting members, whereas a connection is comprised of a joint and portions of adjoining members. Chapter 15 is focused on design requirements for beam-to-column and slab-to-column joints.

For structures assigned to Seismic Design Categories (SDC) B through F, joints may be required to withstand several reversals of loading. Chapter 18 provides requirements for earthquake-resistant structures that are applied in addition to the basic requirements for joints in Chapter 15.

R15.2—General

Tests of joints with extensions of beams with lengths at least equal to their depths have indicated similar joint shear strengths to those of joints with continuous beams. These findings suggest that extensions of beams and columns, when properly dimensioned and reinforced with longitudinal and transverse bars, provide effective confinement to the joint faces (Meinheit and Jirsa 1981). Extensions that provide beam and column continuity through a joint do not contribute to joint shear force if they do not support externally applied loads.

Tests (Hanson and Conner 1967) have shown that beam-column joints laterally supported on four sides by beams of approximately equal depth exhibit superior behavior compared to joints without all four faces confined by beams under reversed cyclic loading.

Corner joints occur where two non-colinear members transfer moment and terminate at the joint. A roof-level exterior joint is an example of a corner joint between two members, also referred to as a knee joint. Corner joints are vulnerable to flexural failure from either closing or opening moments even if flexural strengths at the joint faces are sufficient. Considering transfer of moment across a diagonal section through a corner joint connecting to a cantilevered member is critical because the moment acting through the joint cannot be redistributed.

Chapter 23 provides requirements for design and detailing of corner joints when using the strut-and-tie method. Klein (2008) provides additional guidance on design of frame corners using the strut-and-tie method. The requirements for transverse reinforcement in corner joints are given in 15.3. ACI 352R provides additional guidance on detailing of joints.

For joints in which the beam depth is significantly greater than the column depth a diagonal strut between the joint corners may not be effective. Therefore, the Code requires that joints in which the beam depth exceeds twice the column depth be designed using the strut-and-tie method of Chapter 23.

Transfer of bending through joints between slabs and corner or edge columns is covered in Chapter 8.

In the 2019 Code, classification of beam and column members framing into joint faces was modified to distin-

CODE

(a) The beam extends at least one beam depth h beyond the joint face.
(b) Longitudinal and transverse reinforcement from the beam on the opposite side of the joint is continued through the extension.

15.2.8 A beam-column joint shall be considered to be confined for the direction of joint shear considered if two transverse beams satisfying (a), (b), and (c) are provided:

(a) Width of each transverse beam is at least three-quarters of the width of the column face into which the beam frames
(b) Transverse beams extend at least one beam depth h beyond the joint faces
(c) Transverse beams contain at least two continuous top and bottom bars satisfying 9.6.1.2 and No. 3 or larger stirrups satisfying 9.6.3.4 and 9.7.6.2.2

15.2.9 For slab-column connections transferring moment, strength and detailing requirements shall be in accordance with applicable provisions in Chapter 8 and Sections 15.3.2 and 22.6.

15.3—Detailing of joints

15.3.1 *Beam-column joint transverse reinforcement*

15.3.1.1 Beam-column joints shall satisfy 15.3.1.2 through 15.3.1.4 unless (a) through (c) are satisfied:

(a) Joint is considered confined by transverse beams in accordance with 15.2.8 for all shear directions considered
(b) Joint is not part of a designated seismic-force-resisting system
(c) Joint is not part of a structure assigned to SDC D, E, or F

15.3.1.2 Joint transverse reinforcement shall consist of ties, spirals, or hoops satisfying the requirements of 25.7.2 for ties, 25.7.3 for spirals, and 25.7.4 for hoops.

15.3.1.3 At least two layers of horizontal transverse reinforcement shall be provided within the depth of the shallowest beam framing into the joint.

15.3.1.4 Spacing of joint transverse reinforcement s shall not exceed 8 in. within the depth of the deepest beam framing into the joint.

15.3.2 *Slab-column joint transverse reinforcement*

15.3.2.1 Except where laterally supported on four sides by a slab, column transverse reinforcement shall be continued through a slab-column joint, including column capital, drop panel, and shear cap, in accordance with 25.7.2 for ties, 25.7.3 for spirals, and 25.7.4 for hoops.

COMMENTARY

guish those members contributing to joint shear from those that do not contribute to joint shear but may serve to confine the joint. For a given joint shear direction, lateral confinement is provided by transverse beams while the width of the beams generating joint shear is accounted for through the effective joint width in 15.4.2.4. These classifications are made for the purpose of establishing nominal joint shear strength in Tables 15.4.2.3 and 18.8.4.3. For beam-column joints with circular columns, the column width and depth may be taken as those of a square section of equivalent area.

R15.3—Detailing of joints

R15.3.1 *Beam-column joint transverse reinforcement*

Tests (Hanson and Connor 1967) have shown that the joint region of a beam-to-column connection in the interior of a building does not require shear reinforcement if the joint is laterally supported on four sides by beams of approximately equal depth. However, joints that are not restrained in this manner, such as at the exterior of a building, require shear reinforcement to prevent deterioration due to shear cracking (ACI 352R). These joints may also require transverse reinforcement to prevent buckling of longitudinal column reinforcement.

CODE

15.3.3 *Longitudinal reinforcement*

15.3.3.1 Development of longitudinal reinforcement terminated in the joint or within a column or beam extension, as defined in 15.2.6(a) and 15.2.7(a), shall be in accordance with 25.4.

15.3.3.2 Longitudinal reinforcement terminated in the joint with a standard hook shall have the hook turned toward mid-depth of the beam or column.

15.4—Strength requirements for beam-column joints

15.4.1 *Required shear strength*

15.4.1.1 Joint shear force V_u shall be calculated on a plane at mid-height of the joint using flexural tensile and compressive beam forces and column shear consistent with (a) or (b):

(a) The maximum moment transferred between the beam and column as determined from factored-load analysis for beam-column joints with continuous beams in the direction of joint shear considered

(b) Beam nominal moment strengths M_n

15.4.2 *Design shear strength*

15.4.2.1 Design shear strength of cast-in-place beam-column joints shall satisfy:

$$\phi V_n \geq V_u$$

15.4.2.2 ϕ shall be in accordance with 21.2.1 for shear.

15.4.2.3 V_n of the joint shall be calculated in accordance with Table 15.4.2.3.

COMMENTARY

R15.3.3 *Longitudinal reinforcement*

R15.3.3.1 Where bars are continued through an unloaded extension at the opposite face, the bar length within the extension can be considered as part of the development length.

R15.4—Strength requirements for beam-column joints

Joint shear strength is evaluated separately in each principal direction of loading in accordance with 15.4.

R15.4.2 *Design shear strength*

The effective area of the joint, A_j, is illustrated in Fig. R15.4.2. In no case is A_j greater than the column cross-sectional area. A circular column may be considered as having a square section of equal area. The varied levels of shear strength provided by 15.4.2.3 are based on the recommendations of ACI 352R, although it is noted that the ACI 352R definition of effective cross-sectional joint area is sometimes different than A_j. Values of effective joint width calculated using ACI 352R and ACI 318, however, are the same or similar for many design situations.

CODE

Table 15.4.2.3—Nominal joint shear strength V_n

Column	Beam in direction of V_u	Confinement by transverse beams according to 15.2.8	V_n, lb[1]
Continuous or meets 15.2.6	Continuous or meets 15.2.7	Confined	$24\lambda\sqrt{f_c'}A_j$
		Not confined	$20\lambda\sqrt{f_c'}A_j$
	Other	Confined	$20\lambda\sqrt{f_c'}A_j$
		Not confined	$15\lambda\sqrt{f_c'}A_j$
Other	Continuous or meets 15.2.7	Confined	$20\lambda\sqrt{f_c'}A_j$
		Not confined	$15\lambda\sqrt{f_c'}A_j$
	Other	Confined	$15\lambda\sqrt{f_c'}A_j$
		Not confined	$12\lambda\sqrt{f_c'}A_j$

[1]λ shall be 0.75 for lightweight concrete and 1.0 for normalweight concrete.

15.4.2.4 Effective cross-sectional area within a joint, A_j, shall be calculated as the product of joint depth and effective joint width. Joint depth shall be the overall depth of the column, h, in the direction of joint shear considered. Effective joint width shall be the overall width of the column where the beam is wider than the column. Where the column is wider than the beam, effective joint width shall not exceed the lesser of (a) and (b):

(a) Beam width plus joint depth
(b) Twice the perpendicular distance from longitudinal axis of beam to nearest side face of the column

COMMENTARY

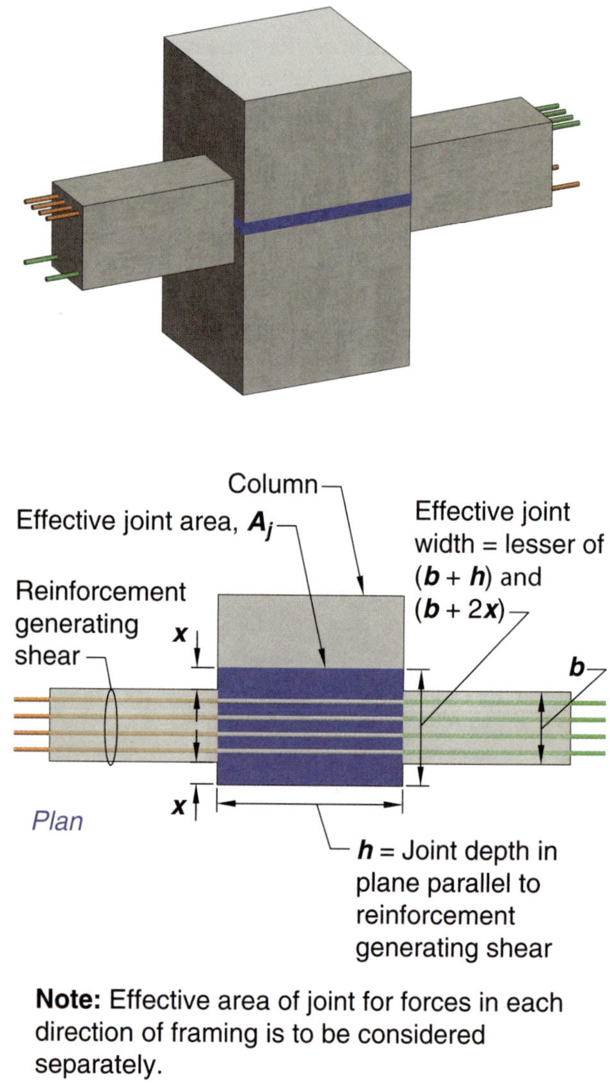

Fig. R15.4.2—Effective joint area.

15.5—Transfer of column axial force through the floor system

15.5.1 If f_c' of a floor system is less than $0.7f_c'$ of a column, transmission of axial force through the floor system shall be in accordance with (a), (b), or (c):

(a) Concrete of compressive strength specified for the column shall be placed in the floor system at the column location. Column concrete shall extend outward at least 2 ft into the floor system from face of column for the full depth of the floor system and be integrated with floor concrete.
(b) Design strength of a column through a floor system shall be calculated using the lower value of concrete strength with vertical dowels and transverse reinforcement as required to achieve design strength.
(c) For beam-column joints laterally supported on four sides by beams of approximately equal depth that satisfy

R15.5—Transfer of column axial force through the floor system

The requirements of this section consider the effect of floor system concrete strength on column axial strength (Bianchini et al. 1960). If floor system concrete strength is less than 70 percent of column concrete strength, methods in 15.5.1(a) or 15.5.1(b) may be applied to corner or edge columns. Methods in 15.5.1(a), (b), or (c) may be applied to interior columns.

Application of the concrete placement procedure described in 15.5.1(a) requires the placing of two different concrete mixtures in the floor system. The Code requires that column concrete be placed through the thickness of the floor system and that mixtures be placed and remain plastic such that the two can be vibrated so they are well integrated. Additional inspection may be required for this process. As required in Chapter 26, it is the responsibility of the licensed design professional to indicate on the construction docu-

CODE

15.2.7 and 15.2.8(a) and for slab-column joints supported on four sides by the slab, it shall be permitted to calculate the design strength of the column using an assumed concrete strength in the column joint equal to 75 percent of column concrete strength plus 35 percent of floor system concrete strength, where the value of column concrete strength shall not exceed 2.5 times the floor system concrete strength.

COMMENTARY

ments where the higher- and lower-strength concretes are to be placed.

Research (Ospina and Alexander 1998) has shown that heavily loaded slabs do not provide as much confinement as lightly loaded slabs when ratios of column concrete strength to slab concrete strength exceed approximately 2.5. Consequently, a limit is given in 15.5.1(c) on the ratio of concrete strengths assumed in design.

As an alternative to 15.5.1(a) or 15.5.1(c), 15.5.1(b) permits the use of dowel bars and confinement reinforcement to increase the effective compressive strength of concrete in the column core (Paultre and Légeron 2008; Richart et al. 1929).

Notes

Standard Spread footing design Example:

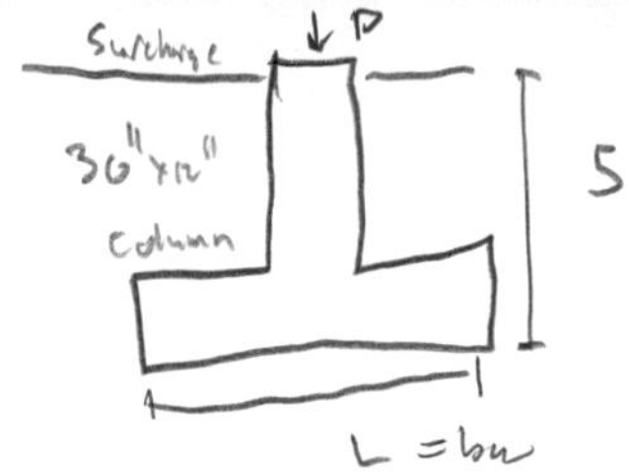

1.) det pressure on soil

$q = (PDL)(H) + \text{Surcharge} + \frac{P_{DL} + P_{LL}}{A}$

$\sigma_{all} = q_a = q$ solve for A

$P_u = 1.2\ P_{DL} + 1.6\ P_{LL}$

$q_{nu} = \frac{P_u}{A}$

Check 1-way Shear

Use tributary shear area

$V_u = q_{nu}\ A$

Capacity Normal weight

$\phi V_c = \phi 2 \lambda \sqrt{f'_c}\ b_w d$

Check $V_c < 5 \lambda \sqrt{f'_c}\ b_w d$

Bearing

Column size

$\phi P_{nb} = \phi (0.85) f'_c\ A_1$

P_u

B on footing

$\phi P_{nb} = \phi (0.85) f'_c\ A_1 \sqrt{\frac{A_2}{A_1}}$

$A_2 =$

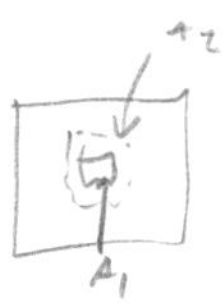

Flexure checks

$M_u = q_{nu}\ b_w f \left(\frac{f}{2}\right)$

Check d

$c = a/\beta_1$ $\quad \varepsilon_t = \frac{d_t - c}{c}(0.003) > 0.005 \quad \phi = 0.9$

$\left[\frac{\phi f_y^2}{2(0.85 f'_c b)}\right](A_s)^2 - (\phi f_y d) A_s + M_u = 0$ solve for A_s

$\phi M_n = \phi A_s f_y \left(d - \frac{\left(\frac{A_s f_y}{0.85 f'_c b}\right)}{2} \right)$

Reinforcing and Spacing Limits

$A_{s,max} = 0.85 \left(\frac{f'_c}{f_y}\right) \beta_1 \left(\frac{3}{7}\right)(d)(b)$

$S_{max} = \min (18 \text{ or } 3(h))$

$S_{min} = \max \begin{cases} 1'' \\ d_b \\ \left(\frac{4}{3}\right) d_{agg} \end{cases}$

longitudinal

$S_{min} \begin{pmatrix} 1.5 \\ 1.5 d_b \\ \frac{4}{3} d_{agg} \end{pmatrix}$ 4h7

CHAPTER 16—CONNECTIONS BETWEEN MEMBERS

CODE

16.1—Scope

16.1.1 This chapter shall apply to the design of joints and connections at the intersection of concrete members and for load transfer between concrete surfaces, including (a) through (d):

(a) Connections of precast members
(b) Connections between foundations and either cast-in-place or precast members
(c) Horizontal shear strength of composite concrete flexural members
(d) Brackets and corbels

16.2—Connections of precast members

16.2.1 *General*

16.2.1.1 Transfer of forces by means of grouted joints, shear keys, bearing, anchors, mechanical connectors, steel reinforcement, reinforced topping, or a combination of these, shall be permitted.

16.2.1.2 Adequacy of connections shall be verified by analysis or test.

16.2.1.3 Connection details that rely solely on friction caused by gravity loads shall not be permitted.

16.2.1.4 Connections, and regions of members adjacent to connections, shall be designed to resist forces and accommodate deformations due to all load effects in the precast structural system.

16.2.1.5 Design of connections shall consider structural effects of restraint of volume change in accordance with 5.3.6.

16.2.1.6 Design of connections shall consider the effects of tolerances specified for fabrication and erection of precast members.

COMMENTARY

R16.2—Connections of precast members

R16.2.1 *General*

Connection details should be arranged to minimize the potential for cracking due to restrained creep, shrinkage, and temperature movements. The Precast/Prestressed Concrete Institute (MNL 123) provides information on recommended connection details for precast concrete structures.

R16.2.1.1 If two or more connection methods are used to satisfy the requirements for force transfer, their individual load-deformation characteristics should be considered to confirm that the mechanisms work together as intended.

R16.2.1.4 The structural behavior of precast members may differ substantially from that of similar members that are cast-in-place. Design of connections to minimize or transmit forces due to shrinkage, creep, temperature change, elastic deformation, differential settlement, wind, and earthquake require particular consideration in precast construction.

R16.2.1.5 Connections should be designed to either permit the displacements or resist the forces induced by lack of fit, volume changes caused by shrinkage, creep, thermal, and other environmental effects. Connections intended to resist the forces should do so without loss of strength. Restraint assumptions should be consistent in all interconnected members. There are also cases in which the intended force may be in one direction, but it may affect the strength of the connection in another. For example, shrinkage-induced longitudinal tension in a precast beam may affect the vertical shear strength on the corbel supporting it.

R16.2.1.6 Refer to R26.9.1(a).

CODE

16.2.1.7 Design of a connection with multiple components shall consider the differences in stiffness, strength, and ductility of the components.

16.2.1.8 Integrity ties shall be provided in the vertical, longitudinal, and transverse directions and around the perimeter of a structure in accordance with 16.2.4 or 16.2.5.

16.2.2 *Required strength*

16.2.2.1 Required strength of connections and adjacent regions shall be calculated in accordance with the factored load combinations in Chapter 5.

16.2.2.2 Required strength of connections and adjacent regions shall be calculated in accordance with the analysis procedures in Chapter 6.

16.2.2.3 For bearing connections, N_{uc} shall be (a) or (b), but need not exceed $N_{uc,max}$, where $N_{uc,max}$ is the maximum restraint force that can be transmitted through the load path of a bearing connection multiplied by the load factor used for live loads in combinations with other factored load effects.

(a) For connections not on bearing pads, N_{uc} shall be calculated simultaneously with V_u using factored load combinations in accordance with 5.3.6. The restraint force shall be treated as a live load.
(b) For connections on bearing pads, N_{uc} shall be 20 percent of the sustained unfactored vertical reaction multiplied by a load factor of 1.6.

16.2.2.4 If the friction coefficient for a bearing material has been determined by results of tests, $N_{uc,max}$ shall be permitted to be determined by multiplying the sustained unfactored vertical reaction by the friction coefficient and a load factor of 1.6.

COMMENTARY

R16.2.1.8 Appendix B of the *PCI Design Handbook* (PCI MNL 120) provides a review of structural integrity and minimum integrity ties for precast concrete bearing wall structures.

R16.2.2 *Required strength*

R16.2.2.3 Bearing connections subjected to sustained loads will experience volume change restraint forces due to the effects of creep, shrinkage, and temperature change. Sustained loads are dead loads and any other permanent loads such as soil loads or equipment loads that may be included with live loads. Section 5.3.6 prescribes the general consideration for restraint of volume change and differential settlement in combination with other loading but does not define a specific load factor for precast concrete bearing conditions. Load factors are provided with these provisions. $N_{uc,max}$ provides a capacity-design limit.

For mechanical connections, steel-to-steel contact, or other high-friction bearings, the horizontal force is usually due to volume change restraint. Such bearing connections will experience volume change restraint forces due to the effects of creep, shrinkage, and temperature change. Because the magnitude of volume change restraint forces acting on bearing connections cannot usually be determined with a high degree of accuracy, it is required to treat the restraint force N_{uc} as a live load in 16.2.2.3(a) when using the factored load combinations of 5.3.6 and multiplied by 1.6 in 16.2.2.3(b).

Common precast concrete bearing connections use elastomeric pads or other structural bearing media that limit transferred forces by pad deformation or slip. The limiting load of such connections can be taken as 20 percent of the sustained unfactored reaction, as recognized by 16.2.2.3(b).

R16.2.2.4 Bearings explicitly designed for low friction, such as polytetrafluoroethylene (PTFE)-faced sliding bearings, may reduce volume change restraint forces. If the friction coefficient has been reliably determined for a bearing material considering service conditions such as temperature, aging, and exposure, that information can be used to calculate the maximum restraint force.

CODE

16.2.3 *Design strength*

16.2.3.1 For each applicable load combination, design strengths of precast member connections shall satisfy

$$\phi S_n \geq U \tag{16.2.3.1}$$

16.2.3.2 ϕ shall be determined in accordance with 21.2.

16.2.3.3 At the contact surface between supported and supporting members, or between a supported or supporting member and an intermediate bearing element, nominal bearing strength for concrete surfaces, $\boldsymbol{B_n}$, shall be calculated in accordance with 22.8. $\boldsymbol{B_n}$ shall be the lesser of the nominal concrete bearing strengths for the supported or supporting member surface, and shall not exceed the strength of intermediate bearing elements, if present.

16.2.3.4 If shear is the primary result of imposed loading and shear transfer occurs across a given plane, it shall be permitted to calculate $\boldsymbol{V_n}$ in accordance with the shear friction provisions in 22.9.

16.2.4 *Minimum connection strength and integrity tie requirements*

16.2.4.1 Except where the provisions of 16.2.5 govern, longitudinal and transverse integrity ties shall connect precast members to a lateral-force-resisting system, and vertical integrity ties shall be provided in accordance with 16.2.4.3 to connect adjacent floor and roof levels.

16.2.4.2 Where precast members form floor or roof diaphragms, the connections between the diaphragm and those members being laterally supported by the diaphragm shall have a nominal tensile strength of not less than 300 lb per linear ft.

16.2.4.3 Vertical integrity ties shall be provided at horizontal joints between all vertical precast structural members, except cladding, and shall satisfy (a) or (b):

(a) Connections between precast columns shall have vertical integrity ties, with a nominal tensile strength of at least $\mathbf{200}\boldsymbol{A_g}$ **lb**, where A_g is the gross area of the column. For columns with a larger cross section than required by consideration of loading, a reduced effective area based on the cross section required shall be permitted. The reduced effective area shall be at least one-half the gross area of the column.

COMMENTARY

R16.2.4 *Minimum connection strength and integrity tie requirements*

R16.2.4.1 It is not intended that these minimum requirements supersede other applicable provisions of the Code for design of precast concrete structures.

The overall integrity of a structure can be substantially enhanced by minor changes in the amount, location, and detailing of member reinforcement and in the detailing of connection hardware. The integrity ties should constitute a complete load path, and load transfers along that load path should be as direct as possible. Eccentricity of the load path, especially within any connection, should be minimized.

R16.2.4.2 The connection between the diaphragm and the member laterally supported by the diaphragm may be direct or indirect. For example, a column may be connected directly to the diaphragm, or it may be connected to a spandrel beam, which is connected to the diaphragm.

R16.2.4.3 Base connections and connections at horizontal joints in precast columns and wall panels, including structural walls, are designed to transfer all design forces and moments. The minimum integrity tie requirements of this provision are not additive to these design requirements. Common practice is to place the wall integrity ties symmetrically about the vertical centerline of the wall panel and within the outer quarters of the panel width, wherever possible.

CODE

(b) Connections between precast wall panels shall have at least two vertical integrity ties, with a nominal tensile strength of at least 10,000 lb per tie.

16.2.5 *Integrity tie requirements for precast concrete bearing wall structures three stories or more in height*

COMMENTARY

R16.2.5 *Integrity tie requirements for precast concrete bearing wall structures three stories or more in height*

Section 16.2.4 gives requirements for integrity ties that apply to all precast concrete structures. The specific requirements in this section apply only to precast concrete bearing wall structures with three or more stories, often called large panel structures. If the requirements of this section conflict with the requirements of 16.2.4, the requirements in this section control.

These minimum provisions for structural integrity ties in large panel bearing wall structures are intended to provide an alternate load path in case of loss of a bearing wall support (Portland Cement Association 1980). Tie requirements calculated for specific load effects may exceed these minimum provisions. The minimum integrity tie requirements are illustrated in Fig. R16.2.5, and are based on PCI's recommendations for design of precast concrete bearing wall buildings (PCI Committee on Precast Concrete Bearing Wall Buildings 1976). Integrity tie strength is based on yield strength. Appendix B of the *PCI Design Handbook* (PCI MNL 120) provides a review of structural integrity and minimum integrity ties for precast concrete bearing wall structures.

CODE | **COMMENTARY**

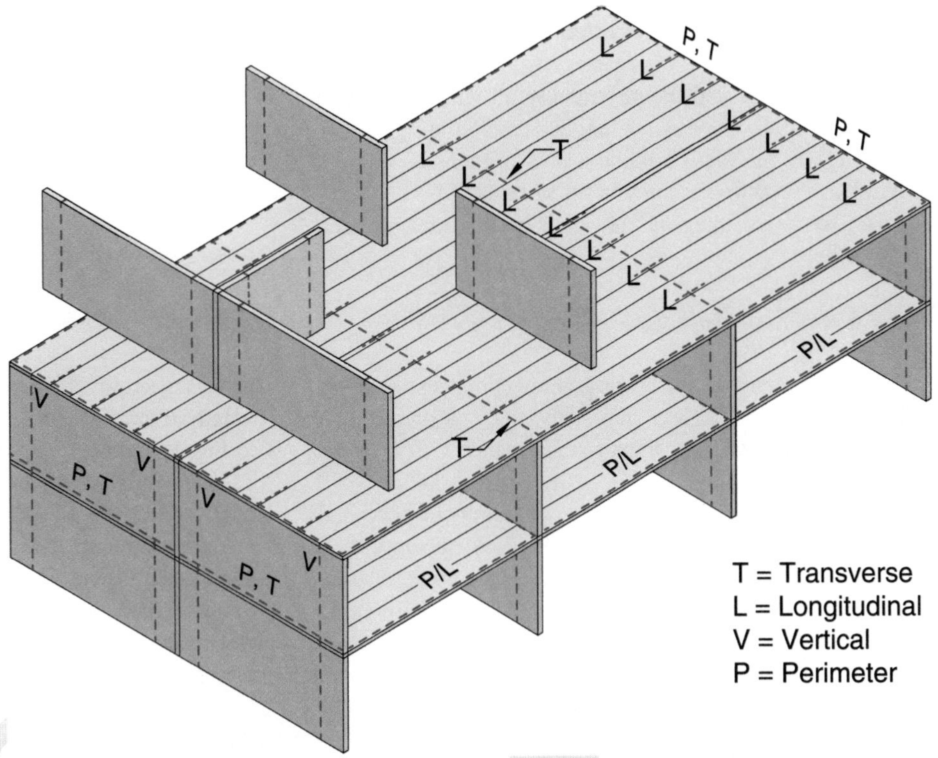

***Fig. R16.2.5**—Typical arrangement of integrity ties in large panel structures.*

16.2.5.1 Integrity ties in floor and roof systems shall satisfy (a) through (f):

(a) Longitudinal and transverse integrity ties shall be provided in floor and roof systems to provide a nominal tensile strength of at least 1500 lb per foot of width or length.

R16.2.5.1(a) Longitudinal integrity ties may project from slabs and be lap spliced, welded, mechanically connected, or embedded in grout joints with sufficient length and cover to develop the required force. Bond length for non-tensioned prestressing reinforcement, if used, should be sufficient to develop the yield strength (Salmons and McCrate 1977; PCA 1980).

(b) Longitudinal and transverse integrity ties shall be provided over interior wall supports and between the floor or roof system and exterior walls.

(c) Longitudinal and transverse integrity ties shall be positioned in or within 2 ft of the plane of the floor or roof system.

R16.2.5.1(c) It is not uncommon to have integrity ties positioned in the walls reasonably close to the plane of the floor or roof system.

(d) Longitudinal integrity ties shall be oriented parallel to floor or roof slab spans and shall be spaced not greater than 10 ft on center. Provisions shall be made to transfer forces around openings.

(e) Transverse integrity ties shall be oriented perpendicular to floor or roof slab spans and shall be spaced not greater than the bearing wall spacing.

R16.2.5.1(e) Transverse integrity ties may be uniformly spaced and either encased in the panels or in a topping, or they may be concentrated at the transverse bearing walls.

(f) Integrity ties at the perimeter of each floor and roof, within 4 ft of the edge, shall provide a nominal tensile strength of at least 16,000 lb.

R16.2.5.1(f) The perimeter integrity tie requirements need not be additive with the longitudinal and transverse integrity tie requirements.

CODE

16.2.5.2 Vertical integrity ties shall satisfy (a) through (c):

(a) Integrity ties shall be provided in all wall panels and shall be continuous over the height of the building.
(b) Integrity ties shall provide a nominal tensile strength of at least 3000 lb per horizontal foot of wall.
(c) At least two integrity ties shall be provided in each wall panel.

16.2.6 *Minimum dimensions at bearing connections*

16.2.6.1 Dimensions of bearing connections shall satisfy 16.2.6.2 or 16.2.6.3 unless shown by analysis or test that lesser dimensions will not impair performance.

16.2.6.2 For precast slabs, beams, or stemmed members, minimum design dimensions from the face of support to end of precast member in the direction of the span, considering specified tolerances, shall be in accordance with Table 16.2.6.2.

Table 16.2.6.2—Minimum design dimensions from face of support to end of precast member

Member type	Minimum distance, in.	
Solid or hollow-core slab	Greater of:	$\ell_n/180$
		2
Beam or stemmed member	Greater of:	$\ell_n/180$
		3

16.2.6.3 Bearing pads adjacent to unarmored faces shall be set back from the face of the support and the end of the supported member a distance not less than 0.5 in. or the chamfer dimension at a chamfered face.

COMMENTARY

R16.2.6 *Minimum dimensions at bearing connections*

This section differentiates between bearing length and length of the end of a precast member over the support (refer to Fig. R16.2.6).

Bearing pads distribute concentrated loads and reactions over the bearing area, and allow limited horizontal and rotational movements for stress relief. To prevent spalling under heavily loaded bearing areas, bearing pads should not extend to the edge of the support unless the edge is armored. Edges can be armored with anchored steel plates or angles. Section 16.5 gives requirements for bearing on brackets or corbels.

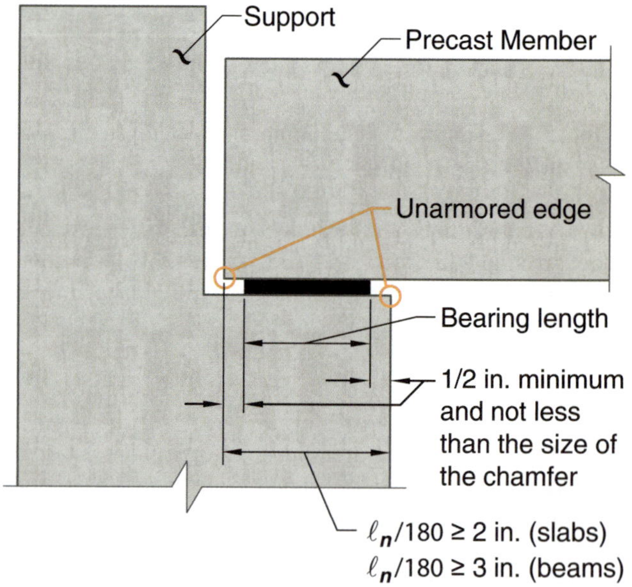

Fig. R16.2.6—Bearing length on support.

16.3—Connections to foundations

16.3.1 *General*

16.3.1.1 Factored forces and moments at base of columns, walls, or pedestals shall be transferred to supporting foundations by bearing on concrete and by reinforcement, dowels, anchor bolts, or mechanical connectors.

16.3.1.2 Reinforcement, dowels, or mechanical connectors between a supported member and foundation shall be designed to transfer (a) and (b):

R16.3—Connections to foundations

The requirements of 16.3.1 through 16.3.3 apply to both cast-in-place and precast construction. Additional requirements for cast-in-place construction are given in 16.3.4 and 16.3.5, while additional requirements for precast construction are given in 16.3.6.

CODE

(a) Compressive forces that exceed the lesser of the concrete bearing strengths of either the supported member or the foundation, calculated in accordance with 22.8
(b) Any calculated tensile force across the interface

16.3.1.3 At the base of a composite column with a structural steel core, (a) or (b) shall be satisfied:

(a) Base of structural steel section shall be designed to transfer the total factored forces from the entire composite member to the foundation.
(b) Base of structural steel section shall be designed to transfer the factored forces from the steel core only, and the remainder of the total factored forces shall be transferred to the foundation by compression in the concrete and by reinforcement.

16.3.2 *Required strength*

16.3.2.1 Factored forces and moments transferred to foundations shall be calculated in accordance with the factored load combinations in Chapter 5 and analysis procedures in Chapter 6.

16.3.3 *Design strength*

16.3.3.1 Design strengths of connections between columns, walls, or pedestals and foundations shall satisfy Eq. (16.3.3.1) for each applicable load combination. For connections between precast members and foundations, requirements for vertical integrity ties in 16.2.4.3 or 16.2.5.2 shall be satisfied.

$$\phi S_n \geq U \qquad (16.3.3.1)$$

where S_n is the nominal flexural, shear, axial, torsional, or bearing strength of the connection.

16.3.3.2 ϕ shall be determined in accordance with 21.2.

16.3.3.3 Combined moment and axial strength of connections shall be calculated in accordance with 22.4.

16.3.3.4 At the contact surface between a supported member and foundation, or between a supported member or foundation and an intermediate bearing element, nominal bearing strength B_n shall be calculated in accordance with 22.8 for concrete surfaces. B_n shall be the lesser of the nominal concrete bearing strengths for the supported member or foundation surface, and shall not exceed the strength of intermediate bearing elements, if present.

16.3.3.5 At the contact surface between supported member and foundation, V_n shall be calculated in accordance with the shear-friction provisions in 22.9 or by other appropriate means.

COMMENTARY

R16.3.3 *Design strength*

R16.3.3.4 In the common case of a column bearing on a footing, where the area of the footing is larger than the area of the column, the bearing strength should be checked at the base of the column and the top of the footing. In the absence of dowels or column reinforcement that continue into the foundation, the strength of the lower part of the column should be checked using the strength of the concrete alone.

R16.3.3.5 Shear-friction may be used to check for transfer of lateral forces to the supporting pedestal or footing. As an alternative to using shear-friction across a shear plane, shear keys may be used, provided that the reinforcement crossing the joint satisfies 16.3.4.1 for cast-in-place construction or 16.3.6.1 for precast construction. In precast construction, resistance to lateral forces may be provided by mechanical or welded connections.

16.3.3.6 At the base of a precast column, pedestal, or wall, anchor bolts and anchors for mechanical connections shall be designed in accordance with Chapter 17. Forces developed during erection shall be considered.

R16.3.3.6 Chapter 17 covers anchor design, including seismic design requirements. In precast concrete construction, erection considerations may control base connection design and need to be considered.

16.3.3.7 At the base of a precast column, pedestal, or wall, mechanical connectors shall be designed to reach their design strength before anchorage failure or failure of surrounding concrete.

16.3.4 *Minimum reinforcement for connections between cast-in-place members and foundation*

R16.3.4 *Minimum reinforcement for connections between cast-in-place members and foundation*

The Code requires a minimum amount of reinforcement between all supported and supporting members to ensure ductile behavior. This reinforcement is required to provide a degree of structural integrity during the construction stage and during the life of the structure.

16.3.4.1 For connections between a cast-in-place column or pedestal and foundation, A_s crossing the interface shall be at least $\mathbf{0.005}A_g$, where A_g is the gross area of the supported member.

R16.3.4.1 The minimum area of reinforcement at the base of a column may be provided by extending the longitudinal bars and anchoring them into the footing or by providing properly anchored dowels.

16.3.4.2 For connections between a cast-in-place wall and foundation, area of vertical reinforcement crossing the interface shall satisfy 11.6.1.

16.3.5 *Details for connections between cast-in-place members and foundation*

R16.3.5 *Details for connections between cast-in-place members and foundation*

16.3.5.1 At the base of a cast-in-place column, pedestal, or wall, reinforcement required to satisfy 16.3.3 and 16.3.4 shall be provided either by extending longitudinal bars into supporting foundation or by dowels.

16.3.5.2 Where continuity is required, splices and mechanical connectors for the longitudinal reinforcement or dowels shall satisfy 10.7.5 and, if applicable, 18.13.2.2.

16.3.5.3 If a pinned or rocker connection is used at the base of a cast-in-place column or pedestal, the connection to foundation shall satisfy 16.3.3.

16.3.5.4 At footings, compression lap splices of No. 14 and No. 18 bars that are in compression for all factored load combinations shall be permitted in accordance with 25.5.5.3.

R16.3.5.4 Satisfying 16.3.3.1 might require that each No. 14 or 18 bar be spliced in compression to more than one No. 11 or smaller dowel bar.

16.3.6 *Details for connections between precast members and foundation*

CODE

16.3.6.1 At the base of a precast column, pedestal, or wall, the connection to the foundation shall satisfy 16.2.4.3 or 16.2.5.2.

16.3.6.2 If the applicable load combinations of 16.3.3 result in no tension at the base of precast walls, vertical integrity ties required by 16.2.4.3(b) shall be permitted to be developed into an adequately reinforced concrete slab-on-ground.

16.4—Horizontal shear transfer in composite concrete flexural members

16.4.1 *General*

16.4.1.1 In a composite concrete flexural member, full transfer of horizontal shear forces shall be provided at contact surfaces of interconnected elements.

16.4.1.2 Where tension exists across any contact surface between interconnected concrete elements, horizontal shear transfer by contact shall be permitted only where transverse reinforcement is provided in accordance with 16.4.6 and 16.4.7.

16.4.1.3 Surface preparation assumed for design shall be specified in the construction documents.

16.4.2 *Required strength*

16.4.2.1 Factored forces transferred along the contact surface in composite concrete flexural members shall be calculated in accordance with the factored load combinations in Chapter 5.

16.4.2.2 Required strength shall be calculated in accordance with the analysis procedures in Chapter 6.

16.4.3 *Design strength*

16.4.3.1 Design strength for horizontal shear transfer shall satisfy Eq. (16.4.3.1) at all locations along the contact surface in a composite concrete flexural member, unless 16.4.5 is satisfied:

$$\phi V_{nh} \geq V_u \qquad (16.4.3.1)$$

where nominal horizontal shear strength V_{nh} is calculated in accordance with 16.4.4.

16.4.3.2 ϕ shall be determined in accordance with 21.2.

16.4.4 *Nominal horizontal shear strength*

COMMENTARY

R16.4—Horizontal shear transfer in composite concrete flexural members

R16.4.1 *General*

R16.4.1.1 Full transfer of horizontal shear forces between segments of composite members can be provided by horizontal shear strength at contact surfaces through interface shear, properly anchored ties, or both.

R16.4.1.3 Section 26.5.6 requires the licensed design professional to specify the surface preparation in the construction documents.

R16.4.4 *Nominal horizontal shear strength*

CODE

16.4.4.1 If $V_u > \phi(500b_vd)$, V_{nh} shall be taken as V_n calculated in accordance with 22.9, where b_v is the width of the contact surface, and d is in accordance with 16.4.4.3.

16.4.4.2 If $V_u \leq \phi(500b_vd)$, V_{nh} shall be calculated in accordance with Table 16.4.4.2, where $A_{v,min}$ is in accordance with 16.4.6, b_v is the width of the contact surface, and d is in accordance with 16.4.4.3.

Table 16.4.4.2—Nominal horizontal shear strength

<table>
<tr><th>Shear transfer reinforcement</th><th>Contact surface preparation[1]</th><th colspan="2">V_{nh}, lb</th><th></th></tr>
<tr><td rowspan="3">$A_v \geq A_{v,min}$</td><td rowspan="2">Concrete placed against hardened concrete intentionally roughened to a full amplitude of approximately 1/4 in.</td><td rowspan="2">Lesser of:</td><td>$\lambda\left(260+0.6\frac{A_v f_{yt}}{b_v s}\right)b_v d$</td><td>(a)</td></tr>
<tr><td>$500b_vd$</td><td>(b)</td></tr>
<tr><td>Concrete placed against hardened concrete not intentionally roughened</td><td colspan="2">$80b_vd$</td><td>(c)</td></tr>
<tr><td>Other cases</td><td>Concrete placed against hardened concrete intentionally roughened</td><td colspan="2">$80b_vd$</td><td>(d)</td></tr>
</table>

[1]Concrete contact surface shall be clean and free of laitance.

16.4.4.3 In Table 16.4.4.2, d shall be the distance from extreme compression fiber for the entire composite section to the centroid of prestressed and nonprestressed longitudinal tension reinforcement, if any, but need not be taken less than $0.80h$ for prestressed concrete members.

16.4.4.4 Transverse reinforcement in the previously cast concrete that extends into the cast-in-place concrete and is anchored on both sides of the interface shall be permitted to be included as ties for calculation of V_{nh}.

16.4.5 *Alternative method for calculating design horizontal shear strength*

16.4.5.1 As an alternative to 16.4.3.1, factored horizontal shear V_{uh} shall be calculated from the change in flexural compressive or tensile force in any segment of the composite concrete member, and Eq. (16.4.5.1) shall be satisfied at all locations along the contact surface:

$$\phi V_{nh} \geq V_{uh} \tag{16.4.5.1}$$

Nominal horizontal shear strength V_{nh} shall be calculated in accordance with 16.4.4.1 or 16.4.4.2, where area of contact surface shall be substituted for b_vd and V_{uh} shall be substituted for V_u. Provisions shall be made to transfer the change in compressive or tensile force as horizontal shear force across the interface.

16.4.5.2 Where shear transfer reinforcement is designed to resist horizontal shear to satisfy Eq. (16.4.5.1), the tie area to tie spacing ratio along the member shall approxi-

COMMENTARY

R16.4.4.2 The permitted horizontal shear strengths and the requirement of 1/4 in. amplitude for intentional roughness are based on tests discussed in Kaar et al. (1960), Saemann and Washa (1964), and Hanson (1960).

R16.4.4.3 In composite prestressed concrete members, the depth of the tension reinforcement may vary along the member. The definition of d used in Chapter 22 for determining the vertical shear strength is also appropriate for determining the horizontal shear strength.

R16.4.5 *Alternative method for calculating design horizontal shear strength*

R16.4.5.2 The distribution of horizontal shear stresses along the contact surface in a composite member will reflect the distribution of shear along the member. Horizontal

CODE

mately reflect the distribution of interface shear forces in the composite concrete flexural member.

16.4.5.3 Transverse reinforcement in a previously cast section that extends into the cast-in-place section and is anchored on both sides of the interface shall be permitted to be included as ties for calculation of V_{nh}.

16.4.6 *Minimum reinforcement for horizontal shear transfer*

16.4.6.1 Where shear transfer reinforcement is designed to resist horizontal shear, $A_{v,min}$ shall be the greater of (a) and (b):

(a) $0.75\sqrt{f_c'}\dfrac{b_w s}{f_y}$

(b) $50\dfrac{b_w s}{f_y}$

16.4.7 *Reinforcement detailing for horizontal shear transfer*

16.4.7.1 Shear transfer reinforcement shall consist of single bars or wire, multiple leg stirrups, or vertical legs of welded wire reinforcement.

16.4.7.2 Where shear transfer reinforcement is designed to resist horizontal shear, longitudinal spacing of shear transfer reinforcement shall not exceed the lesser of 24 in. and four times the least dimension of the supported element.

16.4.7.3 Shear transfer reinforcement shall be developed in interconnected elements in accordance with 25.7.1.

16.5—Brackets and corbels

16.5.1 *General*

COMMENTARY

shear failure will initiate where the horizontal shear stress is a maximum and will spread to regions of lower stress. Because the slip at peak horizontal shear resistance is small for a concrete-to-concrete contact surface, longitudinal redistribution of horizontal shear resistance is very limited. Therefore, the spacing of ties along the contact surface should provide horizontal shear resistance distributed approximately the same as the distribution of shear stress along the contact surface.

R16.4.6 *Minimum reinforcement for horizontal shear transfer*

R16.4.6.1 The requirements for minimum area of shear transfer reinforcement are based on test data given in Kaar et al. (1960), Saemann and Washa (1964), Hanson (1960), Grossfield and Birnstiel (1962), and Mast (1968).

R16.4.7 *Reinforcement detailing for horizontal shear transfer*

R16.4.7.3 Proper anchorage of ties extending across the interface is required to maintain contact along the interface.

R16.5—Brackets and corbels

R16.5.1 *General*

Brackets and corbels are short cantilevers that tend to act as simple trusses or deep beams, rather than beams, which are designed for shear according to 22.5. The corbel shown in Fig. R16.5.1a and Fig. 16.5.1b may fail by shearing along the interface between the column and the corbel, yielding of the tension tie, crushing or splitting of the compression strut, or localized bearing or shearing failure under the loading plate. These failure modes are illustrated and discussed in Elzanaty et al. (1986).

CODE

COMMENTARY

The method of design addressed in this section has only been validated experimentally for $a_v/d \leq 1.0$. In addition, an upper limit is provided for N_{uc} because this method of design has only been validated experimentally for $N_{uc} \leq V_u$.

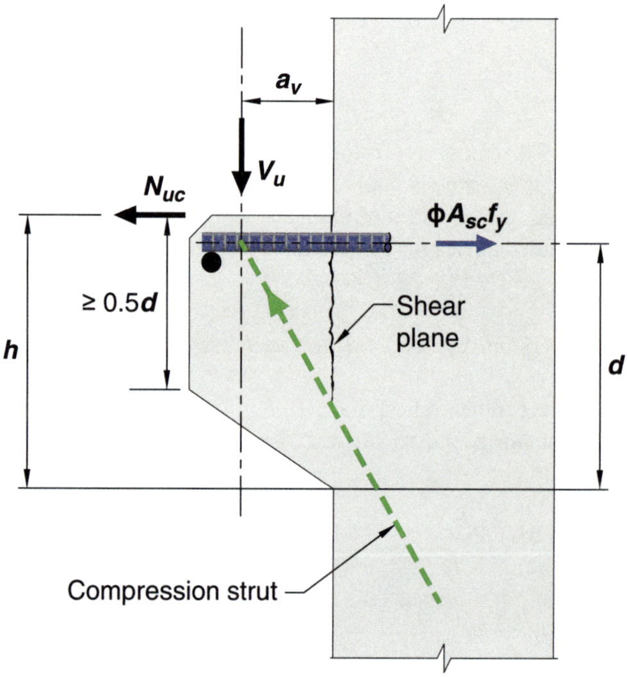

Fig. R16.5.1a—Structural action of a corbel.

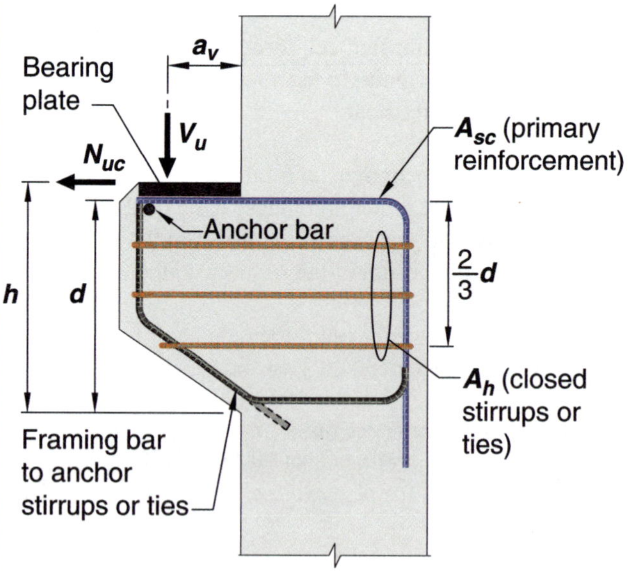

Fig. R16.5.1b—Notation used in Section 16.5.

16.5.1.1 Brackets and corbels with shear span-to-depth ratio $a_v/d \leq 1.0$ and with factored restraint force $N_{uc} \leq V_u$ shall be permitted to be designed in accordance with 16.5.

R16.5.1.1 Design of brackets and corbels in accordance with Chapter 23 is permitted, regardless of shear span.

CODE

COMMENTARY

16.5.2 *Dimensional limits*

16.5.2.1 Effective depth d for a bracket or corbel shall be calculated at the face of the support.

16.5.2.2 Overall depth of bracket or corbel at the outside edge of the bearing area shall be at least $0.5d$.

16.5.2.3 No part of the bearing area on a bracket or corbel shall project farther from the face of support than (a) or (b):

(a) End of the straight portion of the primary tension reinforcement
(b) Interior face of the transverse anchor bar, if one is provided

16.5.2.4 For normalweight concrete, the bracket or corbel dimensions shall be selected such that V_u/ϕ shall not exceed the least of (a) through (c):

(a) $0.2f_c'b_wd$
(b) $(480 + 0.08f_c')b_wd$
(c) $1600b_wd$

16.5.2.5 For lightweight concrete, the bracket or corbel dimensions shall be selected such that V_u/ϕ shall not exceed the lesser of (a) and (b):

(a) $\left(0.2-0.07\dfrac{a_v}{d}\right)f_c'b_wd$

(b) $\left(800-280\dfrac{a_v}{d}\right)b_wd$

R16.5.2 *Dimensional limits*

R16.5.2.2 A minimum depth, as shown in Fig. R16.5.1a and R16.5.1b, is required at the outside edge of the bearing area so that a premature failure will not occur due to a major crack propagating from below the bearing area to the sloping face of the corbel or bracket. Failures of this type have been observed (Kriz and Raths 1965) in corbels having depths at the outside edge of the bearing area less than required in 16.5.2.2.

R16.5.2.3 The restriction on the location of the bearing area is necessary to ensure development of the specified yield strength of the primary tension reinforcement near the load.

If the corbel is designed to resist restraint force N_{uc}, a bearing plate should be provided and fully anchored to the primary tension reinforcement (Fig. R16.5.1b).

R16.5.2.4 These limits impose dimensional restrictions on brackets and corbels necessary to comply with the maximum shear friction strength allowed on the critical section at the face of support.

R16.5.2.5 Tests (Mattock et al. 1976a) have shown that the maximum shear friction strength of lightweight concrete brackets and corbels is a function of both f_c' and a_v/d.

16.5.3 *Required strength*

16.5.3.1 The section at the face of the support shall be designed to resist simultaneously the factored shear V_u, the factored restraint force N_{uc}, and the factored moment M_u.

16.5.3.2 Factored restraint force, N_{uc}, and shear, V_u, shall be the maximum values calculated in accordance with the factored load combinations in Chapter 5. It shall be permitted to calculate N_{uc} in accordance with 16.2.2.3 or 16.2.2.4, as appropriate.

R16.5.3 *Required strength*

R16.5.3.1 Figure R16.5.1b shows the forces applied to the corbel. M_u can be calculated as $[V_ua_v + N_{uc}(h - d)]$.

R16.5.3.2 In editions of the Code prior to ACI 318-19, specific provisions for restraint forces at bearing connections were included only for corbels and brackets. In 2019, 16.2.2.3 and 16.2.2.4 were added to include consideration of restraint forces at all bearing connections. Consequently the provisions applicable only to brackets or corbels were removed and a reference made to 16.2.2.3 or 16.2.2.4.

CODE

16.5.3.3 Required strength shall be calculated in accordance with the analysis procedures in Chapter 6, and the requirements in this section.

16.5.4 *Design strength*

16.5.4.1 Design strength at all sections shall satisfy $\phi S_n \geq U$, including (a) through (c). Interaction between load effects shall be considered.

(a) $\phi N_n \geq N_{uc}$
(b) $\phi V_n \geq V_u$
(c) $\phi M_n \geq M_u$

16.5.4.2 ϕ shall be determined in accordance with 21.2.

16.5.4.3 Nominal tensile strength N_n provided by A_n shall be calculated by

$$N_n = A_n f_y \quad (16.5.4.3)$$

16.5.4.4 Nominal shear strength V_n provided by A_{vf} shall be calculated in accordance with provisions for shear-friction in 22.9, where A_{vf} is the area of reinforcement that crosses the assumed shear plane.

16.5.4.5 Nominal flexural strength M_n provided by A_f shall be calculated in accordance with the design assumptions in 22.2.

16.5.5 *Reinforcement limits*

16.5.5.1 Area of primary tension reinforcement, A_{sc}, shall be at least the greatest of (a) through (c):

(a) $A_f + A_n$
(b) $(2/3)A_{vf} + A_n$
(c) $0.04(f_c'/f_y)(b_w d)$

16.5.5.2 Total area of closed stirrups or ties parallel to primary tension reinforcement, A_h, shall be at least:

$$A_h = 0.5(A_{sc} - A_n) \quad (16.5.5.2)$$

COMMENTARY

R16.5.5 *Reinforcement limits*

R16.5.5.1 Test results (Mattock et al. 1976a) indicate that the total amount of primary tension reinforcement, A_{sc}, required to cross the face of the support should be the greatest of:

(a) The sum of the amount of reinforcement needed to resist demands from flexure, A_f, plus the amount of reinforcement needed to resist the axial force, A_n, as determined by 16.5.4.3.
(b) The sum of two-thirds of the total required shear friction reinforcement, A_{vf}, as determined by 16.5.4.4, plus the amount of reinforcement needed to resist the axial force, A_n, determined by 16.5.4.3. The remaining $A_{vf}/3$ should be provided as closed stirrups parallel to A_{sc} as required by 16.5.5.2.
(c) A minimum amount of reinforcement, multiplied by the ratio of concrete strength to steel strength. This amount is required to prevent the possibility of sudden failure should the bracket or corbel crack under the action of flexure and outward tensile force.

R16.5.5.2 Closed stirrups parallel to the primary tension reinforcement are necessary to prevent a premature diagonal tension failure of the corbel or bracket. Distribution of A_h is required to be in accordance with 16.5.6.6. The total amount

CODE

COMMENTARY

of reinforcement required to cross the face of the support, as shown in Fig. R16.5.1b, is the sum of A_{sc} and A_h.

16.5.6 *Reinforcement detailing*

R16.5.6 *Reinforcement detailing*

16.5.6.1 Concrete cover shall be in accordance with 20.5.1.3.

16.5.6.2 Minimum spacing for deformed reinforcement shall be in accordance with 25.2.

16.5.6.3 At the front face of a bracket or corbel, primary tension reinforcement shall be anchored by (a), (b), or (c):

(a) A weld to a transverse bar of at least equal size that is designed to develop f_y of primary tension reinforcement
(b) Bending the primary tension reinforcement back to form a horizontal loop
(c) Other means of anchorage that develops f_y

R16.5.6.3 For brackets and corbels of variable depth (refer to Fig. R16.5.1a), the stress at ultimate in the reinforcement is almost constant at approximately f_y from the face of support to the load point. This is because the horizontal component of the inclined concrete compression strut is transferred to the primary tension reinforcement at the location of the vertical load. Therefore, reinforcement should be fully anchored at its outer end (refer to 16.5.6.3) and in the supporting column (refer to 16.5.6.4), so as to be able to develop its specified yield strength from the face of support to the vertical load (refer to Fig. R16.5.6.3a). Satisfactory anchorage at the outer end can be obtained by bending the primary tension reinforcement bars in a horizontal loop as specified in 16.5.6.3b, or by welding a bar of equal diameter or a suitably sized angle across the ends of the primary tension reinforcement bars. The weld detail used successfully in the corbel tests reported in Mattock et al. (1976a) is shown in Fig. R16.5.6.3b. Refer to ACI Committee 408 (1966).

An end hook in the vertical plane, with the minimum diameter bend, is not totally effective because a zone of unreinforced concrete beneath the point of loading will exist for loads applied close to the end of the bracket or corbel. For wide brackets (perpendicular to the plane of the figure) and loads not applied close to the end, U-shaped bars in a horizontal plane provide effective end hooks.

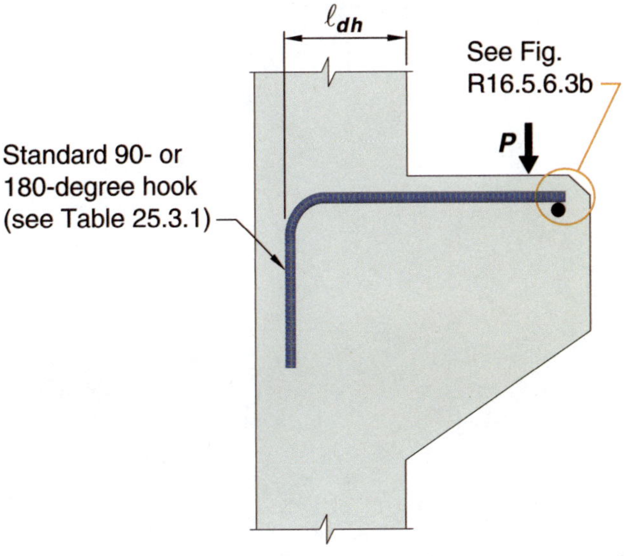

Fig. R16.5.6.3a—Member largely dependent on support and end anchorages.

CODE

COMMENTARY

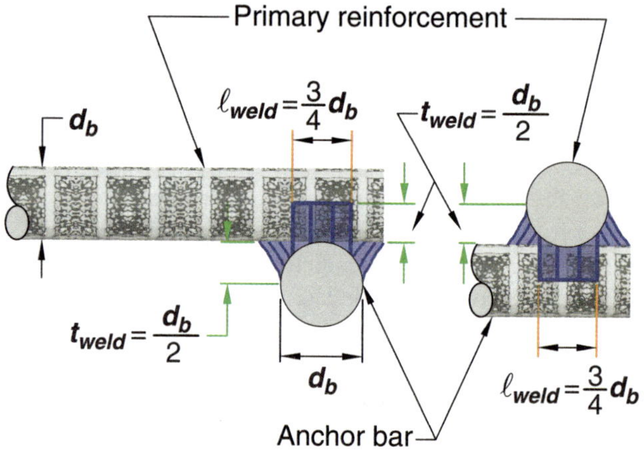

Fig. R16.5.6.3b—Weld details used in tests of Mattock et al. (1976a).

16.5.6.4 Primary tension reinforcement shall be developed at the face of the support.

16.5.6.5 Development of tension reinforcement shall account for distribution of stress in reinforcement that is not directly proportional to the bending moment.

R16.5.6.5 Calculated stress in reinforcement at service loads, f_s, does not decrease linearly in proportion to a decreasing moment in brackets, corbels, and members of variable depth. Additional consideration is required for proper development of the flexural reinforcement.

16.5.6.6 Closed stirrups or ties shall be spaced such that A_h is uniformly distributed within **(2/3)*d*** measured from the primary tension reinforcement.

R16.5.6.6 Refer to R16.5.5.2.

CHAPTER 17—ANCHORING TO CONCRETE

17.1—Scope

17.1.1 This chapter shall apply to the design of anchors in concrete used to transmit loads by means of tension, shear, or a combination of tension and shear between: (a) connected structural elements; or (b) safety-related attachments and structural elements. Safety levels specified are intended for in-service conditions rather than for short-term handling and construction conditions.

17.1.2 Provisions of this chapter shall apply to the following anchor types (a) through (g):

(a) Headed studs and headed bolts having a geometry that has been demonstrated to result in a pullout strength in uncracked concrete equal to or exceeding $\mathbf{1.4}N_p$, where N_p is given in Eq. (17.6.3.2.2a).
(b) Hooked bolts having a geometry that has been demonstrated to result in a pullout strength without the benefit of friction in uncracked concrete equal to or exceeding $\mathbf{1.4}N_p$, where N_p is given in Eq. (17.6.3.2.2b)
(c) Post-installed expansion (torque-controlled and displacement-controlled) anchors that meet the assessment criteria of ACI 355.2.
(d) Post-installed undercut anchors that meet the assessment criteria of ACI 355.2.
(e) Post-installed adhesive anchors that meet the assessment criteria of ACI 355.4.
(f) Post-installed screw anchors that meet the assessment criteria of ACI 355.2.
(g) Attachments with shear lugs.

17.1.3 The removal and resetting of post-installed mechanical anchors is prohibited.

17.1.4 This chapter does not apply for load applications that are predominantly high-cycle fatigue or due to impact.

COMMENTARY

R17.1—Scope

R17.1.1 This chapter is restricted in scope to structural anchors that transmit loads related to strength, stability, or life safety. Two types of applications are envisioned. The first is connections between structural elements where the failure of an anchor or anchor group could result in loss of equilibrium or stability of any portion of the structure. The second is where safety-related attachments that are not part of the structure (such as sprinkler systems, heavy suspended pipes, or barrier rails) are attached to structural elements. The levels of safety defined by the factored load combinations and ϕ-factors are appropriate for structural applications. Other standards may require more stringent safety levels during temporary handling.

The format for this chapter was revised in 2019 to be more consistent with the other chapters of this Code.

R17.1.2 Typical cast-in headed studs and headed bolts with head geometries consistent with ASME B1.1, B18.2.1, and B18.2.6 have been tested and proven to behave predictably; therefore, calculated pullout strengths are acceptable.

Post-installed expansion, screw, and undercut anchors do not have predictable pullout strengths, and therefore qualification tests to establish the pullout strengths according to ACI 355.2 are required. For post-installed expansion, screw, and undercut anchors to be used in conjunction with the requirements of this chapter, the results of the ACI 355.2 tests have to indicate that pullout failures exhibit acceptable load-displacement characteristics or that pullout failures are precluded by another failure mode.

For adhesive anchors, the characteristic bond stress and suitability for structural applications are established by testing in accordance with ACI 355.4. Adhesive anchors are particularly sensitive to a number of factors including installation direction and load type. If adhesive anchors are used to resist sustained tension, the provisions include testing requirements for horizontal or upwardly inclined installations in 17.2.3, design requirements in 17.5.2.2, certification requirements in 26.7, and inspection requirements in 26.13. Adhesive anchors qualified in accordance with ACI 355.4 are tested in concrete with compressive strengths within two ranges: 2500 to 4000 psi and 6500 to 8500 psi. Bond strength is, in general, not highly sensitive to concrete compressive strength.

R17.1.3 ACI 355.2 prohibits reuse of post-installed mechanical anchors.

R17.1.4 The exclusion of load applications producing high-cycle fatigue or extremely short duration impact (such as blast or shock wave) from the scope of this chapter is not meant to exclude earthquake loads. Section 17.10 presents additional requirements for design when earthquake loads are included.

CODE

17.1.5 This chapter does not apply to specialty inserts, through-bolts, multiple anchors connected to a single steel plate at the embedded end of the anchors, grouted anchors, or power driven anchors such as powder or pneumatic actuated fasteners.

17.1.6 Reinforcement used as part of an embedment shall have development length established in accordance with other parts of this Code. If reinforcement is used as anchorage, concrete breakout failure shall be considered. Alternatively, anchor reinforcement in accordance with 17.5.2.1 shall be provided.

17.2—General

17.2.1 Anchors and anchor groups shall be designed for critical effects of factored loads calculated by elastic analysis. If nominal strength is controlled by ductile steel elements, plastic analysis is permitted provided that deformation compatibility is taken into account.

17.2.1.1 Anchor group effects shall be considered if two or more anchors loaded by a common structural element are spaced closer than the spacing required for unreduced breakout strength. If adjacent anchors are not loaded by a common structural element, group effects shall consider simultaneous maximum loading of adjacent anchors.

17.2.2 Adhesive anchors shall be installed in concrete having a minimum age of 21 days at time of anchor installation.

17.2.3 Adhesive anchors installed horizontally or upwardly inclined shall be qualified in accordance with ACI 355.4 requirements for sensitivity to installation direction.

17.2.4 *Lightweight concrete modification factor, λ_a*

17.2.4.1 Modification factor λ_a for lightweight concrete shall be in accordance with Table 17.2.4.1. It shall be

COMMENTARY

R17.1.5 The wide variety of shapes and configurations of specialty inserts precludes prescription of generalized tests and design equations.

R17.1.6 Concrete breakout strength in tension and shear should be considered for reinforcing bars in a group used as anchorage. Concrete breakout behavior can occur even if reinforcement is fully developed in accordance with Chapter 25. Breakout behavior of straight reinforcement as a group is analogous to tension and shear breakout behavior of adhesive anchors whereby h_{ef} is taken as equal to or less than the embedded bar length. Similarly, breakout behavior of hooked and headed reinforcement groups is similar to tension and shear breakout behavior of headed anchors. Consideration should be given to extending bars beyond the development length.

As an alternative to explicit determination of the concrete breakout strength of a group, anchor reinforcement provided in accordance with 17.5.2.1 may be used, or the reinforcement should be extended.

R17.2—General

R17.2.1 If the strength of an anchor group is governed by concrete breakout, the behavior is brittle, and there is limited redistribution of forces between the highly stressed and less stressed anchors. In this case, the theory of elasticity is required to be used, assuming the attachment that distributes loads to the anchors is sufficiently stiff. The forces in the anchors are considered to be proportional to the external load and its distance from the neutral axis of the anchor group.

If anchor strength is governed by ductile yielding of the anchor steel, significant redistribution of anchor forces can occur. In this case, an analysis based on the theory of elasticity will be conservative. Cook and Klingner (1992a,b) and Lotze et al. (2001) discuss nonlinear analysis, using theory of plasticity, for the determination of the strengths of ductile anchor groups.

R17.2.2 The design performance of adhesive anchors cannot be ensured by establishing a minimum concrete compressive strength at the time of installation in early-age concrete. Therefore, a concrete age of at least 21 days at the time of adhesive anchor installation was adopted.

R17.2.3 ACI 355.4 includes optional tests to confirm the suitability of adhesive anchors for horizontal or upwardly inclined installations.

R17.2.4 *Lightweight concrete modification factor, λ_a*

R17.2.4.1 The number of tests available to establish the strength of anchors in lightweight concrete is limited. Tests of headed studs cast in lightweight concrete indicate that the

CODE

permitted to use an alternate value of λ_a if tests are performed and evaluated in accordance with ACI 355.2 or ACI 355.4.

Table 17.2.4.1—Modification factor λ_a for lightweight concrete

Case	λ_a[1]
Cast-in and undercut anchor concrete failure	1.0λ
Expansion, screw, and adhesive anchor concrete failure	0.8λ
Adhesive anchor bond failure per Eq. (17.6.5.2.1)	0.6λ

[1]λ shall be in accordance with 19.2.4

17.2.5 Anchors shall be installed and inspected in accordance with the requirements of 26.7 and 26.13.

17.3—Design Limits

17.3.1 The value of f_c' used for calculation purposes in this chapter shall not exceed 10,000 psi for cast-in anchors and 8000 psi for post-installed anchors. Post-installed anchors shall not be used in concrete with a strength greater than 8000 psi without testing to verify acceptable performance.

17.3.2 For anchors with diameters $d_a \leq 4$ **in.**, concrete breakout strength requirements shall be considered satisfied by the design procedures of 17.6.2 and 17.7.2.

COMMENTARY

present reduction factor λ adequately represents the influence of lightweight concrete (Shaikh and Yi 1985; Anderson and Meinheit 2005). Anchor manufacturer data developed for evaluation reports on post-installed expansion, screw, undercut, and adhesive anchors indicate that a reduced λ is needed to provide the necessary safety factor for the respective design strength. ACI 355.2 and ACI 355.4 provide procedures whereby a specific value of λ_a can be used based on testing, assuming the lightweight concrete is similar to the reference test material.

R17.3—Design limits

R17.3.1 A limited number of tests of cast-in and post-installed anchors in high-strength concrete (Primavera et al. 1997) indicate that the design procedures contained in this chapter become unconservative with increasing concrete strength, particularly for cast-in anchors in concrete with compressive strengths in the range of 11,000 to 12,000 psi. Until further tests are available, an upper limit on f_c' of 10,000 psi has been imposed for the design of cast-in anchors. This limitation is consistent with those for shear strength, torsion strength, and reinforcement development length in this Code (22.5.3.1, 22.6.3.1, 22.7.2.1, 25.4.1.4). For some post-installed anchors, the capacity may be negatively affected by very high-strength concrete. These effects are associated with difficulty in fully expanding expansion anchors, cutting grooves in the sidewall of the predrilled hole by the screw anchor's threads, and reduced bond strength of adhesive anchors. The 8000 psi limit for post-installed anchors reflects the current concrete strength range for testing specified in ACI 355.2 and ACI 355.4. The 8000 psi limit may be exceeded if verified with tests.

R17.3.2 The limitation on anchor diameter is based on the current range of test data. In the 2002 through 2008 editions of the Code, there were limitations on the diameter and embedment of anchors to calculate the concrete breakout strength. These limitations were necessitated by the lack of test results on anchors with diameters larger than 2 in. and embedment lengths longer than 24 in. In 2011, limitations on anchor diameter and embedment length were revised to limit the diameter to 4 in. based on the results of tension and shear tests on large-diameter anchors with deep embedments (Lee et al. 2007, 2010). These tests included 4.25 in. diameter anchors, embedded 45 in., tested in tension and 3 in. diameter anchors tested in shear. The 4 in. diameter limit was selected to maintain consistency with the largest diameter anchor permitted in ASTM F1554. Other ASTM specifications permit up to 8 in. diameter anchors; however, they have not been tested to ensure applicability of the 17.6.2 and 17.7.2 concrete breakout provisions.

CODE

17.3.3 For adhesive anchors with embedment depths $4d_a \leq h_{ef} \leq 20d_a$, bond strength requirements shall be considered satisfied by the design procedure of 17.6.5.

17.3.4 For screw anchors with embedment depths $5d_a \leq h_{ef} \leq 10d_a$, and $h_{ef} \geq$ **1.5 in.**, concrete breakout strength requirements shall be considered satisfied by the design procedures of 17.6.2 and 17.7.2.

17.3.5 Anchors shall satisfy the edge distances, spacings, and thicknesses in 17.9 unless supplementary reinforcement is provided to control splitting failure.

17.4—Required strength

17.4.1 Required strength shall be calculated in accordance with the factored load combinations in Chapter 5.

17.4.2 For anchors in structures assigned to SDC C, D, E, and F, the additional requirements of 17.10 shall apply.

17.5—Design strength

17.5.1 For each applicable factored load combination, design strength of individual anchors and anchor groups shall satisfy $\phi S_n \geq U$. Interaction between load effects shall be considered in accordance with 17.8.1.

17.5.1.1 Strength reduction factor, ϕ, shall be determined in accordance with 17.5.3.

COMMENTARY

R17.3.3 ACI 355.4 limits the embedment depth of adhesive anchors to $4d_a \leq h_{ef} \leq 20d_a$, which represents the theoretical limits of the bond model (Eligehausen et al. 2006a).

R17.3.4 Screw anchor research by Olsen et al. (2012) is based on the nominal screw anchor diameter corresponding to the nominal drill bit size (for example a 5/8 in. screw anchor installs in a hole drilled by a 5/8 in. ANSI drill bit). This definition of screw anchor size is approximately the diameter of the core or shank of the screw rather than the size of the larger external diameter of the thread. This definition differs from the diameter of standard anchors with ASME B1.1 threads that have a reduced shaft area and smaller effective area. The effective area of the screw anchor, as with other post-installed mechanical anchors, is provided by the manufacturer.

The Olsen et al. (2012) empirical design model was derived from a database of tests in cracked and uncracked concrete on metric-sized screw anchors tested in Europe and inch-sized anchors tested by independent laboratories in accordance with ICC-ES AC193.

For concrete screw anchors, the effective embedment depth, h_{ef}, is determined as a reduction from the nominal embedment based on geometric characteristics of the screw. The effective embedment is verified during the qualification testing under ACI 355.2 and provided by the manufacturer for use in design. Using the reduced, effective embedment depth with the concrete capacity design (CCD) method is shown to adequately represent the behavior of concrete screws in the current concrete screw database and also validates the effects and limitations of certain relevant parameters, such as the effective embedment depth and spacing of anchors (17.9).

R17.5—Design strength

CODE

17.5.1.2 Nominal strength for an anchor or anchor groups shall be based on design models that result in predictions of strength in substantial agreement with results of comprehensive tests. The materials used in the tests shall be compatible with the materials used in the structure. The nominal strength shall be based on the 5 percent fractile of the basic individual anchor strength. For nominal strengths related to concrete strength, modifications for size effects, number of anchors, effects of close spacing of anchors, proximity to edges, depth of the concrete member, eccentric loadings of anchor groups, and influence of cracking shall be taken into account. Limits on edge distance and anchor spacing in the design models shall be consistent with the tests that verified the model. Strength of anchors shall be based on design models that satisfy 17.5.1.2 for the following:

(a) Steel strength of anchor in tension
(b) Concrete breakout strength of anchor in tension
(c) Pullout strength of a single cast-in anchor and single post-installed expansion, screw, and undercut anchor in tension
(d) Concrete side-face blowout strength of headed anchor in tension
(e) Bond strength of adhesive anchor in tension
(f) Steel strength of anchor in shear
(g) Concrete breakout strength of anchor in shear
(h) Concrete pryout strength of anchor in shear

COMMENTARY

R17.5.1.2 This section provides requirements for establishing the strength of anchors in concrete. The various types of steel and concrete failure modes for anchors are shown in Fig. R17.5.1.2(a) and R17.5.1.2(b). Comprehensive discussions of anchor failure modes are included in CEB (1997), Fuchs et al. (1995), Eligehausen and Balogh (1995), and Cook et al. (1998). Tension failure modes related to concrete include concrete breakout failure (applicable to all anchor types), pullout failure (applicable to cast-in anchors, post-installed expansion, screw, and undercut anchors), side-face blowout failure (applicable to headed anchors), and bond failure (applicable to adhesive anchors). Shear failure modes related to concrete include concrete breakout failure and concrete pryout (applicable to all anchor types). These failure modes are described in the deemed-to-comply provisions of 17.6.2, 17.6.3, 17.6.4, 17.6.5, 17.7.2, and 17.7.3.

Any model that complies with the requirements of 17.5.1.2 and 17.5.2.3 can be used to establish the concrete-related strengths. Additionally, anchor tensile and shear strengths are limited by the minimum spacings and edge distances of 17.9 to preclude splitting. The design of post-installed anchors recognizes that the strength of anchors is sensitive to appropriate installation; installation requirements are included in Chapter 26. Some post-installed anchors are less sensitive to installation errors and tolerances. This is reflected in various ϕ-factors given in 17.5.3 and based on the assessment criteria of ACI 355.2 and ACI 355.4.

The breakout strength of an unreinforced connection can be taken as an indication of the load at which significant cracking will occur. Such cracking can represent a serviceability problem if not controlled (refer to R17.7.2.1).

CODE | COMMENTARY

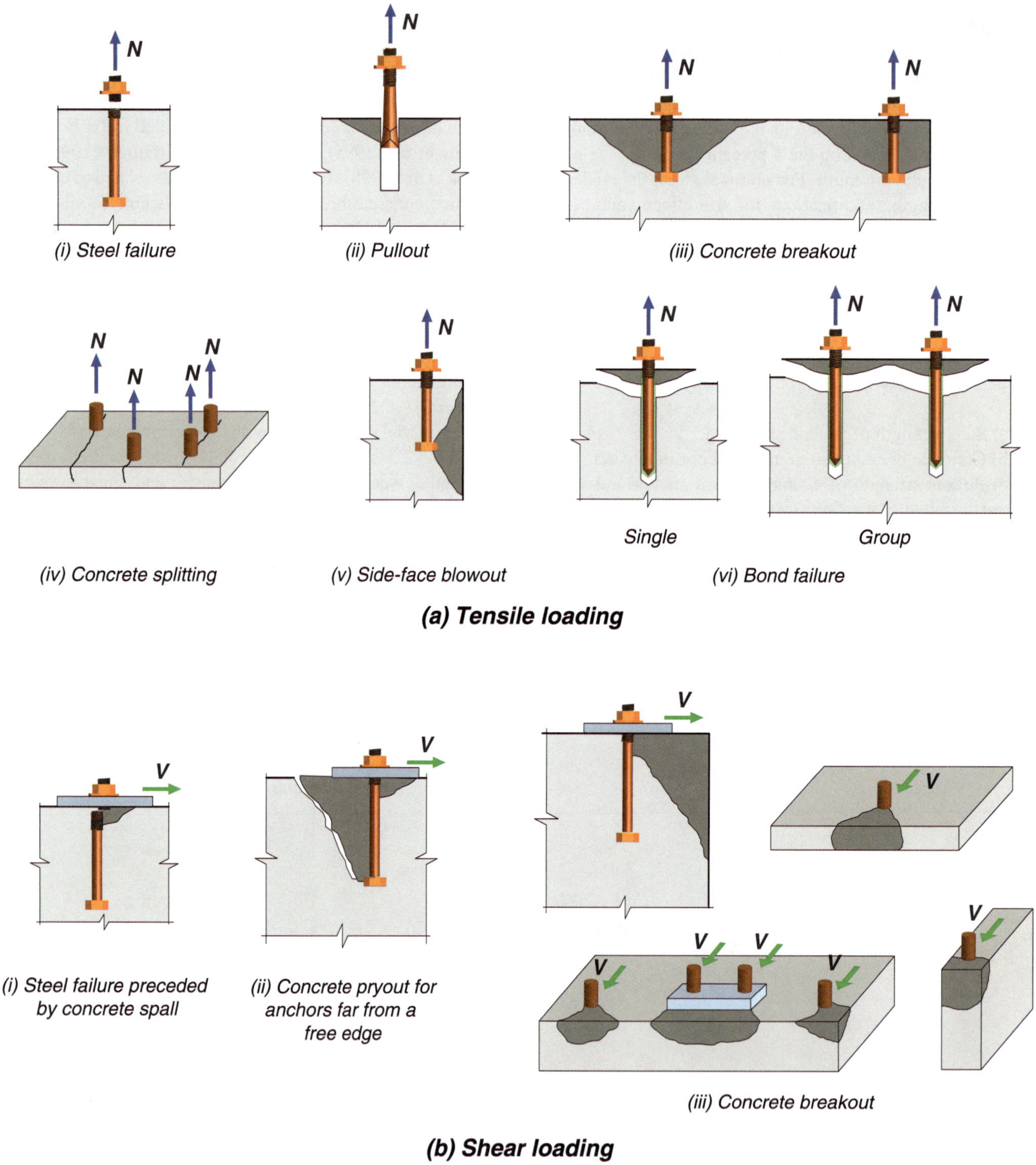

Fig. R17.5.1.2—Failure modes for anchors.

17.5.1.3 Strength of anchors shall be permitted to be determined in accordance with 17.6 for 17.5.1.2(a) through (e), and 17.7 for 17.5.1.2(f) through (h). For adhesive anchors that resist sustained tension, the requirements of 17.5.2.2 shall apply.

R17.5.1.3 The method for concrete breakout design deemed to comply with the requirements of 17.5.1.2 was developed from the concrete capacity design (CCD) Method (Fuchs et al. (1995); Eligehausen and Balogh (1995), which was an adaptation of the Kappa Method (Eligehausen and Fuchs 1988; Eligehausen et al. 2006a) with a breakout failure surface angle of approximately 35 degrees (Fig.

CODE

17.5.1.3.1 Anchor group effects shall be considered wherever two or more anchors have spacing less than the critical spacing in Table 17.5.1.3.1, where only those anchors susceptible to the particular failure mode under investigation shall be included in the group.

Table 17.5.1.3.1—Critical spacing

Failure mode under investigation	Critical spacing
Concrete breakout in tension	$3h_{ef}$
Bond strength in tension	$2c_{Na}$
Concrete breakout in shear	$3c_{a1}$

17.5.1.4 Strength of anchors shall be permitted to be based on test evaluation using the 5 percent fractile of applicable test results for 17.5.1.2 (a) through (h).

COMMENTARY

17.5.1.3a and b). It is considered to be sufficiently accurate, relatively easy to apply, and capable of extension to irregular layouts. The CCD Method predicts the strength of an anchor or anchor group by using a basic equation for tension in cracked concrete, which is multiplied by factors that account for the number of anchors, edge distance, spacing, eccentricity, and absence of cracking. For shear, a similar approach is used. Experimental and numerical investigations have demonstrated the applicability of the CCD Method to adhesive anchors as well (Eligehausen et al. 2006a).

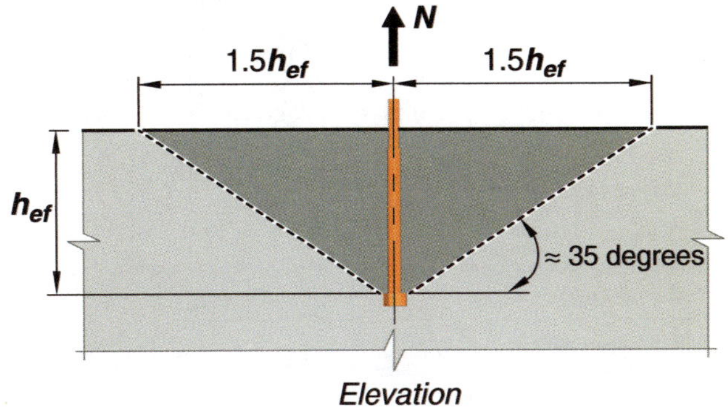

Fig. R17.5.1.3a—Breakout cone for tension.

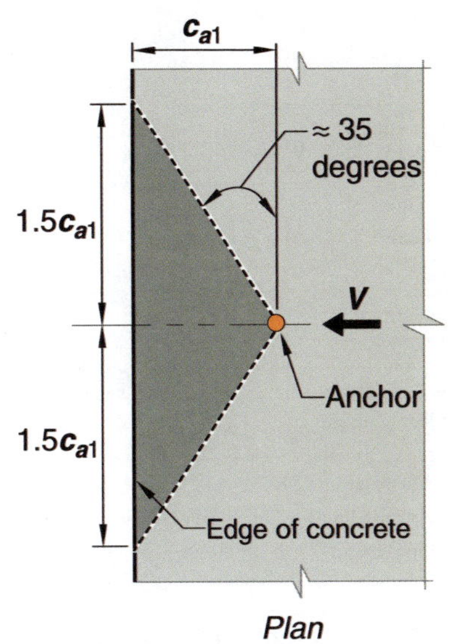

Fig. R17.5.1.3b—Breakout cone for shear.

R17.5.1.4 Sections 17.5.1.2 and 17.5.2.3 establish the performance factors for which anchor design models are required to be verified. Many possible design approaches exist, and the user is always permitted to "design by test" using 17.5.1.4 as long as sufficient data are available to verify the model. Test procedures can be used to determine the single-anchor breakout strength in tension and in shear. The test results, however, are required to be evaluated on a basis statistically equivalent to that used to select the values for the concrete breakout method considered to satisfy provisions of 17.5.1.2. The basic strength cannot be taken

CODE

17.5.2 For each applicable factored load combination, design strength of anchors shall satisfy the criteria in Table 17.5.2.

Table 17.5.2—Design strength requirements of anchors

Failure mode	Single anchor	Anchor group[1]: Individual anchor in a group	Anchor group[1]: Anchors as a group
Steel strength in tension (17.6.1)[2]	$\phi N_{sa} \geq N_{ua}$	$\phi N_{sa} \geq N_{ua,i}$	
Concrete breakout strength in tension[3] (17.6.2)	$\phi N_{cb} \geq N_{ua}$		$\phi N_{cbg} \geq N_{ua,g}$
Pullout strength in tension (17.6.3)	$\phi N_{pn} \geq N_{ua}$	$\phi N_{pn} \geq N_{ua,i}$	
Concrete side-face blowout strength in tension (17.6.4)	$\phi N_{sb} \geq N_{ua}$		$\phi N_{sbg} \geq N_{ua,g}$
Bond strength of adhesive anchor in tension (17.6.5)	$\phi N_{a} \geq N_{ua}$		$\phi N_{ag} \geq N_{ua,g}$
Steel strength in shear (17.7.1)	$\phi V_{sa} \geq V_{ua}$	$\phi V_{sa} \geq V_{ua,i}$	
Concrete breakout strength in shear[3] (17.7.2)	$\phi V_{cb} \geq V_{ua}$		$\phi V_{cbg} \geq V_{ua,g}$
Concrete pryout strength in shear (17.7.3)	$\phi V_{cp} \geq V_{ua}$		$\phi V_{cpg} \geq V_{ua,g}$

[1]Design strengths for steel and pullout failure modes shall be calculated for the most highly stressed anchor in the group.

[2]Sections referenced in parentheses are pointers to models that are permitted to be used to evaluate the nominal strengths.

[3]If anchor reinforcement is provided in accordance with 17.5.2.1, the design strength of the anchor reinforcement shall be permitted to be used instead of the concrete breakout strength

17.5.2.1 The design strength of anchor reinforcement shall be permitted to be used instead of the concrete breakout strength if (a) or (b) is satisfied.

(a) For tension, if anchor reinforcement is developed in accordance with Chapter 25 on both sides of the concrete breakout surface

(b) For shear, if anchor reinforcement is developed in accordance with Chapter 25 on both sides of the concrete breakout surface, or encloses and contacts the anchor and is developed beyond the breakout surface.

17.5.2.1.1 Strength reduction factor ϕ for anchor reinforcement shall be in accordance with 17.5.3.

COMMENTARY

greater than the 5 percent fractile. The number of tests has to be sufficient for statistical validity and should be considered in the determination of the 5 percent fractile.

R17.5.2 Under combined tension and bending, individual anchors in a group may be required to resist different magnitudes of tensile force. Similarly, under combined shear and torsion, individual anchors in a group may be required to resist different magnitudes of shear. Table 17.5.2 includes requirements to design single anchors and individual anchors in a group to safeguard against all potential failure modes. For steel and pullout failure modes, the most highly stressed anchor in the group should be checked to ensure it has sufficient strength to resist its required load. For concrete breakout, the anchors should be checked as a group. Elastic analysis or plastic analysis of ductile anchors as described in 17.2.1 may be used to determine the loads resisted by each anchor.

The addition of reinforcement in the direction of the load to restrain concrete breakout can enhance the strength and deformation capacity of the anchor connection. Such enhancement is practical with cast-in anchors such as those used in precast sections. Klingner et al. (1982), *fib* (2011), ACI 349, and Eligehausen et al. (2006b) provide information regarding the effect of reinforcement on the behavior of anchors. The effect of reinforcement is not included in the ACI 355.2 and ACI 355.4 anchor acceptance tests or in the concrete breakout calculation method of 17.6.2 and 17.7.2. Anchor reinforcement may be provided in accordance with 17.5.2.1 and developed according to Chapter 25 instead of calculating breakout strength.

R17.5.2.1 For conditions where the factored tensile or shear force exceeds the concrete breakout strength of the anchor(s) or if the breakout strength is not evaluated, the nominal strength can be based on properly developed anchor reinforcement as illustrated in Fig. R17.5.2.1a for tension and Fig. R17.5.2.1b(i) and Fig. R17.5.2.1b(ii) for shear. Because anchor reinforcement is placed below where the shear is applied (refer to Fig. R17.5.2.1b), the force in the anchor reinforcement will be larger than the shear force. Anchor reinforcement is distinguished from supplementary reinforcement in that it is designed exclusively for the anchor loads and is intended to preclude concrete breakout. Strut-and-tie models may be used to design anchor reinforcement.

CODE

COMMENTARY

For practical reasons, anchor reinforcement is only used for cast-in anchor applications.

(a) Care needs to be taken in the selection and positioning of anchor reinforcement for tension. Ideally tension anchor reinforcement should consist of stirrups, ties, or hairpins placed as close as practicable to the anchor. It is beneficial for the anchor reinforcement to enclose the surface reinforcement where applicable. Anchor reinforcement spaced less than $\mathbf{0.5}\boldsymbol{h_{ef}}$ from the anchor centerline may be considered as effective. The research (Eligehausen et al. 2006b) on which these provisions are based was limited to anchor reinforcement with maximum diameter equivalent to a No. 5 bar.

(b) To ensure development of anchor reinforcement for shear, the enclosing anchor reinforcement shown in Fig. R17.5.2.1(b)(i) should be in contact with the anchor and placed as close as practicable to the concrete surface. The research (Eligehausen et al. 2006b) on which the provisions for enclosing reinforcement are based was limited to anchor reinforcement with maximum diameter equivalent to a No. 5 bar. The larger bend radii associated with larger bar diameters may significantly reduce the effectiveness of the anchor reinforcement for shear; therefore, anchor reinforcement larger than a No. 6 bar is not recommended. Because development for full $\boldsymbol{f_y}$ is required, the use of excess reinforcement to reduce development length is not permitted for anchor reinforcement.

The anchor reinforcement for shear may also consist of stirrups, ties, hoops, or hairpins enclosing the edge reinforcement embedded in the breakout volume and placed as close to the anchors as practicable (refer to Fig. R17.5.2.1b(ii)). Generally, reinforcement spaced less than the smaller of $\mathbf{0.5}\boldsymbol{c_{a1}}$ and $\mathbf{0.3}\boldsymbol{c_{a2}}$ from the anchor centerline should be included as anchor reinforcement. In this case, the anchor reinforcement must be developed on both sides of the breakout surface. For equilibrium, edge reinforcement is required. The research on which these provisions are based was limited to anchor reinforcement with maximum diameter equivalent to a No. 6 bar.

CODE

COMMENTARY

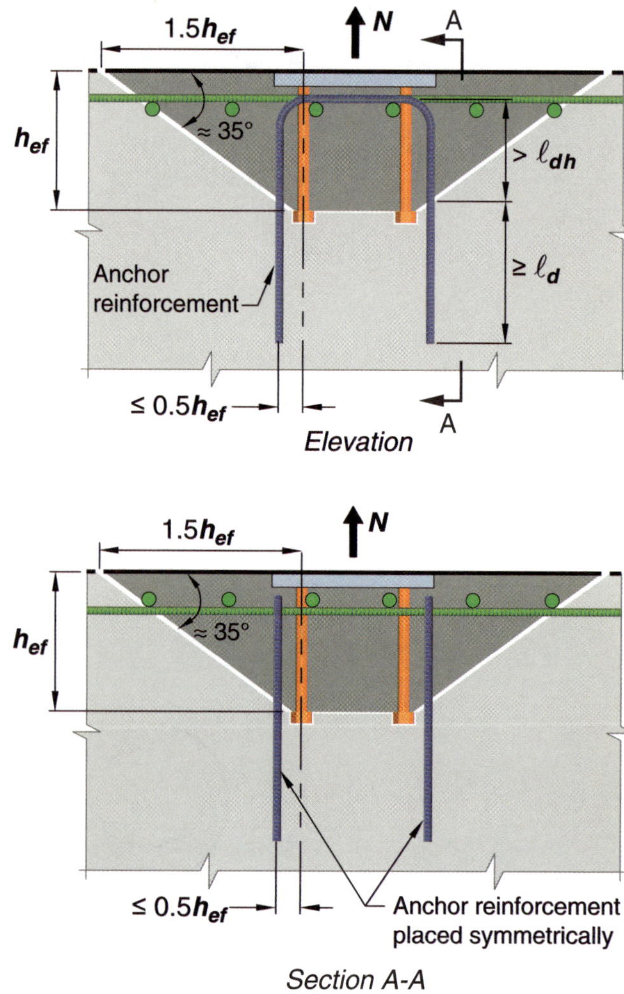

Fig. R17.5.2.1a*—Anchor reinforcement for tension.*

CODE

COMMENTARY

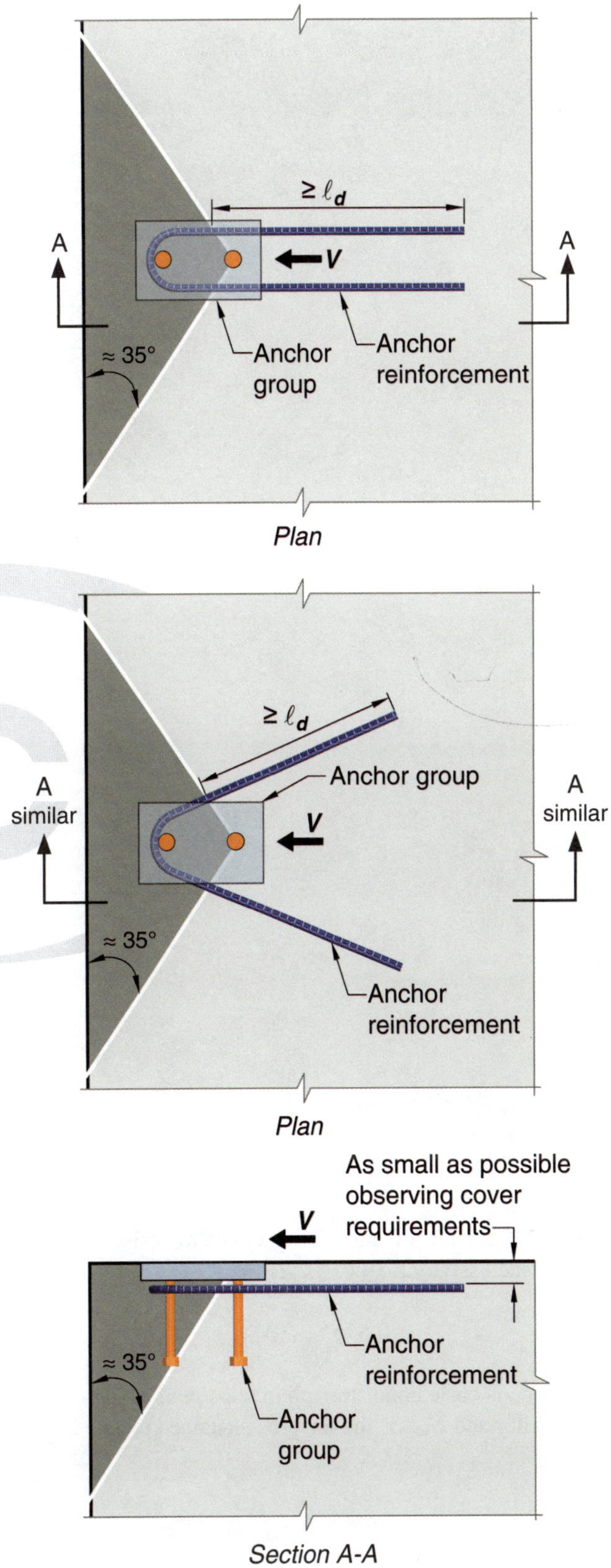

Fig. R17.5.2.1b(i)—*Hairpin anchor reinforcement for shear.*

CODE

COMMENTARY

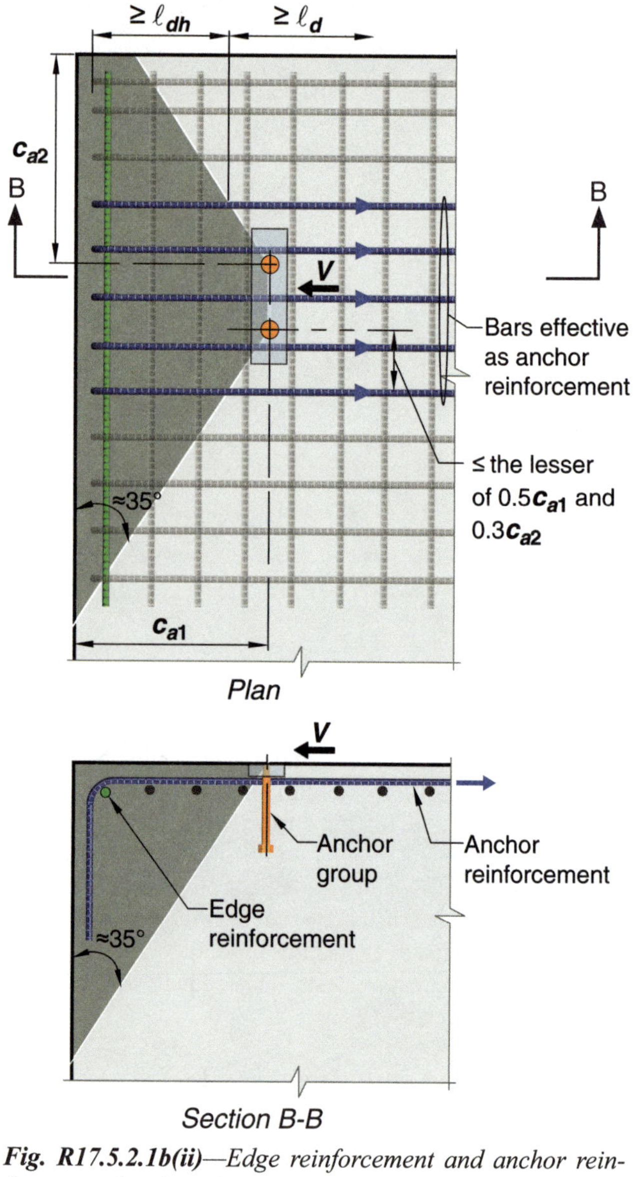

***Fig. R17.5.2.1b(ii)**—Edge reinforcement and anchor reinforcement for shear.*

17.5.2.2 Design of adhesive anchors to resist sustained tension shall satisfy Eq. (17.5.2.2)

$$0.55\phi N_{ba} \geq N_{ua,s} \quad (17.5.2.2)$$

where N_{ba} is basic bond strength in tension of a single adhesive anchor and $N_{ua,s}$ is the factored sustained tensile load.

R17.5.2.2 For adhesive anchors that resist sustained tensile load, an additional calculation for the sustained portion of the factored load for a reduced bond resistance is required to account for possible bond strength reductions under sustained tension. The resistance of adhesive anchors to sustained tension is particularly dependent on correct installation, including hole cleaning, adhesive metering and mixing, and prevention of voids in the adhesive bond line (annular gap). In addition, care should be taken in the selection of the correct adhesive and bond strength for the expected on-site conditions such as the concrete condition during installation (dry or saturated, cold or hot), the drilling method used (rotary impact drill, rock drill, or core drill), and anticipated in-service temperature variations in the concrete.

CODE

COMMENTARY

The 0.55 factor used for the additional calculation for sustained tension is correlated with ACI 355.4 test requirements and provides satisfactory performance of adhesive anchors under sustained tensile loads in accordance with ACI 355.4. Product evaluation according to ACI 355.4 is based on sustained tensile loads being present for 50 years at a standard temperature of 70°F and 10 years at a temperature of 110°F. For longer life spans (for example, greater than 50 years) or higher temperatures, lower factors should be considered. Additional information on use of adhesive anchors for such conditions can be found by consulting with the adhesive manufacturer.

Adhesive anchors are particularly sensitive to installation direction and load type. Adhesive anchors installed overhead that resist sustained tension are of concern because previous applications of this type have led to failures (National Transportation Safety Board 2007). Other anchor types may be more appropriate for such cases. For adhesive anchors that resist sustained tension in horizontal or upwardly inclined orientations, it is essential to meet test requirements of ACI 355.4 for sensitivity to installation direction, use certified installers, and require special inspection. Inspection and installation requirements are provided in Chapter 26.

17.5.2.2.1 For groups of adhesive anchors subject to sustained tension, Eq. (17.5.2.2) shall be satisfied for the anchor that resists the highest sustained tension.

R17.5.2.2.1 The check for anchor groups is limited to the highest loaded anchor in the group, analogous to the design for pullout.

17.5.2.3 If both N_{ua} and V_{ua} are present, interaction effects shall be considered using an interaction expression that results in calculated strengths in substantial agreement with results of comprehensive tests. This requirement shall be considered satisfied by 17.8.

17.5.2.4 Anchors shall satisfy the edge distances, spacings, and thicknesses in 17.9 to preclude splitting failure.

17.5.2.5 Anchors in structures assigned to Seismic Design Category C, D, E, or F shall satisfy the additional requirements of 17.10.

17.5.2.6 Attachments with shear lugs used to transfer structural loads shall satisfy the requirements of 17.11.

17.5.3 Strength reduction factor ϕ for anchors in concrete shall be in accordance with Tables 17.5.3(a), 17.5.3(b), and 17.5.3(c). Strength reduction factor ϕ for anchor reinforcement shall be 0.75.

R17.5.3 The ϕ-factors for the anchor steel strength in Table 17.5.3(a) are based on using f_{uta} to determine the nominal strength of the anchor (refer to 17.6.1 and 17.7.1) rather than f_{ya}, as used in the design of reinforced concrete members. Although the ϕ-factors for use with f_{uta} appear low, they result in a level of safety consistent with the use of higher ϕ-factors applied to f_{ya}. The ϕ-factors for shear, which are smaller than for tension, do not reflect basic material differences but rather account for the possibility of a non-uniform distribution of shear in connections with multiple anchors.

CODE

Table 17.5.3(a)—Anchor strength governed by steel

Type of steel element	Strength reduction factor ϕ	
	Tension (steel)	Shear (steel)
Ductile	0.75	0.65
Brittle	0.65	0.60

Table 17.5.3(b)—Anchor strength governed by concrete breakout, bond, and side-face blowout

Supplementary reinforcement	Type of anchor installation	Anchor Category[1] from ACI 355.2 or ACI 355.4	Strength reduction factor ϕ	
			Tension (concrete breakout, bond, or side-face blowout)	Shear (concrete breakout)
Supplementary reinforcement present	Cast-in anchors	Not applicable	0.75	0.75
	Post-installed anchors	1	0.75	
		2	0.65	
		3	0.55	
Supplementary reinforcement not present	Cast-in Anchors	Not applicable	0.70	0.70
	Post-installed anchors	1	0.65	
		2	0.55	
		3	0.45	

[1]Anchor Category 1 indicates low sensitivity to installation and high reliability; Anchor Category 2 indicates medium sensitivity and medium reliability; Anchor Category 3 indicates high sensitivity and lower reliability.

Table 17.5.3(c)—Anchor strength governed by concrete pullout, or pryout strength

Type of anchor installation	Anchor Category[1] from ACI 355.2 or ACI 355.4	Strength reduction factor ϕ	
		Tension (concrete pullout)	Shear (concrete pryout)
Cast-in anchors	Not applicable	0.70	0.70
Post-installed anchors	1	0.65	
	2	0.55	
	3	0.45	

[1]Anchor Category 1 indicates low sensitivity to installation and high reliability; Anchor Category 2 indicates medium sensitivity and medium reliability; and Anchor Category 3 indicates high sensitivity and lower reliability.

COMMENTARY

The ϕ-factors for anchor strength governed by concrete breakout, bond, and side-face blowout in Table 17.5.3(b) are separated into two groups based on the presence or absence of supplementary reinforcement. The supplementary reinforcement classifications of this table replace the "Condition A" and "Condition B" designations in previous Codes. Applications with supplementary reinforcement provide more deformation capacity, permitting the ϕ-factors to be increased. An explicit design of supplementary reinforcement for anchor-related forces is not required; however, the arrangement of supplementary reinforcement should generally conform to that of the anchor reinforcement shown in Fig. R17.5.2.1(a) and R17.5.2.1(b)(i) and (ii). Unlike anchor reinforcement, full development of supplementary reinforcement beyond the assumed breakout failure plane is not required.

For concrete breakout in shear for all anchor types and for brittle concrete failure modes for cast-in anchors, the basic strength reduction factor for brittle concrete failures (**ϕ = 0.70**) was chosen based on results of probabilistic studies. While this factor is greater than the strength reduction factor of structural plain concrete (**ϕ = 0.60**), the nominal resistance expressions used in this chapter and in the test requirements are based on the 5 percent fractiles; therefore, **ϕ = 0.60** would be overly conservative. Comparison with other design procedures and probabilistic studies (Farrow and Klingner 1995) indicated that the choice of **ϕ = 0.70** is justified. For the same cases with supplementary reinforcement, the value of **ϕ = 0.75** is compatible with the level of safety for shear failures in concrete beams; and has been recommended in the *PCI Design Handbook* (MNL 120) and by ACI 349.

Tests included in ACI 355.2 and ACI 355.4 to assess sensitivity to installation procedures determine the Anchor Categories as given in Table 17.5.3(b) for proprietary post-installed expansion, screw, undercut, and adhesive anchors. ACI 355.2 tests for installation sensitivity measure effects of variability in anchor torque during installation, tolerance on drilled hole size, and energy level used in setting anchors; for expansion, screw, and undercut anchors intended for use in cracked concrete, increased crack widths are considered. ACI 355.4 tests for installation sensitivity assess the influence of adhesive mixing and the influence of hole cleaning in dry, saturated, and water-filled/underwater bore holes.

17.6—Tensile strength

17.6.1 *Steel strength of anchors in tension,* N_{sa}

17.6.1.1 Nominal steel strength of anchors in tension as governed by the steel, N_{sa}, shall be evaluated based on the

R17.6—Tensile strength

R17.6.1 *Steel strength of anchors in tension,* N_{sa}

CODE

properties of the anchor material and the physical dimensions of the anchors.

17.6.1.2 Nominal steel strength of an anchor in tension, N_{sa}, shall be calculated by:

$$N_{sa} = A_{se,N} f_{uta} \quad (17.6.1.2)$$

where $A_{se,N}$ is the effective cross-sectional area of an anchor in tension, in.^{2}, and f_{uta} used for calculations shall not exceed either $1.9f_{ya}$ or 125,000 psi.

17.6.2 *Concrete breakout strength of anchors in tension,* N_{cb}

17.6.2.1 Nominal concrete breakout strength in tension, N_{cb} of a single anchor or N_{cbg} of an anchor group satisfying 17.5.1.3.1, shall be calculated by (a) or (b), respectively:

(a) For a single anchor

$$N_{cb} = \frac{A_{Nc}}{A_{Nco}} \psi_{ed,N} \psi_{c,N} \psi_{cp,N} N_b \quad (17.6.2.1a)$$

(b) For an anchor group

$$N_{cbg} = \frac{A_{Nc}}{A_{Nco}} \psi_{ec,N} \psi_{ed,N} \psi_{c,N} \psi_{cp,N} N_b \quad (17.6.2.1b)$$

where $\psi_{ec,N}$, $\psi_{ed,N}$, $\psi_{c,N}$, and $\psi_{cp,N}$ are given in 17.6.2.3, 17.6.2.4, 17.6.2.5, and 17.6.2.6, respectively.

COMMENTARY

R17.6.1.2 The nominal strength of anchors in tension is best represented as a function of f_{uta} rather than f_{ya} because the large majority of anchor materials do not exhibit a well-defined yield point. AISC has based tension strength of anchors on $A_{se,N} f_{uta}$ since the 1986 edition of their specifications. The use of Eq. (17.6.1.2) with the load factors provided in 5.3 and the ϕ-factors provided in 17.5.3 result in design strengths consistent with AISC 360.

The limitation of $1.9f_{ya}$ on f_{uta} is to ensure that, under service load conditions, the anchor does not exceed f_{ya}. Although not a concern for standard structural steel anchors (maximum value of f_{uta}/f_{ya} is 1.6 for ASTM A307), the limitation is applicable to some stainless steels. The limit on f_{uta} of $1.9f_{ya}$ was determined by converting the LRFD provisions to corresponding service level conditions. From 5.3, the average load factor of 1.4 (from $1.2D + 1.6L$) divided by the highest ϕ-factor (0.75 for tension) results in a limit of f_{uta}/f_{ya} of 1.4/0.75 = 1.87.

For post-installed anchors having a reduced cross-sectional area anywhere along the anchor length, such as wedge-type anchors, the effective cross-sectional area of the anchor should be provided by the manufacturer. For threaded rods and headed bolts, ASME B1.1 defines $A_{se,N}$ as

$$A_{se,N} = \frac{\pi}{4}\left(d_a - \frac{0.9743}{n_t}\right)^2$$

where n_t is the number of threads per inch.

R17.6.2 *Concrete breakout strength of anchors in tension,* N_{cb}

R17.6.2.1 The effects of multiple anchors, spacing of anchors, and edge distance on the nominal concrete breakout strength in tension are included by applying the modification factors A_{Nc}/A_{Nco} and $\psi_{ed,N}$ in Eq. (17.6.2.1a) and (17.6.2.1b).

Figure R17.6.2.1(a) shows A_{Nco} and the development of Eq. (17.6.2.1.4). A_{Nco} is the maximum projected area for a single anchor. Figure R17.6.2.1(b) shows examples of the projected areas for various single-anchor and multiple-anchor arrangements. Because A_{Nc} is the total projected area for an anchor group, and A_{Nco} is the area for a single anchor, there is no need to include n, the number of anchors, in Eq. (17.6.2.1b). If anchor groups are positioned in such a way that their projected areas overlap, the value of A_{Nc} is required to be reduced accordingly.

CODE

COMMENTARY

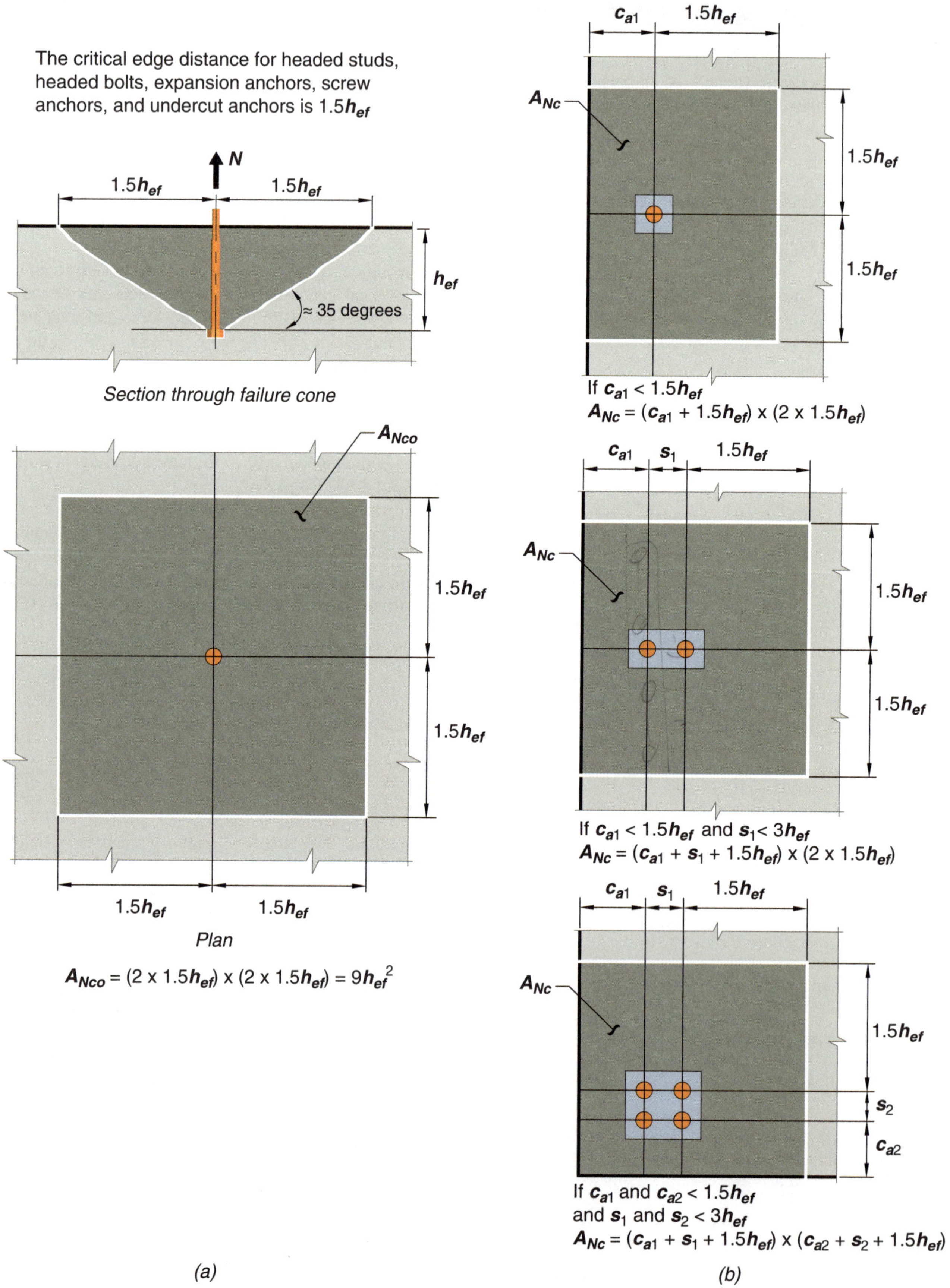

Fig. R17.6.2.1—(a) Calculation of A_{Nco} *and (b) calculation of* A_{Nc} *for single anchors and anchor groups.*

CODE

17.6.2.1.1 A_{Nc} is the projected concrete failure area of a single anchor or of an anchor group that is approximated as the base of the rectilinear geometrical shape that results from projecting the failure surface outward $1.5h_{ef}$ from the centerlines of the anchor, or in the case of an anchor group, from a line through a row of adjacent anchors. A_{Nc} shall not exceed nA_{Nco}, where n is the number of anchors in the group that resist tension.

17.6.2.1.2 If anchors are located less than $1.5h_{ef}$ from three or more edges, the value of h_{ef} used to calculate A_{Nc} in accordance with 17.6.2.1.1, as well as for the equations in 17.6.2.1 through 17.6.2.4, shall be the greater of (a) and (b):

(a) $c_{a,max}/1.5$
(b) $s/3$, where s is the maximum spacing between anchors within the group.

COMMENTARY

R17.6.2.1.2 For anchors located less than $1.5h_{ef}$ from three or more edges, the CCD Method (refer to R17.5.1.3), which is the basis for the equations in 17.6.2.1 through 17.6.2.4, gives overly conservative results for the tensile breakout strength (Lutz 1995). This occurs because the ordinary definitions of A_{Nc}/A_{Nco} do not correctly reflect the edge effects. This problem is corrected by limiting the value of h_{ef} used in the equations in 17.6.2.1 through 17.6.2.4 to $(c_{a,max})/1.5$, where $c_{a,max}$ is the greatest of the influencing edge distances that do not exceed the actual $1.5h_{ef}$. In no case should $(c_{a,max})/1.5$ be taken less than one-third of the maximum spacing between anchors within the group. The limit on h_{ef} of at least one-third of the maximum spacing between anchors within the group prevents the use of a calculated strength based on individual breakout volumes for an anchor group configuration. This approach is illustrated in Fig. R17.6.2.1.2. In this example, the proposed limit on the value of h_{ef} to be used in calculations where $h_{ef} = (c_{a,max})/1.5$, results in $h_{ef} = h'_{ef} =$ **4 in.** For this example, this would be the proper value to be used for h_{ef} in calculating the resistance even if the actual embedment depth is greater.

The requirement of 17.6.2.1.2 may be visualized by moving the actual concrete breakout surface, which originates at the actual h_{ef}, toward the surface of the concrete parallel to the applied tensile load. The value of h_{ef} used in 17.6.2.1 through 17.6.2.4 is determined when (a) the outer boundaries of the failure surface first intersect a free edge; or (b) the intersection of the breakout surface between anchors within the group first intersects the surface of the concrete. For the example shown in Fig. R17.6.2.1.2, point "A" shows the intersection of the assumed failure surface for limiting h_{ef} with the concrete surface.

CODE

COMMENTARY

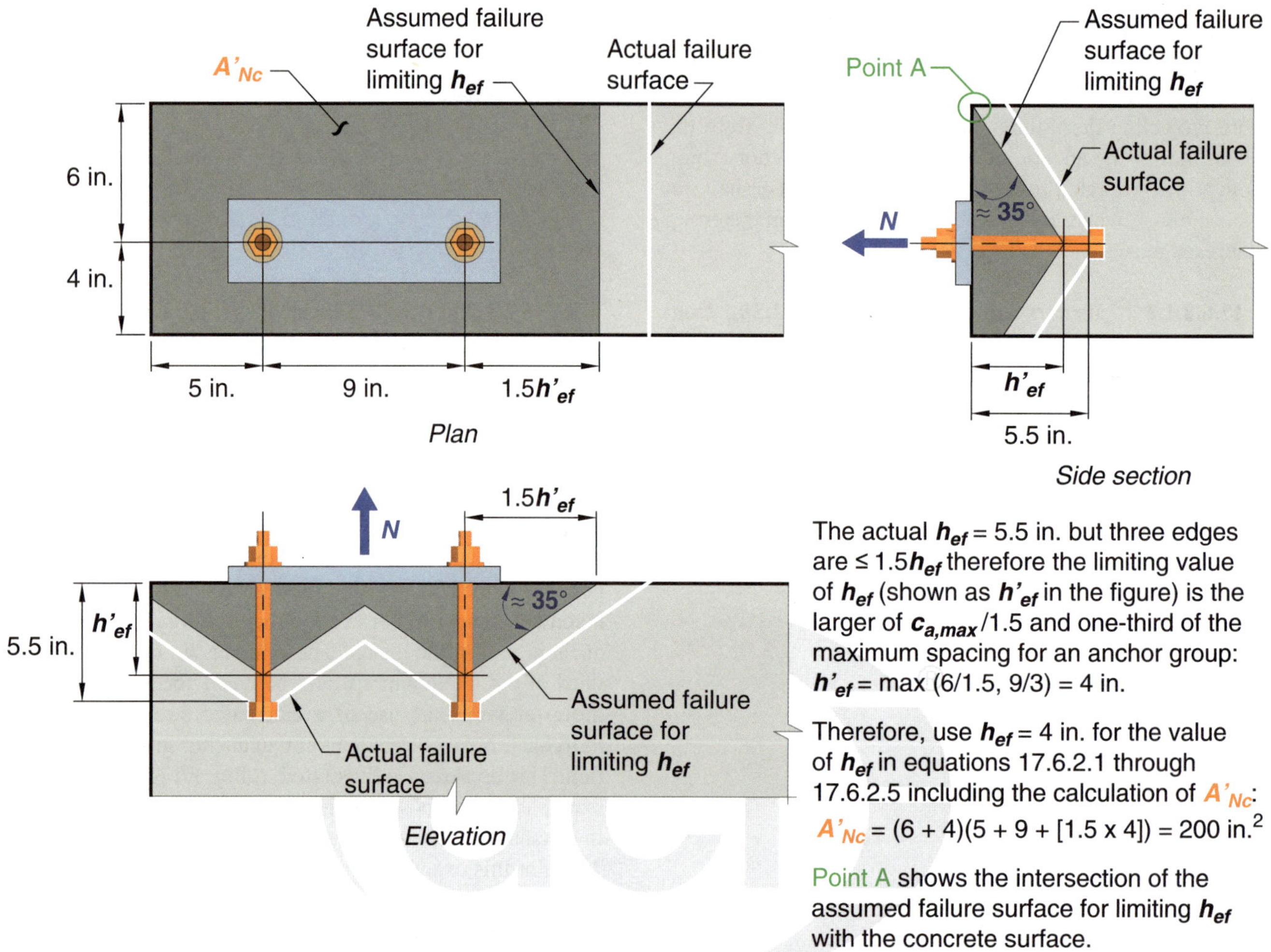

***Fig. R17.6.2.1.2**—Example of tension where anchors are located in narrow members.*

17.6.2.1.3 If an additional plate or washer is added at the head of the anchor, it shall be permitted to calculate the projected area of the failure surface by projecting the failure surface outward $1.5h_{ef}$ from the effective perimeter of the plate or washer. The effective perimeter shall not exceed the value at a section projected outward more than the thickness of the washer or plate from the outer edge of the head of the anchor.

17.6.2.1.4 A_{Nco} is the projected concrete failure area of a single anchor with an edge distance of at least $1.5h_{ef}$ and shall be calculated by Eq. (17.6.2.1.4).

$$A_{Nco} = 9h_{ef}^{\ 2} \qquad (17.6.2.1.4)$$

17.6.2.2 *Basic single anchor breakout strength,* N_b

17.6.2.2.1 Basic concrete breakout strength of a single anchor in tension in cracked concrete, N_b, shall be calculated by Eq. (17.6.2.2.1), except as permitted in 17.6.2.2.3

$$N_b = k_c\lambda_a\sqrt{f_c'}\,h_{ef}^{\ 1.5} \qquad (17.6.2.2.1)$$

R17.6.2.2 *Basic single anchor breakout strength,* N_b

R17.6.2.2.1 The equation for the basic concrete breakout strength was derived assuming concrete breakout with an angle of approximately 35 degrees, considering fracture mechanics concepts (Fuchs et al. 1995; Eligehausen and Balogh 1995; Eligehausen and Fuchs 1988; *fib* 2011).

CODE

where k_c = **24** for cast-in anchors and **17** for post-installed anchors.

COMMENTARY

The values of k_c in Eq. (17.6.2.2.1) were determined from a large database of test results in uncracked concrete at the 5 percent fractile (Fuchs et al. 1995). The values were adjusted to corresponding k_c values for cracked concrete (Eligehausen and Balogh 1995; Goto 1971). Tests have shown that the values of k_c applicable to adhesive anchors are approximately equal to those derived for expansion anchors (Eligehausen et al. 2006a; Zhang et al. 2001).

17.6.2.2.2 k_c for post-installed anchors shall be permitted to be increased based on ACI 355.2 or ACI 355.4 product-specific tests, but shall not exceed 24.

17.6.2.2.3 For single cast-in headed studs and headed bolts with **11 in.** $\leq h_{ef} \leq$ **25 in.**, N_b shall be calculated by:

$$N_b = 16\lambda_a\sqrt{f_c'}\,h_{ef}^{5/3} \qquad (17.6.2.2.3)$$

R17.6.2.2.3 For anchors with a deeper embedment (h_{ef} > **11 in.**), test evidence indicates the use of $h_{ef}^{1.5}$ can be overly conservative for some cases. An alternative expression (Eq. (17.6.2.2.3)) is provided using $h_{ef}^{5/3}$ for evaluation of cast-in headed studs and headed bolts with **11 in.** $\leq h_{ef} \leq$ **25 in.** This expression can also be appropriate for some undercut post-installed anchors. However, for such anchors, the use of Eq. (17.6.2.2.3) should be justified by test results in accordance with 17.5.1.4. Experimental and numerical investigations indicate that Eq. (17.6.2.2.3) may be unconservative for h_{ef} > **25 in.** if bearing pressure on the anchor head is at or near the limit permitted by Eq. (17.6.3.2.2a) (Ožbolt et al. 2007).

17.6.2.3 *Breakout eccentricity factor,* $\psi_{ec,N}$

17.6.2.3.1 Modification factor for anchor groups loaded eccentrically in tension, $\psi_{ec,N}$, shall be calculated by Eq. (17.6.2.3.1).

$$\psi_{ec,N} = \frac{1}{\left(1+\dfrac{e_N'}{1.5h_{ef}}\right)} \leq 1.0 \qquad (17.6.2.3.1)$$

R17.6.2.3 *Breakout eccentricity factor,* $\psi_{ec,N}$

R17.6.2.3.1 Figure 17.6.2.3.1(a) shows an anchor group where all anchors are in tension but the resultant force is eccentric with respect to the centroid of the anchor group. Anchors can also be loaded in such a way that only some of the anchors are in tension (Fig. 17.6.2.3.1(b)). In this case, only the anchors in tension are to be considered for the calculation of e_N'. The eccentricity e_N' of the resultant tensile force is determined with respect to the center of gravity of the anchors in tension.

CODE

COMMENTARY

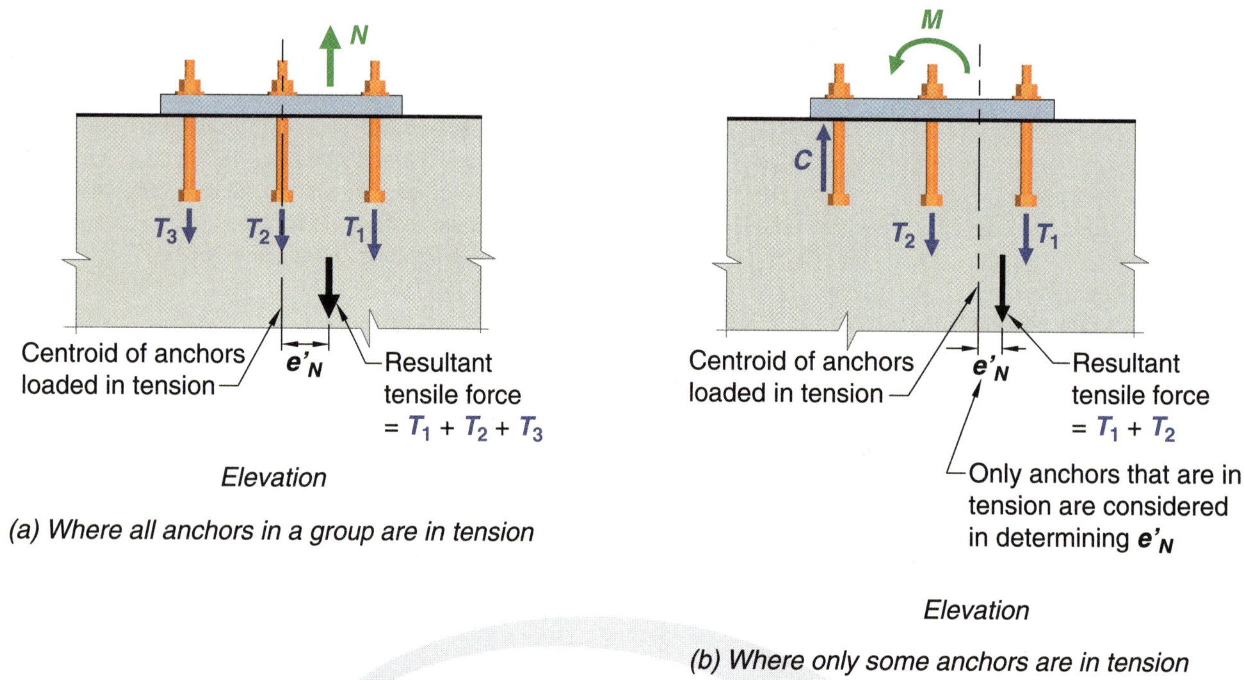

Fig. R17.6.2.3.1—Definition of e_N' *for an anchor group.*

17.6.2.3.2 If the loading on an anchor group is such that only some of the anchors in the group are in tension, only those anchors that are in tension shall be considered for determining eccentricity e'_N in Eq. (17.6.2.3.1) and for the calculation of N_{cbg} according to Eq. (17.6.2.1b).

17.6.2.3.3 If the loading is eccentric with respect to two orthogonal axes, $\psi_{ec,N}$ shall be calculated for each axis individually, and the product of these factors shall be used as $\psi_{ec,N}$ in Eq. (17.6.2.1b).

17.6.2.4 *Breakout edge effect factor,* $\psi_{ed,N}$

17.6.2.4.1 Modification factor for edge effects for single anchors or anchor groups loaded in tension, $\psi_{ed,N}$, shall be determined by (a) or (b).

(a) If $c_{a,min} \geq 1.5h_{ef}$, then $\psi_{ed,N} = 1.0$ (17.6.2.4.1a)

(b) If $c_{a,min} < 1.5h_{ef}$, then $\psi_{ed,N} = 0.7 + 0.3\dfrac{c_{a,min}}{1.5h_{ef}}$ (17.6.2.4.1b)

R17.6.2.4 *Breakout edge effect factor,* $\psi_{ed,N}$

R17.6.2.4.1 If anchors are located close to an edge such that there is insufficient space for a complete breakout volume to develop, the strength of the anchor is further reduced beyond that reflected in A_{Nc}/A_{Nco}. If the smallest side cover distance is at least $1.5h_{ef}$, the design model assumes a complete breakout volume can form, and there is no reduction ($\psi_{ed,N} = 1$). If the side cover is less than $1.5h_{ef}$, the factor $\psi_{ed,N}$ is required to adjust for the edge effect (Fuchs et al. 1995).

17.6.2.5 *Breakout cracking factor,* $\psi_{c,N}$

17.6.2.5.1 Modification factor for the influence of cracking in anchor regions at service load levels, $\psi_{c,N}$, shall be determined by (a) or (b):

(a) For anchors located in a region of a concrete member where analysis indicates no cracking at service load levels, $\psi_{c,N}$ shall be permitted to be:

R17.6.2.5 *Breakout cracking factor,* $\psi_{c,N}$

R17.6.2.5.1 Post-installed anchors that do not meet the requirements for use in cracked concrete according to ACI 355.2 or ACI 355.4 should be used only in regions that will remain uncracked. The analysis for the determination of crack formation should include the effects of restrained shrinkage (refer to 24.4.2). The anchor qualification tests of ACI 355.2 or ACI 355.4 require that anchors in cracked

CODE

$\psi_{c,N}$ = **1.25** for cast-in anchors

$\psi_{c,N}$ = **1.4** for post-installed anchors, if the value of k_c used in Eq. (17.6.2.2.1) is 17. If the value of k_c used in Eq. (17.6.2.2.1) is taken from the ACI 355.2 or ACI 355.4 product evaluation report for post-installed anchors:

(i) $\psi_{c,N}$ shall be based on the ACI 355.2 or ACI 355.4 product evaluation report for anchors qualified for use in both cracked and uncracked concrete

(ii) $\psi_{c,N}$ shall be taken as 1.0 for anchors qualified for use in uncracked concrete.

(b) For anchors located in a region of a concrete member where analysis indicates cracking at service load levels, $\psi_{c,N}$ shall be taken as 1.0 for both cast-in anchors and post-installed anchors, and 17.6.2.6 shall be satisfied.

17.6.2.5.2 Post-installed anchors shall be qualified for use in cracked concrete in accordance with ACI 355.2 or ACI 355.4. Cracking in the concrete shall be controlled by flexural reinforcement distributed in accordance with 24.3.2, or equivalent crack control shall be provided by confining reinforcement.

17.6.2.6 *Breakout splitting factor,* $\psi_{cp,N}$

17.6.2.6.1 Modification factor for post-installed anchors designed for uncracked concrete in accordance with 17.6.2.5 without supplementary reinforcement to control splitting, $\psi_{cp,N}$, shall be determined by (a) or (b) using the critical distance c_{ac} as defined in 17.9.5.

(a) If $c_{a,min} \geq c_{ac}$, then $\psi_{cp,N} = 1.0$ (17.6.2.6.1a)

(b) If $c_{a,min} < c_{ac}$, then $\psi_{cp,N} = \dfrac{c_{a,min}}{c_{ac}} \geq \dfrac{1.5h_{ef}}{c_{ac}}$ (17.6.2.6.1b)

17.6.2.6.2 For all other cases, including cast-in anchors, $\psi_{cp,N}$ shall be taken as 1.0.

17.6.3 *Pullout strength of a single cast-in anchor or a single post-installed expansion, screw, or undercut anchor in tension,* N_{pn}

17.6.3.1 Nominal pullout strength of a single cast-in anchor or a single-post-installed expansion, screw, or undercut anchor in tension, N_{pn}, shall be calculated by:

$$N_{pn} = \psi_{c,P}N_p \qquad (17.6.3.1)$$

COMMENTARY

concrete zones perform well in a crack that is 0.012-in. wide. If wider cracks are expected, reinforcement to control the crack width to approximately 0.012 in. should be provided. Refer to ACI 224R for more information.

The concrete breakout strengths given by Eq. (17.6.2.2.1) and (17.6.2.2.3) assume cracked concrete ($\psi_{c,N}$ = **1.0**) with $\psi_{c,N}k_c$ = **24** for cast-in anchors and 17 for post-installed anchors. If the uncracked concrete $\psi_{c,N}$ factors are applied (1.25 for cast-in and 1.4 for post-installed), $\psi_{c,N}k_c$ factors become 30 for cast-in anchors and 24 for post-installed anchors. This agrees with field observations and tests demonstrating cast-in anchor strength exceeds that of post-installed for both cracked and uncracked concrete.

R17.6.2.6 *Breakout splitting factor,* $\psi_{cp,N}$

R17.6.2.6.1 The design provisions in 17.6 are based on the assumption that the basic concrete breakout strength can be achieved if the minimum edge distance $c_{a,min}$ equals $1.5h_{ef}$. Test results (Asmus 1999), however, indicate that many torque-controlled and displacement-controlled expansion anchors and some undercut anchors require edge distances exceeding $1.5h_{ef}$ to achieve the basic concrete breakout strength if tested in uncracked concrete without supplementary reinforcement to control splitting. When a tensile load is applied, the resulting tensile stresses at the embedded end of the anchor are added to the tensile stresses induced due to anchor installation, and splitting failure may occur before reaching the concrete breakout strength given in 17.6.2.1. To account for this potential splitting mode of failure, the basic concrete breakout strength is reduced by a factor $\psi_{cp,N}$ if $c_{a,min}$ is less than the critical edge distance c_{ac}.

R17.6.2.6.2 If supplementary reinforcement to control splitting is present or if the anchors are located in a region where analysis indicates cracking of the concrete at service loads, the reduction factor $\psi_{cp,N}$ is taken as 1.0.

R17.6.3 *Pullout strength of a single cast-in anchor or a single post-installed expansion, screw, or undercut anchor in tension,* N_{pn}

R17.6.3.1 The design requirements for pullout are applicable to cast-in anchors and post-installed expansion, screw, and undercut anchors. They are not applicable to adhesive anchors, which are instead evaluated for bond failure in accordance with 17.6.5.

CODE

where $\psi_{c,P}$ is given in 17.6.3.3.

17.6.3.2 *Basic single anchor pullout strength,* $\mathbf{N_p}$

17.6.3.2.1 For post-installed expansion, screw, and undercut anchors, the values of N_p shall be based on the 5 percent fractile of results of tests performed and evaluated according to ACI 355.2. It is not permissible to calculate the pullout strength in tension for such anchors.

17.6.3.2.2 For single anchors, it shall be permitted to evaluate the pullout strength in tension, N_p, for use in Eq. (17.6.3.1) in accordance with (a) or (b). Alternatively, it shall be permitted to use values of N_p based on the 5 percent fractile of tests performed and evaluated in the same manner as the ACI 355.2 procedures but without the benefit of friction.

(a) For cast-in headed studs and headed bolts, N_p shall be calculated by:

$$N_p = 8A_{brg}f_c' \quad (17.6.3.2.2a)$$

(b) For J- or L-bolts, Np shall be calculated by:

$$N_p = 0.9f_c'e_hd_a \quad (17.6.3.2.2b)$$

where $\mathbf{3d_a \le e_h \le 4.5d_a}$.

17.6.3.3 *Pullout cracking factor,* $\psi_{c,P}$

17.6.3.3.1 Modification factor to account for the influence of cracking in anchor regions at service load levels, $\psi_{c,P}$, shall be determined by (a) or (b):

(a) For anchors located in a region of a concrete member where analysis indicates no cracking at service load levels, $\psi_{c,P}$ shall be permitted to be 1.4.

(b) For anchors located in a region of a concrete member where analysis indicates cracking at service load levels, $\psi_{c,P}$, shall be taken as 1.0.

COMMENTARY

R17.6.3.2 *Basic single anchor pullout strength,* $\mathbf{N_p}$

R17.6.3.2.2 The pullout strength equations given in 17.6.3.2.2(a) and 17.6.3.2.2(b) are only applicable to cast-in headed and hooked anchors (Kuhn and Shaikh 1996; *fib* 2011); they are not applicable to post-installed expansion, screw, and undercut anchors that use various mechanisms for end anchorage unless the validity of the pullout strength equations is verified by tests.

The value calculated from Eq. (17.6.3.2.2a) corresponds to the force at which crushing of the concrete occurs due to bearing of the anchor head (*fib* 2011; ACI 349). It is not the force required to pull the anchor completely out of the concrete; therefore, the equation does not contain a term relating to embedment depth. Local crushing of the concrete greatly reduces the stiffness of the connection, and generally will be the beginning of a pullout failure. The pullout strength in tension of headed studs or headed bolts can be increased by providing reinforcement, such as closely spaced spirals, throughout the head region. This increase can be demonstrated by tests, as required by the Licensed Design Professional for the specific application.

Equation (17.6.3.2.2b) for hooked bolts was developed by Lutz based on the results of Kuhn and Shaikh (1996). Reliance is placed on the bearing component only, neglecting any frictional component, because crushing inside the hook will greatly reduce the stiffness of the connection and generally will be the beginning of a pullout failure. The limits on e_h are based on the range of variables used in the three test programs reported in Kuhn and Shaikh (1996).

CODE

17.6.4 *Concrete side-face blowout strength of headed anchors in tension,* $\mathbf{N_{sb}}$

17.6.4.1 For a single headed anchor with deep embedment close to an edge ($h_{ef} > 2.5c_{a1}$), the nominal side-face blowout strength, N_{sb}, shall be calculated by:

$$N_{sb} = 160c_{a1}\sqrt{A_{brg}}\lambda_a\sqrt{f'_c} \quad (17.6.4.1)$$

17.6.4.1.1 If c_{a2} for the single headed anchor is less than $3c_{a1}$, the value of N_{sb} shall be multiplied by the factor $(1 + c_{a2}/c_{a1})/4$, where $1.0 \leq c_{a2}/c_{a1} \leq 3.0$.

17.6.4.2 For multiple headed anchors with deep embedment close to an edge ($h_{ef} > 2.5c_{a1}$) and anchor spacing less than $6c_{a1}$, the nominal strength of those anchors susceptible to a side-face blowout failure, N_{sbg}, shall be calculated by:

$$N_{sbg} = \left(1 + \frac{s}{6c_{a1}}\right)N_{sb} \quad (17.6.4.2)$$

where s is the distance between the outer anchors along the edge, and N_{sb} is obtained from Eq. (17.6.4.1) without modification for a perpendicular edge distance.

17.6.5 *Bond strength of adhesive anchors in tension,* $\mathbf{N_a}$ *or* $\mathbf{N_{ag}}$

17.6.5.1 Nominal bond strength in tension, N_a of a single adhesive anchor or N_{ag} of an adhesive anchor group satisfying 17.5.1.3.1, shall be calculated by (a) or (b), respectively.

(a) For a single adhesive anchor:

$$N_a = \frac{A_{Na}}{A_{Nao}}\psi_{ed,Na}\psi_{cp,Na}N_{ba} \quad (17.6.5.1a)$$

(b) For an adhesive anchor group:

$$N_{ag} = \frac{A_{Na}}{A_{Nao}}\psi_{ec,Na}\psi_{ed,Na}\psi_{cp,Na}N_{ba} \quad (17.6.5.1b)$$

where $\psi_{ec,Na}$, $\psi_{ed,Na}$, and $\psi_{cp,Na}$ are given in 17.6.5.3, 17.6.5.4, and 17.6.5.5, respectively.

17.6.5.1.1 A_{Na} is the projected influence area of a single adhesive anchor or an adhesive anchor group that is approximated as a rectilinear area that projects outward a distance c_{Na} from the centerline of the adhesive anchor, or in the case of an adhesive anchor group, from a line through a row of adjacent adhesive anchors. A_{Na} shall not exceed nA_{Nao}, where n is the number of adhesive anchors in the group that resist tension.

COMMENTARY

R17.6.4 *Concrete side-face blowout strength of headed anchors in tension,* $\mathbf{N_{sb}}$

R17.6.4.1 The design requirements for side-face blowout are based on the recommendations of Furche and Eligehausen (1991) and are applicable to headed anchors that usually are cast-in. Splitting during installation rather than side-face blowout generally governs post-installed anchors and is evaluated by ACI 355.2 requirements.

R17.6.4.2 To calculate nominal side-face blowout strength for multiple headed anchors, only those anchors close to an edge ($c_{a1} < 0.4h_{ef}$) that are loaded in tension should be considered. Their strength is compared to the portion of the tensile load applied to those anchors.

R17.6.5 *Bond strength of adhesive anchors in tension,* $\mathbf{N_a}$ *or* $\mathbf{N_{ag}}$

R17.6.5.1 Evaluation of bond strength applies only to adhesive anchors. Single anchors with small embedment loaded to failure in tension may exhibit concrete breakout failures, while deeper embedments produce bond failures. Adhesive anchors that exhibit bond failures when loaded individually may exhibit concrete failures in a group or in a near-edge condition. In all cases, the strength in tension of adhesive anchors is limited by concrete breakout strength as given by Eq. (17.6.2.1a) and (17.6.2.1b) (Eligehausen et al. 2006a).

The influence of anchor spacing and edge distance on both bond strength and concrete breakout strength must be evaluated for adhesive anchors. The influence of anchor spacing and edge distance on the nominal bond strength of adhesive anchors in tension are included in the modification factors A_{Na}/A_{Nao} and $\psi_{ed,Na}$ in Eq. (17.6.5.1a) and (17.6.5.1b).

The influence of nearby edges and adjacent loaded anchors on bond strength is dependent on the volume of concrete mobilized by a single adhesive anchor. In contrast to the projected concrete failure area concept used in Eq. (17.6.2.1a) and (17.6.2.1b) to calculate the breakout strength of an adhesive anchor, the influence area associated with the bond strength of an adhesive anchor used in Eq. (17.6.5.1a) and (17.6.5.1b) is not a function of the embedment depth, but rather a function of the anchor diameter and characteristic bond stress. The critical distance c_{Na} is assumed the same whether the concrete is cracked or uncracked. For simplicity,

CODE

17.6.5.1.2 A_{Nao} is the projected influence area of a single adhesive anchor with an edge distance of at least c_{Na}:

$$A_{Nao} = (2c_{Na})^2 \quad (17.6.5.1.2a)$$

where

$$c_{Na} = 10d_a\sqrt{\frac{\tau_{uncr}}{1100}} \quad (17.6.5.1.2b)$$

COMMENTARY

the relationship for c_{Na} in Eq. (17.6.5.1.2b) uses τ_{uncr}, the characteristic bond stress in uncracked concrete. This has been verified by experimental and numerical studies (Eligehausen et al. 2006a). Figure R17.6.5.1(a) shows A_{Nao} and the development of Eq. (17.6.5.1.2a). A_{Nao} is the projected influence area for the bond strength of a single adhesive anchor. Figure R17.6.5.1(b) shows an example of the projected influence area for an anchor group. Because, in this case, A_{Na} is the projected influence area for an anchor group, and A_{Nao} is the projected influence area for a single anchor, there is no need to include n, the number of anchors, in Eq. (17.6.5.1b). If individual anchors in a group (anchors loaded by a common base plate or attachment) are positioned in such a way that the projected influence areas of the individual anchors overlap, the value of A_{Na} is less than nA_{Nao}.

The tensile strength of closely spaced adhesive anchors with low bond strength may significantly exceed the value given by Eq. (17.6.5.1b). A correction factor is given in the literature (Eligehausen et al. 2006a) to address this issue, but for simplicity, this factor is not included in the Code.

CODE

COMMENTARY

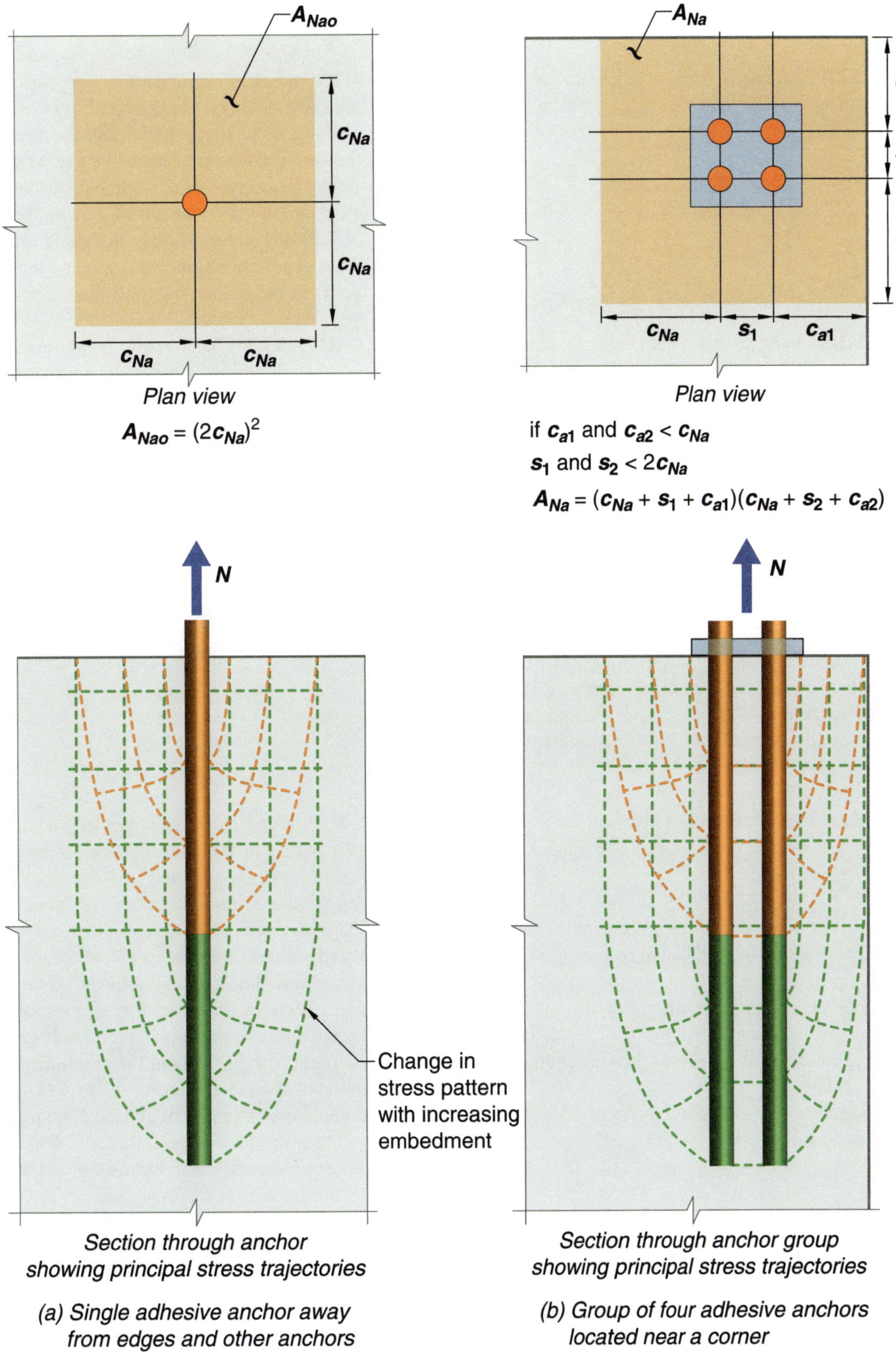

***Fig. R17.6.5.1**—Calculation of influence areas* $\mathbf{A_{Nao}}$ *and* $\mathbf{A_{Na}}$.

CODE

17.6.5.2 *Basic single anchor bond strength,* N_{ba}

17.6.5.2.1 Basic bond strength of a single adhesive anchor in tension in cracked concrete, N_{ba}, shall be calculated by Eq. (17.6.5.2.1)

$$N_{ba} = \lambda_a \tau_{cr} \pi d_a h_{ef} \tag{17.6.5.2.1}$$

17.6.5.2.2 Characteristic bond stress, τ_{cr}, shall be taken as the 5 percent fractile of results of tests performed and evaluated in accordance with ACI 355.4.

17.6.5.2.3 If analysis indicates cracking at service load levels, adhesive anchors shall be qualified for use in cracked concrete in accordance with ACI 355.4.

17.6.5.2.4 For adhesive anchors located in a region of a concrete member where analysis indicates no cracking at service load levels, τ_{uncr} shall be permitted to be used in place of τ_{cr} in Eq. (17.6.5.2.1) and shall be taken as the 5 percent fractile of results of tests performed and evaluated according to ACI 355.4.

17.6.5.2.5 It shall be permitted to use the minimum characteristic bond stress values in Table 17.6.5.2.5, provided (a) through (e) are satisfied:

(a) Anchors shall meet the requirements of ACI 355.4
(b) Anchors shall be installed in holes drilled with a rotary impact drill or rock drill
(c) Concrete compressive strength at time of anchor installation shall be at least 2500 psi
(d) Concrete age at time of anchor installation shall be at least 21 days
(e) Concrete temperature at time of anchor installation shall be at least 50°F

COMMENTARY

R17.6.5.2 *Basic single anchor bond strength,* N_{ba}

R17.6.5.2.1 The equation for basic bond strength of adhesive anchors as given in Eq. (17.6.5.2.1) represents a uniform bond stress model that has been shown to provide the best prediction of adhesive anchor bond strength based on numerical studies and comparisons of different models to an international database of experimental results (Cook et al. 1998). The basic bond strength is valid for bond failures that occur between the concrete and the adhesive as well as between the anchor and the adhesive.

R17.6.5.2.2 Characteristic bond stresses should be based on tests performed in accordance with ACI 355.4 and should reflect the particular combination of installation and use conditions anticipated during construction and during anchor service life. If product-specific information is unavailable at the time of design, Table 17.6.5.2.5 provides lower-bound default values.

R17.6.5.2.5 The characteristic bond stresses in Table 17.6.5.2.5 are the minimum values permitted for adhesive anchor systems qualified in accordance with ACI 355.4 for the tabulated installation and use conditions. Use of these values is restricted to the combinations of specific conditions listed; values for other combinations of installation and use conditions should not be inferred. If both sustained tension and earthquake-induced forces are required to be resisted by the anchors, the applicable factors given in the footnotes of Table 17.6.5.2.5 should be multiplied together. The table assumes a concrete age of at least 21 days and a concrete compressive strength of at least 2500 psi.

The terms "indoor" and "outdoor" as used in Table 17.6.5.2.5 refer to a specific set of installation and service environments. Indoor conditions represent anchors installed in dry concrete with a rotary impact drill or rock drill and subjected to limited concrete temperature variations over the service life of the anchor. Outdoor conditions are assumed to occur if, at the time of installation, the concrete is exposed to weather that might leave the concrete wet. Anchors installed in outdoor conditions are also assumed to be subject to greater concrete temperature variations such as might be associated with freezing and thawing or elevated temperatures resulting from direct sun exposure. While the indoor/outdoor characterization is useful for many applications, there may be situations in which a literal

CODE

Table 17.6.5.2.5—Minimum characteristic bond stresses[1][2]

Installation and service environment	Moisture content of concrete at time of anchor installation	Peak in-service temperature of concrete, °F	τ_{cr}, psi	τ_{uncr}, psi
Outdoor	Dry to fully saturated	175	200	650
Indoor	Dry	110	300	1000

[1]If anchor design includes sustained tension, multiply values of τ_{cr} and τ_{uncr} by 0.4.
[2]If anchor design includes earthquake-induced forces for structures assigned to SDC C, D, E, or F, multiply values of τ_{cr} by 0.8 and τ_{uncr} by 0.4.

COMMENTARY

interpretation of the terms "indoor" and "outdoor" do not apply. For example, anchors installed before the building envelope is completed may involve drilling in saturated concrete. As such, the minimum characteristic bond stress associated with the outdoor condition in Table 17.6.5.2.5 applies, regardless of whether the service environment is "indoor" or "outdoor."

Rotary impact drills and rock drills produce non-uniform hole geometries that are generally favorable for bond. Installation of adhesive anchors in core-drilled holes may result in substantially lower characteristic bond stresses. Because this effect is highly product dependent, design of anchors to be installed in core-drilled holes should adhere to the product-specific characteristic bond stresses established through testing in accordance with ACI 355.4.

The characteristic bond stresses associated with specific adhesive anchor systems are dependent on a number of parameters. Consequently, care should be taken to include all parameters relevant to the value of characteristic bond stress used in the design. These parameters include but are not limited to:

(a) Type and duration of loading—bond strength is reduced for sustained tension
(b) Concrete cracking—bond strength is higher in uncracked concrete
(c) Anchor size—bond strength is generally inversely proportional to anchor diameter
(d) Drilling method—bond strength may be lower for anchors installed in core-drilled holes
(e) Degree of concrete saturation at time of hole drilling and anchor installation—bond strength may be reduced due to concrete saturation
(f) Concrete temperature at time of installation—installation of anchors in cold conditions may result in retarded adhesive cure and reduced bond strength
(g) Concrete age at time of installation—installation in early-age concrete may result in reduced bond strength (refer to R17.2.2)
(h) Peak concrete temperatures during anchor service life—under specific conditions (for example, anchors in thin concrete members exposed to direct sunlight), elevated concrete temperatures can result in reduced bond strength
(i) Chemical exposure—anchors used in industrial environments may be exposed to increased levels of contaminants that can reduce bond strength over time

Anchors tested and assessed under ACI 355.4 may in some cases not be qualified for all of the installation and service environments represented in Table 17.6.5.2.5. Therefore, where the minimum values given in Table 17.6.5.2.5 are used for design, the relevant installation and service environments should be specified in accordance with 26.7.1(i), (j), (k), and (l), and only anchors that have been qualified under ACI 355.4 for the installation and service environments corresponding to the characteristic bond stress taken from Table 17.6.5.2.5 should be specified.

CODE

COMMENTARY

Characteristic bond stresses associated with qualified adhesive anchor systems for a specific set of installation and use conditions may substantially exceed the minimum values provided in Table 17.6.5.2.5. For example, 1/2-in. to 3/4-in. diameter anchors installed in impact-drilled holes in dry concrete where use is limited to indoor conditions in uncracked concrete as described above may exhibit characteristic bond stresses τ_{uncr} in the range of 2000 to 2500 psi.

17.6.5.3 *Bond eccentricity factor,* $\psi_{ec,Na}$

17.6.5.3.1 Modification factor for adhesive anchor groups loaded eccentrically in tension, $\psi_{ec,Na}$, shall be calculated by Eq (17.6.5.3.1).

$$\psi_{ec,Na} = \frac{1}{\left(1+\frac{e'_N}{c_{Na}}\right)} \leq 1.0 \qquad (17.6.5.3.1)$$

17.6.5.3.2 If the loading on an adhesive anchor group is such that only some of the adhesive anchors are in tension, only those adhesive anchors that are in tension shall be considered for determining eccentricity e'_N in Eq. (17.6.5.3.1) and for the calculation of N_{ag} according to Eq. (17.6.5.1b).

17.6.5.3.3 If a load is eccentric about two orthogonal axes, $\psi_{ec,Na}$ shall be calculated for each axis individually, and the product of these factors shall be used as $\psi_{ec,Na}$ in Eq. (17.6.5.1b).

R17.6.5.3 *Bond eccentricity factor,* $\psi_{ec,Na}$

R17.6.5.3.1 Refer to R17.6.2.3.1.

17.6.5.4 *Bond edge effect factor,* $\psi_{ed,Na}$

17.6.5.4.1 Modification factor for edge effects for single adhesive anchors or adhesive anchor groups in tension, $\psi_{ed,Na}$, shall be determined by (a) or (b) using the critical distance c_{Na} as defined in Eq. (17.6.5.1.2b).

(a) If $c_{a,min} \geq c_{Na}$, then $\psi_{ed,Na} = 1.0$ (17.6.5.4.1a)

(b) If $c_{a,min} < c_{Na}$, then $\psi_{ed,Na} = 0.7 + 0.3\frac{c_{a,min}}{c_{Na}}$ (17.6.5.4.1b)

R17.6.5.4 *Bond edge effect factor,* $\psi_{ed,Na}$

R17.6.5.4.1 If anchors are located close to an edge, their strength is further reduced beyond that reflected in A_{Na}/A_{Nao}. The factor $\psi_{ed,Na}$ accounts for the edge effect (Fuchs et al. 1995; Eligehausen et al. 2006a).

17.6.5.5 *Bond splitting factor,* $\psi_{cp,Na}$

17.6.5.5.1 Modification factor for adhesive anchors designed for uncracked concrete in accordance with 17.6.5.1 without supplementary reinforcement to control splitting, $\psi_{cp,Na}$, shall be determined by (a) or (b) where c_{ac} is defined in 17.9.5

(a) If $c_{a,min} \geq c_{ac}$, then $\psi_{cp,Na} = 1.0$ (17.6.5.5.1a)

(b) If $c_{a,min} < c_{ac}$, then $\psi_{cp,Na} = \frac{c_{a,min}}{c_{ac}} \geq \frac{c_{Na}}{c_{ac}}$ (17.6.5.5.1b)

CODE

17.6.5.5.2 For all other cases, $\psi_{cp,Na}$ shall be taken as 1.0.

17.7—Shear strength

17.7.1 *Steel strength of anchors in shear,* V_{sa}

17.7.1.1 Nominal steel strength of anchors in shear as governed by the steel, V_{sa}, shall be evaluated based on the properties of the anchor material and the physical dimensions of the anchors. If concrete breakout is a potential failure mode, the required steel shear strength shall be consistent with the assumed breakout surface.

17.7.1.2 Nominal strength of an anchor in shear, V_{sa}, shall not exceed (a) through (c):

(a) For cast-in headed stud anchor

$$V_{sa} = A_{se,V} f_{uta} \qquad (17.7.1.2a)$$

where $A_{se,V}$ is the effective cross-sectional area of an anchor in shear, in.^{2}, and f_{uta} used for calculations shall not exceed either $\mathbf{1.9}f_{ya}$ or 125,000 psi.

(b) For cast-in headed bolt and hooked bolt anchors and for post-installed anchors where sleeves do not extend through the shear plane

$$V_{sa} = 0.6A_{se,V} f_{uta} \qquad (17.7.1.2b)$$

where $A_{se,V}$ is the effective cross-sectional area of an anchor in shear, in.^{2}, and the value of f_{uta} shall not exceed either $\mathbf{1.9}f_{ya}$ or 125,000 psi.

(c) For post-installed anchors where sleeves extend through the shear plane, V_{sa} shall be based on the 5 percent fractile of results of tests performed and evaluated in accordance with ACI 355.2. Alternatively, Eq. (17.7.1.2b) shall be permitted to be used.

17.7.1.2.1 If anchors are used with built-up grout pads, nominal strength V_{sa} calculated in accordance with 17.7.1.2 shall be multiplied by 0.80.

17.7.2 *Concrete breakout strength of anchors in shear,* V_{cb}

17.7.2.1 Nominal concrete breakout strength in shear, V_{cb} of a single anchor or V_{cbg} of an anchor group satisfying 17.5.1.3.1, shall be calculated in accordance with (a) through (d):

(a) For shear perpendicular to the edge on a single anchor

$$V_{cb} = \frac{A_{Vc}}{A_{Vco}} \psi_{ed,V} \psi_{c,V} \psi_{h,V} V_b \qquad (17.7.2.1a)$$

(b) For shear perpendicular to the edge on an anchor group

$$V_{cbg} = \frac{A_{Vc}}{A_{Vco}} \psi_{ec,V} \psi_{ed,V} \psi_{c,V} \psi_{h,V} V_b \qquad (17.7.2.1b)$$

COMMENTARY

R17.7—Shear strength

R17.7.1 *Steel strength of anchors in shear,* V_{sa}

R17.7.1.1 The shear applied to each anchor in an anchor group may vary depending on assumptions for the concrete breakout surface and load redistribution (refer to R17.7.2.1).

R17.7.1.2 The nominal shear strength of anchors is best represented as a function of f_{uta} rather than f_{ya} because the large majority of anchor materials do not exhibit a well-defined yield point. Welded studs develop a higher steel shear strength than headed anchors due to the fixity provided by the weld between the studs and the base plate. The use of Eq. (17.7.1.2a) and (17.7.1.2b) with the load factors of 5.3 and the ϕ-factors of 17.5.3 result in design strengths consistent with AISC 360.

The limitation of $\mathbf{1.9}f_{ya}$ on f_{uta} is to ensure that, under service load conditions, the anchor stress does not exceed f_{ya}. The limit on f_{uta} of $\mathbf{1.9}f_{ya}$ was determined by converting the LRFD provisions to corresponding service-level conditions, as discussed in R17.6.1.2.

For post-installed anchors having a reduced cross-sectional area anywhere along the anchor length, the effective cross-sectional area of the anchor should be provided by the manufacturer. For threaded rods and headed bolts, ASME B1.1 defines $A_{se,V}$ as

$$A_{se,V} = \frac{\pi}{4}\left(d_a - \frac{0.9743}{n_t}\right)^2$$

where n_t is the number of threads per inch.

R17.7.2 *Concrete breakout strength of anchors in shear,* V_{cb}

R17.7.2.1 The shear strength equations were developed from the CCD Method (refer to R17.5.1.3). They assume a breakout angle of approximately 35 degrees (refer to Fig. R17.5.1.3b) and consider fracture mechanics theory. The effects of multiple anchors, spacing of anchors, edge distance, and thickness of the concrete member on nominal concrete breakout strength in shear are included by applying the reduction factor of A_{Vc}/A_{Vco} in Eq. (17.7.2.1a) and (17.7.2.1b), and $\psi_{ec,V}$ in Eq. (17.7.2.1b). For anchors far from the edge, 17.7.2 usually will not govern. For these cases, 17.7.1 and 17.7.3 often govern.

CODE

(c) For shear parallel to an edge, V_{cb} or V_{cbg} shall be permitted to be twice the value of the shear calculated by Eq. (17.7.2.1a) or (17.7.2.1b), respectively, with the shear assumed to act perpendicular to the edge and $\psi_{ed,V}$ taken equal to 1.0.

(d) For anchors located at a corner, the limiting nominal concrete breakout strength shall be calculated for each edge, and the lesser value shall be used.

where $\psi_{ec,V}$, $\psi_{ed,V}$, $\psi_{c,V}$, and $\psi_{h,V}$ are given in 17.7.2.3, 17.7.2.4, 17.7.2.5, and 17.7.2.6, respectively.

17.7.2.1.1 A_{Vc} is the projected area of the failure surface on the side of the concrete member at its edge for a single anchor or an anchor group. It shall be permitted to evaluate A_{Vc} as the base of a truncated half-pyramid projected on the side face of the member where the top of the half-pyramid is given by the axis of the anchor row selected as critical. The value of c_{a1} shall be taken as the distance from the edge to this axis. A_{Vc} shall not exceed nA_{Vco}, where n is the number of anchors in the group.

COMMENTARY

Figure R17.7.2.1a shows A_{Vco} and the development of Eq. (17.7.2.1.3). A_{Vco} is the maximum projected area for a single anchor that approximates the surface area of the full breakout volume for an anchor unaffected by edge distance, spacing, or depth of member. Figure R17.7.2.1b shows examples of the projected areas for various single-anchor and multiple-anchor arrangements. A_{Vc} approximates the full surface area of the breakout for the particular arrangement of anchors. Because A_{Vc} is the total projected area for an anchor group, and A_{Vco} is the area for a single anchor, there is no need to include the number of anchors in the equation.

As shown in the examples in Fig. R17.7.2.1b of two-anchor groups loaded in shear, when using Eq. (17.7.2.1b) for cases where the anchor spacing s is greater than the edge distance to the near-edge anchor $c_{a1,1}$, both assumptions for load distribution illustrated in Cases 1 and 2 should be considered. This is because the anchors nearest to the free edge could fail first or the entire group could fail as a unit with the failure surface originating from the anchors farthest from the edge. For Case 1, the steel shear strength is provided by both anchors. For Case 2, the steel shear strength is provided entirely by the anchor farthest from the edge; no contribution of the anchor near the edge is considered. In addition, checking the near-edge anchor for concrete breakout under service loads is advisable to preclude undesirable cracking at service conditions. If the anchor spacing s is less than the edge distance to the near-edge anchor, the failure surfaces may merge (Eligehausen et al. 2006b) and Case 3 of Fig. R17.7.2.1b may be taken as a conservative approach.

If the anchors are welded to a common plate (regardless of anchor spacing s), when the anchor nearest the front edge begins to form a breakout failure, shear is transferred to the stiffer and stronger rear anchor. For this reason, only Case 2 need be considered, which is consistent with Section 6.5.5 of the *PCI Design Handbook* (PCI MNL 120). For determination of steel shear strength, it is conservative to consider only the anchor farthest from the edge. However, for anchors having a ratio of $s/c_{a1,1}$ less than 0.6, both the front and rear anchors may be assumed to resist the shear (Anderson and Meinheit 2007). For ratios of $s/c_{a1,1}$ greater than 1, it is advisable to check concrete breakout of the near-edge anchor to preclude undesirable cracking at service conditions.

Further discussion of design for multiple anchors is given in Primavera et al. (1997).

For anchors near a corner required to resist a shear force with components normal to each edge, a satisfactory solution is to check the connection independently for each component of the shear force. Other specialized cases, such as the shear resistance of anchor groups where all anchors do not have the same edge distance, are treated in Eligehausen et al. (2006a).

The detailed provisions of 17.7.2.1(a) apply to the case of shear directed toward an edge. If the shear is directed away from the edge, the strength will usually be governed by 17.7.1 or 17.7.3. The case of shear parallel to an edge

CODE

COMMENTARY

is shown in Fig. R17.7.2.1c. The maximum shear that can be applied parallel to the edge, $V_{||}$, as governed by concrete breakout, is twice the maximum shear that can be applied perpendicular to the edge, $V_{\perp}$. For a single anchor required to resist shear near a corner (refer to Fig. R17.7.2.1d), the provisions for shear applied perpendicular to the edge should be checked in addition to the provisions for shear applied parallel to the edge.

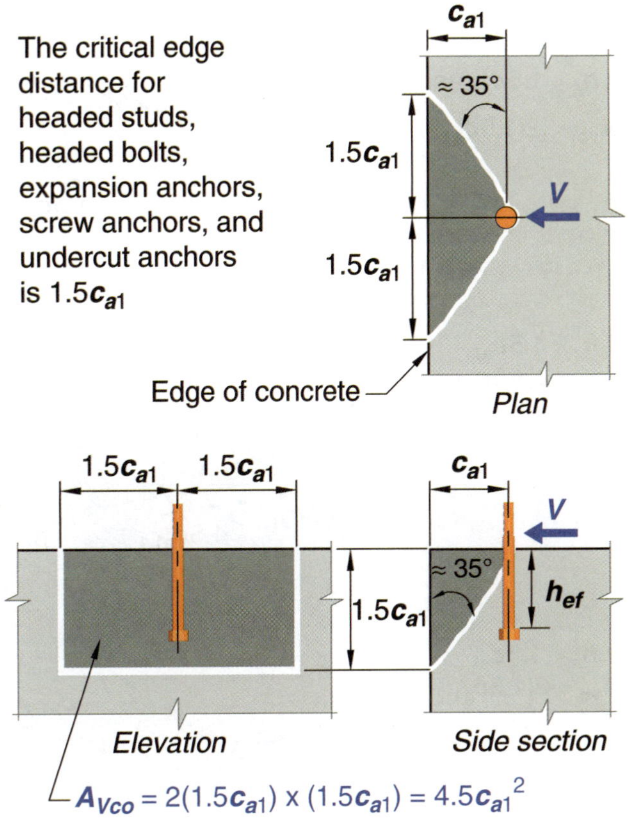

Fig. R17.7.2.1a—Calculation of A_{Vco}.

CODE

COMMENTARY

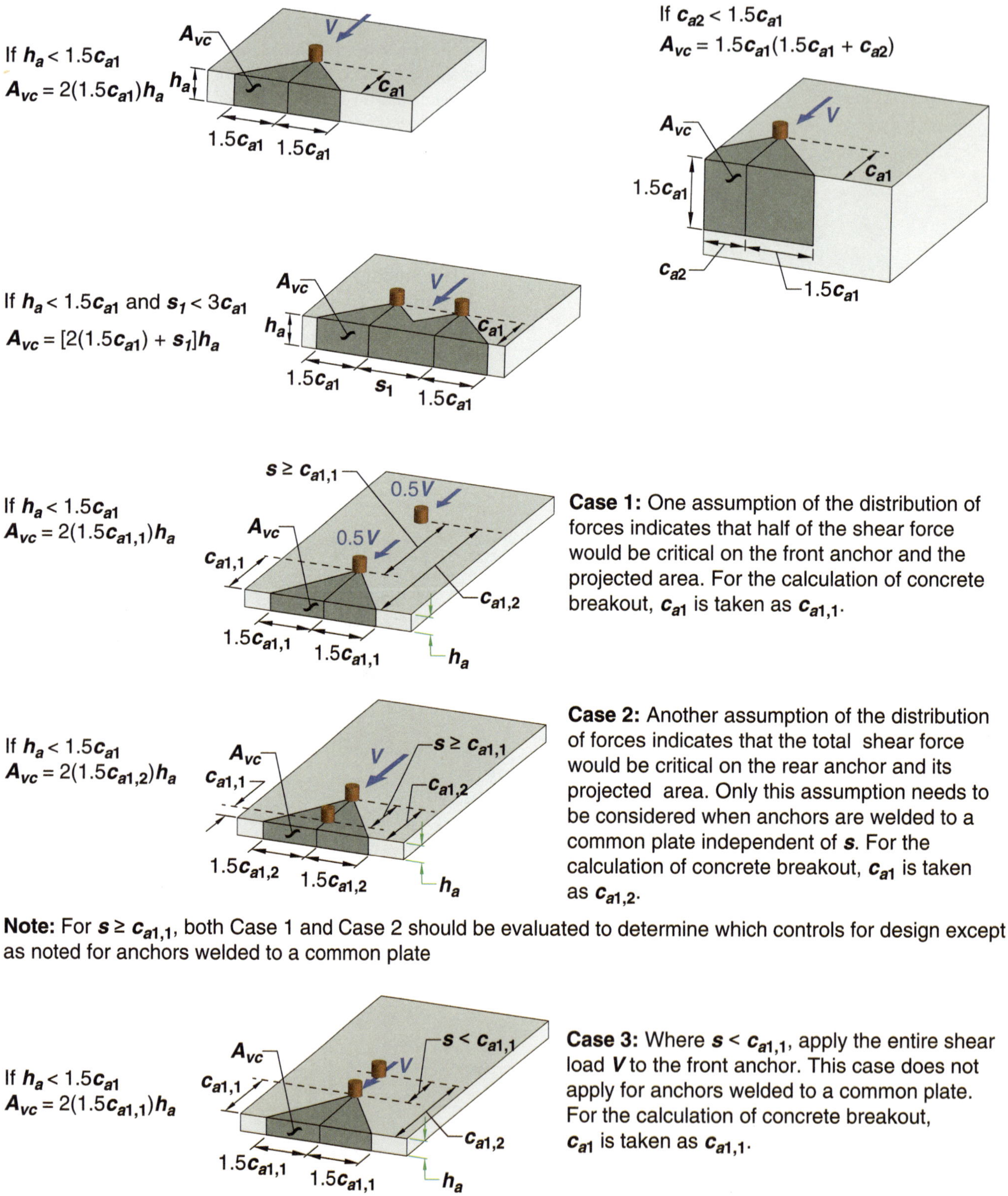

Fig. R17.7.2.1b—Calculation of A_{vc} for single anchors and anchor groups.

CODE

COMMENTARY

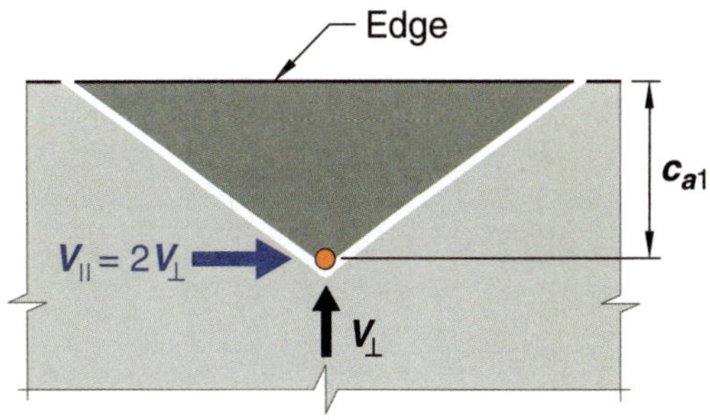

Fig. R17.7.2.1c—Shear force parallel to an edge.

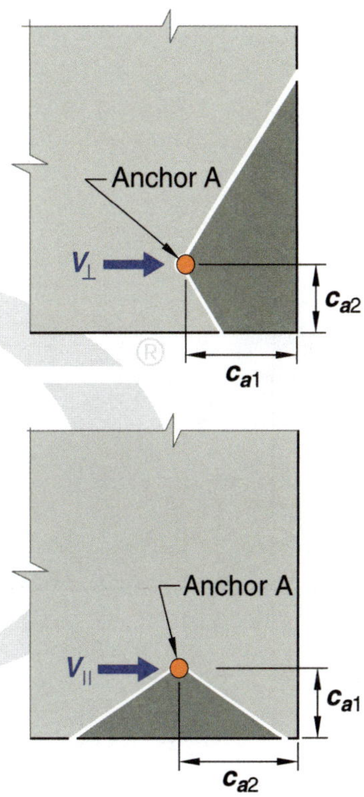

Fig. R17.7.2.1d—Shear near a corner.

17.7.2.1.2 If anchors are located in narrow sections of limited thickness such that both edge distances c_{a2} and thickness h_a are less than $1.5c_{a1}$, the value of c_{a1} used to calculate A_{Vc} in accordance with 17.7.2.1.1 as well as for the equations in 17.7.2.1 through 17.7.2.6 shall not exceed the greatest of (a) through (c).

(a) $c_{a2}/1.5$, where c_{a2} is the greatest edge distance
(b) $h_a/1.5$
(c) $s/3$, where s is the maximum spacing perpendicular to direction of shear, between anchors within a group

R17.7.2.1.2 For anchors located in narrow sections of limited thickness where the edge distances perpendicular to the direction of load and the member thickness are less than $1.5c_{a1}$, the shear breakout strength calculated by the CCD Method (refer to R17.5.1.3) is overly conservative. These cases were studied for the Kappa Method (Eligehausen and Fuchs 1988), and the problem was pointed out by Lutz (1995). Similar to the approach used for concrete breakout strength in tension in 17.6.2.1.2, the concrete breakout strength in shear for this case is more accurately evaluated if the value of c_{a1} used in 17.7.2.1 through 17.7.2.6 and in the calculation of A_{Vc} is limited to the maximum of two-thirds of the greater of the two edge distances perpendicular to the direction of shear, two-thirds of the member thickness, and one-third of the maximum spacing between individual anchors within the group, measured perpendicular to the direction of shear. The limit on c_{a1} of at least one-third of the maximum spacing between anchors within the group

CODE

COMMENTARY

prevents the use of a calculated strength based on individual breakout volumes for an anchor group configuration.

This approach is illustrated in Fig. R17.7.2.1.2. In this example, the limiting value of c_{a1} is denoted as c'_{a1} and is used to calculate A_{Vc}, A_{Vco}, $\psi_{ed,V}$, and $\psi_{h,V}$ as well as V_b (not shown). The requirement of 17.7.2.1.2 may be visualized by moving the actual concrete breakout surface originating at the actual c_{a1} toward the surface of the concrete in the direction of the applied shear. The value of c_{a1} used to calculate A_{Vc} and to be used in 17.7.2.1 through 17.7.2.6 is determined when (a) an outer boundary of the failure surface first intersects the concrete surface;, or (b) the intersection of the breakout surface between individual anchors within the group first intersects the concrete surface. For the example shown in Fig. R17.7.2.1.2, point "A" shows the intersection of the assumed failure surface for limiting c_{a1} with the concrete surface.

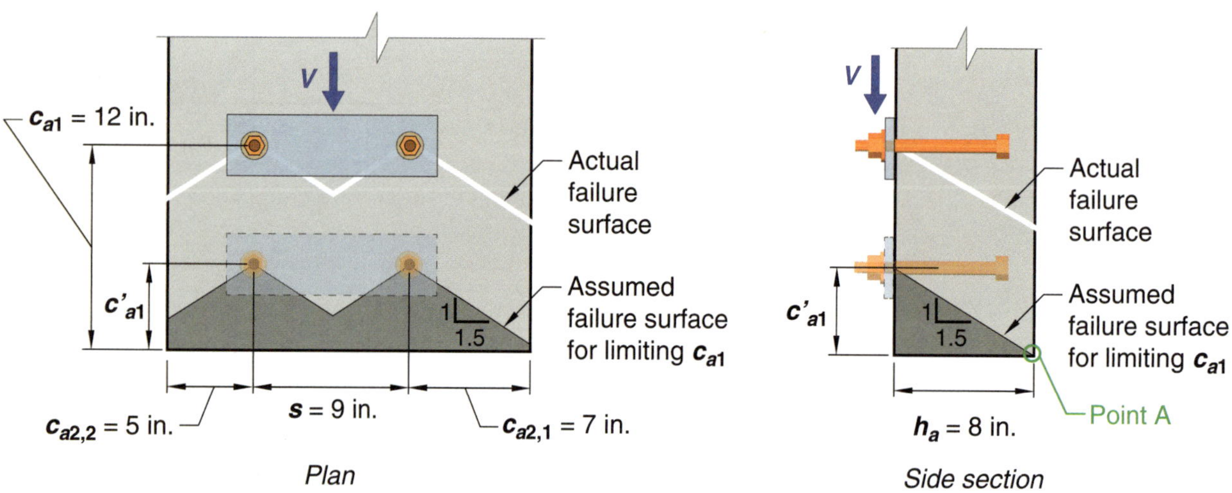

1. The actual c_{a1} = 12 in.
2. The two edge distances c_{a2} as well as h_a are all less than $1.5c_{a1}$.
3. The limiting value of c_{a1} (shown as c'_{a1} in the figure) to be used to calculate A_{Vc} and to be used in 17.7.2.1 through 17.7.2.6 is the largest of the following:

   $(c_{a2,max})/1.5 = (7)/1.5 = 4.67$ in.

   $(h_a)/1.5 = (8)/1.5 = 5.33$ in. (controls)

   $s/3 = 9/3 = 3$ in.
4. For this case, A_{Vc}, A_{Vco}, $\psi_{ed,V}$, and $\psi_{h,V}$ are:

   $A_{Vc} = (5 + 9 + 7)(1.5 \times 5.33) = 168$ in.^{2}

   $A_{Vco} = 4.5(5.33)^2 = 128$ in.^{2}

   $\psi_{ed,V} = 0.7 + 0.3(5)/5.33 = 0.98$

   $\psi_{h,V}$ = 1.0 because $c_{a1} = (h_a)/1.5$. Point A shows the intersection of the assumed failure surface with the concrete surface that establishes the limiting value of c_{a1}.

Fig. R17.7.2.1.2—Example of shear where anchors are located in narrow members of limited thickness.

CODE

17.7.2.1.3 A_{Vco} is the projected area for a single anchor in a deep member with a distance from edges of at least $\mathbf{1.5}c_{a1}$ in the direction perpendicular to the shear. It shall be permitted to calculate A_{Vco} by Eq. (17.7.2.1.3), which gives the area of the base of a half-pyramid with a side length parallel to the edge of $\mathbf{3}c_{a1}$ and a depth of $\mathbf{1.5}c_{a1}$.

$$A_{Vco} = 4.5(c_{a1})^2 \qquad (17.7.2.1.3)$$

17.7.2.1.4 If anchors are located at varying distances from the edge and the anchors are welded to the attachment so as to distribute the force to all anchors, it shall be permitted to evaluate the strength based on the distance to the farthest row of anchors from the edge. In this case, it shall be permitted to base the value of c_{a1} on the distance from the edge to the axis of the farthest anchor row that is selected as critical, and all of the shear shall be assumed to be resisted by this critical anchor row alone.

17.7.2.2 *Basic single anchor breakout strength,* V_b

17.7.2.2.1 Basic concrete breakout strength of a single anchor in shear in cracked concrete, V_b, shall not exceed the lesser of (a) and (b):

(a) $$V_b = \left(7\left(\frac{\ell_e}{d_a}\right)^{0.2}\sqrt{d_a}\right)\lambda_a\sqrt{f_c'}(c_{a1})^{1.5} \qquad (17.7.2.2.1a)$$

where ℓ_e is the load-bearing length of the anchor for shear:
$\ell_e = h_{ef}$ for anchors with a constant stiffness over the full length of embedded section, such as headed studs and post-installed anchors with one tubular shell over full length of the embedment depth;
$\ell_e = 2d_a$ for torque-controlled expansion anchors with a distance sleeve separated from expansion sleeve;
$\ell_e \leq 8d_a$ in all cases.

(b) $$V_b = 9\lambda_a\sqrt{f_c'}(c_{a1})^{1.5} \qquad (17.7.2.2.1b)$$

17.7.2.2.2 For cast-in headed studs, headed bolts, or hooked bolts that are continuously welded to steel attachments, basic concrete breakout strength of a single anchor in shear in cracked concrete, V_b, shall be the lesser of Eq. (17.7.2.2.1b) and Eq. (17.7.2.2.2) provided that (a) through (d) are satisfied.

$$V_b = \left(8\left(\frac{\ell_e}{d_a}\right)^{0.2}\sqrt{d_a}\right)\lambda_a\sqrt{f_c'}(c_{a1})^{1.5} \qquad (17.7.2.2.2)$$

where ℓ_e is defined in 17.7.2.2.1.

(a) Steel attachment thickness is the greater of $\mathbf{0.5}d_a$ and 3/8 in.

COMMENTARY

R17.7.2.2 *Basic single anchor breakout strength,* V_b

R17.7.2.2.1 Like the concrete breakout tensile strength, the concrete breakout shear strength does not increase with the failure surface, which is proportional to $(c_{a1})^2$. Instead, the strength increases proportionally to $(c_{a1})^{1.5}$ due to the size effect. The constant, 7, in the shear strength equation (17.7.2.2.1a) was determined from test data reported in Fuchs et al. (1995) at the 5 percent fractile adjusted for cracking.

The strength is also influenced by the anchor stiffness and the anchor diameter (Fuchs et al. 1995; Eligehausen and Balogh 1995; Eligehausen et al. 1987, 2006b; Eligehausen and Fuchs 1988). The influence of anchor stiffness and diameter is not apparent in large-diameter anchors (Lee et al. 2010), resulting in a limitation on the shear breakout strength provided by Eq. (17.7.2.2.1b).

R17.7.2.2.2 For cast-in headed bolts continuously welded to an attachment, test data (Shaikh and Yi 1985) show that somewhat higher shear strength exists, possibly due to the stiff welded connection clamping the bolt more effectively than an attachment with an anchor gap. Because of this, the basic shear breakout strength for such anchors is increased, but the upper limit of Eq. (17.7.2.2.1b) is imposed because tests on large-diameter anchors welded to steel attachments are not available to justify a higher value than Eq. (17.7.2.2.1b). The design of supplementary reinforcement is discussed in *fib* (2011), Eligehausen et al. (1987, 2006b), and Eligehausen and Fuchs (1988).

CODE

(b) Anchor spacing s is at least 2.5 in.
(c) Reinforcement is provided at the corners if $c_{a2} \leq 1.5h_{ef}$
(d) For anchor groups, the strength is calculated based on the strength of the row of anchors farthest from the edge.

17.7.2.3 *Breakout eccentricity factor, $\psi_{ec,V}$*

17.7.2.3.1 Modification factor for anchor groups loaded eccentrically in shear, $\psi_{ec,V}$, shall be calculated by Eq. (17.7.2.3.1).

$$\psi_{ec,V} = \frac{1}{\left(1+\frac{e'_V}{1.5c_{a1}}\right)} \leq 1.0 \qquad (17.7.2.3.1)$$

17.7.2.3.2 If the loading on an anchor group is such that only some of the anchors in the group are in shear, only those anchors that are in shear in the same direction shall be considered for determining the eccentricity e'_V in Eq. (17.7.2.3.1) and for the calculation of V_{cbg} according to Eq. (17.7.2.1b).

17.7.2.4 *Breakout edge effect factor, $\psi_{ed,V}$*

17.7.2.4.1 Modification factor for edge effects for single anchors or anchor groups loaded in shear, $\psi_{ed,V}$, shall be determined by (a) or (b) using the lesser value of c_{a2}.

(a) If $c_{a2} \geq 1.5c_{a1}$, then $\psi_{ed,V} = 1.0$ (17.7.2.4.1a)

COMMENTARY

R17.7.2.3 *Breakout eccentricity factor, $\psi_{ec,V}$*

R17.7.2.3.1 This section provides a modification factor for an eccentric shear toward an edge on an anchor group. If the shear originates above the plane of the concrete surface, the shear should first be resolved as a shear in the plane of the concrete surface, acting in combination with a moment that may or may not also cause tension in the anchors, depending on the normal force. Figure R17.7.2.3.1 defines the term e'_V for calculating the $\psi_{ec,V}$ modification factor that accounts for the fact that more shear is applied to one anchor than others, tending to split the concrete near an edge.

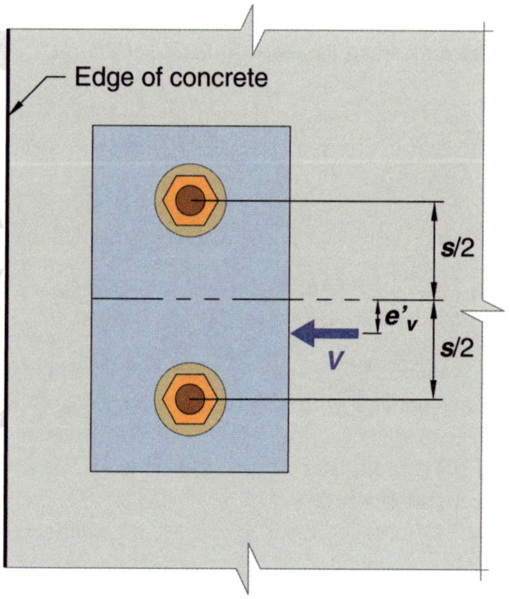

Fig. R17.7.2.3.1—Definition of e'_V for an anchor group.

CODE

(b) If $c_{a2} < 1.5c_{a1}$, then $\psi_{ed,V} = 0.7 + 0.3\frac{c_{a2}}{1.5c_{a1}}$ (17.7.2.4.1b)

17.7.2.5 *Breakout cracking factor, $\psi_{c,V}$*

17.7.2.5.1 Modification factor for the influence of cracking in anchor regions at service load levels and presence or absence of supplementary reinforcement, $\psi_{c,V}$, shall be determined as follows:

(a) For anchors located in a region of a concrete member where analysis indicates no cracking at service load levels, $\psi_{c,V}$ shall be permitted to be 1.4.

(b) For anchors located in a region of a concrete member where analysis indicates cracking at service load levels, $\psi_{c,V}$ shall be in accordance with Table 17.7.2.5.1.

Table 17.7.2.5.1—Modification factor where analysis indicates cracking at service load levels, $\psi_{c,V}$

Condition	$\psi_{c,V}$
Anchors without supplementary reinforcement or with edge reinforcement smaller than a No. 4 bar	1.0
Anchors with reinforcement of at least a No. 4 bar or greater between the anchor and the edge	1.2
Anchors with reinforcement of at least a No. 4 bar or greater between the anchor and the edge, and with the reinforcement enclosed within stirrups spaced at not more than 4 in.	1.4

17.7.2.6 *Breakout thickness factor, $\psi_{h,V}$*

17.7.2.6.1 Modification factor for anchors located in a concrete member where $h_a < 1.5c_{a1}$, $\psi_{h,V}$ shall be calculated by Eq. (17.7.2.6.1)

$$\psi_{h,V} = \sqrt{\frac{1.5c_{a1}}{h_a}} \geq 1.0 \qquad (17.7.2.6.1)$$

17.7.3 *Concrete pryout strength of anchors in shear,* V_{cp} *or* V_{cpg}

17.7.3.1 Nominal pryout strength, V_{cp} of a single anchor or V_{cpg} of an anchor group satisfying 17.5.1.3.1, shall not exceed (a) or (b), respectively.

(a) For a single anchor

$$V_{cp} = k_{cp}N_{cp} \qquad (17.7.3.1a)$$

(b) For an anchor group

$$V_{cpg} = k_{cp}N_{cpg} \qquad (17.7.3.1b)$$

where

COMMENTARY

R17.7.2.6 *Breakout thickness factor, $\psi_{h,V}$*

R17.7.2.6.1 For anchors located in a concrete member where $h_a < 1.5c_{a1}$, tests (*fib* 2011; Eligehausen et al. 2006b) have shown that the concrete breakout strength in shear is not directly proportional to the member thickness h_a. The factor $\psi_{h,V}$ accounts for this effect.

R17.7.3 *Concrete pryout strength of anchors in shear,* V_{cp} *or* V_{cpg}

R17.7.3.1 Fuchs et al. (1995) indicates that the pryout shear resistance can be approximated as one to two times the anchor tensile resistance with the lower value appropriate for h_{ef} less than 2.5 in. Because it is possible that the bond strength of adhesive anchors could be less than the concrete breakout strength, it is necessary to consider both 17.6.2.1 and 17.6.5.1 to calculate pryout strength.

CODE

k_{cp} = **1.0** for h_{ef} < **2.5 in.**
k_{cp} = **2.0** for $h_{ef} \geq$ **2.5 in.**

17.7.3.1.1 For cast-in anchors and post-installed expansion, screw, and undercut anchors, N_{cp} shall be taken as N_{cb} calculated by Eq. (17.6.2.1a), and for adhesive anchors, N_{cp} shall be the lesser of N_a calculated by Eq. (17.6.5.1a) and N_{cb} calculated by Eq. (17.6.2.1a).

17.7.3.1.2 For cast-in anchors and post-installed expansion, screw, and undercut anchors, N_{cpg} shall be taken as N_{cbg} calculated by Eq. (17.6.2.1b), and for adhesive anchors, N_{cpg} shall be the lesser of N_{ag} calculated by Eq. (17.6.5.1b) and N_{cbg} calculated by Eq. (17.6.2.1b).

17.8—Tension and shear interaction

17.8.1 Unless tension and shear interaction effects are considered in accordance with 17.5.2.3, anchors or anchor groups that resist both tension and shear shall satisfy 17.8.2 and 17.8.3. The values of ϕN_n and ϕV_n shall be in accordance with 17.5.2 or 17.10.

17.8.2 It shall be permitted to neglect the interaction between tension and shear if (a) or (b) is satisfied.

(a) $N_{ua}/(\phi N_n) \leq 0.2$ (17.8.2a)
(b) $V_{ua}/(\phi V_n) \leq 0.2$ (17.8.2b)

17.8.3 If $N_{ua}/(\phi N_n) > 0.2$ for the governing strength in tension and $V_{ua}/(\phi V_n) > 0.2$ for the governing strength in shear, then Eq. (17.8.3) shall be satisfied.

$$\frac{N_{ua}}{\phi N_n} + \frac{V_{ua}}{\phi V_n} \leq 1.2 \quad (17.8.3)$$

17.9—Edge distances, spacings, and thicknesses to preclude splitting failure

17.9.1 Minimum spacings and edge distances for anchors and minimum thicknesses of members shall conform to this section, unless supplementary reinforcement is provided to control splitting. Lesser values from product-specific tests

COMMENTARY

R17.8—Tension and shear interaction

The tension-shear interaction expression has traditionally been expressed as

$$\left(\frac{N_{ua}}{N_n}\right)^{\varsigma} + \left(\frac{V_{ua}}{V_n}\right)^{\varsigma} \leq 1.0$$

where ς varies from 1 to 2. The current trilinear recommendation is a simplification of the expression where ς = **5/3** (Fig. R17.8). The limits were chosen to eliminate the requirement for calculation of interaction effects where very small values of the second force are present. Any other interaction expression that is verified by test data, however, can be used to satisfy 17.5.2.3.

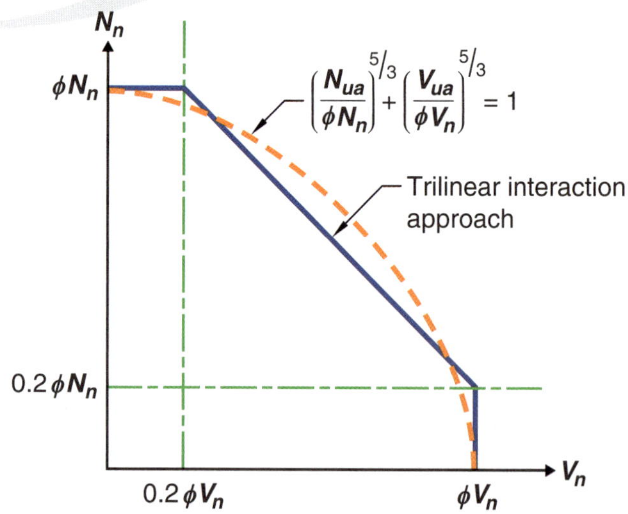

***Fig. R17.8**—Shear and tensile load interaction equation.*

R17.9—Edge distances, spacings, and thicknesses to preclude splitting failure

R17.9.1 Minimum spacings, edge distances, and thicknesses are dependent on the anchor characteristics. Installation forces and torques in post-installed anchors can cause splitting of the surrounding concrete. Such splitting also can

CODE

performed in accordance with ACI 355.2 or ACI 355.4 shall be permitted.

COMMENTARY

be produced in subsequent torquing during connection of attachments to anchors including cast-in anchors. The primary source of values for minimum spacings, edge distances, and thicknesses of post-installed anchors should be the product-specific tests of ACI 355.2 and ACI 355.4. In some cases, however, specific products are not known in the design stage. Approximate values are provided for use in design.

17.9.2 Unless determined in accordance with 17.9.3, minimum spacing parameters shall conform to Table 17.9.2(a).

Table 17.9.2(a)—Minimum spacing and edge distance requirements

Spacing parameter	Anchor type: Cast-in anchors — Not torqued	Cast-in anchors — Torqued	Post-installed expansion and undercut anchors	Post-installed screw anchors
Minimum anchor spacing	$4d_a$	$6d_a$	$6d_a$	Greater of $0.6h_{ef}$ and $6d_a$
Minimum edge distance	Specified cover requirements for reinforcement according to 20.5.1.3	$6d_a$	Greatest of (a), (b), and (c): (a) Specified cover requirements for reinforcement according to 20.5.1.3 (b) Twice the maximum aggregate size (c) Minimum edge distance requirements according to ACI 355.2 or 355.4, or Table 17.9.2(b) when product information is absent	

Table 17.9.2(b)—Minimum edge distance in absence of product-specific ACI 355.2 or ACI 355.4 test information

Post-installed anchor type	Minimum edge distance
Torque-controlled	$8d_a$
Displacement-controlled	$10d_a$
Screw	$6d_a$
Undercut	$6d_a$
Adhesive	$6d_a$

R17.9.2 Edge cover for anchors with deep embedments can have a significant effect on the side face blowout strength provided in 17.6.4. It is therefore advantageous to increase edge cover beyond that required in 20.5.1.3 to increase side-face blowout strength.

Drilling holes for post-installed anchors can cause micro-cracking. The requirement for edge distance to be at least twice the maximum aggregate size is to reduce effects of such microcracking.

17.9.3 For anchors where installation does not produce a splitting force and that will not be torqued, if the edge distance or spacing is less than those given in 17.9.2, calculations shall be performed by substituting for d_a a lesser value d_a' that meets the requirements of 17.9.2. Calculated forces applied to the anchor shall be limited to the values corresponding to an anchor having a diameter of d_a'.

R17.9.3 In some cases, it may be desirable to use a larger-diameter anchor than the requirements of 17.9.2 permit. In these cases, it is permissible to use a larger-diameter anchor, provided the design strength of the anchor is based on a smaller assumed anchor diameter d_a'.

17.9.4 Value of h_{ef} for a post-installed expansion, screw, or undercut post-installed anchor shall not exceed the greater

R17.9.4 Splitting failures are caused by load transfer between the bolt and the concrete. The limitations on the

CODE

of 2/3 of the member thickness, h_a, and the member thickness minus 4 in., unless determined from tests in accordance with ACI 355.2.

17.9.5 Critical edge distance c_{ac} shall be in accordance with Table 17.9.5 unless determined from tension tests in accordance with ACI 355.2 or ACI 355.4.

Table 17.9.5—Critical edge distance

Post-installed anchor type	Critical edge distance c_{ac}
Torque-controlled	$4h_{ef}$
Displacement-controlled	$4h_{ef}$
Screw	$4h_{ef}$
Undercut	$2.5h_{ef}$
Adhesive	$2h_{ef}$

17.10—Earthquake-resistant anchor design requirements

17.10.1 Anchors in structures assigned to Seismic Design Category (SDC) C, D, E, or F shall satisfy the additional requirements of this section.

COMMENTARY

value of h_{ef} do not apply to cast-in and adhesive anchors because the splitting forces associated with these anchor types are less than for expansion, screw, and undercut anchors.

For all post-installed anchors, the embedment depth for a given member thickness should be limited to avoid back-face blowout on the opposite side of the concrete member during hole drilling and anchor setting. This depth limit is dependent on many variables, including anchor type, drilling method, drilling technique, type and size of drilling equipment, presence of reinforcement, and strength and condition of the concrete.

R17.9.5 The critical edge distance c_{ac} is required for design of post-installed anchors for use in uncracked concrete where no supplemental reinforcement is available to restrain splitting cracks. To permit the design of these types of anchors if product-specific information is not available, conservative default values for c_{ac} are provided. Alternately, product-specific values of c_{ac} may be determined in accordance with ACI 355.2 or ACI 355.4. Corner-test requirements in the aforementioned qualification standards may not be satisfied with $c_{a,min} = 1.5h_{ef}$ for many expansion, screw, undercut, and adhesive anchors due to tensile and flexural stresses associated with anchor installation and loading, which may result in a premature splitting failure.

R17.10—Earthquake-resistant anchor design requirements

R17.10.1 Unless 17.10.5.1 or 17.10.6.1 apply, all anchors in structures assigned to Seismic Design Categories (SDC) C, D, E, or F are required to satisfy the additional requirements of 17.10.2 through 17.10.7, regardless of whether earthquake-induced forces are included in the controlling load combination for the anchor design. In addition, all post-installed anchors in structures assigned to SDC C, D, E, or F must meet the requirements of ACI 355.2 or ACI 355.4 for prequalification of anchors to resist earthquake-induced forces. Ideally, for tension, anchor strength should be governed by yielding of the ductile steel element of the anchor. If the anchor cannot meet the specified ductility requirements of 17.10.5.3(a), then the attachment should be designed to yield if it is structural or light gauge steel, or designed to crush if it is wood. If ductility requirements of 17.10.5.3(a) are satisfied, then any attachments to the anchor should be designed not to yield. In designing attachments using yield mechanisms to provide adequate ductility, as permitted by 17.10.5.3(b) and 17.10.6.3(a), the ratio of specified yield strength to expected strength for the material of the attachment should be considered in determining the design force. The value used for the expected strength should consider both material overstrength and strain hardening effects. For example, the material in a connection element could yield and, due to an increase in its strength with strain hardening, cause a secondary failure of a sub-element or place extra force or deformation demands on the anchors.

CODE	COMMENTARY

For a structural steel attachment, if only the specified yield strength of the steel is known, the expected strength should be taken as approximately 1.5 times the specified yield strength. If the actual yield strength of the steel is known, the expected strength should be taken as approximately 1.25 times the actual yield strength.

Under earthquake conditions, the direction of shear may not be predictable. The full shear should be assumed in any direction for a safe design.

17.10.2 Provisions of this chapter shall not apply to the design of anchors in plastic hinge zones of concrete structures resisting earthquake-induced forces.

R17.10.2 The possible higher levels of cracking and spalling in plastic hinge zones are beyond the conditions for which the nominal concrete-governed strength values in this chapter are applicable. Plastic hinge zones are considered to extend a distance equal to twice the member depth from any column or beam face, and also include any other sections in walls, frames, and slabs where yielding of reinforcement is likely to occur as a result of lateral displacements.

If anchors must be located in plastic hinge regions, they should be detailed so that the anchor forces are transferred directly to anchor reinforcement that is designed to transmit the anchor forces into the body of the member beyond the anchorage region. Configurations that rely on concrete tensile strength should not be used.

17.10.3 Post-installed anchors shall be qualified for earthquake-induced forces in accordance with ACI 355.2 or ACI 355.4. The pullout strength, N_p, and steel strength in shear, V_{sa}, of post-installed expansion, screw, and undercut anchors shall be based on the results of the ACI 355.2 Simulated Seismic Tests. For adhesive anchors, the steel strength in shear, V_{sa}, and the characteristic bond stresses, τ_{uncr} and τ_{cr}, shall be based on results of the ACI 355.4 Simulated Seismic Tests.

R17.10.3 Anchors that are not suitable for use in cracked concrete should not be used to resist earthquake-induced forces. Qualification of post-installed anchors for use in cracked concrete is an integral part of the qualification for resisting earthquake-induced forces in ACI 355.2 and ACI 355.4. The design values obtained from the Simulated Seismic Tests of ACI 355.2 and ACI 355.4 are expected to be less than those for static load applications.

17.10.4 Anchor reinforcement used in structures assigned to SDC C, D, E, or F shall be deformed reinforcement and shall be in accordance with the anchor reinforcement requirements of 20.2.2.

17.10.5 *Tensile loading design requirements*

17.10.5.1 If the tensile component of the strength-level earthquake-induced force applied to a single anchor or anchor group does not exceed 20 percent of the total factored anchor tensile force associated with the same load combination, it shall be permitted to design a single anchor or anchor group in accordance with 17.6 and the tensile strength requirements of Table 17.5.2.

17.10.5.2 If the tensile component of the strength-level earthquake-induced force applied to anchors exceeds 20 percent of the total factored anchor tensile force associated with the same load combination, anchors and their attachments shall be designed in accordance with 17.10.5.3. The

R17.10.5 *Tensile loading design requirements*

R17.10.5.1 The requirements of 17.10.5.3 need not apply if the applied earthquake-induced tensile force is a small fraction of the total factored tensile force.

R17.10.5.2 If the ductile steel element is ASTM A36 or ASTM A307 steel, the f_{uta}/f_{ya} value is typically approximately 1.5, and the anchor can stretch considerably before rupturing at the threads. For other steels, calculations may need to be made to ensure that similar behavior can occur. Section R17.6.1.2 provides additional information on the

CODE

anchor design tensile strength shall be determined in accordance with 17.10.5.4.

17.10.5.3 Anchors and their attachments shall satisfy (a), (b), (c), or (d).

(a) For single anchors, the concrete-governed strength shall be greater than the steel strength of the anchor. For anchor groups, the ratio of the tensile load on the most highly stressed anchor to the steel strength of that anchor shall be equal to or greater than the ratio of the tensile load on anchors loaded in tension to the concrete-governed strength of those anchors. In each case:

(i) The steel strength shall be taken as 1.2 times the nominal steel strength of the anchor.

(ii) The concrete-governed strength shall be taken as the nominal strength considering pullout, side-face blowout, concrete breakout, and bond strength as applicable. For consideration of pullout in groups, the ratio shall be calculated for the most highly stressed anchor.

In addition, the following shall be satisfied:

(iii) Anchors shall transmit tensile loads via a ductile steel element with a stretch length of at least $\mathbf{8d_a}$ unless otherwise determined by analysis.

(iv) Anchors that resist load reversals shall be protected against buckling.

(v) If connections are threaded and the ductile steel elements are not threaded over their entire length, the ratio of f_{uta}/f_{ya} shall be at least 1.3 unless the threaded portions are upset. The upset portions shall not be included in the stretch length.

(vi) Deformed reinforcing bars used as ductile steel elements to resist earthquake-induced forces shall be in accordance with the anchor reinforcement requirements of 20.2.2.

(b) Anchor or anchor groups shall be designed for the maximum tension that can be transmitted to the anchor or group of anchors based on the development of a ductile yield mechanism in the attachment in tension, flexure, shear, or bearing, or a combination of those conditions, considering both material overstrength and strain-hardening effects for the attachment. The anchor design tensile strength shall be calculated in accordance with 17.10.5.4.

(c) Anchor or anchor groups shall be designed for the maximum tension that can be transmitted to the anchors by a non-yielding attachment. The anchor design tensile strength shall be calculated in accordance with 17.10.5.4.

(d) Anchor or anchor groups shall be designed for the maximum tension obtained from factored load combinations that include $\boldsymbol{E}$, with $\boldsymbol{E_h}$ increased by $\boldsymbol{\Omega_o}$. The anchor design tensile strength shall be calculated in accordance with 17.10.5.4.

COMMENTARY

steel properties of anchors. Use of upset threaded ends, whereby the threaded end of the anchor is enlarged to compensate for the area reduction associated with threading, can ensure that yielding occurs over the stretch length regardless of the tensile to yield strength ratio.

R17.10.5.3 Four options are provided for determining the required anchor or attachment strength to protect against nonductile tensile failure:

In option (a), anchor ductility requirements are imposed, and the required anchor strength is that determined using strength-level earthquake-induced forces acting on the structure. Research (Hoehler and Eligehausen 2008; Vintzileou and Eligehausen 1992) has shown that if the steel of the anchor yields before the concrete anchorage fails, no reduction in the anchor tensile strength is needed for earthquake–induced forces. Ductile steel anchors should satisfy the definition for **steel element, ductile** in Chapter 2. To facilitate comparison between steel strength, which is based on the most highly-stressed anchor, and concrete strength based on group behavior, the design is performed on the basis of the ratio of applied load to strength for the steel and concrete, respectively.

For some structures, anchors provide the best locations for energy dissipation in the nonlinear range of response. The stretch length of the anchor, shown in Fig. R17.10.5.3, affects the lateral displacement capacity of the structure; therefore, that length needs to be sufficient such that the displacement associated with the design-basis earthquake can be achieved (FEMA P750). Observations from earthquakes indicate that the provision of a stretch length of $\mathbf{8d_a}$ results in good structural performance. If the required stretch length is calculated, the relative stiffness of the connected elements needs to be considered. When an anchor is subject to load reversals, and its yielding length outside the concrete exceeds $\mathbf{6d_a}$, buckling of the anchor in compression is likely. Buckling can be restrained by placing the anchor in a tube. However, care must be taken that the tube does not share in resisting the tensile load assumed to act on the anchor. For anchor bolts that are not threaded over their length, it is important to ensure that yielding occurs over the unthreaded portion of the bolt within the stretch length before failure in the threads. This is accomplished by maintaining sufficient margin between the specified yield and tensile strengths of the bolt. It should be noted that the available stretch length may be adversely influenced by construction techniques (for example, the addition of leveling nuts to the examples illustrated in Fig. R17.10.5.3).

In option (b), the anchor is designed for the tensile force associated with the expected strength of the attachment. Care must be taken in design to consider the consequences of potential differences between the specified yield strength and the expected strength of the attachment. An example is the design of connections of intermediate precast walls where a connection not designed to yield should develop at

CODE

COMMENTARY

least $\mathbf{1.5}\boldsymbol{S_y}$, where $\boldsymbol{S_y}$ is the nominal strength of the yielding element based on its specified yield strength (refer to 18.5.2.2). Similarly, steel design manuals require structural steel connections that are designated nonyielding and part of the seismic load path to have design strengths that exceed a multiple of the nominal strength. That multiple depends on a factor relating the likely actual to specified yield strength of the material and an additional factor exceeding unity to account for material strain hardening. For attachments of cold-formed steel or wood, similar principles should be used to determine the expected strength of the attachment in order to determine the required strength of the anchors.

Additional guidance on the use of options (a) through (d) is provided in the 2009 edition of the NEHRP Recommended Seismic Provisions for New Buildings and Other Structures (FEMA P750). The design of anchors in accordance with option (a) should be used only if the anchor yield behavior is well defined and if the interaction of the yielding anchor with other elements in the load path has been adequately addressed. For the design of anchors in accordance with option (b), the force associated with yield of a steel attachment, such as an angle, baseplate, or web tab, should be the expected strength rather than the specified yield strength of the steel. Option (c) may apply to cases, such as the design of sill bolts where crushing of the wood limits the force that can be transferred to the bolt, or where the provisions of the American National Standards Institute/American Institute of Steel Construction (AISC) Code Seismic Provisions for Structural Steel Buildings (AISC 341) specify design loads based on member strengths.

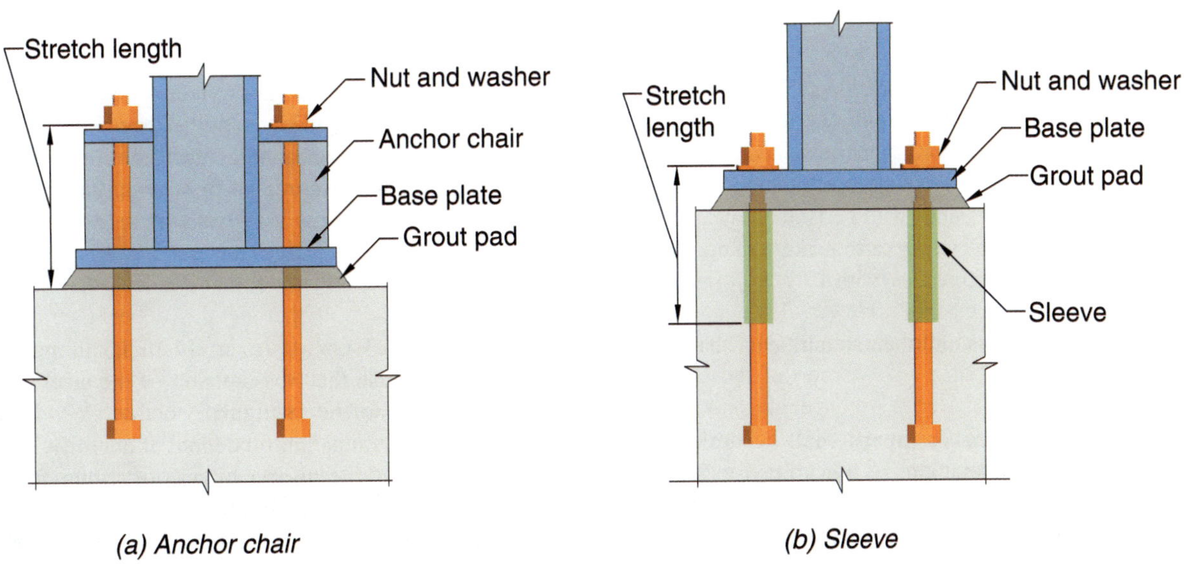

Fig. R17.10.5.3—Illustrations of stretch length.

CODE

17.10.5.4 The anchor design tensile strength shall be calculated from (a) through (e) for the failure modes given in Table 17.5.2 assuming the concrete is cracked unless it can be demonstrated that the concrete remains uncracked.

(a) ϕN_{sa} for a single anchor, or for the most highly stressed individual anchor in an anchor group
(b) $0.75\phi N_{cb}$ or $0.75\phi N_{cbg}$, except that N_{cb} or N_{cbg} need not be calculated if anchor reinforcement satisfying 17.5.2.1(a) is provided
(c) $0.75\phi N_{pn}$ for a single anchor or for the most highly stressed individual anchor in an anchor group
(d) $0.75\phi N_{sb}$ or $0.75\phi N_{sbg}$
(e) $0.75\phi N_a$ or $0.75\phi N_{ag}$

where ϕ is in accordance with 17.5.3.

17.10.5.5 If anchor reinforcement is provided in accordance with 17.5.2.1(a), no reduction in design tensile strength beyond that given in 17.5.2.1 shall be required.

17.10.6 *Shear design requirements*

17.10.6.1 If the shear component of the strength-level earthquake-induced force applied to a single anchor or anchor group does not exceed 20 percent of the total factored anchor shear associated with the same load combination, it shall be permitted to design a single anchor or anchor group in accordance with 17.7 and the shear strength requirements of 17.5.2.

17.10.6.2 If the shear component of the strength-level earthquake-induced force applied to anchors exceeds 20 percent of the total factored anchor shear associated with the same load combination, anchors and their attachments shall be designed in accordance with 17.10.6.3. The anchor design shear strength for resisting earthquake-induced forces shall be determined in accordance with 17.7.

17.10.6.3 Anchors and their attachments shall satisfy (a), (b) or (c).

(a) Anchor or anchor groups shall be designed for the maximum shear that can be transmitted to the anchor or anchor groups based on the development of a ductile yield mechanism in the attachment in tension, flexure, shear, or bearing, or a combination of those conditions, and considering both material overstrength and strain-hardening effects in the attachment.
(b) Anchor or anchor groups shall be designed for the maximum shear that can be transmitted to the anchors by a non-yielding attachment.

COMMENTARY

R17.10.5.4 The reduced anchor nominal tensile strengths associated with concrete failure modes is to account for increased cracking and spalling in the concrete resulting from earthquake effects. Because earthquake-resistant design generally assumes that all or portions of the structure are loaded beyond yield, it is likely that the concrete is cracked throughout for the purpose of calculating anchor strength. In locations where it can be demonstrated that the concrete does not crack, uncracked concrete may be assumed in calculating anchor strength as governed by concrete failure modes.

R17.10.5.5 If anchor reinforcement conforming to 17.5.2.1a is used, with the properties as defined in 20.2.2.5, separation of the potential breakout from the substrate is unlikely to occur provided the anchor reinforcement is designed for a force exceeding the concrete breakout strength.

R17.10.6 *Shear design requirements*

R17.10.6.1 The requirements of 17.10.6.3 need not apply if the applied earthquake-induced shear is a small fraction of the total factored shear.

R17.10.6.2 If the shear component of the earthquake-induced force applied to the anchor exceeds 20 percent of the total anchor shear force, three options are recognized to determine the required shear strength to protect the anchor or anchor group against premature shear failure.

R17.10.6.3 Option (a) of 17.10.5.3 is not permitted for shear because the cross section of the steel element of the anchor cannot be configured so that steel failure in shear provides any meaningful degree of ductility.

Design of the anchor or anchor group for the strength associated with force-limiting mechanisms under option (b), such as the bearing strength at holes in a steel attachment or the combined crushing and bearing strength for wood members, may be particularly relevant. Tests on typical anchor bolt connections for wood-framed structural walls (Fennel et al. 2009) demonstrated that wood components attached to concrete with minimum edge distances exhibited ductile behavior. Wood "yield" (crushing) was the first limiting state and resulted in nail slippage in shear. Nail

CODE

(c) Anchor or anchor groups shall be designed for the maximum shear obtained from factored load combinations that include $\boldsymbol{E}$, with $\boldsymbol{E_h}$ increased by $\boldsymbol{\Omega_o}$.

17.10.6.4 If anchor reinforcement is provided in accordance with 17.5.2.1(b), no reduction in design shear strength beyond that given in 17.5.2.1 shall be required.

17.10.7 *Tension and shear interaction*

17.10.7.1 Single anchors or anchor groups that resist both tensile and shear forces shall be designed in accordance with 17.8, and the anchor design tensile strength calculated in accordance with 17.10.5.4.

17.11—Attachments with shear lugs

17.11.1 *General*

17.11.1.1 It is permitted to design attachments with shear lugs in accordance with 17.11.1.1.1 through 17.11.1.1.9. Alternatively, it is permitted to design using alternative methods if adequate strength and load transfer can be demonstrated by analysis or tests.

17.11.1.1.1 Shear lugs shall be constructed of rectangular plates, or steel shapes composed of plate-like elements, welded to an attachment base plate.

COMMENTARY

slippage combined with bolt bending provided the required ductility and toughness for the structural walls and limited the loads acting on the bolts. Procedures for defining bearing and shear limit states for connections to cold-formed steel are described in AISI S100, and examples of strength calculations are provided in the AISI manual (AISI D100). In such cases, exceeding the bearing strength may lead to tearing and an unacceptable loss of connectivity. If anchors are located far from edges, it may not be possible to design such that anchor reinforcement controls the anchor strength. In such cases, anchors should be designed for overstrength in accordance with option (c).

R17.10.6.4 If anchor reinforcement conforming to 17.5.2.1b is used, with the properties as defined in 20.2.2.5, separation of the potential breakout from the substrate is unlikely to occur provided the anchor reinforcement is designed for a force exceeding the concrete breakout strength.

R17.11—Attachments with shear lugs

R17.11.1 *General*

R17.11.1.1 The provisions of 17.11 cover concrete failure modes of attachments with shear lugs. These provisions do not cover the steel or welding design of the attachment base plate or shear lugs.

Attachments with shear lugs may be embedded in cast-in-place or precast concrete, or post-installed by using a blockout in the concrete that receives the shear lug and is then filled with a fluid, non-shrink grout as shown in Fig. R17.11.1.1a. Base plates with anchors provide moment resistance, which prevents pryout action on the shear lugs. Attachments with embedded shapes and without base plates and anchors, which must resist moment by pryout action on the embedment, are not covered in this section.

Bearing strength in shear refers to the strength prior to concrete fracture in front of the shear lug. Bearing failure occurs at small displacements (Cook and Michler 2017). Following bearing failure, there is a significant decrease in strength and increase in lateral displacement leading eventually to steel failure of the anchors (Fig. R17.11.1.1b) at lateral displacements at least an order of magnitude greater than that corresponding to bearing failure.

Types of attachments with shear lugs that satisfy 17.11.1.1.1 through 17.11.1.1.9 are shown in Fig. R17.11.1.1a. Shear lugs that are different than those covered in 17.11.1.1.1 through 17.11.1.1.9, such as shear lugs composed of steel

CODE

COMMENTARY

pipe or attachments with shear lugs where the top of plate is located below the concrete surface, can be used provided adequate strength and load transfer can be demonstrated by analysis or tests.

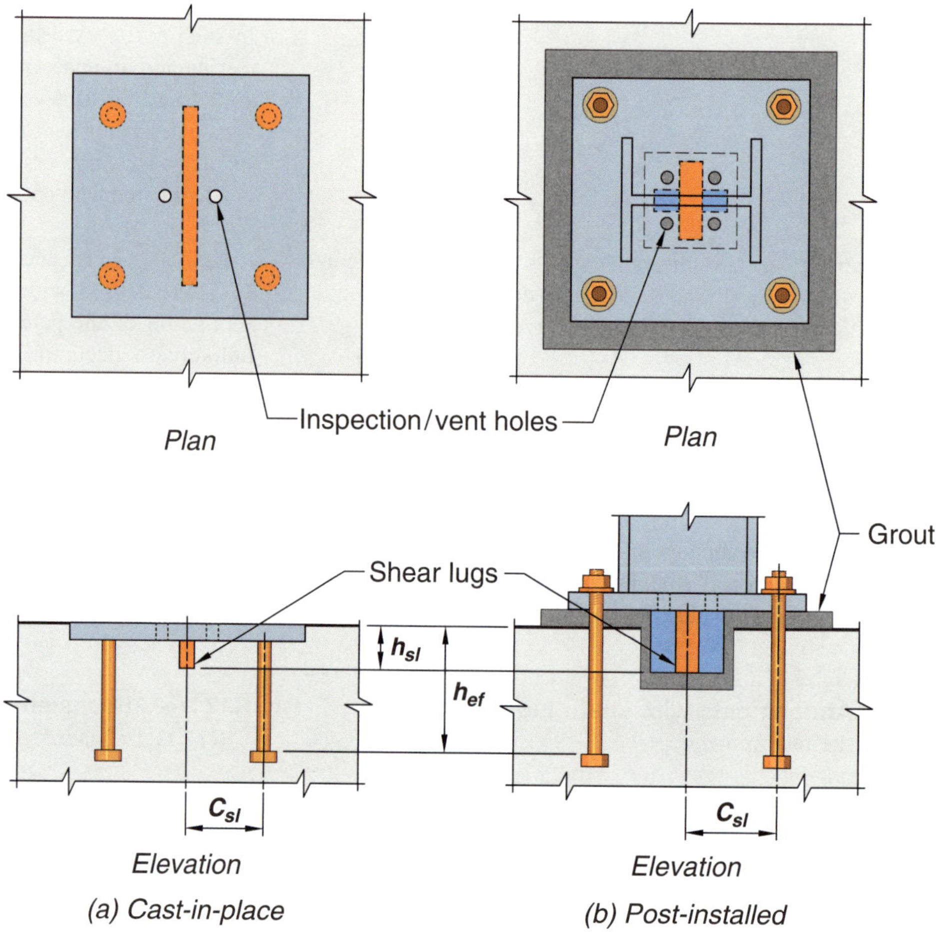

***Fig. R17.11.1.1a**—Examples of attachments with shear lugs.*

CODE

COMMENTARY

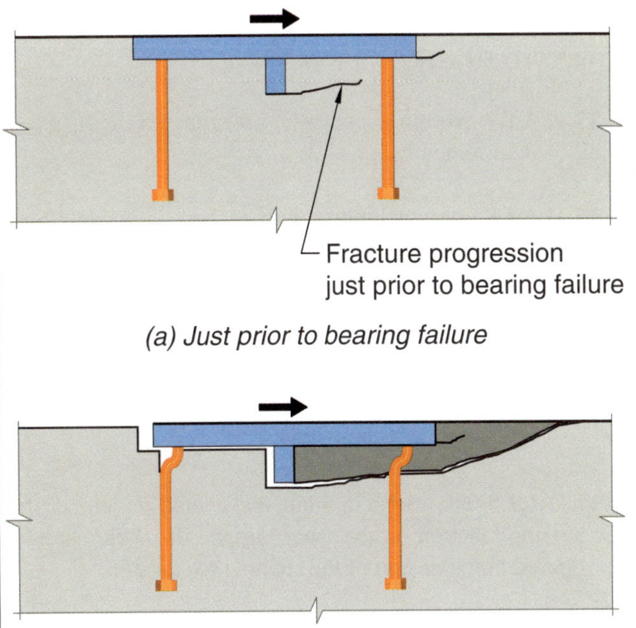

***Fig. R17.11.1.1b**—Bearing failure and subsequent anchor steel failure for embedded plate with shear lug (if concrete breakout is not applicable)*

17.11.1.1.2 A minimum of four anchors shall be provided that satisfy the requirements of Chapter 17 with the exception of the requirements of 17.5.1.2(f), (g), and (h) and the corresponding requirements of Table 17.5.2 for steel strength of anchors in shear, concrete breakout strength of anchors in shear, and concrete pryout strength of anchors in shear.

17.11.1.1.3 For anchors welded to the attachment base plate, tension and shear interaction requirements of 17.8 shall include a portion of the total shear on the anchor.

R17.11.1.1.3 Although neglected in the bearing strength evaluation in 17.11.2, welded anchors resist a portion of the shear load because they displace the same as the shear lug. The portion of the applied shear, V_u, that each anchor carries, $V_{ua,i}$, is given by

$$V_{ua,i} = V_u\left(\frac{2d_a^2}{A_{ef,sl} + n2d_a^2}\right)$$

The effective bearing area of an anchor is assumed to be the diameter of the anchor multiplied by an effective bearing depth of twice its diameter (Cook and Michler 2017). The bearing reaction on the anchor is not large enough to fail the anchor in shear alone but does need to be considered in tension and shear interaction for steel failure (refer to 17.8).

17.11.1.1.4 Bearing strength in shear shall satisfy $\phi V_{brg,sl} \geq V_u$ with ϕ **= 0.65**.

17.11.1.1.5 Nominal bearing strength in shear, $V_{brg,sl}$, shall be determined by 17.11.2.

CODE

17.11.1.1.6 Concrete breakout strength of the shear lug shall satisfy $\phi V_{cb,sl} \geq V_u$ with $\phi = 0.65$.

17.11.1.1.7 Nominal concrete breakout strength, $V_{cb,sl}$, shall be determined by 17.11.3.

17.11.1.1.8 For attachments with anchors in tension, both (a) and (b) shall be satisfied:

(a) $h_{ef}/h_{sl} \geq 2.5$
(b) $h_{ef}/c_{sl} \geq 2.5$

17.11.1.1.9 The moment from the couple developed by the bearing reaction on the shear lug and the shear shall be considered in the design of the anchors for tension.

17.11.1.2 Horizontally installed steel base plates with shear lugs shall have a minimum 1 in. diameter hole along each of the long sides of the shear lug.

17.11.2 *Bearing strength in shear of attachments with shear lugs,* $V_{brg,sl}$

17.11.2.1 Nominal bearing strength in shear of a shear lug, $V_{brg,sl}$, shall be calculated as:

$$V_{brg,sl} = 1.7 f_c' A_{ef,sl} \psi_{brg,sl} \quad (17.11.2.1)$$

where $\psi_{brg,sl}$ is given in 17.11.2.2.

COMMENTARY

R17.11.1.1.8 The lower bound limitations on the ratios of anchor embedment depth to shear lug embedment depth and anchor embedment depth to the distance between the centerline of the anchors in tension and the centerline of the shear lug in the direction of shear are based on available test data. The required lower limits reduce potential interaction between concrete breakout of the anchors in tension and bearing failure in shear of the shear lug.

R17.11.1.1.9 The bearing reaction on shear lugs occurs further below the surface of the concrete than the bearing reaction on anchors and embedded plates. As a result, the couple caused by the bearing reaction and the shear load needs to be considered when determining anchor tension.

R17.11.1.2 Base plate holes are necessary to verify proper concrete or grout consolidation around the shear lug and to avoid trapping air immediately below a horizontal plate. Holes in the base plate should be placed close to each face of the shear lug. For a single shear lug, place at least one inspection hole near the center of each long side of the shear lug. For a cruciform-shaped shear lug, four inspection holes are recommended, one per quadrant. For other configurations or long shear lug lengths, the licensed design professional should specify inspection hole locations that will permit adequate observation and allow trapped air to escape.

R17.11.2 *Bearing strength in shear of attachments with shear lugs,* $V_{brg,sl}$

R17.11.2.1 The nominal bearing strength in shear of a shear lug, $V_{brg,sl}$, given by Eq. (17.11.2.1) is based on a uniform bearing stress of $1.7f_c'$ acting over the effective area of the shear lug as discussed in Cook and Michler (2017). Although the bearing strength in shear of attachments with shear lugs is a function of bearing on the shear lug, embedded plate (if present), and welded anchors (if present), the method presented in 17.11.2 only includes the contribution of shear lugs. Cook and Michler (2017) discuss development of the method and a less conservative procedure to include bearing on the embedded plate and welded anchors.

CODE

17.11.2.1.1 The effective bearing area, $A_{ef,sl}$, shall be below the surface of the concrete, perpendicular to the applied shear, and composed of areas according to (a) through (d):

(a) Bearing area of shear lugs located within $2t_{sl}$ of the bottom surface of the base plate if the top or bottom surface of the base plate is flush with the surface of the concrete

(b) Bearing area of shear lugs located within $2t_{sl}$ of the surface of the concrete if the base plate is above the surface of the concrete

(c) Bearing area of shear lugs located within $2t_{sl}$ of the interface with stiffeners

(d) Bearing area on the leading edge of stiffeners below the surface of the concrete

COMMENTARY

R17.11.2.1.1 Figure R17.11.2.1.1 shows examples of effective bearing areas. The effective bearing area for stiffened shear lugs is applicable to both welded plates and steel shapes composed of plate-like elements in which case the web would be the stiffening element. The limit of a distance of $2t_{sl}$ in determining the effective bearing area is described in Cook and Michler (2017).

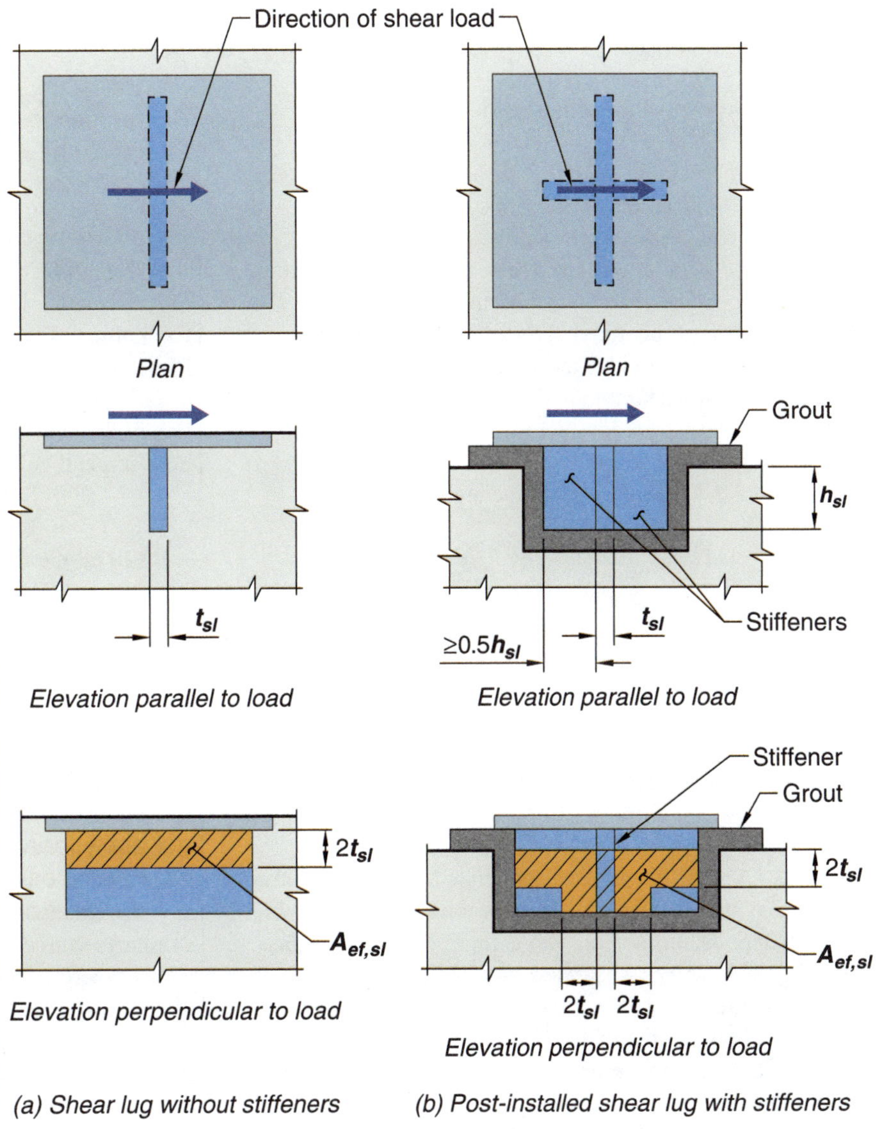

Note: Anchors and inspection holes not shown for clarity.

***Fig. R17.11.2.1.1**—Examples of effective bearing areas for attachments with shear lugs.*

CODE

17.11.2.2 *Bearing factor*, $\psi_{brg,sl}$

17.11.2.2.1 Modification factor, $\psi_{brg,sl}$, for the effects of axial load, P_u, on bearing strength in shear, shall be determined by (a), (b), or (c):

(a) For applied axial tension:

$$\psi_{brg,sl} = 1 + \frac{P_u}{nN_{sa}} \leq 1.0 \qquad (17.11.2.2.1a)$$

where P_u is negative for tension and n is the number of anchors in tension.

(b) For no applied axial load:

$$\psi_{brg,sl} = 1 \qquad (17.11.2.2.1b)$$

(c) For applied axial compression:

$$\psi_{brg,sl} = 1 + 4\frac{P_u}{A_{bp}f_c'} \leq 2.0 \qquad (17.11.2.2.1c)$$

where P_u is positive for compression.

17.11.2.3 If used, the length of shear lug stiffeners in the direction of the shear load shall not be less than $0.5h_{sl}$.

17.11.2.4 For attachments with multiple shear lugs arranged perpendicular to the direction of applied shear, the bearing strength of the individual shear lugs may be considered to be additive provided the shear stress on a shear plane in the concrete at the bottom of the shear lugs, and extending between the shear lugs, does not exceed $0.2f_c'$. The nominal bearing strength of each individual lug shall be determined by Eq. (17.11.2.1) using the effective area of the lug.

17.11.3 *Concrete breakout strength of shear lug*, $V_{cb,sl}$

17.11.3.1 Nominal concrete breakout strength of a shear lug for shear perpendicular to the edge, $V_{cb,sl}$, shall be determined from 17.7.2 using Eq. (17.7.2.1a), where V_b is calculated using Eq. (17.7.2.2.1b) with c_{a1} taken as the distance from the bearing surface of the shear lug to the free edge and where A_{vc} is the projected area of the failure surface on the side of the concrete member.

17.11.3.1.1 A_{vc} is the projected concrete failure area on the side face of the concrete that is approximated as the rectangular shape resulting from projecting horizontally $1.5c_{a1}$ from the edge of the shear lug and projecting vertically $1.5c_{a1}$ from the edge of the effective depth of the shear lug, $h_{ef,sl}$. The effective area of the shear lug, $A_{ef,sl}$, shall not be included. The effective embedment depth of the shear lug, $h_{ef,sl}$, shall be taken as the distance from the concrete surface to the bottom of the effective bearing area, $A_{ef,sl}$.

COMMENTARY

R17.11.2.4 The limitation for considering multiple shear lugs to be effective is based on the maximum limits for shear friction in Table 22.9.4.4 and two tests reported in Rotz and Reifschneider (1984). The area of the shear plane is the clear distance between adjacent shear lugs measured in the direction of the applied shear multiplied by the width of the shear lugs perpendicular to the applied shear.

R17.11.3 *Concrete breakout strength of shear lug*, $V_{cb,sl}$

R17.11.3.1 The method for evaluating concrete breakout strength where shear is perpendicular to an edge is similar to that used in 17.7.2 for anchors. The difference is in the determination of A_{Vc}, which is illustrated in Fig. R17.11.3.1. The method has been confirmed by tests where the shear lug is concentrically loaded in shear (Gomez et al. 2009; Cook and Michler 2017). With shear transferred by the shear lug, embedded plate (if present), and welded anchors (if present), the bearing surfaces all displace the same amount with any incremental change in applied shear. This behavior is similar to connections with anchors welded to steel attachments where concrete edge failure originates from the row of anchors farthest from the edge. In anchorages with shear lugs, the effective contributions to concrete breakout strength from the bearing areas of the shear lug and embedded plate (if present) dominate over the contribution from the effective bearing area of anchors farther from the edge than the shear lug. As a result, concrete breakout strength for the anchorage

CODE

COMMENTARY

should be determined based on the concrete breakout surface originating at the shear lug (Fig. R17.11.3.1).

The nominal concrete breakout strength of a shear lug is based on Eq. (17.7.2.2.1b) for V_b that applies to concrete edge failure in shear for large diameter anchors.

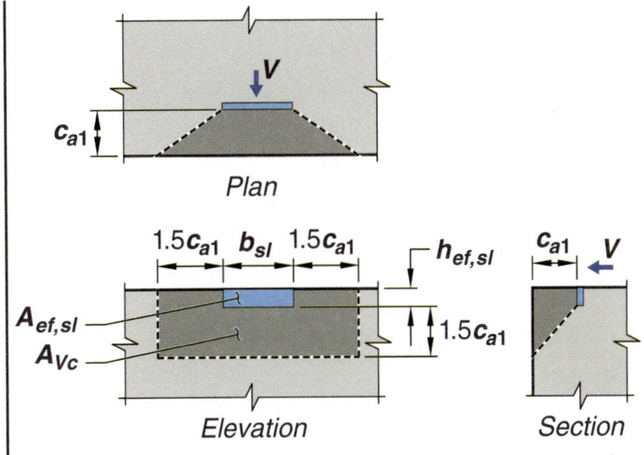

***Fig. R17.11.3.1**—Example of* A_{Vc} *for a shear lug near an edge.*

17.11.3.2 Nominal concrete breakout strength of a shear lug for shear parallel to the edge shall be permitted to be determined in accordance with 17.7.2.1(c) using Eq. (17.7.2.1(a)) with c_{a1} taken as the distance from the edge to the center of the shear lug and with $\psi_{ec,V}$ taken as 1.0.

R17.11.3.2 The concrete breakout strength for shear lugs loaded parallel to the edge is based on 17.7.2.1(c) for concrete failure with load applied parallel to the free edge, assuming shear lug breakout behavior is similar to that of a single anchor.

17.11.3.3 For shear lugs located at a corner, the limiting concrete breakout strength shall be determined for each edge, and the minimum value shall be used.

R17.11.3.3 The concrete breakout strength for shear lugs located near a corner is based on 17.7.2.1(d) for anchors.

17.11.3.4 For cases with multiple shear lugs, the concrete breakout strength shall be determined for each potential breakout surface.

R17.11.3.4 The concrete breakout strength for multiple shear lugs is based on R17.7.2.1 and shown in Fig. R17.7.2.1b Case 1 and Case 2.

Notes

Bar Development, hooking

Tension development length

R25.4.22

CHAPTER 18—EARTHQUAKE-RESISTANT STRUCTURES

CODE

18.1—Scope

18.1.1 This chapter shall apply to the design of nonprestressed and prestressed concrete structures assigned to Seismic Design Categories (SDC) B through F, including, where applicable:

(a) Structural systems designated as part of the seismic-force-resisting system, including diaphragms, moment frames, structural walls, and foundations
(b) Members not designated as part of the seismic-force-resisting system but required to support other loads while undergoing deformations associated with earthquake effects

18.1.2 Structures designed according to the provisions of this chapter are intended to resist earthquake motions through ductile inelastic response of selected members.

18.2—General

18.2.1 *Structural systems*

18.2.1.1 All structures shall be assigned to a SDC in accordance with 4.4.6.1.

COMMENTARY

R18.1—Scope

Chapter 18 does not apply to structures assigned to Seismic Design Category (SDC) A. For structures assigned to SDC B and C, Chapter 18 applies to structural systems designated as part of the seismic-force-resisting system. For structures assigned to SDC D through F, Chapter 18 applies to both structural systems designated as part of the seismic-force-resisting system and structural systems not designated as part of the seismic-force-resisting system.

Chapter 18 contains provisions considered to be the minimum requirements for a cast-in-place or precast concrete structure capable of sustaining a series of oscillations into the inelastic range of response without critical deterioration in strength. The integrity of the structure in the inelastic range of response should be maintained because the design earthquake forces defined in documents such as ASCE/SEI 7, the 2018 IBC, the UBC (ICBO 1997), and the NEHRP (FEMA P749) provisions are considered less than those corresponding to linear response at the anticipated earthquake intensity (FEMA P749; Blume et al. 1961; Clough 1960; Gulkan and Sozen 1974).

The design philosophy in Chapter 18 is for cast-in-place concrete structures to respond in the nonlinear range when subjected to design-level ground motions, with decreased stiffness and increased energy dissipation but without critical strength decay. Precast concrete structures designed in accordance with Chapter 18 are intended to emulate cast-in-place construction, except 18.5, 18.9.2.3, and 18.11.2.2, which permit precast construction with alternative yielding mechanisms. The combination of reduced stiffness and increased energy dissipation tends to reduce the response accelerations and lateral inertia forces relative to values that would occur were the structure to remain linearly elastic and lightly damped (Gulkan and Sozen 1974). Thus, the use of design forces representing earthquake effects such as those in ASCE/SEI 7 requires that the seismic-force-resisting system retain a substantial portion of its strength into the inelastic range under displacement reversals.

The provisions of Chapter 18 relate detailing requirements to type of structural framing and SDC. Seismic design categories are adopted directly from ASCE/SEI 7, and relate to considerations of seismic hazard level, soil type, occupancy, and use. Before the 2008 Code, low, intermediate, and high seismic risk designations were used to delineate detailing requirements. For a qualitative comparison of seismic design categories and seismic risk designations, refer to Table R5.2.2. The assignment of a structure to a SDC is regulated by the general building code (refer to 4.4.6.1).

R18.2—General

Structures assigned to SDC A need not satisfy requirements of Chapter 18 but must satisfy all other applicable requirements of this Code. Structures assigned to Seismic Design Categories B through F must satisfy requirements of

CODE

18.2.1.2 All members shall satisfy Chapters 1 to 17 and 19 to 26. Structures assigned to SDC B, C, D, E, or F also shall satisfy 18.2.1.3 through 18.2.1.7, as applicable. Where Chapter 18 conflicts with other chapters of this Code, Chapter 18 shall govern.

18.2.1.3 Structures assigned to SDC B shall satisfy 18.2.2.

18.2.1.4 Structures assigned to SDC C shall satisfy 18.2.2, 18.2.3, and 18.13.

18.2.1.5 Structures assigned to SDC D, E, or F shall satisfy 18.2.2 through 18.2.8 and 18.12 through 18.14.

18.2.1.6 Structural systems designated as part of the seismic-force-resisting system shall be restricted to those designated by the general building code, or determined by other authority having jurisdiction in areas without a legally adopted building code. Except for SDC A, for which Chapter 18 does not apply, (a) through (h) shall be satisfied for each structural system designated as part of the seismic-force-resisting system, in addition to 18.2.1.3 through 18.2.1.5:

(a) Ordinary moment frames shall satisfy 18.3

(b) Ordinary reinforced concrete structural walls need not satisfy any detailing provisions in Chapter 18, unless required by 18.2.1.3 or 18.2.1.4

(c) Intermediate moment frames shall satisfy 18.4

(d) Intermediate precast walls shall satisfy 18.5

(e) Special moment frames shall satisfy 18.2.3 through 18.2.8 and 18.6 through 18.8

(f) Special moment frames constructed using precast concrete shall satisfy 18.2.3 through 18.2.8 and 18.9

(g) Special structural walls shall satisfy 18.2.3 through 18.2.8 and 18.10

(h) Special structural walls constructed using precast concrete shall satisfy 18.2.3 through 18.2.8 and 18.11

18.2.1.7 A reinforced concrete structural system not satisfying this chapter shall be permitted if it is demonstrated by experimental evidence and analysis that the proposed system will have strength and toughness equal to or exceeding those provided by a comparable reinforced concrete structure satisfying this chapter.

COMMENTARY

Chapter 18 in addition to all other applicable requirements of this Code.

Sections 18.2.1.3 through 18.2.1.5 identify those parts of Chapter 18 that apply to the building based on its assigned SDC, regardless of the vertical elements of the seismic-force-resisting system. ASCE/SEI 7 defines the permissible vertical elements of the seismic-force-resisting system and applies where adopted. The remaining commentary of R18.2 summarizes the intent of ACI 318 regarding which vertical elements should be permissible in a building considering its SDC. Section 18.2.1.6 defines the requirements for the vertical elements of the seismic-force-resisting system.

The design and detailing requirements should be compatible with the level of inelastic response assumed in the calculation of the design earthquake forces. The terms "ordinary," "intermediate," and "special" are used to facilitate this compatibility. For any given structural element or system, the terms "ordinary," "intermediate," and "special," refer to increasing requirements for detailing and proportioning, with expectations of increased deformation capacity. Structures assigned to SDC B are not expected to be subjected to strong ground motion, but instead are expected to experience low levels of ground motion at long time intervals. This Code provides some requirements for beam-column ordinary moment frames to improve deformation capacity.

Structures assigned to SDC C may be subjected to moderately strong ground motion. The designated seismic-force-resisting system typically comprises some combination of ordinary cast-in-place structural walls, intermediate precast structural walls, and intermediate moment frames. The general building code also may contain provisions for use of other seismic-force-resisting systems in SDC C. Provision 18.2.1.6 defines requirements for whatever system is selected.

Structures assigned to SDC D, E, or F may be subjected to strong ground motion. It is the intent of ACI Committee 318 that the seismic-force-resisting system of structural concrete buildings assigned to SDC D, E, or F be provided by special moment frames, special structural walls, or a combination of the two. In addition to 18.2.2 through 18.2.8, these structures also are required to satisfy requirements for continuous inspection (26.13.1.3), diaphragms and trusses (18.12), foundations (18.13), and gravity-load-resisting elements that are not designated as part of the seismic-force-resisting system (18.14). These provisions have been developed to provide the structure with adequate deformation capacity for the high demands expected for these seismic design categories.

The general building code may also permit the use of intermediate moment frames as part of dual systems for some buildings assigned to SDC D, E, or F. It is not the intent of ACI Committee 318 to recommend the use of intermediate moment frames as part of moment-resisting frame or dual systems in SDC D, E, or F. The general building code may also permit substantiated alternative or nonprescriptive designs or, with various supplementary provisions, the use

CODE

COMMENTARY

of ordinary or intermediate systems for nonbuilding structures in the higher seismic design categories. These are not the typical applications that were considered in the writing of this chapter, but wherever the term "ordinary or intermediate moment frame" is used in reference to reinforced concrete, 18.3 or 18.4 apply.

Table R18.2 summarizes the applicability of the provisions of Chapter 18 as they are typically applied when using the minimum requirements in the various seismic design categories. Where special systems are used for structures in SDC B or C, it is not required to satisfy the requirements of 18.14, although it should be verified that members not designated as part of the seismic-force-resisting system will be stable under design displacements.

Table R18.2—Sections of Chapter 18 to be satisfied in typical applications[1]

<table>
<tr><th rowspan="2">Component resisting earthquake effect, unless otherwise noted</th><th colspan="4">SDC</th></tr>
<tr><th>A (None)</th><th>B (18.2.1.3)</th><th>C (18.2.1.4)</th><th>D, E, F (18.2.1.5)</th></tr>
<tr><td>Analysis and design requirements</td><td rowspan="10">None</td><td>18.2.2</td><td>18.2.2</td><td>18.2.2, 18.2.4</td></tr>
<tr><td>Materials</td><td>None</td><td>None</td><td>18.2.5 through 18.2.8</td></tr>
<tr><td>Frame members</td><td>18.3</td><td>18.4</td><td>18.6 through 18.9</td></tr>
<tr><td>Structural walls and coupling beams</td><td>None</td><td>None</td><td>18.10</td></tr>
<tr><td>Precast structural walls</td><td>None</td><td>18.5</td><td>18.5[2], 18.11</td></tr>
<tr><td>Diaphragms and trusses</td><td>None</td><td>18.12</td><td>18.12</td></tr>
<tr><td>Foundations</td><td>None</td><td>18.13</td><td>18.13</td></tr>
<tr><td>Frame members not designated as part of the seismic-force-resisting system</td><td>None</td><td>None</td><td>18.14</td></tr>
<tr><td>Anchors</td><td>None</td><td>18.2.3</td><td>18.2.3</td></tr>
</table>

[1]In addition to requirements of Chapters 1 through 17, 19 through 26, and ACI 318.2, except as modified by Chapter 18. Section 14.1.4 also applies in SDC D, E, and F.

[2]As permitted by the general building code.

The proportioning and detailing requirements in Chapter 18 are based predominantly on field and laboratory experience with monolithic reinforced concrete building structures and precast concrete building structures designed and detailed to behave like monolithic building structures. Extrapolation of these requirements to other types of cast-in-place or precast concrete structures should be based on evidence provided by field experience, tests, or analysis. The acceptance criteria for moment frames given in ACI 374.1 can be used in conjunction with Chapter 18 to demonstrate that the strength, energy dissipation capacity, and deformation capacity of a proposed frame system equals or exceeds that provided by a comparable monolithic concrete system. ACI ITG-5.1 provides similar information for precast wall systems.

The toughness requirement in 18.2.1.7 refers to the requirement to maintain structural integrity of the entire seismic-force-resisting system at lateral displacements anticipated for the maximum considered earthquake motion. Depending on the energy-dissipation characteristics of the structural system used, such displacements may be larger than for a monolithic reinforced concrete structure satisfying the prescriptive provisions of other parts of this Code.

CODE

18.2.2 *Analysis and proportioning of structural members*

18.2.2.1 The interaction of all structural and nonstructural members that affect the linear and nonlinear response of the structure to earthquake motions shall be considered in the analysis.

18.2.2.2 Rigid members assumed not to be a part of the seismic-force-resisting system shall be permitted provided their effect on the response of the system is considered in the structural design. Consequences of failure of structural and nonstructural members that are not a part of the seismic-force-resisting system shall be considered.

18.2.2.3 Structural members extending below the base of structure that are required to transmit forces resulting from earthquake effects to the foundation shall comply with the requirements of Chapter 18 that are consistent with the seismic-force-resisting system above the base of structure.

COMMENTARY

R18.2.2 *Analysis and proportioning of structural members*

It is assumed that the distribution of required strength to the various components of a seismic-force-resisting system will be determined from the analysis of a linearly elastic model of the system acted upon by the factored forces, as required by the general building code. If nonlinear response history analyses are to be used, base motions should be selected after a detailed study of the site conditions and local seismic history.

Because the basis for earthquake-resistant design admits nonlinear response, it is necessary to investigate the stability of the seismic-force-resisting system, as well as its interaction with other structural and nonstructural members, under expected lateral displacements corresponding to maximum considered earthquake ground motion. For lateral displacement calculations, assuming all the structural members to be fully cracked is likely to lead to better estimates of the possible drift than using uncracked stiffness for all members. The analysis assumptions described in 6.6.3.1 may be used to estimate lateral deflections of reinforced concrete building systems.

The main objective of Chapter 18 is the safety of the structure. The intent of 18.2.2.1 and 18.2.2.2 is to draw attention to the influence of nonstructural members on structural response and to hazards from falling objects.

Section 18.2.2.3 serves as an alert that the base of structure as defined in analysis may not necessarily correspond to the foundation or ground level. Details of columns and walls extending below the base of structure to the foundation are required to be consistent with those above the base of structure.

In selecting member sizes for earthquake-resistant structures, it is important to consider constructibility problems related to congestion of reinforcement. The design should be such that all reinforcement can be assembled and placed in the proper location and that concrete can be cast and consolidated properly. Using the upper limits of permitted reinforcement ratios may lead to construction problems.

CODE

18.2.3 *Anchoring to concrete*

18.2.3.1 Anchors resisting earthquake-induced forces in structures assigned to SDC C, D, E, or F shall be in accordance with 17.10.

CODE

18.2.4 *Strength reduction factors*

18.2.4.1 Strength reduction factors shall be in accordance with Chapter 21.

18.2.5 *Concrete in special moment frames and special structural walls*

18.2.5.1 Specified compressive strength of concrete in special moment frames and special structural walls shall be in accordance with the special seismic systems requirements of Table 19.2.1.1.

18.2.6 *Reinforcement in special moment frames and special structural walls*

18.2.6.1 Reinforcement in special moment frames and special structural walls shall be in accordance with the special seismic systems requirements of 20.2.2.

COMMENTARY

R18.2.4 *Strength reduction factors*

R18.2.4.1 Chapter 21 contains strength reduction factors for all members, joints, and connections of earthquake-resistant structures, including specific provisions in 21.2.4 for buildings that use special moment frames, special structural walls, and intermediate precast walls.

R18.2.5 *Concrete in special moment frames and special structural walls*

Requirements of this section refer to concrete quality in frames and walls that resist earthquake-induced forces. The maximum specified compressive strength of lightweight concrete to be used in structural design calculations is limited to 5000 psi, primarily because of paucity of experimental and field data on the behavior of members made with lightweight concrete subjected to displacement reversals in the nonlinear range. If convincing evidence is developed for a specific application, the limit on maximum specified compressive strength of lightweight concrete may be increased to a level justified by the evidence.

R18.2.6 *Reinforcement in special moment frames and special structural walls*

R18.2.6.1 Nonprestressed reinforcement for seismic systems is required to meet 20.2.2.4 and 20.2.2.5. Starting with ACI 318-19, ASTM A706 Grades 80 and 100 reinforcement is permitted to resist moments, axial, and shear forces in special structural walls and all components of special structural walls, including coupling beams and wall piers. ASTM A706 Grade 80 reinforcement is also permitted in special moment frames. Results of tests and analytical studies presented in NIST (2014) and Sokoli and Ghannoum (2016) indicate that properly detailed beams and columns of special moment frames with ASTM A706 Grade 80 reinforcement exhibit strength and deformation capacities similar to those of members reinforced with Grade 60 reinforcement. The use of Grade 100 reinforcement is not allowed in special moment frames because there is insufficient data to demonstrate satisfactory seismic performance.

To allow the use of ASTM A706 Grade 80 and 100 reinforcement, the 2019 Code includes limits for spacing of transverse reinforcement to provide adequate longitudinal bar support to control longitudinal bar buckling. In special moment frames, the use of Grade 80 reinforcement requires increased joint depths to prevent excessive slip of beam bars passing through beam-column joints (18.8.2.3).

The requirement for a tensile strength greater than the yield strength of the reinforcement (20.2.2.5, Table 20.2.1.3(b)) is based on the assumption that the capability of a structural member to develop inelastic rotation capacity is a function of the length of the yield region along the axis of the member. In interpreting experimental results, the length of

CODE

COMMENTARY

the yield region has been related to the relative magnitudes of nominal and yield moments (ACI 352R). According to this interpretation, the greater the ratio of nominal to yield moment, the longer the yield region. Chapter 20 requires that the ratio of actual tensile strength to actual yield strength be at least 1.25 for ASTM A615 Grade 60.

The restrictions on the value of f_{yt} apply to all types of transverse reinforcement, including spirals, circular hoops, rectilinear hoops, and crossties. Research results (Budek et al. 2002; Muguruma and Watanabe 1990; Sugano et al. 1990) indicate that higher yield strengths can be used effectively as confinement reinforcement as specified in 18.7.5.4. The increases to 80,000 psi and 100,000 psi for shear design of some special seismic system members is based on research indicating the design shear strength can be developed (Wallace 1998; Aoyama 2001; Budek et al. 2002; Sokoli and Ghannoum 2016; Cheng et al. 2016; Huq et al. 2018; Weber-Kamin et al. 2019). The 60,000 psi restriction on the value of f_{yt} for deformed bar in 20.2.2.4 for calculating nominal shear strength is intended to limit the width of shear cracks at service-level loads. Service-level cracking is not a concern in members of the seismic-force-resisting system subjected to design-level earthquake forces.

18.2.7 *Mechanical splices in special moment frames and special structural walls*

R18.2.7 *Mechanical splices in special moment frames and special structural walls*

In a structure undergoing inelastic deformations during an earthquake, the tensile stresses in reinforcement may approach the tensile strength of the reinforcement. The requirements for Type 2 mechanical splices are intended to avoid a splice failure when the reinforcement is subjected to expected stress levels in yielding regions. Type 1 mechanical splices on any grade of reinforcement and Type 2 mechanical splices on Grade 80 and Grade 100 reinforcement may not be capable of resisting the stress levels expected in yielding regions. The locations of these mechanical splices are restricted because tensile stresses in reinforcement in yielding regions can exceed the strength requirements of 18.2.7.1. The restriction on all Type 1 mechanical splices and on Type 2 mechanical splices on Grade 80 and Grade 100 reinforcement applies to all reinforcement resisting earthquake effects, including transverse reinforcement.

Recommended detailing practice would preclude the use of splices in regions of potential yielding in members resisting earthquake effects. If use of mechanical splices in regions of potential yielding cannot be avoided, there should be documentation on the actual strength characteristics of the bars to be spliced, on the force-deformation characteristics of the spliced bar, and on the ability of the mechanical splice to be used to meet the specified performance requirements.

Although mechanical splices as defined by 18.2.7 need not be staggered, staggering is encouraged and may be necessary for constructibility or provide enough space around the splice for installation or to meet the clear spacing requirements.

CODE

18.2.7.1 Mechanical splices shall be classified as (a) or (b):

(a) Type 1 – Mechanical splice conforming to 25.5.7

(b) Type 2 – Mechanical splice conforming to 25.5.7 and capable of developing the specified tensile strength of the spliced bars

18.2.7.2 Except for Type 2 mechanical splices on Grade 60 reinforcement, mechanical splices shall not be located within a distance equal to twice the member depth from the column or beam face for special moment frames or from critical sections where yielding of the reinforcement is likely to occur as a result of lateral displacements beyond the linear range of behavior. Type 2 mechanical splices on Grade 60 reinforcement shall be permitted at any location, except as noted in 18.9.2.1(c).

18.2.8 *Welded splices in special moment frames and special structural walls*

18.2.8.1 Welded splices in reinforcement resisting earthquake-induced forces shall conform to 25.5.7 and shall not be located within a distance equal to twice the member depth from the column or beam face for special moment frames or from critical sections where yielding of the reinforcement is likely to occur as a result of lateral displacements beyond the linear range of behavior.

18.2.8.2 Welding of stirrups, ties, inserts, or other similar elements to longitudinal reinforcement required by design shall not be permitted.

18.3—Ordinary moment frames

18.3.1 *Scope*

18.3.1.1 This section shall apply to ordinary moment frames forming part of the seismic-force-resisting system.

18.3.2 Beams shall have at least two continuous bars at both top and bottom faces. Continuous bottom bars shall have area not less than one-fourth the maximum area of bottom bars along the span. These bars shall be anchored to develop f_y in tension at the face of support.

18.3.3 Columns having unsupported length $\ell_u \leq 5c_1$ shall have ϕV_n at least the lesser of (a) and (b):

(a) The shear associated with development of nominal moment strengths of the column at each restrained end of the unsupported length due to reverse curvature bending. Column flexural strength shall be calculated for the factored

COMMENTARY

R18.2.8 *Welded splices in special moment frames and special structural walls*

R18.2.8.1 Welding of reinforcement should be in accordance with AWS D1.4 as required in Chapter 26. The locations of welded splices are restricted because reinforcement tension stresses in yielding regions can exceed the strength requirements of 25.5.7. The restriction on welded splices applies to all reinforcement resisting earthquake effects, including transverse reinforcement.

R18.2.8.2 Welding of crossing reinforcing bars can lead to local embrittlement of the steel. If welding of crossing bars is used to facilitate fabrication or placement of reinforcement, it should be done only on bars added for such purposes. The prohibition of welding crossing reinforcing bars does not apply to bars that are welded with welding operations under continuous, competent control, as in the manufacture of welded-wire reinforcement.

R18.3—Ordinary moment frames

This section applies only to ordinary moment frames assigned to SDC B. The provisions for beam reinforcement are intended to improve continuity in the framing members and thereby improve lateral force resistance and structural integrity; these provisions do not apply to slab-column moment frames. The provisions for columns are intended to provide additional capacity to resist shear for columns with proportions that would otherwise make them more susceptible to shear failure under earthquake loading.

CODE

axial force, consistent with the direction of the lateral forces considered, resulting in the highest flexural strength.

(b) The maximum shear obtained from design load combinations that include $\boldsymbol{E}$, with $\boldsymbol{\Omega_o E}$ substituted for $\boldsymbol{E}$.

18.3.4 Beam-column joints shall satisfy Chapter 15 with joint shear V_u calculated on a plane at mid-height of the joint using tensile and compressive beam forces and column shear consistent with beam nominal moment strengths M_n.

18.4—Intermediate moment frames

18.4.1 *Scope*

18.4.1.1 This section shall apply to intermediate moment frames including two-way slabs without beams forming part of the seismic-force-resisting system.

18.4.2 *Beams*

18.4.2.1 Beams shall have at least two continuous bars at both top and bottom faces. Continuous bottom bars shall have area not less than one-fourth the maximum area of bottom bars along the span. These bars shall be anchored to develop f_y in tension at the face of support.

18.4.2.2 The positive moment strength at the face of the joint shall be at least one-third the negative moment strength provided at that face of the joint. Neither the negative nor the positive moment strength at any section along the length of the beam shall be less than one-fifth the maximum moment strength provided at the face of either joint.

18.4.2.3 ϕV_n shall be at least the lesser of (a) and (b):

(a) The sum of the shear associated with development of nominal moment strengths of the beam at each restrained end of the clear span due to reverse curvature bending and the shear calculated for factored gravity and vertical earthquake loads
(b) The maximum shear obtained from design load combinations that include $\boldsymbol{E}$, with $\boldsymbol{E}$ taken as twice that prescribed by the general building code

18.4.2.4 At both ends of the beam, hoops shall be provided over a length of at least $2h$ measured from the face of the supporting member toward midspan. The first hoop shall be located not more than 2 in. from the face of the supporting member. Spacing of hoops shall not exceed the smallest of (a) through (d):

(a) $d/4$
(b) Eight times the diameter of the smallest longitudinal bar enclosed
(c) 24 times the diameter of the hoop bar

COMMENTARY

R18.4—Intermediate moment frames

The objective of the requirements in 18.4.2.3 and 18.4.3.1 is to reduce the risk of failure in shear in beams and columns during an earthquake. Two options are provided to determine the factored shear force.

R18.4.2 *Beams*

According to 18.4.2.3(a), the factored shear force is determined from a free-body diagram obtained by cutting through the beam ends, with end moments assumed equal to the nominal moment strengths acting in reverse curvature bending, both clockwise and counterclockwise. Figure R18.4.2 demonstrates only one of the two options that are to be considered for every beam. To determine the maximum beam shear, it is assumed that its nominal moment strengths (ϕ = **1.0** for moment) are developed simultaneously at both ends of its clear span. As indicated in Fig. R18.4.2, the shear associated with this condition $[(M_{n\ell} + M_{nr})/\ell_n]$ is added algebraically to the shear due to the factored gravity loads and vertical earthquake effects to obtain the design shear for the beam. For the example shown, dead load, live load, and snow load have been assumed to be uniformly distributed. The figure also shows that vertical earthquake effects are to be included, as is typically required by the general building code. For example, ASCE/SEI 7 requires vertical earthquake effects, $\boldsymbol{0.2S_{DS}}$, to be included.

Provision 18.4.2.3(b) bases V_u on the load combination including the earthquake effect $\boldsymbol{E}$, which should be doubled. For example, the load combination defined by Eq. (5.3.1.e) would be

$$U = 1.2D + 2.0E + 1.0L + 0.2S$$

where $\boldsymbol{E}$ is the value specified by the general building code. The factor of 1.0 applied to $\boldsymbol{L}$ is allowed to be reduced to 0.5 in accordance with 5.3.3.

Transverse reinforcement at the ends of the beam is required to be hoops. In most cases, transverse reinforcement required by 18.4.2.3 for the design shear force will be more than those required by 18.4.2.4.

Beams may be subjected to axial compressive force due to prestressing or applied loads. The additional requirements

CODE

(d) 12 in.

18.4.2.5 Transverse reinforcement spacing shall not exceed $d/2$ throughout the length of the beam.

18.4.2.6 In beams having factored axial compressive force exceeding $A_g f_c'/10$, transverse reinforcement required by 18.4.2.5 shall conform to 25.7.2.2 and either 25.7.2.3 or 25.7.2.4.

COMMENTARY

in 18.4.2.6 are intended to provide lateral support for beam longitudinal reinforcement.

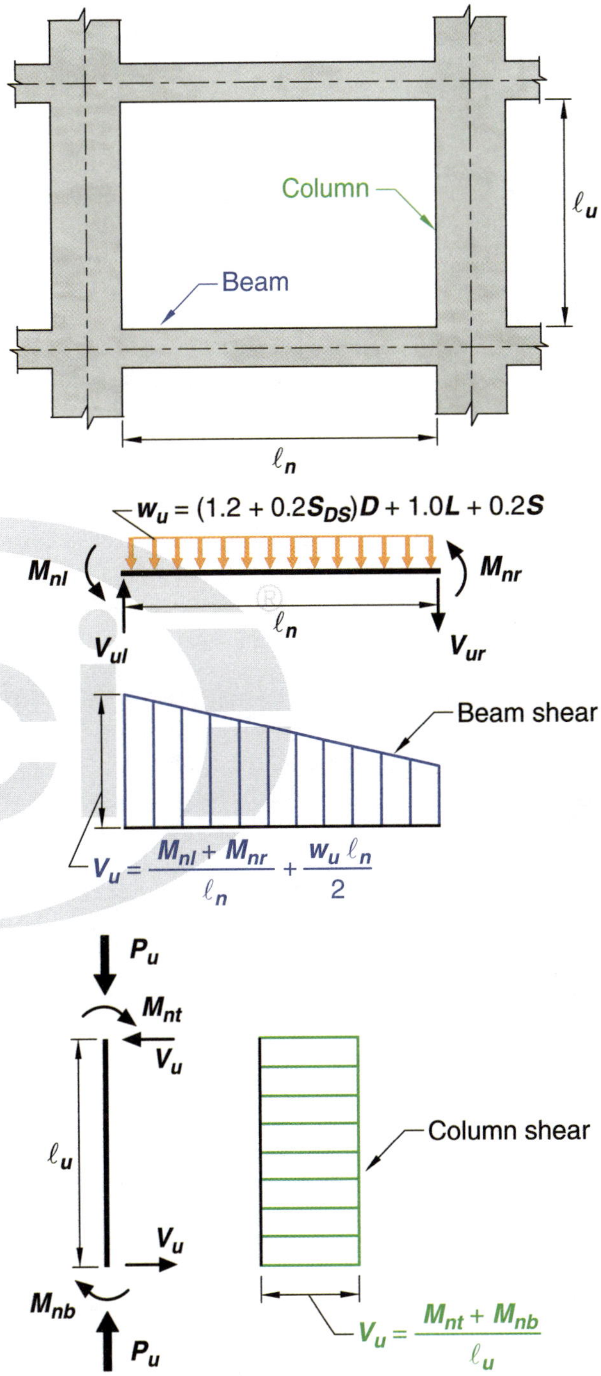

Fig. R18.4.2—Design shears for intermediate moment frames.

18.4.3 *Columns*

18.4.3.1 ϕV_n shall be at least the lesser of (a) and (b):

(a) The shear associated with development of nominal moment strengths of the column at each restrained end of

R18.4.3 *Columns*

According to 18.4.3.1(a), the factored shear force is determined from a free-body diagram obtained by cutting through the column ends, with end moments assumed equal to the nominal moment strengths acting in reverse curva-

CODE

the unsupported length due to reverse curvature bending. Column flexural strength shall be calculated for the factored axial force, consistent with the direction of the lateral forces considered, resulting in the highest flexural strength

(b) The maximum shear obtained from factored load combinations that include $\boldsymbol{E}$, with $\boldsymbol{\Omega_o E}$ substituted for $\boldsymbol{E}$

18.4.3.2 Columns shall be spirally reinforced in accordance with Chapter 10 or shall be in accordance with 18.4.3.3 through 18.4.3.5. Provision 18.4.3.6 shall apply to all columns supporting discontinuous stiff members.

18.4.3.3 At both ends of the column, hoops shall be provided at spacing s_o over a length ℓ_o measured from the joint face. Spacing s_o shall not exceed the least of (a) through (c):

(a) For Grade 60, the smaller of $\boldsymbol{8d_b}$ of the smallest longitudinal bar enclosed and 8 in.

(b) For Grade 80, the smaller of $\boldsymbol{6d_b}$ of the smallest longitudinal bar enclosed and 6 in.

(c) One-half of the smallest cross-sectional dimension of the column

Length ℓ_o shall not be less than the longest of (d), (e), and (f):

(d) One-sixth of the clear span of the column

(e) Maximum cross-sectional dimension of the column

(f) 18 in.

18.4.3.4 The first hoop shall be located not more than $s_o/2$ from the joint face.

18.4.3.5 Outside of length ℓ_o, spacing of transverse reinforcement shall be in accordance with 10.7.6.5.2.

18.4.3.6 Columns supporting reactions from discontinuous stiff members, such as walls, shall be provided with transverse reinforcement at the spacing s_o in accordance with 18.4.3.3 over the full height beneath the level at which the discontinuity occurs if the portion of factored axial compressive force in these members related to earthquake effects exceeds $\boldsymbol{A_g f_c'/10}$. If design forces have been magnified to account for the overstrength of the vertical elements of the seismic-force-resisting system, the limit of $\boldsymbol{A_g f_c'/10}$ shall be increased to $\boldsymbol{A_g f_c'/4}$. Transverse reinforcement shall extend above and below the column in accordance with 18.7.5.6(b).

18.4.4 *Joints*

18.4.4.1 Beam-column joints shall satisfy the detailing requirements of 15.3.1.2, 15.3.1.3, and 18.4.4.2 through 18.4.4.5.

18.4.4.2 If a beam framing into the joint and generating joint shear has depth exceeding twice the column depth,

COMMENTARY

ture bending, both clockwise and counterclockwise. Figure R18.4.2 demonstrates only one of the two options that are to be considered for every column. The factored axial force $\boldsymbol{P_u}$ should be chosen to develop the largest moment strength of the column within the range of design axial forces. Provision 18.4.3.1(b) for columns is similar to 18.4.2.3(b) for beams except it bases $\boldsymbol{V_u}$ on load combinations including the earthquake effect $\boldsymbol{E}$, with $\boldsymbol{E}$ increased by the overstrength factor $\boldsymbol{\Omega_o}$ rather than the factor 2.0. In ASCE/SEI 7, $\boldsymbol{\Omega_o = 3.0}$ for intermediate moment frames. The higher factor for columns relative to beams is because of greater concerns about shear failures in columns.

Transverse reinforcement at the ends of columns is required to be spirals or hoops. The amount of transverse reinforcement at the ends must satisfy both 18.4.3.1 and 18.4.3.2. Note that hoops require seismic hooks at both ends. The maximum spacing allowed for hoops is intended to inhibit or delay buckling of longitudinal reinforcement.

Discontinuous structural walls and other stiff members can impose large axial forces on supporting columns during earthquakes. The required transverse reinforcement in 18.4.3.6 is to improve column toughness under anticipated demands. The factored axial compressive force related to earthquake effect should include the factor $\boldsymbol{\Omega_o}$ if required by the general building code.

R18.4.4 *Joints*

R18.4.4.2 For joints in which the beam depth is significantly greater than the column depth, a diagonal strut between

CODE

analysis and design of the joint shall be based on the strut-and-tie method in accordance with Chapter 23 and (a) and (b) shall be satisfied:

(a) Design joint shear strength determined in accordance with Chapter 23 shall not exceed ϕV_n calculated in accordance with 15.4.2.
(b) Detailing requirements of 18.4.4.3 through 18.4.4.5 shall be satisfied.

18.4.4.3 Longitudinal reinforcement terminated in a joint shall extend to the far face of the joint core and shall be developed in tension in accordance with 18.8.5 and in compression in accordance with 25.4.9.

18.4.4.4 Spacing of joint transverse reinforcement s shall not exceed the lesser of 18.4.3.3(a) through (c) within the height of the deepest beam framing into the joint.

18.4.4.5 Where the top beam longitudinal reinforcement consists of headed deformed bars that terminate in the joint, the column shall extend above the top of the joint a distance at least the depth h of the joint. Alternatively, the beam reinforcement shall be enclosed by additional vertical joint reinforcement providing equivalent confinement to the top face of the joint.

18.4.4.6 Slab-column joints shall satisfy transverse reinforcement requirements of 15.3.2. Where slab-column joint transverse reinforcement is required, at least one layer of joint transverse reinforcement shall be placed between the top and bottom slab reinforcement.

18.4.4.7 *Shear strength requirements for beam-column joints*

18.4.4.7.1 Design shear strength of cast-in-place beam-column joints shall satisfy:

$$\phi V_n \geq V_u$$

18.4.4.7.2 V_u of the joint shall be determined in accordance with 18.3.4.

18.4.4.7.3 ϕ shall be in accordance with 21.2.1 for shear.

COMMENTARY

the joint corners may not be effective. Therefore, the Code requires that joints in which the beam depth exceeds twice the column depth be designed using the strut-and-tie method of Chapter 23.

R18.4.4.3 Refer to R18.8.2.2.

R18.4.4.4 The maximum spacing of transverse reinforcement within a joint is consistent with the spacing limits for reinforcement in columns of intermediate moment frames.

R18.4.4.5 This provision refers to a knee joint in which beam reinforcement terminates with headed deformed bars. Such joints require confinement of the headed beam bars along the top face of the joint. This confinement can be provided by either (a) a column that extends above the top of the joint or (b) vertical reinforcement hooked around the beam top reinforcing bars and extending downward into the joint in addition to the column longitudinal reinforcement. Detailing guidance and design recommendations for vertical joint reinforcement may be found in ACI 352R.

18.4.4.7 *Shear strength requirements for beam-column joints*

R18.4.4.7.2 Factored joint shear force is determined assuming that beams framing into the joint develop end moments equal to their nominal moment strengths. Consequently, joint shear force generated by the flexural reinforcement is calculated for a stress of f_y in the reinforcement. This is consistent with 18.4.2 and 18.4.3 for determination of minimum design shear strength in beams and columns of intermediate moment frames.

Seismic 18

CODE

18.4.4.7.4 V_n of the joint shall be in accordance with 18.8.4.3.

18.4.5 *Two-way slabs without beams*

18.4.5.1 Factored slab moment at the support including earthquake effects, $\boldsymbol{E}$, shall be calculated for load combinations given in Eq. (5.3.1e) and (5.3.1g). Reinforcement to resist $\boldsymbol{M_{sc}}$ shall be placed within the column strip defined in 8.4.1.5.

18.4.5.2 Reinforcement placed within the effective width given in 8.4.2.2.3 shall be designed to resist $\gamma_f \boldsymbol{M_{sc}}$. Effective slab width for exterior and corner connections shall not extend beyond the column face a distance greater than $\boldsymbol{c_t}$ measured perpendicular to the slab span.

18.4.5.3 At least one-half of the reinforcement in the column strip at the support shall be placed within the effective slab width given in 8.4.2.2.3.

18.4.5.4 At least one-fourth of the top reinforcement at the support in the column strip shall be continuous throughout the span.

18.4.5.5 Continuous bottom reinforcement in the column strip shall be at least one-third of the top reinforcement at the support in the column strip.

18.4.5.6 At least one-half of all bottom middle strip reinforcement and all bottom column strip reinforcement at midspan shall be continuous and shall develop $\boldsymbol{f_y}$ at the face of columns, capitals, brackets, or walls.

18.4.5.7 At discontinuous edges of the slab, all top and bottom reinforcement at the support shall be developed at the face of columns, capitals, brackets, or walls.

COMMENTARY

R18.4.5 *Two-way slabs without beams*

Section 18.4.5 applies to two-way slabs without beams, such as flat plates.

Using load combinations of Eq. (5.3.1e) and (5.3.1g) may result in moments requiring top and bottom reinforcement at the supports.

The moment $\boldsymbol{M_{sc}}$ refers, for a given design load combination with $\boldsymbol{E}$ acting in one horizontal direction, to that portion of the factored slab moment that is balanced by the supporting members at a joint. It is not necessarily equal to the total design moment at the support for a load combination including earthquake effect. In accordance with 8.4.2.2.3, only a fraction of the moment $\boldsymbol{M_{sc}}$ is assigned to the slab effective width. For edge and corner connections, flexural reinforcement perpendicular to the edge is not considered fully effective unless it is placed within the effective slab width (ACI 352.1R; Pan and Moehle 1989). Refer to Fig. R18.4.5.1.

Application of the provisions of 18.4.5 is illustrated in Fig. R18.4.5.2 and R18.4.5.3.

CODE

COMMENTARY

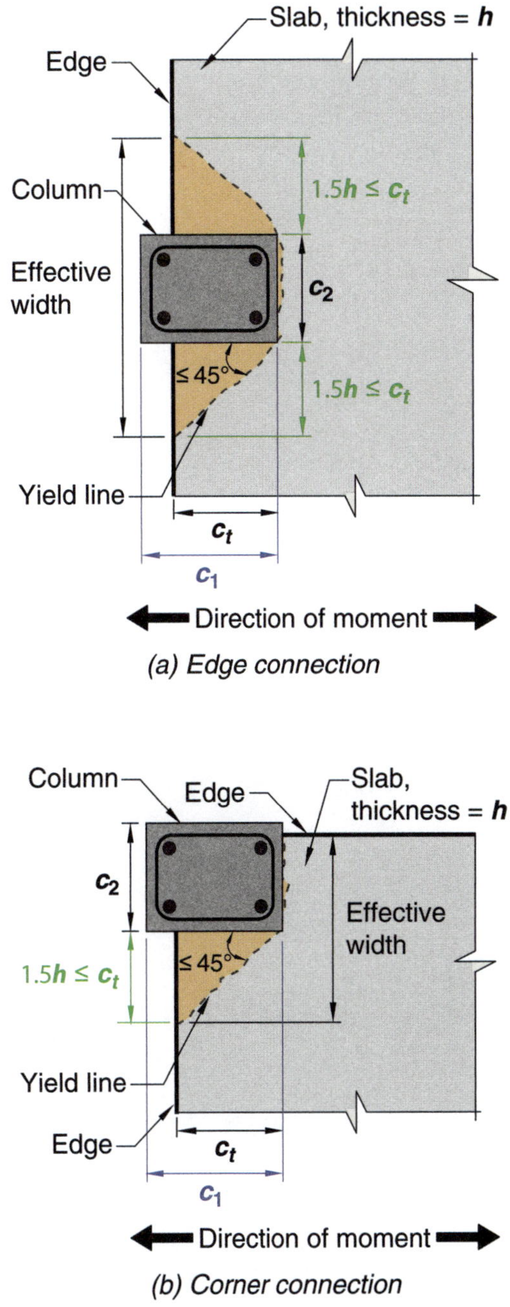

***Fig. R18.4.5.1**—Effective width for reinforcement placement in edge and corner connections.*

CODE

COMMENTARY

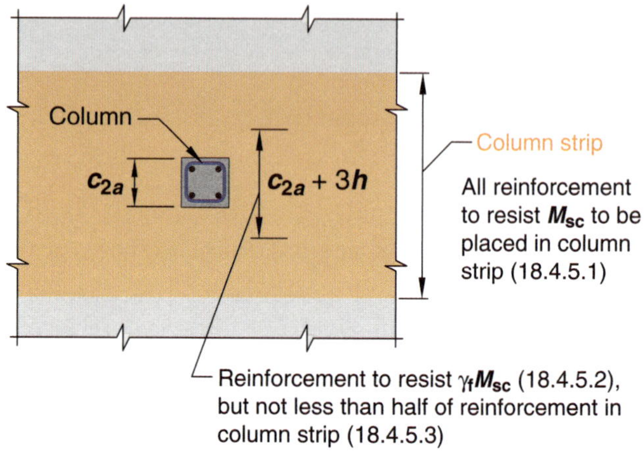

Fig. R18.4.5.2—Location of reinforcement in slabs.

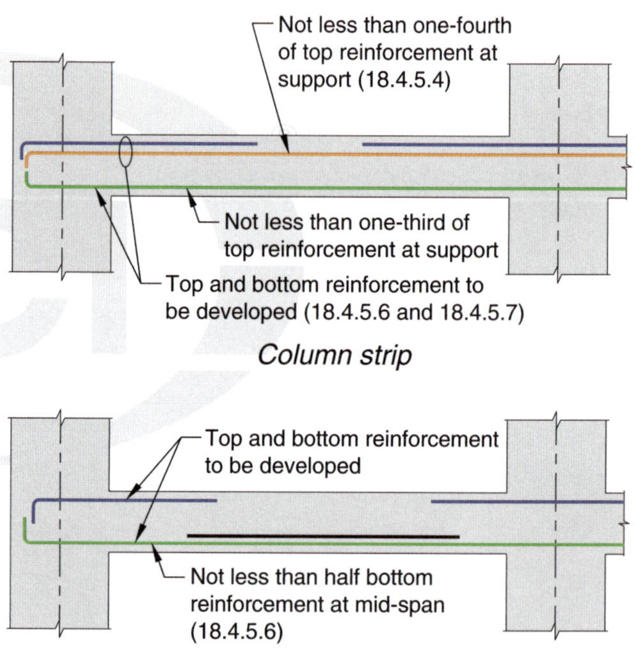

Fig. R18.4.5.3—Arrangement of reinforcement in slabs.

18.4.5.8 At the critical sections for columns defined in 22.6.4.1, two-way shear stress caused by factored gravity loads without moment transfer shall not exceed $0.4\phi v_c$ for nonprestressed slab-column connections and $0.5\phi v_c$ for unbonded post-tensioned slab-column connections with f_{pc} in each direction meeting the requirements of 8.6.2.1, where v_c shall be calculated in accordance with 22.6.5. This requirement need not be satisfied if the slab-column connection satisfies 18.14.5.

R18.4.5.8 The requirements apply to two-way slabs that are designated part of the seismic-force-resisting system. Nonprestressed slab-column connections in laboratory tests (Pan and Moehle 1989) exhibited reduced lateral displacement ductility when the shear stress at the column connection exceeded the recommended limit of $0.4\phi v_c$. Based on laboratory test data (Kang and Wallace 2006; Kang et al. 2007), a higher maximum factored gravity shear stress of $0.5\phi v_c$ is allowed for unbonded post-tensioned slab-column connections with f_{pc} in each direction meeting the requirements of 8.6.2.1. Post-tensioned slab-column connections with f_{pc} in each direction not meeting the requirements of 8.6.2.1 can be designed as nonprestressed slab-column connections in accordance with 8.2.3. Slab-column connections also must

CODE | COMMENTARY

satisfy shear and moment strength requirements of Chapter 8 under load combinations including earthquake effect.

18.5—Intermediate precast structural walls

18.5.1 *Scope*

18.5.1.1 This section shall apply to intermediate precast structural walls forming part of the seismic-force-resisting system.

18.5.2 *General*

18.5.2.1 In connections between wall panels, or between wall panels and the foundation, yielding shall be restricted to steel elements or reinforcement.

18.5.2.2 For elements of the connection that are not designed to yield, the required strength shall be based on $1.5S_y$ of the yielding portion of the connection.

18.5.2.3 In structures assigned to SDC D, E, or F, wall piers shall be designed in accordance with 18.10.8 or 18.14.

R18.5—Intermediate precast structural walls

Connections between precast wall panels or between wall panels and the foundation are required to resist forces induced by earthquake motions and to provide for yielding in the vicinity of connections. If mechanical splices are used to directly connect primary reinforcement, the probable strength of the splice should be at least 1.5 times the specified yield strength of the reinforcement.

18.6—Beams of special moment frames

18.6.1 *Scope*

18.6.1.1 This section shall apply to beams of special moment frames that form part of the seismic-force-resisting system and are proportioned primarily to resist flexure and shear.

18.6.1.2 Beams of special moment frames shall frame into columns of special moment frames satisfying 18.7.

R18.6—Beams of special moment frames

R18.6.1 *Scope*

This section applies to beams of special moment frames resisting lateral loads induced by earthquake motions. In previous Codes, any frame member subjected to a factored axial compressive force exceeding $(A_g f_c'/10)$ under any load combination was to be proportioned and detailed as described in 18.7. In the 2014 Code, all requirements for beams are contained in 18.6 regardless of the magnitude of axial compressive force.

This Code is written with the assumption that special moment frames comprise horizontal beams and vertical columns interconnected by beam-column joints. It is acceptable for beams and columns to be inclined provided the resulting system behaves as a frame—that is, lateral resistance is provided primarily by moment transfer between beams and columns rather than by strut or brace action. In special moment frames, it is acceptable to design beams to resist combined moment and axial force as occurs in beams that act both as moment frame members and as chords or collectors of a diaphragm. It is acceptable for beams of special moment frames to cantilever beyond columns, but such cantilevers are not part of the special moment frame that forms part of the seismic-force-resisting system. It is acceptable for beams of a special moment frame to connect into a wall boundary if the boundary is reinforced as a special moment frame column in accordance with 18.7. A concrete braced frame, in which lateral resistance is provided primarily by axial forces in beams and columns, is not a recognized seismic-force-resisting system.

CODE

18.6.2 *Dimensional limits*

18.6.2.1 Beams shall satisfy (a) through (c):

(a) Clear span ℓ_n shall be at least $4d$
(b) Width b_w shall be at least the lesser of $0.3h$ and 10 in.
(c) Projection of the beam width beyond the width of the supporting column on each side shall not exceed the lesser of c_2 and $0.75c_1$.

COMMENTARY

R18.6.2 *Dimensional limits*

Experimental evidence (Hirosawa 1977) indicates that, under reversals of displacement into the nonlinear range, behavior of continuous members having length-to-depth ratios of less than 4 is significantly different from the behavior of relatively slender members. Design rules derived from experience with relatively slender members do not apply directly to members with length-to-depth ratios less than 4, especially with respect to shear strength.

Geometric constraints indicated in 18.6.2.1(b) and (c) were derived from practice and research (ACI 352R) on reinforced concrete frames resisting earthquake-induced forces. The limits in 18.6.2.1(c) define the maximum beam width that can effectively transfer forces into the beam-column joint. An example of maximum effective beam width is shown in Fig. R18.6.2.

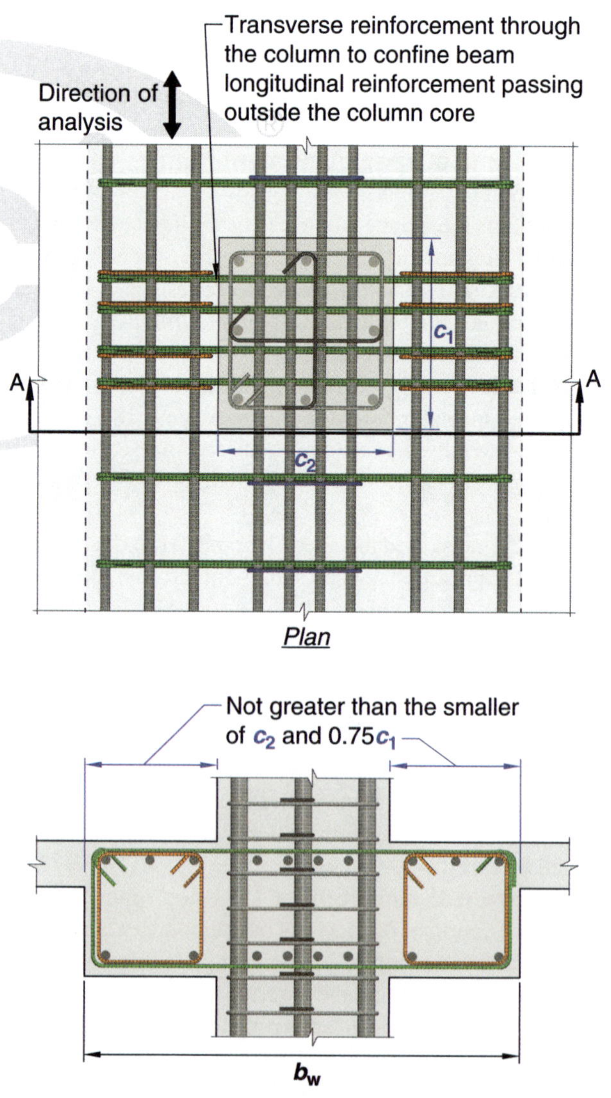

Fig. R18.6.2—Maximum effective width of wide beam and required transverse reinforcement.

CODE

18.6.3 *Longitudinal reinforcement*

18.6.3.1 Beams shall have at least two continuous bars at both top and bottom faces. At any section, for top as well as for bottom reinforcement, the amount of reinforcement shall be at least that required by 9.6.1.2, and the reinforcement ratio **ρ** shall not exceed 0.025 for Grade 60 reinforcement and 0.02 for Grade 80 reinforcement.

18.6.3.2 Positive moment strength at joint face shall be at least one-half the negative moment strength provided at that face of the joint. Both the negative and the positive moment strength at any section along member length shall be at least one-fourth the maximum moment strength provided at face of either joint.

18.6.3.3 Lap splices of deformed longitudinal reinforcement shall be permitted if hoop or spiral reinforcement is provided over the lap length. Spacing of the transverse reinforcement enclosing the lap-spliced bars shall not exceed the lesser of $d/4$ and 4 in. Lap splices shall not be used in locations (a) through (c):

(a) Within the joints
(b) Within a distance of twice the beam depth from the face of the joint
(c) Within a distance of twice the beam depth from critical sections where flexural yielding is likely to occur as a result of lateral displacements beyond the elastic range of behavior

18.6.3.4 Mechanical splices shall conform to 18.2.7 and welded splices shall conform to 18.2.8.

18.6.3.5 Unless used in a special moment frame as permitted by 18.9.2.3, prestressing shall satisfy (a) through (d):

(a) The average prestress f_{pc} calculated for an area equal to the least cross-sectional dimension of the beam multiplied by the perpendicular cross-sectional dimension shall not exceed the lesser of 500 psi and $f_c'/10$.
(b) Prestressed reinforcement shall be unbonded in potential plastic hinge regions, and the calculated strains in prestressed reinforcement under the design displacement shall be less than 0.01.
(c) Prestressed reinforcement shall not contribute more than one-fourth of the positive or negative flexural strength at the critical section in a plastic hinge region and shall be anchored at or beyond the exterior face of the joint.
(d) Anchorages of post-tensioning tendons resisting earthquake-induced forces shall be capable of allowing tendons to withstand 50 cycles of loading, with prestressed reinforcement forces bounded by 40 and 85 percent of the specified tensile strength of the prestressing reinforcement.

COMMENTARY

R18.6.3 *Longitudinal reinforcement*

R18.6.3.1 The limiting reinforcement ratios of 0.025 and 0.02 are based primarily on considerations of providing adequate deformation capacity, avoiding reinforcement congestion, and, indirectly, on limiting shear stresses in beams of typical proportions.

R18.6.3.3 Lap splices of reinforcement are prohibited along lengths where flexural yielding is anticipated because such splices are not reliable under conditions of cyclic loading into the inelastic range. Transverse reinforcement for lap splices at any location is mandatory because of the potential of concrete cover spalling and the need to confine the splice.

R18.6.3.5 These provisions were developed, in part, based on observations of building performance in earthquakes (ACI 423.3R). For calculating the average prestress, the least cross-sectional dimension in a beam normally is the web dimension, and is not intended to refer to the flange thickness. In a potential plastic hinge region, the limitation on strain and the requirement for unbonded tendons are intended to prevent fracture of tendons under inelastic earthquake deformation. Calculation of strain in the prestressed reinforcement is required considering the anticipated inelastic mechanism of the structure. For prestressed reinforcement unbonded along the full beam span, strains generally will be well below the specified limit. For prestressed reinforcement with short unbonded length through or adjacent to the joint, the additional strain due to earthquake deformation is calculated as the product of the depth to the neutral axis and the sum of plastic hinge rotations at the joint, divided by the unbonded length.

The restrictions on the flexural strength provided by the tendons are based on the results of analytical and experimental studies (Ishizuka and Hawkins 1987; Park and

Thompson 1977). Although satisfactory seismic performance can be obtained with greater amounts of prestressed reinforcement, this restriction is needed to allow the use of the same response modification and deflection amplification factors as those specified in model codes for special moment frames without prestressed reinforcement. Prestressed special moment frames will generally contain continuous prestressed reinforcement that is anchored with adequate cover at or beyond the exterior face of each beam-column connection located at the ends of the moment frame.

Fatigue testing for 50 cycles of loading between 40 and 80 percent of the specified tensile strength of the prestressed reinforcement has been a long-standing industry practice (ACI 423.3R; ACI 423.7). The 80 percent limit was increased to 85 percent to correspond to the 1 percent limit on the strain in prestressed reinforcement. Testing over this range of stress is intended to conservatively simulate the effect of a severe earthquake. Additional details on testing procedures are provided in ACI 423.7.

CODE

18.6.4 *Transverse reinforcement*

18.6.4.1 Hoops shall be provided in the following regions of a beam:

(a) Over a length equal to twice the beam depth measured from the face of the supporting column toward midspan, at both ends of the beam
(b) Over lengths equal to twice the beam depth on both sides of a section where flexural yielding is likely to occur as a result of lateral displacements beyond the elastic range of behavior.

18.6.4.2 Where hoops are required, primary longitudinal reinforcing bars closest to the tension and compression faces shall have lateral support in accordance with 25.7.2.3 and 25.7.2.4. The spacing of transversely supported flexural reinforcing bars shall not exceed 14 in. Skin reinforcement required by 9.7.2.3 need not be laterally supported.

18.6.4.3 Hoops in beams shall be permitted to be made up of two pieces of reinforcement: a stirrup having seismic hooks at both ends and closed by a crosstie. Consecutive crossties engaging the same longitudinal bar shall have their 90-degree hooks at opposite sides of the flexural member. If the longitudinal reinforcing bars secured by the crossties are confined by a slab on only one side of the beam, the 90-degree hooks of the crossties shall be placed on that side.

18.6.4.4 The first hoop shall be located not more than 2 in. from the face of a supporting column. Spacing of the hoops shall not exceed the least of (a) through (d):

(a) $d/4$
(b) 6 in.

COMMENTARY

R18.6.4 *Transverse reinforcement*

Transverse reinforcement is required primarily to confine the concrete and maintain lateral support for the reinforcing bars in regions where yielding is expected. Examples of hoops suitable for beams are shown in Fig. R18.6.4.

In earlier Code editions, the upper limit on hoop spacing was the least of $d/4$, eight longitudinal bar diameters, 24 tie bar diameters, and 12 in. The upper limits were changed in the 2011 edition because of concerns about adequacy of longitudinal bar buckling restraint and confinement in large beams.

In the case of members with varying strength along the span or members for which the permanent load represents a large proportion of the total design load, concentrations of inelastic rotation may occur within the span. If such a condition is anticipated, transverse reinforcement is also required in regions where yielding is expected. Because spalling of the concrete shell might occur, especially at and near regions of flexural yielding, all web reinforcement is required to be provided in the form of closed hoops.

Seismic 18

CODE

(c) For Grade 60, $6d_b$ of the smallest primary flexural reinforcing bar excluding longitudinal skin reinforcement required by 9.7.2.3

(d) For Grade 80, $5d_b$ of the smallest primary flexural reinforcing bar excluding longitudinal skin reinforcement required by 9.7.2.3

18.6.4.5 Where hoops are required, they shall be designed to resist shear according to 18.6.5.

18.6.4.6 Where hoops are not required, stirrups with seismic hooks at both ends shall be spaced at a distance not more than $d/2$ throughout the length of the beam.

18.6.4.7 In beams having factored axial compressive force exceeding $A_g f_c'/10$, hoops satisfying 18.7.5.2 through 18.7.5.4 shall be provided along lengths given in 18.6.4.1. Along the remaining length, hoops satisfying 18.7.5.2 shall have spacing s not exceeding the least of 6 in., $6d_b$ of the smallest Grade 60 enclosed longitudinal beam bar, and $5d_b$ of the smallest Grade 80 enclosed longitudinal beam bar. Where concrete cover over transverse reinforcement exceeds 4 in., additional transverse reinforcement having cover not exceeding 4 in. and spacing not exceeding 12 in. shall be provided.

COMMENTARY

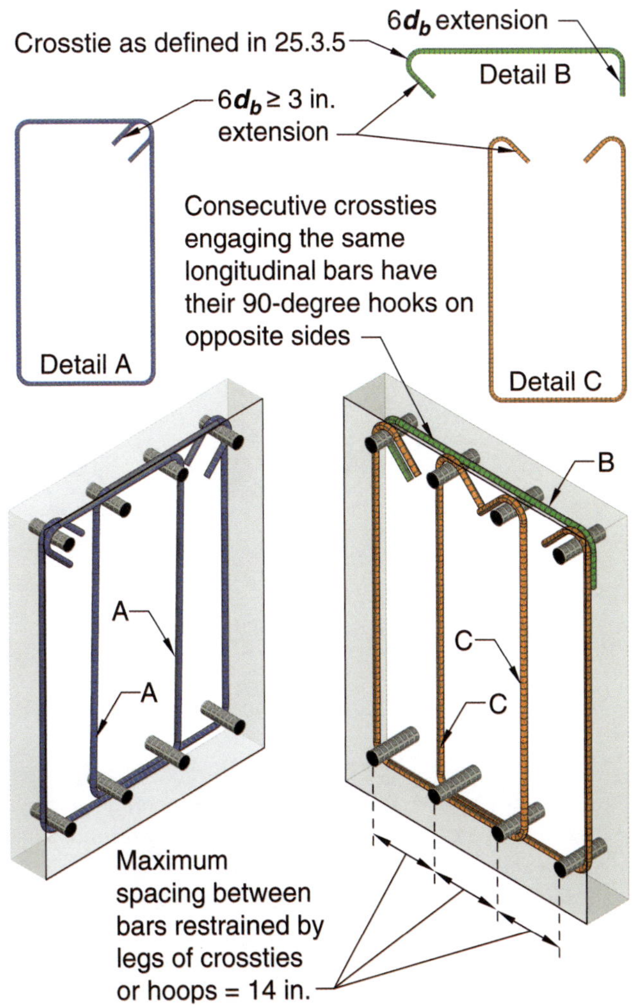

Fig. R18.6.4— Examples of overlapping hoops and illustration of limit on maximum horizontal spacing of supported longitudinal bars.

18.6.5 *Shear strength*

18.6.5.1 *Design forces*

The design shear force V_e shall be calculated from consideration of the forces on the portion of the beam between faces of the joints. It shall be assumed that moments of opposite sign corresponding to probable flexural strength, M_{pr}, act at the joint faces and that the beam is loaded with the factored gravity and vertical earthquake loads along its span.

18.6.5.2 *Transverse reinforcement*

Transverse reinforcement over the lengths identified in 18.6.4.1 shall be designed to resist shear assuming $V_c = 0$ when both (a) and (b) occur:

R18.6.5 *Shear strength*

Unless a beam possesses a moment strength that is on the order of 3 or 4 times the design moment, it should be assumed that it will yield in flexure in the event of a major earthquake. The design shear force should be selected so as to be a good approximation of the maximum shear that may develop in a member. Therefore, required shear strength for frame members is related to flexural strengths of the designed member rather than to factored shear forces indicated by lateral load analysis. The conditions described by 18.6.5.1 are illustrated in Fig. R18.6.5. The figure also shows that vertical earthquake effects are to be included, as is typically required by the general building code. For example, ASCE/SEI 7 requires vertical earthquake effects, $0.2S_{DS}$, to be included.

Because the actual yield strength of the longitudinal reinforcement may exceed the specified yield strength and because strain hardening of the reinforcement is likely to

CODE

(a) The earthquake-induced shear force calculated in accordance with 18.6.5.1 represents at least one-half of the maximum required shear strength within those lengths.

(b) The factored axial compressive force $\boldsymbol{P_u}$ including earthquake effects is less than $\boldsymbol{A_g f_c'/20}$.

COMMENTARY

take place at a joint subjected to large rotations, required shear strengths are determined using a stress of at least $\mathbf{1.25}\boldsymbol{f_y}$ in the longitudinal reinforcement.

Experimental studies (Popov et al. 1972) of reinforced concrete members subjected to cyclic loading have demonstrated that more shear reinforcement is required to ensure a flexural failure if the member is subjected to alternating nonlinear displacements than if the member is loaded in only one direction: the necessary increase of shear reinforcement being higher in the case of no axial load. This observation is reflected in the Code (refer to 18.6.5.2) by eliminating the term representing the contribution of concrete to shear strength. The added conservatism on shear is deemed necessary in locations where potential flexural hinging may occur. However, this stratagem, chosen for its relative simplicity, should not be interpreted to mean that no concrete is required to resist shear. On the contrary, it may be argued that the concrete core resists all the shear with the shear (transverse) reinforcement confining and strengthening the concrete. The confined concrete core plays an important role in the behavior of the beam and should not be reduced to a minimum just because the design expression does not explicitly recognize it.

CODE | **COMMENTARY**

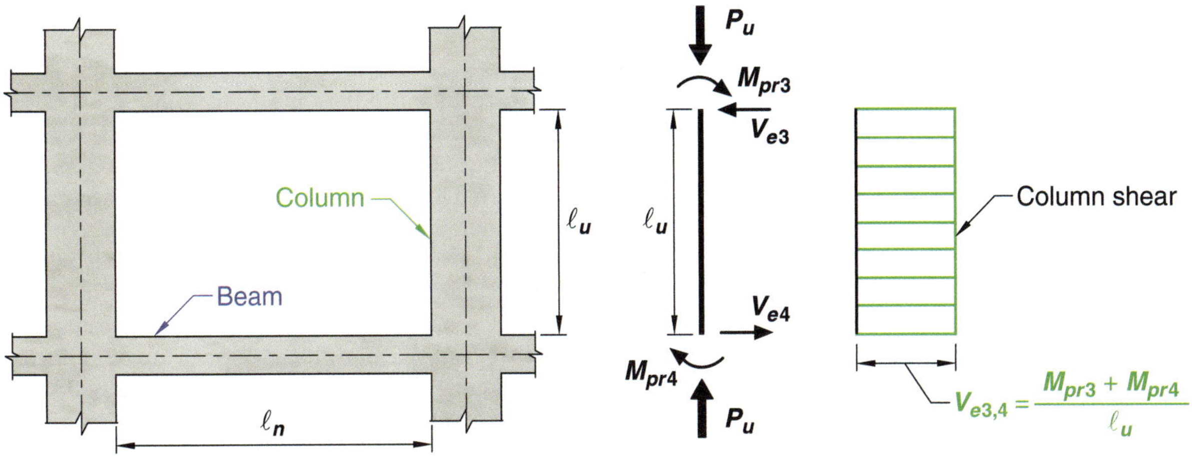

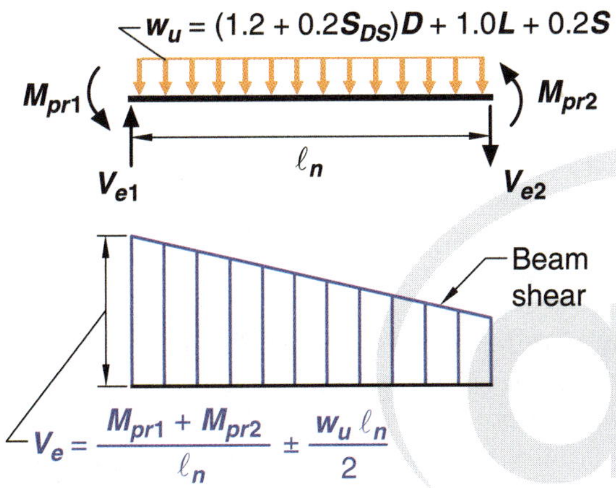

Notes on Fig. R18.6.5:

1. Direction of shear force V_e depends on relative magnitudes of gravity loads and shear generated by end moments.
2. End moments M_{pr} based on steel tensile stress of $1.25f_y$, where f_y is specified yield strength. (Both end moments should be considered in both directions, clockwise and counter-clockwise).
3. End moment M_{pr} for columns need not be greater than moments generated by the M_{pr} of the beams framing into the beam-column joints. V_e should not be less than that required by analysis of the structure.

Fig. R18.6.5—Design shears for beams and columns.

18.7—Columns of special moment frames

18.7.1 *Scope*

18.7.1.1 This section shall apply to columns of special moment frames that form part of the seismic-force-resisting system and are proportioned primarily to resist flexure, shear, and axial forces.

18.7.2 *Dimensional limits*

18.7.2.1 Columns shall satisfy (a) and (b):

(a) The shortest cross-sectional dimension, measured on a straight line passing through the geometric centroid, shall be at least 12 in.
(b) The ratio of the shortest cross-sectional dimension to the perpendicular dimension shall be at least 0.4.

18.7.3 *Minimum flexural strength of columns*

18.7.3.1 Columns shall satisfy 18.7.3.2 or 18.7.3.3, except at connections where the column is discontinuous above the connection and the column factored axial compressive force

R18.7—Columns of special moment frames

R18.7.1 *Scope*

This section applies to columns of special moment frames regardless of the magnitude of axial force. Before 2014, the Code permitted columns with low levels of axial stress to be detailed as beams.

R18.7.2 *Dimensional limits*

The geometric constraints in this provision follow from previous practice (Seismology Committee of SEAOC 1996).

R18.7.3 *Minimum flexural strength of columns*

The intent of 18.7.3.2 is to reduce the likelihood of yielding in columns that are considered as part of the seismic-force-resisting system. If columns are not stronger than beams framing into a joint, there is increased likelihood of inelastic

CODE

P_u under load combinations including earthquake effect, E, are less than $A_g f_c'/10$.

18.7.3.2 The flexural strengths of the columns shall satisfy

$$\sum M_{nc} \geq (6/5)\sum M_{nb} \qquad (18.7.3.2)$$

where

$\sum M_{nc}$ is sum of nominal flexural strengths of columns framing into the joint, evaluated at the faces of the joint. Column flexural strength shall be calculated for the factored axial force, consistent with the direction of the lateral forces considered, resulting in the lowest flexural strength.

$\sum M_{nb}$ is sum of nominal flexural strengths of the beams framing into the joint, evaluated at the faces of the joint. In T-beam construction, where the slab is in tension under moments at the face of the joint, slab reinforcement within an effective slab width defined in accordance with 6.3.2 shall be assumed to contribute to M_{nb} if the slab reinforcement is developed at the critical section for flexure.

Flexural strengths shall be summed such that the column moments oppose the beam moments. Equation (18.7.3.2) shall be satisfied for beam moments acting in both directions in the vertical plane of the frame considered.

18.7.3.3 If 18.7.3.2 is not satisfied at a joint, the lateral strength and stiffness of the columns framing into that joint shall be ignored when calculating strength and stiffness of the structure. These columns shall conform to 18.14.

COMMENTARY

action. In the worst case of weak columns, flexural yielding can occur at both ends of all columns in a given story, resulting in a column failure mechanism that can lead to collapse. Connections with discontinuous columns above the connection, such as roof-level connections, are exempted if the column axial load is low, because special moment frame columns with low axial stress are inherently ductile and column yielding at such levels is unlikely to create a column failure mechanism that can lead to collapse.

In 18.7.3.2, the nominal strengths of the beams and columns are calculated at the joint faces, and those strengths are compared directly using Eq. (18.7.3.2). The 1995 and earlier Codes required design strengths to be compared at the center of the joint, which typically produced similar results but with added calculation effort.

In determining the nominal moment strength of a beam section in negative bending (top in tension), longitudinal reinforcement contained within an effective flange width of a top slab that acts monolithically with the beam increases the beam strength. French and Moehle (1991), on beam-column subassemblies under lateral loading, indicates that using the effective flange widths defined in 6.3.2 gives reasonable estimates of beam negative moment strengths of interior connections at story displacements approaching 2 percent of story height. This effective width is conservative where the slab terminates in a weak spandrel.

If 18.7.3.2 cannot be satisfied at a joint, 18.7.3.3 requires that any positive contribution of the column or columns involved to the lateral strength and stiffness of the structure is to be ignored. Negative contributions of the column or columns should not be ignored. For example, ignoring the stiffness of the columns ought not to be used as a justification for reducing the design base shear. If inclusion of those columns in the analytical model of the building results in an increase in torsional effects, the increase should be considered as required by the general building code. Furthermore, the column must be provided with transverse reinforcement to increase its resistance to shear and axial forces.

18.7.4 *Longitudinal reinforcement*

18.7.4.1 Area of longitudinal reinforcement, A_{st}, shall be at least $0.01A_g$ and shall not exceed $0.06A_g$.

18.7.4.2 In columns with circular hoops, there shall be at least six longitudinal bars.

R18.7.4 *Longitudinal reinforcement*

The lower limit of the area of longitudinal reinforcement is to control time-dependent deformations and to have the yield moment exceed the cracking moment. The upper limit of the area reflects concern for reinforcement congestion, load transfer from floor elements to column (especially in low-rise construction) and the development of high shear stresses.

Spalling of the shell concrete, which is likely to occur near the ends of the column in frames of typical configuration, makes lap splices in these locations vulnerable. If lap splices are to be used at all, they should be located near the midheight where stress reversal is likely to be limited to a smaller stress range than at locations near the joints. Transverse reinforcement is required along the lap-splice length

CODE

COMMENTARY

because of the uncertainty in moment distributions along the height and the need for confinement of lap splices subjected to stress reversals (Sivakumar et al. 1983).

18.7.4.3 Over column clear height, longitudinal reinforcement shall be selected such that $\mathbf{1.25}\boldsymbol{\ell}_d \leq \boldsymbol{\ell}_u\mathbf{/2}$.

R18.7.4.3 Bond splitting failure along longitudinal bars within the clear column height may occur under earthquake demands (Ichinose 1995; Sokoli and Ghannoum 2016).

Splitting can be controlled by restricting longitudinal bar size, increasing the amount of transverse reinforcement, or increasing concrete strength, all of which reduce the development length of longitudinal bars (ℓ_d) over column clear height (ℓ_u). Increasing the ratio of column-to-beam moment strength at joints can reduce the inelastic demands on longitudinal bars in columns under earthquake demands.

18.7.4.4 Mechanical splices shall conform to 18.2.7 and welded splices shall conform to 18.2.8. Lap splices shall be permitted only within the center half of the member length, shall be designed as tension lap splices, and shall be enclosed within transverse reinforcement in accordance with 18.7.5.2 and 18.7.5.3.

18.7.5 *Transverse reinforcement*

R18.7.5 *Transverse reinforcement*

This section is concerned with confining the concrete and providing lateral support to the longitudinal reinforcement.

18.7.5.1 Transverse reinforcement required in 18.7.5.2 through 18.7.5.4 shall be provided over a length ℓ_o from each joint face and on both sides of any section where flexural yielding is likely to occur as a result of lateral displacements beyond the elastic range of behavior. Length ℓ_o shall be at least the greatest of (a) through (c):

(a) The depth of the column at the joint face or at the section where flexural yielding is likely to occur
(b) One-sixth of the clear span of the column
(c) 18 in.

R18.7.5.1 This section stipulates a minimum length over which to provide closely-spaced transverse reinforcement at the column ends, where flexural yielding normally occurs. Research results indicate that the length should be increased by 50 percent or more in locations, such as the base of a building, where axial loads and flexural demands may be especially high (Watson et al. 1994).

18.7.5.2 Transverse reinforcement shall be in accordance with (a) through (f):

(a) Transverse reinforcement shall comprise either single or overlapping spirals, circular hoops, or single or overlapping rectilinear hoops with or without crossties.
(b) Bends of rectilinear hoops and crossties shall engage peripheral longitudinal reinforcing bars.
(c) Crossties of the same or smaller bar size as the hoops shall be permitted, subject to the limitation of 25.7.2.2. Consecutive crossties shall be alternated end for end along the longitudinal reinforcement and around the perimeter of the cross section.

R18.7.5.2 Sections 18.7.5.2 and 18.7.5.3 provide requirements for configuration of transverse reinforcement for columns and joints of special moment frames. Figure R18.7.5.2 shows an example of transverse reinforcement provided by one hoop and three crossties. Crossties with a 90-degree hook are not as effective as either crossties with 135-degree hooks or hoops in providing confinement. For lower values of P_u/A_gf_c' and lower concrete compressive strengths, crossties with 90-degree hooks are adequate if the ends are alternated along the length and around the perimeter of the column. For higher values of P_u/A_gf_c', for which compression-controlled behavior is expected, and for higher compressive strengths, for which behavior tends to be more brittle, the improved confinement provided by having corners of hoops or seismic hooks supporting all

CODE

(d) Where rectilinear hoops or crossties are used, they shall provide lateral support to longitudinal reinforcement in accordance with 25.7.2.2 and 25.7.2.3.

(e) Reinforcement shall be arranged such that the spacing h_x of longitudinal bars laterally supported by the corner of a crosstie or hoop leg shall not exceed 14 in. around the perimeter of the column.

(f) Where $P_u > 0.3A_g f_c'$ or $f_c' > 10{,}000$ **psi** in columns with rectilinear hoops, every longitudinal bar or bundle of bars around the perimeter of the column core shall have lateral support provided by the corner of a hoop or by a seismic hook, and the value of h_x shall not exceed 8 in. P_u shall be the largest value in compression consistent with factored load combinations including E.

COMMENTARY

longitudinal bars is important to achieving intended performance. Where these conditions apply, crossties with seismic hooks at both ends are required. The 8 in. limit on h_x is also intended to improve performance under these critical conditions. For bundled bars, bends or hooks of hoops and crossties need to enclose the bundle, and longer extensions on hooks should be considered. Column axial load P_u should reflect factored compressive demands from both earthquake and gravity loads.

In past editions of the Code, the requirements for transverse reinforcement in columns, walls, beam-column joints, and diagonally reinforced coupling beams referred to the same equations. In the 2014 edition of the Code, the equations and detailing requirements differ among the member types based on consideration of their loadings, deformations, and performance requirements. Additionally, h_x previously referred to the distance between legs of hoops or crossties. In the 2014 edition of the Code, h_x refers to the distance between longitudinal bars supported by those hoops or crossties.

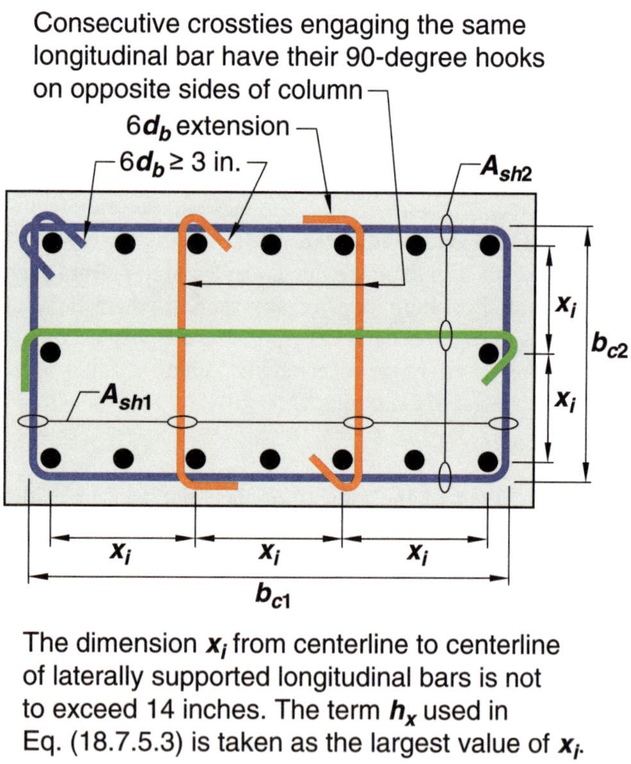

Fig. R18.7.5.2—Example of transverse reinforcement in columns.

18.7.5.3 Spacing of transverse reinforcement shall not exceed the least of (a) through (d):

(a) One-fourth of the minimum column dimension

(b) For Grade 60, $6d_b$ of the smallest longitudinal bar

(c) For Grade 80, $5d_b$ of the smallest longitudinal bar

(d) s_o, as calculated by:

R18.7.5.3 The requirement that spacing not exceed one-fourth of the minimum member dimension or 6 in. is for concrete confinement. If the maximum spacing of crossties or legs of overlapping hoops within the section is less than 14 in., then the 4 in. limit can be increased as permitted by Eq. (18.7.5.3). The spacing limit as a function of the longitudinal bar diameter is intended to provide adequate longitudinal bar restraint to control buckling after spalling.

CODE

$$s_o = 4 + \left(\frac{14 - h_x}{3}\right) \quad (18.7.5.3)$$

The value of s_o from Eq. (18.7.5.3) shall not exceed 6 in. and need not be taken less than 4 in.

18.7.5.4 Amount of transverse reinforcement shall be in accordance with Table 18.7.5.4.

The concrete strength factor k_f and confinement effectiveness factor k_n are calculated according to Eq. (18.7.5.4a) and (18.7.5.4b).

$$\text{(a) } k_f = \frac{f_c'}{25{,}000} + 0.6 \geq 1.0 \quad (18.7.5.4a)$$

$$\text{(b) } k_n = \frac{n_l}{n_l - 2} \quad (18.7.5.4b)$$

where n_l is the number of longitudinal bars or bar bundles around the perimeter of a column core with rectilinear hoops that are laterally supported by the corner of hoops or by seismic hooks.

Table 18.7.5.4—Transverse reinforcement for columns of special moment frames

Transverse reinforcement	Conditions		Applicable expressions
A_{sh}/sb_c for rectilinear hoop	$P_u \leq 0.3A_gf_c'$ and $f_c' \leq 10{,}000$ psi	Greater of (a) and (b)	$0.3\left(\frac{A_g}{A_{ch}} - 1\right)\frac{f_c'}{f_{yt}}$ (a)
	$P_u > 0.3A_gf_c'$ or $f_c' > 10{,}000$ psi	Greatest of (a), (b), and (c)	$0.09\frac{f_c'}{f_{yt}}$ (b) $0.2k_fk_n\frac{P_u}{f_{yt}A_{ch}}$ (c)
ρ_s for spiral or circular hoop	$P_u \leq 0.3A_gf_c'$ and $f_c' \leq 10{,}000$ psi	Greater of (d) and (e)	$0.45\left(\frac{A_g}{A_{ch}} - 1\right)\frac{f_c'}{f_{yt}}$ (d)
	$P_u > 0.3A_gf_c'$ or $f_c' > 10{,}000$ psi	Greatest of (d), (e), and (f)	$0.12\frac{f_c'}{f_{yt}}$ (e) $0.35k_f\frac{P_u}{f_{yt}A_{ch}}$ (f)

18.7.5.5 Beyond the length ℓ_o given in 18.7.5.1, the column shall contain spiral reinforcement satisfying 25.7.3 or hoop and crosstie reinforcement satisfying 25.7.2 and 25.7.4 with spacing s not exceeding the least of 6 in., $6d_b$ of the smallest Grade 60 longitudinal column bar, and $5d_b$ of the smallest Grade 80 longitudinal column bar, unless a greater amount of transverse reinforcement is required by 18.7.4.4 or 18.7.6.

18.7.5.6 Columns supporting reactions from discontinued stiff members, such as walls, shall satisfy (a) and (b):

COMMENTARY

R18.7.5.4 The effect of helical (spiral) reinforcement and adequately configured rectilinear hoop reinforcement on deformation capacity of columns is well established (Sakai and Sheikh 1989). Expressions (a), (b), (d), and (e) in Table 18.7.5.4 have historically been used in ACI 318 to calculate the required confinement reinforcement to ensure that spalling of shell concrete does not result in a loss of column axial load strength. Expressions (c) and (f) were developed from a review of column test data (Elwood et al. 2009) and are intended to result in columns capable of sustaining a drift ratio of 0.03 with limited strength degradation. Expressions (c) and (f) are triggered for axial load greater than $0.3A_gf_c'$, which corresponds approximately to the onset of compression-controlled behavior for symmetrically reinforced columns. The k_n term (Paultre and Légeron 2008) decreases the required confinement for columns with closely spaced, laterally supported longitudinal reinforcement because such columns are more effectively confined than columns with more widely spaced longitudinal reinforcement. The k_f term increases the required confinement for columns with $f_c' >$ **10,000 psi** because such columns can experience brittle failure if not well confined. Concrete strengths greater than 15,000 psi should be used with caution given the limited test data for such columns. The concrete strength used to determine the confinement reinforcement is required to be the same as that specified in the construction documents.

Expressions (a), (b), and (c) in Table 18.7.5.4 are to be satisfied in both cross-sectional directions of the rectangular core. For each direction, b_c is the core dimension perpendicular to the tie legs that constitute A_{sh}, as shown in Fig. R18.7.5.2.

Research results indicate that high strength reinforcement can be used effectively as confinement reinforcement. Section 20.2.2.4 permits a value of f_{yt} as high as 100,000 psi to be used in Table 18.7.5.4.

R18.7.5.5 This provision is intended to provide reasonable protection to the midheight of columns outside the length ℓ_o. Observations after earthquakes have shown significant damage to columns in this region, and the minimum hoops or spirals required should provide more uniform strength of the column along its length.

R18.7.5.6 Columns supporting discontinued stiff members, such as walls or trusses, may develop considerable inelastic response. Therefore, it is required that these

CODE

(a) Transverse reinforcement required by 18.7.5.2 through 18.7.5.4 shall be provided over the full height at all levels beneath the discontinuity if the factored axial compressive force in these columns, related to earthquake effect, exceeds $A_g f_c'/10$. Where design forces have been magnified to account for the overstrength of the vertical elements of the seismic-force-resisting system, the limit of $A_g f_c'/10$ shall be increased to $A_g f_c'/4$.
(b) Transverse reinforcement shall extend into the discontinued member at least ℓ_d of the largest longitudinal column bar, where ℓ_d is in accordance with 18.8.5. Where the lower end of the column terminates on a wall, the required transverse reinforcement shall extend into the wall at least ℓ_d of the largest longitudinal column bar at the point of termination. Where the column terminates on a footing or mat, the required transverse reinforcement shall extend at least 12 in. into the footing or mat.

18.7.5.7 If the concrete cover outside the confining transverse reinforcement required by 18.7.5.1, 18.7.5.5, and 18.7.5.6 exceeds 4 in., additional transverse reinforcement having cover not exceeding 4 in. and spacing not exceeding 12 in. shall be provided.

18.7.6 *Shear strength*

18.7.6.1 *Design forces*

18.7.6.1.1 The design shear force V_e shall be calculated from considering the maximum forces that can be generated at the faces of the joints at each end of the column. These joint forces shall be calculated using the maximum probable flexural strengths, M_{pr}, at each end of the column associated with the range of factored axial forces, P_u, acting on the column. The column shears need not exceed those calculated from joint strengths based on M_{pr} of the beams framing into the joint. In no case shall V_e be less than the factored shear calculated by analysis of the structure.

18.7.6.2 *Transverse reinforcement*

18.7.6.2.1 Transverse reinforcement over the lengths ℓ_o, given in 18.7.5.1, shall be designed to resist shear assuming $V_c = 0$ when both (a) and (b) occur:

(a) The earthquake-induced shear force, calculated in accordance with 18.7.6.1, is at least one-half of the maximum required shear strength within ℓ_o.
(b) The factored axial compressive force P_u including earthquake effects is less than $A_g f_c'/20$.

COMMENTARY

columns have the specified reinforcement throughout their length. This covers all columns beneath the level at which the stiff member has been discontinued, unless the factored forces corresponding to earthquake effect are low. Refer to R18.12.7.6 for discussion of the overstrength factor Ω_o.

R18.7.5.7 The unreinforced shell may spall as the column deforms to resist earthquake effects. Separation of portions of the shell from the core caused by local spalling creates a falling hazard. The additional reinforcement is required to reduce the risk of portions of the shell falling away from the column.

R18.7.6 *Shear strength*

R18.7.6.1 *Design forces*

R18.7.6.1.1 The procedures of 18.6.5.1 also apply to columns. Above the ground floor, the moment at a joint may be limited by the flexural strength of the beams framing into the joint. Where beams frame into opposite sides of a joint, the combined strength is the sum of the negative moment strength of the beam on one side of the joint and the positive moment strength of the beam on the other side of the joint. Moment strengths are to be determined using a strength reduction factor of 1.0 and reinforcement with an effective yield stress equal to at least $1.25f_y$. Distribution of the combined moment strength of the beams to the columns above and below the joint should be based on analysis.

CODE

18.8—Joints of special moment frames

18.8.1 *Scope*

18.8.1.1 This section shall apply to beam-column joints of special moment frames forming part of the seismic-force-resisting system.

18.8.2 *General*

18.8.2.1 Forces in longitudinal beam reinforcement at the joint face shall be calculated assuming that the stress in the flexural tensile reinforcement is $1.25f_y$.

18.8.2.2 Longitudinal reinforcement terminated in a joint shall extend to the far face of the joint core and shall be developed in tension in accordance with 18.8.5 and in compression in accordance with 25.4.9.

18.8.2.3 Where longitudinal beam reinforcement extends through a beam-column joint, the depth h of the joint parallel to the beam longitudinal reinforcement shall be at least the greatest of (a) through (c):

(a) $\frac{20}{\lambda}d_b$ of the largest Grade 60 longitudinal bar, where

$\lambda = 0.75$ for lightweight concrete and 1.0 for all other cases

(b) $26d_b$ of the largest Grade 80 longitudinal bar

(c) $h/2$ of any beam framing into the joint and generating joint shear as part of the seismic-force-resisting system in the direction under consideration

COMMENTARY

R18.8—Joints of special moment frames

R18.8.2 *General*

Development of inelastic rotations at the faces of joints of reinforced concrete frames is associated with strains in the flexural reinforcement well in excess of the yield strain. Consequently, joint shear force generated by the flexural reinforcement is calculated for a stress of $1.25f_y$ in the reinforcement (refer to 18.8.2.1). A detailed explanation of the reasons for the possible development of stresses in excess of the yield strength in beam tensile reinforcement is provided in ACI 352R.

R18.8.2.2 The design provisions for hooked bars are based mainly on research and experience for joints with standard 90-degree hooks. Therefore, standard 90-degree hooks generally are preferred to standard 180-degree hooks unless unusual considerations dictate use of 180-degree hooks. For bars in compression, the development length corresponds to the straight portion of a hooked or headed bar measured from the critical section to the onset of the bend for hooked bars and from the critical section to the head for headed bars.

R18.8.2.3 Depth h of the joint is defined in Fig. R15.4.2. The column dimension parallel to the beam reinforcement in joints with circular columns may be taken as that of a square section of equivalent area. Research (Meinheit and Jirsa 1977; Briss et al. 1978; Ehsani 1982; Durrani and Wight 1982; Leon 1989; Aoyama 2001; Lin et al. 2000) has shown that straight longitudinal beam bars may slip within the beam-column joint during a series of large moment reversals. The bond stresses on these straight bars may be very large. To reduce slip substantially during the formation of adjacent beam hinging, it would be necessary to have a ratio of column dimension to bar diameter of approximately 32 for Grade 60 bars, which would result in very large joints. Tests demonstrate adequate behavior if the ratio of joint depth to maximum beam longitudinal bar diameter for Grade 60 reinforcement is at least 20 for normalweight concrete and 26 for lightweight concrete. A joint depth of $26d_b$ for Grade 80 reinforcement is intended to achieve similar performance to that of a joint depth of $20d_b$ for Grade 60 reinforcement and normalweight concrete. The limits on joint depth provide reasonable control on the amount of slip of the beam bars in a beam-column joint, considering the number of anticipated inelastic excursions of the building frame during a major earthquake. A thorough treatment of this topic is given in Zhu and Jirsa (1983).

CODE

COMMENTARY

Requirement (c) on joint aspect ratio applies only to beams that are designated as part of the seismic-force-resisting system. Joints having depth less than half the beam depth require a steep diagonal compression strut across the joint, which may be less effective in resisting joint shear. Tests to demonstrate performance of such joints have not been reported in the literature.

18.8.2.3.1 Concrete used in joints with Grade 80 longitudinal reinforcement shall be normalweight concrete.

R18.8.2.3.1 Test data justifying the combination of lightweight concrete and Grade 80 longitudinal reinforcement in joints are not available.

18.8.3 *Transverse reinforcement*

18.8.3.1 Joint transverse reinforcement shall satisfy 18.7.5.2, 18.7.5.3, 18.7.5.4, and 18.7.5.7, except as permitted in 18.8.3.2.

18.8.3.2 Where beams frame into all four sides of the joint and where each beam width is at least three-fourths the column width, the amount of reinforcement required by 18.7.5.4 shall be permitted to be reduced by one-half, and the spacing required by 18.7.5.3 shall be permitted to be increased to 6 in. within the overall depth h of the shallowest framing beam.

18.8.3.3 Longitudinal beam reinforcement outside the column core shall be confined by transverse reinforcement passing through the column that satisfies spacing requirements of 18.6.4.4, and requirements of 18.6.4.2, and 18.6.4.3, if such confinement is not provided by a beam framing into the joint.

R18.8.3 *Transverse reinforcement*

The Code requires transverse reinforcement in a joint regardless of the magnitude of the calculated shear force.

R18.8.3.2 The amount of confining reinforcement may be reduced and the spacing may be increased if beams of adequate dimensions frame into all four sides of the joint.

R18.8.3.3 The required transverse reinforcement, or transverse beam if present, is intended to confine the beam longitudinal reinforcement and improve force transfer to the beam-column joint.

An example of transverse reinforcement through the column provided to confine the beam reinforcement passing outside the column core is shown in Fig. R18.6.2. Additional detailing guidance and design recommendations for both interior and exterior wide-beam connections with beam reinforcement passing outside the column core may be found in ACI 352R.

18.8.4 *Shear strength*

18.8.4.1 Joint shear force V_u shall be calculated on a plane at mid-height of the joint from calculated forces at the joint faces using tensile and compressive beam forces determined in accordance with 18.8.2.1 and column shear consistent with beam probable flexural strengths M_{pr}.

18.8.4.2 ϕ shall be in accordance with 21.2.4.4.

18.8.4.3 V_n of the joint shall be in accordance with Table 18.8.4.3.

R18.8.4 *Shear strength*

The shear strength values given in 18.8.4.3 are based on the recommendation in ACI 352R for joints with members that are expected to undergo reversals of deformation into the inelastic range, although the ACI 352R definition of effective cross-sectional joint area is sometimes different. The given nominal joint shear strengths do not explicitly consider transverse reinforcement in the joint because tests of joints (Meinheit and Jirsa 1977) and deep beams (Hirosawa 1977) have indicated that joint shear strength is not sensitive to transverse reinforcement if at least the required minimum amount is provided in the joint.

Cyclic loading tests of joints with extensions of beams with lengths at least equal to their depths have indicated similar joint shear strengths to those of joints with continuous beams. These findings suggest that extensions of beams and

CODE

Table 18.8.4.3—Nominal joint shear strength V_n

Column	Beam in direction of V_u	Confinement by transverse beams according to 15.2.8	V_n, lb[1]
Continuous or meets 15.2.6	Continuous or meets 15.2.7	Confined	$20\lambda\sqrt{f'_c}A_j$
		Not confined	$15\lambda\sqrt{f'_c}A_j$
	Other	Confined	$15\lambda\sqrt{f'_c}A_j$
		Not confined	$12\lambda\sqrt{f'_c}A_j$
Other	Continuous or meets 15.2.7	Confined	$15\lambda\sqrt{f'_c}A_j$
		Not confined	$12\lambda\sqrt{f'_c}A_j$
	Other	Confined	$12\lambda\sqrt{f'_c}A_j$
		Not confined	$8\lambda\sqrt{f'_c}A_j$

[1]λ shall be 0.75 for lightweight concrete and 1.0 for normalweight concrete. A_j shall be calculated in accordance with 15.4.2.4.

COMMENTARY

columns, when properly dimensioned and reinforced with longitudinal and transverse bars, provide effective confinement to the joint faces, thus delaying joint strength deterioration at large deformations (Meinheit and Jirsa 1981).

Seismic 18

18.8.5 *Development length of bars in tension*

18.8.5.1 For bar sizes No. 3 through No. 11 terminating in a standard hook, ℓ_{dh} shall be calculated by Eq. (18.8.5.1), but ℓ_{dh} shall be at least the greater of $8d_b$ and 6 in. for normalweight concrete and at least the greater of $10d_b$ and 7-1/2 in. for lightweight concrete.

$$\ell_{dh} = f_y d_b/(65\lambda\sqrt{f'_c}) \quad (18.8.5.1)$$

The value of λ shall be 0.75 for concrete containing lightweight aggregate and 1.0 otherwise.

The hook shall be located within the confined core of a column or of a boundary element, with the hook bent into the joint.

R18.8.5 *Development length of bars in tension*

R18.8.5.1 Minimum embedment length in tension for deformed bars with standard hooks is determined using Eq. (18.8.5.1), which is based on the requirements of 25.4.3. The embedment length of a bar with a standard hook is the distance, parallel to the bar, from the critical section (where the bar is to be developed) to a tangent drawn to the outside edge of the hook. The tangent is to be drawn perpendicular to the axis of the bar (refer to Table 25.3.1).

Because Chapter 18 stipulates that the hook is to be embedded in confined concrete, the coefficients 0.7 (for concrete cover) and 0.8 (for ties) have been incorporated in the constant used in Eq. (18.8.5.1). The development length that would be derived directly from 25.4.3 is increased to reflect the effect of load reversals. Factors such as the actual stress in the reinforcement being more than the yield strength and the effective development length not necessarily starting at the face of the joint were implicitly considered in the formulation of the expression for basic development length that has been used as the basis for Eq. (18.8.5.1).

The requirement for the hook to project into the joint is to improve development of a diagonal compression strut across the joint. The requirement applies to beam and column bars terminated at a joint with a standard hook.

18.8.5.2 For headed deformed bars satisfying 20.2.1.6, development in tension shall be in accordance with 25.4.4, by substituting a bar stress of $1.25f_y$ for f_y.

R18.8.5.2 The factor 1.25 is intended to represent the potential increase in stresses due to inelastic response, including strain hardening that may occur in beams of special moment frames.

CODE

18.8.5.3 For bar sizes No. 3 through No. 11, ℓ_d, the development length in tension for a straight bar, shall be at least the greater of (a) and (b):

(a) 2.5 times the length in accordance with 18.8.5.1 if the depth of the concrete cast in one lift beneath the bar does not exceed 12 in.
(b) 3.25 times the length in accordance with 18.8.5.1 if the depth of the concrete cast in one lift beneath the bar exceeds 12 in.

18.8.5.4 Straight bars terminated at a joint shall pass through the confined core of a column or a boundary element. Any portion of ℓ_d not within the confined core shall be increased by a factor of 1.6.

18.8.5.5 If epoxy-coated reinforcement is used, the development lengths in 18.8.5.1, 18.8.5.3, and 18.8.5.4 shall be multiplied by applicable factors in 25.4.2.5 or 25.4.3.2.

18.9—Special moment frames constructed using precast concrete

18.9.1 *Scope*

18.9.1.1 This section shall apply to special moment frames constructed using precast concrete forming part of the seismic-force-resisting system.

COMMENTARY

R18.8.5.3 Minimum development length in tension for straight bars is a multiple of the length indicated by 18.8.5.1. Section 18.8.5.3(b) refers to top bars. Lack of reference to No. 14 and No. 18 bars in 18.8.5 is due to the paucity of information on anchorage of such bars subjected to load reversals simulating earthquake effects.

R18.8.5.4 If the required straight embedment length of a reinforcing bar extends beyond the confined volume of concrete (as defined in 18.6.4, 18.7.5, or 18.8.3), the required development length is increased on the premise that the limiting bond stress outside the confined region is less than that inside.

$$\ell_{dm} = 1.6(\ell_d - \ell_{dc}) + \ell_{dc}$$

or

$$\ell_{dm} = 1.6\ell_d - 0.6\ell_{dc}$$

where ℓ_{dm} is the required development length if bar is not entirely embedded in confined concrete; ℓ_d is the required development length in tension for straight bar as defined in 18.8.5.3; and ℓ_{dc} is the length of bar embedded in confined concrete.

R18.9—Special moment frames constructed using precast concrete

The detailing provisions in 18.9.2.1 and 18.9.2.2 are intended to produce frames that respond to design displacements essentially like monolithic special moment frames.

Precast frame systems composed of concrete elements with ductile connections are expected to experience flexural yielding in connection regions. Reinforcement in ductile connections can be made continuous by using mechanical splices or any other technique that provides development in tension or compression of at least the specified tensile strength of bars (Yoshioka and Sekine 1991; Kurose et al. 1991; Restrepo et al. 1995a,b). Requirements for mechanical splices are in addition to those in 18.2.7 and are intended to avoid strain concentrations over a short length of reinforcement adjacent to a splice device. Additional requirements for shear strength are provided in 18.9.2.1 to prevent sliding on connection faces. Precast frames composed of elements with ductile connections may be designed to promote yielding at locations not adjacent to the joints. Therefore, design shear V_e, as calculated according to 18.6.5.1 or 18.7.6.1, may not be conservative.

CODE

COMMENTARY

Precast concrete frame systems composed of elements joined using strong connections are intended to experience flexural yielding outside the connections. Strong connections include the length of the mechanical splice hardware as shown in Fig. R18.9.2.2. Capacity-design techniques are used in 18.9.2.2(c) to ensure the strong connection remains elastic following formation of plastic hinges. Additional column requirements are provided to avoid hinging and strength deterioration of column-to-column connections.

Strain concentrations have been observed to cause brittle fracture of reinforcing bars at the face of mechanical splices in laboratory tests of precast beam-column connections (Palmieri et al. 1996). Locations of strong connections should be selected carefully or other measures should be taken, such as debonding of reinforcing bars in highly stressed regions, to avoid strain concentrations that can result in premature fracture of reinforcement.

18.9.2 *General*

18.9.2.1 Special moment frames with ductile connections constructed using precast concrete shall satisfy (a) through (c):

(a) Requirements of 18.6 through 18.8 for special moment frames constructed with cast-in-place concrete

(b) V_n for connections calculated according to 22.9 shall be at least $2V_e$, where V_e is in accordance with 18.6.5.1 or 18.7.6.1

(c) Mechanical splices of beam reinforcement shall be located not closer than $h/2$ from the joint face and shall satisfy 18.2.7

18.9.2.2 Special moment frames with strong connections constructed using precast concrete shall satisfy (a) through (e):

(a) Requirements of 18.6 through 18.8 for special moment frames constructed with cast-in-place concrete

(b) Provision 18.6.2.1(a) shall apply to segments between locations where flexural yielding is intended to occur due to design displacements

(c) Design strength of the strong connection, ϕS_n, shall be at least S_e

(d) Primary longitudinal reinforcement shall be made continuous across connections and shall be developed outside both the strong connection and the plastic hinge region

(e) For column-to-column connections, ϕS_n shall be at least $1.4S_e$, ϕM_n shall be at least $0.4M_{pr}$ for the column within the story height, and ϕV_n shall be at least V_e in accordance with 18.7.6.1

R18.9.2 *General*

CODE

COMMENTARY

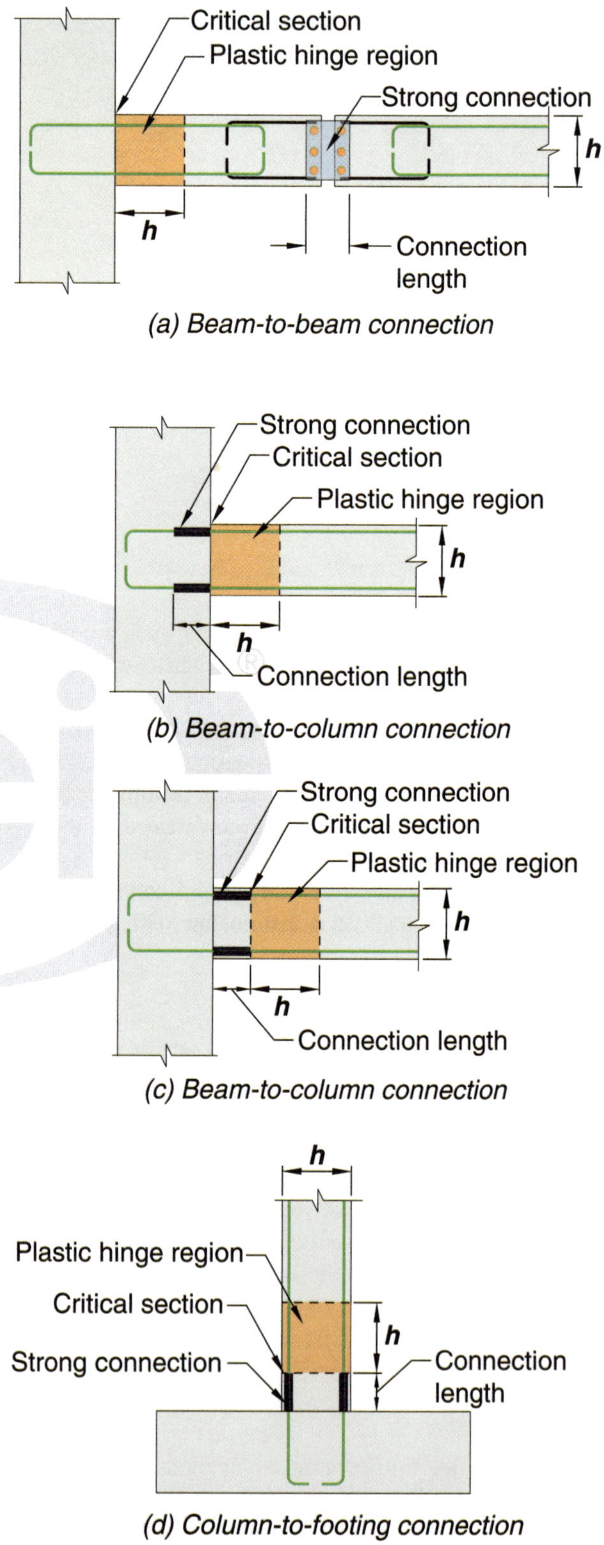

***Fig. R18.9.2.2**—Strong connection examples.*

18.9.2.3 Special moment frames constructed using precast concrete and not satisfying 18.9.2.1 or 18.9.2.2 shall satisfy (a) through (c):

R18.9.2.3 Precast frame systems not satisfying the prescriptive requirements of Chapter 18 have been demonstrated in experimental studies to provide satisfactory seismic performance characteristics (Stone et al. 1995; Nakaki et al. 1995).

Seismic 18

CODE

(a) ACI 374.1
(b) Details and materials used in the test specimens shall be representative of those used in the structure
(c) The design procedure used to proportion the test specimens shall define the mechanism by which the frame resists gravity and earthquake effects, and shall establish acceptance values for sustaining that mechanism. Portions of the mechanism that deviate from Code requirements shall be contained in the test specimens and shall be tested to determine upper bounds for acceptance values.

COMMENTARY

ACI 374.1 defines a protocol for establishing a design procedure, validated by analysis and laboratory tests, for such frames. The design procedure should identify the load path or mechanism by which the frame resists gravity and earthquake effects. The tests should be configured to investigate critical behaviors, and the measured quantities should establish upper-bound acceptance values for components of the load path, which may be in terms of limiting stresses, forces, strains, or other quantities. The design procedure used for the structure should not deviate from that used to design the test specimens, and acceptance values should not exceed values that were demonstrated by the tests to be acceptable. Materials and components used in the structure should be similar to those used in the tests. Deviations may be acceptable if the licensed design professional can demonstrate that those deviations do not adversely affect the behavior of the framing system.

ACI 550.3 defines design requirements for one type of special precast concrete moment frame for use in accordance with 18.9.2.3.

18.10—Special structural walls

18.10.1 *Scope*

18.10.1.1 This section shall apply to special structural walls, including ductile coupled walls, and all components of special structural walls including coupling beams and wall piers forming part of the seismic-force-resisting system.

18.10.1.2 Special structural walls constructed using precast concrete shall be in accordance with 18.11 in addition to 18.10.

R18.10—Special structural walls

R18.10.1 *Scope*

This section contains requirements for the dimensions and details of special structural walls and all components including coupling beams and wall piers. Wall piers are defined in Chapter 2. Design provisions for vertical wall segments depend on the aspect ratio of the wall segment in the plane of the wall (h_w/ℓ_w), and the aspect ratio of the horizontal cross section (ℓ_w/b_w), and generally follow the descriptions in Table R18.10.1. The limiting aspect ratios for wall piers are based on engineering judgment. It is intended that flexural yielding of the vertical reinforcement in the pier should limit shear demand on the pier.

Table R18.10.1—Governing design provisions for vertical wall segments[1]

Clear height of vertical wall segment/length of vertical wall segment, (h_w/ℓ_w)	Length of vertical wall segment/wall thickness (ℓ_w/b_w)		
	(ℓ_w/b_w) ≤ 2.5	2.5 < (ℓ_w/b_w) ≤ 6.0	(ℓ_w/b_w) > 6.0
$h_w/\ell_w < 2.0$	Wall	Wall	Wall
$h_w/\ell_w \geq 2.0$	Wall pier required to satisfy specified column design requirements; refer to 18.10.8.1	Wall pier required to satisfy specified column design requirements or alternative requirements; refer to 18.10.8.1	Wall

[1]h_w is the clear height, ℓ_w is the horizontal length, and b_w is the width of the web of the wall segment.

18.10.2 *Reinforcement*

R18.10.2 *Reinforcement*

CODE

18.10.2.1 The distributed web reinforcement ratios, ρ_ℓ and ρ_t, for structural walls shall be at least 0.0025, except that if V_u does not exceed $\lambda\sqrt{f_c'}A_{cv}$, ρ_t shall be permitted to be reduced to the values in 11.6. Reinforcement spacing each way in structural walls shall not exceed 18 in. Reinforcement contributing to V_n shall be continuous and shall be distributed across the shear plane.

18.10.2.2 At least two curtains of reinforcement shall be used in a wall if $V_u > 2\lambda\sqrt{f_c'}A_{cv}$ or $h_w/\ell_w \geq 2.0$, in which h_w and ℓ_w refer to height and length of entire wall, respectively.

18.10.2.3 Reinforcement in structural walls shall be developed or spliced for f_y in tension in accordance with 25.4, 25.5, and (a) through (d):

(a) Except at the top of a wall, longitudinal reinforcement shall extend at least 12 ft above the point at which it is no longer required to resist flexure but need not extend more than ℓ_d above the next floor level.

(b) At locations where yielding of longitudinal reinforcement is likely to occur as a result of lateral displacements, development lengths of longitudinal reinforcement shall be 1.25 times the values calculated for f_y in tension.

(c) Lap splices of longitudinal reinforcement within boundary regions shall not be permitted over a height equal to h_{sx} above, and ℓ_d below, critical sections where yielding of longitudinal reinforcement is likely to occur as a result of lateral displacements. The value of h_{sx} need not exceed 20 ft. Boundary regions include those within lengths specified in 18.10.6.4(a) and within a length equal to the wall thickness measured beyond the intersecting region(s) of connected walls.

(d) Mechanical splices of reinforcement shall conform to 18.2.7 and welded splices of reinforcement shall conform to 18.2.8.

COMMENTARY

Minimum reinforcement requirements in 18.10.2.1 follow from preceding Codes. The requirement for distributed shear reinforcement is related to the intent to control the width of inclined cracks. The requirement for two layers of reinforcement in walls resisting substantial design shears in 18.10.2.2 is based on the observation that, under ordinary construction conditions, the probability of maintaining a single layer of reinforcement near the middle of the wall section is quite low. Furthermore, presence of reinforcement close to the surface tends to inhibit fragmentation of the concrete in the event of severe cracking during an earthquake. The requirement for two layers of vertical reinforcement in more slender walls is to improve lateral stability of the compression zone under cyclic loads following yielding of vertical reinforcement in tension.

R18.10.2.3 Requirements are based on provisions in Chapter 25, with modifications to address issues specific to structural walls, as well as to the use of high-strength reinforcement. Because actual forces in longitudinal reinforcement of structural walls may exceed calculated forces, reinforcement should be developed or spliced to reach the yield strength of the bar in tension. Termination of longitudinal (vertical) reinforcement in structural walls should be specified so that bars extend above elevations where they are no longer required to resist design flexure and axial force; extending bars ℓ_d above the next floor level is a practical approach to achieving this requirement. A limit of 12 ft is included for cases with large story heights. Bar terminations should be accomplished gradually over a wall height and should not be located close to critical sections where yielding of longitudinal reinforcement is expected, which typically occurs at the base of a wall with a uniform, or nearly uniform, cross section over the building height. Strain hardening of reinforcement results in spread of plasticity away from critical sections as lateral deformations increase. Research (Aaletti et al. 2012; Hardisty et al. 2015) shows that lap splices should be avoided in walls where flexural yielding is anticipated, for example at the base of walls, because they may lead to large localized strains and bar fractures. Figure R18.10.2.3 illustrates boundary regions where lap splices are not permitted.

At locations where yielding of longitudinal reinforcement is expected, a 1.25 multiplier is applied to account for the likelihood that the actual yield strength exceeds the specified yield strength of the bar, as well as the influence of strain hardening and cyclic load reversals. Where transverse reinforcement is used, development lengths for straight and hooked bars may be reduced as permitted in 25.4.2 and 25.4.3, respectively, because closely spaced transverse reinforcement improves the performance of splices and hooks subjected to repeated inelastic demands (ACI 408.2R).

CODE | COMMENTARY

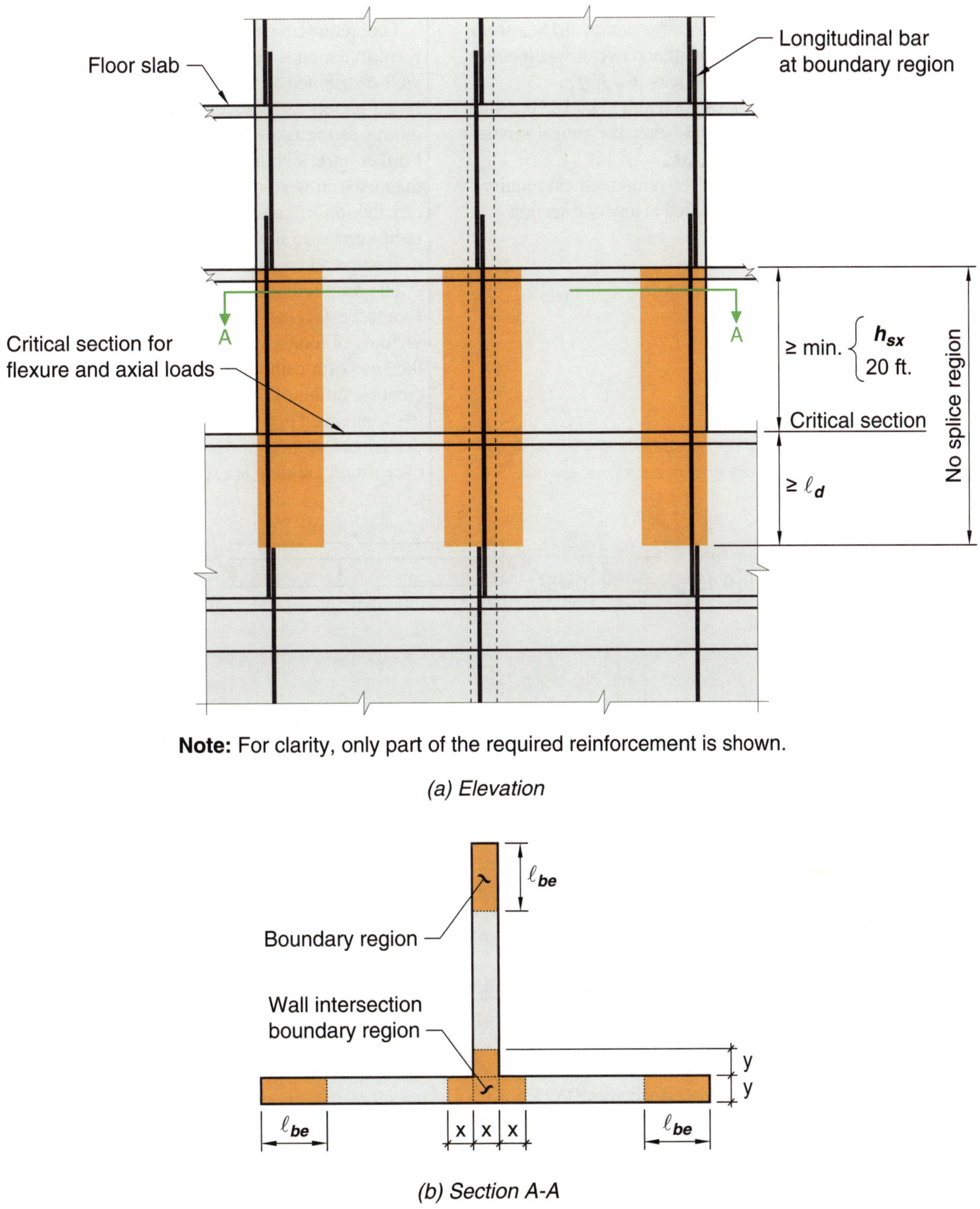

Fig. R18.10.2.3—Wall boundary regions within heights where lap splices are not permitted.

18.10.2.4 Walls or wall piers with $h_w/\ell_w \geq 2.0$ that are effectively continuous from the base of structure to top of wall and are designed to have a single critical section for flexure and axial loads shall have longitudinal reinforcement at the ends of a vertical wall segment that satisfies (a) through (c).

R18.10.2.4 This provision is based on the assumption that inelastic response of the wall is dominated by flexural action at a critical, yielding section. The wall should be proportioned so that the critical section occurs where intended. If there is potential for more than one critical section, it is prudent to provide the minimum boundary reinforcement at all such sections.

CODE

(a) Longitudinal reinforcement ratio within $0.15\ell_w$ from the end of a vertical wall segment, and over a width equal to the wall thickness, shall be at least $6\sqrt{f_c'}/f_y$.

(b) The longitudinal reinforcement required by 18.10.2.4(a) shall extend vertically above and below the critical section at least the greater of ℓ_w and $M_u/3V_u$.

(c) No more than 50 percent of the reinforcement required by 18.10.2.4(a) shall be terminated at any one section.

COMMENTARY

The requirement for minimum longitudinal reinforcement in the ends of the wall is to promote the formation of well-distributed secondary flexural cracks in the wall plastic hinge region to achieve the required deformation capacity during earthquakes (Lu et al. 2017; Sritharan et al. 2014). Furthermore, significantly higher in-place concrete strengths than used in design calculations may be detrimental to the distribution of cracking. 18.10.2.4(a) specifies the required reinforcement ratio in the end tension zones, as shown for different wall sections in Fig. R18.10.2.4.

The longitudinal reinforcement required by 18.10.2.4(a) should be located at a critical section where concentrated yielding of longitudinal reinforcement is expected (typically the base of a cantilever wall) and must continue to a sufficient elevation of the wall to avoid a weak section adjacent to the intended plastic hinge region. A height above or below the critical section of $M_u/3V_u$ is used to identify the length over which yielding is expected.

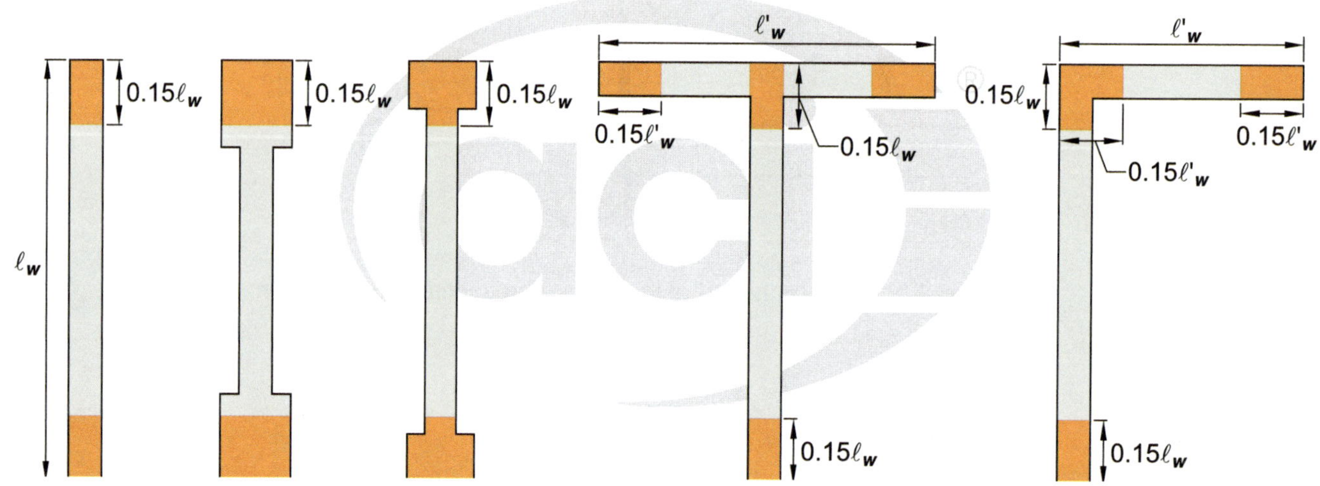

Fig. R18.10.2.4—Locations of longitudinal reinforcement required by 18.10.2.4(a) in different configurations of wall sections.

18.10.2.5 Reinforcement in coupling beams shall be developed for f_y in tension in accordance with 25.4, 25.5, and (a) and (b):

(a) If coupling beams are reinforced according to 18.6.3.1, the development length of longitudinal reinforcement shall be 1.25 times the values calculated for f_y in tension.

(b) If coupling beams are reinforced according to 18.10.7.4, the development length of diagonal reinforcement shall be 1.25 times the values calculated for f_y in tension.

18.10.3 *Design forces*

R18.10.3 *Design forces*

The possibility of yielding in components of structural walls should be considered, as in the portion of a wall between two window openings, in which case the actual shear may be in excess of the shear indicated by lateral load analysis based on factored design forces.

CODE

18.10.3.1 The design shear force V_e shall be calculated by:

$$V_e = \Omega_v \omega_v V_u \leq 3V_u \quad (18.10.3.1)$$

where V_u, Ω_v, and ω_v are defined in 18.10.3.1.1, 18.10.3.1.2, and 18.10.3.1.3, respectively.

18.10.3.1.1 V_u is the shear force obtained from code lateral load analysis with factored load combinations.

18.10.3.1.2 Ω_v shall be in accordance with Table 18.10.3.1.2.

Table 18.10.3.1.2—Overstrength factor Ω_v at critical section

Condition	Ω_v	
$h_{wcs}/\ell_w > 1.5$	Greater of	M_{pr}/M_u [1]
		1.5[2]
$h_{wcs}/\ell_w \leq 1.5$	1.0	

[1] For the load combination producing the largest value of Ω_v.

[2] Unless a more detailed analysis demonstrated a smaller value, but not less than 1.0.

18.10.3.1.3 For walls with $h_{wcs}/\ell_w < 2.0$, ω_v shall be taken as 1.0. Otherwise, ω_v shall be calculated as:

$$\omega_v = 0.9 + \frac{n_s}{10} \quad n_s \leq 6$$
$$\omega_v = 1.3 + \frac{n_s}{30} \leq 1.8 \quad n_s > 6 \quad (18.10.3.1.3)$$

where n_s shall not be taken less than the quantity $0.007h_{wcs}$.

COMMENTARY

R18.10.3.1 Design shears for structural walls are obtained from lateral load analysis with appropriate load factors increased to account for: (i) flexural overstrength at critical sections where yielding of longitudinal reinforcement is expected; and (ii) dynamic amplification due to higher mode effects, as illustrated in Fig. R18.10.3.1. The approach used to determine the amplified shear forces is similar to that used in New Zealand Standard 3101 (2006). Because M_n and M_{pr} depend on axial force, which varies for different load combinations, and loading direction for flanged and coupled walls, the condition producing the largest value of Ω_v should be used. Although the value of 1.5 in 18.10.3.1.2 is greater than the minimum value obtained for the governing load combination with a ϕ factor of 0.9 and a tensile stress of at least $1.25f_y$ in the longitudinal reinforcement, a value greater than 1.5 may be appropriate if provided longitudinal reinforcement exceeds that required. Dynamic amplification is not significant in walls with $h_w/\ell_w < 2$. A limit of $0.007h_{wcs}$ is imposed on n_s to account for buildings with large story heights. The application of Ω_V to V_u does not preclude the application of a redundancy factor if required by the general building code.

CODE

COMMENTARY

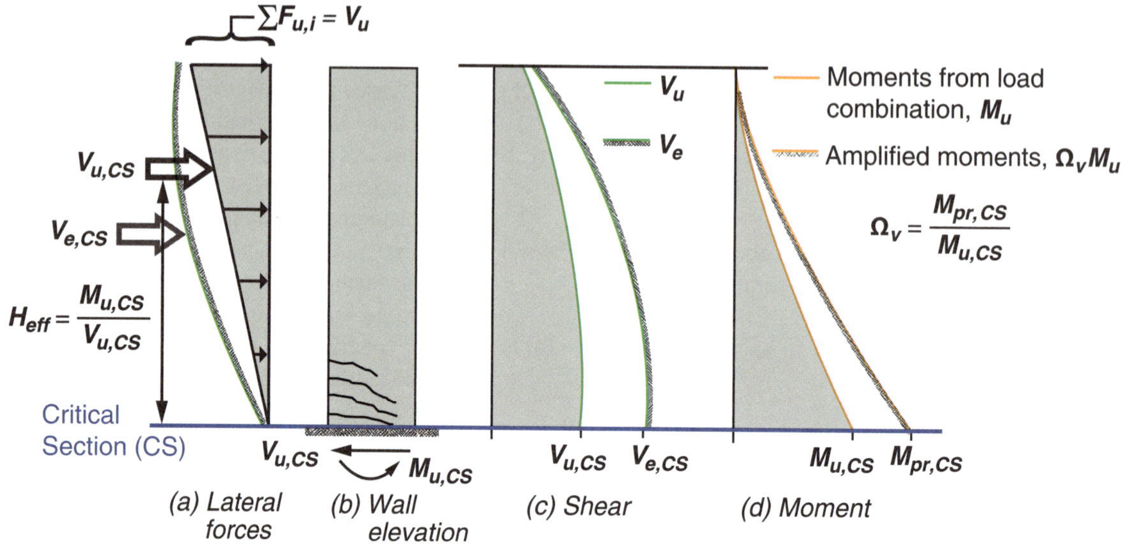

Fig. R18.10.3.1—Determination of shear demand for walls with $h_w/\ell_w \geq 2.0$ *(Moehle et al 2011).*

18.10.4 *Shear strength*

18.10.4.1 V_n shall be calculated by:

$$V_n = (\alpha_c \lambda \sqrt{f_c'} + \rho_t f_{yt}) A_{cv} \tag{18.10.4.1}$$

where:
$\alpha_c = 3$ for $h_w/\ell_w \leq 1.5$
$\alpha_c = 2$ for $h_w/\ell_w \geq 2.0$
It shall be permitted to linearly interpolate the value of α_c between 3 and 2 for $1.5 < h_w/\ell_w < 2.0$.

18.10.4.2 In 18.10.4.1, the value of ratio h_w/ℓ_w used to calculate V_n for segments of a wall shall be the greater of the ratios for the entire wall and the segment of wall considered.

18.10.4.3 Walls shall have distributed shear reinforcement in two orthogonal directions in the plane of the wall. If h_w/ℓ_w does not exceed 2.0, reinforcement ratio ρ_ℓ shall be at least the reinforcement ratio ρ_t.

18.10.4.4 For all vertical wall segments sharing a common lateral force, V_n shall not be taken greater than $8\sqrt{f_c'}A_{cv}$. For any one of the individual vertical wall segments, V_n shall not be taken greater than $10\sqrt{f_c'}A_{cw}$, where A_{cw} is the area of concrete section of the individual vertical wall segment considered.

18.10.4.5 For horizontal wall segments and coupling beams, V_n shall not be taken greater than $10\sqrt{f_c'}A_{cw}$, where A_{cw} is the area of concrete section of a horizontal wall segment or coupling beam.

R18.10.4 *Shear strength*

Equation (18.10.4.1) recognizes the higher shear strength of walls with high shear-to-moment ratios (Hirosawa 1977; Joint ACI-ASCE Committee 326 1962; Barda et al. 1977). The nominal shear strength is given in terms of the gross area of the section resisting shear, A_{cv}. For a rectangular section without openings, the term A_{cv} refers to the gross area of the cross section rather than to the product of the width and the effective depth.

A vertical wall segment refers to a part of a wall bounded horizontally by openings or by an opening and an edge. For an isolated wall or a vertical wall segment, ρ_t refers to horizontal reinforcement and ρ_ℓ refers to vertical reinforcement.

The ratio h_w/ℓ_w may refer to overall dimensions of a wall, or of a segment of the wall bounded by two openings, or an opening and an edge. The intent of 18.10.4.2 is to make certain that any segment of a wall is not assigned a unit strength greater than that for the entire wall. However, a wall segment with a ratio of h_w/ℓ_w higher than that of the entire wall should be proportioned for the unit strength associated with the ratio h_w/ℓ_w based on the dimensions for that segment.

To restrain the inclined cracks effectively, reinforcement included in ρ_t and ρ_ℓ should be appropriately distributed along the length and height of the wall (refer to 18.10.4.3). Chord reinforcement provided near wall edges in concentrated amounts for resisting bending moment is not to be included in determining ρ_t and ρ_ℓ. Within practical limits, shear reinforcement distribution should be uniform and at a small spacing.

If the factored shear force at a given level in a structure is resisted by several walls or several vertical wall segments of a perforated wall, the average unit shear strength assumed for the total available cross-sectional area is limited to $8\sqrt{f_c'}$ with the additional requirement that the unit shear strength assigned to any single vertical wall segment does not exceed $10\sqrt{f_c'}$. The upper limit of strength to be assigned to any

CODE / COMMENTARY

one member is imposed to limit the degree of redistribution of shear force.

Horizontal wall segments in 18.10.4.5 refer to wall sections between two vertically aligned openings (refer to Fig. R18.10.4.5). It is, in effect, a vertical wall segment rotated through 90 degrees. A horizontal wall segment is also referred to as a coupling beam when the openings are aligned vertically over the building height. When designing a horizontal wall segment or coupling beam, ρ_t refers to vertical reinforcement and ρ_ℓ refers to horizontal reinforcement.

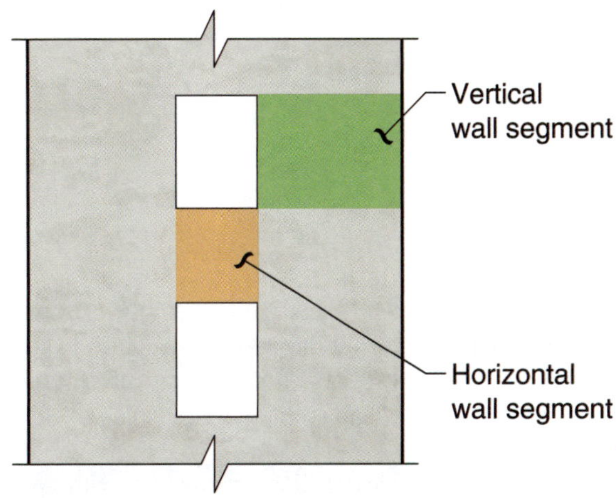

***Fig. R18.10.4.5**—Wall with openings.*

18.10.4.6 The requirements of 21.2.4.1 shall not apply to walls or wall piers designed according to 18.10.6.2.

R18.10.4.6 Section 21.2.4.1 does not apply because walls designed according to 18.10.6.2 are controlled by flexural yielding, and code level shear forces have been amplified.

18.10.5 *Design for flexure and axial force*

R18.10.5 *Design for flexure and axial force*

18.10.5.1 Structural walls and portions of such walls subject to combined flexure and axial loads shall be designed in accordance with 22.4. Concrete and developed longitudinal reinforcement within effective flange widths, boundary elements, and the wall web shall be considered effective. The effects of openings shall be considered.

R18.10.5.1 Flexural strength of a wall or wall segment is determined according to procedures commonly used for columns. Strength should be determined considering the applied axial and lateral forces. Reinforcement concentrated in boundary elements and distributed in flanges and webs should be included in the strength calculations based on a strain compatibility analysis. The foundation supporting the wall should be designed to resist the wall boundary and web forces. For walls with openings, the influence of the opening or openings on flexural and shear strengths is to be considered and a load path around the opening or openings should be verified. Capacity-design concepts and the strut-and-tie method may be useful for this purpose (Taylor et al. 1998).

18.10.5.2 Unless a more detailed analysis is performed, effective flange widths of flanged sections shall extend from the face of the web a distance equal to the lesser of one-half the distance to an adjacent wall web and 25 percent of the total wall height above the section under consideration.

R18.10.5.2 Where wall sections intersect to form L-, T-, C-, or other cross-sectional shapes, the influence of the flange on the behavior of the wall should be considered by selecting appropriate flange widths. Tests (Wallace 1996) show that effective flange width increases with increasing drift level and the effectiveness of a flange in compression differs from that for a flange in tension. The value used for the effective compression flange width has little effect on

CODE

18.10.6 *Boundary elements of special structural walls*

18.10.6.1 The need for special boundary elements at the edges of structural walls shall be evaluated in accordance with 18.10.6.2 or 18.10.6.3. The requirements of 18.10.6.4 and 18.10.6.5 shall also be satisfied.

18.10.6.2 Walls or wall piers with $\boldsymbol{h_{wcs}/\ell_w \geq 2.0}$ that are effectively continuous from the base of structure to top of wall and are designed to have a single critical section for flexure and axial loads shall satisfy (a) and (b):

(a) Compression zones shall be reinforced with special boundary elements where

$$\frac{1.5\delta_u}{h_{wcs}} \geq \frac{\ell_w}{600c} \qquad (18.10.6.2a)$$

and $\boldsymbol{c}$ corresponds to the largest neutral axis depth calculated for the factored axial force and nominal moment strength consistent with the direction of the design displacement $\boldsymbol{\delta_u}$. Ratio $\boldsymbol{\delta_u/h_{wcs}}$ shall not be taken less than 0.005.

(b) If special boundary elements are required by (a), then (i) and either (ii) or (iii) shall be satisfied.

(i) Special boundary element transverse reinforcement shall extend vertically above and below the critical section a least the greater of $\boldsymbol{\ell_w}$ and $\boldsymbol{M_u/4V_u}$, except as permitted in 18.10.6.4(i).

(ii) $\boldsymbol{b \geq \sqrt{0.025\ell c_w}}$

(iii) $\boldsymbol{\delta_c/h_{wcs} \geq 1.5\delta_u/h_{wcs}}$, where:

$$\frac{\delta_c}{h_{wcs}} = \frac{1}{100}\left(4 - \frac{1}{50}\left(\frac{\ell_w}{b}\right)\left(\frac{c}{b}\right) - \frac{V_e}{8\sqrt{f_c'}A_{cv}}\right) \qquad (18.10.6.2b)$$

The value of $\boldsymbol{\delta_c/h_{wcs}}$ in Eq. (18.10.6.2b) need not be taken less than 0.015.

COMMENTARY

the strength and deformation capacity of the wall; therefore, to simplify design, a single value of effective flange width based on an estimate of the effective tension flange width is used in both tension and compression.

R18.10.6 *Boundary elements of special structural walls*

R18.10.6.1 Two design approaches for evaluating detailing requirements at wall boundaries are included in 18.10.6.1. Provision 18.10.6.2 allows the use of displacement-based design of walls, in which the structural details are determined directly on the basis of the expected lateral displacements of the wall. The provisions of 18.10.6.3 are similar to those of the 1995 Code, and have been retained because they are conservative for assessing required transverse reinforcement at wall boundaries for many walls. Provisions 18.10.6.4 and 18.10.6.5 apply to structural walls designed by either 18.10.6.2 or 18.10.6.3.

R18.10.6.2 This section is based on the assumption that inelastic response of the wall is dominated by flexural action at a critical, yielding section. The wall should be proportioned and reinforced so that the critical section occurs where intended.

Equation (18.10.6.2a) follows from a displacement-based approach (Moehle 1992; Wallace and Orakcal 2002). The approach assumes that special boundary elements are required to confine the concrete where the strain at the extreme compression fiber of the wall exceeds a critical value when the wall is displaced to 1.5 times the design displacement. Consistent with a displacement-based design approach, the design displacement in Eq. (18.10.6.2a) is taken at the top of the wall, and the wall height is taken as the height above the critical section. The multiplier of 1.5 on design displacement was added to Eq. (18.10.6.2) in the 2014 Code to produce detailing requirements more consistent with the building code performance intent of a low probability of collapse in Maximum Considered Earthquake level shaking. The lower limit of 0.005 on the quantity $\boldsymbol{\delta_u/h_{wcs}}$ requires special boundary elements if wall boundary longitudinal reinforcement tensile strain does not reach approximately twice the limit used to define tension-controlled beam sections according to 21.2.2. The lower limit of 0.005 on the quantity $\boldsymbol{\delta_u/h_{wcs}}$ requires moderate wall deformation capacity for stiff buildings.

The neutral axis depth $\boldsymbol{c}$ in Eq. (18.10.6.2) is the depth calculated according to 22.2 corresponding to development of nominal flexural strength of the wall when displaced in the same direction as $\boldsymbol{\delta_u}$. The axial load is the factored axial load that is consistent with the design load combination that produces the design displacement $\boldsymbol{\delta_u}$.

The height of the special boundary element is based on estimates of plastic hinge length and extends beyond the zone over which yielding of tension reinforcement and spalling of concrete are likely to occur.

CODE

COMMENTARY

Equation (18.10.6.2b) is based on the mean top-of-wall drift capacity at 20 percent loss of lateral strength proposed by Abdullah and Wallace (2019). The requirement that drift capacity exceed 1.5 times the drift demand results in a low probability of strength loss for the design earthquake. The expression for b in (ii) is derived from Eq. (18.10.6.2b), assuming values of $V_u/(8A_{cv}\sqrt{f_c'})$ and δ_u/h_{wcs} of approximately 1.0 and 0.015, respectively. If b varies over c, an average or representative value of b should be used. For example, at the flanged end of a wall, b should be taken equal to the effective flange width defined in 18.10.5.2, unless c extends into the web, then a weighted average should be used for b. At the end of a wall without a flange, b should be taken equal to the wall thickness. If the drift capacity does not exceed the drift demand for a trial design, then changes to the design are required to increase wall drift capacity, reduces wall drift demand, or both, such that drift capacity exceeds drift demand for each wall in a given building.

18.10.6.3 Structural walls not designed in accordance with 18.10.6.2 shall have special boundary elements at boundaries and edges around openings of structural walls where the maximum extreme fiber compressive stress, corresponding to load combinations including earthquake effects E, exceeds $0.2f_c'$. The special boundary element shall be permitted to be discontinued where the calculated compressive stress is less than $0.15f_c'$. Stresses shall be calculated for the factored loads using a linearly elastic model and gross section properties. For walls with flanges, an effective flange width as given in 18.10.5.2 shall be used.

R18.10.6.3 By this procedure, the wall is considered to be acted on by gravity loads and the maximum shear and moment induced by earthquake in a given direction. Under this loading, the compressed boundary at the critical section resists the tributary gravity load plus the compressive resultant associated with the bending moment.

Recognizing that this loading condition may be repeated many times during the strong motion, the concrete is to be confined where the calculated compressive stresses exceed a nominal critical value equal to $0.2f_c'$. The stress is to be calculated for the factored forces on the section assuming linear response of the gross concrete section. The compressive stress of $0.2f_c'$ is used as an index value and does not necessarily describe the actual state of stress that may develop at the critical section under the influence of the actual inertia forces for the anticipated earthquake intensity.

18.10.6.4 If special boundary elements are required by 18.10.6.2 or 18.10.6.3, (a) through (k) shall be satisfied:

(a) The boundary element shall extend horizontally from the extreme compression fiber a distance at least the greater of $c - 0.1\ell_w$ and $c/2$, where c is the largest neutral axis depth calculated for the factored axial force and nominal moment strength consistent with δ_u.

(b) Width of the flexural compression zone, b, over the horizontal distance calculated by 18.10.6.4(a), including flange if present, shall be at least $h_u/16$.

(c) For walls or wall piers with $h_w/\ell_w \geq 2.0$ that are effectively continuous from the base of structure to top of wall, designed to have a single critical section for flexure and axial loads, and with $c/\ell_w \geq 3/8$, width of the flexural compression zone b over the length calculated in 18.10.6.4(a) shall be greater than or equal to 12 in.

(d) In flanged sections, the boundary element shall include the effective flange width in compression and shall extend at least 12 in. into the web.

R18.10.6.4 The horizontal dimension of the special boundary element is intended to extend at least over the length where the concrete compressive strain exceeds the critical value. For flanged wall sections, including box shapes, L-shapes, and C-shapes, the calculation to determine the need for special boundary elements should include a direction of lateral load consistent with the orthogonal combinations defined in ASCE/SEI 7. The value of $c/2$ in 18.10.6.4(a) is to provide a minimum length of the special boundary element. Good detailing practice is to arrange the longitudinal reinforcement and the confinement reinforcement such that all primary longitudinal reinforcement at the wall boundary is supported by transverse reinforcement.

A slenderness limit is introduced into the 2014 edition of this Code based on lateral instability failures of slender wall boundaries observed in recent earthquakes and tests (Wallace 2012; Wallace et al. 2012). For walls with large cover, where spalling of cover concrete would lead to a

CODE

(e) The boundary element transverse reinforcement shall satisfy 18.7.5.2(a) through (d) and 18.7.5.3, except the transverse reinforcement spacing limit of 18.7.5.3(a) shall be one-third of the least dimension of the boundary element. The maximum vertical spacing of transverse reinforcement in the boundary element shall also not exceed that in Table 18.10.6.5(b).

(f) Transverse reinforcement shall be arranged such that the spacing h_x between laterally supported longitudinal bars around the perimeter of the boundary element shall not exceed the lesser of 14 in. and two-thirds of the boundary element thickness. Lateral support shall be provided by a seismic hook of a crosstie or corner of a hoop. The length of a hoop leg shall not exceed two times the boundary element thickness, and adjacent hoops shall overlap at least the lesser of 6 in. and two-thirds the boundary element thickness.

(g) The amount of transverse reinforcement shall be in accordance with Table 18.10.6.4(g).

Table 18.10.6.4(g)—Transverse reinforcement for special boundary elements

Transverse reinforcement	Applicable expressions		
A_{sh}/sb_c for rectilinear hoop	Greater of	$0.3\left(\frac{A_g}{A_{ch}}-1\right)\frac{f_c'}{f_{yt}}$	(a)
		$0.09\frac{f_c'}{f_{yt}}$	(b)
ρ_s for spiral or circular hoop	Greater of	$0.45\left(\frac{A_g}{A_{ch}}-1\right)\frac{f_c'}{f_{yt}}$	(c)
		$0.12\frac{f_c'}{f_{yt}}$	(d)

(h) Concrete within the thickness of the floor system at the special boundary element location shall have specified compressive strength at least 0.7 times f_c' of the wall.

(i) For a distance above and below the critical section specified in 18.10.6.2(b), web vertical reinforcement shall have lateral support provided by the corner of a hoop or by a crosstie with seismic hooks at each end. Transverse reinforcement shall have a vertical spacing not to exceed 12 in. and diameter satisfying 25.7.2.2.

(j) Where the critical section occurs at the wall base, the boundary element transverse reinforcement at the wall base shall extend into the support at least ℓ_d, in accordance with 18.10.2.3, of the largest longitudinal reinforcement in the special boundary element. Where the special boundary element terminates on a footing, mat, or pile cap, special boundary element transverse reinforcement shall extend at least 12 in. into the footing, mat, or pile cap, unless a greater extension is required by 18.13.2.4.

COMMENTARY

significantly reduced section, increased boundary element thickness should be considered.

A value of $c/\ell_w \geq 3/8$ is used to define a wall critical section that is not tension-controlled according to 21.2.2. A minimum wall thickness of 12 in. is imposed to reduce the likelihood of lateral instability of the compression zone after spalling of cover concrete.

Where flanges are highly stressed in compression, the web-to-flange interface is likely to be highly stressed and may sustain local crushing failure unless special boundary element reinforcement extends into the web.

Required transverse reinforcement at wall boundaries is based on column provisions. Expression (a) of Table 18.10.6.4(g) was applied to wall special boundary elements prior to the 1999 edition of this Code. It is reinstated in the 2014 edition of this Code due to concerns that expression (b) of Table 18.10.6.4(g) by itself does not provide adequate transverse reinforcement for thin walls where concrete cover accounts for a significant portion of the wall thickness. For wall special boundary elements having rectangular cross section, A_g and A_{ch} in expressions (a) and (c) in Table 18.10.6.4(g) are defined as $A_g = \ell_{be}b$ and $A_{ch} = b_{c1}b_{c2}$, where dimensions are shown in Fig. R18.10.6.4b. This considers that concrete spalling is likely to occur only on the exposed faces of the confined boundary element. Tests (Thomsen and Wallace 2004) show that adequate performance can be achieved using vertical spacing greater than that permitted by 18.7.5.3(a). The limits on spacing between laterally supported longitudinal bars are intended to provide more uniform spacing of hoops and crossties for thin walls.

Configuration requirements for boundary element transverse reinforcement and crossties for web longitudinal reinforcement are summarized in Fig. R18.10.6.4a. A limit is placed on the relative lengths of boundary element hoop legs because tests (Segura and Wallace 2018; Welt et al. 2017; Arteta 2015) show that a single perimeter hoop with supplemental crossties that have alternating 90-degree and 135-degree hooks are not as effective as overlapping hoops and crossties with seismic hooks at both ends if ℓ_{be} exceeds approximately $2b$.

These tests also show that loss of axial load-carrying capacity of a wall can occur immediately following damage to the wall boundary elements if web vertical reinforcement within the plastic hinge region is not restrained. Use of web crossties outside of boundary elements also results in a less abrupt transition in transverse reinforcement used to provide concrete confinement and restrain buckling of longitudinal reinforcement, which addresses potential increases in the neutral axis depth due to shear (diagonal compression) and uncertainties in axial load.

Requirements for vertical extensions of boundary elements are summarized in Fig. R18.10.6.4c (Moehle et al. 2011).

The horizontal reinforcement in a structural wall with low shear-to-moment ratio resists shear through truss action, with the horizontal bars acting like the stirrups in a beam.

CODE

(k) Horizontal reinforcement in the wall web shall extend to within 6 in. of the end of the wall. Reinforcement shall be anchored to develop f_y within the confined core of the boundary element using standard hooks or heads. Where the confined boundary element has sufficient length to develop the horizontal web reinforcement, and $A_s f_y/s$ of the horizontal web reinforcement does not exceed $A_s f_{yt}/s$ of the boundary element transverse reinforcement parallel to the horizontal web reinforcement, it shall be permitted to terminate the horizontal web reinforcement without a standard hook or head.

COMMENTARY

Thus, the horizontal bars provided for shear reinforcement must be developed within the confined core of the boundary element and extended as close to the end of the wall as cover requirements and proximity of other reinforcement permit. The requirement that the horizontal web reinforcement be anchored within the confined core of the boundary element and extended to within 6 in. from the end of the wall applies to all horizontal bars whether straight, hooked, or headed, as illustrated in Fig. R18.10.6.4c.

The requirements in 18.10.2.4 apply to the minimum longitudinal reinforcement in the ends of walls, including those with special boundary elements.

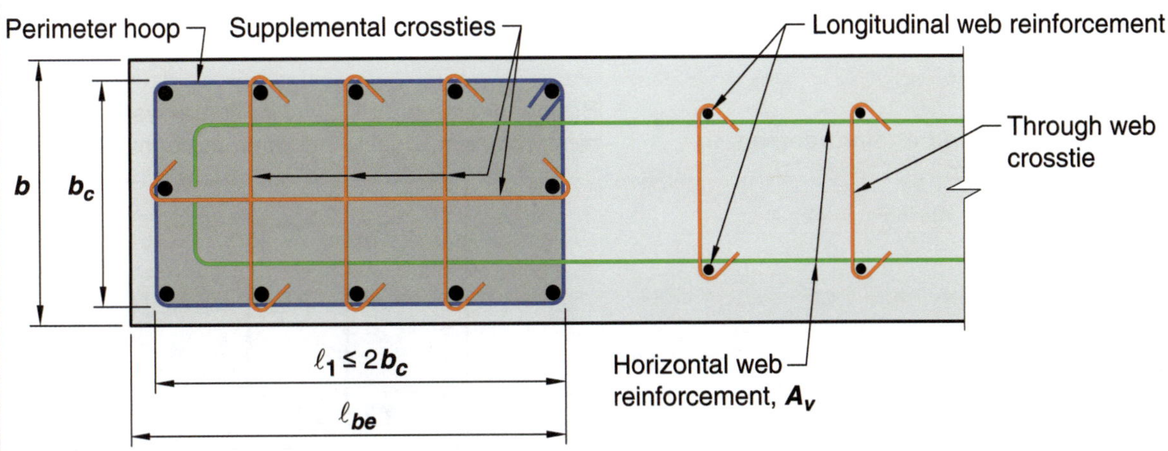

(a) Perimeter hoop with supplemental 135-degree crossties and 135-degree crossties supporting distributed web longitudinal reinforcement

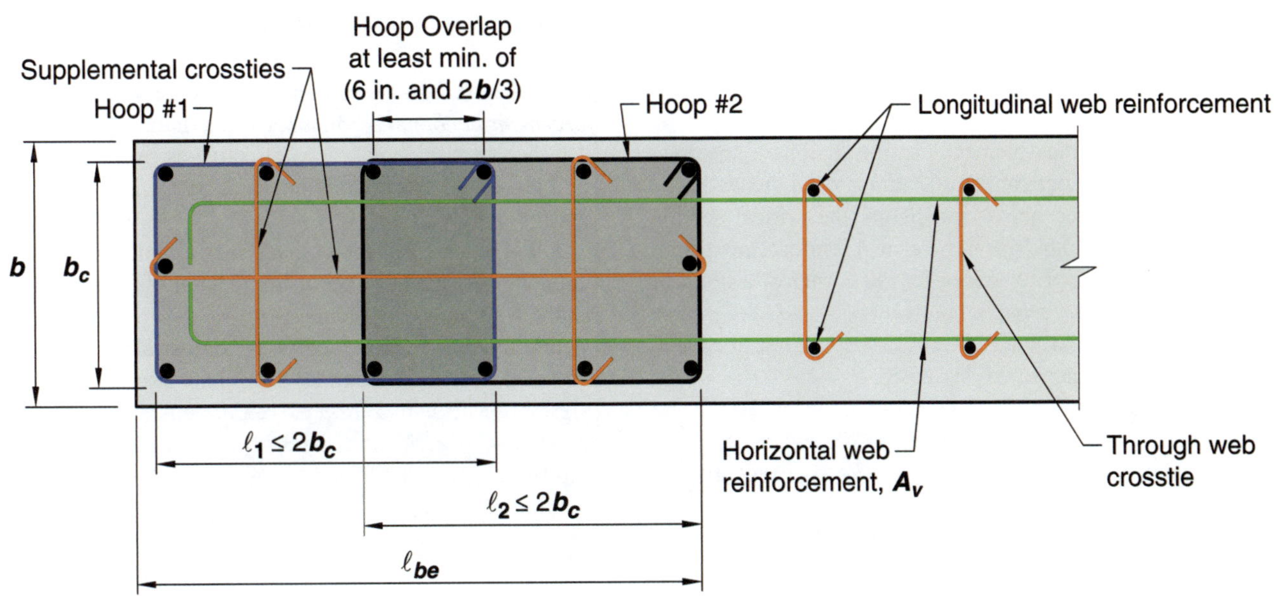

(b) Overlapping hoops with supplemental 135-degree crossties and 135-degree crossties supporting distributed web longitudinal reinforcement

Fig. R18.10.6.4a*—Configurations of boundary transverse reinforcement and web crossties.*

CODE

COMMENTARY

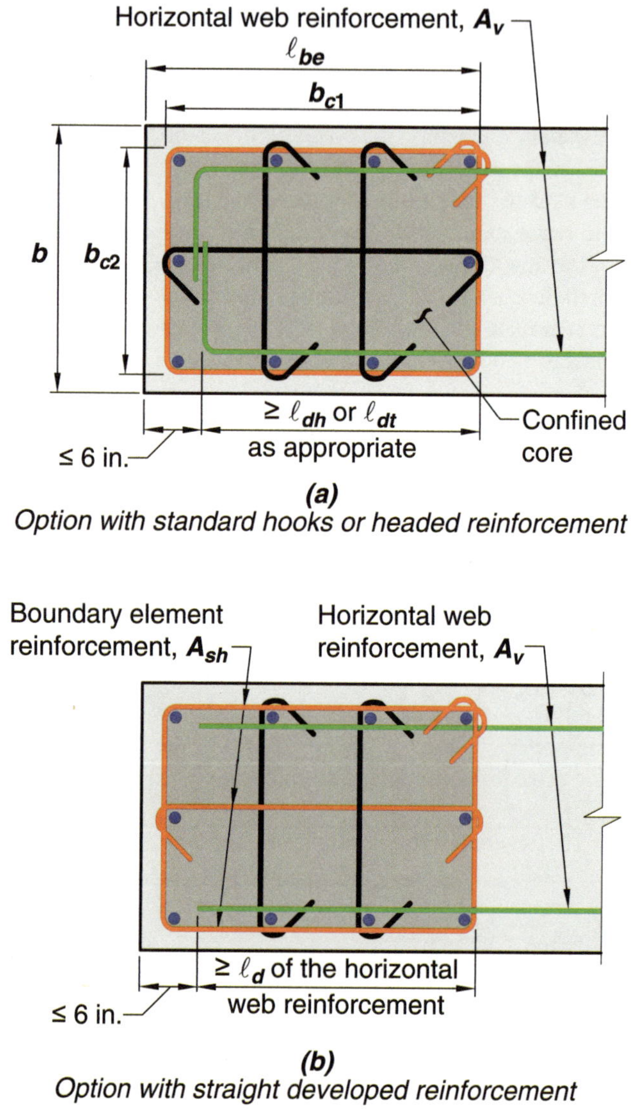

Fig. R18.10.6.4b—Development of wall horizontal reinforcement in confined boundary element.

CODE | COMMENTARY

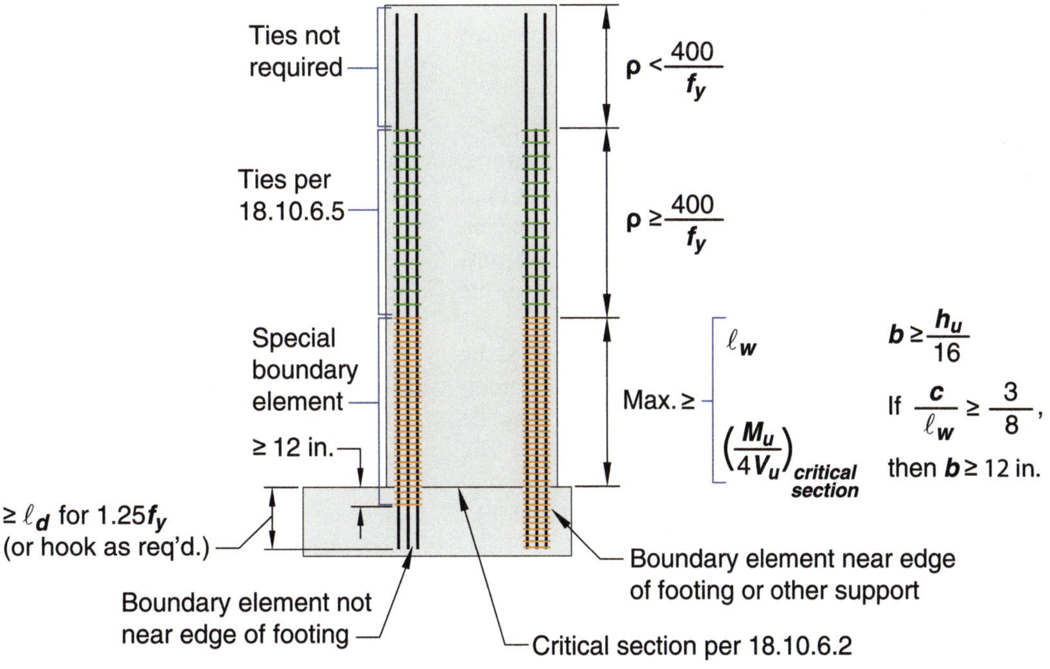

(a) Wall with $h_w/\ell_w \geq 2.0$ and a single critical section controlled by flexure and axial load designed using 18.10.6.2, 18.10.6.4, and 18.10.6.5

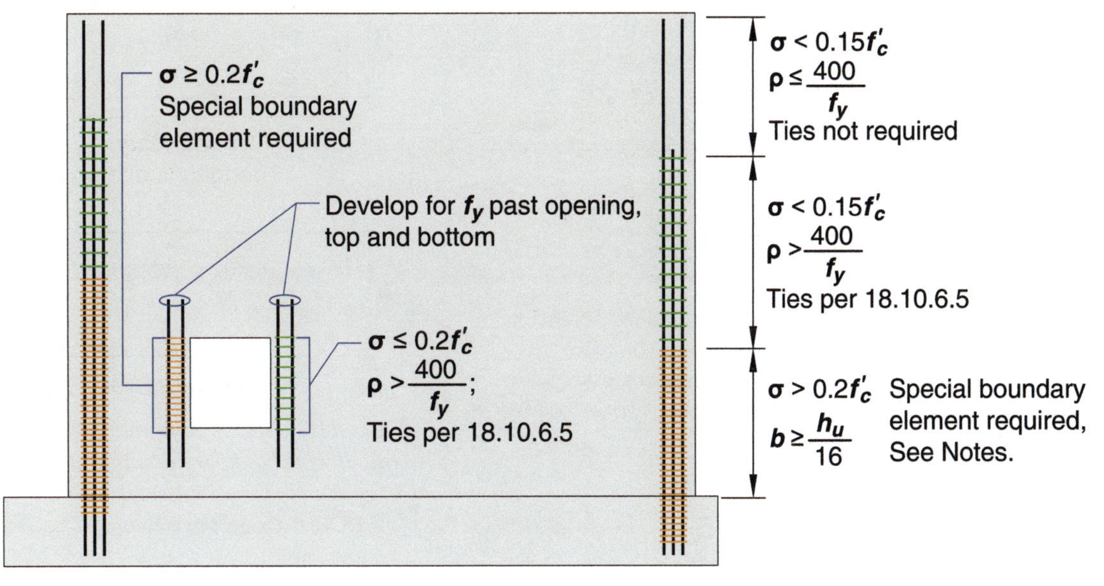

Notes: Requirement for special boundary element is triggered if maximum extreme fiber compressive stress $\sigma \geq 0.2f'_c$. Once triggered, the special boundary element extends until $\sigma < 0.15f'_c$. Since $h_w/\ell_w \leq 2.0$, 18.10.6.4(c) does not apply.

(b) Wall and wall pier designed using 18.10.6.3, 18.10.6.4, and 18.10.6.5.

***Fig. R18.10.6.4c**—Summary of boundary element requirements for special walls.*

CODE

18.10.6.5 Where special boundary elements are not required by 18.10.6.2 or 18.10.6.3, (a) and (b) shall be satisfied:

(a) Except where V_u in the plane of the wall is less than $\lambda\sqrt{f_c'}A_{cv}$, horizontal reinforcement terminating at the edges of structural walls without boundary elements shall have a standard hook engaging the edge reinforcement or the edge reinforcement shall be enclosed in U-stirrups having the same size and spacing as, and spliced to, the horizontal reinforcement.

(b) If the maximum longitudinal reinforcement ratio at the wall boundary exceeds $\mathbf{400/f_y}$, boundary transverse reinforcement shall satisfy 18.7.5.2(a) through (e) over the distance calculated in accordance with 18.10.6.4(a). The vertical spacing of transverse reinforcement at the wall boundary shall be in accordance with Table 18.10.6.5(b).

Table 18.10.6.5(b)—Maximum vertical spacing of transverse reinforcement at wall boundary

Grade of primary flexural reinforcing bar	Transverse reinforcement required	Maximum vertical spacing of transverse reinforcement[1]	
60	Within the greater of ℓ_w and $M_u/4V_u$ above and below critical sections[2]	Lesser of:	$6d_b$
			6 in.
	Other locations	Lesser of:	$8d_b$
			8 in.
80	Within the greater of ℓ_w and $M_u/4V_u$ above and below critical sections[2]	Lesser of:	$5d_b$
			6 in.
	Other locations	Lesser of:	$6d_b$
			6 in.
100	Within the greater of ℓ_w and $M_u/4V_u$ above and below critical sections[2]	Lesser of:	$4d_b$
			6 in.
	Other locations	Lesser of:	$6d_b$
			6 in.

[1]In this table, d_b is the diameter of the smallest primary flexural reinforcing bar.

[2]Critical sections are defined as locations where yielding of longitudinal reinforcement is likely to occur as a result of lateral displacements.

COMMENTARY

R18.10.6.5 Cyclic load reversals may lead to buckling of boundary longitudinal reinforcement even in cases where the demands on the boundary of the wall do not require special boundary elements. For walls with moderate amounts of boundary longitudinal reinforcement, ties are required to inhibit buckling. The longitudinal reinforcement ratio is intended to include only the reinforcement at the wall boundary, as indicated in Fig. R18.10.6.5. A greater spacing of ties relative to 18.10.6.4(e) is allowed due to the lower deformation demands on the walls. Requirements of 18.10.6.5 apply over the entire wall height and are summarized in Fig. R18.10.6.4c for cases where special boundary elements are required (Moehle et al. 2011).

The addition of hooks or U-stirrups at the ends of horizontal wall reinforcement provides anchorage so that the reinforcement will be effective in resisting shear forces. It will also tend to inhibit the buckling of the vertical edge reinforcement. In walls with low in-plane shear, the development of horizontal reinforcement is not necessary.

Limits on spacing of transverse reinforcement are intended to prevent bar buckling until reversed cyclic strains extend well into the inelastic range. To achieve similar performance capability, smaller spacing is required for higher-strength longitudinal reinforcement.

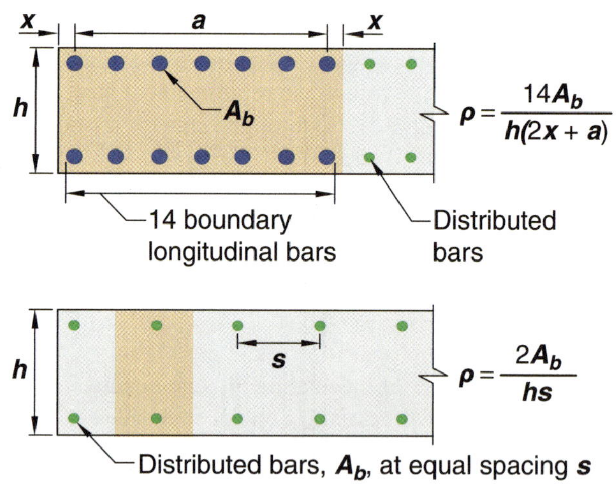

Fig. R18.10.6.5—Longitudinal reinforcement ratios for typical wall boundary conditions.

18.10.7 *Coupling beams*

18.10.7.1 Coupling beams with $(\ell_n/h) \geq 4$ shall satisfy the requirements of 18.6, with the wall boundary interpreted as being a column. The provisions of 18.6.2.1(b) and (c) need not be satisfied if it can be shown by analysis that the beam has adequate lateral stability.

18.10.7.2 Coupling beams with $(\ell_n/h) < 2$ and with $V_u \geq 4\lambda\sqrt{f_c'}A_{cw}$ shall be reinforced with two intersecting groups of diagonally placed bars symmetrical about the midspan, unless it can be shown that loss of stiffness and strength of the coupling

R18.10.7 *Coupling beams*

Coupling beams connecting structural walls can provide stiffness and energy dissipation. In many cases, geometric limits result in coupling beams that are deep in relation to their clear span. Deep coupling beams may be controlled by shear and may be susceptible to strength and stiffness deterioration under earthquake loading. Test results (Paulay and Binney 1974; Barney et al. 1980) have shown that confined diagonal reinforcement provides adequate resistance in deep coupling beams.

CODE

beams will not impair the vertical load-carrying ability of the structure, the egress from the structure, or the integrity of nonstructural components and their connections to the structure.

18.10.7.3 Coupling beams not governed by 18.10.7.1 or 18.10.7.2 shall be permitted to be reinforced either with two intersecting groups of diagonally placed bars symmetrical about the midspan or according to 18.6.3 through 18.6.5, with the wall boundary interpreted as being a column.

18.10.7.4 Coupling beams reinforced with two intersecting groups of diagonally placed bars symmetrical about the midspan shall satisfy (a), (b), and either (c) or (d), and the requirements of 9.9 need not be satisfied:

(a) V_n shall be calculated by

$$V_n = 2A_{vd}f_y\sin\alpha \leq 10\sqrt{f_c'}A_{cw} \qquad (18.10.7.4)$$

where α is the angle between the diagonal bars and the longitudinal axis of the coupling beam.

(b) Each group of diagonal bars shall consist of a minimum of four bars provided in two or more layers.

(c) Each group of diagonal bars shall be enclosed by rectilinear transverse reinforcement having out-to-out dimensions of at least $b_w/2$ in the direction parallel to b_w and $b_w/5$ along the other sides, where b_w is the web width of the coupling beam. The transverse reinforcement shall be in accordance with 18.7.5.2(a) through (e), with A_{sh} not less than the greater of (i) and (ii):

(i) $0.09sb_c\dfrac{f_c'}{f_{yt}}$

(ii) $0.3sb_c\left(\dfrac{A_g}{A_{ch}}-1\right)\dfrac{f_c'}{f_{yt}}$

For the purpose of calculating A_g, the concrete cover in 20.5.1 shall be assumed on all four sides of each group of diagonal bars. The transverse reinforcement shall have spacing measured parallel to the diagonal bars satisfying 18.7.5.3(d) and not exceeding $6d_b$ of the smallest diagonal bars, and shall have spacing of crossties or legs of hoops measured perpendicular to the diagonal bars not exceeding 14 in. The transverse reinforcement shall continue through the intersection of the diagonal bars. At the intersection, it is permitted to modify the arrangement of the transverse reinforcement provided the spacing and volume ratio requirements are satisfied. Additional longitudinal and transverse reinforcement shall be distributed around the beam perimeter with total area in each direction of at least $0.002b_ws$ and spacing not exceeding 12 in.

(d) Transverse reinforcement shall be provided for the entire beam cross section in accordance with 18.7.5.2(a) through (e) with A_{sh} not less than the greater of (i) and (ii):

COMMENTARY

Experiments show that diagonally oriented reinforcement is effective only if the bars are placed with a large inclination. Therefore, diagonally reinforced coupling beams are restricted to beams having aspect ratio $\ell_n/h < 4$. The 2008 edition of this Code was changed to clarify that coupling beams of intermediate aspect ratio can be reinforced according to 18.6.3 through 18.6.5.

Diagonal bars should be placed approximately symmetrically in the beam cross section, in two or more layers. The diagonally placed bars are intended to provide the entire shear and corresponding moment strength of the beam. Designs deriving their moment strength from combinations of diagonal and longitudinal bars are not covered by these provisions.

Two confinement options are described. According to 18.10.7.4(c), each diagonal element consists of a cage of longitudinal and transverse reinforcement, as shown in Fig. R18.10.7a. Each cage contains at least four diagonal bars and confines a concrete core. The requirement on side dimensions of the cage and its core is to provide adequate stability to the cross section when the bars are loaded beyond yielding. The minimum dimensions and required reinforcement clearances may control the wall width. Revisions were made in the 2008 Code to relax spacing of transverse reinforcement confining the diagonal bars, to clarify that confinement is required at the intersection of the diagonals, and to simplify design of the longitudinal and transverse reinforcement around the beam perimeter; beams with these new details are expected to perform acceptably. The expressions for transverse reinforcement A_{sh} are based on ensuring compression capacity of an equivalent column section is maintained after spalling of cover concrete.

Section 18.10.7.4(d) describes a second option for confinement of the diagonals introduced in the 2008 Code (refer to Fig. R18.10.7b). This second option is to confine the entire beam cross section instead of confining the individual diagonals. This option can considerably simplify field placement of hoops, which can otherwise be especially challenging where diagonal bars intersect each other or enter the wall boundary.

For coupling beams not used as part of the lateral-force-resisting system, the requirements for diagonal reinforcement may be waived.

Test results (Barney et al. 1980) demonstrate that beams reinforced as described in 18.10.7 have adequate ductility at shear forces exceeding $10\sqrt{f_c'}b_wd$. Consequently, the use of a limit of $10\sqrt{f_c'}A_{cw}$ provides an acceptable upper limit.

18 Seismic

CODE

(i) $0.09sb_c \dfrac{f_c'}{f_{yt}}$

(ii) $0.3sb_c \left(\dfrac{A_g}{A_{ch}} - 1\right)\dfrac{f_c'}{f_{yt}}$

Longitudinal spacing of transverse reinforcement shall not exceed the lesser of 6 in. and $6d_b$ of the smallest diagonal bars. Spacing of crossties or legs of hoops both vertically and horizontally in the plane of the beam cross section shall not exceed 8 in. Each crosstie and each hoop leg shall engage a longitudinal bar of equal or greater diameter. It shall be permitted to configure hoops as specified in 18.6.4.3.

COMMENTARY

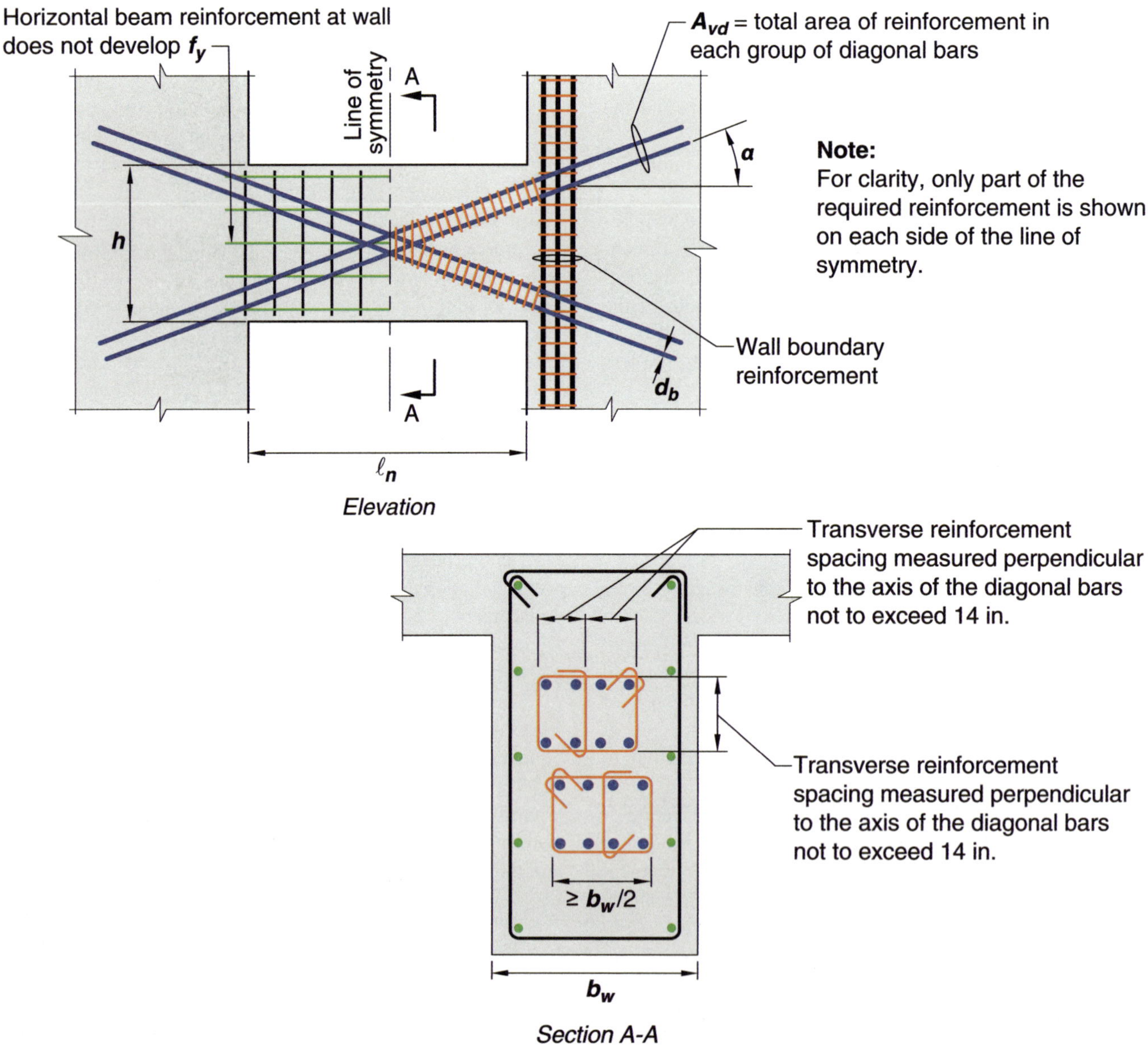

***Fig. R18.10.7a**—Confinement of individual diagonals in coupling beams with diagonally oriented reinforcement. Wall boundary reinforcement shown on one side only for clarity.*

CODE | COMMENTARY

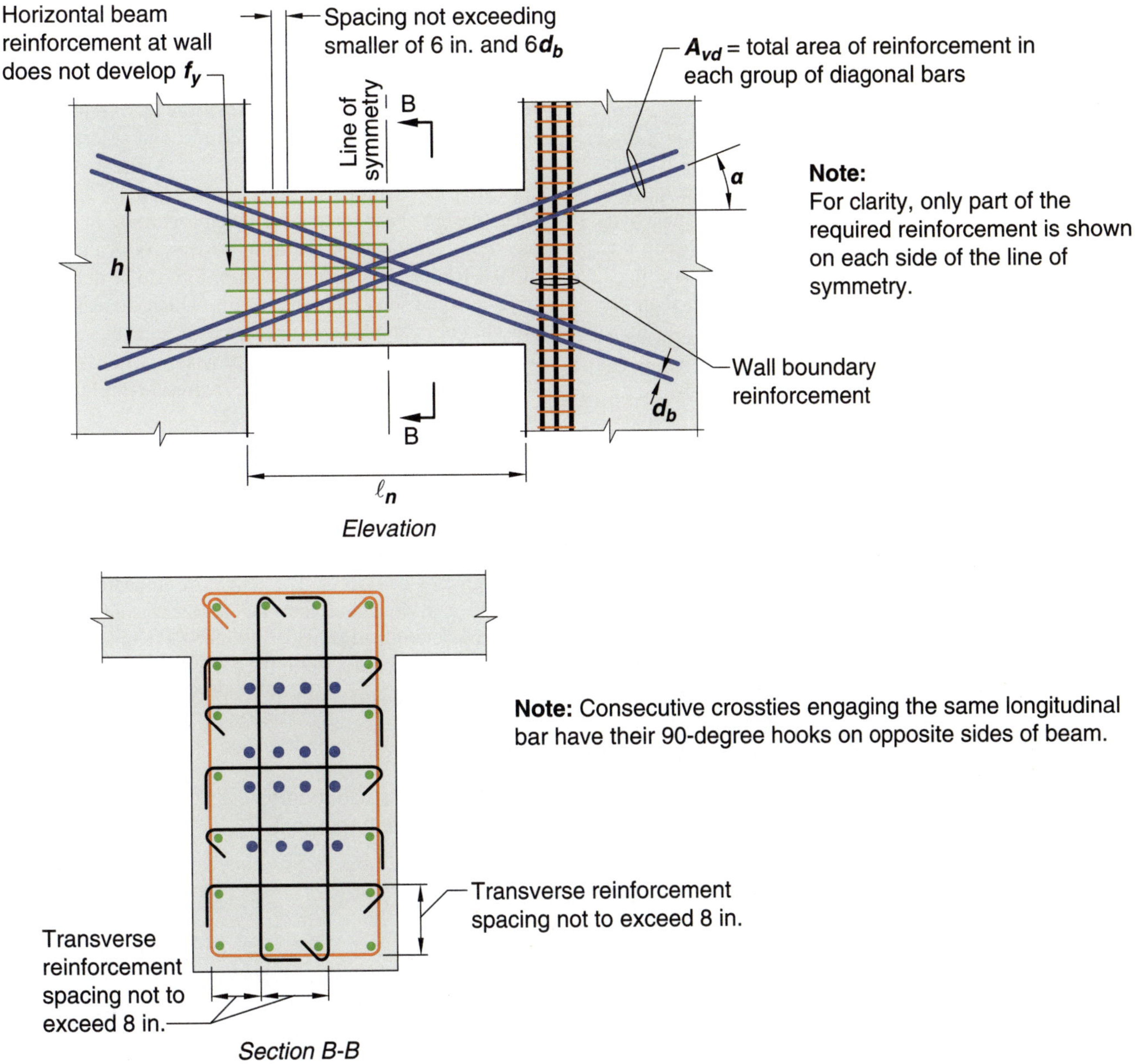

Fig. R18.10.7b*—Full confinement of diagonally reinforced concrete beam section in coupling beams with diagonally oriented reinforcement. Wall boundary reinforcement shown on one side only for clarity.*

18.10.8 *Wall piers*

18.10.8.1 Wall piers shall satisfy the special moment frame requirements for columns of 18.7.4, 18.7.5, and 18.7.6, with joint faces taken as the top and bottom of the clear height of the wall pier. Alternatively, wall piers with $(\ell_w/b_w) > 2.5$ shall satisfy (a) through (f):

(a) Design shear force shall be calculated in accordance with 18.7.6.1 with joint faces taken as the top and bottom of the clear height of the wall pier. If the general building code includes provisions to account for overstrength of the seismic-force-resisting system, the design shear force

R18.10.8 *Wall piers*

Door and window placements in structural walls sometimes lead to narrow vertical wall segments that are considered to be wall piers. The dimensions defining wall piers are given in Chapter 2. Shear failures of wall piers have been observed in previous earthquakes. The intent of this section is to provide sufficient shear strength to wall piers such that inelastic response, if it occurs, will be primarily in flexure. The provisions apply to wall piers designated as part of the seismic-force-resisting system. Provisions for wall piers not designated as part of the seismic-force-resisting system are given in 18.14. The effect of all vertical wall segments on the

CODE

need not exceed Ω_o times the factored shear calculated by analysis of the structure for earthquake load effects.

(b) V_n and distributed shear reinforcement shall satisfy 18.10.4.

(c) Transverse reinforcement shall be hoops except it shall be permitted to use single-leg horizontal reinforcement parallel to ℓ_w where only one curtain of distributed shear reinforcement is provided. Single-leg horizontal reinforcement shall have 180-degree bends at each end that engage wall pier boundary longitudinal reinforcement.

(d) Vertical spacing of transverse reinforcement shall not exceed 6 in.

(e) Transverse reinforcement shall extend at least 12 in. above and below the clear height of the wall pier.

(f) Special boundary elements shall be provided if required by 18.10.6.3.

18.10.8.2 For wall piers at the edge of a wall, horizontal reinforcement shall be provided in adjacent wall segments above and below the wall pier and be designed to transfer the design shear force from the wall pier into the adjacent wall segments.

COMMENTARY

response of the structural system, whether designated as part of the seismic-force-resisting system or not, should be considered as required by 18.2.2. Wall piers having $(\ell_w/b_w) \leq 2.5$ behave essentially as columns. Provision 18.10.8.1 requires that such members satisfy reinforcement and shear strength requirements of 18.7.4 through 18.7.6. Alternative provisions are provided for wall piers having $(\ell_w/b_w) > 2.5$.

The design shear force determined according to 18.7.6.1 may be unrealistically large in some cases. As an alternative, 18.10.8.1(a) permits the design shear force to be determined using factored load combinations in which the earthquake effect has been amplified to account for system overstrength. Documents such as the NEHRP provisions (FEMA P749), ASCE/SEI 7, and the 2018 IBC represent the amplified earthquake effect using the factor Ω_o.

Section 18.10.8.2 addresses wall piers at the edge of a wall. Under in-plane shear, inclined cracks can propagate into segments of the wall directly above and below the wall pier. Unless there is sufficient reinforcement in the adjacent wall segments, shear failure within the adjacent wall segments can occur. The length of embedment of the provided reinforcement into the adjacent wall segments should be determined considering both development length requirements and shear strength of the wall segments (refer to Fig. R18.10.8).

CODE

COMMENTARY

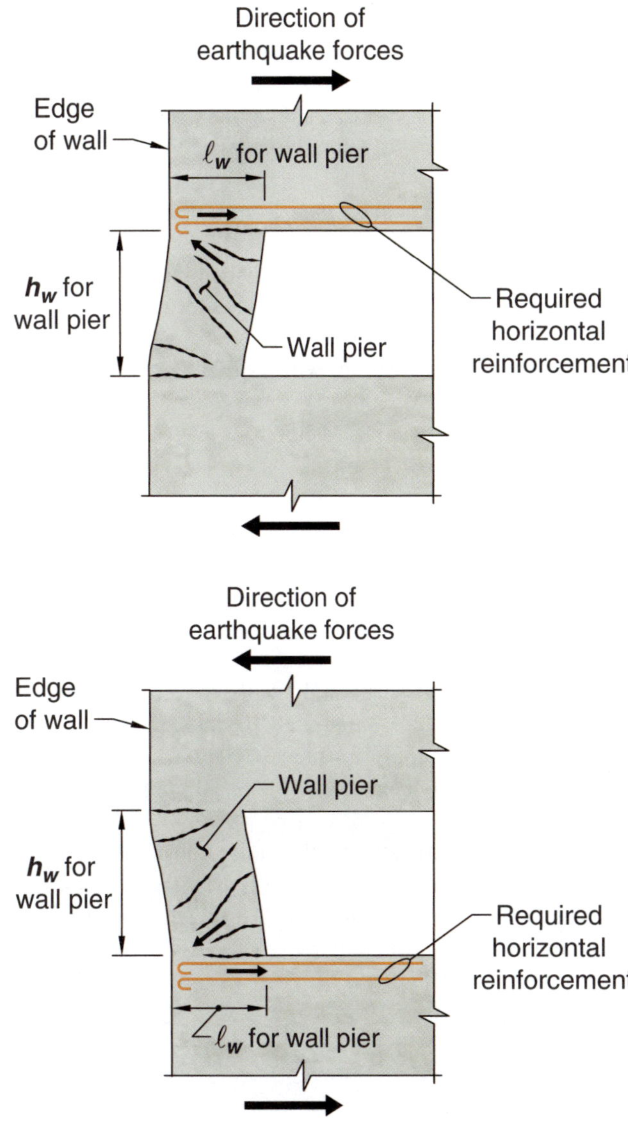

***Fig. R18.10.8**—Required horizontal reinforcement in wall segments above and below wall piers at the edge of a wall.*

18.10.9 *Ductile coupled walls*

18.10.9.1 Ductile coupled walls shall satisfy the requirements of this section.

18.10.9.2 Individual walls shall satisfy $h_{wcs}/\ell_w \geq 2$ and the applicable provisions of 18.10 for special structural walls.

18.10.9.3 Coupling beams shall satisfy 18.10.7 and (a) through (c) in the direction considered.

(a) Coupling beams shall have $\ell_n/h \geq 2$ at all levels of the building.
(b) All coupling beams at a floor level shall have $\ell_n/h \leq 5$ in at least 90 percent of the levels of the building.
(c) The requirements of 18.10.2.5 shall be satisfied at both ends of all coupling beams.

R18.10.9 *Ductile coupled walls*

The aspect ratio limits and development length requirements for ductile coupled walls are intended to induce an energy dissipation mechanism associated with inelastic deformation reversal of coupling beams. Wall stiffness and strength at each end of coupling beams should be sufficient to develop this intended behavior.

CODE

18.10.10 *Construction joints*

18.10.10.1 Construction joints in structural walls shall be specified according to 26.5.6, and contact surfaces shall be roughened consistent with condition (b) of Table 22.9.4.2.

18.10.11 *Discontinuous walls*

18.10.11.1 Columns supporting discontinuous structural walls shall be reinforced in accordance with 18.7.5.6.

18.11—Special structural walls constructed using precast concrete

18.11.1 *Scope*

18.11.1.1 This section shall apply to special structural walls constructed using precast concrete forming part of the seismic-force-resisting system.

18.11.2 *General*

18.11.2.1 Special structural walls constructed using precast concrete shall satisfy 18.10 and 18.5.2, except 18.10.2.4 shall not apply for precast walls where deformation demands are concentrated at the panel joints.

18.11.2.2 Special structural walls constructed using precast concrete and unbonded post-tensioning tendons and not satisfying the requirements of 18.11.2.1 are permitted provided they satisfy the requirements of ACI ITG-5.1.

18.12—Diaphragms and trusses

18.12.1 *Scope*

18.12.1.1 This section shall apply to diaphragms and collectors forming part of the seismic-force-resisting system in structures assigned to SDC D, E, or F and to SDC C if 18.12.1.2 applies.

18.12.1.2 Section 18.12.11 shall apply to diaphragms constructed using precast concrete members and forming part of the seismic-force-resisting system for structures assigned to SDC C, D, E, or F.

COMMENTARY

R18.11—Special structural walls constructed using precast concrete

R18.11.2 *General*

R18.11.2.2 Experimental and analytical studies (Priestley et al. 1999; Perez et al. 2003; Restrepo 2002) have demonstrated that some types of precast structural walls post-tensioned with unbonded tendons, and not satisfying the prescriptive requirements of Chapter 18, provide satisfactory seismic performance characteristics. ACI ITG-5.1 defines a protocol for establishing a design procedure, validated by analysis and laboratory tests, for such walls, with or without coupling beams.

ACI ITG-5.2 defines design requirements for one type of special structural wall constructed using precast concrete and unbonded post-tensioning tendons, and validated for use in accordance with 18.11.2.2.

R18.12—Diaphragms and trusses

R18.12.1 *Scope*

Diaphragms as used in building construction are structural elements (such as a floor or roof) that provide some or all of the following functions:

(a) Support for building elements (such as walls, partitions, and cladding) resisting horizontal forces but not acting as part of the seismic-force-resisting system
(b) Transfer of lateral forces from the point of application to the vertical elements of the seismic-force-resisting system
(c) Connection of various components of the vertical seismic-force-resisting system with appropriate strength,

CODE

18.12.1.3 Section 18.12.12 shall apply to structural trusses forming part of the seismic-force-resisting system in structures assigned to SDC D, E, or F.

18.12.2 *Design forces*

18.12.2.1 The earthquake design forces for diaphragms shall be obtained from the general building code using the applicable provisions and load combinations.

18.12.3 *Seismic load path*

18.12.3.1 All diaphragms and their connections shall be designed and detailed to provide for transfer of forces to collector elements and to the vertical elements of the seismic-force-resisting system.

18.12.3.2 Elements of a structural diaphragm system that are subjected primarily to axial forces and used to transfer

COMMENTARY

stiffness, and ductility so the building responds as intended in the design (Wyllie 1987).

R18.12.2 *Design forces*

R18.12.2.1 In the general building code, earthquake design forces for floor and roof diaphragms typically are not calculated directly during the lateral-force analysis that provides story forces and story shears. Instead, diaphragm design forces at each level are calculated by a formula that amplifies the story forces recognizing dynamic effects and includes minimum and maximum limits. These forces are used with the governing load combinations to design diaphragms for shear and moment.

For collector elements, the general building code in the United States specifies load combinations that amplify earthquake forces by a factor $\mathbf{\Omega}_o$. The forces amplified by $\mathbf{\Omega}_o$ are also used for the local diaphragm shear forces resulting from the transfer of collector forces, and for local diaphragm flexural moments resulting from any eccentricity of collector forces. The specific requirements for earthquake design forces for diaphragms and collectors depend on which edition of the general building code is used. The requirements may also vary according to the SDC.

For most concrete buildings subjected to inelastic earthquake demands, it is desirable to limit inelastic behavior of floor and roof diaphragms under the imposed earthquake forces and deformations. It is preferable for inelastic behavior to occur only in the intended locations of the vertical seismic-force-resisting system that are detailed for ductile response, such as in beam plastic hinges of special moment frames, or in flexural plastic hinges at the base of structural walls or in coupling beams. For buildings without long diaphragm spans between lateral-force-resisting elements, elastic diaphragm behavior is typically not difficult to achieve. For buildings where diaphragms could reach their flexural or shear strength before yielding occurs in the vertical seismic-force-resisting system, the licensed design professional should consider providing increased diaphragm strength.

For reinforced concrete diaphragms, ASCE/SEI 7 Sections 12.10.1 and 12.10.2 provide requirements to determine design forces for reinforced concrete diaphragms. For precast concrete diaphragms in buildings assigned to SDC C, D, E, or F, the provisions of ASCE/SEI 7 Section 12.10.3 apply.

R18.12.3 *Seismic load path*

R18.12.3.2 This provision applies to strut-like elements that occur around openings, diaphragm edges, or other

CODE

diaphragm shear or flexural forces around openings or other discontinuities shall satisfy the requirements for collectors in 18.12.7.6 and 18.12.7.7.

COMMENTARY

discontinuities in diaphragms. Figure R18.12.3.2 shows an example. Such elements can be subjected to earthquake axial forces in combination with bending and shear from earthquake or gravity loads.

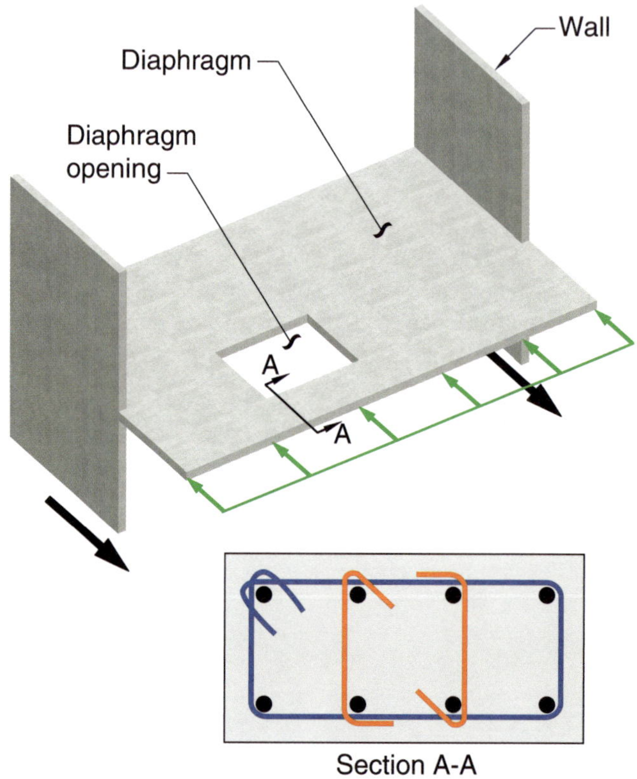

Fig. R18.12.3.2—Example of diaphragm subject to the requirements of 18.12.3.2 and showing an element having confinement as required by 18.12.7.6.

18.12.4 *Cast-in-place composite topping slab diaphragms*

18.12.4.1 A cast-in-place composite topping slab on a precast floor or roof shall be permitted as a structural diaphragm, provided the cast-in-place topping slab is reinforced and the surface of the previously hardened concrete on which the topping slab is placed is clean, free of laitance, and intentionally roughened.

R18.12.4 *Cast-in-place composite topping slab diaphragms*

R18.12.4.1 A bonded topping slab is required so that the floor or roof system can provide restraint against slab buckling. Reinforcement is required to ensure the continuity of the shear transfer across precast joints. The connection requirements are introduced to promote a complete system with necessary shear transfers.

18.12.5 *Cast-in-place noncomposite topping slab diaphragms*

18.12.5.1 A cast-in-place noncomposite topping on a precast floor or roof shall be permitted as a structural diaphragm, provided the cast-in-place topping slab acting alone is designed and detailed to resist the design earthquake forces.

R18.12.5 *Cast-in-place noncomposite topping slab diaphragms*

R18.12.5.1 Composite action between the topping slab and the precast floor elements is not required, provided that the topping slab is designed to resist the design earthquake forces.

18.12.6 *Minimum thickness of diaphragms*

18.12.6.1 Concrete slabs and composite topping slabs serving as diaphragms used to transmit earthquake forces shall be at least 2 in. thick. Topping slabs placed over precast

R18.12.6 *Minimum thickness of diaphragms*

R18.12.6.1 The minimum thickness of concrete diaphragms reflects current practice in joist and waffle systems and composite topping slabs on precast floor and

CODE

floor or roof elements, acting as diaphragms and not relying on composite action with the precast elements to resist the design earthquake forces, shall be at least 2-1/2 in. thick.

18.12.7 *Reinforcement*

18.12.7.1 The minimum reinforcement ratio for diaphragms shall be in conformance with 24.4. Except for post-tensioned slabs, reinforcement spacing each way in floor or roof systems shall not exceed 18 in. Where welded wire reinforcement is used as the distributed reinforcement to resist shear in topping slabs placed over precast floor and roof elements, the wires parallel to the joints between the precast elements shall be spaced not less than 10 in. on center. Reinforcement provided for shear strength shall be continuous and shall be distributed uniformly across the shear plane.

18.12.7.2 Bonded tendons used as reinforcement to resist collector forces, diaphragm shear, or flexural tension shall be designed such that the stress due to design earthquake forces does not exceed 60,000 psi. Precompression from unbonded tendons shall be permitted to resist diaphragm design forces if a seismic load path is provided.

18.12.7.3 All reinforcement used to resist collector forces, diaphragm shear, or flexural tension shall be developed or spliced for f_y in tension.

18.12.7.4 Type 2 splices are required where mechanical splices on Grade 60 reinforcement are used to transfer forces between the diaphragm and the vertical elements of the seismic-force-resisting system. Grade 80 and Grade 100 reinforcement shall not be mechanically spliced for this application.

18.12.7.5 Longitudinal reinforcement for collectors shall be proportioned such that the average tensile stress over length (a) or (b) does not exceed ϕf_y where the value of f_y is limited to 60,000 psi.

COMMENTARY

roof systems. Thicker slabs are required if the topping slab is not designed to act compositely with the precast system to resist the design earthquake forces.

R18.12.7 *Reinforcement*

R18.12.7.1 Minimum reinforcement ratios for diaphragms correspond to the required amount of temperature and shrinkage reinforcement (refer to 24.4). The maximum spacing for reinforcement is intended to control the width of inclined cracks. Minimum average prestress requirements (refer to 24.4.4.1) are considered to be adequate to limit the crack widths in post-tensioned floor systems; therefore, the maximum spacing requirements do not apply to these systems.

The minimum spacing requirement for welded wire reinforcement in topping slabs on precast floor systems is to avoid fracture of the distributed reinforcement during an earthquake. Cracks in the topping slab open immediately above the boundary between the flanges of adjacent precast members, and the wires crossing those cracks are restrained by the transverse wires (Wood et al. 2000). Therefore, all the deformation associated with cracking should be accommodated in a distance not greater than the spacing of the transverse wires. A minimum spacing of 10 in. for the transverse wires is required to reduce the likelihood of fracture of the wires crossing the critical cracks during a design earthquake. The minimum spacing requirements do not apply to diaphragms reinforced with individual bars, because strains are distributed over a longer length.

R18.12.7.3 Bar development and lap splices are designed according to requirements of Chapter 25 for reinforcement in tension. Reductions in development or splice length for calculated stresses less than f_y are not permitted, as indicated in 25.4.10.2.

R18.12.7.5 Table 20.2.2.4(a) permits the maximum design yield strength to be 80,000 psi for portions of a collector, for example, at and near critical sections. The average stress in the collector is limited to control diaphragm cracking over the length of the collector. The calculation of average stress along the length is not necessary if the collector is

CODE

(a) Length between the end of a collector and location at which transfer of load to a vertical element begins
(b) Length between two vertical elements

18.12.7.6 Collector elements with compressive stresses exceeding $\mathbf{0.2}f_c'$ at any section shall have transverse reinforcement satisfying 18.7.5.2(a) through (e) and 18.7.5.3, except the spacing limit of 18.7.5.3(a) shall be one-third of the least dimension of the collector. The amount of transverse reinforcement shall be in accordance with Table 18.12.7.6. The specified transverse reinforcement is permitted to be discontinued at a section where the calculated compressive stress is less than $\mathbf{0.15}f_c'$.

If design forces have been amplified to account for the overstrength of the vertical elements of the seismic-force-resisting system, the limit of $\mathbf{0.2}f_c'$ shall be increased to $\mathbf{0.5}f_c'$, and the limit of $\mathbf{0.15}f_c'$ shall be increased to $\mathbf{0.4}f_c'$.

Table 18.12.7.6—Transverse reinforcement for collector elements

Transverse reinforcement	Applicable expressions		
A_{sh}/sb_c for rectilinear hoop	$0.09\dfrac{f_c'}{f_{yt}}$		(a)
ρ_s for spiral or circular hoop	Greater of:	$0.45\left(\dfrac{A_g}{A_{ch}}-1\right)\dfrac{f_c'}{f_{yt}}$	(b)
		$0.12\dfrac{f_c'}{f_{yt}}$	(c)

18.12.7.7 Longitudinal reinforcement detailing for collector elements at splices and anchorage zones shall satisfy (a) or (b):

(a) Center-to-center spacing of at least three longitudinal bar diameters, but not less than 1-1/2 in., and concrete clear cover of at least two and one-half longitudinal bar diameters, but not less than 2 in.
(b) Area of transverse reinforcement, providing A_v at least the greater of $\mathbf{0.75}\sqrt{f_c'}(b_w s/f_{yt})$ and $\mathbf{50}b_w s/f_{yt}$, except as required in 18.12.7.6

18.12.8 *Flexural strength*

18.12.8.1 Diaphragms and portions of diaphragms shall be designed for flexure in accordance with Chapter 12. The effects of openings shall be considered.

COMMENTARY

designed for f_y of 60,000 psi even if Grade 80 reinforcement is specified.

R18.12.7.6 In documents such as the NEHRP Provisions (FEMA P750), ASCE/SEI 7, the 2018 IBC, and the Uniform Building Code (ICBO 1997), collector elements of diaphragms are designed for forces amplified by a factor Ω_o to account for the overstrength in the vertical elements of the seismic-force-resisting systems. The amplification factor Ω_o ranges between 2 and 3 for most concrete structures, depending on the document selected and on the type of seismic-force-resisting system. In some documents, the factor can be calculated based on the maximum forces that can be developed by the elements of the vertical seismic-force-resisting system.

Compressive stress calculated for the factored forces on a linearly elastic model based on gross section of the structural diaphragm is used as an index value to determine whether confining reinforcement is required. A calculated compressive stress of $\mathbf{0.2}f_c'$, or $\mathbf{0.5}f_c'$ for forces amplified by Ω_o, is assumed to indicate that integrity of the entire structure depends on the ability of that member to resist substantial compressive force under severe cyclic loading. Transverse reinforcement is required at such locations to provide confinement for the concrete and the reinforcement.

R18.12.7.7 This section is intended to reduce the possibility of bar buckling and provide adequate bar development conditions in the vicinity of splices and anchorage zones.

R18.12.8 *Flexural strength*

R18.12.8.1 Flexural strength for diaphragms is calculated using the same assumptions as for walls, columns, or beams. The design of diaphragms for flexure and other actions uses the applicable load combinations of 5.3.1 to consider earthquake forces acting concurrently with gravity or other loads.

The influence of slab openings on flexural and shear strength is to be considered, including evaluating the potential critical sections created by the openings. The strut-and-tie method is potentially useful for designing diaphragms with openings.

CODE

COMMENTARY

Earlier design practice assumed design moments for diaphragms were resisted entirely by chord forces acting at opposite edges of the diaphragm. This idealization was implicit in earlier versions of the Code, but has been replaced by an approach in which all longitudinal reinforcement, within the limits of 18.12.7, is assumed to contribute to the flexural strength of the diaphragm. This change reduces the required area of longitudinal reinforcement concentrated near the edge of the diaphragm, but should not be interpreted as a requirement to eliminate all boundary reinforcement.

18.12.9 *Shear strength*

18.12.9.1 V_n of diaphragms shall not exceed:

$$V_n = A_{cv}\left(2\lambda\sqrt{f_c'} + \rho_t f_y\right) \quad (18.12.9.1)$$

For cast-in-place topping slab diaphragms on precast floor or roof members, A_{cv} shall be calculated using only the thickness of topping slab for noncomposite topping slab diaphragms and the combined thickness of cast-in-place and precast elements for composite topping slab diaphragms. For composite topping slab diaphragms, the value of f_c' used to calculate V_n shall not exceed the lesser of f_c' for the precast members and f_c' for the topping slab.

18.12.9.2 V_n of diaphragms shall not exceed $8\sqrt{f_c'}A_{cv}$.

18.12.9.3 Above joints between precast elements in noncomposite and composite cast-in-place topping slab diaphragms, V_n shall not exceed:

$$V_n = A_{vf} f_y \mu \quad (18.12.9.3)$$

where A_{vf} is the total area of shear friction reinforcement within the topping slab, including both distributed and boundary reinforcement, that is oriented perpendicular to joints in the precast system and coefficient of friction, μ, is 1.0λ, where λ is given in 19.2.4. At least one-half of A_{vf} shall be uniformly distributed along the length of the potential shear plane. The area of distributed reinforcement in the topping slab shall satisfy 24.4.3.2 in each direction.

18.12.9.4 Above joints between precast elements in noncomposite and composite cast-in-place topping slab diaphragms, V_n shall not exceed the limits in 22.9.4.4, where A_c is calculated using only the thickness of the topping slab.

R18.12.9 *Shear strength*

The shear strength requirements for diaphragms are similar to those for slender structural walls and are based on the shear provisions for beams. The term A_{cv} refers to the gross area of the diaphragm, but may not exceed the thickness times the width of the diaphragm. This corresponds to the gross area of the effective deep beam that forms the diaphragm. Distributed slab reinforcement ρ_t used to calculate shear strength of a diaphragm in Eq. (18.12.9.1) is positioned perpendicular to the diaphragm flexural reinforcement. Provision 18.12.9.2 limits the maximum shear strength of the diaphragm.

In addition to satisfying 18.12.9.1 and 18.12.9.2, cast-in-place topping slab diaphragms must also satisfy 18.12.9.3 and 18.12.9.4. Cast-in-place topping slabs on a precast floor or roof system tend to have shrinkage cracks that are aligned with the joints between adjacent precast members. Therefore, the additional shear strength requirements for topping slab diaphragms in 18.12.9.3 are based on a shear friction model (Wood et al. 2000), and the assumed crack plane corresponds to joints in the precast system along the direction of the applied shear, as shown in Fig. R22.9.4.3a. The coefficient of friction, μ, in the shear friction model is taken equal to 1.0 for normalweight concrete due to the presence of these shrinkage cracks.

Both distributed and boundary reinforcement in the topping slab may be considered as shear friction reinforcement A_{vf}. Boundary reinforcement within the diaphragm was called chord reinforcement in ACI 318 before 2008. Although the boundary reinforcement also resists forces due to moment and axial force in the diaphragm, the reduction in the shear friction resistance in the tension zone is offset by the increase in shear friction resistance in the compression zone. Therefore, the area of boundary reinforcement used to resist shear friction need not be added to the area of boundary reinforcement used to resist moment and axial force. The distributed topping slab reinforcement must contribute at least one-half of the nominal shear strength. It is assumed that connections between the precast elements do not contribute to the shear strength of the topping slab diaphragm.

Provision 18.12.9.4 limits the maximum shear that may be transmitted by shear friction within a topping slab diaphragm.

CODE

18.12.10 *Construction joints*

18.12.10.1 Construction joints in diaphragms shall be specified according to 26.5.6, and contact surfaces shall be roughened consistent with condition (b) of Table 22.9.4.2.

18.12.11 *Precast concrete diaphragms*

18.12.11.1 Diaphragms and collectors constructed using precast concrete members with composite topping slab and not satisfying 18.12.4, and untopped precast concrete diaphragms, are permitted provided they satisfy the requirements of ACI 550.5. Cast-in-place noncomposite topping slab diaphragms shall satisfy 18.12.5 and 18.12.6.

18.12.11.2 Connections and reinforcement at joints used in the construction of precast concrete diaphragms satisfying 18.12.11.1 shall have been tested in accordance with ACI 550.4.

18.12.11.3 Extrapolation of data on connections and reinforcement at joints to project details that result in larger construction tolerances than those used to qualify connections in accordance with ACI 550.4 shall not be permitted.

18.12.12 *Structural trusses*

18.12.12.1 Structural truss elements with compressive stresses exceeding $\mathbf{0.2}\boldsymbol{f_c'}$ at any section shall have transverse reinforcement, in accordance with 18.7.5.2, 18.7.5.3, 18.7.5.7, and Table 18.12.12.1, over the length of the element.

COMMENTARY

R18.12.11 *Precast concrete diaphragms*

R18.12.11.1 ACI 550.5 provides requirements for the design of precast concrete diaphragms with connections whose performance has been validated by ACI 550.4 testing. ACI 550.5 permits a maximum tolerance for positioning and completion of connections of 1/2 in., which can be difficult to achieve with normal construction practices. Section 26.13.1.3 requires continuous inspection of precast concrete diaphragm connections to verify that construction is performed properly and tolerances not greater than 1/2 in. for all connections are achieved. Results from ACI 550.4 testing are not to be extrapolated to allow greater tolerances.

Topped precast concrete floors designed in accordance with Chapter 18 need careful consideration of support conditions to verify precast concrete members have sufficient seating for anticipated displacements and ability to accommodate relative rotations between beam supports and the member (Henry et al. 2017).

R18.12.12 *Structural trusses*

R18.12.12.1 The expressions for transverse reinforcement A_{sh} are based on ensuring compression capacity of an equivalent column section is maintained after spalling of cover concrete.

CODE

Table 18.12.12.1—Transverse reinforcement for structural trusses

Transverse reinforcement	Applicable expressions		
A_{sh}/sb_c for rectilinear hoop	Greater of:	$0.3\left(\frac{A_g}{A_{ch}}-1\right)\frac{f'_c}{f_{yt}}$	(a)
		$0.09\frac{f'_c}{f_{yt}}$	(b)
ρ_s for spiral or circular hoop	Greater of:	$0.45\left(\frac{A_g}{A_{ch}}-1\right)\frac{f'_c}{f_{yt}}$	(c)
		$0.12\frac{f'_c}{f_{yt}}$	(d)

18.12.12.2 All continuous reinforcement in structural truss elements shall be developed or spliced for f_y in tension.

18.13—Foundations

18.13.1 *Scope*

18.13.1.1 This section shall apply to foundations resisting earthquake-induced forces or transferring earthquake-induced forces between structure and ground.

18.13.1.2 The provisions in this section for piles, drilled piers, caissons, and slabs-on-ground shall supplement other applicable Code design and construction criteria, including 1.4.6 and 1.4.7.

18.13.2 *Footings, foundation mats, and pile caps*

18.13.2.1 The provisions of this section shall apply to structures assigned to SDC D, E, or F.

18.13.2.2 Longitudinal reinforcement of columns and structural walls resisting forces induced by earthquake effects shall extend into the footing, mat, or pile cap, and shall be fully developed for tension at the interface.

18.13.2.3 Columns designed assuming fixed-end conditions at the foundation shall comply with 18.13.2.2 and, if hooks are required, longitudinal reinforcement resisting flexure shall have 90-degree hooks near the bottom of the foundation with the free end of the bars oriented toward the center of the column.

COMMENTARY

R18.13—Foundations

R18.13.1 *Scope*

Requirements for foundations supporting buildings assigned to SDC C, D, E, or F represent a consensus of a minimum level of good practice in designing and detailing concrete foundations. However, because repairs to foundations can be extremely difficult and expensive, it may be desirable that the elements of the foundation remain essentially elastic during strong ground motions. Methods to achieve this goal include designing the foundation to include an overstrength factor or an increased seismic demand level when compared to the superstructure, or comparing strengths to demands predicted by nonlinear response history analyses with appropriate consideration of uncertainty in demands (Klemencic et al. 2012).

R18.13.2 *Footings, foundation mats, and pile caps*

R18.13.2.3 Tests (Nilsson and Losberg 1976) have demonstrated that flexural members terminating in a footing, slab, or beam (a T-joint or L-joint) should have their hooks turned inward toward the axis of the member for the joint to be able to resist the flexure in the member forming the stem of the T or L.

CODE

18.13.2.4 Columns or boundary elements of special structural walls that have an edge within one-half the footing depth from an edge of the footing shall have transverse reinforcement in accordance with 18.7.5.2 through 18.7.5.4 provided below the top of the footing. This reinforcement shall extend into the footing, mat, or pile cap a length equal to the development length, calculated for f_y in tension, of the column or boundary element longitudinal reinforcement.

18.13.2.5 Where earthquake effects create uplift forces in boundary elements of special structural walls or columns, flexural reinforcement shall be provided in the top of the footing, mat, or pile cap to resist actions resulting from the factored load combinations, and shall be at least that required by 7.6.1 or 9.6.1.

18.13.2.6 Structural plain concrete in footings and basement walls shall be in accordance with 14.1.4.

18.13.2.7 Pile caps incorporating batter piles shall be designed to resist the full compressive strength of the batter piles acting as short columns. The slenderness effects of batter piles shall be considered for the portion of the piles in soil that is not capable of providing lateral support, or in air or water.

18.13.3 *Grade beams and slabs-on-ground*

18.13.3.1 For structures assigned to SDC D, E, or F, grade beams and beams that are part of a mat foundation subjected to flexure from columns that are part of the seismic-force-resisting system shall be in accordance with 18.6.

18.13.3.2 For structures assigned to SDC C, D, E, or F, slabs-on-ground that resist in-plane earthquake forces from walls or columns that are part of the seismic-force-resisting system shall be designed as diaphragms in accordance with 18.12. The construction documents shall clearly indicate that the slab-on-ground is a structural diaphragm and part of the seismic-force-resisting system.

18.13.4 *Foundation seismic ties*

18.13.4.1 For structures assigned to SDC C, D, E, or F, individual pile caps, piers, or caissons shall be interconnected by foundation seismic ties in orthogonal directions, unless it can be demonstrated that equivalent restraint is provided by other means.

COMMENTARY

R18.13.2.4 Columns or boundary members supported close to the edge of the foundation, as often occurs near property lines, should be detailed to prevent an edge failure of the footing, pile cap, or mat.

R18.13.2.5 The purpose of this section is to emphasize that top reinforcement in footings, mats, and pile caps may be required, in addition to other required reinforcement.

R18.13.2.6 Foundation and basement walls should be reinforced in buildings assigned to SDC D, E, or F.

R18.13.2.7 Batter piles typically attract higher lateral forces during earthquakes than vertical piles. Extensive structural damage has been observed at the junction of batter piles and building foundations. The pile cap and surrounding structure should be designed for the potentially large forces that can be developed in batter piles.

R18.13.3 *Grade beams and slabs-on-ground*

For earthquake conditions, slabs-on-ground (soil-supported slabs) are often part of the lateral-force-resisting system and should be designed in accordance with this Code as well as other appropriate standards or guidelines (refer to 1.4.8).

R18.13.3.1 Grade beams resisting flexural stresses from column moments should have reinforcement details similar to the beams of the frame above the foundation.

R18.13.3.2 Slabs-on-ground often act as a diaphragm to tie the building together at the ground level and minimize the effects of out-of-phase ground motion that may occur over the footprint of the building. The construction documents should clearly state that these slabs-on-ground are structural members so as to prohibit saw cutting of the slab.

R18.13.4 *Foundation seismic ties*

R18.13.4.1 The foundation seismic ties should sufficiently interconnect foundations to act as a unit and be designed to minimize the relative movement of an individual column or tie relative to the foundation. This is essential where surface soils are soft enough to require deep foundations or where the site soils are susceptible to liquefaction.

CODE

18.13.4.2 For structures assigned to SDC D, E, or F, individual spread footings founded on soil defined in ASCE/SEI 7 as Site Class E or F shall be interconnected by foundation seismic ties.

18.13.4.3 Where required, foundation seismic ties shall have a design strength in tension and compression at least equal to $0.1S_{DS}$ times the greater of the pile cap or column factored dead load plus factored live load unless it is demonstrated that equivalent restraint will be provided by (a), (b), (c), or (d):

(a) Reinforced concrete beams within the slab-on-ground
(b) Reinforced concrete slabs-on-ground
(c) Confinement by competent rock, hard cohesive soils, or very dense granular soils
(d) Other means approved by the building official

18.13.4.4 For structures assigned to SDC D, E, or F, grade beams designed to act as horizontal foundation seismic ties between pile caps or footings shall have continuous longitudinal reinforcement that shall be developed within or beyond the supported column or anchored within the pile cap or footing at all discontinuities and shall satisfy (a) and (b):

(a) The smallest cross-sectional dimension of the grade beam shall be at least equal to the clear spacing between connected columns divided by 20, but need not exceed 18 in.
(b) Closed tie transverse reinforcement shall be provided at a spacing not to exceed the lesser of 0.5 times the smallest orthogonal cross-sectional dimension and 12 in.

18.13.5 *Deep foundations*

18.13.5.1 This section shall apply to the following types of deep foundations

(a) Uncased cast-in-place concrete drilled or augered piles
(b) Metal cased concrete piles
(c) Concrete filled pipe piles
(d) Precast concrete piles

18.13.5.2 For structures assigned to SDC C, D, E, or F, piles, piers, or caissons resisting tension loads shall have continuous longitudinal reinforcement over their length to resist design tension forces.

18.13.5.3 For structures assigned to SDC C, D, E, or F, the minimum longitudinal and transverse reinforcement required by 18.13.5.7 through 18.13.5.10 shall be extended over the entire unsupported length for the portion of deep foundation member in air or water, or in soil that is not capable of providing adequate lateral restraint to prevent buckling throughout this length.

COMMENTARY

R18.13.4.2 The ties between footings should have the same characteristics as the ties between pile caps in R18.13.4.1.

R18.13.4.3 The minimum foundation seismic tie design strength requirement based on a percentage of the factored dead plus live load provides a minimum connection between foundation elements. Other types of restraint can be used if substantiated as equivalent to the minimum tie design strength. The required design strength for the tie beam must be at least equal to $\mathbf{0.1}\boldsymbol{S_{DS}}$ times the larger force on either end of the tie beam, and that force is from the column or pile cap, whichever applies.

R18.13.5 *Deep foundations*

Adequate performance of piles and caissons for earthquake effects requires that these provisions be met in addition to other applicable standards or guidelines (refer to R1.4.7).

R18.13.5.3 Minimum reinforcement lengths for both longitudinal and transverse reinforcement are based on the assumption that soil is capable of providing lateral support. For portions of the pile above ground, typically in air or water, or where soil is not capable of providing this lateral restraint, the minimum reinforced lengths should be increased, and the member should be designed as a column.

CODE

18.13.5.4 For structures assigned to SDC C, D, E, or F, hoops, spirals, and ties in deep foundation members shall be terminated with seismic hooks.

18.13.5.5 For structures assigned to SDC D, E, or F or located in Site Class E or F, concrete deep foundation members shall have transverse reinforcement in accordance with 18.7.5.2, 18.7.5.3, and Table 18.7.5.4 Item (e) within seven member diameters above and below the interfaces between strata that are hard or stiff and strata that are liquefiable or soft.

18.13.5.6 For structures assigned to SDC D, E, or F, in foundations supporting one- and two-story stud bearing wall construction, concrete piles, piers or caissons, and foundation ties are exempt from the transverse reinforcement requirements of 18.13.5.3 through 18.13.5.5.

18.13.5.7 *Uncased cast-in-place drilled or augered concrete piles or piers*

18.13.5.7.1 For structures assigned to SDC C, D, E, or F, reinforcement shall be provided in uncased cast-in-place drilled or augered concrete piles where required by analysis and in accordance with the requirements in Table 18.13.5.7.1.

COMMENTARY

R18.13.5.5 During earthquakes, piles can be subjected to high flexural and shear demands at points of discontinuity, such as at interfaces between stiff and soft soil strata. ASCE/SEI 7 defines limits for soil strata. Transverse reinforcement is required in these regions to provide ductile behavior. In determining the portions of a pile with increased transverse reinforcement, accommodations are often made to the length of the reinforced zone for transverse reinforcement to account for variations in the driven pile tip elevations and variations in the interface elevations between stiff and soft soil strata.

R18.13.5.7 *Uncased cast-in-place drilled or augered concrete piles or piers*

R18.13.5.7.1 Longitudinal and transverse reinforcement requirements prescribed by this section result in ductility consistent with the applicable Seismic Design Category (SDC) to withstand ground deformation that occurs during earthquakes.

Where piles are subjected to significant uplift forces, the longitudinal reinforcement length required by analysis may exceed the minimum reinforcement length requirements.

Transverse reinforcement is required at the top of the pile to provide ductile performance where flexural yielding can potentially occur. For SDC D, E, and F and Site Classes A, B, C, and D, one-half of the transverse reinforcement for special moment frame columns is acceptable because some level of confinement is attributed to competent soils. For Site Class E and F, full column confinement is required because the soils are either liquefiable or not considered competent enough to provide confinement.

CODE | COMMENTARY

Table 18.13.5.7.1—Minimum reinforcement for uncased cast-in-place or augered concrete piles or piers

<table>
<tr><th colspan="2">Minimum reinforcement</th><th>SDC C –
All Site Classes</th><th>SDC D, E, and F –
Site Class A, B, C, and D</th><th>SDC D, E, and F –
Site Class E and F</th></tr>
<tr><td colspan="2">Minimum longitudinal reinforcement ratio (minimum number of bars)</td><td>0.0025
(minimum number of bars in accordance with 10.7.3.1)</td><td>0.005
(minimum number of bars in accordance with 10.7.3.1)</td><td>0.005
(minimum number of bars in accordance with 10.7.3.1)</td></tr>
<tr><td colspan="2">Minimum reinforced pile length</td><td>Longest of (a) through (d):
(a) 1/3 pile length
(b) 10 ft
(c) 3 times the pile diameter
(d) Flexural length of pile - distance from bottom of pile cap to where $0.4M_{cr}$ exceeds M_u</td><td>Longest of (a) through (d):
(a) 1/2 pile length
(b) 10 ft
(c) 3 times the pile diameter
(d) Flexural length of pile - distance from bottom of pile cap to where $0.4M_{cr}$ exceeds M_u</td><td>Full length of pile except in accordance with [1] or [2].</td></tr>
<tr><td rowspan="4">Transverse confinement reinforcement zone</td><td>Length of reinforcement zone</td><td>3 times the pile diameter from the bottom of the pile cap</td><td>3 times the pile diameter from the bottom of the pile cap</td><td>7 times the pile diameter from the bottom of the pile cap</td></tr>
<tr><td rowspan="2">Type of transverse reinforcement</td><td rowspan="2">Closed ties or spirals with a minimum 3/8 in. diameter</td><td colspan="2">Minimum of No. 3 closed tie or 3/8 in. diameter spiral for piles $\leq$ 20 in. diameter
Minimum No. 4 closed tie or 1/2 in. diameter spiral for piles $>$ 20 in. diameter</td></tr>
<tr><td colspan="2">In accordance with 18.7.5.2</td></tr>
<tr><td>Spacing and amount of transverse reinforcement</td><td>Spacing shall not exceed lesser of 6 in. or 8 longitudinal bar diameters</td><td>In accordance with 18.7.5.3 and not less than one-half the requirement of Table 18.7.5.4 Item (e)</td><td>In accordance with 18.7.5.3 and not less than the requirement of Table 18.7.5.4 Item (e).</td></tr>
<tr><td rowspan="3">Transverse reinforcement in remainder of reinforced pile length</td><td rowspan="2">Type of transverse reinforcement</td><td rowspan="2">Closed ties or spirals with minimum 3/8 in. diameter</td><td colspan="2">Minimum of No. 3 closed tie or 3/8 in. diameter spiral for piles $\leq$ 20 in. diameter
Minimum of No. 4 closed tie or 1/2 in. diameter spiral for piles $>$ 20 in. diameter</td></tr>
<tr><td colspan="2">In accordance with 18.7.5.2</td></tr>
<tr><td>Spacing and amount of transverse reinforcement</td><td>Maximum spacing of 16 longitudinal bar diameters</td><td colspan="2">Spacing shall not exceed the least of (a) through (c):
(a) 12 longitudinal bar diameters
(b) 1/2 the pile diameter
(c) 12 in.</td></tr>
</table>

[1] For piles sufficiently embedded in firm soil or rock, reinforcement shall be permitted to be terminated a length above the tip equal to the lesser of 5 percent of the pile length and 33 percent of the length of the pile within rock or firm soil.

[2] In lieu of providing full length minimum flexural reinforcement, the deep foundation element shall be designed to withstand maximum imposed curvatures from the earthquake ground motions and structural response. Curvatures shall include free-field soil strains modified for soil-foundation-structure interaction coupled with foundation element deformations associated with earthquake loads imparted to the foundation by the structure. Minimum reinforced length shall not be less than the requirement for SDC D, E, or F; Site Class D.

18.13.5.7.2 Minimum longitudinal and transverse reinforcement shall be provided along minimum reinforced lengths measured from the top of the pile in accordance with Table 18.13.5.7.1.

18.13.5.7.3 Longitudinal reinforcement shall extend at least the development length in tension beyond the flexural length of the pile, which is defined in Table 18.13.5.7.1 as the distance from the bottom of the pile cap to where $\mathbf{0.4}\boldsymbol{M_{cr}} > \boldsymbol{M_u}$.

R18.13.5.7.3 Reinforcement should extend ℓ_d beyond the point where plain concrete is no longer adequate to resist the factored moment.

18.13.5.8 *Metal-cased concrete piles*

R18.13.5.8 *Metal-cased concrete piles*

18.13.5.8.1 For structures assigned to SDC C, D, E, or F, longitudinal reinforcement requirements and minimum reinforced lengths for metal-cased concrete piles shall be the same as for uncased concrete piles in 18.13.5.7.

18.13.5.8.2 Metal-cased concrete piles shall have a spiral-welded metal casing of a thickness not less than 0.0747 in.

R18.13.5.8.2 Spiral-welded metal casing with the specified wall thickness provides confinement equivalent to

CODE

(No. 14 gauge) that is adequately protected from possible deleterious action due to soil constituents, changing water levels, or other factors indicated by boring records of site conditions.

18.13.5.9 *Concrete-filled pipe piles*

18.13.5.9.1 For structures assigned to SDC C, D, E or F, concrete-filled pipe piles shall have longitudinal reinforcement in the top of the pile with a total area of at least $\mathbf{0.01}A_g$ and with a minimum length within the pile equal to two times the required embedment length into the pile cap, but not less than the development length in tension of the reinforcement.

18.13.5.10 *Precast concrete piles*

18.13.5.10.1 For precast concrete driven piles, the length of transverse reinforcement provided shall be sufficient to account for potential variations in the elevation of pile tips.

18.13.5.10.2 Precast nonprestressed concrete piles for structures assigned to SDC C shall satisfy (a) through (d):

(a) Minimum longitudinal steel reinforcement ratio shall be 0.01.
(b) Longitudinal reinforcement shall be enclosed within a minimum of No. 3 closed ties or 3/8 in. diameter spirals, for up to 20 in. diameter piles, and No. 4 closed ties or 1/2 in. diameter spirals, for larger diameter piles.
(c) Spacing of transverse reinforcement within a distance of 3 times the least cross-sectional dimension of the pile from the bottom of the pile cap shall not exceed the lesser of 8 times the diameter of the smallest longitudinal bar and 6 in.
(d) Transverse reinforcement shall be provided throughout the length of the pile at a spacing not exceeding 6 in.

18.13.5.10.3 For structures assigned to SDC D, E, or F, precast nonprestressed concrete piles shall satisfy the requirements of 18.13.5.10.2 and the requirements for uncased cast-in-place or augered concrete piles in SDC D, E, or F in Table 18.13.5.7.1.

18.13.5.10.4 For structures assigned to SDC C, precast-prestressed concrete piles shall satisfy (a) and (b):

(a) If the transverse reinforcement consists of spirals or circular hoops, the volumetric ratio of transverse rein-

COMMENTARY

closed ties or spirals required in an uncased concrete pile and eliminates the need for confinement ties.

R18.13.5.9 *Concrete-filled pipe piles*

R18.13.5.9.1 For resistance to uplift forces, concrete bond to the steel pipe is to be ignored in determining anchorage of the pile. Concrete shrinkage can be detrimental to bond, therefore shrinkage should be controlled, or force transfer via other methods such as headed studs or surface irregularities on the pipe should be considered. Reinforcement at the top of the pile is extended into the pile cap to tie the elements together and assist transfer of force to the pile cap.

R18.13.5.10 *Precast concrete piles*

R18.13.5.10.1 The potential for driving precast piles to a tip elevation different than that specified in the construction documents should be considered when detailing the pile. If the pile reaches refusal at a shallower depth, a longer length of pile will need to be cut off. If this possibility is not foreseen, the length of transverse reinforcement required by these provisions may not be provided after the excess pile length is cut off.

R18.13.5.10.4(a) In a study of minimum confinement reinforcement for prestressed concrete piles (Sritharan et al.

CODE

forcement, ρ_s, in the upper 20 ft shall not be less than that calculated by Eq. (18.13.5.10.4a) or calculated from a more detailed analysis by Eq. (18.13.5.10.4b):

$$0.15\left(\frac{f_c'}{f_{yt}}\right) \quad (18.13.5.10.4a)$$

$$0.04\left(\frac{f_c'}{f_{yt}}\right)\left(2.8+\frac{2.3P_u}{f_c'A_g}\right) \quad (18.13.5.10.4b)$$

and f_{yt} shall not be taken greater than 100,000 psi.

(b) A minimum of one-half of the volumetric ratio of spiral reinforcement required by Eq. (18.13.5.10.4a) or Eq. (18.13.5.10.4b) shall be provided for the remaining length of the pile.

18.13.5.10.5 For structures assigned to SDC D, E, or F, precast-prestressed concrete piles shall satisfy (a) through (e) and the ductile pile region shall be defined as the length of pile measured from the bottom of the pile cap to the point of zero curvature plus 3 times the least pile dimension, but not less than 35 ft. If the total pile length in the soil is 35 ft or less, the ductile pile region shall be taken as the entire length of the pile:

(a) In the ductile pile region, the center-to-center spacing of spirals or hoop reinforcement shall not exceed the least of 0.2 times the least pile dimension, 6 times the diameter of the longitudinal strand, and 6 in.

(b) Spiral reinforcement shall be spliced by lapping one full turn, by welding, or by the use of a mechanical splice. If spiral reinforcement is lap spliced, the ends of the spiral shall terminate in a seismic hook. Mechanical and welded splices of deformed bars shall comply with 25.5.7.

(c) If the transverse reinforcement consists of spirals, or circular hoops, the volumetric ratio of transverse reinforcement, ρ_s, in the ductile pile region shall not be less than that calculated by Eq. (18.13.5.10.5a) or calculated from a more detailed analysis by Eq. (18.13.5.10.5b), and the required volumetric ratio shall be permitted to be obtained by providing an inner and outer spiral.

$$0.2\left(\frac{f_c'}{f_{yt}}\right) \quad (18.13.5.10.5a)$$

$$0.06\left(\frac{f_c'}{f_{yt}}\right)\left(2.8+\frac{2.3P_u}{f_c'A_g}\right) \quad (18.13.5.10.5b)$$

and f_{yt} shall not be taken as greater than 100,000 psi.

(d) Outside of the ductile pile region, spiral or hoop reinforcement shall be provided with a volumetric ratio not

COMMENTARY

2016), the relationship between curvature ductility demand on prestressed piles and overall system ductility demand was considered in the context of all soil profiles identified in ASCE/SEI 7. It was concluded that Eq. (18.13.5.10.4b) results in adequate deformation capacity for structures assigned to SDC C. The factored axial force on a pile should be determined from Eq. (5.3.1c) and Eq. (5.3.1g) with 5.3.7 and 5.3.8 as applicable.

R18.13.5.10.5 Observed damage from earthquakes and concerns about the accuracy of calculated pile demands have led to prescriptive requirements for confinement of potential yielding regions of piles. The required confinement is intended to provide adequate ductility capacity for structures assigned to SDC D, E, and F (Sritharan et al. 2016).

CODE

less than one-half of that required within the ductile pile region, and the maximum spacing shall be in accordance with Table 13.4.5.6(b).

(e) If transverse reinforcement consists of rectangular hoops and crossties, the total cross-sectional area of lateral transverse reinforcement in the ductile region shall be the greater of Eq. (18.13.5.10.5c) and Eq. (18.13.5.10.5d). The hoops and crossties shall be equivalent to deformed bars not less than No. 3 in size, and rectangular hoop ends shall terminate at a corner with seismic hooks.

$$A_{sh} = 0.3sb_c\left(\frac{f_c'}{f_{yt}}\right)\left(\frac{A_g}{A_{ch}} - 1.0\right)\left(0.5 + \frac{1.4P_u}{f_c'A_g}\right) \quad (18.13.5.10.5c)$$

$$A_{sh} = 0.12sb_c\left(\frac{f_c'}{f_{yt}}\right)\left(0.5 + \frac{1.4P_u}{f_c'A_g}\right) \quad (18.13.5.10.5d)$$

and f_{yt} shall not be taken as greater than 100,000 psi.

18.13.5.10.6 For structures assigned to SDC C, D, E, or F, the maximum factored axial load for precast prestressed piles subjected to a combination of earthquake lateral force and axial load shall not exceed the following values:

(a) $\mathbf{0.2}f_c'A_g$ for square piles
(b) $\mathbf{0.4}f_c'A_g$ for circular or octagonal piles

COMMENTARY

R18.13.5.10.6 The axial load in precast prestressed piles is limited to preclude spalling of the concrete cover prior to the pile section experiencing flexural cracking, as this will result in a significant loss in pile resistance (Sritharan et al. 2016).

18.13.6 *Anchorage of piles, piers, and caissons*

18.13.6.1 For structures assigned to SDC C, D, E, or F, the longitudinal reinforcement in piles, piers, or caissons resisting tension loads shall be detailed to transfer tension forces within the pile cap to supported structural members.

18.13.6.2 For structures assigned to SDC C, D, E, or F, concrete piles and concrete filled pipe piles shall be connected to the pile cap by embedding the pile reinforcement in the pile cap a distance equal to the development length or by the use of field-placed dowels anchored in the concrete pile. For deformed bars, the compression development length is used if the pile is in compression. In the case of uplift, the tension development length is used without reduction in length for excess reinforcement.

18.13.6.3 For structures assigned to SDC D, E, or F, if tension forces induced by earthquake effects are transferred between pile cap or mat foundation and precast pile by reinforcing bars grouted or post-installed in the top of the pile, the grouting system shall have been demonstrated by testing to develop at least $\mathbf{1.25}f_y$ of the bar.

R18.13.6 *Anchorage of piles, piers, and caissons*

R18.13.6.1 A load path is necessary at pile caps to transfer tension forces from the reinforcing bars in the column or boundary element through the pile cap to the reinforcement of the pile or caisson. Examples of different types of pile connections to pile caps are available in ASCE/COPRI Standard for the Seismic Design of Piers and Wharves (61-14).

R18.13.6.2 Development length is determined according to requirements of Chapter 25. Reductions in development length for calculated stresses less than f_y are not permitted, as indicated in 25.4.10.2. Full development of the pile longitudinal reinforcement into the pile cap is intended to enable the capacity of the pile to pile cap connection to meet or exceed the pile section strength.

R18.13.6.3 Grouted dowels in a blockout in the top of a precast concrete pile need to be developed, and testing is a practical means of demonstrating strength. Alternatively, reinforcing bars can be cast in the upper portion of the pile, exposed by chipping of concrete and mechanically spliced or welded to an extension.

CODE

18.14—Members not designated as part of the seismic-force-resisting system

18.14.1 *Scope*

18.14.1.1 This section shall apply to members not designated as part of the seismic-force-resisting system in structures assigned to SDC D, E, and F.

18.14.2 *Design actions*

18.14.2.1 Members not designated as part of the seismic-force-resisting system shall be evaluated for gravity load combinations of 5.3 including the effect of vertical ground motion acting simultaneously with the design displacement δ_u.

18.14.3 *Cast-in-place beams, columns, and joints*

18.14.3.1 Cast-in-place beams, columns, and joints shall be detailed in accordance with 18.14.3.2 or 18.14.3.3 depending on the magnitude of moments and shears induced in those members when subjected to the design displacement δ_u. If effects of δ_u are not explicitly checked, the provisions of 18.14.3.3 shall be satisfied.

18.14.3.2 Where the induced moments and shears do not exceed the design moment and shear strength of the frame member, (a) through (d) shall be satisfied:

(a) Beams shall satisfy 18.6.3.1. Transverse reinforcement shall be provided throughout the length of the beam at spacing not to exceed $d/2$. Where factored axial force exceeds $A_g f_c'/10$, transverse reinforcement shall be hoops satisfying 18.7.5.2 at a spacing not to exceed the lesser of $6d_b$ of the smallest enclosed longitudinal bar and 6 in.
(b) Columns shall satisfy 18.7.4.1 and 18.7.6. Spiral reinforcement satisfying 25.7.3 or hoop reinforcement satisfying 25.7.4 shall be provided over the full length of the column with spacing not to exceed the lesser of $6d_b$ of the smallest enclosed longitudinal bar and 6 in. Transverse reinforcement satisfying 18.7.5.2(a) through (e) shall be provided over a length ℓ_o, as defined in 18.7.5.1, from each joint face.
(c) Columns with factored gravity axial forces exceeding $0.35P_o$ shall satisfy 18.14.3.2(b) and 18.7.5.7. The minimum amount of transverse reinforcement provided shall be, for rectilinear hoops, one-half the greater of Table 18.7.5.4 parts (a) and (b) and, for spiral or circular hoops, one-half the greater of Table 18.7.5.4 parts (d) and (e). This transverse

COMMENTARY

R18.14—Members not designated as part of the seismic-force-resisting system

This section applies only to structures assigned to SDC D, E, or F. For those SDCs, all structural members not designated as a part of the seismic-force-resisting system are required to be designed to support gravity loads and the load effects of vertical ground motion, while subjected to the design displacement. For concrete structures, the provisions of this section satisfy this requirement for columns, beams, slabs, and wall piers of the gravity system.

Design displacement is defined in Chapter 2. Models used to determine design displacement of buildings should be chosen to produce results that conservatively bound the values expected during the design earthquake and should include, as appropriate, effects of concrete cracking, foundation flexibility, and deformation of floor and roof diaphragms.

The provisions of 18.14 are intended to enable ductile flexural yielding of columns, beams, slabs, and wall piers under the design displacement, by providing sufficient confinement and shear strength in elements that yield.

R18.14.3 *Cast-in-place beams, columns, and joints*

R18.14.3.1 Cast-in-place columns and beams are assumed to yield if the combined effects of factored gravity loads and design displacements exceed the strengths specified, or if the effects of design displacements are not calculated. Requirements for transverse reinforcement and shear strength vary with member type and whether the member yields under the design displacement.

CODE

reinforcement shall be provided over a length ℓ_o, as defined in 18.7.5.1, from each joint face.
(d) Joints shall satisfy Chapter 15.

18.14.3.3 Where the induced moments or shears exceed ϕM_n or ϕV_n of the frame member, or if induced moments or shears are not calculated, (a) through (d) shall be satisfied:

(a) Materials, mechanical splices, and welded splices shall satisfy the requirements for special moment frames in 18.2.5 through 18.2.8.
(b) Beams shall satisfy 18.14.3.2(a) and 18.6.5.
(c) Columns shall satisfy 18.7.4, 18.7.5, and 18.7.6.
(d) Joints shall satisfy 18.4.4.1.

18.14.4 *Precast beams and columns*

18.14.4.1 Precast concrete frame members assumed not to contribute to lateral resistance, including their connections, shall satisfy (a) through (d):

(a) Requirements of 18.14.3
(b) Ties specified in 18.14.3.2(b) over the entire column height, including the depth of the beams
(c) Structural integrity reinforcement, in accordance with 4.10
(d) Bearing length at the support of a beam shall be at least 2 in. longer than determined from 16.2.6

18.14.5 *Slab-column connections*

18.14.5.1 For slab-column connections of two-way slabs without beams, slab shear reinforcement satisfying the requirements of 18.14.5.3 and either 8.7.6 or 8.7.7 shall be provided at any slab critical section defined in 22.6.4.1 for the following conditions:

(a) Nonprestressed slabs where $\Delta_x/h_{sx} \geq 0.035 - (1/20)(v_{uv}/\phi v_c)$
(b) Unbonded post-tensioned slabs with f_{pc} in each direction meeting the requirements of 8.6.2.1, where $\Delta_x/h_{sx} \geq 0.040 - (1/20)(v_{uv}/\phi v_c)$

The load combinations to be evaluated for v_{uv} shall only include those with E. The value of (Δ_x/h_{sx}) shall be taken as the greater of the values of the adjacent stories above and below the slab-column connection, v_c shall be calculated in accordance with 22.6.5; and, for unbonded post-tensioned slabs, the value of V_p shall be taken as zero when calculating v_c.

COMMENTARY

R18.14.4 *Precast beams and columns*

R18.14.4.1 Damage to some buildings with precast concrete gravity systems during the 1994 Northridge earthquake was attributed to several factors addressed in this section. Columns should contain ties over their entire height, frame members not proportioned to resist earthquake forces should be tied together, and longer bearing lengths should be used to maintain integrity of the gravity system during ground motion. The 2 in. increase in bearing length is based on an assumed 4 percent story drift ratio and 50 in. beam depth, and is considered to be conservative for the ground motions expected for structures assigned to SDC D, E, or F. In addition to this provision, precast frame members assumed not to contribute to lateral resistance should also satisfy the requirements for cast-in-place construction addressed in 18.14.3, as applicable.

R18.14.5 *Slab-column connections*

R18.14.5.1 Provisions for shear reinforcement at slab-column connections are intended to reduce the likelihood of slab punching shear failure if the design story drift ratio exceeds the value specified.

No calculation of induced moments is required, based on research (Megally and Ghali 2002; Moehle 1996; Kang and Wallace 2006; Kang et al. 2007) that identifies the likelihood of punching shear failure considering the story drift ratio and shear stress v_{uv} due to gravity loads and the vertical component of earthquake loads, without moment transfer, about the slab critical section. Figure R18.14.5.1 illustrates the requirement for nonprestressed and unbonded post-tensioned slab-column connections. The requirement can be satisfied by adding slab shear reinforcement, increasing slab thickness, changing the design to reduce the design story drift ratio, or a combination of these.

If column capitals, drop panels, shear caps, or other changes in slab thickness are used, the requirements of 18.14.5 are evaluated at all potential critical sections, as required by 22.6.5.1.

CODE

COMMENTARY

Post-tensioned slab-column connections with f_{pc} in each direction not meeting the requirements of 8.6.2.1 can be designed as nonprestressed slab-column connections in accordance with 8.2.3.

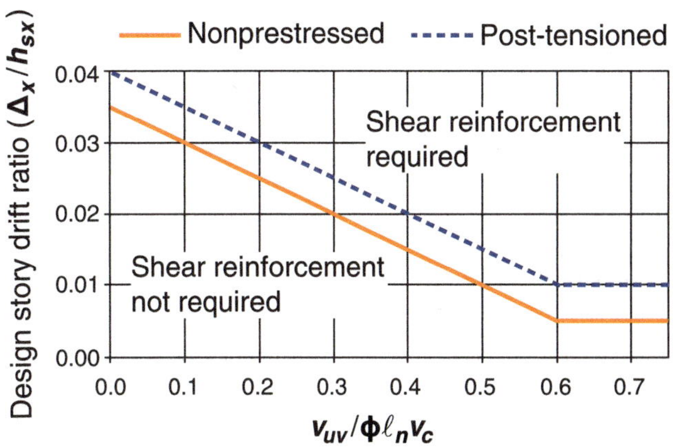

Fig. R18.14.5.1—Illustration of the criteria of 18.14.5.1.

18.14.5.2 The shear reinforcement requirements of 18.14.5.1 need not be satisfied if (a) or (b) is met:

(a) $\Delta_x/h_{sx} \leq 0.005$ for nonprestressed slabs
(b) $\Delta_x/h_{sx} \leq 0.01$ for unbonded post-tensioned slabs with f_{pc} in each direction meeting the requirements of 8.6.2.1

18.14.5.3 Required slab shear reinforcement shall provide $v_s \geq 3.5\sqrt{f_c'}$ at the slab critical section and shall extend at least four times the slab thickness from the face of the support adjacent to the slab critical section.

18.14.6 *Wall piers*

18.14.6.1 Wall piers not designated as part of the seismic-force-resisting system shall satisfy the requirements of 18.10.8. Where the general building code includes provisions to account for overstrength of the seismic-force-resisting system, it shall be permitted to calculate the design shear force as Ω_o times the shear induced under design displacements, δ_u.

R18.14.6 *Wall piers*

R18.14.6.1 Section 18.10.8 requires that the design shear force be determined according to 18.7.6.1, which in some cases may result in unrealistically large forces. As an alternative, the design shear force can be determined as the product of an overstrength factor and the shear induced when the wall pier is displaced by δ_u. The overstrength factor Ω_o included in FEMA P749, ASCE/SEI 7, and the 2018 IBC can be used for this purpose.

Notes

Factors of Safety for Retaining Walls Example:

$\phi = 37^\circ$ $\gamma = 120$ pcf $q_{ult} = 9000$ psf $\delta = 24^\circ$

$C_A = 0$ psf

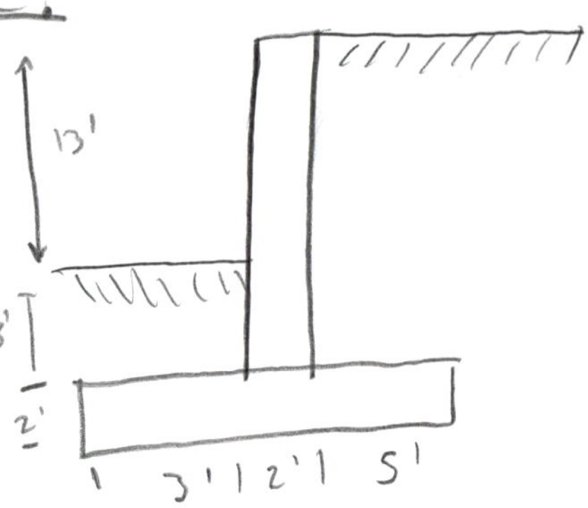

$$k_p = \frac{1+\sin(\phi')}{1-\sin(\phi')} = \frac{1+\sin(37)}{1-\sin(37)} = 4.02$$

$$k_a = \frac{1}{k_p} = 0.25$$

Normal force

Shape	L	H	W	V	γ	W
1	2	16	1	32	120	4800
2	10	2	1	20	120	3000
3	5	16	1	40	120	9600
4	3	3	1	9	120	1040
Sum						14,440

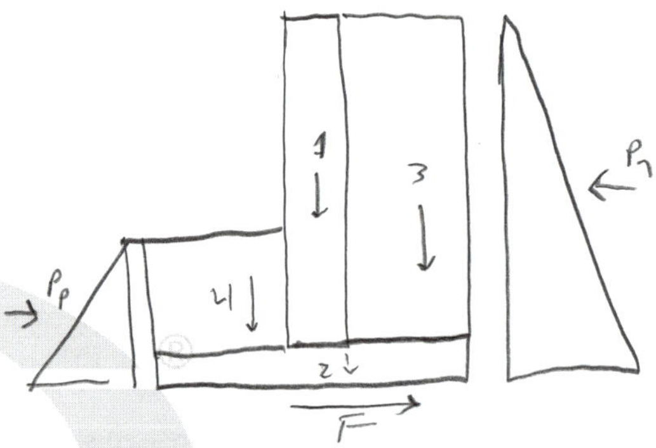

$P_a = \frac{1}{2}k_a \gamma H^2 = \frac{1}{2}(0.25)(125)(18')^2$ ← active p, goes to base

$P_p = \frac{1}{2}(4.02)(120)(5')^2 = 6030$

$F = C_A A + \tan\delta(\Sigma W) = 0[(10)(1)] + \tan(24)(14,440) = 8228$

$FS_{P_p} = \frac{F+P_P}{P_a}$ $FS_{nPP} = \frac{F}{P_a}$

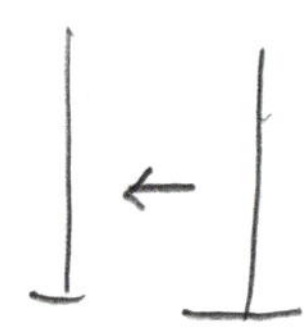

sliding

M_o

Force	F (lbs)	M arm (ft)	M
Pa	4860	$6 = \frac{1}{3}(18)$	29,160

Overturning

M_r

Shape	F (lbs)	Moment arm (ft)	Moment
1	4800	4	19,200
2	3000	5	15,000
3	9600	2.5	72,000
4	1080	1.5	1,620
Sum			107,820

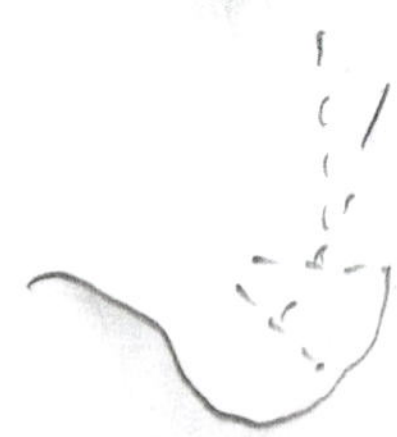

Bearing capacity

$FS = \frac{M_r}{M_o}$

Bearing: $\Sigma M_{@\,toes} = 0 = M_r - M_o - Q\bar{x} \Rightarrow \bar{x} = \frac{M_r - M_o}{\Sigma W} = \frac{107,820 - 29160}{18,480} = 4.25$

$e = \frac{B}{2} - \bar{x} = \frac{10}{2} - 4.25 = 0.75'$

Check kern: $\frac{B}{6} = \frac{10}{6} = 1.67 > e$ ∴ OK

$$q = \frac{\Sigma W}{B(1')} \pm \frac{[\Sigma W(e)](\frac{B}{2})}{\frac{1}{12}(1)(B)^3}$$ take max $FS = \frac{q_{ult}}{q_{max}}$

CHAPTER 19—CONCRETE: DESIGN AND DURABILITY REQUIREMENTS

CODE

19.1—Scope

19.1.1 This chapter shall apply to concrete, including:

(a) Properties to be used for design
(b) Durability requirements

19.1.2 This chapter shall apply to durability requirements for grout used for bonded tendons in accordance with 19.4.

19.2—Concrete design properties

19.2.1 *Specified compressive strength*

19.2.1.1 The value of f_c' shall be in accordance with (a) through (d):

(a) Limits for f_c' in Table 19.2.1.1. Limits apply to both normalweight and lightweight concrete.
(b) Durability requirements in Table 19.3.2.1
(c) Structural strength requirements
(d) f_c' for lightweight concrete in special moment frames and special structural walls, and their foundations, shall not exceed 5000 psi, unless demonstrated by experimental evidence that members made with lightweight concrete provide strength and toughness equal to or exceeding those of comparable members made with normalweight concrete of the same strength.

Table 19.2.1.1—Limits for f_c'

Application	Minimum f_c', psi
General	2500
Foundations for structures assigned to SDC A, B, or C	2500
Foundations for Residential and Utility use and occupancy classification with stud bearing wall construction two stories or less assigned to SDC D, E, or F	2500
Foundations for structures assigned to SDC D, E, or F other than Residential and Utility use and occupancy classification with stud bearing wall construction two stories or less	3000
Special moment frames Special structural walls with Grade 60 or 80 reinforcement	3000
Special structural walls with Grade 100 reinforcement	5000
Precast-nonprestressed driven piles Drilled shafts	4000
Precast-prestressed driven piles	5000

19.2.1.2 The specified compressive strength shall be used for proportioning of concrete mixtures in 26.4.3 and for testing and acceptance of concrete in 26.12.3.

19.2.1.3 Unless otherwise specified, f_c' shall be based on 28-day tests. If other than 28 days, test age for f_c' shall be indicated in the construction documents.

19.2.2 *Modulus of elasticity*

COMMENTARY

R19.2—Concrete design properties

R19.2.1 *Specified compressive strength*

Requirements for concrete mixtures are based on the philosophy that concrete should provide both adequate strength and durability. The Code defines a minimum value of f_c' for structural concrete. There is no limit on the maximum value of f_c' except as required by specific Code provisions.

Concrete mixtures proportioned in accordance with 26.4.3 should achieve an average compressive strength that exceeds the value of f_c' used in the structural design calculations. The amount by which the average strength of concrete exceeds f_c' is based on statistical concepts. When concrete is designed to achieve a strength level greater than f_c', it ensures that the concrete strength tests will have a high probability of meeting the strength acceptance criteria in 26.12.3. The durability requirements prescribed in Table 19.3.2.1 are to be satisfied in addition to meeting the minimum f_c' of 19.2.1. Under some circumstances, durability requirements may dictate a higher f_c' than that required for structural purposes.

Available test data do not include lower strength concrete with Grade 100 reinforcement in special structural walls (refer to R18.2.6).

For design of special moment frames and special structural walls used to resist earthquake forces, the Code limits the maximum f_c' of lightweight concrete to 5000 psi. This limit is imposed primarily because of a paucity of experimental and field data on the behavior of members made with lightweight concrete subjected to displacement reversals in the nonlinear range.

Minimum concrete strengths are increased for special seismic systems with f_y > **80,000 psi** to enhance bar anchorage and reduce the neutral axis depth for improved performance.

The Code also limits f_c' for design of anchors to concrete. The requirements are in 17.3.1.

R19.2.2 *Modulus of elasticity*

Concrete

19

CODE

19.2.2.1 It shall be permitted to calculate E_c in accordance with (a) or (b):

(a) For values of w_c between 90 and 160 lb/ft^3

$$E_c = w_c^{1.5} 33 \sqrt{f_c'} \text{ (in psi)} \quad (19.2.2.1.a)$$

(b) For normalweight concrete

$$E_c = 57{,}000 \sqrt{f_c'} \text{ (in psi)} \quad (19.2.2.1.b)$$

COMMENTARY

R19.2.2.1 Equations in 19.2.2.1 provide an estimate of E_c for general design use. Studies leading to the expression for E_c of concrete are summarized in Pauw (1960), where E_c is defined as the slope of the line drawn from a stress of zero to 45 percent of the compressive strength using the stress-strain curve of the concrete. This definition is slightly different than the definition in ASTM C469. ASTM C469 defines E_c using 40 percent of the compressive strength.

The modulus of elasticity is sensitive to a number of variables including aggregate type, concrete constituents, mixture proportions, bond between paste and aggregate, and the age of the concrete. This sensitivity, coupled with the inherent variability in the properties of the constituent materials and quality control exercised during construction, can result in differences between measured and calculated values for deflection, drift, periods of vibration, and other quantities that depend on E_c. Refer to ACI 435R for more information on the use of E_c, especially when used in deflection calculations.

Modulus of elasticity determined by calculation using the Code equations has been shown to be appropriate for most applications based on many years of use. For some applications, however, these equations may not provide sufficiently accurate estimates of actual values. Larger differences between measured and calculated values of E_c have been observed for high-strength concrete (**$f_c' >$ 8000 psi**), lightweight concrete, and for mixtures with low coarse aggregate volume, as can occur with self-consolidating concrete. Refer to ACI 363R, ACI 213R, and ACI 237R for more information.

19.2.2.2 It shall be permitted to specify E_c based upon testing of concrete mixtures to be used in the Work in accordance with (a) through (c):

(a) Specified E_c shall be used for proportioning concrete mixtures in accordance with 26.4.3.
(b) Testing to verify that the specified E_c has been achieved shall be conducted, and results shall be provided with the mixture submittal.
(c) Test age of measurement of E_c shall be 28 days or as indicated in the construction documents.

R19.2.2.2 For any project, E_c used for design may be specified and verified by testing. Design conditions that are sensitive to the value of E_c may warrant testing. Examples include applications where deflections are critical, tall buildings or similar structures for which axial deformation or lateral stiffness impact performance, and where estimation of E_c is important to acceptable vibration or seismic performance.

In cases where an unintended change of stiffness may have an adverse effect on the design, such as for some seismic applications, the licensed design professional may choose to specify a range of acceptable values of E_c at a specified test age. If a range of values of E_c is specified, details of a testing program and acceptance criteria should be provided in the construction documents.

The licensed design professional may choose to specify laboratory testing of E_c at multiple ages. It should be recognized that the development of E_c over time cannot be controlled with precision.

19.2.3 *Modulus of rupture*

19.2.3.1 Modulus of rupture, f_r, for concrete shall be calculated by:

$$f_r = 7.5\lambda \sqrt{f_c'} \quad (19.2.3.1)$$

where the value of λ is in accordance with 19.2.4.

CODE

19.2.4 *Lightweight concrete*

19.2.4.1 Except as required in Table 25.4.2.5, the value of λ shall be determined using Table 19.2.4.1(a) based on the equilibrium density, w_c, of the concrete mixture used in design, or Table 19.2.4.1(b) based on the composition of the aggregate in the concrete mixture assumed in the design.

Table 19.2.4.1(a)—Values of λ for lightweight concrete based on equilibrium density

w_c, lb/ft³	λ	
≤ 100	0.75	(a)
100 < w_c ≤ 135	0.0075w_c ≤ 1.0	(b)
> 135	1.0	(c)

Table 19.2.4.1(b)—Values of λ for lightweight concrete based on composition of aggregates

Concrete	Composition of aggregates	λ
All-lightweight	Fine: ASTM C330 Coarse: ASTM C330	0.75
Lightweight, fine blend	Fine: Combination of ASTM C330 and C33 Coarse: ASTM C330	0.75 to 0.85[1]
Sand-lightweight	Fine: ASTM C33 Coarse: ASTM C330	0.85
Sand-lightweight, coarse blend	Fine: ASTM C33 Coarse: Combination of ASTM C330 and C33	0.85 to 1[2]

[1]Linear interpolation from 0.75 to 0.85 is permitted based on the absolute volume of normalweight fine aggregate as a fraction of the total absolute volume of fine aggregate.

[2]Linear interpolation from 0.85 to 1 is permitted based on the absolute volume of normalweight coarse aggregate as a fraction of the total absolute volume of aggregate.

19.2.4.2 It shall be permitted to take λ as 0.75 for lightweight concrete.

19.2.4.3 The value of λ shall be taken as 1.0 for normalweight concrete.

COMMENTARY

R19.2.4 *Lightweight concrete*

The modification factor λ is used to account for the reduced mechanical properties of lightweight concrete compared with normalweight concrete of the same compressive strength. For design using lightweight concrete, shear strength, friction properties, splitting resistance, bond between concrete and reinforcement, and development length requirements are not taken as equivalent to normalweight concrete of the same compressive strength.

The methodology for determining λ was changed in the 2019 Code to include a new method that is based on the equilibrium density of the lightweight concrete. The new method allows the designer to select a value for λ based on the equilibrium density of the lightweight concrete that is used in design. Laboratory testing on the specific mixture to be used in the structure can be accomplished if the designer desires to determine a more accurate value of λ (Ivey and Buth 1967; Hanson 1961). Table 19.2.4.1 is based on data from tests (Graybeal 2014; Greene and Graybeal 2013, 2015) of concrete made with many types of structural lightweight aggregate and having a wide range of mixture proportions that resulted in equilibrium densities over a range of 90 to 135 lb/ft³.

The second method for determining λ, which is retained from the previous code, is based on the composition of aggregates. In most cases, local concrete and aggregate suppliers have standard lightweight concrete mixtures and can provide the volumetric fractions to determine the value of λ. In the absence of such data, it is permissible to use the lower-bound value of λ for the type of lightweight concrete specified. This method is based on the assumption that, for equivalent compressive strength levels, the tensile strength of lightweight concrete is a fixed fraction of the tensile strength of normalweight concrete (Ivey and Buth 1967). The multipliers used for λ are based on data from tests on concrete made with many types of structural lightweight aggregate.

A previously included method to calculate λ based on splitting tensile strength and the corresponding value of measured compressive strength was removed from the Code in 2019.

In editions of the Code prior to 2019, the upper limit on the equilibrium density for lightweight concrete was 115 lb/ft³. With the lower limit for normalweight concrete established at 135 lb/ft³, a 20 lb/ft³ range remained that was undefined. In practice, to achieve an equilibrium density in the range of 115 to 135 lb/ft³, the use of some amount of lightweight aggregate is required. The 2019 Code removes this undefined range by defining lightweight concrete as having an equilibrium density from 90 to 135 lb/ft³.

CODE

19.3—Concrete durability requirements

COMMENTARY

R19.3—Concrete durability requirements

The Code addresses concrete durability on the basis of exposure categories and exposure classes as defined in Table 19.3.1.1. The licensed design professional assigns members in the structure to the appropriate exposure category and class. The assigned exposure classes, which are based on the severity of exposure, are used to establish the appropriate

CODE

19.3.1 *Exposure categories and classes*

19.3.1.1 The licensed design professional shall assign exposure classes in accordance with the severity of the anticipated exposure of members for each exposure category in Table 19.3.1.1.

Table 19.3.1.1—Exposure categories and classes

Category	Class	Condition	
Freezing and thawing (F)	F0	Concrete not exposed to freezing-and-thawing cycles	
	F1	Concrete exposed to freezing-and-thawing cycles with limited exposure to water	
	F2	Concrete exposed to freezing-and-thawing cycles with frequent exposure to water	
	F3	Concrete exposed to freezing-and-thawing cycles with frequent exposure to water and exposure to deicing chemicals	
Sulfate (S)		Water-soluble sulfate (SO_4^{2-}) in soil, percent by mass[1]	Dissolved sulfate (SO_4^{2-}) in water, ppm[2]
	S0	$SO_4^{2-} < 0.10$	$SO_4^{2-} < 150$
	S1	$0.10 \leq SO_4^{2-} < 0.20$	$150 \leq SO_4^{2-} < 1500$ or seawater
	S2	$0.20 \leq SO_4^{2-} \leq 2.00$	$1500 \leq SO_4^{2-} \leq 10,000$
	S3	$SO_4^{2-} > 2.00$	$SO_4^{2-} > 10,000$
In contact with water (W)	W0	Concrete dry in service	
	W1	Concrete in contact with water where low permeability is not required	
	W2	Concrete in contact with water where low permeability is required	
Corrosion protection of reinforcement (C)	C0	Concrete dry or protected from moisture	
	C1	Concrete exposed to moisture but not to an external source of chlorides	
	C2	Concrete exposed to moisture and an external source of chlorides from deicing chemicals, salt, brackish water, seawater, or spray from these sources	

[1]Percent sulfate by mass in soil shall be determined by ASTM C1580.

[2]Concentration of dissolved sulfates in water, in ppm, shall be determined by ASTM D516 or ASTM D4130.

COMMENTARY

concrete properties from Table 19.3.2.1 to include in the construction documents.

The Code does not include provisions for especially severe exposures, such as acids or high temperatures.

R19.3.1 *Exposure categories and classes*

The Code addresses four exposure categories that affect the requirements for concrete to ensure adequate durability:

Exposure Category F applies to concrete exposed to moisture and cycles of freezing and thawing, with or without deicing chemicals.

Exposure Category S applies to concrete in contact with soil or water containing deleterious amounts of water-soluble sulfate ions.

Exposure Category W applies to concrete in contact with water.

Exposure Category C applies to nonprestressed and prestressed concrete exposed to conditions that require additional protection against corrosion of reinforcement.

Severity of exposure within each category is defined by classes with increasing numerical values representing increasingly severe exposure conditions. A classification of 0 is assigned if the exposure severity has negligible effect (is benign) or the exposure category does not apply to the member.

The following discussion provides assistance for selecting the appropriate exposure class for each of the exposure categories. Members are required to be assigned to four exposure classes, one for each exposure category, and are also required to meet the most restrictive requirements of all of these exposures. For example, the slabs of a parking garage in a cold climate might be assigned to Exposure Classes F3, S0, W2, and C2, and a potable water tank inside a heated building might be assigned to Exposure Classes F0, S0, W2, and C1.

Exposure Category F: Whether concrete is damaged by cycles of freezing and thawing depends on the amount of water in the pores of the concrete at the time of freezing (Powers 1975). The amount of water present may be described in terms of the degree of saturation of the concrete. If the degree of saturation is high enough, there will be sufficient water in the concrete pores to produce internal tensile stresses large enough to cause cracking when the water freezes and expands. The entire member need not be saturated to be susceptible to damage. For example, if the top 3/8 in. of a slab or outer 1/4 in. of a wall is saturated, those portions are vulnerable to damage from freezing and thawing, regardless of how dry the interior may be.

For any portion of a member to be resistant to freezing and thawing, that portion of the concrete needs to have sufficient entrained air and adequate strength. Adequate strength is obtained by requiring a low ***w/cm***, which also reduces the pore volume and increases resistance to water penetration. Entrained air makes it more difficult for the

CODE

COMMENTARY

concrete to become saturated and allows for expansion of the water when it freezes.

Exposure class varies with degree of exposure to water, as this will influence the likelihood that any portion of the concrete will be saturated when exposed to cyclic freezing and thawing. Conditions that increase the potential for saturation include longer-duration or more-frequent contact with water without intervening drainage or drying periods. The likelihood that concrete in a member will be saturated depends on project location, member location and orientation in the structure, and climate. Records of performance of similar members in existing structures in the same general location can also provide guidance in assigning exposure classes.

Exposure Category F is subdivided into four exposure classes:

(a) Exposure Class F0 is assigned to concrete that will not be exposed to cycles of freezing and thawing.

(b) Exposure Class F1 is assigned to concrete that will be exposed to cycles of freezing and thawing and that will have limited exposure to water. Limited exposure to water implies some contact with water and water absorption; however, it is not anticipated that the concrete will absorb sufficient water to become saturated. The licensed design professional should review the exposure conditions carefully to support the decision that the concrete is not anticipated to become saturated before freezing. Even though concrete in this exposure class is not expected to become saturated, a minimum entrained air content of 3.5 to 6 percent is required to reduce the potential for damage in case portions of the concrete member become saturated.

(c) Exposure Class F2 is assigned to concrete that will be exposed to cycles of freezing and thawing and that will have frequent exposure to water. Frequent exposure to water implies that some portions of the concrete will absorb sufficient water such that over time they will have the potential to be saturated before freezing. If there is doubt about whether to assign Exposure Classes F1 or F2 to a member, the more conservative choice, F2, should be selected. Exposure Classes F1 and F2 are conditions where exposure to deicing chemicals is not anticipated.

(d) Exposure Class F3 is assigned to concrete that will be exposed to cycles of freezing and thawing with the same degree of exposure to water as Exposure Class F2. Additionally, concrete in Exposure Class F3 is anticipated to be exposed to deicing chemicals. Deicing chemicals can increase water absorption and retention (Spragg et al. 2011), which would enable the concrete to become saturated more readily.

Table R19.3.1 provides examples of concrete members for each of these exposure classes.

CODE

COMMENTARY

Table R19.3.1—Examples of structural members in Exposure Category F

Exposure class	Examples
F0	• Members in climates where freezing temperatures will not be encountered • Members that are inside structures and will not be exposed to freezing • Foundations not exposed to freezing • Members that are buried in soil below the frost line
F1	• Members that will not be subject to snow and ice accumulation, such as exterior walls, beams, girders, and slabs not in direct contact with soil • Foundation walls may be in this class depending upon their likelihood of being saturated
F2	• Members that will be subject to snow and ice accumulation, such as exterior elevated slabs • Foundation or basement walls extending above grade that have snow and ice buildup against them • Horizontal and vertical members in contact with soil
F3	• Members exposed to deicing chemicals, such as horizontal members in parking structures • Foundation or basement walls extending above grade that can experience accumulation of snow and ice with deicing chemicals

Exposure Category S is subdivided into four exposure classes:

(a) Exposure Class S0 is assigned for conditions where the water-soluble sulfate concentration in contact with concrete is low and injurious sulfate attack is not a concern.

(b) Exposure Classes S1, S2, and S3 are assigned for structural concrete members in direct contact with soluble sulfates in soil or water. The severity of exposure increases from Exposure Class S1 to S3 based on the more critical value of measured water-soluble sulfate concentration in soil or the concentration of dissolved sulfate in water. Seawater exposure is classified as Exposure Class S1.

Exposure Category W is subdivided into three exposure classes:

(a) Members are assigned to Exposure Class W0 if they are dry in service.

(b) Members are assigned to Exposure Class W1 if they may be exposed to continuous contact with water, to intermittent sources of water, or can absorb water from surrounding soil. Members assigned to W1 do not require concrete with low permeability.

(c) Members are assigned to Exposure Class W2 if they may be exposed to continuous contact with water, to intermittent sources of water, or can absorb water from surrounding soil, and if the penetration of water through the concrete might reduce durability or serviceability. Members assigned to W2 require concrete with low permeability.

Exposure Category C is subdivided into three exposure classes:

CODE

COMMENTARY

(a) Exposure Class C0 is assigned if exposure conditions do not require additional protection against the initiation of reinforcement corrosion.

(b) Exposure Classes C1 and C2 are assigned to nonprestressed and prestressed concrete members, depending on the degree of exposure to external sources of moisture and chlorides in service. Examples of exposures to external sources of chlorides include concrete in direct contact with deicing chemicals, salt, salt water, brackish water, seawater, or spray from these sources.

19.3.2 *Requirements for concrete mixtures*

19.3.2.1 Based on the exposure classes assigned from Table 19.3.1.1, concrete mixtures shall conform to the most restrictive requirements in Table 19.3.2.1.

R19.3.2 *Requirements for concrete mixtures*

Durability of concrete is impacted by the resistance of the concrete to fluid penetration. This is primarily affected by the ***w/cm*** and the composition of cementitious materials used in concrete. For a given ***w/cm***, the use of fly ash, slag cement, silica fume, or a combination of these materials will typically increase the resistance of concrete to fluid penetration and thus improve concrete durability. The Code provides limits on ***w/cm*** in Table 19.3.2.1 to achieve low permeability and the intended durability. ASTM C1202 can be used to provide an indication of concrete's resistance to fluid penetration.

Because ***w/cm*** of concrete cannot be accurately verified in the field using standard test methods, strength tests are used as a surrogate. Representative values for minimum f_c' have been assigned to each ***w/cm*** limit in Table 19.3.2.1. The acceptance criteria for strength tests in 26.12 establish a basis to indicate that the maximum ***w/cm*** has not been exceeded. For this approach to be reliable, the values of f_c' specified in construction documents should be consistent with the maximum ***w/cm***. Considering the wide range of materials and concrete mixtures possible, including regional variations, the minimum f_c' limit in Table 19.3.2.1 associated with the maximum ***w/cm*** should not be considered absolute. The average strength of concrete mixtures for a given ***w/cm*** can in some cases be considerably higher than the average strength expected for the representative value of f_c'. For a given exposure class, the licensed design professional may choose to specify a higher value of f_c' than listed in the table to obtain better consistency between the maximum ***w/cm*** and f_c'. This improves the confidence that concrete complies with the ***w/cm*** limit if the strength acceptance criteria are satisfied.

As stated in the footnote to Table 19.3.2.1, maximum ***w/cm*** limits are not specified for lightweight concrete because the amount of mixing water that is absorbed by the lightweight aggregates makes calculation of ***w/cm*** uncertain. Therefore, only a minimum f_c' is specified to achieve the required durability.

Table 19.3.2.1 provides the requirements for concrete on the basis of the assigned exposure classes. The most restrictive requirements are applicable. For example, a member assigned to Exposure Class W1 and Exposure Class S2 would require concrete to comply with a maximum ***w/cm*** of

CODE

COMMENTARY

0.45 and a minimum f_c' of 4500 psi because the requirement for Exposure Class S2 is more restrictive than the requirement for Exposure Class W1.

Exposure Classes F1, F2, and F3: In addition to complying with a maximum ***w/cm*** limit and a minimum f_c', concrete for members subject to freezing-and-thawing exposures is required to be air entrained in accordance with 19.3.3.1. Members assigned to Exposure Class F3 are additionally required to comply with the limitations on the quantity of pozzolans and slag cement in the composition of the cementitious materials as given in 26.4.2.2(b).

The requirements for plain concrete members in Exposure Class F3 are less restrictive because of the reduced likelihood of problems caused by reinforcement corrosion. The licensed design professional should consider the details of the minimal reinforcement to be included in plain concrete members to ensure that the less restrictive requirements are appropriate for the specific project.

Exposure Classes S1, S2, and S3: Table 19.3.2.1 lists the appropriate types of cement and the maximum ***w/cm*** and minimum f_c' for various sulfate exposure conditions. In selecting cement for sulfate resistance, the principal consideration is its tricalcium aluminate (C_3A) content.

The use of fly ash (ASTM C618, Class F), natural pozzolans (ASTM C618, Class N), silica fume (ASTM C1240), or slag cement (ASTM C989) has been shown to improve the sulfate resistance of concrete (Li and Roy 1986; ACI 233R; ACI 234R). Therefore, Footnote [7] to Table 19.3.2.1 provides a performance option to determine the appropriate amounts of these materials to use in combination with the specific cement types listed. ASTM C1012 is permitted to be used to evaluate the sulfate resistance of mixtures using combinations of cementitious materials in accordance with 26.4.2.2(c).

Some ASTM C595 and ASTM C1157 blended cements can meet the testing requirements of 26.4.2.2.(c) without addition of pozzolans or slag cement to the blended cement as manufactured.

Note that sulfate-resisting cement will not increase resistance of concrete to some chemically aggressive solutions—for example, sulfuric acid. The construction documents should explicitly cover such cases.

In addition to the proper selection of cementitious materials, other requirements for durable concrete exposed to water-soluble sulfates are essential, such as ***w/cm***, strength, consolidation, uniformity, cover of reinforcement, and moist curing to develop the potential properties of the concrete.

Exposure Class S1: ASTM C150 Type II cement is limited to a maximum C_3A content of 8 percent and is acceptable for use in Exposure Class S1. Blended cements under ASTM C595 with the MS designation, which indicates the cement meets requirements for moderate sulfate resistance, are also appropriate for use. Under ASTM C1157, the appropriate designation for moderate sulfate exposure is Type MS.

CODE

COMMENTARY

Seawater is listed under Exposure Class S1 (moderate exposure) in Table 19.3.1.1, even though it generally contains more than 1500 ppm SO_4^{2-}. Less expansion is produced by a given cement in seawater compared with freshwater with the same sulfate content (ACI 201.2R). Therefore, seawater is included in the same exposure class as solutions with lower sulfate concentrations. Portland cement with C_3A up to 10 percent is allowed in concrete mixtures exposed to seawater if the maximum ***w/cm*** is limited to 0.40 (refer to the footnote to Table 19.3.2.1).

Exposure Class S2: ASTM C150 Type V cement is limited to a maximum C_3A content of 5 percent and is acceptable for use in Exposure Class S2. The appropriate binary and ternary blended cements under ASTM C595 include the suffix (HS) as part of their designation, which indicates the cement conforms to requirements for high sulfate resistance. Under ASTM C1157, the appropriate designation for severe sulfate exposure is Type HS.

Exposure Class S3 (Option 1): The benefit of the addition of pozzolan or slag cement allows for a greater ***w/cm*** than required for Option 2. The amounts of supplementary cementitious materials are based on records of successful service or testing in accordance with 26.4.2.2(c).

Exposure Class S3 (Option 2): This option allows the use of ASTM C150 Type V portland cement meeting the optional limit of 0.040 percent maximum expansion, ASTM C595 binary and ternary blended cements with the (HS) suffix in their designation, and ASTM C1157 Type HS cements without the use of additional pozzolan or slag cement, but it instead requires a lower ***w/cm*** than that required for Option 1. This lower ***w/cm*** reduces the permeability of the concrete and thus increases sulfate resistance (Lenz 1992). Use of this lower ***w/cm*** permits a shorter testing period to qualify the sulfate resistance of a cementitious system in accordance with 26.4.2.2(c).

In addition to the proper selection of cementitious materials, other requirements for durable concrete exposed to water-soluble sulfates are essential, such as low ***w/cm***, strength, adequate consolidation, uniformity, adequate cover of reinforcement, and sufficient moist curing to develop the potential properties of the concrete.

Exposure Class W1: This exposure class does not have specific requirements for low permeability. However, because of the exposure to water, the Code (26.4.2.2(d)) has a requirement to demonstrate that aggregates used in concrete are not alkali reactive according to ASTM C1778. If the aggregates are alkali-silica reactive, the Code (26.4.2.2(d)) also requires submission of proposed mitigation measures. The Code (26.4.2.2(d)) prohibits the use of aggregates that are alkali-carbonate reactive.

Exposure Class W2: This exposure class requires low concrete permeability. The primary means to obtain a concrete with low permeability is to reduce ***w/cm***. For a given *w/cm*, permeability can be reduced by optimizing the cementitious materials used in the concrete mixture.

CODE

COMMENTARY

In addition, because of the exposure to water, the Code (26.4.2.2(d)) has a requirement to demonstrate that aggregates used in concrete are not alkali reactive according to ASTM C1778. If the aggregates are alkali-silica reactive, the Code (26.4.2.2(d)) also requires submission of proposed mitigation measures. The Code (26.4.2.2(d)) prohibits the use of aggregates that are alkali-carbonate reactive.

Exposure Class C2: For nonprestressed and prestressed concrete in Exposure Class C2, the maximum ***w/cm***, minimum specified compressive strength, and minimum cover are the basic requirements to be considered. Conditions should be evaluated for structures exposed to chlorides, such as in parking structures where chlorides may be tracked in by vehicles, or in structures near seawater. Coated reinforcement, corrosion-resistant steel reinforcement, and cover greater than the minimum required in 20.5 can provide additional protection under such conditions. Use of slag cement meeting ASTM C989 or fly ash meeting ASTM C618 and increased levels of specified compressive strength provide increased protection. Use of silica fume meeting ASTM C1240 with an appropriate high-range water reducer, ASTM C494, Types F and G, or ASTM C1017 can also provide additional protection (Ozyildirim and Halstead 1988). The use of ASTM C1202 to test concrete mixtures proposed for use will provide additional information on the performance of the mixtures.

Chloride limits for Exposure Category C: For Exposure Classes C0, C1, and C2, the chloride ion limits apply to the chlorides contributed from the concrete materials, not from the environment surrounding the concrete. Even for Exposure Class C0, water-soluble chlorides introduced from the concrete materials can potentially induce corrosion of the reinforcement and must be limited for both nonprestressed and prestressed concrete, regardless of external exposure. For nonprestressed concrete, the permitted maximum amount of water-soluble chloride ions incorporated into the concrete, depends on the degree of exposure to an anticipated external source of moisture and chlorides. For prestressed concrete, the same limit of 0.06 percent chloride ion by mass of cementitious material applies regardless of exposure. The limits on chloride ion content for prestressed concrete are reduced from those for nonprestressed concrete because corrosion of prestressed reinforcement generally has more severe consequences than corrosion of nonprestressed reinforcement. Corrosion-induced reduction in the cross-sectional area of the prestressed reinforcement may result in fracture of the steel (ACI 222R). The presence of chloride ions may cause corrosion of embedded aluminum such as conduits, especially if the aluminum is in contact with embedded steel and the concrete is in a humid environment. Requirements for protecting aluminum embedments from corrosion are given in 20.6.3 and 26.8.2.

Allowable chloride limits are based on the mass of total cementitious materials rather than portland cement alone. This change was made in ACI 318-19 to reflect findings that

CODE

COMMENTARY

demonstrate the beneficial effects of supplementary cementitious materials (SCMs) in reducing permeability and binding chlorides, thus helping to inhibit corrosion (Kosmatka and Wilson 2016). Because there are diminishing effects with increasing amounts of SCMs, the Code limits the mass of SCMs to 50 percent of the total cementitious materials that can be used to calculate the allowable amount of chloride ions in concrete (Tepke et al. 2016).

Additional information on the effects of chlorides on the corrosion of steel reinforcement is given in ACI 201.2R, which provides guidance on concrete durability, and ACI 222R, which provides guidance on factors that impact corrosion of metals in concrete. Requirements for the evaluation of chloride ion content are provided in 26.4.2.2.

CODE

Table 19.3.2.1—Requirements for concrete by exposure class

Exposure class		Maximum *w/cm*[1,2]	Minimum f_c', psi	Additional requirements — Air content			Limits on cementitious materials
F0		N/A	2500	N/A			N/A
F1		0.55	3500	Table 19.3.3.1 for concrete or Table 19.3.3.3 for shotcrete			N/A
F2		0.45	4500	Table 19.3.3.1 for concrete or Table 19.3.3.3 for shotcrete			N/A
F3		0.40[3]	5000[3]	Table 19.3.3.1 for concrete or Table 19.3.3.3 for shotcrete			26.4.2.2(b)
				Cementitious materials[4] — Types			Calcium chloride admixture
				ASTM C150	ASTM C595	ASTM C1157	
S0		N/A	2500	No type restriction	No type restriction	No type restriction	No restriction
S1		0.50	4000	II[5][6]	Types with (MS) designation	MS	No restriction
S2		0.45	4500	V[6]	Types with (HS) designation	HS	Not permitted
S3	Option 1	0.45	4500	V plus pozzolan or slag cement[7]	Types with (HS) designation plus pozzolan or slag cement[7]	HS plus pozzolan or slag cement[7]	Not permitted
S3	Option 2	0.40	5000	V[8]	Types with (HS) designation	HS	Not permitted
W0		N/A	2500	None			
W1		N/A	2500	26.4.2.2(d)			
W2		0.50	4000	26.4.2.2(d)			
				Maximum water-soluble chloride ion (Cl^-) content in concrete, percent by mass of cementitious materials[9,10] — Nonprestressed concrete	Maximum water-soluble chloride ion (Cl^-) content in concrete, percent by mass of cementitious materials[9,10] — Prestressed concrete	Additional provisions	
C0		N/A	2500	1.00	0.06	None	
C1		N/A	2500	0.30	0.06		
C2		0.40	5000	0.15	0.06	Concrete cover[11]	

[1]The *w/cm* is based on all cementitious and supplementary cementitious materials in the concrete mixture.

[2]The maximum *w/cm* limits do not apply to lightweight concrete.

[3]For plain concrete, the maximum *w/cm* shall be 0.45 and the minimum f_c' shall be 4500 psi.

[4]Alternative combinations of cementitious materials to those listed are permitted for all sulfate exposure classes when tested for sulfate resistance and meeting the criteria in 26.4.2.2(c).

[5]For seawater exposure, other types of portland cements with tricalcium aluminate (C_3A) contents up to 10 percent are permitted if the *w/cm* does not exceed 0.40.

[6]Other available types of cement such as Type I or Type III are permitted in Exposure Classes S1 or S2 if the C_3A contents are less than 8 percent for Exposure Class S1 or less than 5 percent for Exposure Class S2.

[7]The amount of the specific source of the pozzolan or slag cement to be used shall be at least the amount that has been determined by service record to improve sulfate resistance when used in concrete containing Type V cement. Alternatively, the amount of the specific source of the pozzolan or slag cement to be used shall be at least the amount tested in accordance with ASTM C1012 and meeting the criteria in 26.4.2.2(c).

[8]If Type V cement is used as the sole cementitious material, the optional sulfate resistance requirement of 0.040 percent maximum expansion in ASTM C150 shall be specified.

[9]The mass of supplementary cementitious materials used in determining the chloride content shall not exceed the mass of the portland cement.

[10]Criteria for determination of chloride content are in 26.4.2.2.

[11]Concrete cover shall be in accordance with 20.5.

19.3.3 *Additional requirements for freezing-and-thawing exposure*

19.3.3.1 Concrete subject to freezing-and-thawing Exposure Classes F1, F2, or F3 shall be air entrained. Except as

COMMENTARY

R19.3.3 *Additional requirements for freezing-and-thawing exposure*

R19.3.3.1 A table of required air contents for concrete to resist damage from cycles of freezing and thawing is

CODE

permitted in 19.3.3.6, air content shall conform to Table 19.3.3.1.

Table 19.3.3.1—Total air content for concrete exposed to cycles of freezing and thawing

Nominal maximum aggregate size, in.	Target air content, percent	
	F1	F2 and F3
3/8	6.0	7.5
1/2	5.5	7.0
3/4	5.0	6.0
1	4.5	6.0
1-1/2	4.5	5.5
2	4.0	5.0
3	3.5	4.5

19.3.3.2 Concrete shall be sampled in accordance with ASTM C172, and air content shall be measured in accordance with ASTM C231 or ASTM C173.

19.3.3.3 Wet-mix shotcrete subject to freezing-and-thawing Exposure Classes F1, F2, or F3 shall be air entrained. Dry-mix shotcrete subject to freezing-and-thawing Exposure Class F3 shall be air entrained. Except as permitted in 19.3.3.6, air content shall conform to Table 19.3.3.3.

Table 19.3.3.3—Total air content for shotcrete exposed to cycles of freezing and thawing

Mixture type	Sampling location	Target air content, percent		
		F1	F2	F3
Wet-mix shotcrete	Before placement	5.0	6.0	6.0
Dry-mix shotcrete	In-place	N/A[1]	N/A[1]	4.5

[1]Entrained air is not required in dry-mix shotcrete for these exposure classes.

COMMENTARY

included in the Code, based on guidance provided for proportioning concrete mixtures in ACI 211.1. Entrained air will not protect concrete containing coarse aggregates that undergo disruptive volume changes when frozen in a saturated condition.

R19.3.3.2 The sampling of fresh concrete for acceptance based on air content is usually performed as the concrete is discharged from a mixer or a transportation unit (for example, a ready mixed concrete truck) to the conveying equipment used to transfer the concrete to the forms. ASTM C172 primarily covers sampling of concrete as it is discharged from a mixer or a transportation unit, but recognizes that specifications may require sampling at other points such as discharge from a pump. Table 19.3.3.1 was developed for testing as-delivered concrete. ASTM C231 is applicable to normalweight concrete and ASTM C173 is applicable to normalweight or lightweight concrete.

If the licensed design professional requires measurement of air content of fresh concrete at additional sampling locations, such requirements should be stated in the construction documents, including the sampling protocol, test methods to be used, and the criteria for acceptance.

R19.3.3.3 Adding air-entraining admixtures improves freezing-and-thawing resistance of wet-mix shotcrete (ACI 506R). Having air contents before placement as specified in Table 19.3.3.3 will provide required performance in freezing and thawing. Air contents greater than those specified will not improve shotcrete performance because once adequate air content for durability is achieved, there is no further benefit. As in all concrete, too much in-place air will reduce strength.

Dry-mix shotcrete without air entrainment has performed well in freezing-and-thawing environments with no exposure to saltwater or deicing salts (ACI 506R; Seegebrecht et al. 1989). For exposure to saltwater or deicing salts, air-entraining admixtures, in either a wet or dry form, can be added to dry-mix shotcrete to provide the required air content for durability in these exposures (Bertrand and Vezina 1994). The higher air content of wet-mix shotcrete sampled at the point of delivery accounts for expected air losses during shooting.

CODE

19.3.3.4 Wet-mix shotcrete shall be sampled in accordance with ASTM C172, and air content shall be measured in accordance with ASTM C231 or ASTM C173.

19.3.3.5 Dry-mix shotcrete shall be sampled and air content shall be measured as directed by the licensed design professional.

19.3.3.6 For $f_c' \geq$ **5000 psi**, reduction of air content indicated in Table 19.3.3.1 and 19.3.3.3 by 1.0 percentage point is permitted.

19.3.3.7 The maximum percentage of pozzolans, including fly ash and silica fume, and slag cement in concrete assigned to Exposure Class F3, shall be in accordance with 26.4.2.2(b).

19.3.4 *Additional requirements for chloride ion content*

19.3.4.1 Nonprestressed concrete that will be cast against stay-in-place galvanized steel forms shall comply with the chloride ion limits for Exposure Class C1 unless a more stringent limit is required by other project conditions.

COMMENTARY

R19.3.3.5 If the licensed design professional requires measurement of air content of fresh dry-mix shotcrete, such requirements are to be stated in the construction documents, including the sampling frequency, sampling protocol, test methods to be used, and the criteria for acceptance.

The air content required for dry-mix shotcrete is for sampling of in-place shotcrete. This air content can be verified by taking cores from shotcrete test panels for analysis in accordance with ASTM C457. During the mixture development process, shotcrete test panels may be prepared with different amounts of air-entraining admixture and cored to determine a dosage that will provide the required amount of air after placement.

The use of ASTM C457 for quality control during construction is not practical. Although there are no standard tests for air content of dry-mix shotcrete during construction, there are industry accepted methods for testing. These methods involve obtaining samples of dry-mix shotcrete and performing standard tests such as ASTM C231 to determine air content.

Field measurements of air content of dry-mix shotcrete have been obtained by shooting the material directly into a bowl of an air meter (Betrand and Vezina 1994). Samples for air content testing can also be taken from material shot into test panels, into a wheelbarrow, or onto the ground. These samples can then be used for testing in accordance with ASTM C231 (Zhang 2015).

R19.3.3.6 This section permits a 1.0 percentage point lower air content for concrete with f_c' equal to or greater than 5000 psi. Such higher-strength concretes, which have a lower ***w/cm*** and porosity, have greater resistance to cycles of freezing and thawing.

R19.3.3.7 This provision is intended for application during concrete mixture proportioning. The provision has been duplicated in 26.4.2.2(b). Additional commentary information is presented in Chapter 26.

R19.3.4 *Additional requirements for chloride ion content*

R19.3.4.1 Corrosion of galvanized steel sheet or stay-in-place galvanized steel forms may occur, especially in humid environments or where drying is inhibited by the thickness of the concrete, coatings, or impermeable coverings. If stay-in-place galvanized steel forms are used, the maximum chloride limit of 0.30 percent is required. For more severe environments, such as for concrete in Exposure Class C2, a more stringent limit of 0.15 percent would be required.

At the time of design, the licensed design professional may not know if aluminum embedments or stay-in-place galva-

CODE

COMMENTARY

nized steel forms will be used. Use of aluminum embedments is covered in 26.8.2. Use of stay-in-place galvanized steel forms is covered in 26.4.2.2.

19.4—Grout durability requirements

19.4.1 Water-soluble chloride ion content of grout for bonded tendons shall not exceed 0.06 percent when tested in accordance with ASTM C1218, measured by mass of chloride ion to mass of cementitious materials.

Notes

Retaining Wall Continued

There for cut ever other #4 bar at 2.6' + 1.8' = 4.4' for simplicity do 5'

Horizontal steel min

$A_{s,min} = 0.0020\, A_g$ Page 171, need two layer split and ple #4 at 8" each f

Heel design

$R_u = 1.5\,[5'(16')(120\,pcf)] + 1.2\,[5'(2')(150\,pcf)] = 13{,}320\ lb$

$V_u = R_u$

$\phi V_n = \phi 2\sqrt{f'_c}\, bd = (0.75)(2)\sqrt{3000\ psi}\,(12'')(21.5)'' = 21{,}197\ lbs$

$M_u = R_u\left(\tfrac{1}{2}(x)\right) = (13{,}320)\left(\tfrac{1}{2}(5)\right) = 33{,}300\ ft.lbs = 399{,}600\ in\_lbs$

x = length of heel

get As, spacing and min same as before

Toe design

$\phi V_n = \phi 2\sqrt{f'_c}\, bd = 0.75\,(2)\sqrt{3000}\,(12'')(20.5'') = 20{,}211\ lbs$

$V_u = [(2894)(3')] + [\tfrac{1}{2}(3908 - 2894)(3')] - [1.2(3')(2')(150)]$

$V_u = 9123$

$M_u = [(2894)(3')(1.5')] + [(\tfrac{1}{2}(3908\ psf - 2894\ psf)(3')(2')] -$

$[1.2(3')(2')(150)(1.5')]] = 14{,}445 = 173{,}340\ in.lbs$

Find As and Spacing $A_s = \frac{total}{A_b} \times 12$

$A_{s\,min}$ transverse $= 0.0018\,A_g\,(12)(24) = 0.518\ in^2/ft$

A_s total $= 0.4 + 0.16 = 0.56\ in^2/ft$ ✓

Check development lengths from next example / hooking

Temperature and shrinkage steel in footing

$A_{s\,min} = 0.0018\,A_g = 0.0018\,(24'')(120'') - 5.18\ in^2$

CHAPTER 20—STEEL REINFORCEMENT PROPERTIES, DURABILITY, & EMBEDMENTS

20.1—Scope

20.1.1 This chapter shall apply to steel reinforcement, and shall govern (a) through (c):

(a) Material properties
(b) Properties to be used for design
(c) Durability requirements, including minimum specified cover requirements

20.1.2 Provisions of 20.6 shall apply to embedments.

R20.1—Scope

R20.1.1 Materials permitted for use as reinforcement are specified. Other metal elements, such as inserts, anchor bolts, or plain bars for dowels at isolation or contraction joints, are not normally considered reinforcement under the provisions of this Code. Fiber-reinforced polymer (FRP) reinforcement is not addressed in this Code. ACI Committee 440 has developed guidelines for the use of FRP reinforcement (ACI 440.1R and ACI 440.2R).

20.2—Nonprestressed bars and wires

20.2.1 *Material properties*

20.2.1.1 Nonprestressed bars and wires shall be deformed, except plain bars or wires are permitted for use in spirals.

20.2.1.2 Yield strength of nonprestressed bars and wires shall be determined by either (a) or (b):

(a) The offset method, using an offset of 0.2 percent in accordance with ASTM A370
(b) The yield point by the halt-of-force method, provided the nonprestressed bar or wire has a sharp-kneed or well-defined yield point

20.2.1.3 Deformed bars shall conform to (a), (b), (c), (d), or (e), except bar sizes larger than No. 18 shall not be permitted:

(a) ASTM A615 – carbon steel, including requirements specified in Table 20.2.1.3(a)
(b) ASTM A706 – low-alloy steel, including requirements specified in (i), (ii), and (iii):
(i) Tensile property requirements for ASTM A706 Grade 100 reinforcement shall be as specified in Table 20.2.1.3(b), and bend test requirements for ASTM A706 Grade 100 reinforcement shall be the same as the bend test requirements for ASTM A706 Grade 80 reinforcement.

R20.2—Nonprestressed bars and wires

R20.2.1 *Material properties*

R20.2.1.2 The majority of nonprestressed steel bar reinforcement exhibits actual stress-strain behavior that is sharply yielding or sharp-kneed (elasto-plastic stress-strain behavior). However, reinforcement products such as bars of higher strength grade, steel wire, coiled steel bar, and stainless steel bars and wire generally do not exhibit sharply-yielding stress-strain behavior, but instead are gradually-yielding. The method used to measure yield strength of reinforcement needs to provide for both types of reinforcement stress-strain relationships.

A study (Paulson et al. 2013) considering reinforcement manufactured during 2008 through 2012 found that the offset method, using an offset of 0.2 percent, provides for a reasonable estimate of the strength of reinforced concrete structures.

The yield strength is determined by the manufacturer during tensile tests performed at the mill on samples of reinforcement. Test methods for determining yield strength of steel, including the offset method and yield point by halt-of-force method, are referenced either in the ASTM standards for nonprestressed bars and wire or in ASTM A370 Test Methods and Definitions.

R20.2.1.3 The requirements specified in 20.2.1.3(a) and (b), and in Tables 20.2.1.3(a) through (c), are necessary because the referenced standards in Chapter 3, ASTM A615-18$^{\varepsilon 1}$ and ASTM A706-16, do not include these requirements. For project specifications, these requirements should be specified along with the corresponding ASTM requirements. The requirements provide for harmonization of minimum tensile strength requirements between ASTM A615 and ASTM A706, add new ductility requirements to both ASTM A615 and ASTM A706, and introduce Grade 100 reinforcement for ASTM A706. These requirements accommodate the introduction of higher strength reinforcement into the Code for special seismic applications and have been developed considering both structural safety and

CODE

(ii) Uniform elongation requirements for all grades of ASTM A706 reinforcement shall be as specified in Table 20.2.1.3(c), and uniform elongation shall be determined as the elongation at the maximum force sustained by the reinforcing bar test piece.

(iii) For all grades of ASTM A706 reinforcement, the radius at the base of each deformation shall be at least 1.5 times the height of the deformation. This requirement applies to all deformations, including transverse lugs, longitudinal ribs, grade ribs, grade marks, and intersections between deformations. Conformance shall be assessed by measurements taken on newly-machined rolls used to manufacture reinforcing bars, instead of measurements taken on bar samples.

Table 20.2.1.3(a)—Modified tensile strength and additional tensile property requirements for ASTM A615 reinforcement

	Grade 40	Grade 60	Grade 80	Grade 100
Tensile strength, minimum, psi	60,000	80,000	100,000	115,000
Ratio of actual tensile strength to actual yield strength, minimum	1.10	1.10	1.10	1.10

Table 20.2.1.3(b)—Tensile property requirements for ASTM A706 Grade 100 reinforcement

	Grade 100
Tensile strength, minimum, psi	117,000
Ratio of actual tensile strength to actual yield strength, minimum	1.17
Yield strength, minimum, psi	100,000
Yield strength, maximum, psi	118,000
Fracture elongation in 8 in., minimum, %	10

Table 20.2.1.3(c)—Uniform elongation requirements for ASTM A706 reinforcement

	Grade 60	Grade 80	Grade 100
Uniform elongation, minimum, percent			
Bar designation No.			
3, 4, 5, 6, 7, 8, 9, 10	9	7	6
11, 14, 18	6	6	6

(c) ASTM A996 – axle steel and rail steel; bars from rail steel shall be Type R

(d) ASTM A955 – stainless steel

(e) ASTM A1035 – low-carbon chromium steel

20.2.1.4 Plain bars for spiral reinforcement shall conform to ASTM A615, A706, A955, or A1035.

COMMENTARY

production of reinforcement. The method for determination of uniform elongation specified in 20.2.1.3(b)(ii) is taken from ASTM E8.

Low-alloy steel deformed bars conforming to ASTM A706 are intended for applications where controlled tensile properties are required. ASTM A706 also includes restrictions on chemical composition to enhance weldability for Grades 60 and 80.

Rail-steel deformed bars used with this Code are required to conform to ASTM A996, including the provisions for Type R bars. Type R bars are required to meet more restrictive provisions for bend tests than other types of rail steel.

Stainless steel deformed bars are used in applications where high corrosion resistance or controlled magnetic permeability are required.

Low-carbon chromium steel is a high-strength material that is permitted for use as transverse reinforcement for confinement in special earthquake-resistant structural systems and spirals in columns. Refer to Tables 20.2.2.4(a) and (b). ASTM A1035 provides requirements for bars of two minimum yield strength levels—100,000 psi and 120,000 psi—designated as Grade 100 and Grade 120, respectively, but the maximum f_{yt} permitted for design calculations in this Code is limited in accordance with 20.2.2.3.

In 2015, ASTM A615 included bar sizes larger than No. 18, and in 2016, ASTM A1035 also included bar sizes larger than No. 18. Bar sizes larger than No. 18 are not permitted by this Code due to the lack of information on their performance including bar bends and development lengths.

R20.2.1.4 Plain bars are permitted only for spiral reinforcement used as transverse reinforcement for columns,

CODE

COMMENTARY

transverse reinforcement for shear and torsion, or confining reinforcement for splices.

20.2.1.5 Welded deformed bar mats shall conform to ASTM A184. Deformed bars used in welded deformed bar mats shall conform to ASTM A615 or A706.

20.2.1.6 Headed deformed bars shall conform to ASTM A970, including Annex A1 requirements for Class HA head dimensions.

R20.2.1.6 The limitation to Class HA head dimensions from Annex A1 of ASTM A970 is due to a lack of test data for headed deformed bars that do not meet Class HA dimensional requirements. Heads not conforming to Class HA limits on bar deformation obstructions and bearing face features have been shown to provide lower anchorage strength than the heads used in the tests that serve as the basis for 25.4.4 (Shao et al. 2016).

20.2.1.7 Deformed wire, plain wire, welded deformed wire reinforcement, and welded plain wire reinforcement shall conform to (a) or (b), except that yield strength shall be determined in accordance with 20.2.1.2:

(a) ASTM A1064 – carbon steel
(b) ASTM A1022 – stainless steel

R20.2.1.7 Plain wire is permitted only for spiral reinforcement and in welded plain wire reinforcement, the latter of which is considered deformed. Stainless steel wire and stainless steel welded wire reinforcement are used in applications where high corrosion resistance or controlled magnetic permeability is required. The physical and mechanical property requirements for deformed stainless steel wire and deformed and plain welded wire reinforcement under ASTM A1022 are the same as those for deformed wire, deformed welded wire reinforcement, and plain welded wire reinforcement under ASTM A1064.

20.2.1.7.1 Deformed wire sizes D4 through D31 shall be permitted.

R20.2.1.7.1 An upper limit is placed on the size of deformed wire because tests (Rutledge and Devries 2002) have shown that D45 wire will achieve only approximately 60 percent of the bond strength in tension given by Eq. (25.4.2.4a).

20.2.1.7.2 Deformed wire sizes larger than D31 shall be permitted in welded wire reinforcement if treated as plain wire for calculation of development and splice lengths in accordance with 25.4.7 and 25.5.4, respectively.

20.2.1.7.3 Except as permitted for welded wire reinforcement used as stirrups in accordance with 25.7.1, spacing of welded intersections in welded wire reinforcement in the direction of calculated stress shall not exceed (a) or (b):

(a) 16 in. for welded deformed wire reinforcement
(b) 12 in. for welded plain wire reinforcement

20.2.2 *Design properties*

20.2.2.1 For nonprestressed bars and wires, the stress below f_y shall be E_s times steel strain. For strains greater than that corresponding to f_y, stress shall be considered independent of strain and equal to f_y.

R20.2.2 *Design properties*

R20.2.2.1 For deformed reinforcement, it is reasonably accurate to assume that the stress in reinforcement is proportional to strain below the specified yield strength f_y. The increase in strength due to the effect of strain hardening of the reinforcement is neglected for nominal strength calculations. In nominal strength calculations, the force developed in tension or compression reinforcement is calculated as:

CODE

20.2.2.2 Modulus of elasticity, E_s, for nonprestressed bars and wires shall be permitted to be taken as 29,000,000 psi.

20.2.2.3 Yield strength for nonprestressed bars and wires shall be based on the specified grade of reinforcement and shall not exceed the values given in 20.2.2.4 for the associated applications.

20.2.2.4 Types of nonprestressed bars and wires to be specified for particular structural applications shall be in accordance with Table 20.2.2.4(a) for deformed reinforcement and Table 20.2.2.4(b) for plain reinforcement.

COMMENTARY

if $\varepsilon_s < \varepsilon_y$ (yield strain)

$$A_s f_s = A_s E_s \varepsilon_s$$

if $\varepsilon_s \geq \varepsilon_y$

$$A_s f_s = A_s f_y$$

where ε_s is the value from the strain diagram at the location of the reinforcement.

R20.2.2.4 Tables 20.2.2.4(a) and 20.2.2.4(b) limit the maximum values of yield strength to be used in design calculations for nonprestressed deformed reinforcement and nonprestressed plain spiral reinforcement, respectively.

Grade 100 reinforcement is now permitted to resist tension and compression in some applications. For reinforcement resisting compression, strain compatibility calculations indicate that stresses are not likely to exceed 80,000 psi before strain in unconfined concrete reaches the strain limit of 0.003 unless special confinement reinforcement is provided to increase the limiting concrete compressive strain. For beams, the deflection provisions of 24.2 and the limitations on distribution of flexural reinforcement of 24.3 become increasingly critical as f_y increases.

In Table 20.2.2.4(a), for deformed reinforcement in special moment frames and special structural walls, the use of longitudinal reinforcement with strength substantially higher than that assumed in design will lead to higher shear and bond stresses at the time of development of yield moments. These conditions may lead to brittle failures in shear or bond and should be avoided even if such failures may occur at higher loads than those anticipated in design. Therefore, ASTM A706 specifies both a lower and an upper limit on the actual yield strength of the steel and requires a minimum tensile-to-yield strength ratio. ASTM A615 Grade 60 reinforcement in special seismic systems is permitted only if the requirements of 20.2.2.5(b) are satisfied. ASTM A706 Grade 80 and Grade 100 are now permitted to resist tension and compression in some applications. ASTM A706 Grade 80 and Grade 100 are now permitted to resist moments, axial forces, and shear forces in special structural walls and all components of structural walls, including coupling beams and wall piers. ASTM A706 Grade 80 is also permitted in special moment frames. For reinforcement resisting compression, strain compatibility calculations indicate that stresses are not likely to exceed 80,000 psi before strain in unconfined concrete reaches the strain limit of

CODE

COMMENTARY

0.003 unless special confinement reinforcement is provided to increase the limiting concrete compressive strain.

The maximum value of yield strength for calculation purposes is limited to 100,000 psi for both nonprestressed deformed reinforcement and plain spiral reinforcement in Tables 20.2.2.4(a) and (b), respectively, when used for lateral support of longitudinal bars or for concrete confinement. The research that supports this limit for confinement is given in Saatcioglu and Razvi (2002), Pessiki et al. (2001), and Richart et al. (1929). For reinforcement in special moment frames and special structural walls, the research that indicated that higher yield strengths can be used effectively for confinement reinforcement is given in Budek et al. (2002), Muguruma and Watanabe (1990), and Sugano et al. (1990).

The limit of 60,000 psi on the values of f_y and f_{yt} used in design for most shear and torsional reinforcement is intended to control the width of inclined cracks under service-level gravity loads. The higher yield strength of 80,000 psi permitted in shear design for welded deformed wire reinforcement is also intended to control width of inclined cracks and is based on Guimares et al. (1992), Griezic et al. (1994), and Furlong et al. (1991). In particular, full-scale beam tests described in Griezic et al. (1994) indicated that the widths of inclined shear cracks at service load levels were less for beams reinforced with smaller diameter welded deformed wire reinforcement cages designed on the basis of a yield strength of 75,000 psi than beams reinforced with deformed Grade 60 stirrups.

For strength-level earthquake load effects, tests of members using higher strength reinforcement have shown acceptable behavior (Wallace 1998; Aoyama 2001; Budek et al. 2002; Sokoli and Ghannoum 2016; Cheng et al. 2016; Huq et al. 2018; Weber-Kamin et al. 2019), leading to the allowance of ASTM A706 Grade 80 reinforcement for special seismic systems and ASTM A706 Grade 100 for special structural walls in the 2019 Code, as indicated in Table 20.2.2.4(a).

Footnote [6] of Table 20.2.2.4(a) is provided because ASTM A1064 and A1022 only require the welds to develop 35,000 psi in the interconnected wires. Hoops, stirrups, and other elements used in special seismic systems should have anchorages that are capable of developing **1.25**f_y or **1.25**f_{yt}, as applicable, or tensile strength of the bar or wire, whichever is less, so that moderate ductility capacity can be achieved. A welded product that is capable of developing these stress limits could be approved for use through Code Section 1.10.

Footnote [3] of Table 20.2.2.4(a) limiting slab and beam bars passing through or extending from special structural walls to reinforcement meeting 20.2.2.5 provides for greater ductility of these members that are not designated as part of the seismic-force-resisting system but are likely to undergo large nonlinear rotational demands.

The 80,000 psi limit on f_y for ties of members or regions of members designed using the strut-and-tie method is imposed because of scarcity of test data justifying a higher limit. The yield strength f_y of "other" ties is limited to 60,000 psi for consistency with the usage "shear."

CODE

COMMENTARY

Table 20.2.2.4(a)—Nonprestressed deformed reinforcement

Usage	Application		Maximum value of f_y or f_{yt} permitted for design calculations, psi	Applicable ASTM specification: Deformed bars	Deformed wires	Welded wire reinforcement	Welded deformed bar mats
Flexure; axial force; and shrinkage and temperature	Special seismic systems	Special moment frames	80,000	A706[2]	Not permitted	Not permitted	Not permitted
		Special structural walls[1]	100,000				
	Other		100,000[3] [4]	A615, A706, A955, A996, A1035	A1064, A1022	A1064, A1022	A184[5]
Lateral support of longitudinal bars; or concrete confinement	Special seismic systems		100,000	A615, A706, A955, A996, A1035	A1064, A1022	A1064[6], A1022[6]	Not permitted
	Spirals		100,000	A615, A706, A955, A996, A1035	A1064, A1022	Not permitted	Not permitted
	Other		80,000	A615, A706, A955, A996	A1064, A1022	A1064, A1022	Not permitted
Shear	Special seismic systems[7]	Special moment frames[8]	80,000	A615, A706, A955, A996	A1064, A1022	A1064[6], A1022[6]	Not permitted
		Special structural walls[9]	100,000				
	Spirals		60,000	A615, A706, A955, A996	A1064, A1022	Not permitted	Not permitted
	Shear friction		60,000	A615, A706, A955, A996	A1064, A1022	A1064, A1022	Not permitted
	Stirrups, ties, hoops		60,000	A615, A706, A955, A996, A1035	A1064, A1022	A1064 and A1022 welded plain wire	Not permitted
			80,000	Not permitted	Not permitted	A1064 and A1022 welded deformed wire	Not permitted
Torsion	Longitudinal and transverse		60,000	A615, A706, A955, A996	A1064, A1022	A1064, A1022	Not permitted
Anchor reinforcement	Special seismic systems		80,000	A706[2]	Not permitted	Not permitted	Not permitted
	Other		80,000	A615, A706, A955, A996	A1064, A1022	A1064, A1022	A184[5]
Regions designed using strut-and-tie method	Longitudinal ties		80,000	A615, A706, A955, A996	A1064, A1022	A1064, A1022	Not permitted
	Other		60,000				

[1]All components of special structural walls, including coupling beams and wall piers.

[2]ASTM A615 Grade 60 shall be permitted if requirements of 20.2.2.5(b) are satisfied.

[3]In slabs and beams not part of a special seismic system, bars that pass through or extend from special structural walls shall satisfy 20.2.2.5.

[4]Longitudinal reinforcement with f_y > 80,000 psi is not permitted for intermediate moment frames and ordinary moment frames resisting earthquake demands E.

[5]Welded deformed bar mats shall be permitted to be assembled using only ASTM A615 or A706 deformed bars of Grade 60 or Grade 80.

[6]ASTM A1064 and A1022 are not permitted in special seismic systems if the weld is required to resist stresses in response to confinement, lateral support of longitudinal bars, shear, or other actions.

[7]This application also includes shear reinforcement with a maximum value of 80,000 psi f_y or f_{yt} permitted for design calculations for diaphragms and foundations for load combinations including earthquake forces if part of a building with a special seismic system.

[8]Shear reinforcement in this application includes stirrups, ties, hoops, and spirals in special moment frames.

[9]Shear reinforcement in this application includes all transverse reinforcement in special structural walls, coupling beams, and wall piers. Diagonal bars in coupling beams shall comply with ASTM A706 or Footnote [2].

CODE

Table 20.2.2.4(b)—Nonprestressed plain spiral reinforcement

Usage	Application	Maximum value of f_y or f_{yt} permitted for design calculations, psi	Applicable ASTM specification	
			Plain bars	Plain wires
Lateral support of longitudinal bars; or concrete confinement	Spirals in special seismic systems	100,000	A615, A706, A955, A1035	A1064, A1022
	Spirals	100,000	A615, A706, A955, A1035	A1064, A1022
Shear	Spirals	60,000	A615, A706, A955, A1035	A1064, A1022
Torsion in nonprestressed beams	Spirals	60,000	A615, A706, A955, A1035	A1064, A1022

20.2.2.5 Deformed nonprestressed longitudinal reinforcement resisting earthquake-induced moment, axial force, or both, in special seismic systems and anchor reinforcement in Seismic Design Categories (SDC) C, D, E, and F shall be in accordance with (a) or (b):

(a) ASTM A706, Grade 60, 80, or 100 for special structural walls and Grade 60 and 80 for special moment frames.
(b) ASTM A615 Grade 60 if (i) through (iv) are satisfied. ASTM A615 Grade 80 and Grade 100 are not permitted in special seismic systems.
 (i) Actual yield strength based on mill tests does not exceed f_y by more than 18,000 psi
 (ii) Ratio of the actual tensile strength to the actual yield strength is at least 1.25
 (iii) Minimum fracture elongation in 8 in. shall be at least 14 percent for bar sizes No. 3 through No. 6, at least 12 percent for bar sizes No. 7 through No. 11, and at least 10 percent for bar sizes No. 14 and No. 18.
 (iv) Minimum uniform elongation shall be at least 9 percent for bar sizes No. 3 through No. 10, and at least 6 percent for bar sizes No. 11, No. 14, and No. 18.

20.3—Prestressing strands, wires, and bars

20.3.1 *Material properties*

20.3.1.1 Except as required in 20.3.1.3 for special moment frames and special structural walls, prestressing reinforcement shall conform to (a), (b), (c), or (d):

(a) ASTM A416 – strand
(b) ASTM A421 – wire
(c) ASTM A421 – low-relaxation wire including Supplementary Requirement S1, "Low-Relaxation Wire and Relaxation Testing"
(d) ASTM A722 – high-strength bar

20.3.1.2 Prestressing strands, wires, and bars not listed in ASTM A416, A421, or A722 are permitted provided they conform to minimum requirements of these specifications and are shown by test or analysis not to impair the performance of the member.

COMMENTARY

R20.2.2.5 The requirement for the tensile strength to be greater than the yield strength of the reinforcement by a factor of 1.25 is based on the assumption that the capability of a structural member to develop inelastic rotation capacity is a function of the length of the yield region along the axis of the member. In interpreting experimental results, the length of the yield region has been related to the relative magnitudes of probable and yield moments (ACI 352R). According to this interpretation, the greater the ratio of probable-to-yield moment, the longer the yield region. Members with reinforcement not satisfying this condition can also develop inelastic rotation, but their behavior is sufficiently different to exclude them from direct consideration on the basis of rules derived from experience with members reinforced with strain-hardening steel.

The required minimum elongations in 20.2.2.5(b) for ASTM A615 Grade 60 are the same as the values in ASTM A706 for Grade 60 deformed reinforcement.

ASTM A615 Grade 80 and Grade 100 are not permitted to resist moments and axial forces in special seismic systems because of concern associated with low-cycle fatigue behavior (Slavin and Ghannoum 2015).

R20.3—Prestressing strands, wires, and bars

R20.3.1 *Material properties*

R20.3.1.1 Because low-relaxation prestressing reinforcement is addressed in a supplementary requirement to ASTM A421, which applies only if low-relaxation material is specified, the appropriate ASTM reference is listed as a separate entity.

20 Reinforcement

CODE

20.3.1.3 Prestressing reinforcement resisting earthquake-induced moment, axial force, or both, in special moment frames, special structural walls, and all components of special structural walls including coupling beams and wall piers, cast using precast concrete shall comply with ASTM A416 or ASTM A722.

20.3.2 *Design properties*

20.3.2.1 Modulus of elasticity, E_p, for prestressing reinforcement shall be determined from tests or as reported by the manufacturer.

20.3.2.2 Tensile strength, f_{pu}, shall be based on the specified grade or type of prestressing reinforcement and shall not exceed the values given in Table 20.3.2.2.

Table 20.3.2.2—Prestressing strands, wires, and bars

Type	Maximum value of f_{pu} permitted for design calculations, psi	Applicable ASTM Specification
Strand (stress-relieved and low-relaxation)	270,000	A416
Wire (stress-relieved and low-relaxation)	250,000	A421
		A421, including Supplementary Requirement S1 "Low-Relaxation Wire and Relaxation Testing"
High-strength bar	150,000	A722

20.3.2.3 *Stress in bonded prestressed reinforcement at nominal flexural strength,* f_{ps}

20.3.2.3.1 As an alternative to a more accurate calculation of f_{ps} based on strain compatibility, values of f_{ps} calculated in accordance with Eq. (20.3.2.3.1) shall be permitted for members with bonded prestressed reinforcement if all prestressed reinforcement is in the tension zone and $f_{se} \geq 0.5 f_{pu}$.

$$f_{ps} = f_{pu}\left\{1 - \frac{\gamma_p}{\beta_1}\left[\rho_p \frac{f_{pu}}{f'_c} + \frac{d}{d_p}\frac{f_y}{f'_c}(\rho - \rho')\right]\right\} \qquad (20.3.2.3.1)$$

where γ_p is in accordance with Table 20.3.2.3.1.

If compression reinforcement is considered for the calculation of f_{ps} by Eq. (20.3.2.3.1), (a) and (b) shall be satisfied.

(a) If d' exceeds $0.15d_p$, the compression reinforcement shall be neglected in Eq. (20.3.2.3.1).

COMMENTARY

R20.3.2 *Design properties*

R20.3.2.1 Default values of E_p between 28,500,000 and 29,000,000 psi are commonly used for design purposes. More accurate values based on tests or the manufacturer's reports may be needed for elongation checks during stressing.

R20.3.2.2 ASTM A416 specifies two grades of strand tensile strength: 250,000 and 270,000 psi.

ASTM A421 specifies tensile strengths of 235,000, 240,000, and 250,000 psi, depending on the diameter and type of wire. For the most common diameter, 0.25 in., ASTM A421 specifies a tensile strength of 240,000 psi.

R20.3.2.3 *Stress in bonded prestressed reinforcement at nominal flexural strength,* f_{ps}

R20.3.2.3.1 Use of Eq. (20.3.2.3.1) may underestimate the strength of beams with high percentages of reinforcement and, for more accurate evaluations of their strength, the strain compatibility and equilibrium method should be used. If part of the prestressed reinforcement is in the compression zone, a strain compatibility and equilibrium method should be used.

The γ_p term in Eq. (20.3.2.3.1) and Table 20.3.2.3.1 reflects the influence of different types of prestressing reinforcement on the value of f_{ps}. Table R20.3.2.3.1 shows prestressing reinforcement type and the associated ratio f_{py}/f_{pu}.

R20.3.2.3.1(a) If d' is large, the strain in compression reinforcement can be considerably less than its yield strain. In such a case, the compression reinforcement does not influence f_{ps} as favorably as implied by Eq. (20.3.2.3.1). For this

CODE

(b) If compression reinforcement is included in Eq. (20.3.2.3.1), the term

$$\left[\rho_p \frac{f_{pu}}{f_c'} + \frac{d}{d_p}\frac{f_y}{f_c'}(\rho - \rho')\right]$$

shall not be taken less than 0.17.

Table 20.3.2.3.1—Values of γ_p for use in Eq. (20.3.2.3.1)

f_{py}/f_{pu}	γ_p
≥ 0.80	0.55
≥ 0.85	0.40
≥ 0.90	0.28

20.3.2.3.2 For pretensioned strands, the strand design stress at sections of members located within ℓ_d from the free end of strand shall not exceed that calculated in accordance with 25.4.8.3.

20.3.2.4 *Stress in unbonded prestressed reinforcement at nominal flexural strength,* f_{ps}

20.3.2.4.1 As an alternative to a more accurate calculation of f_{ps}, values of f_{ps} calculated in accordance with Table 20.3.2.4.1 shall be permitted for members prestressed with unbonded tendons if $f_{se} \geq 0.5f_{pu}$.

Table 20.3.2.4.1—Approximate values of f_{ps} at nominal flexural strength for unbonded tendons

ℓ_n/h	f_{ps}	
≤ 35	The least of:	$f_{se} + 10{,}000 + f_c'/(100\rho_p)$
		$f_{se} + 60{,}000$
		f_{py}
> 35	The least of:	$f_{se} + 10{,}000 + f_c'/(300\rho_p)$
		$f_{se} + 30{,}000$
		f_{py}

COMMENTARY

reason, if d' exceeds $0.15d_p$, Eq. (20.3.2.3.1) is applicable only if the compression reinforcement is neglected.

R20.3.2.3.1(b) The ρ' term in Eq. (20.3.2.3.1) reflects the increased value of f_{ps} obtained when compression reinforcement is provided in a beam with a large reinforcement index. If the term $[\rho_p(f_{pu}/f_c') + (d/d_p)(f_y/f_c')(\rho - \rho')]$ is small, the neutral axis depth is small, the compressive reinforcement does not develop its yield strength, and Eq. (20.3.2.3.1) becomes unconservative. For this reason, the term $[\rho_p(f_{pu}/f_c') + (d/d_p)(f_y/f_c')(\rho - \rho')]$ may not be taken less than 0.17 if compression reinforcement is taken into account when calculating f_{ps}. The compression reinforcement may be conservatively neglected when using Eq. (20.3.2.3.1) by taking ρ' as zero, in which case the term $[\rho_p(f_{pu}/f_c') + (d/d_p)(f_y/f_c')(\rho)]$ may be less than 0.17 and an acceptable value of f_{ps} is obtained.

Table R20.3.2.3.1—Ratio of f_{py}/f_{pu} associated with reinforcement type

Prestressing reinforcement type		f_{py}/f_{pu}
High-strength prestressing bars	ASTM A722 Type I (Plain)	≥ 0.85
	ASTM A722 Type II (Deformed)	≥ 0.80
Stress-relieved strand and wire	ASTM A416 ASTM A421	≥ 0.85
Low-relaxation strand and wire	ASTM A416 ASTM A421	≥ 0.90

R20.3.2.4 *Stress in unbonded prestressed reinforcement at nominal flexural strength,* f_{ps}

R20.3.2.4.1 The term $[f_{se} + 10{,}000 + f_c'/(300\rho_p)]$ reflects results of tests on members with unbonded tendons and span-to-depth ratios greater than 35 (one-way slabs, flat plates, and flat slabs) (Mojtahedi and Gamble 1978). These tests also indicate that the term $[f_{se} + 10{,}000 + f_c'/(100\rho_p)]$, formerly used for all span-depth ratios, overestimates the amount of stress increase in such members. Although these same tests indicate that the moment strength of those shallow members designed using $[f_{se} + 10{,}000 + f_c'/(100\rho_p)]$ meets the factored load strength requirements, this reflects the effect of the Code requirements for minimum bonded reinforcement as well as the limitation on concrete tensile stress that often control the amount of prestressing force provided.

CODE

20.3.2.5 *Permissible tensile stresses in prestressed reinforcement*

20.3.2.5.1 The tensile stress in prestressed reinforcement shall not exceed the limits in Table 20.3.2.5.1.

Table 20.3.2.5.1—Maximum permissible tensile stresses in prestressed reinforcement

Stage	Location	Maximum tensile stress	
During stressing	At jacking end	Least of:	$0.94f_{py}$
			$0.80f_{pu}$
			Maximum jacking force recommended by the supplier of anchorage device
Immediately after force transfer	At post-tensioning anchorage devices and couplers	$0.70f_{pu}$	

20.3.2.6 *Prestress losses*

20.3.2.6.1 Prestress losses shall be considered in the calculation of the effective tensile stress in the prestressed reinforcement, f_{se}, and shall include (a) through (f):

(a) Prestressed reinforcement seating at transfer
(b) Elastic shortening of concrete
(c) Creep of concrete
(d) Shrinkage of concrete
(e) Relaxation of prestressed reinforcement
(f) Friction loss due to intended or unintended curvature in post-tensioning tendons

20.3.2.6.2 Calculated friction loss in post-tensioning tendons shall be based on experimentally determined wobble and curvature friction coefficients.

20.3.2.6.3 Where loss of prestress in a member is anticipated due to connection of the member to adjoining

COMMENTARY

R20.3.2.5 *Permissible tensile stresses in prestressed reinforcement*

R20.3.2.5.1 Because of the high yield strength of low-relaxation strand and wire meeting the requirements of ASTM A416 and ASTM A421 including Supplementary Requirement S1 "Low-Relaxation Wire and Relaxation Testing," it is appropriate to specify permissible stresses in terms of specified minimum ASTM yield strength along with the specified minimum ASTM tensile strength. Because of the higher allowable initial prestressed reinforcement stresses permitted since the 1983 Code, final stresses can be greater. For structures subject to corrosive conditions or repeated loadings, consideration should be given to limiting the final stress.

R20.3.2.6 *Prestress losses*

R20.3.2.6.1 ACI 423.10R provides a comprehensive treatment of the estimation of prestress losses.

Actual losses, greater or smaller than the calculated values, have little effect on the design strength of the member, but affect service load behavior (deflections, camber, cracking load) and connections. At service loads, overestimation of prestress losses can be almost as detrimental as underestimation because the former can result in excessive camber and horizontal movement.

R20.3.2.6.2 Estimation of friction losses in post-tensioned tendons is addressed in the *Post-Tensioning Manual* (TAB.1). Values of the wobble and curvature friction coefficients to be used for the particular types of prestressing reinforcement and particular types of ducts should be obtained from the manufacturers of the tendons. An unrealistically low estimate of the friction loss can lead to improper camber, or potential deflection, of the member and inadequate prestress. Overestimation of the friction may result in extra prestressing force. This could lead to excessive camber and excessive shortening of a member. If the friction factors are determined to be less than those assumed in the design, the tendon stressing should be adjusted to provide only that prestressing force in the critical portions of the structure required by the design.

When safety or serviceability of the structure may be involved, the acceptable range of prestressing reinforcement jacking forces or other limiting requirements should either be given or approved by the licensed design professional in conformance with the permissible stresses of 20.3.2.5 and 24.5.

CODE

construction, such loss of prestress shall be included in design calculations.

20.4—Headed shear stud reinforcement

20.4.1 Headed shear stud reinforcement and stud assemblies shall conform to ASTM A1044.

20.5—Provisions for durability of steel reinforcement

20.5.1 *Specified concrete cover*

20.5.1.1 Unless the general building code requires a greater concrete cover for fire protection, the minimum specified concrete cover shall be in accordance with 20.5.1.2 through 20.5.1.4.

COMMENTARY

R20.4—Headed shear stud reinforcement

R20.4.1 The configuration of the studs for headed shear stud reinforcement differs from the configuration of the headed-type shear studs prescribed in Section 7 of AWS D1.1 (2015) and referenced for use in Chapter 17 of this Code (Fig. R20.4.1). Ratios of the head to shank cross-sectional areas of the AWS D1.1 studs range from approximately 2.5 to 4. In contrast, ASTM A1044 requires the area of the head of headed shear stud reinforcement to be at least 10 times the area of the shank. Thus, the AWS D1.1 headed studs are not suitable for use as headed shear stud reinforcement. The base rail, where provided, anchors one end of the studs; ASTM A1044 specifies material width and thickness of the base rail that are sufficient to provide the required anchorage without yielding for stud shank diameters of 0.375, 0.500, 0.625, and 0.750 in. In ASTM A1044, the minimum specified yield strength of headed shear studs is 51,000 psi.

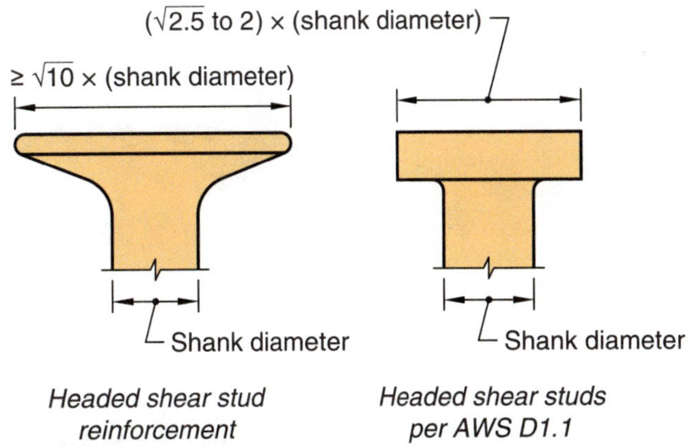

***Fig. R20.4.1**—Configurations of stud heads.*

R20.5—Provisions for durability of steel reinforcement

R20.5.1 *Specified concrete cover*

This section addresses concrete cover over reinforcement and does not include requirements for concrete cover over embedments such as pipes, conduits, and fittings, which are addressed in 20.6.5.

R20.5.1.1 Concrete cover as protection of reinforcement from weather and other effects is measured from the concrete surface to the outermost surface of the reinforcement to which the cover requirement applies. Where concrete cover is prescribed for a class of structural members, it is measured to the outer edge of stirrups, ties, or spirals if transverse reinforcement encloses main bars; to the outermost layer of bars if more than one layer is used without stirrups or ties; to the metal end fitting or duct of post-tensioning tendons; or to the outermost part of the head on headed bars.

CODE

COMMENTARY

The condition "exposed to weather or in contact with ground" refers to direct exposure to moisture changes and not just to temperature changes. Slab soffits are not usually considered directly exposed unless subject to alternate wetting and drying, including that due to condensation conditions or direct leakage from exposed top surface, run off, or similar effects.

Alternative methods of protecting the reinforcement from weather may be provided if they are equivalent to the additional concrete cover required by the Code. When approved by the building official under the provisions of 1.10, reinforcement with alternative protection from weather may not have concrete cover less than the cover required for reinforcement not exposed to weather.

Development length provisions given in Chapter 25 are a function of cover over the reinforcement. To meet requirements for development length, it may be necessary to use cover greater than the minimums specified in 20.5.1.

20.5.1.2 It shall be permitted to consider concrete floor finishes as part of required cover for nonstructural purposes.

R20.5.1.2 Concrete floor finishes may be considered for nonstructural purposes such as cover for reinforcement and fire protection. Provisions should be made, however, to ensure that the concrete finish will not spall off, thus resulting in decreased cover. Furthermore, considerations for development of reinforcement require minimum monolithic concrete cover in accordance with 20.5.1.3.

20.5.1.3 *Specified concrete cover requirements*

R20.5.1.3 *Specified concrete cover requirements*

20.5.1.3.1 Nonprestressed cast-in-place concrete members shall have specified concrete cover for reinforcement at least that given in Table 20.5.1.3.1.

Table 20.5.1.3.1—Specified concrete cover for cast-in-place nonprestressed concrete members

Concrete exposure	Member	Reinforcement	Specified cover, in.
Cast against and permanently in contact with ground	All	All	3
Exposed to weather or in contact with ground	All	No. 6 through No. 18 bars	2
		No. 5 bar, W31 or D31 wire, and smaller	1-1/2
Not exposed to weather or in contact with ground	Slabs, joists, and walls	No. 14 and No. 18 bars	1-1/2
		No. 11 bar and smaller	3/4
	Beams, columns, pedestals, and tension ties	Primary reinforcement, stirrups, ties, spirals, and hoops	1-1/2

CODE

20.5.1.3.2 Cast-in-place prestressed concrete members shall have specified concrete cover for reinforcement, ducts, and end fittings at least that given in Table 20.5.1.3.2.

Table 20.5.1.3.2—Specified concrete cover for cast-in-place prestressed concrete members

Concrete exposure	Member	Reinforcement	Specified cover, in.
Cast against and permanently in contact with ground	All	All	3
Exposed to weather or in contact with ground	Slabs, joists, and walls	All	1
	All other	All	1-1/2
Not exposed to weather or in contact with ground	Slabs, joists, and walls	All	3/4
	Beams, columns, and tension ties	Primary reinforcement	1-1/2
		Stirrups, ties, spirals, and hoops	1

20.5.1.3.3 Precast nonprestressed or prestressed concrete members manufactured under plant conditions shall have specified concrete cover for reinforcement, ducts, and end fittings at least that given in Table 20.5.1.3.3.

COMMENTARY

R20.5.1.3.3 The lesser cover thicknesses for precast construction reflect the greater control for proportioning, placing, and curing inherent in precasting. Manufactured under plant conditions does not imply that precast members should be manufactured in a plant. Structural elements precast at the job site will also qualify under this section if the control of form dimensions, placing of reinforcement, quality control of concrete, and curing procedures are equal to that normally expected in a plant.

Concrete cover to pretensioned strand as described in this section is intended to provide minimum protection from weather and other effects. Such cover may not be sufficient to transfer or develop the stress in the strand, and it may be necessary to increase the cover accordingly.

CODE

Table 20.5.1.3.3—Specified concrete cover for precast-nonprestressed or prestressed concrete members manufactured under plant conditions

Concrete exposure	Member	Reinforcement	Specified cover, in.
Exposed to weather or in contact with ground	Walls	No. 14 and No. 18 bars; tendons larger than 1-1/2 in. diameter	1-1/2
		No. 11 bars and smaller; W31 and D31 wire and smaller; tendons and strands 1-1/2 in. diameter and smaller	3/4
	All other	No. 14 and No. 18 bars; tendons larger than 1-1/2 in. diameter	2
		No. 6 through No. 11 bars; tendons and strands larger than 5/8 in. diameter through 1-1/2 in. diameter	1-1/2
		No. 5 bar, W31 or D31 wire, and smaller; tendons and strands 5/8 in. diameter and smaller	1-1/4
Not exposed to weather or in contact with ground	Slabs, joists, and walls	No. 14 and No. 18 bars; tendons larger than 1-1/2 in. diameter	1-1/4
		Tendons and strands 1-1/2 in. diameter and smaller	3/4
		No. 11 bar, W31 or D31 wire, and smaller	5/8
	Beams, columns, pedestals, and tension ties	Primary reinforcement	Greater of d_b and 5/8 and need not exceed 1-1/2
		Stirrups, ties, spirals, and hoops	3/8

20.5.1.3.4 Deep foundation members shall have specified concrete cover for reinforcement at least that given in Table 20.5.1.3.4.

COMMENTARY

CODE

Table 20.5.1.3.4—Specified concrete cover for deep foundation members

Concrete exposure	Deep foundation member type	Reinforcement	Specified cover, in.
Cast against and permanently in contact with ground, not enclosed by steel pipe, tube permanent casing, or stable rock socket	Cast-in-place	All	3
Enclosed by steel pipe, tube, permanent casing, or stable rock socket	Cast-in-place	All	1-1/2
Permanently in contact with ground	Precast-nonprestressed	All	1-1/2
	Precast-prestressed		
Exposed to seawater	Precast-nonprestressed	All	2-1/2
	Precast-prestressed	All	2

20.5.1.3.5 For bundled bars, specified concrete cover shall be at least the smaller of (a) and (b):

(a) The equivalent diameter of the bundle
(b) 2 in.

and for concrete cast against and permanently in contact with ground, the specified cover shall be 3 in.

20.5.1.3.6 For headed shear stud reinforcement, specified concrete cover for the heads and base rails shall be at least that required for the reinforcement in the member.

COMMENTARY

R20.5.1.3.6 Concrete cover requirements for headed shear stud reinforcement are illustrated in Fig. R20.5.1.3.6.

CODE | **COMMENTARY**

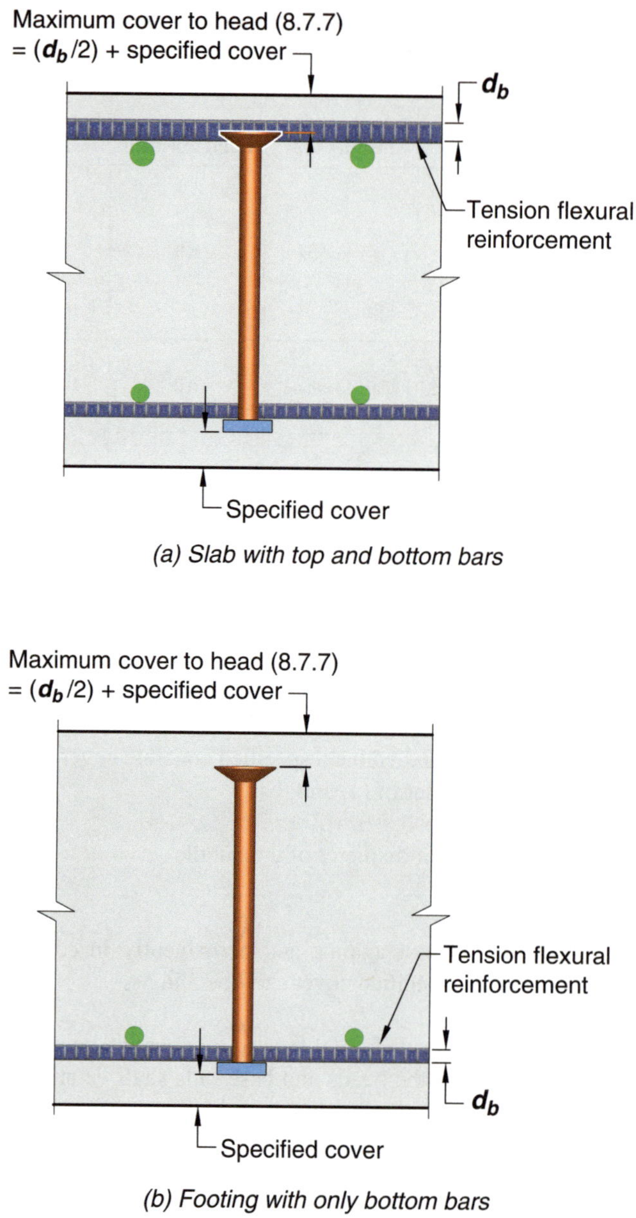

Fig. R20.5.1.3.6*—Concrete cover requirements for headed shear stud reinforcement.*

20.5.1.4 *Specified concrete cover requirements for corrosive environments*

20.5.1.4.1 In corrosive environments or other severe exposure conditions, the specified concrete cover shall be increased as deemed necessary. The applicable requirements for concrete based on exposure categories in 19.3 shall be satisfied, or other protection shall be provided.

R20.5.1.4 *Specified concrete cover requirements for corrosive environments*

Corrosive environments are defined in 19.3.1, R19.3.1, and R19.3.2. Additional information on corrosion in parking structures is given in ACI 362.1R.

R20.5.1.4.1 Where concrete will be exposed to external sources of chlorides in service, such as deicing salts, brackish water, seawater, or spray from these sources, concrete should be proportioned to satisfy the requirements for the applicable exposure class in Chapter 19. These include maximum ***w/cm***, minimum strength for normalweight and lightweight concrete, and maximum chloride ion in the concrete. Addi-

CODE

COMMENTARY

tionally, for corrosion protection, a specified concrete cover for reinforcement not less than 2 in. for walls and slabs and not less than 2-1/2 in. for other members is recommended. For precast concrete members manufactured under plant control conditions, a specified concrete cover not less than 1-1/2 in. for walls and slabs and not less than 2 in. for other members is recommended.

20.5.1.4.2 For prestressed concrete members classified as Class T or C in 24.5.2 and exposed to corrosive environments or other severe exposure categories such as those given in 19.3, the specified concrete cover for prestressed reinforcement shall be at least one and one-half times the cover in 20.5.1.3.2 for cast-in-place members and in 20.5.1.3.3 for precast members.

20.5.1.4.3 If the precompressed tension zone is not in tension under sustained loads, 20.5.1.4.2 need not be satisfied.

20.5.2 *Nonprestressed coated reinforcement*

20.5.2.1 Nonprestressed coated reinforcement shall conform to Table 20.5.2.1.

Table 20.5.2.1—Nonprestressed coated reinforcement

Type of coating	Applicable ASTM specifications		
	Bar	Wire	Welded wire
Zinc-coated	A767	Not permitted	A1060
Epoxy-coated	A775 or A934	A884	A884
Zinc and epoxy dual-coated	A1055	Not permitted	Not permitted

20.5.2.2 Deformed bars to be zinc-coated, epoxy-coated, or zinc and epoxy dual-coated shall conform to 20.2.1.3(a), (b), or (c).

20.5.2.3 Wire and welded wire reinforcement to be epoxy-coated shall conform to 20.2.1.7(a).

R20.5.2 *Nonprestressed coated reinforcement*

R20.5.2.1 Zinc-coated (hot-dipped galvanized) bars (ASTM A767), epoxy-coated bars (ASTM A775 and A934), and zinc and epoxy dual-coated bars (ASTM A1055) are used in applications where corrosion resistance of reinforcement is of particular concern such as in parking structures, bridge structures, and other highly corrosive environments.

20.5.3 *Corrosion protection for unbonded prestressing reinforcement*

20.5.3.1 Unbonded prestressing reinforcement shall be encased in sheathing, and the space between the prestressing reinforcement and the sheathing shall be completely filled with a material formulated to inhibit corrosion. Sheathing shall be watertight and continuous over the unbonded length.

R20.5.3 *Corrosion protection for unbonded prestressing reinforcement*

R20.5.3.1 Material for corrosion protection of unbonded prestressing reinforcement should have the properties identified in 19.1 of Breen et al. (1994).

Typically, sheathing is a continuous, seamless, high-density polyethylene material that is extruded directly onto the coated prestressing reinforcement.

CODE

20.5.3.2 The sheathing shall be connected to all stressing, intermediate, and fixed anchorages in a watertight fashion.

20.5.3.3 Unbonded single-strand tendons shall be protected to provide resistance to corrosion in accordance with ACI 423.7.

20.5.4 *Corrosion protection for grouted tendons*

20.5.4.1 Ducts for grouted tendons shall be grout-tight and nonreactive with concrete, prestressing reinforcement, grout, and corrosion inhibitor admixtures.

20.5.4.2 Ducts shall be maintained free of water.

20.5.4.3 Ducts for grouted single-wire, single-strand, or single-bar tendons shall have an inside diameter at least 1/4 in. larger than the diameter of the prestressing reinforcement.

20.5.4.4 Ducts for grouted multiple wire, multiple strand, or multiple bar tendons shall have an inside cross-sectional area at least two times the cross-sectional area of the prestressing reinforcement.

20.5.5 *Corrosion protection for post-tensioning anchorages, couplers, and end fittings*

20.5.5.1 Anchorages, couplers, and end fittings shall be protected to provide long-term resistance to corrosion.

20.5.6 *Corrosion protection for external post-tensioning*

20.5.6.1 External tendons and tendon anchorage regions shall be protected to provide resistance to corrosion.

20.6—Embedments

20.6.1 Embedments shall not significantly impair the strength of the structure and shall not reduce fire protection.

COMMENTARY

R20.5.4 *Corrosion protection for grouted tendons*

R20.5.4.2 Water in ducts may cause corrosion of the prestressing reinforcement, may lead to bleeding and segregation of grout, and may cause distress to the surrounding concrete if subjected to freezing conditions. A corrosion inhibitor should be used to provide temporary corrosion protection if prestressing reinforcement is exposed in the ducts for prolonged periods of time before grouting (ACI 423.7).

R20.5.5 *Corrosion protection for post-tensioning anchorages, couplers, and end fittings*

R20.5.5.1 For recommendations regarding protection, refer to 4.2 and 4.3 of Mojtahedi and Gamble (1978) and 3.4, 3.6, 5, 6, and 6.3 of Breen et al. (1994).

R20.5.6 *Corrosion protection for external post-tensioning*

R20.5.6.1 Corrosion protection can be achieved by a variety of methods. The corrosion protection provided should be suitable to the environment in which the tendons are located. Some conditions will require that the prestressing reinforcement be protected by concrete cover or by cement grout in polyethylene or metal tubing; other conditions will permit the protection provided by coatings such as paint or grease. Corrosion protection methods should meet the fire protection requirements of the general building code, unless the installation of external post-tensioning is to only improve serviceability.

R20.6—Embedments

R20.6.1 Any embedments not harmful to concrete or reinforcement can be placed in the concrete, but the work should be done in such a manner that the structure will not

CODE

20.6.2 Embedment materials shall not be harmful to concrete or reinforcement.

20.6.3 Aluminum embedments shall be coated or covered to prevent aluminum-concrete reaction and electrolytic action between aluminum and steel.

20.6.4 Reinforcement with an area at least 0.002 times the area of the concrete section shall be provided perpendicular to pipe embedments.

20.6.5 Specified concrete cover for pipe embedments with their fittings shall be at least 1-1/2 in. for concrete exposed to earth or weather, and at least 3/4 in. for concrete not exposed to weather, or not in contact with ground.

COMMENTARY

be endangered. Many general building codes have adopted ASME Piping Code B31.1 for power piping and B31.3 for chemical and petroleum piping. The licensed design professional should be sure that the appropriate piping codes are used in the design and testing of the system. The contractor should not be permitted to install conduits, pipes, ducts, or sleeves that are not shown in the construction documents or not approved by the licensed design professional.

R20.6.3 The Code prohibits the use of aluminum in structural concrete unless it is effectively coated or covered. Aluminum reacts with concrete and, in the presence of chloride ions, may also react electrolytically with steel, causing cracking, spalling, or both. Aluminum electrical conduits present a special problem because stray electric current accelerates the adverse reaction. Provision 26.4.2.2(f) prohibits calcium chloride or any admixture containing chloride from being used in concrete with aluminum embedments.

Notes

Structural Design Example

Start off with Retaining wall design find M_O M_R, W_u, $\bar{x}$, e, kern, qu and calc slope. Same diagram too

$$\text{slope} = \frac{q_{max} - q_{min}}{B} = \frac{3408-928}{10} = 334 \text{ psf/ft}$$

$d = 24'' - 2'' - 0.5'' = 21.5''$

Stem design

$P_u = 1.6\, k_a \gamma h = 1.6(0.25)(120\text{ pcf})(16') = 768$ psf

$R_u = \frac{1}{2} p_u h = \frac{1}{2}(768)(16') = 6144$ lbs

$V_u = 6144$ lbs

Capacity

$\phi 2 \sqrt{f'_c}\, b d = (0.75)(2)\sqrt{3000 \text{ psi}}\,(12'')(24.5'') = 21,152$ lbs

$M_u = [\frac{1}{2} P_u h][\frac{1}{3} h] = [6144][\frac{1}{3}(16)] = 32,768$ ft-lbs $= 393,216$ in-lbs

Use quadratic from strip footing example to get $A_s = 0.344$

$\text{Spacing} = \frac{A_b}{A_s}(12'')$

$A_{s,min} = 0.0012 A_g = 0.0012(12)(24)$ check Page 171 to make sure

- #4 @ 6 → $A_s = 0.4$ in²/ft
- #5 @ 6 → $A_s = 0.37$ in²/ft

$$a = \frac{A_s f_y}{0.85 f'_c b} = \frac{0.2 \text{ in}^2 (60{,}000 \text{ psf})}{0.85(3000 \text{ psi})(12'')} = 0.392''$$

$$\phi M_n = \phi A_s f_y (d - \tfrac{a}{2}) = (0.9)(0.2)(60{,}000)\left[21.5'' - \frac{0.392}{2}\right]$$

$\phi M_n = 230,083$ in.lbs $= 19,174$ ft.lbs

$P_u(x) = 1.6\, k_a \gamma x = 48x$

$M_u = \frac{1}{2} p_u(x)(x)[\frac{1}{3}(x)] = 8x^3$

$\phi M_n = M_u$ solve for $x = 13.4'$ cut from top, every other vertical #4 on interior face.

$$l_d = \frac{f_y \psi_t \psi_e}{25 \lambda \sqrt{f'_c}} d_b = \frac{(60{,}000)(1)(1)}{25(1)\sqrt{3000}}(0.5)'' = 21.9'' = 1.8'$$

CHAPTER 21—STRENGTH REDUCTION FACTORS

CODE

21.1—Scope

21.1.1 This chapter shall apply to the selection of strength reduction factors used in design, except as permitted by Chapter 27.

21.2—Strength reduction factors for structural concrete members and connections

21.2.1 Strength reduction factors ϕ shall be in accordance with Table 21.2.1, except as modified by 21.2.2, 21.2.3, and 21.2.4.

Table 21.2.1—Strength reduction factors ϕ

	Action or structural element	ϕ	Exceptions
(a)	Moment, axial force, or combined moment and axial force	0.65 to 0.90 in accordance with 21.2.2	Near ends of pretensioned members where strands are not fully developed, ϕ shall be in accordance with 21.2.3.
(b)	Shear	0.75	Additional requirements are given in 21.2.4 for structures designed to resist earthquake effects.
(c)	Torsion	0.75	—
(d)	Bearing	0.65	—
(e)	Post-tensioned anchorage zones	0.85	—
(f)	Brackets and corbels	0.75	—
(g)	Struts, ties, nodal zones, and bearing areas designed in accordance with strut-and-tie method in Chapter 23	0.75	—
(h)	Components of connections of precast members controlled by yielding of steel elements in tension	0.90	—
(i)	Plain concrete elements	0.60	—
(j)	Anchors in concrete elements	0.45 to 0.75 in accordance with Chapter 17	—

21.2.2 Strength reduction factor for moment, axial force, or combined moment and axial force shall be in accordance with Table 21.2.2.

COMMENTARY

R21.1—Scope

R21.1.1 The purposes of strength reduction factors ϕ are: (1) to account for the probability of under-strength members due to variations in material strengths and dimensions; (2) to account for inaccuracies in the design equations; (3) to reflect the available ductility and required reliability of the member under the load effects being considered; and (4) to reflect the importance of the member in the structure (MacGregor 1976; Winter 1979).

R21.2—Strength reduction factors for structural concrete members and connections

R21.2.1 The strength reduction factors in this Code are compatible with the ASCE/SEI 7 load combinations, which are the basis for the required factored load combinations in Chapter 5:

(e) Laboratory tests of post-tensioned anchorage zones (Breen et al. 1994) indicate a wide range of scatter in the results. This observation is addressed with a ϕ-factor of 0.85 and by limiting the nominal compressive strength of unconfined concrete in the general zone to $\mathbf{0.7\lambda f_{ci}'}$ in 25.9.4.5.2, where λ is defined in 19.2.4. Thus, the effective design strength of unconfined concrete is $\mathbf{0.85 \times 0.7\lambda f_{ci}' = 0.6\lambda f_{ci}'}$ in the general zone.

(f) Bracket and corbel behavior is predominantly controlled by shear; therefore, a single value of $\phi = \mathbf{0.75}$ is used for all potential modes of failure.

(i) The strength reduction factor ϕ for plain concrete members is the same for all potential modes of failure. Because both the flexural tension strength and shear strength for plain concrete depend on the tensile strength of the concrete, without the reserve strength or ductility that might otherwise be provided by reinforcement, equal strength reduction factors for moment and shear are considered to be appropriate.

R21.2.2 The nominal strength of a member that is subjected to moment or combined moment and axial force is determined for the condition where the strain in the extreme compression fiber is equal to the assumed strain limit of 0.003. The net tensile strain ε_t is the tensile strain calculated in the extreme tension reinforcement at nominal strength,

CODE

21.2.2.1 For deformed reinforcement, ε_{ty} shall be f_y/E_s. For Grade 60 deformed reinforcement, it shall be permitted to take ε_{ty} equal to 0.002.

21.2.2.2 For all prestressed reinforcement, ε_{ty} shall be taken as 0.002.

COMMENTARY

exclusive of strains due to prestress, creep, shrinkage, and temperature. The net tensile strain in the extreme tension reinforcement is determined from a linear strain distribution at nominal strength, shown in Fig. R21.2.2a for a nonprestressed member.

Members subjected to only axial compression are considered to be compression-controlled and members subjected to only axial tension are considered to be tension-controlled.

If the net tensile strain in the extreme tension reinforcement is sufficiently large ($\geq \varepsilon_{ty}$ + **0.003**), the section is defined as tension-controlled, for which warning of failure by excessive deflection and cracking may be expected. The limit $\geq \varepsilon_{ty}$ + **0.003** provides sufficient ductility for most applications. Before the 2019 Code, the tension-controlled limit on ε_t was defined as 0.005 established primarily on the basis of Grade 60 nonprestressed reinforcement and prestressed reinforcement, with some consideration given to higher grades of nonprestressed reinforcement (Mast 1992). Beginning with the 2019 Code, to accommodate nonprestressed reinforcement of higher grades, the tension-controlled limit on ε_t in Table 21.2.2 is defined as ε_{ty} + **0.003**. This expression is consistent with the recommendations of Mast (1992) for the general case of reinforcement other than Grade 60, and test data show that the expression leads to elements with adequate ductility.

One condition where greater ductile behavior is required is in design for redistribution of moments in continuous members and frames, which is addressed in 6.6.5. Because redistribution of moment depends on the ductility available in the hinge regions, redistribution of moment is limited to sections that have a net tensile strain of at least 0.0075.

If the net tensile strain in the extreme tension reinforcement is small ($\leq \varepsilon_{ty}$), a brittle compression failure condition is expected, with little warning of impending failure. Before ACI 318-14, the compression-controlled strain limit was defined as 0.002 for Grade 60 reinforcement and all prestressed reinforcement, but it was not explicitly defined for other types of reinforcement. The compression-controlled strain limit ε_{ty} is defined in 21.2.2.1 and 21.2.2.2 for deformed and prestressed reinforcement, respectively.

Beams and slabs are usually tension-controlled, whereas columns may be compression-controlled. Some members, such as those with small axial forces and large bending moments, experience net tensile strain in the extreme tension reinforcement between the limits of ε_{ty} and (ε_{ty} + **0.003**). These sections are in a transition region between compression-controlled and tension-controlled.

This section specifies the appropriate strength reduction factors for tension-controlled and compression-controlled sections, and for intermediate cases in the transition region. Beginning with the 2019 Code, the expression (ε_{ty} + **0.003**) defines the limit on ε_t for tension-controlled behavior in Table 21.2.2. For sections subjected to combined axial force and moment, design strengths are determined by multiplying both P_n and M_n by the appropriate single value of ϕ.

CODE

COMMENTARY

A lower ϕ-factor is used for compression-controlled sections than for tension-controlled sections because compression-controlled sections have less ductility, are more sensitive to variations in concrete strength, and generally occur in members that support larger loaded areas than members with tension-controlled sections. Columns with spiral reinforcement are assigned a higher ϕ-factor than columns with other types of transverse reinforcement because spiral columns have greater ductility and toughness. For sections within the transition region, the value of ϕ may be determined by linear interpolation, as shown in Fig. R21.2.2b.

Table 21.2.2—Strength reduction factor ϕ for moment, axial force, or combined moment and axial force

Net tensile stain ε_t	Classification	ϕ			
		Type of transverse reinforcement			
		Spirals conforming to 25.7.3		Other	
$\varepsilon_t \leq \varepsilon_{ty}$	Compression-controlled	0.75	(a)	0.65	(b)
$\varepsilon_{ty} < \varepsilon_t < \varepsilon_{ty} + 0.003$	Transition[1]	$0.75 + 0.15\dfrac{(\varepsilon_t - \varepsilon_{ty})}{(0.003)}$	(c)	$0.65 + 0.25\dfrac{(\varepsilon_t - \varepsilon_{ty})}{(0.003)}$	(d)
$\varepsilon_t \geq \varepsilon_{ty} + 0.003$	Tension-controlled	0.90	(e)	0.90	(f)

[1]For sections classified as transition, it shall be permitted to use ϕ corresponding to compression-controlled sections.

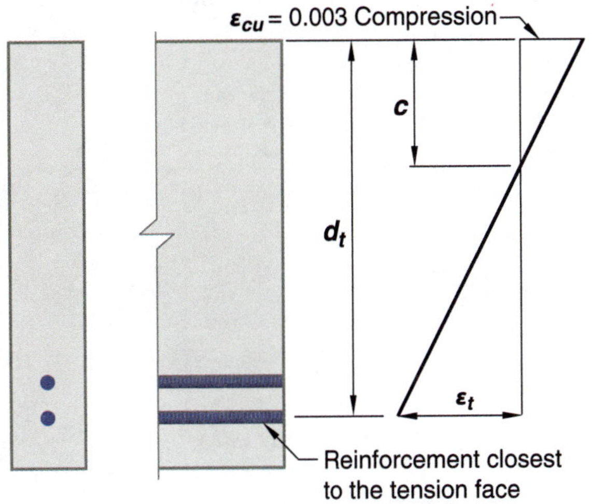

***Fig. R21.2.2a**—Strain distribution and net tensile strain in a nonprestressed member.*

CODE

COMMENTARY

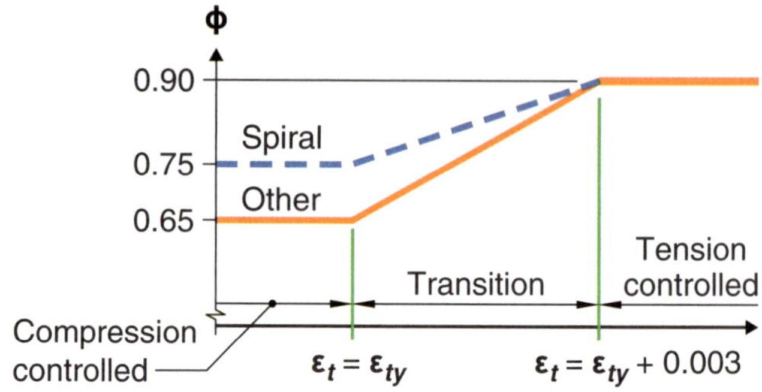

***Fig. R21.2.2b**—Variation of ϕ with net tensile strain in extreme tension reinforcement, ε_t.*

21.2.3 For sections in pretensioned flexural members where all strands are not fully developed, ϕ for moment shall be calculated in accordance with Table 21.2.3, where ℓ_{tr} is calculated using Eq. (21.2.3), ϕ_p is the value of ϕ determined in accordance with Table 21.2.2 at the cross section closest to the end of member where all strands are developed, and ℓ_d is given in 25.4.8.1.

$$\ell_{tr} = \left(\frac{f_{se}}{3000}\right) d_b \qquad (21.2.3)$$

Table 21.2.3—Strength reduction factor ϕ for sections near the end of pretensioned members

Condition near end of member	Stress in concrete under service load[1]	Distance from end of member to section under consideration	ϕ	
All strands bonded	Not applicable	$\leq \ell_{tr}$	0.75	(a)
		ℓ_{tr} to ℓ_d	Linear interpolation from 0.75 to ϕ_p[2]	(b)
One or more strands debonded	No tension calculated	$\leq (\ell_{db} + \ell_{tr})$	0.75	(c)
		$(\ell_{db} + \ell_{tr})$ to $(\ell_{db} + \ell_d)$	Linear interpolation from 0.75 to ϕ_p[2]	(d)
	Tension calculated	$\leq (\ell_{db} + \ell_{tr})$	0.75	(e)
		$(\ell_{db} + \ell_{tr})$ to $(\ell_{db} + 2\ell_d)$	Linear interpolation from 0.75 to ϕ_p[2]	(f)

[1]Stress calculated using gross cross-sectional properties in extreme concrete fiber of precompressed tension zone under service loads after allowance for all prestress losses at section under consideration.

[2]It shall be permitted to use a strength reduction factor of 0.75.

R21.2.3 If a critical section along a pretensioned member occurs in a region where not all the strands are fully developed, bond slip failure may occur. This mode of failure resembles a brittle shear failure; hence, ϕ values for flexure are reduced relative to the value of ϕ at the cross section where all strands are fully developed. For sections between the end of the transfer length and the end of the development length, the value of ϕ may be determined by linear interpolation, as shown in Fig. R21.2.3a, where ϕ_p corresponds to the value of ϕ at the cross section closest to the end of the member where all strands are fully developed.

Where bonding of one or more strands does not extend to the end of the member, instead of more rigorous analysis, ϕ should be taken as 0.75 from the end of the member to the end of the transfer length of the strand with the longest debonded length. Beyond this point, ϕ may be varied linearly to ϕ_p at the cross section where all strands are developed, as shown in Fig. R21.2.3b. Alternatively, the value of ϕ may be taken as 0.75 until all strands are fully developed. Embedment of debonded strand is considered to begin at the termination of the debonding sleeves. Beyond this point, the provisions of 25.4.8.1 are used to determine whether the strands develop over a length of ℓ_d or $2\ell_d$, depending on the calculated stress in the precompressed tension zone under service loads (Fig. R21.2.3b).

CODE

COMMENTARY

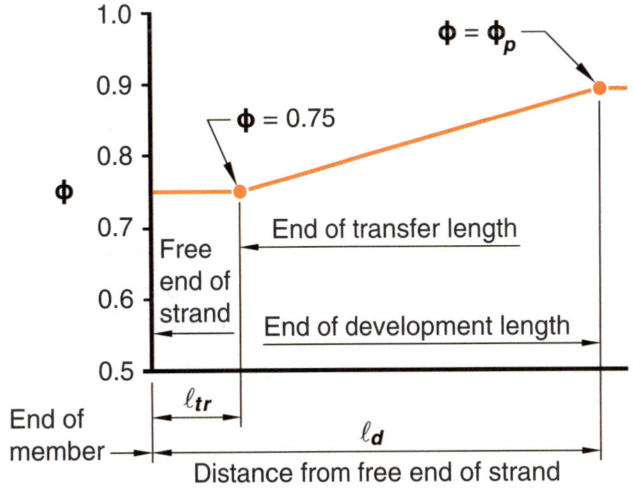

Fig. R21.2.3a*—Variation of* ϕ *with distance from the free end of strand in pretensioned member with fully bonded strands.*

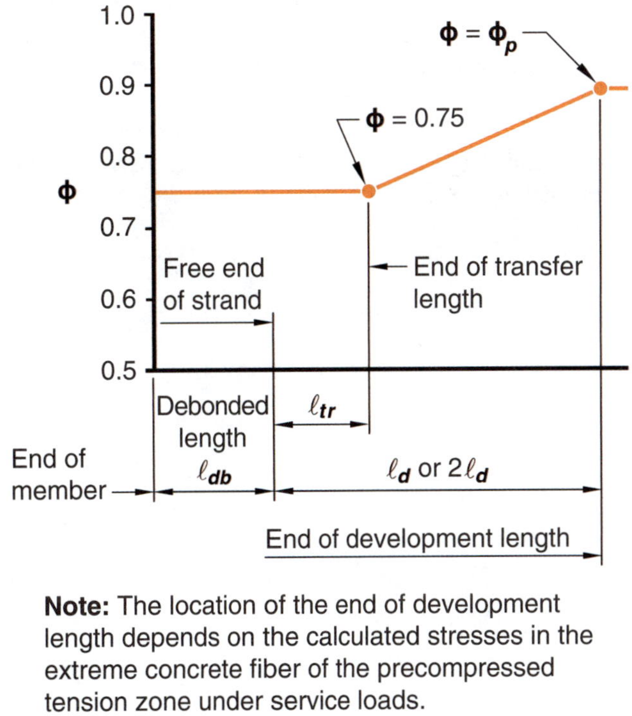

Fig. R21.2.3b*—Variation of* ϕ *with distance from the free end of strand in pretensioned member with debonded strands.*

21.2.4 For structures that rely on elements in (a), (b), or (c) to resist earthquake effects $\boldsymbol{E}$, the value of ϕ for shear shall be modified in accordance with 21.2.4.1 through 21.2.4.4:

(a) Special moment frames
(b) Special structural walls
(c) Intermediate precast structural walls in structures assigned to Seismic Design Category D, E, or F

CODE

21.2.4.1 For any member designed to resist ***E***, ϕ for shear shall be 0.60 if the nominal shear strength of the member is less than the shear corresponding to the development of the nominal moment strength of the member. The nominal moment strength shall be the maximum value calculated considering factored axial loads from load combinations that include ***E***.

21.2.4.2 For diaphragms, ϕ for shear shall not exceed the least value of ϕ for shear used for the vertical components of the primary seismic-force-resisting system.

21.2.4.3 For foundation elements supporting the primary seismic-force-resisting system, ϕ for shear shall not exceed the least value of ϕ for shear used for the vertical components of the primary seismic-force-resisting system.

21.2.4.4 For beam-column joints of special moment frames and diagonally reinforced coupling beams, ϕ for shear shall be 0.85.

COMMENTARY

R21.2.4.1 This provision addresses shear-controlled members, such as low-rise walls, portions of walls between openings, or diaphragms, for which nominal shear strength is less than the shear corresponding to development of nominal flexural strength for the pertinent loading conditions.

R21.2.4.2 Short structural walls were the primary vertical elements of the lateral-force-resisting system in many of the parking structures that sustained damage during the 1994 Northridge earthquake. In some cases, walls remained essentially linear elastic, while diaphragms responded inelastically. This provision is intended to increase strength of the diaphragm and its connections in buildings for which the shear strength reduction factor for walls is 0.60, as those structures tend to have relatively high overstrength.

R21.2.4.3 This provision is intended to provide consistent reliability for shear in foundation elements that support shear-controlled walls designed with a strength reduction factor of 0.6.

CHAPTER 22—SECTIONAL STRENGTH

22.1—Scope

22.1.1 This chapter shall apply to calculating nominal strength at sections of members, including (a) through (g):

(a) Flexural strength
(b) Axial strength or combined flexural and axial strength
(c) One-way shear strength
(d) Two-way shear strength
(e) Torsional strength
(f) Bearing
(g) Shear friction

22.1.2 Sectional strength requirements of this chapter shall be satisfied unless the member or region of the member is designed in accordance with Chapter 23.

22.1.3 Design strength at a section shall be taken as the nominal strength multiplied by the applicable strength reduction factor ϕ given in Chapter 21.

R22.1—Scope

R22.1.1 The provisions in this chapter apply where the strength of the member is evaluated at critical sections.

R22.1.2 Chapter 23 provides methods for designing discontinuity regions where section-based methods do not apply.

22.2—Design assumptions for moment and axial strength

22.2.1 *Equilibrium and strain compatibility*

22.2.1.1 Equilibrium shall be satisfied at each section.

22.2.1.2 Strain in concrete and nonprestressed reinforcement shall be assumed proportional to the distance from neutral axis.

22.2.1.3 Strain in prestressed concrete and in bonded and unbonded prestressed reinforcement shall include the strain due to effective prestress.

22.2.1.4 Changes in strain for bonded prestressed reinforcement shall be assumed proportional to the distance from neutral axis.

R22.2—Design assumptions for moment and axial strength

R22.2.1 *Equilibrium and strain compatibility*

The flexural and axial strength of a member calculated by the strength design method of the Code requires that two basic conditions be satisfied: 1) equilibrium; and 2) compatibility of strains. Equilibrium refers to the balancing of forces acting on the cross section at nominal strength. The relationship between the stress and strain for the concrete and the reinforcement at nominal strength is established within the design assumptions allowed by 22.2.

R22.2.1.2 It is reasonable to assume a linear distribution of strain across a reinforced concrete cross section (plane sections remain plane), even near nominal strength except in cases as described in Chapter 23.

The strain in both nonprestressed reinforcement and in concrete is assumed to be directly proportional to the distance from the neutral axis. This assumption is of primary importance in design for determining the strain and corresponding stress in the reinforcement.

R22.2.1.4 The change in strain for bonded prestressed reinforcement is influenced by the change in strain at the section under consideration. For unbonded prestressed reinforcement, the change in strain is influenced by external load, reinforcement location, and boundary conditions along the length of the reinforcement. Current Code equations for calculating f_{ps} for unbonded tendons, as provided in 20.3.2.4, have been correlated with test results.

CODE

22.2.2 *Design assumptions for concrete*

22.2.2.1 Maximum strain at the extreme concrete compression fiber shall be assumed equal to 0.003.

22.2.2.2 Tensile strength of concrete shall be neglected in flexural and axial strength calculations.

22.2.2.3 The relationship between concrete compressive stress and strain shall be represented by a rectangular, trapezoidal, parabolic, or other shape that results in prediction of strength in substantial agreement with results of comprehensive tests.

22.2.2.4 The equivalent rectangular concrete stress distribution in accordance with 22.2.2.4.1 through 22.2.2.4.3 satisfies 22.2.2.3.

22.2.2.4.1 Concrete stress of $\mathbf{0.85}\boldsymbol{f_c'}$ shall be assumed uniformly distributed over an equivalent compression zone bounded by edges of the cross section and a line parallel to the neutral axis located a distance $\boldsymbol{a}$ from the fiber of maximum compressive strain, as calculated by:

$$a = \beta_1 c \qquad (22.2.2.4.1)$$

22.2.2.4.2 Distance from the fiber of maximum compressive strain to the neutral axis, $\boldsymbol{c}$, shall be measured perpendicular to the neutral axis.

22.2.2.4.3 Values of $\boldsymbol{\beta_1}$ shall be in accordance with Table 22.2.2.4.3.

COMMENTARY

R22.2.2 *Design assumptions for concrete*

R22.2.2.1 The maximum concrete compressive strain at crushing of the concrete has been observed in tests of various kinds to vary from 0.003 to higher than 0.008 under special conditions. However, the strain at which strength of the member is developed is usually 0.003 to 0.004 for members of normal proportions, materials, and strength.

R22.2.2.2 The tensile strength of concrete in flexure (modulus of rupture) is a more variable property than the compressive strength and is approximately 10 to 15 percent of the compressive strength. Tensile strength of concrete in flexure is conservatively neglected in calculating the nominal flexural strength. The strength of concrete in tension, however, is important in evaluating cracking and deflections at service loads.

R22.2.2.3 At high strain levels, the stress-strain relationship for concrete is nonlinear (stress is not proportional to strain). As stated in 22.2.2.1, the maximum usable strain is set at 0.003 for design.

The actual distribution of concrete compressive stress within a cross section is complex and usually not known explicitly. The important properties of the concrete stress distribution can be approximated closely using any one of several different assumptions for the shape of the stress distribution.

R22.2.2.4 For design, the Code allows the use of an equivalent rectangular compressive stress distribution (stress block) to replace the more detailed approximation of the concrete stress distribution.

R22.2.2.4.1 The equivalent rectangular stress distribution does not represent the actual stress distribution in the compression zone at nominal strength, but does provide essentially the same nominal combined flexural and axial compressive strength as obtained in tests (Mattock et al. 1961).

R22.2.2.4.3 The values for $\boldsymbol{\beta_1}$ were determined experimentally. The lower limit of $\boldsymbol{\beta_1}$ is based on experimental data from beams constructed with concrete strengths greater than 8000 psi (Leslie et al. 1976; Karr et al. 1978).

CODE

Table 22.2.2.4.3—Values of β_1 for equivalent rectangular concrete stress distribution

f_c', psi	β_1	
$2500 \leq f_c' \leq 4000$	0.85	(a)
$4000 < f_c' < 8000$	$0.85 - \frac{0.05(f_c' - 4000)}{1000}$	(b)
$f_c' \geq 8000$	0.65	(c)

22.2.3 *Design assumptions for nonprestressed reinforcement*

22.2.3.1 Deformed reinforcement used to resist tensile or compressive forces shall conform to 20.2.1.

22.2.3.2 Stress-strain relationship and modulus of elasticity for deformed reinforcement shall be idealized in accordance with 20.2.2.1 and 20.2.2.2.

22.2.4 *Design assumptions for prestressed reinforcement*

22.2.4.1 For members with bonded prestressed reinforcement conforming to 20.3.1, stress at nominal flexural strength, f_{ps}, shall be calculated in accordance with 20.3.2.3.

22.2.4.2 For members with unbonded prestressed reinforcement conforming to 20.3.1, f_{ps} shall be calculated in accordance with 20.3.2.4.

22.2.4.3 If the embedded length of the prestressed strand is less than ℓ_d, the design stress of the prestressed strand shall not exceed the value given in 25.4.8.3, as modified by 25.4.8.1(b).

COMMENTARY

22.3—Flexural strength

R22.3—Flexural strength

22.3.1 *General*

22.3.1.1 Nominal flexural strength M_n shall be calculated in accordance with the assumptions of 22.2.

22.3.2 *Prestressed concrete members*

R22.3.2 *Prestressed concrete members*

22.3.2.1 Deformed reinforcement conforming to 20.2.1, provided in conjunction with prestressed reinforcement, shall be permitted to be considered to contribute to the tensile force and be included in flexural strength calculations at a stress equal to f_y.

22.3.2.2 Other nonprestressed reinforcement shall be permitted to be considered to contribute to the flexural strength if a strain compatibility analysis is performed to calculate stresses in such reinforcement.

R22.3.2.2 Bond length for nontensioned prestressing strand (Salmons and McCrate 1977; PCA 1980) should be sufficient to develop the stress consistent with strain compatibility analysis at the critical section.

22.3.3 *Composite concrete members*

R22.3.3 *Composite concrete members*

CODE

22.3.3.1 Provisions of 22.3.3 apply to members constructed in separate placements but connected so that all elements resist loads as a unit.

22.3.3.2 For calculation of M_n for composite slabs and beams, use of the entire composite section shall be permitted.

22.3.3.3 For calculation of M_n for composite slabs and beams, no distinction shall be made between shored and unshored members.

22.3.3.4 For calculation of M_n for composite members where the specified concrete compressive strength of different elements varies, properties of the individual elements shall be used in design. Alternatively, it shall be permitted to use the value of f_c' for the element that results in the most critical value of M_n.

22.4—Axial strength or combined flexural and axial strength

22.4.1 *General*

22.4.1.1 Nominal flexural and axial strength shall be calculated in accordance with the assumptions of 22.2.

22.4.2 *Maximum axial compressive strength*

22.4.2.1 Nominal axial compressive strength P_n shall not exceed $P_{n,max}$ in accordance with Table 22.4.2.1, where P_o is calculated by Eq. (22.4.2.2) for nonprestressed members and by Eq. (22.4.2.3) for prestressed members. The value of f_y shall be limited to a maximum of 80,000 psi.

Table 22.4.2.1—Maximum axial strength

Member	Transverse reinforcement	$P_{n,max}$	
Nonprestressed	Ties conforming to 22.4.2.4	$0.80P_o$	(a)
	Spirals conforming to 22.4.2.5	$0.85P_o$	(b)
Prestressed	Ties	$0.80P_o$	(c)
	Spirals	$0.85P_o$	(d)
Deep foundation member	Ties conforming to Ch. 13	$0.80P_o$	(e)

22.4.2.2 For nonprestressed members, P_o shall be calculated by:

$$P_o = 0.85f_c'(A_g - A_{st}) + f_yA_{st} \qquad (22.4.2.2)$$

COMMENTARY

R22.3.3.1 The scope of Chapter 22 is intended to include composite concrete flexural members. In some cases with cast-in-place concrete, separate placements of concrete may be designed to act as a unit. In these cases, the interface is designed for the loads that will be transferred across the interface. Composite structural steel-concrete beams are not covered in this chapter. Design provisions for these types of composite members are covered in AISC 360.

R22.4—Axial strength or combined flexural and axial strength

R22.4.2 *Maximum axial compressive strength*

R22.4.2.1 To account for accidental eccentricity, the design axial strength of a section in pure compression is limited to 80 to 85 percent of the nominal axial strength. These percentage values approximate the axial strengths at eccentricity-to-depth ratios of 0.10 and 0.05 for tied and spirally reinforced members conforming to 22.4.2.4 and 22.4.2.5, respectively. The same axial load limitation applies to both cast-in-place and precast compression members. The value of f_y is limited to 80,000 psi because the compression capacity of the concrete is likely to be reached before this stress is exceeded. The transverse reinforcement requirements for columns do not apply to deep foundation members. Chapter 13 provides the detailing requirements for these members.

CODE

where A_{st} is the total area of nonprestressed longitudinal reinforcement.

22.4.2.3 For prestressed members, P_o shall be calculated by:

$$P_o = 0.85f_c'(A_g - A_{st} - A_{pd}) + f_yA_{st} - (f_{se} - 0.003E_p)A_{pt} \quad (22.4.2.3)$$

where A_{pt} is the total area of prestressing reinforcement, and A_{pd} is the total area occupied by duct, sheathing, and prestressing reinforcement; the value of f_{se} shall be at least $\mathbf{0.003E_p}$. For grouted, post-tensioned tendons, it shall be permitted to assume A_{pd} equals A_{pt}.

22.4.2.4 Tie reinforcement for lateral support of longitudinal reinforcement in compression members shall satisfy 10.7.6.2 and 25.7.2.

22.4.2.5 Spiral reinforcement for lateral support of longitudinal reinforcement in compression members shall satisfy 10.7.6.3 and 25.7.3.

22.4.3 *Maximum axial tensile strength*

22.4.3.1 Nominal axial tensile strength of a nonprestressed, composite, or prestressed member, P_{nt}, shall not be taken greater than $P_{nt,max}$, calculated by:

$$P_{nt,max} = f_yA_{st} + (f_{se} + \Delta f_p)A_{pt} \quad (22.4.3.1)$$

where $(f_{se} + \Delta f_p)$ shall not exceed f_{py}, and A_{pt} is zero for nonprestressed members.

22.5—One-way shear strength

22.5.1 *General*

22.5.1.1 Nominal one-way shear strength at a section, V_n, shall be calculated by:

$$V_n = V_c + V_s \quad (22.5.1.1)$$

COMMENTARY

R22.4.2.3 The effects of prestressing on the axial strength of compression members are taken into account in Eq. (22.4.2.3). Equation (22.4.2.3) is similar to Eq. (22.4.2.2) for nonprestressed compression members. The effective area of concrete subjected to the limiting stress of $\mathbf{0.85f_c'}$ is reduced by the term A_{pd} to account for the area of ducts, sheathing, and prestressing reinforcement. A third term is added to account for the reduction of column capacity due to the prestress force. At nominal strength, the stress in the prestressed reinforcement, f_{se}, is decreased by $\mathbf{0.003E_p}$, where 0.003 is the assumed compressive strain at the axial capacity of the member.

R22.5—One-way shear strength

R22.5.1 *General*

R22.5.1.1 In a member without shear reinforcement, shear is assumed to be resisted by the concrete. In a member with shear reinforcement, a portion of the shear strength is assumed to be provided by the concrete and the remainder by the shear reinforcement.

The one-way shear equations for nonprestressed concrete were changed in the 2019 Code with the primary objectives of including effect of member depth, commonly referred to as the "size effect," and the effects of the longitudinal reinforcement ratio on shear strength.

The shear strength provided by concrete, V_c, is taken as the shear causing inclined cracking (Joint ACI-ASCE Committee 426 1973; MacGregor and Hanson 1969; Joint ACI-ASCE Committee 326 1962). After cracking, V_c is attributed to aggregate interlock, dowel action, and the shear transmitted across the concrete compression zone.

Sect. Strength 22

CODE

COMMENTARY

The shear strength is based on an average shear stress over the effective cross section, $b_w d$.

Chapter 23 allows the use of the strut-and-tie method in the shear design of any structural concrete member, or discontinuity region in a member.

22.5.1.2 Cross-sectional dimensions shall be selected to satisfy Eq. (22.5.1.2).

$$V_u \leq \phi(V_c + 8\sqrt{f_c'}b_w d) \quad (22.5.1.2)$$

R22.5.1.2 The limit on cross-sectional dimensions in 22.5.1.2 is intended to minimize the likelihood of diagonal compression failure in the concrete and limit the extent of cracking.

22.5.1.3 For nonprestressed members, V_c shall be calculated in accordance with 22.5.5.

22.5.1.4 For prestressed members, V_c, V_{ci}, and V_{cw} shall be calculated in accordance with 22.5.6 or 22.5.7.

22.5.1.5 For calculation of V_c, V_{ci}, and V_{cw}, λ shall be in accordance with 19.2.4.

22.5.1.6 V_s shall be calculated in accordance with 22.5.8.

22.5.1.7 Effect of any openings in members shall be considered in calculating V_n.

R22.5.1.7 Openings in the web of a member can reduce its shear strength. The effects of openings are discussed in Section 4.7 of Joint ACI-ASCE Committee 426 (1973), Barney et al. (1977), and Schlaich et al. (1987). The strut-and-tie method as addressed in Chapter 23 can be used to design members with openings.

22.5.1.8 Effect of axial tension due to creep and shrinkage in members shall be considered in calculating V_c.

R22.5.1.8 Consideration of axial tension requires engineering judgment. Axial tension often occurs due to volume changes, but it may be low enough not to be detrimental to the performance of a structure with adequate expansion joints and satisfying minimum longitudinal reinforcement requirements. It may be desirable to design shear reinforcement to resist the total shear if there is uncertainty about the magnitude of axial tension.

22.5.1.9 Effect of inclined flexural compression in variable depth members shall be permitted to be considered in calculating V_c.

R22.5.1.9 In a member of variable depth, the internal shear at any section is increased or decreased by the vertical component of the inclined flexural stresses.

22.5.1.10 The interaction of shear forces acting along orthogonal axes shall be permitted to be neglected if (a) or (b) is satisfied.

(a) $$\frac{v_{u,x}}{\phi v_{n,x}} \leq 0.5 \quad (22.5.1.10a)$$

(b) $$\frac{v_{u,y}}{\phi v_{n,y}} \leq 0.5 \quad (22.5.1.10b)$$

R22.5.1.10 and **R.22.5.1.11** Reinforced concrete members, such as columns and beams, may be subjected to biaxial shear. For symmetrically reinforced circular sections, nominal one-way shear strength about any axis is the same. Therefore, when a circular section is subjected to shear along two centroidal axes, shear strength can be evaluated using the resultant shear. However, for rectangular and other cross sections, calculating nominal one-way shear strength along the axis of the resultant shear is not practical. Tests and analytical results for columns have indicated that for biaxial shear loading, the shear strength follows an elliptical interaction diagram that requires calculating nominal one-way shear strength along two orthogonal directions (Umehara

CODE

22.5.1.11 If $\frac{v_{u,x}}{\phi v_{n,x}} > 0.5$ and $\frac{v_{u,y}}{\phi v_{n,y}} > 0.5$ then Eq. (22.5.1.11) shall be satisfied.

$$\frac{v_{u,x}}{\phi v_{n,x}} + \frac{v_{u,y}}{\phi v_{n,y}} \leq 1.5 \qquad (22.5.1.11)$$

COMMENTARY

and Jirsa 1984). Considering shear along each centroidal axis independently can be unconservative. Thus, linear interaction accounts for biaxial shear.

22.5.2 *Geometric assumptions*

22.5.2.1 For calculation of V_c and V_s in prestressed members, d shall be taken as the distance from the extreme compression fiber to the centroid of prestressed and any nonprestressed longitudinal reinforcement but need not be taken less than $0.8h$.

22.5.2.2 For calculation of V_c and V_s, it shall be permitted to assume (a) through (c):

(a) d equal to 0.8 times the diameter for circular sections
(b) b_w equal to the diameter for solid circular sections
(c) b_w equal to twice the wall thickness for hollow circular sections

R22.5.2 *Geometric assumptions*

R22.5.2.1 Although the value of d may vary along the span of a prestressed beam, studies (MacGregor and Hanson 1969) have shown that, for prestressed concrete members, d need not be taken less than $0.8h$. The beams considered had some straight prestressed reinforcement or reinforcing bars at the bottom of the section and had stirrups that enclosed the longitudinal reinforcement.

R22.5.2.2 Shear tests of members with circular sections indicate that the effective area can be taken as the gross area of the section or as an equivalent rectangular area (Joint ACI-ASCE Committee 426 1973; Faradji and Diaz de Cossio 1965; Khalifa and Collins 1981).

Although the transverse reinforcement in a circular section may not consist of straight legs, tests indicate that Eq. (22.5.8.5.3) is conservative if d is taken as defined in 22.5.2.2 (Faradji and Diaz de Cossio 1965; Khalifa and Collins 1981).

22.5.3 *Limiting material strengths*

22.5.3.1 The value of $\sqrt{f_c'}$ used to calculate V_c, V_{ci}, and V_{cw} for one-way shear shall not exceed 100 psi, unless allowed in 22.5.3.2.

22.5.3.2 Values of $\sqrt{f_c'}$ greater than 100 psi shall be permitted in calculating V_c, V_{ci}, and V_{cw} for reinforced or prestressed concrete beams and concrete joist construction having minimum web reinforcement in accordance with 9.6.3.4 or 9.6.4.2.

22.5.3.3 The values of f_y and f_{yt} used to calculate V_s shall not exceed the limits in 20.2.2.4.

R22.5.3 *Limiting material strengths*

R22.5.3.1 Because of a lack of test data and practical experience with concretes having compressive strengths greater than 10,000 psi, the Code imposes a maximum value of 100 psi on $\sqrt{f_c'}$ for use in the calculation of shear strength of concrete members. Exceptions to this limit are permitted in beams and joists if the transverse reinforcement satisfies the requirements in 22.5.3.2.

R22.5.3.2 Based on the beam test results in Mphonde and Frantz (1984), Elzanaty et al. (1986), Roller and Russell (1990), Johnson and Ramirez (1989), and Ozcebe et al. (1999), an increase in the minimum amount of transverse reinforcement is required for high-strength concrete. These tests indicate a reduction in reserve shear strength occurs as f_c' increases in beams reinforced with transverse reinforcement providing an effective shear stress of 50 psi. By providing minimum transverse reinforcement, which increases as f_c' increases, the reduction in shear strength is offset.

R22.5.3.3 The upper limit of 60,000 psi on the value of f_y and f_{yt} used in design is intended to control diagonal crack widths.

CODE

22.5.4 *Composite concrete members*

22.5.4.1 This section shall apply to members constructed in separate placements but connected so that all elements resist loads as a unit.

22.5.4.2 For calculation of V_n for composite members, no distinction shall be made between shored and unshored members.

22.5.4.3 For calculation of V_n for composite members where the specified concrete compressive strength, unit weight, or other properties of different elements vary, properties of the individual elements shall be used in design. Alternatively, it shall be permitted to use the properties of the element that results in the most critical value of V_n.

22.5.4.4 If an entire composite member is assumed to resist vertical shear, it shall be permitted to calculate V_c assuming a monolithically cast member of the same cross-sectional shape.

22.5.4.5 If an entire composite member is assumed to resist vertical shear, it shall be permitted to calculate V_s assuming a monolithically cast member of the same cross-sectional shape if shear reinforcement is fully anchored into the interconnected elements in accordance with 25.7.

22.5.5 V_c *for nonprestressed members*

22.5.5.1 For nonprestressed members, V_c shall be calculated in accordance with Table 22.5.5.1 and 22.5.5.1.1 through 22.5.5.1.3.

Table 22.5.5.1—V_c for nonprestressed members

Criteria	V_c		
$A_v \geq A_{v,min}$	Either of:	$\left[2\lambda\sqrt{f_c'}+\frac{N_u}{6A_g}\right]b_w d$	(a)
		$\left[8\lambda(\rho_w)^{1/3}\sqrt{f_c'}+\frac{N_u}{6A_g}\right]b_w d$	(b)
$A_v < A_{v,min}$	$\left[8\lambda_s\lambda(\rho_w)^{1/3}\sqrt{f_c'}+\frac{N_u}{6A_g}\right]b_w d$		(c)

Notes:
1. Axial load, N_u, is positive for compression and negative for tension.
2. V_c shall not be taken less than zero.

COMMENTARY

R22.5.4 *Composite concrete members*

R22.5.4.1 The scope of Chapter 22 includes composite concrete members. In some cases with cast-in-place concrete, separate placements of concrete may be designed to act as a unit. In these cases, the interface is designed for the loads that will be transferred across the interface. Composite structural steel-concrete beams are not covered in this Code. Design provisions for such composite members are covered in AISC 360.

R22.5.5 V_c *for nonprestressed members*

R22.5.5.1 Test results (Kuchma et al. 2019) for nonprestressed members without shear reinforcement indicate that measured shear strength, attributed to concrete, does not increase in direct proportion with member depth. This phenomenon is often referred to as the "size effect." For example, if the member depth doubles, the shear at failure for the deeper beam may be less than twice the shear at failure of the shallower beam (Sneed and Ramirez 2010). $A_{v,min}$ for beams and one-way slabs is defined in 9.6.3.4.

Research (Angelakos et al. 2001; Lubell et al. 2004; Brown et al. 2006; Becker and Buettner 1985; Anderson 1978; Bažant et al. 2007) has shown that shear stress at failure is lower for beams with increased depth and a reduced area of longitudinal reinforcement.

In Table 22.5.5.1, for $A_v > A_{v,min}$, either equation for V_c may be used. Equation (a) is provided as a simpler option.

When calculating V_c by Table 22.5.5.1, an axial tension force can cause V_c to have a negative value. In those cases, the Code specifies that V_c should be taken equal to zero.

CODE | **COMMENTARY**

The criteria column in Table 22.5.5.1 references $A_{v,min}$, which is defined in Table 9.6.3.4 and 10.6.2.2 and referenced throughout the Code.

When applying equations in Table 22.5.5.1, the value of A_s to be used in the calculation of ρ_w may be taken as the sum of the areas of longitudinal bars located more than two-thirds of the overall member depth away from the extreme compression fiber. Definitions for b_w and d to be used with circular sections are given in 22.5.2.2.

22.5.5.1.1 V_c shall not be taken greater than $5\lambda\sqrt{f_c'}\,b_wd$.

22.5.5.1.2 In Table 22.5.5.1, the value of $N_u/6A_g$ shall not be taken greater than $0.05f_c'$.

22.5.5.1.3 The size effect modification factor, λ_s, shall be determined by

$$\lambda_s = \sqrt{\frac{2}{1+\frac{d}{10}}} \leq 1 \qquad (22.5.5.1.3)$$

R22.5.5.1.3 The parameters within the size effect modification factor, λ_s, are consistent with fracture mechanics theory for reinforced concrete (Bažant et al. 2007; Frosch et al. 2017).

22.5.6 V$_c$ *for prestressed members*

R22.5.6 V$_c$ *for prestressed members*

22.5.6.1 This section shall apply to the calculation of V_c for post-tensioned and pretensioned members in regions where the effective force in the prestressed reinforcement is fully transferred to the concrete. For regions of pretensioned members where the effective force in the prestressed reinforcement is not fully transferred to the concrete, 22.5.7 shall govern the calculation of V_c.

22.5.6.2 For prestressed flexural members with $A_{ps}f_{se} \geq 0.4(A_{ps}f_{pu} + A_sf_y)$, V_c shall be calculated in accordance with Table 22.5.6.2, but need not be less than $2\lambda\sqrt{f_c'}\,b_wd$. Alternatively, it shall be permitted to calculate V_c in accordance with 22.5.6.3.

Table 22.5.6.2—Approximate method for calculating V_c

	V_c	
Least of (a), (b), and (c):	$\left(0.6\lambda\sqrt{f_c'}+700\frac{V_ud_p}{M_u}\right)b_wd$ [1],[2]	(a)
	$(0.6\lambda\sqrt{f_c'}+700)b_wd$	(b)
	$5\lambda\sqrt{f_c'}b_wd$	(c)

[1] M_u occurs simultaneously with V_u at the section considered.

[2] When calculating the V_ud_p/M_u term in Eq. 22.5.6.2(a), d_p is the distance from the extreme compression fiber to the centroid of prestressed reinforcement. It shall not be permitted to take d_p as $0.80h$ as in 22.5.2.1.

R22.5.6.2 This provision offers a simple means of calculating V_c for prestressed concrete beams (MacGregor and Hanson 1969). This provision may be applied to beams having prestressed reinforcement only, or to members reinforced with a combination of prestressed and nonprestressed reinforcement. Expression (a) in Table 22.5.6.2 is most applicable to members subject to uniform loading.

In applying the expression in row (a) to simply-supported members subject to uniform loads, Eq. (R22.5.6.2) can be used:

$$\frac{V_ud_p}{M_u} = \frac{d_p(\ell-2x)}{x(\ell-x)} \qquad (R22.5.6.2)$$

where ℓ is the span length, and x is the distance from the section being investigated to the support. For concrete with f_c' equal to 5000 psi, V_c from 22.5.6.2 varies, as shown in Fig. R22.5.6.2. Design aids based on this equation are given in ASCE Joint Committee (1940).

CODE

COMMENTARY

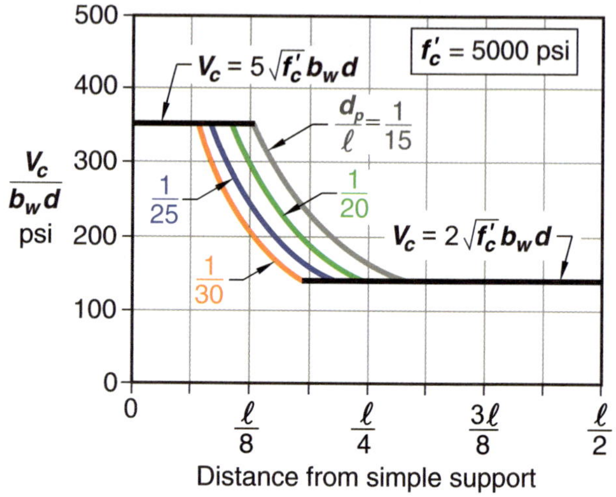

Fig. R22.5.6.2—Application of Table 22.5.6.2 to uniformly loaded prestressed members with f'_c *= 5000 psi.*

22.5.6.3 For prestressed members, V_c shall be permitted to be the lesser of V_{ci} calculated in accordance with 22.5.6.3.1 and V_{cw} calculated in accordance with 22.5.6.3.2 or 22.5.6.3.3.

R22.5.6.3 Two types of inclined cracking occur in concrete beams: web-shear cracking and flexure-shear cracking. These two types of inclined cracking are illustrated in Fig. R22.5.6.3.

Web-shear cracking begins from an interior point in a member when the principal tensile stresses exceed the tensile strength of the concrete. Flexure-shear cracking is initiated by flexural cracking. When flexural cracking occurs, the shear stresses in the concrete above the crack are increased. The flexure-shear crack develops when the combined shear and flexural-tensile stress exceeds the tensile strength of the concrete.

The nominal shear strength provided by the concrete, V_c, is assumed equal to the lesser of V_{ci} and V_{cw}. The derivations of Eq. (22.5.6.3.1a) and Eq. (22.5.6.3.2) are summarized in ACI Committee 318 (1965).

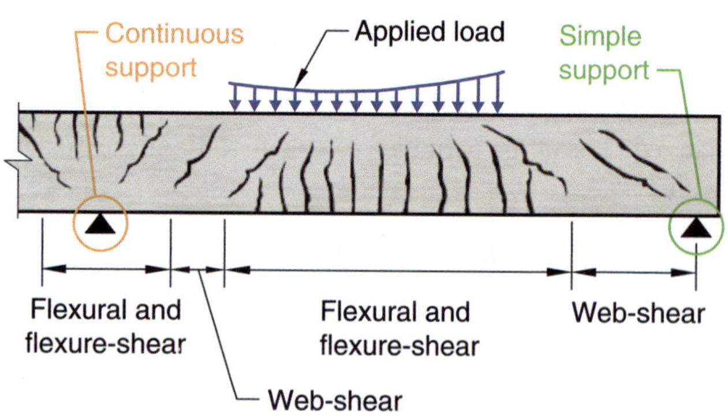

Fig. R22.5.6.3—Types of cracking in concrete beams.

22.5.6.3.1 The flexure-shear strength V_{ci} shall be calculated by (a) but need not be taken less than (b) or (c):

(a) $$V_{ci} = 0.6\lambda\sqrt{f'_c}b_wd_p + V_d + \frac{V_iM_{cre}}{M_{max}} \qquad (22.5.6.3.1a)$$

(b) For members with $A_{ps}f_{se} < 0.4(A_{ps}f_{pu} + A_sf_y)$,

$$V_{ci} = 1.7\lambda\sqrt{f'_c}b_wd \qquad (22.5.6.3.1b)$$

R22.5.6.3.1 In deriving Eq. (22.5.6.3.1a), it was assumed that V_{ci} is the sum of the shear required to cause a flexural crack at the section in question given by:

$$V = \frac{V_iM_{cre}}{M_{max}} \qquad (R22.5.6.3.1a)$$

CODE

(c) For members with $A_{ps}f_{se} \geq 0.4(A_{ps}f_{pu} + A_sf_y)$,

$$V_{ci} = 2\lambda\sqrt{f_c'}b_wd \quad (22.5.6.3.1c)$$

where d_p need not be taken less than **0.80*h***, the values of M_{max} and V_i shall be calculated from the load combinations causing maximum factored moment to occur at section considered, and M_{cre} shall be calculated by:

$$M_{cre} = \left(\frac{I}{y_t}\right)(6\lambda\sqrt{f_c'} + f_{pe} - f_d) \quad (22.5.6.3.1d)$$

COMMENTARY

plus an additional increment of shear required to change the flexural crack to a flexure-shear crack. The externally applied factored loads, from which V_i and M_{max} are determined, include superimposed dead load and live load. In calculating M_{cre} for substitution into Eq. (22.5.6.3.1a), I and y_t are the properties of the section resisting the externally applied loads.

For a composite member, where part of the dead load is resisted by only a part of the section, appropriate section properties should be used to calculate f_d. The shear due to dead loads, V_d, and that due to other loads, V_i, are separated in this case. V_d is then the total shear force due to unfactored dead load acting on that part of the section resisting the dead loads acting prior to composite action plus the unfactored superimposed dead load acting on the composite member. The terms V_i and M_{max} may be taken as

$$V_i = V_u - V_d \quad (R22.5.6.3.1b)$$

$$M_{max} = M_u - M_d \quad (R22.5.6.3.1c)$$

where V_u and M_u are the factored shear and moment due to the total factored loads, and M_d is the moment due to unfactored dead load (the moment corresponding to f_d).

For noncomposite, uniformly loaded beams, the total cross section resists all the shear, and the live and dead load shear force diagrams are similar. In this case, Eq. (22.5.6.3.1a) and Eq. (22.5.6.3.1d) reduce to

$$V_{ci} = 0.6\lambda\sqrt{f_c'}b_wd + \frac{V_uM_{ct}}{M_u} \quad (R22.5.6.3.1d)$$

where

$$M_{ct} = (I/y_t)(6\lambda\sqrt{f_c'} + f_{pe}) \quad (R22.5.6.3.1e)$$

The cracking moment M_{ct} in the two preceding equations represents the total moment, including dead load, required to cause cracking at the extreme fiber in tension. This is not the same as M_{cre} in Eq. (22.5.6.3.1a) where the cracking moment is that due to all loads except the dead load. In Eq. (22.5.6.3.1a), the dead load shear is added as a separate term.

M_u is the factored moment on the beam at the section under consideration, and V_u is the factored shear force occurring simultaneously with M_u. Because the same section properties apply to both dead and live load stresses, there is no need to calculate dead load stresses and shears separately. M_{ct} reflects the total stress change from effective prestress to a tension of $6\lambda\sqrt{f_c'}$, assumed to cause flexural cracking.

22.5.6.3.2 The web-shear strength V_{cw} shall be calculated by:

$$V_{cw} = (3.5\lambda\sqrt{f_c'} + 0.3f_{pc})b_wd_p + V_p \quad (22.5.6.3.2)$$

R22.5.6.3.2 Equation (22.5.6.3.2) is based on the assumption that web-shear cracking occurs at a shear level causing a principal tensile stress of approximately $4\lambda\sqrt{f_c'}$ at the centroidal axis of the cross section. V_p is calculated from the effective prestress force without load factors.

CODE

where d_p need not be taken less than $0.80h$, and V_p is the vertical component of the effective prestress.

22.5.6.3.3 As an alternative to 22.5.6.3.2, it shall be permitted to calculate V_{cw} as the shear force corresponding to dead load plus live load that results in a principal tensile stress of $4\lambda\sqrt{f_c'}$ at location (a) or (b):

(a) Where the centroidal axis of the prestressed cross section is in the web, the principal tensile stress shall be calculated at the centroidal axis.
(b) Where the centroidal axis of the prestressed cross section is in the flange, the principal tensile stress shall be calculated at the intersection of the flange and the web.

22.5.6.3.4 In composite members, the principal tensile stress shall be calculated at the location specified in 22.5.6.3.3 for the composite section, considering superposition of stresses calculated cross sections that resist the corresponding loads.

22.5.7 V$_c$ *for pretensioned members in regions of reduced prestress force*

22.5.7.1 When calculating V_c, the transfer length of prestressed reinforcement, ℓ_{tr}, shall be assumed to be $50d_b$ for strand and $100d_b$ for wire.

22.5.7.2 If bonding of strands extends to the end of the member, the effective prestress force shall be assumed to vary linearly from zero at the end of the prestressed reinforcement to a maximum at a distance ℓ_{tr} from the end of the prestressed reinforcement.

22.5.7.3 At locations corresponding to a reduced effective prestress force in 22.5.7.2, V_c shall be calculated in accordance with (a) through (c):

(a) The reduced effective prestress force shall be used to determine the applicability of 22.5.6.2.
(b) The reduced effective prestress force shall be used to calculate V_{cw} in 22.5.6.3.
(c) The value of V_c calculated using 22.5.6.2 shall not exceed the value of V_{cw} calculated using the reduced effective prestress force.

22.5.7.4 If bonding of strands does not extend to the end of the member, the effective prestress force shall be assumed to vary linearly from zero at the point where bonding commences to a maximum at a distance ℓ_{tr} from that point.

COMMENTARY

R22.5.6.3.4 Generally, in unshored construction the principal tensile stresses due to dead load are caused before composite action and principal tensile stresses due to live load are caused after composite action is developed in a member. In shored construction the principal tensile stresses due to both the dead load and live load are caused after composite action is developed.

R22.5.7 V$_c$ *for pretensioned members in regions of reduced prestress force*

The effect of the reduced prestress near the ends of pretensioned beams on the shear strength should be taken into account. Provisions 22.5.7.2 and 22.5.7.3 relate to the reduced shear strength at sections within the transfer length of prestressed reinforcement when bonding of prestressed reinforcement extends to the end of the member. Provisions 22.5.7.4 and 22.5.7.5 relate to the reduced shear strength at sections within the length over which some of the prestressed reinforcement is not bonded to the concrete, or within the transfer length of the prestressed reinforcement for which bonding does not extend to the end of the beam.

CODE

22.5.7.5 At locations corresponding to a reduced effective prestress force according to 22.5.7.4, V_c shall be calculated in accordance with (a) through (c):

(a) The reduced effective prestress force shall be used to determine the applicability of 22.5.6.2.
(b) The reduced effective prestress force shall be used to calculate V_c in accordance with 22.5.6.3.
(c) The value of V_c calculated using 22.5.6.2 shall not exceed the value of V_{cw} calculated using the reduced effective prestress force.

22.5.8 *One-way shear reinforcement*

22.5.8.1 At each section where $V_u > \phi V_c$, transverse reinforcement shall be provided such that Eq. (22.5.8.1) is satisfied.

$$V_s \geq \frac{V_u}{\phi} - V_c \qquad (22.5.8.1)$$

22.5.8.2 For one-way members reinforced with transverse reinforcement, V_s shall be calculated in accordance with 22.5.8.5.

22.5.8.3 For one-way members reinforced with bent-up longitudinal bars, V_s shall be calculated in accordance with 22.5.8.6.

22.5.8.4 If more than one type of shear reinforcement is provided to reinforce the same portion of a member, V_s shall be the sum of the V_s values for the various types of shear reinforcement.

22.5.8.5 *One-way shear strength provided by transverse reinforcement*

22.5.8.5.1 In nonprestressed and prestressed members, shear reinforcement satisfying (a), (b), or (c) shall be permitted:

(a) Stirrups, ties, or hoops perpendicular to longitudinal axis of member
(b) Welded wire reinforcement with wires located perpendicular to longitudinal axis of member
(c) Spiral reinforcement

COMMENTARY

R22.5.8 *One-way shear reinforcement*

R22.5.8.2 Provisions of 22.5.8.5 apply to all types of transverse reinforcement, including stirrups, ties, hoops, crossties, and spirals.

R22.5.8.5 *One-way shear strength provided by transverse reinforcement*

Design of shear reinforcement is based on a modified truss analogy. In the truss analogy, the force in vertical ties is resisted by shear reinforcement. Shear reinforcement needs to be designed to resist only the shear exceeding that which causes inclined cracking, provided the diagonal members in the truss are assumed to be inclined at 45 degrees. The concrete is assumed to contribute to the shear capacity through resistance across the concrete compressive zone, aggregate interlock, and dowel action in an amount equivalent to that which caused inclined cracking.

Equations (22.5.8.5.3), (22.5.8.5.4), and (22.5.8.6.2a) are presented in terms of nominal shear strength provided by shear reinforcement, V_s. Where shear reinforcement perpendicular to the axis of the member is used, the required area of shear reinforcement, A_v, and its spacing, s, are calculated by

Sect. Strength 22

CODE

22.5.8.5.2 Inclined stirrups making an angle of at least 45 degrees with the longitudinal axis of the member and crossing the plane of the potential shear crack shall be permitted to be used as shear reinforcement in nonprestressed members.

22.5.8.5.3 V_s for shear reinforcement in 22.5.8.5.1 shall be calculated by:

$$V_s = \frac{A_v f_{yt} d}{s} \quad (22.5.8.5.3)$$

where s is the spiral pitch or the longitudinal spacing of the shear reinforcement, and A_v is given in 22.5.8.5.5 or 22.5.8.5.6.

22.5.8.5.4 V_s for shear reinforcement in 22.5.8.5.2 shall be calculated by:

$$V_s = \frac{A_v f_{yt}(\sin\alpha + \cos\alpha)d}{s} \quad (22.5.8.5.4)$$

where α is the angle between the inclined stirrups and the longitudinal axis of the member, s is measured parallel to the longitudinal reinforcement, and A_v is given in 22.5.8.5.5.

22.5.8.5.5 For each rectangular tie, stirrup, hoop, or crosstie, A_v shall be the effective area of all bar legs or wires within spacing s.

22.5.8.5.6 For each circular tie or spiral, A_v shall be two times the area of the bar or wire within spacing s.

22.5.8.6 *One-way shear strength provided by bent-up longitudinal bars*

22.5.8.6.1 The center three-fourths of the inclined portion of bent-up longitudinal bars shall be permitted to be used as shear reinforcement in nonprestressed members if the angle α between the bent-up bars and the longitudinal axis of the member is at least 30 degrees.

22.5.8.6.2 If shear reinforcement consists of a single bar or a single group of parallel bars having an area A_v, all bent the same distance from the support, V_s shall be the lesser of (a) and (b):

(a) $V_s = A_v f_y \sin\alpha$ (22.5.8.6.2a)

(b) $V_s = 3\sqrt{f'_c} b_w d$ (22.5.8.6.2b)

COMMENTARY

$$\frac{A_v}{s} = \frac{(V_u - \phi V_c)}{\phi f_{yt} d} \quad (R22.5.8.5)$$

R22.5.8.5.2 Although inclined stirrups crossing the plane of the potential shear cracks are permitted, their use is not appropriate where the direction of net shear reverses due to changes in transient load.

R22.5.8.5.4 To be effective, it is critical that inclined stirrups cross potential shear cracks. If the inclined stirrups are generally oriented parallel to the potential shear cracks, the stirrups provide no shear strength.

R22.5.8.5.6 Although the transverse reinforcement in a circular section may not consist of straight legs, tests indicate that Eq. (22.5.8.5.3) is conservative if d is taken as defined in 22.5.2.2 (Faradji and Diaz de Cossio 1965; Khalifa and Collins 1981).

R22.5.8.6 *One-way shear strength provided by bent-up longitudinal bars*

To be effective, it is critical that the inclined portion of the bent-up longitudinal bar cross potential shear cracks. If the inclined bars are generally oriented parallel to the potential shear cracks, the bars provide no shear strength.

CODE

where α is the angle between bent-up reinforcement and longitudinal axis of the member.

22.5.8.6.3 If shear reinforcement consists of a series of parallel bent-up bars or groups of parallel bent-up bars at different distances from the support, V_s shall be calculated by Eq. (22.5.8.5.4).

22.6—Two-way shear strength

22.6.1 *General*

22.6.1.1 Provisions 22.6.1 through 22.6.8 apply to the nominal shear strength of two-way members with and without shear reinforcement.

22.6.1.2 Nominal shear strength for two-way members without shear reinforcement shall be calculated by

$$v_n = v_c \tag{22.6.1.2}$$

22.6.1.3 Nominal shear strength for two-way members with shear reinforcement shall be calculated by

$$v_n = v_c + v_s \tag{22.6.1.3}$$

22.6.1.4 Two-way shear shall be resisted by a section with a depth d and an assumed critical perimeter b_o as defined in 22.6.4.

22.6.1.5 v_c for two-way shear shall be calculated in accordance with 22.6.5. For two-way members with shear reinforcement, v_c shall not exceed the limits in 22.6.6.1.

22.6.1.6 For calculation of v_c, λ shall be in accordance with 19.2.4.

22.6.1.7 For two-way members reinforced with single- or multiple-leg stirrups, v_s shall be calculated in accordance with 22.6.7.

22.6.1.8 For two-way members reinforced with headed shear stud reinforcement, v_s shall be calculated in accordance with 22.6.8.

COMMENTARY

R22.6—Two-way shear strength

Factored shear stress in two-way members due to shear and moment transfer is calculated in accordance with the requirements of 8.4.4. Section 22.6 provides requirements for determining nominal shear strength, either without shear reinforcement or with shear reinforcement in the form of stirrups or headed shear studs. Factored shear demand and strength are calculated in terms of stress, permitting superposition of effects from direct shear and moment transfer.

Design provisions for shearheads have been eliminated from the Code because this type of shear reinforcement is seldom used in current practice. Shearheads may be designed following the provisions of ACI 318-14.

R22.6.1 *General*

R22.6.1.4 The critical section perimeter b_o is defined in 22.6.4.

CODE

22.6.2 *Effective depth*

22.6.2.1 For calculation of v_c and v_s for two-way shear, d shall be the average of the effective depths in the two orthogonal directions.

22.6.2.2 For prestressed, two-way members, d need not be taken less than $0.8h$.

22.6.3 *Limiting material strengths*

22.6.3.1 The value of $\sqrt{f_c'}$ used to calculate v_c for two-way shear shall not exceed 100 psi.

22.6.3.2 The value of f_{yt} used to calculate v_s shall not exceed the limits in 20.2.2.4.

22.6.4 *Critical sections for two-way members*

22.6.4.1 For two-way shear, critical sections shall be located so that the perimeter b_o is a minimum but need not be closer than $d/2$ to (a) and (b):

(a) Edges or corners of columns, concentrated loads, or reaction areas

(b) Changes in slab or footing thickness, such as edges of capitals, drop panels, or shear caps

22.6.4.1.1 For square or rectangular columns, concentrated loads, or reaction areas, critical sections for two-way shear in accordance with 22.6.4.1(a) and (b) shall be permitted to be defined assuming straight sides.

22.6.4.1.2 For a circular or regular polygon-shaped column, critical sections for two-way shear in accordance with 22.6.4.1(a) and (b) shall be permitted to be defined assuming a square column of equivalent area.

22.6.4.2 For two-way members reinforced with headed shear reinforcement or single- or multi-leg stirrups, a critical section with perimeter b_o located $d/2$ beyond the outermost peripheral line of shear reinforcement shall also be considered. The shape of this critical section shall be a polygon selected to minimize b_o.

COMMENTARY

R22.6.3 *Limiting material strengths*

R22.6.3.1 There are limited test data on the two-way shear strength of high-strength concrete slabs. Until more experience is obtained for two-way slabs constructed with concretes that have compressive strengths greater than 10,000 psi, it is prudent to limit $\sqrt{f_c'}$ to 100 psi for the calculation of shear strength.

R22.6.3.2 The upper limit of 60,000 psi on the value of f_{yt} used in design is intended to control cracking.

R22.6.4 *Critical sections for two-way members*

The critical section defined in 22.6.4.1(a) for shear in slabs and footings subjected to bending in two directions follows the perimeter at the edge of the loaded area (Joint ACI-ASCE Committee 326 1962). Loaded area for shear in two-way slabs and footings includes columns, concentrated loads, and reaction areas. An idealized critical section located a distance $d/2$ from the periphery of the loaded area is considered.

For members of uniform thickness without shear reinforcement, it is sufficient to check shear using one section. For slabs with changes in thickness or with shear reinforcement, it is necessary to check shear at multiple sections as defined in 22.6.4.1(a) and (b) and 22.6.4.2.

For columns near an edge or corner, the critical perimeter may extend to the edge of the slab.

R22.6.4.2 For two-way members with stirrup or headed stud shear reinforcement, it is required to check shear stress in concrete at a critical section located a distance $d/2$ beyond the point where shear reinforcement is discontinued. Calculated shear stress at this section must not exceed the limits given in expressions (b) and (d) in Table 22.6.6.1. The shape of this outermost critical section should correspond to the minimum value of b_o, as depicted in Fig. R22.6.4.2a, b, and c. Note that these figures depict slabs reinforced with stirrups. The shape of the outermost critical section is similar for slabs with headed shear reinforcement. The square or rectangular critical sections described in 22.6.4.1.1 will not result in the minimum value of b_o for the cases depicted in these figures. Additional

CODE

COMMENTARY

critical section checks are required at a distance ***d*/2** beyond any point where variations in shear reinforcement occur, such as changes in size, spacing, or configuration.

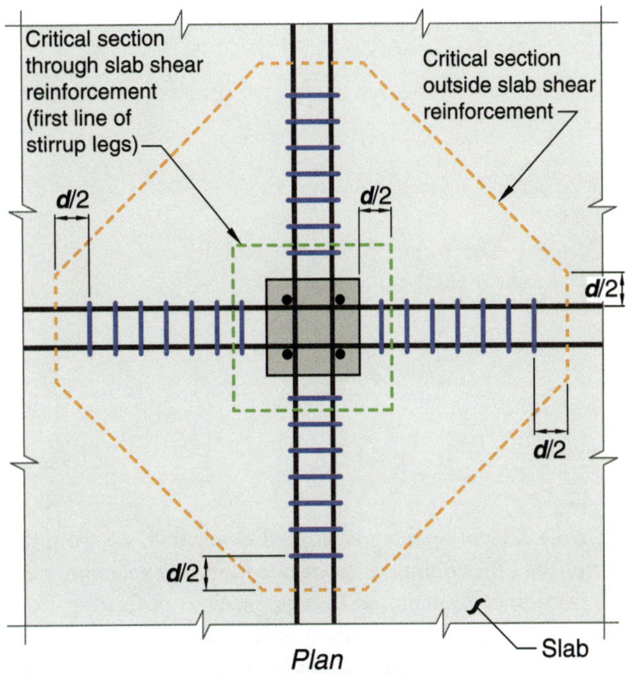

***Fig. R22.6.4.2a**—Critical sections for two-way shear in slab with shear reinforcement at interior column.*

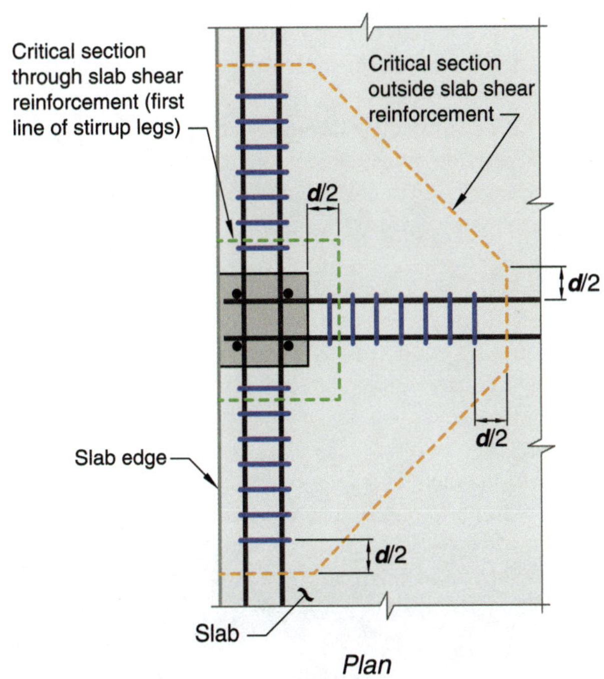

***Fig. R22.6.4.2b**—Critical sections for two-way shear in slab with shear reinforcement at edge column.*

CODE

COMMENTARY

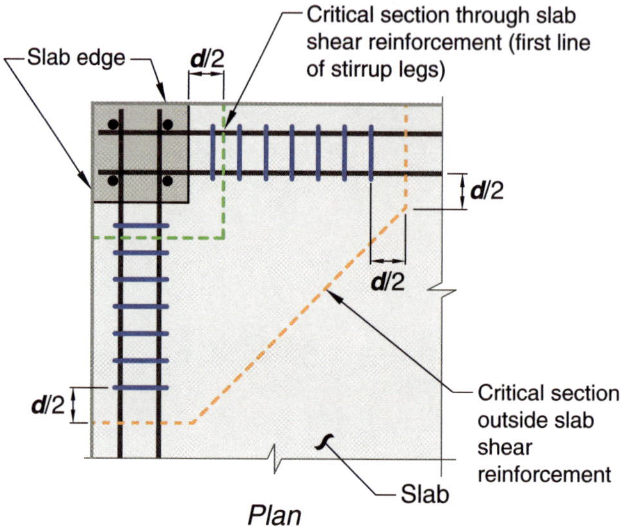

Fig. R22.6.4.2c—Critical sections for two-way shear in slab with shear reinforcement at corner column.

22.6.4.3 If an opening is located closer than $4h$ from the periphery of a column, concentrated load, or reaction area, the portion of b_o enclosed by straight lines projecting from the centroid of the column, concentrated load or reaction area and tangent to the boundaries of the opening shall be considered ineffective.

R22.6.4.3 Provisions for design of openings in slabs (and footings) were developed in Joint ACI-ASCE Committee 326 (1962). The locations of the effective portions of the critical section near typical openings and free edges are shown by the dashed lines in Fig. R22.6.4.3. Research (Joint ACI-ASCE Committee 426 1974) has confirmed that these provisions are conservative.

Research (Genikomsou and Polak 2017) has shown that when openings are located at distances greater than $4d$ from the periphery of a column, the punching shear strength is the same as that for a slab without openings.

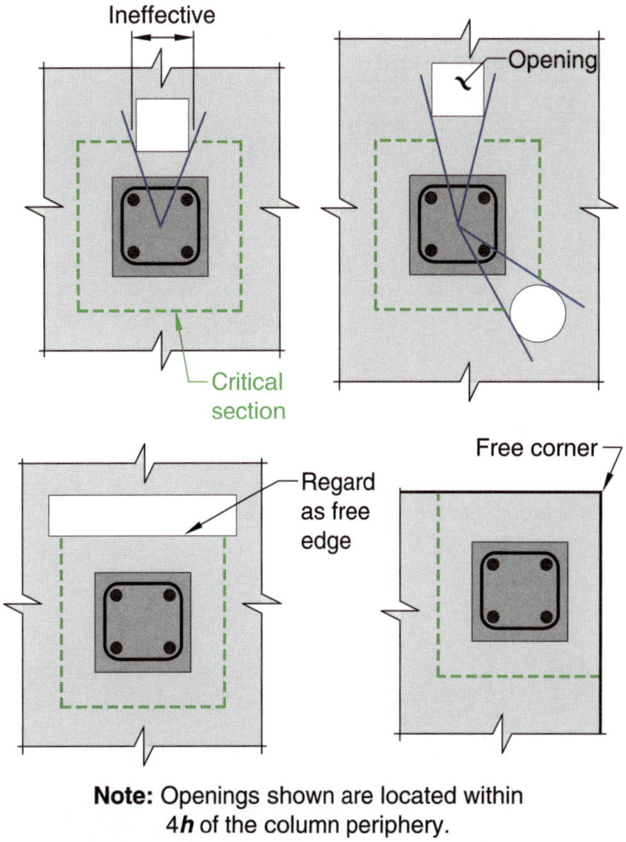

Fig. R22.6.4.3—Effect of openings and free edges (effective perimeter shown with dashed lines).

22.6.5 *Two-way shear strength provided by concrete in members without shear reinforcement*

22.6.5.1 For nonprestressed two-way members, v_c shall be calculated in accordance with 22.6.5.2. For prestressed two-way members, v_c shall be calculated in accordance with (a) or (b):

(a) 22.6.5.2

(b) 22.6.5.5, if the conditions of 22.6.5.4 are satisfied

22.6.5.2 v_c shall be calculated in accordance with Table 22.6.5.2.

R22.6.5 *Two-way shear strength provided by concrete in members without shear reinforcement*

R22.6.5.2 Experimental evidence indicates that the measured concrete shear strength of two-way members without shear reinforcement does not increase in direct proportion with member depth. This phenomenon is referred to as the "size effect." The modification factor λ_s accounts for the dependence of two-way shear strength of slabs on effective depth.

For nonprestressed two-way slabs without a minimum amount of shear reinforcement and with $d > 10$ **in.**, the size effect specified in 22.5.5.1.3 reduces the shear strength of two-way slabs below $4\sqrt{f_c'}\,b_o d$ (Hawkins and Ospina 2017; Dönmez and Bažant 2017).

For square columns, the stress corresponding to the nominal two-way shear strength provided by concrete in

CODE

Table 22.6.5.2—v_c for two-way members without shear reinforcement

v_c		
Least of (a), (b), and (c):	$4\lambda_s\lambda\sqrt{f_c'}$	(a)
	$\left(2+\dfrac{4}{\beta}\right)\lambda_s\lambda\sqrt{f_c'}$	(b)
	$\left(2+\dfrac{\alpha_s d}{b_o}\right)\lambda_s\lambda\sqrt{f_c'}$	(c)

Notes:

(i) λ_s is the size effect factor given in 22.5.5.1.3.

(ii) β is the ratio of long to short sides of the column, concentrated load, or reaction area.

(iii) α_s is given in 22.6.5.3.

COMMENTARY

slabs subjected to bending in two directions is limited to $4\lambda_s\sqrt{f_c'}$. However, tests (Joint ACI-ASCE Committee 426 1974) have indicated that the value of $4\lambda_s\sqrt{f_c'}$ is unconservative when the ratio β of the lengths of the long and short sides of a rectangular column or loaded area is larger than 2.0. In such cases, the actual shear stress on the critical section at punching shear failure varies from a maximum of approximately $4\lambda_s\sqrt{f_c'}$ around the corners of the column or loaded area, down to $2\lambda_s\sqrt{f_c'}$ or less along the long sides between the two end sections. Other tests (Vanderbilt 1972) indicate that v_c decreases as the ratio b_o/d increases. Expressions (b) and (c) in Table 22.6.5.2 were developed to account for these two effects.

For shapes other than rectangular, β is taken to be the ratio of the longest overall dimension of the effective loaded area to the largest overall perpendicular dimension of the effective loaded area, as illustrated for an L-shaped reaction area in Fig. R22.6.5.2. The effective loaded area is that area totally enclosing the actual loaded area, for which the perimeter is a minimum.

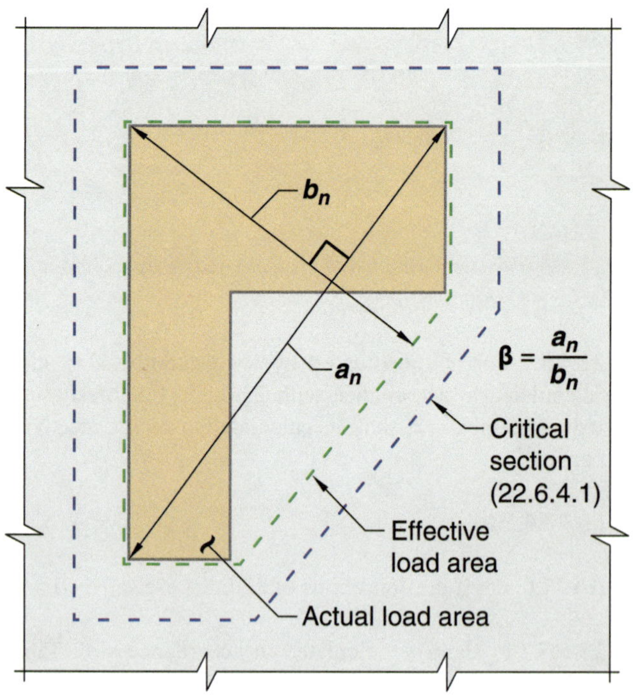

Fig. R22.6.5.2—Value of β for a nonrectangular loaded area.

22.6.5.3 The value of α_s is 40 for interior columns, 30 for edge columns, and 20 for corner columns.

R22.6.5.3 The terms "interior columns," "edge columns," and "corner columns" in this provision refer to critical sections with a continuous slab on four, three, and two sides, respectively.

22.6.5.4 For prestressed, two-way members, it shall be permitted to calculate v_c using 22.6.5.5, provided that (a) through (c) are satisfied:

R22.6.5.4 For prestressed two-way members, modified forms of expressions (b) and (c) in Table 22.6.5.2 are specified. Research (ACI 423.3R) indicates that the shear strength of two-way prestressed slabs around interior columns is

CODE

(a) Bonded reinforcement is provided in accordance with 8.6.2.3 and 8.7.5.3

(b) No portion of the column cross section is closer to a discontinuous edge than four times the slab thickness h

(c) Effective prestress f_{pc} in each direction is not less than 125 psi

22.6.5.5 For prestressed, two-way members conforming to 22.6.5.4, v_c shall be permitted to be the lesser of (a) and (b)

(a) $$v_c = 3.5\lambda\sqrt{f_c'} + 0.3f_{pc} + \frac{V_p}{b_o d} \quad (22.6.5.5a)$$

(b) $$v_c = \left(1.5 + \frac{\alpha_s d}{b_o}\right)\lambda\sqrt{f_c'} + 0.3f_{pc} + \frac{V_p}{b_o d} \quad (22.6.5.5b)$$

where α_s is given in 22.6.5.3; the value of f_{pc} is the average of f_{pc} in the two directions and shall not exceed 500 psi; V_p is the vertical component of all effective prestress forces crossing the critical section; and the value of $\sqrt{f_c'}$ shall not exceed 70 psi.

22.6.6 *Two-way shear strength provided by concrete in members with shear reinforcement*

22.6.6.1 For two-way members with shear reinforcement, v_c at critical sections shall be calculated in accordance with Table 22.6.6.1.

COMMENTARY

conservatively calculated by the expressions in 22.6.5.5, where v_c corresponds to a diagonal tension failure of the concrete initiating at the critical section defined in 22.6.4.1. The mode of failure differs from a punching shear failure around the perimeter of the loaded area of a nonprestressed slab calculated using expression (b) in Table 22.6.5.2. Consequently, the expressions in 22.6.5.5 differ from those for nonprestressed slabs. Values for $\sqrt{f_c'}$ and f_{pc} are restricted in design due to limited test data available beyond the specified limits. When calculating f_{pc}, loss of prestress due to restraint of the slab by structural walls and other structural elements should be taken into account.

R22.6.6 *Two-way shear strength provided by concrete in members with shear reinforcement*

Critical sections for two-way members with shear reinforcement are defined in 22.6.4.1 for the sections adjacent to the column, concentrated load, or reaction area, and 22.6.4.2 for the section located just beyond the outermost peripheral line of stirrup or headed shear stud reinforcement. Values of maximum v_c for these critical sections are given in Table 22.6.6.1. Limiting values of v_u for the critical sections defined in 22.6.4.1 are given in Table 22.6.6.3.

The maximum v_c and limiting value of v_u at the innermost critical section (defined in 22.6.4.1) are higher where headed shear stud reinforcement is provided than the case where stirrups are provided (refer to R8.7.7). Maximum v_c values at the critical sections defined in 22.6.4.2 beyond the outermost peripheral line of shear reinforcement are independent of the type of shear reinforcement provided.

R22.6.6.1 For two-way slabs with stirrups, the maximum value of v_c is taken as $2\lambda_s\lambda\sqrt{f_c'}$ because the stirrups resist all the shear beyond that at inclined cracking (which occurs at approximately half the capacity of a slab without shear reinforcement (that is, $0.5 \times 4\lambda_s\lambda\sqrt{f_c'} = 2\lambda_s\lambda\sqrt{f_c'}$) (Hawkins 1974). The higher value of v_c for two-way slabs with headed shear stud reinforcement is based on research (Elgabry and Ghali 1987).

Sect. Strength 22

CODE

Table 22.6.6.1—v_c for two-way members with shear reinforcement

Type of shear reinforcement	Critical sections	v_c		
Stirrups	All	$2\lambda_s\lambda\sqrt{f_c'}$		(a)
Headed shear stud reinforcement	According to 22.6.4.1	Least of (b), (c), and (d):	$3\lambda_s\lambda\sqrt{f_c'}$	(b)
			$\left(2+\frac{4}{\beta}\right)\lambda_s\lambda\sqrt{f_c'}$	(c)
			$\left(2+\frac{\alpha_s d}{b_o}\right)\lambda_s\lambda\sqrt{f_c'}$	(d)
	According to 22.6.4.2		$2\lambda_s\lambda\sqrt{f_c'}$	(e)

Notes:

(i) λ_s is the size effect factor given in 22.5.5.1.3.

(ii) β is the ratio of long to short sides of the column, concentrated load, or reaction area.

(iii) α_s is given in 22.6.5.3.

22.6.6.2 It shall be permitted to take λ_s as 1.0 if (a) or (b) is satisfied:

(a) Stirrups are designed and detailed in accordance with 8.7.6 and $A_v/s \geq 2\sqrt{f_c'}\,b_o/f_{yt}$.

(b) Smooth headed shear stud reinforcement with stud shaft length not exceeding 10 in. is designed and detailed in accordance with 8.7.7 and $A_v/s \geq 2\sqrt{f_c'}\,b_o/f_{yt}$.

COMMENTARY

R22.6.6.2 The size effect in slabs with ***d* > 10 in.** can be mitigated if a minimum amount of shear reinforcement is provided. The ability of ordinary (smooth) headed shear stud reinforcement to effectively mitigate the size effect on the two-way shear strength of slabs may be compromised if studs longer than 10 in. are used. Until experimental evidence becomes available, it is not permitted to use λ_s equal to 1.0 for slabs with ***d* > 10 in.** without headed shear stud reinforcement with stud shaft length not exceeding 10 in. Stacking or "piggybacking" of headed shear studs, as shown in Fig. R22.6.6.2, introduces an intermediate head that contributes to further anchor the stacked stud.

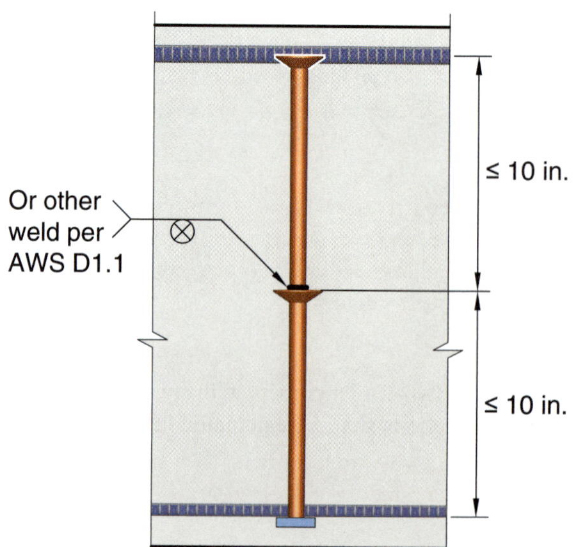

Fig. R22.6.6.2—Stacking (piggybacking) of headed shear stud reinforcement.

CODE

22.6.6.3 For two-way members with shear reinforcement, effective depth shall be selected such that v_u calculated at critical sections does not exceed the values in Table 22.6.6.3.

Table 22.6.6.3—Maximum v_u for two-way members with shear reinforcement

Type of shear reinforcement	Maximum v_u at critical sections defined in 22.6.4.1	
Stirrups	$\phi 6\sqrt{f_c'}$	(a)
Headed shear stud reinforcement	$\phi 8\sqrt{f_c'}$	(b)

22.6.7 *Two-way shear strength provided by single- or multiple-leg stirrups*

22.6.7.1 Single- or multiple-leg stirrups fabricated from bars or wires shall be permitted to be used as shear reinforcement in slabs and footings satisfying (a) and (b):

(a) d is at least 6 in.

(b) d is at least $16d_b$, where d_b is the diameter of the stirrups

22.6.7.2 For two-way members with stirrups, v_s shall be calculated by:

$$v_s = \frac{A_v f_{yt}}{b_o s} \quad (22.6.7.2)$$

where A_v is the sum of the area of all legs of reinforcement on one peripheral line that is geometrically similar to the perimeter of the column section, and s is the spacing of the peripheral lines of shear reinforcement in the direction perpendicular to the column face.

22.6.8 *Two-way shear strength provided by headed shear stud reinforcement*

22.6.8.1 Headed shear stud reinforcement shall be permitted to be used as shear reinforcement in slabs and footings if the placement and geometry of the headed shear stud reinforcement satisfies 8.7.7.

22.6.8.2 For two-way members with headed shear stud reinforcement, v_s shall be calculated by:

$$v_s = \frac{A_v f_{yt}}{b_o s} \quad (22.6.8.2)$$

COMMENTARY

R22.6.7 *Two-way shear strength provided by single- or multiple-leg stirrups*

R22.6.7.2 Because shear stresses are used for two-way shear in this chapter, shear strength provided by transverse reinforcement is averaged over the cross-sectional area of the critical section.

R22.6.8 *Two-way shear strength provided by headed shear stud reinforcement*

Tests (ACI 421.1R) show that headed shear stud reinforcement mechanically anchored as close as practicable to the top and bottom of slabs is effective in resisting punching shear. The critical section beyond the shear reinforcement is generally assumed to have a polygonal shape (refer to Fig. R22.6.4.2a, R22.6.4.2b, and R22.6.4.2c). Equations for calculating shear stresses on such sections are given in ACI 421.1R.

R22.6.8.2 Because shear stresses are used for two-way shear in this chapter, shear strength provided by transverse reinforcement is averaged over the cross-sectional area of the critical section.

Sect. Strength 22

CODE

where A_v is the sum of the area of all shear studs on one peripheral line that is geometrically similar to the perimeter of the column section, and s is the spacing of the peripheral lines of headed shear stud reinforcement in the direction perpendicular to the column face.

22.6.8.3 If headed shear stud reinforcement is provided, A_v/s shall satisfy:

$$\frac{A_v}{s} \geq 2\sqrt{f_c'}\frac{b_o}{f_{yt}} \quad (22.6.8.3)$$

22.7—Torsional strength

COMMENTARY

R22.7—Torsional strength

The design for torsion in this section is based on a thin-walled tube space truss analogy. A beam subjected to torsion is idealized as a thin-walled tube with the core concrete cross section in a solid beam neglected as shown in Fig. R22.7(a). Once a reinforced concrete beam has cracked in torsion, its torsional strength is provided primarily by closed stirrups and longitudinal bars located near the surface of the member. In the thin-walled tube analogy, the strength is assumed to be provided by the outer skin of the cross section roughly centered on the closed stirrups. Both hollow and solid sections are idealized as thin-walled tubes both before and after cracking.

In a closed thin-walled tube, the product of the shear stress τ and the wall thickness t at any point in the perimeter is known as the shear flow, $q = \tau t$. The shear flow q due to torsion acts as shown in Fig. R22.7(a) and is constant at all points around the perimeter of the tube. The path along which it acts extends around the tube at midthickness of the walls of the tube. At any point along the perimeter of the tube, the shear stress due to torsion is $\tau = T/(2A_o t)$, where A_o is the gross area enclosed by the shear flow path, shown shaded in Fig. R22.7(b), and t is the thickness of the wall at the point where τ is being calculated. For a hollow member with continuous walls, A_o includes the area of the hole.

The concrete contribution to torsional strength is ignored, and in cases of combined shear and torsion, the concrete contribution to shear strength does not need to be reduced. The design procedure is derived and compared with test results in MacGregor and Ghoneim (1995) and Hsu (1997).

CODE

COMMENTARY

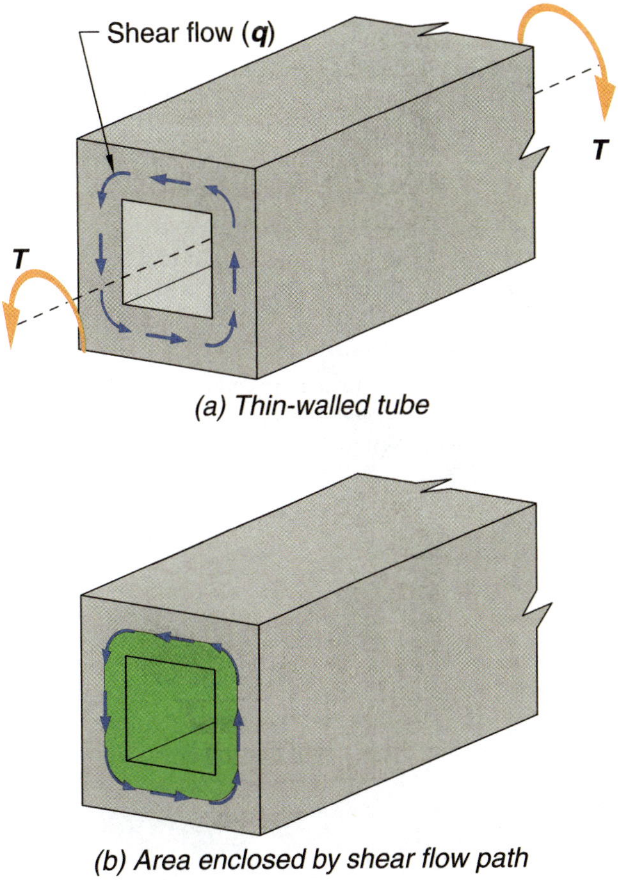

***Fig. R22.7**—(a) Thin-walled tube; and (b) area enclosed by shear flow path.*

22.7.1 *General*

22.7.1.1 This section shall apply to members if $T_u \geq \phi T_{th}$, where ϕ is given in Chapter 21 and threshold torsion T_{th} is given in 22.7.4. If $T_u < \phi T_{th}$, it shall be permitted to neglect torsional effects.

22.7.1.2 Nominal torsional strength shall be calculated in accordance with 22.7.6.

22.7.1.3 For calculation of T_{th} and T_{cr}, λ shall be in accordance with 19.2.4.

R22.7.1 *General*

R22.7.1.1 Torsional moments that do not exceed the threshold torsion T_{th} will not cause a structurally significant reduction in either flexural or shear strength and can be ignored.

22.7.2 *Limiting material strengths*

22.7.2.1 The value of $\sqrt{f_c'}$ used to calculate T_{th} and T_{cr} shall not exceed 100 psi.

22.7.2.2 The values of f_y and f_{yt} for longitudinal and transverse torsional reinforcement shall not exceed the limits in 20.2.2.4.

R22.7.2 *Limiting material strengths*

R22.7.2.1 Because of a lack of test data and practical experience with concretes having compressive strengths greater than 10,000 psi, the Code imposes a maximum value of 100 psi on $\sqrt{f_c'}$ for use in the calculation of torsional strength.

R22.7.2.2 The upper limit of 60,000 psi on the value of f_y and f_{yt} used in design is intended to control diagonal crack width.

CODE

22.7.3 *Factored design torsion*

22.7.3.1 If $T_u \geq \phi T_{cr}$ and T_u is required to maintain equilibrium, the member shall be designed to resist T_u.

22.7.3.2 In a statically indeterminate structure where $T_u \geq \phi T_{cr}$ and a reduction of T_u can occur due to redistribution of internal forces after torsional cracking, it shall be permitted to reduce T_u to ϕT_{cr}, where the cracking torsion T_{cr} is calculated in accordance with 22.7.5.

22.7.3.3 If T_u is redistributed in accordance with 22.7.3.2, the factored moments and shears used for design of the adjoining members shall be in equilibrium with the reduced torsion.

COMMENTARY

R22.7.3 *Factored design torsion*

In designing for torsion in reinforced concrete structures, two conditions may be identified (Collins and Lampert 1973; Hsu and Burton 1974):

(a) The torsional moment cannot be reduced by redistribution of internal forces (22.7.3.1). This type of torsion is referred to as equilibrium torsion because the torsional moment is required for the structure to be in equilibrium. For this condition, illustrated in Fig. R22.7.3(a), torsional reinforcement must be provided to resist the total design torsional moments.

(b) The torsional moment can be reduced by redistribution of internal forces after cracking (22.7.3.2) if the torsion results from the member twisting to maintain compatibility of deformations. This type of torsion is referred to as compatibility torsion.

For this condition, illustrated in Fig. R22.7.3(b), the torsional stiffness before cracking corresponds to that of the uncracked section according to St. Venant's theory. At torsional cracking, however, a large twist occurs under an essentially constant torsional moment, resulting in a large redistribution of forces in the structure (Collins and Lampert 1973; Hsu and Burton 1974). The cracking torsional moment under combined shear, moment, and torsion corresponds to a principal tensile stress somewhat less than the $4\lambda\sqrt{f_c'}$ used in R22.7.5.

If the torsional moment exceeds the cracking torsional moment (22.7.3.2), a maximum factored torsional moment equal to the cracking torsional moment may be assumed to occur at the critical sections near the faces of the supports. The maximum factored torsional moment has been established to limit the width of torsional cracks.

Provision 22.7.3.2 applies to typical and regular framing conditions. With layouts that impose significant torsional rotations within a limited length of the member, such as a large torsional moment located close to a stiff column, or a column that rotates in the reverse directions because of other loading, a more detailed analysis is advisable.

If the factored torsional moment from an elastic analysis based on uncracked section properties is between ϕT_{th} and ϕT_{cr}, torsional reinforcement should be designed to resist the calculated torsional moments.

CODE | COMMENTARY

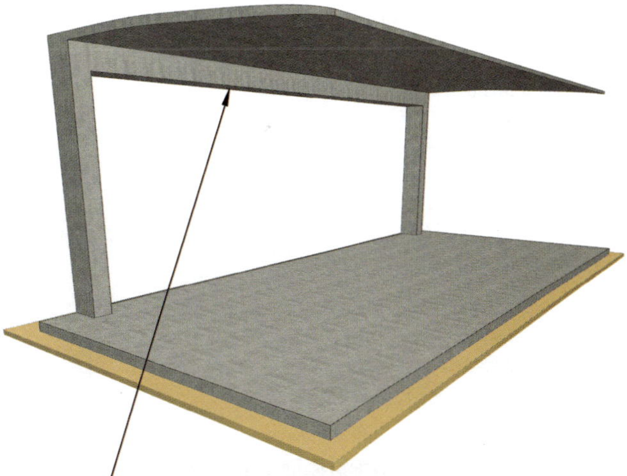

Fig. R22.7.3a—Equilibrium torsion, the design torsional moment may not be reduced (22.7.3.1).

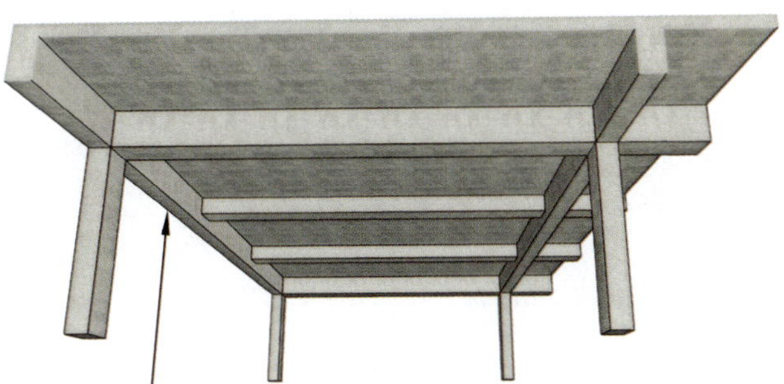

Fig. R22.7.3b—Compatibility torsion, the design torsional moment may be reduced (22.7.3.2).

22.7.4 *Threshold torsion*

22.7.4.1 Threshold torsion T_{th} shall be calculated in accordance with Table 22.7.4.1(a) for solid cross sections and Table 22.7.4.1(b) for hollow cross sections, where N_u is positive for compression and negative for tension.

R22.7.4 *Threshold torsion*

The threshold torsion is defined as one-fourth the cracking torsional moment T_{cr}. For sections of solid members, the interaction between the cracking torsional moment and the inclined cracking shear is approximately circular or elliptical. For such a relationship, a threshold torsional moment of T_{th}, as used in 22.7.4.1, corresponds to a reduction of less than 5 percent in the inclined cracking shear, which is considered negligible.

For torsion, a hollow section is defined as having one or more longitudinal voids, such as a single-cell or multiple-cell box girder. Small longitudinal voids, such as ungrouted post-tensioning ducts that result in $A_g/A_{cp} \geq 0.95$, can be ignored when calculating T_{th}. The interaction between torsional cracking and shear cracking for hollow sections is assumed to vary from the elliptical relationship for members with small voids, to a straight-line relationship for thin-walled

Sect. Strength 22

CODE

Table 22.7.4.1(a)—Threshold torsion for solid cross sections

Type of member	T_{th}	
Nonprestressed member	$\lambda\sqrt{f'_c}\left(\frac{A_{cp}^2}{p_{cp}}\right)$	(a)
Prestressed member	$\lambda\sqrt{f'_c}\left(\frac{A_{cp}^2}{p_{cp}}\right)\sqrt{1+\frac{f_{pc}}{4\lambda\sqrt{f'_c}}}$	(b)
Nonprestressed member subjected to axial force	$\lambda\sqrt{f'_c}\left(\frac{A_{cp}^2}{p_{cp}}\right)\sqrt{1+\frac{N_u}{4A_g\lambda\sqrt{f'_c}}}$	(c)

Table 22.7.4.1(b)—Threshold torsion for hollow cross sections

Type of member	T_{th}	
Nonprestressed member	$\lambda\sqrt{f'_c}\left(\frac{A_g^2}{p_{cp}}\right)$	(a)
Prestressed member	$\lambda\sqrt{f'_c}\left(\frac{A_g^2}{p_{cp}}\right)\sqrt{1+\frac{f_{pc}}{4\lambda\sqrt{f'_c}}}$	(b)
Nonprestressed member subjected to axial force	$\lambda\sqrt{f'_c}\left(\frac{A_g^2}{p_{cp}}\right)\sqrt{1+\frac{N_u}{4A_g\lambda\sqrt{f'_c}}}$	(c)

22.7.5 *Cracking torsion*

22.7.5.1 Cracking torsion T_{cr} shall be calculated in accordance with Table 22.7.5.1 for solid and hollow cross sections, where N_u is positive for compression and negative for tension.

Table 22.7.5.1—Cracking torsion

Type of member	T_{cr}	
Nonprestressed member	$4\lambda\sqrt{f'_c}\left(\frac{A_{cp}^2}{p_{cp}}\right)$	(a)
Prestressed member	$4\lambda\sqrt{f'_c}\left(\frac{A_{cp}^2}{p_{cp}}\right)\sqrt{1+\frac{f_{pc}}{4\lambda\sqrt{f'_c}}}$	(b)
Nonprestressed member subjected to axial force	$4\lambda\sqrt{f'_c}\left(\frac{A_{cp}^2}{p_{cp}}\right)\sqrt{1+\frac{N_u}{4A_g\lambda\sqrt{f'_c}}}$	(c)

COMMENTARY

sections with large voids. For a straight-line interaction, a torsional moment of T_{th} would cause a reduction in the inclined cracking shear of approximately 25 percent, which was considered to be significant. Therefore, the expressions for solid sections are modified by the factor $(A_g/A_{cp})^2$ to develop the expressions for hollow sections. Tests of solid and hollow beams (Hsu 1968) indicate that the cracking torsional moment of a hollow section is approximately (A_g/A_{cp}) times the cracking torsional moment of a solid section with the same outside dimensions. An additional multiplier of (A_g/A_{cp}) reflects the transition from the circular interaction between the inclined cracking loads in shear and torsion for solid members, to the approximately linear interaction for thin-walled hollow sections.

R22.7.5 *Cracking torsion*

The cracking torsional moment under pure torsion, T_{cr}, is derived by replacing the actual section with an equivalent thin-walled tube with a wall thickness t prior to cracking of $0.75A_{cp}/p_{cp}$ and an area enclosed by the wall centerline A_o equal to $2A_{cp}/3$. Cracking is assumed to occur when the principal tensile stress reaches $4\lambda\sqrt{f'_c}$. The stress at cracking, $4\lambda\sqrt{f'_c}$, has purposely been taken as a lower bound value. In a nonprestressed beam loaded with torsion alone, the principal tensile stress is equated to the torsional shear stress, $\tau = T/(2A_ot)$. Thus, cracking occurs when τ reaches $4\lambda\sqrt{f'_c}$, giving the cracking torsional moment T_{cr} as defined by expression (a) in Table 22.7.5.1.

For prestressed members, the torsional cracking load is increased by the prestress given by expression (b) in Table 22.7.5.1. A Mohr's Circle analysis based on average stresses indicates the torsional moment required to cause a principal tensile stress equal to $4\lambda\sqrt{f'_c}$ is $\sqrt{1+f_{pc}/(4\lambda\sqrt{f'_c})}$ times the corresponding torsional cracking moment in a nonprestressed beam. A similar modification is made in expression (c) in Table 22.7.5.1 for members subjected to axial force and torsion.

If the factored torsional moment exceeds ϕT_{cr} in a statically indeterminate structure, a maximum factored torsional moment equal to ϕT_{cr} may be assumed to occur at critical sections near the faces of the supports. This limit has been

CODE / **COMMENTARY**

established to control the width of the torsional cracks. The replacement of A_{cp} with A_g, as in the calculation of T_{th} for hollow sections in 22.7.4.1, is not applied here. Thus, the torsional moment after redistribution is larger and, hence, more conservative.

22.7.6 *Torsional strength*

R22.7.6 *Torsional strength*

The torsional design strength ϕT_n must equal or exceed the torsional moment T_u due to factored loads. In the calculation of T_n, all the torsion is assumed to be resisted by stirrups and longitudinal reinforcement, neglecting any concrete contribution to torsional strength. At the same time, the nominal shear strength provided by concrete, V_c, is assumed to be unchanged by the presence of torsion.

22.7.6.1 For nonprestressed and prestressed members, T_n shall be the lesser of (a) and (b):

(a) $$T_n = \frac{2A_oA_tf_{yt}}{s}\cot\theta \quad (22.7.6.1a)$$

(b) $$T_n = \frac{2A_oA_\ell f_y}{p_h}\tan\theta \quad (22.7.6.1b)$$

where A_o shall be determined by analysis, θ shall not be taken less than 30 degrees nor greater than 60 degrees; A_t is the area of one leg of a closed stirrup resisting torsion; A_ℓ is the area of longitudinal torsional reinforcement; and p_h is the perimeter of the centerline of the outermost closed stirrup.

R22.7.6.1 Equation (22.7.6.1a) is based on the space truss analogy shown in Fig. R22.7.6.1a with compression diagonals at an angle θ, assuming the concrete resists no tension and the reinforcement yields. After torsional cracking develops, the torsional strength is provided mainly by closed stirrups, longitudinal reinforcement, and compression diagonals. The concrete outside these stirrups is relatively ineffective. For this reason A_o, the gross area enclosed by the shear flow path around the perimeter of the tube, is defined after cracking in terms of A_{oh}, the area enclosed by the centerline of the outermost closed transverse torsional reinforcement.

The shear flow q in the walls of the tube, discussed in R22.7, can be resolved into the shear forces V_1 to V_4 acting in the individual sides of the tube or space truss, as shown in Fig. R22.7.6.1a.

As shown in Figure R22.7.6.1b, on a given wall of the tube, the shear flow V_i is resisted by a diagonal compression component, $D_i = V_i/\sin\theta$, in the concrete. An axial tension force, $N_i = V_i(\cot\theta)$, is required in the longitudinal reinforcement to complete the resolution of V_i.

Because the shear flow due to torsion is constant at all points around the perimeter of the tube, the resultants of D_i and N_i act through the midheight of side i. As a result, half of N_i can be assumed to be resisted by each of the top and bottom chords as shown. Longitudinal reinforcement with a strength $A_\ell f_y$ is required to resist the sum of the N_i forces, $\sum N_i$, acting in all of the walls of the tube.

In the derivation of Eq. (22.7.6.1b), axial tension forces are summed along the sides of the area A_o. These sides form a perimeter length p_o approximately equal to the length of the line joining the centers of the bars in the corners of the tube. For ease in calculation, this has been replaced with the perimeter of the closed stirrups, p_h.

Sect. Strength 22

CODE

COMMENTARY

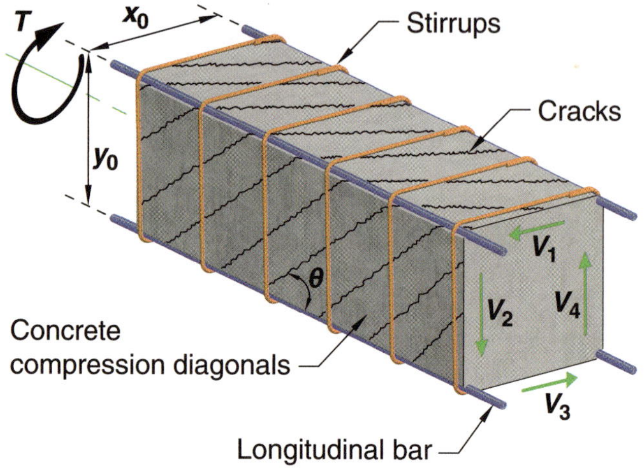

Fig. R22.7.6.1a—Space truss analogy.

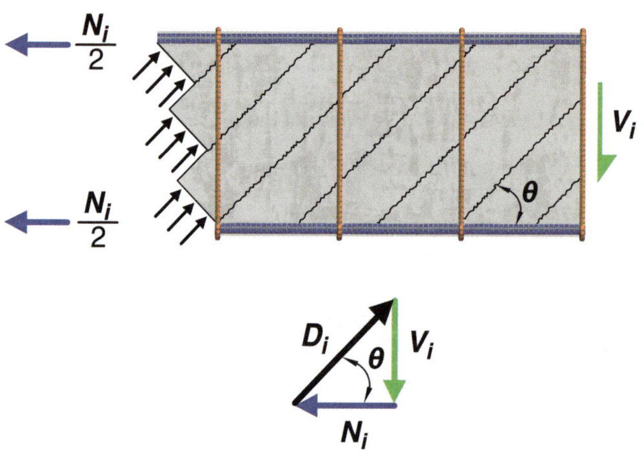

Fig. R22.7.6.1b—Resolution of shear force $\mathbf{V_i}$ *into diagonal compression force* $\mathbf{D_i}$ *and axial tension force* $\mathbf{N_i}$ *in one wall of tube.*

22.7.6.1.1 In Eq. (22.7.6.1a) and (22.7.6.1b), it shall be permitted to take A_o equal to $0.85A_{oh}$.

R22.7.6.1.1 The area A_{oh} is shown in Fig. R22.7.6.1.1 for various cross sections. In I-, T-, L-shaped, or circular sections, A_{oh} is taken as that area enclosed by the outermost transverse reinforcement.

CODE

COMMENTARY

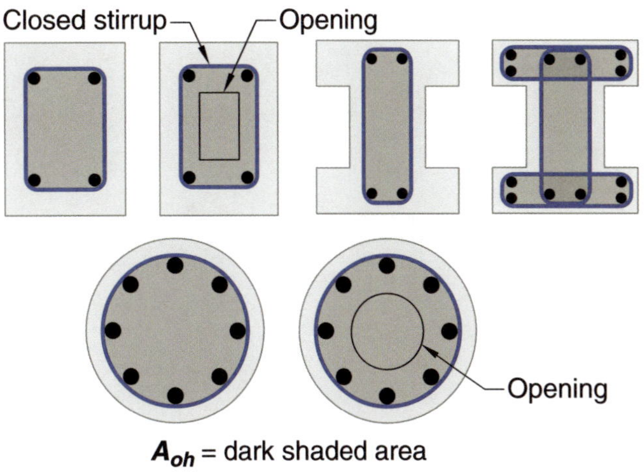

Fig. R22.7.6.1.1—Definition of A_{oh}.

22.7.6.1.2 In Eq. (22.7.6.1a) and (22.7.6.1b), it shall be permitted to take θ equal to (a) or (b):

(a) 45 degrees for nonprestressed members or members with $A_{ps}f_{se} < 0.4(A_{ps}f_{pu} + A_sf_y)$

(b) 37.5 degrees for prestressed members with $A_{ps}f_{se} \geq 0.4(A_{ps}f_{pu} + A_sf_y)$

R22.7.6.1.2 The angle θ can be obtained by analysis (Hsu 1990) or may be taken equal to the values given in 22.7.6.1.2(a) or (b). The same value of θ is required to be used in both Eq. (22.7.6.1a) and (22.7.6.1b). With smaller values of θ, the amount of stirrups required by Eq. (22.7.6.1a) decreases. At the same time, the amount of longitudinal reinforcement required by Eq. (22.7.6.1b) increases.

22.7.7 *Cross-sectional limits*

22.7.7.1 Cross-sectional dimensions shall be selected such that (a) or (b) is satisfied:

(a) For solid sections

$$\sqrt{\left(\frac{V_u}{b_wd}\right)^2+\left(\frac{T_up_h}{1.7A_{oh}^2}\right)^2} \leq \phi\left(\frac{V_c}{b_wd}+8\sqrt{f_c'}\right) \quad (22.7.7.1a)$$

(b) For hollow sections

$$\left(\frac{V_u}{b_wd}\right)+\left(\frac{T_up_h}{1.7A_{oh}^2}\right) \leq \phi\left(\frac{V_c}{b_wd}+8\sqrt{f_c'}\right) \quad (22.7.7.1b)$$

R22.7.7 *Cross-sectional limits*

R22.7.7.1 The size of a cross section is limited for two reasons: first, to reduce excessive cracking, and second, to minimize the potential for crushing of the surface concrete due to inclined compressive stresses due to shear and torsion. In Eq. (22.7.7.1a) and (22.7.7.1b), the two terms on the left-hand side are the shear stresses due to shear and torsion. The sum of these stresses may not exceed the stress causing shear cracking plus $8\sqrt{f_c'}$, similar to the limiting strength given in 22.5.1.2 for shear without torsion. The limit is expressed in terms of V_c to allow its use for nonprestressed or prestressed concrete. It was originally derived on the basis of crack control. It is not necessary to check against crushing of the web because crushing occurs at higher shear stresses.

In a hollow section, the shear stresses due to shear and torsion both occur in the walls of the box as shown in Fig. R22.7.7.1(a) and hence are directly additive at Point A as given in Eq. (22.7.7.1b). In a solid section, the shear stresses due to torsion act in the tubular outside section while the shear stresses due to V_u are spread across the width of the section, as shown in Fig. R22.7.7.1(b). For this reason, stresses are combined in Eq. (22.7.7.1a) using the square root of the sum of the squares rather than by direct addition.

CODE

COMMENTARY

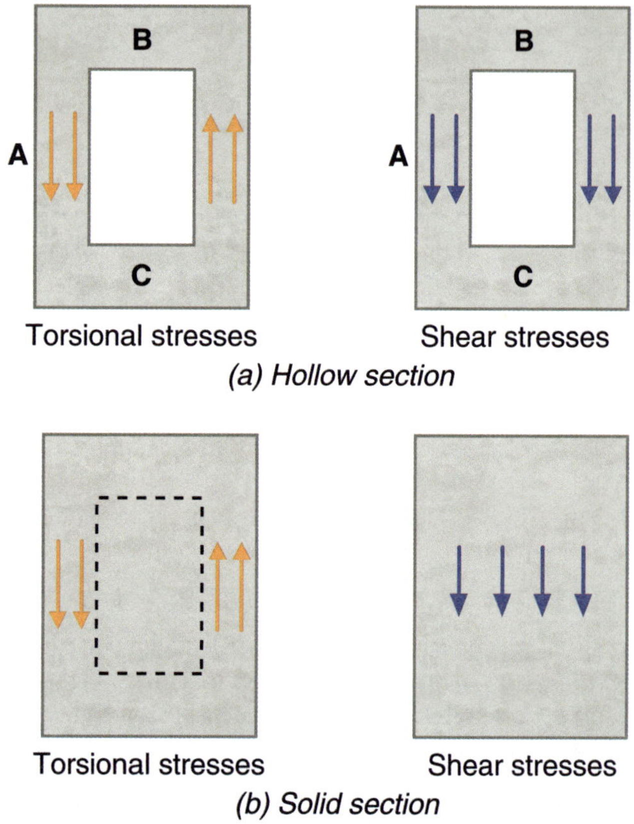

***Fig. R22.7.7.1**—Addition of torsional and shear stresses.*

22.7.7.1.1 For prestressed members, the value of d used in 22.7.7.1 need not be taken less than $0.8h$.

R22.7.7.1.1 Although the value of d may vary along the span of a prestressed beam, studies (MacGregor and Hanson 1969) have shown that, for prestressed concrete members, d need not be taken less than $0.8h$. The beams considered had some straight prestressed reinforcement or reinforcing bars at the bottom of the section and had stirrups that enclosed the longitudinal reinforcement.

22.7.7.1.2 For hollow sections where the wall thickness varies around the perimeter, Eq. (22.7.7.1b) shall be evaluated at the location where the term

$$\left(\frac{V_u}{b_w d}\right)+\left(\frac{T_u p_h}{1.7A_{oh}^2}\right)$$

is a maximum.

R22.7.7.1.2 Generally, the maximum torsional stress will be on the wall where the torsional and shearing stresses are additive (Point A in Fig. R22.7.7.1(a)). If the top or bottom flanges are thinner than the vertical webs, it may be necessary to evaluate Eq. (22.7.7.1b) at Points B and C in Fig. R22.7.7.1(a). At these points, the stresses due to the shear are usually negligible.

22.7.7.2 For hollow sections where the wall thickness is less than A_{oh}/p_h, the term $(T_u p_h/1.7A_{oh}^2)$ in Eq. (22.7.7.1b) shall be taken as $(T_u/1.7A_{oh}t)$, where t is the thickness of the wall of the hollow section at the location where the stresses are being checked.

22.8—Bearing

22.8.1 *General*

R22.8—Bearing

R22.8.1 *General*

CODE

22.8.1.1 Section 22.8 shall apply to the calculation of bearing strength of concrete members.

22.8.1.2 Bearing strength provisions in 22.8 shall not apply to post-tensioned anchorage zones.

22.8.2 *Required strength*

22.8.2.1 Factored compressive force transferred through bearing shall be calculated in accordance with the factored load combinations defined in Chapter 5 and analysis procedures defined in Chapter 6.

22.8.3 *Design strength*

22.8.3.1 Design bearing strength shall satisfy:

$$\phi B_n \geq B_u \qquad (22.8.3.1)$$

for each applicable factored load combination.

22.8.3.2 Nominal bearing strength B_n shall be calculated in accordance with Table 22.8.3.2, where A_1 is the loaded area, and A_2 is the area of the lower base of the largest frustum of a pyramid, cone, or tapered wedge contained wholly within the support and having its upper base equal to the loaded area. The sides of the pyramid, cone, or tapered wedge shall be sloped 1 vertical to 2 horizontal.

Table 22.8.3.2—Nominal bearing strength

Geometry of bearing area	B_n		
Supporting surface is wider on all sides than the loaded area	Lesser of (a) and (b)	$\sqrt{A_2/A_1}(0.85f_c'A_1)$	(a)
		$2(0.85f_c'A_1)$	(b)
Other cases	$0.85f_c'A_1$		(c)

COMMENTARY

R22.8.1.2 Because post-tensioned anchorage zones are usually designed in accordance with 25.9, the bearing strength provisions in 22.8 are not applicable.

R22.8.3 *Design strength*

R22.8.3.2 The permissible bearing stress of $\mathbf{0.85}f_c'$ is based on tests reported in Hawkins (1968). Where the supporting area is wider than the loaded area on all sides, the surrounding concrete confines the bearing area, resulting in an increase in bearing strength. No minimum depth is given for the support, which will most likely be controlled by the punching shear requirements of 22.6.

A_1 is the loaded area but not greater than the bearing plate or bearing cross-sectional area.

Where the top of the support is sloped or stepped, advantage may still be taken of the condition that the supporting member is larger than the loaded area, provided the supporting member does not slope at too great an angle. Figure R22.8.3.2 illustrates the application of the frustum to find A_2 for a support under vertical load transfer.

Adequate bearing strength needs to be provided for cases where the compression force transfer is in a direction other than normal to the bearing surface. For such cases, this section applies to the normal component and the tangential component needs to be transferred by other methods, such as by anchor bolts or shear lugs.

The frustum should not be confused with the path by which a load spreads out as it progresses downward through the support. Such a load path would have steeper sides. However, the frustum described has somewhat flat side slopes to ensure that there is concrete immediately surrounding the zone of high stress at the bearing.

Where tensile forces occur in the plane of bearing, it may be desirable to reduce the allowable bearing stress, provide confinement reinforcement, or both. Guidelines are provided in the *PCI Design Handbook* for precast and prestressed concrete (PCI MNL 120).

CODE

COMMENTARY

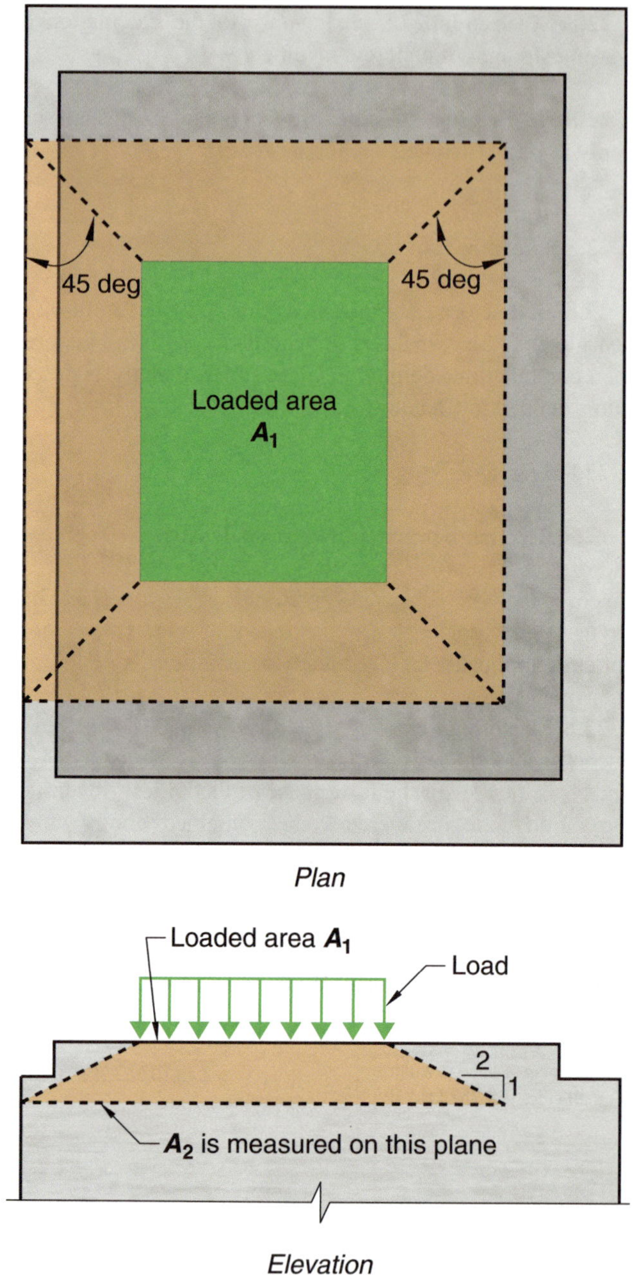

Fig. R22.8.3.2—Application of frustum to find $\mathbf{A_2}$ *in stepped or sloped supports.*

22.9—Shear friction

22.9.1 *General*

22.9.1.1 This section shall apply where it is appropriate to consider shear transfer across any given plane, such as an existing or potential crack, an interface between dissimilar materials, or an interface between two concretes cast at different times.

R22.9—Shear friction

R22.9.1 *General*

R22.9.1.1 The purpose of this section is to provide a design method to address possible failure by shear sliding on a plane. Such conditions include a plane formed by a crack in monolithic concrete, an interface between concrete and steel, and an interface between concretes cast at different times (Birkeland and Birkeland 1966; Mattock and Hawkins 1972).

Although uncracked concrete is relatively strong in direct shear, there is always the possibility that a crack will form in an unfavorable location. The shear-friction concept assumes that such a crack will form, and that reinforcement is provided across the crack to resist relative displacement along it. When shear acts along a crack, one crack face slips relative to the other. If the crack faces are rough and irregular, this slip is accompanied by separation of the crack faces. At nominal strength, the separation is sufficient to stress, in tension, the reinforcement crossing the crack to its specified yield strength. The reinforcement in tension provides a clamping force $A_{vf}f_y$ across the crack faces. The applied shear is then resisted by friction between the crack faces, by resistance to the shearing off of protrusions on the crack faces, and by dowel action of the reinforcement crossing the crack. Successful application of this section depends on proper selection of the location of an assumed crack (PCI MNL 120; Birkeland and Birkeland 1966).

22.9.1.2 The required area of shear-friction reinforcement across the assumed shear plane, A_{vf}, shall be calculated in accordance with 22.9.4. Alternatively, it shall be permitted to use shear transfer design methods that result in prediction of strength in substantial agreement with results of comprehensive tests.

R22.9.1.2 The relationship between shear-transfer strength and the reinforcement crossing the shear plane can be expressed in various ways. Equations (22.9.4.2) and (22.9.4.3) are based on the shear-friction model and provide a conservative estimate of the shear-transfer strength.

Other relationships that provide a more accurate estimate of shear-transfer strength can be used under the requirements of this section. Examples of such procedures can be found in the *PCI Design Handbook* (PCI MNL 120), Mattock et al. (1976b), and Mattock (1974).

22.9.1.3 The value of f_y used to calculate V_n for shear friction shall not exceed the limit in 20.2.2.4.

22.9.1.4 Surface preparation of the shear plane assumed for design shall be specified in the construction documents.

R22.9.1.4 For concrete cast against hardened concrete or structural steel, 26.5.6.1 requires the licensed design professional to specify the surface preparation in the construction documents.

22.9.2 *Required strength*

22.9.2.1 Factored forces across the assumed shear plane shall be calculated in accordance with the factored load combinations defined in Chapter 5 and analysis procedures defined in Chapter 6.

22.9.3 *Design strength*

22.9.3.1 Design shear strength across the assumed shear plane shall satisfy:

$$\phi V_n \geq V_u \qquad (22.9.3.1)$$

for each applicable factored load combination.

22.9.4 *Nominal shear strength*

R22.9.4 *Nominal shear strength*

22.9.4.1 Value of V_n across the assumed shear plane shall be calculated in accordance with 22.9.4.2 or 22.9.4.3. V_n

CODE

shall not exceed the value calculated in accordance with 22.9.4.4.

22.9.4.2 If shear-friction reinforcement is perpendicular to the shear plane, nominal shear strength across the assumed shear plane shall be calculated by:

$$V_n = \mu A_{vf} f_y \quad (22.9.4.2)$$

where A_{vf} is the area of reinforcement crossing the assumed shear plane to resist shear, and μ is the coefficient of friction in accordance with Table 22.9.4.2.

Table 22.9.4.2—Coefficients of friction

Contact surface condition	Coefficient of friction μ[1]	
Concrete placed monolithically	1.4λ	(a)
Concrete placed against hardened concrete that is clean, free of laitance, and intentionally roughened to a full amplitude of approximately 1/4 in.	1.0λ	(b)
Concrete placed against hardened concrete that is clean, free of laitance, and not intentionally roughened	0.6λ	(c)
Concrete placed against as-rolled structural steel that is clean, free of paint, and with shear transferred across the contact surface by headed studs or by welded deformed bars or wires.	0.7λ	(d)

[1]λ = 1.0 for normalweight concrete. For lightweight concrete, λ is calculated as given in 19.2.4, but shall not exceed 0.85.

22.9.4.3 If shear-friction reinforcement is inclined to the shear plane and the shear force induces tension in the shear-friction reinforcement, nominal shear strength across the assumed shear plane shall be calculated by:

$$V_n = A_{vf} f_y(\mu \sin\alpha + \cos\alpha) \quad (22.9.4.3)$$

where **α** is the angle between shear-friction reinforcement and assumed shear plane, and μ is the coefficient of friction in accordance with Table 22.9.4.2.

COMMENTARY

R22.9.4.2 The required area of shear-friction reinforcement, A_{vf}, is calculated using:

$$A_{vf} = \frac{V_u}{\phi f_y \mu} \quad (R22.9.4.2)$$

The upper limit on shear strength that can be achieved using Eq. (22.9.4.2) is given in 22.9.4.4.

In the shear-friction method of calculation, it is assumed that all the shear resistance is due to the friction between the crack faces. It is therefore necessary to use artificially high values of the coefficient of friction in the shear-friction equations so that the calculated shear strength will be in reasonable agreement with test results.

For concrete cast against hardened concrete not roughened in accordance with 22.9.4.2, shear resistance is primarily due to dowel action of the reinforcement. Test results (Mattock 1977) indicate that the reduced value of **μ = 0.6λ** specified for this case is appropriate.

For concrete placed against as-rolled structural steel, the shear-transfer reinforcement may be either reinforcing bars or headed studs. The design of shear connectors for composite action of concrete slabs and steel beams is not covered by these provisions. AISC 360 contains design provisions for these systems.

R22.9.4.3 Inclined shear-friction reinforcement is illustrated in Fig. R22.9.4.3a (Mattock 1974), where α is the acute angle between the bar and the shear plane. Equation (22.9.4.3) applies only when the shear force component parallel to the reinforcement produces tension in the reinforcement and the force component parallel to the shear plane resists part of the shear, as shown in Fig. R22.9.4.3a.

If the shear-friction reinforcement is inclined such that the shear force component parallel to the reinforcement produces compression in the reinforcement, as shown in Fig. R22.9.4.3b, then shear friction does not apply (V_n = **0**).

CODE

COMMENTARY

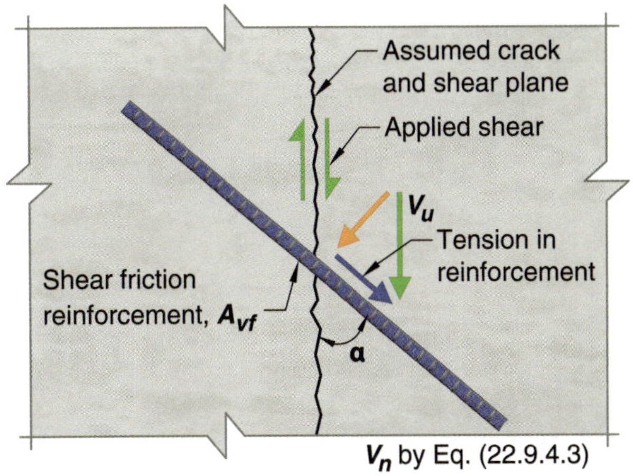

***Fig. R22.9.4.3a**—Tension in shear friction reinforcement.*

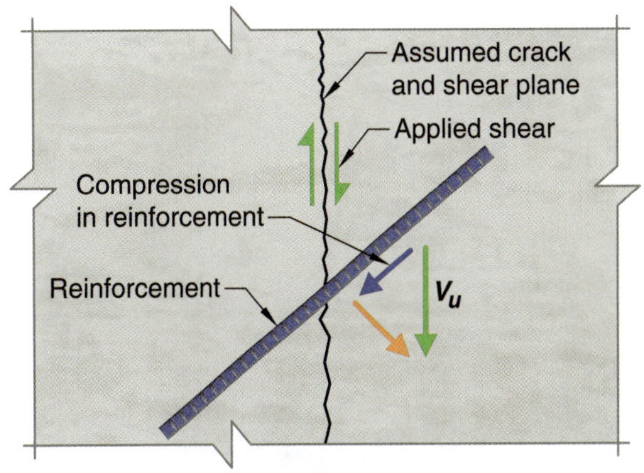

***Fig. R22.9.4.3b**—Compression in reinforcement.*

22.9.4.4 The value of V_n across the assumed shear plane shall not exceed the limits in Table 22.9.4.4. Where concretes of different strengths are cast against each other, the lesser value of f_c' shall be used in Table 22.9.4.4.

R22.9.4.4 Upper limits on shear friction strength are necessary, as Eq. (22.9.4.2) and (22.9.4.3) may become unconservative for some cases (Kahn and Mitchell 2002; Mattock 2001).

Table 22.9.4.4—Maximum V_n across the assumed shear plane

Condition	Maximum V_n		
Normalweight concrete placed monolithically or placed against hardened concrete intentionally roughened to a full amplitude of approximately 1/4 in.	Least of (a), (b), and (c)	$0.2f_c'A_c$	(a)
		$(480 + 0.08f_c')A_c$	(b)
		$1600A_c$	(c)
Other cases	Lesser of (d) and (e)	$0.2f_c'A_c$	(d)
		$800A_c$	(e)

22.9.4.5 Permanent net compression across the shear plane shall be permitted to be added to $A_{vf}f_y$, the force in the shear-friction reinforcement, to calculate required A_{vf}.

R22.9.4.5 This provision is supported by test data (Mattock and Hawkins 1972) and should be used to reduce the amount of shear-friction reinforcement required only if the compressive force across the shear plane is permanent.

22.9.4.6 Area of reinforcement required to resist a net factored tension across an assumed shear plane shall be added to the area of reinforcement required for shear friction crossing the assumed shear plane.

R22.9.4.6 Tension across the shear plane may be caused by restraint of deformations due to temperature change, creep, and shrinkage.

Where moment acts on a shear plane, the flexural compression and tension forces are in equilibrium and do not change the resultant compression $A_{vf}f_y$ acting across the shear plane or the shear-friction resistance. It is therefore not necessary to provide additional reinforcement to resist the flexural tension stresses, unless the required flexural tension reinforcement exceeds the amount of shear-transfer reinforcement provided in the flexural tension zone (Mattock et al. 1975).

22.9.5 *Detailing for shear-friction reinforcement*

R22.9.5 *Detailing for shear-friction reinforcement*

22.9.5.1 Reinforcement crossing the shear plane to satisfy 22.9.4 shall be anchored to develop f_y on both sides of the shear plane.

R22.9.5.1 If no moment acts across the shear plane, reinforcement should be uniformly distributed along the shear plane to minimize crack widths. If a moment acts across the shear plane, the shear-transfer reinforcement should be placed primarily in the flexural tension zone.

Anchorage may be developed by bond, by a mechanical device, or by threaded dowels and screw inserts. Space limitations often require the use of mechanical anchorage devices. For anchorage of headed studs in concrete, refer to *PCI Design Handbook* for precast and prestressed concrete (PCI MNL 120).

The shear-friction reinforcement anchorage should engage the primary reinforcement; otherwise, a potential crack may pass between the shear-friction reinforcement and the body of the concrete. This requirement applies particularly to welded headed studs used with steel inserts.

Retain wall Design Example: Rule of thumb

Height of wall: Below frost depth, Spokane 2'

Length of base: 40-60% height

Thickness of footing 7-10% h, generally same thickness as stem w/in R is

Thick Stem: 10in min over 12-14% base width or 10-12% wall height vs 0.25 in/ft taper of height

Surcharge = 200psf × (total depth) + ...)

H wall: bottom of footing will be 2' below grade, assume a footing thickness of 14" so the BoF will be 26" below grade hwall = 9'-0"

Length of base: 40-60% h

9×0.6 = 5.4' = 6'

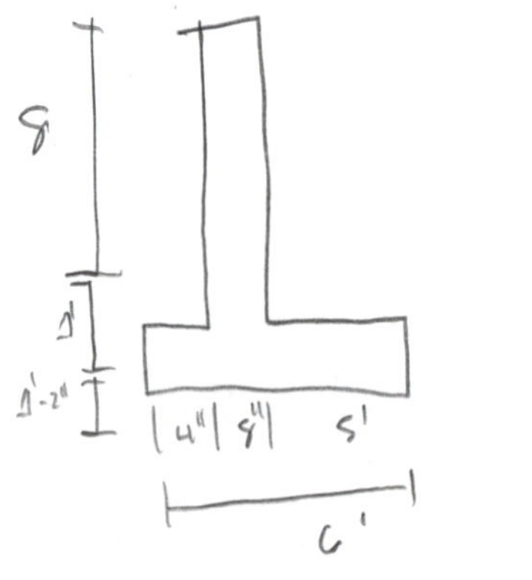

CHAPTER 23—STRUT-AND-TIE METHOD

CODE

23.1—Scope

23.1.1 This chapter shall apply to the design of structural concrete members, or regions of members, where load or geometric discontinuities cause a nonlinear distribution of longitudinal strains within the cross section.

23.1.2 Any structural concrete member, or discontinuity region in a member, shall be permitted to be designed by modeling the member or region as an idealized truss in accordance with this chapter.

COMMENTARY

R23.1—Scope

A discontinuity in the stress distribution occurs at a change in the geometry of a structural element or at a concentrated load or reaction. St. Venant's principle indicates that the stresses due to axial force and bending approach a linear distribution at a distance approximately equal to the overall depth of the member, h, away from the discontinuity. For this reason, discontinuity regions are assumed to extend a distance h from the section where the load or change in geometry occurs.

The shaded regions in Fig. R23.1(a) and (b) show typical D-regions (Schlaich et al. 1987). The plane sections assumption of 9.2.1 is not applicable in such regions. In general, any portion of a member outside a D-region is a B-region where the plane sections assumptions of flexural theory can be applied. The strut-and-tie design method, as described in this chapter, is based on the assumption that D-regions can be analyzed and designed using hypothetical pin-jointed trusses consisting of struts and ties connected at nodes.

The idealized truss specified in 23.2.1, which forms the basis of the strut-and-tie method, is not intended to apply to structural systems configured as actual trusses because secondary effects, such as moments, are not included in the model.

Thickness of footing = 7-10% h

9 x 0.1 = 10.8'' → stay with 14''

Thickness of stem: wall thickness at base is determined either by 12-16% of the base width or as 10-12% of the wall height.

6 x 0.12 = 8.64 → use 8''

CODE

COMMENTARY

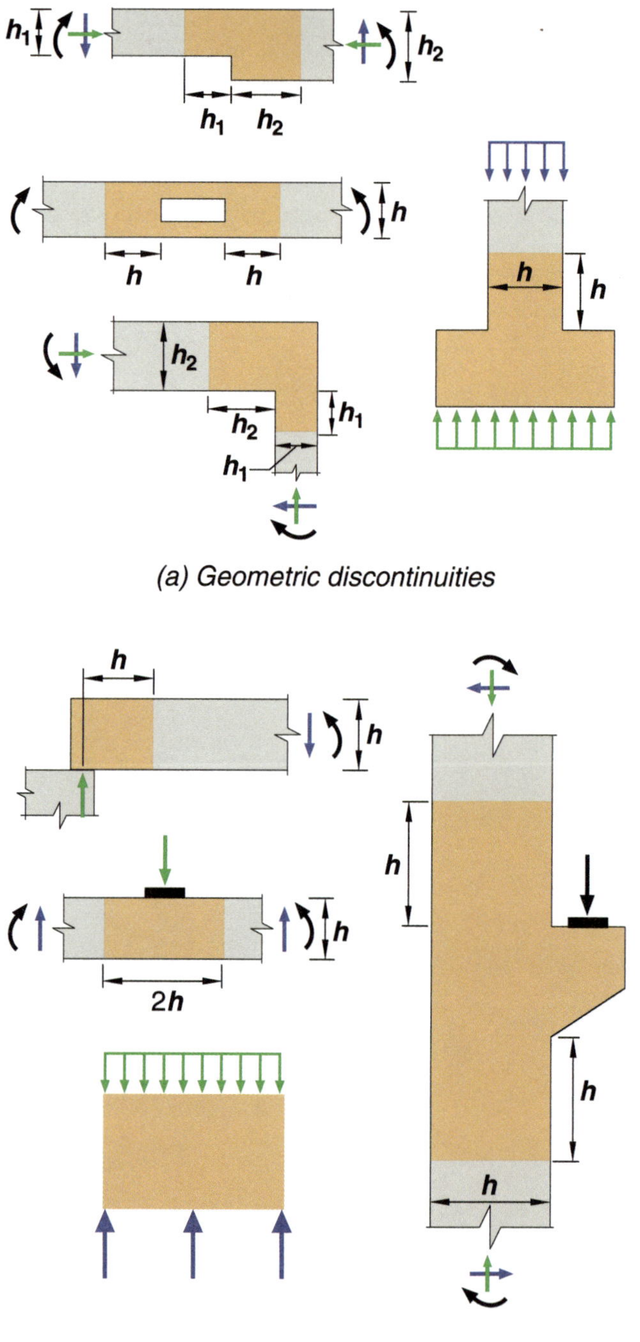

(a) Geometric discontinuities

(b) Loading and geometric discontinuities

Fig. R23.1—D-regions and discontinuities.

23.2—General

23.2.1 Strut-and-tie models shall consist of struts and ties connected at nodes to form an idealized truss in two or three dimensions.

R23.2—General

R23.2.1 For the idealized truss, struts are the compression members, ties are the tension members, and nodes are the joints. Uniformly distributed loads are usually idealized as a series of concentrated loads applied at nodes. Similarly, distributed reinforcement is usually modeled as discrete ties representing groups of individual bars or wires. Details of the use of the strut-and-tie method are given in Schlaich et al. (1987), Collins and Mitchell (1991), MacGregor (1997), FIP (1999), Menn (1986), Muttoni et al. (1997), and ACI 445R.

CODE

COMMENTARY

Design examples for the strut-and-tie method are given in ACI SP-208 (Reineck 2002) and ACI SP-273 (Reineck and Novak 2010). The process of designing by the strut-and-tie method to support the imposed forces acting on and within a D-region is referred to as the strut-and-tie method, and it includes the following four steps:

(1) Define and isolate each D-region.

(2) Calculate resultant forces on each D-region boundary.

(3) Select the model and calculate the forces in the struts and ties to transfer the resultant forces across the D-region. The axes of the struts and ties are chosen to approximately coincide with the axes of the compression and tension fields, respectively.

(4) Design the struts, ties, and nodal zones so that they have sufficient strength. Widths of struts and nodal zones are determined considering the effective concrete strengths defined in 23.4.3 and 23.9.2. Reinforcement is provided for the ties considering the steel strengths defined in 23.7.2. The reinforcement should be anchored in or beyond the nodal zones.

The components of a strut-and-tie model of a single-span deep beam loaded with two concentrated loads are identified in Fig. R23.2.1. The cross-sectional dimensions of a strut or tie are designated as thickness and width, and both directions are perpendicular to the axis of the strut or tie. Thickness is perpendicular to the plane, and width is in the plane of the strut-and-tie model. A tie consists of nonprestressed or prestressed reinforcement plus a portion of the surrounding concrete that is concentric with the axis of the tie. The surrounding concrete is included to define the zone in which the forces in the ties are to be anchored. The concrete in a tie is not used to resist the axial force in the tie. Although not explicitly considered in design, the surrounding concrete will reduce the elongations of the tie, especially at service loads.

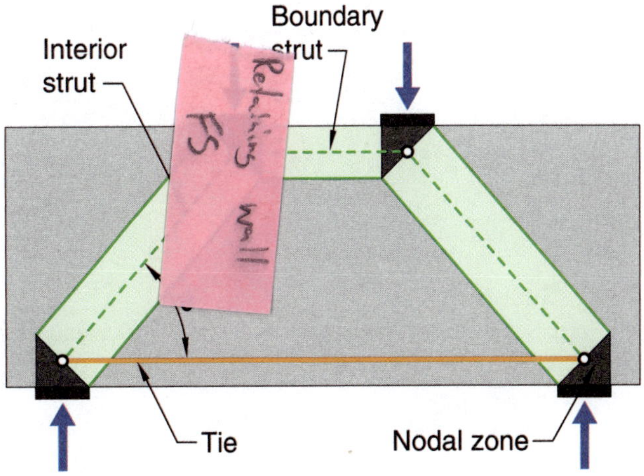

Fig. R23.2.1—Description of strut-and-tie model.

Strut-and-Tie 23

CODE

23.2.2 Geometry of the idealized truss shall be consistent with the dimensions of the struts, ties, nodal zones, bearing areas, and supports.

COMMENTARY

R23.2.2 The struts, ties, and nodal zones making up the strut-and-tie model all have finite widths, typically in the plane of the model, and thicknesses, typically the out-of-plane dimension of the structure, which should be taken into account in selecting the dimensions of the truss. Figures R23.2.2(a) and (b) show a node and the corresponding nodal zone. The vertical and horizontal forces equilibrate the forces in the inclined strut.

If more than three forces act on a nodal zone in a two-dimensional strut-and-tie model, as shown in Fig. R23.2.2b, it is suggested to resolve some of the forces to form three intersecting forces. The strut forces acting on Faces A-E and C-E in Fig. R23.2.2(a) can be replaced with one force acting on Face A-C as shown in Fig. R23.2.2(b). This force passes through the node at D.

Alternatively, the strut-and-tie model can be analyzed assuming all the strut forces act through the node at D, as shown in Fig. R23.2.2(c). In this case, the forces in the two struts on the right side of Node D can be resolved into a single force acting through Point D, as shown in Fig. R23.2.2(d).

If the width of the support in the direction perpendicular to the member is less than the width of the member, transverse reinforcement may be required to restrain vertical splitting in the plane of the node. This can be modeled using a transverse strut-and-tie model.

R23.2.3 The analysis results from the strut-and-tie method represent lower-bound strength limit states. Section 23.5.1 requires distributed reinforcement in D-regions designed by this chapter unless struts are laterally restrained. Distributed reinforcement in D-regions will improve serviceability performance. In addition, crack widths in a tie can be controlled using 24.3.2, assuming the tie is encased in a prism of concrete corresponding to the area of the tie from R23.8.1.

CODE

COMMENTARY

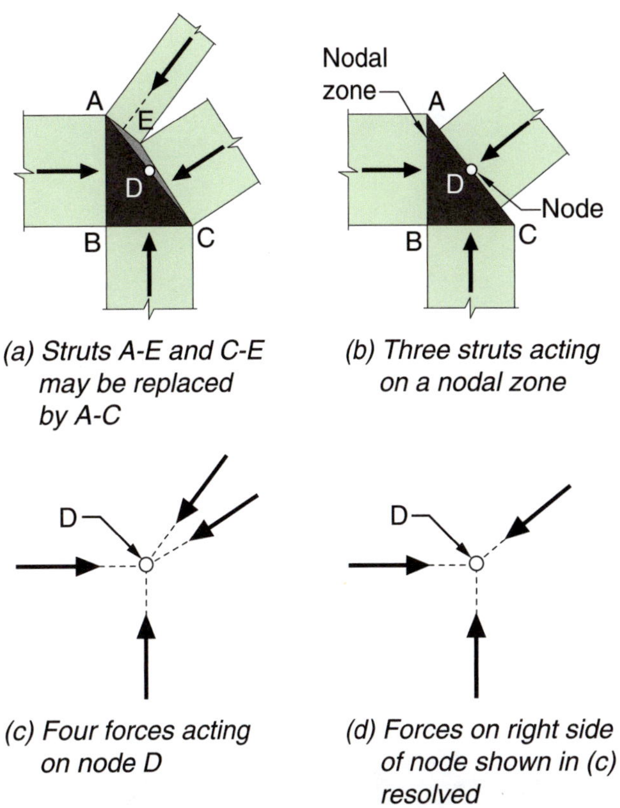

***Fig. R23.2.2**—Resolution of forces on a nodal zone.*

23.2.3 Strut-and-tie models shall be capable of transferring all factored loads to supports or adjacent B-regions.

23.2.4 The internal forces in strut-and-tie models shall be in equilibrium with the applied loads and reactions.

23.2.5 Ties shall be permitted to cross struts and other ties.

23.2.6 Struts shall intersect or overlap only at nodes.

R23.2.6 A hydrostatic nodal zone, by definition, has equal stresses on the loaded faces; these faces are perpendicular to the axes of the struts and ties that act on the node. This type of node is considered a hydrostatic nodal zone because the in-plane stresses are the same in all directions. Strictly speaking, this terminology is incorrect because the in-plane stresses are not equal to the out-of-plane stresses.

Figure R23.2.6a(i) shows a C-C-C nodal zone. If the stresses on the face of the nodal zone are the same in all three struts, the ratios of the lengths of the sides of the nodal zone, w_{n1}:w_{n2}:w_{n3}, are in the same proportions as the three forces, C_1:C_2:C_3.

A C-C-T nodal zone can be represented as a hydrostatic nodal zone if the tie is assumed to extend through the node and is anchored by a plate on the far side of the node, as shown in Fig. R23.2.6a(ii), provided that the size of the plate results in bearing stresses that are equal to the stresses in the struts. The bearing plate on the left side of Fig. R23.2.6a(ii) is used to represent an actual tie anchorage. The tie force can

CODE

COMMENTARY

be anchored by a plate or through embedment of straight bars (Fig. R23.2.6a(iii)), headed bars, or hooked bars. For non-hydrostatic nodes, the face with the highest stress will control the dimensions of the node.

The lightly shaded area in Fig. R23.2.6b is an extended nodal zone. An extended nodal zone is that portion of a member bounded by the intersection of the effective strut width w_s and the effective tie width w_t.

For equilibrium, at least three forces should act on each node in a strut-and-tie model, as shown in Fig. R23.2.6c. Nodes are classified according to the signs of these forces. A C-C-C node resists three compressive forces, a C-C-T node resists two compressive forces and one tensile force, and a C-T-T node resists one compressive force and two tensile forces.

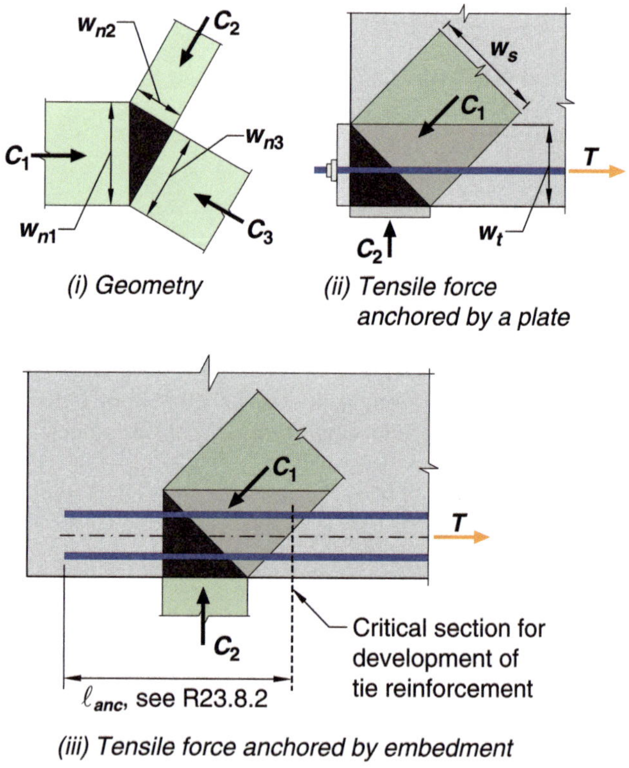

***Fig. R23.2.6a**—Hydrostatic nodes.*

CODE

COMMENTARY

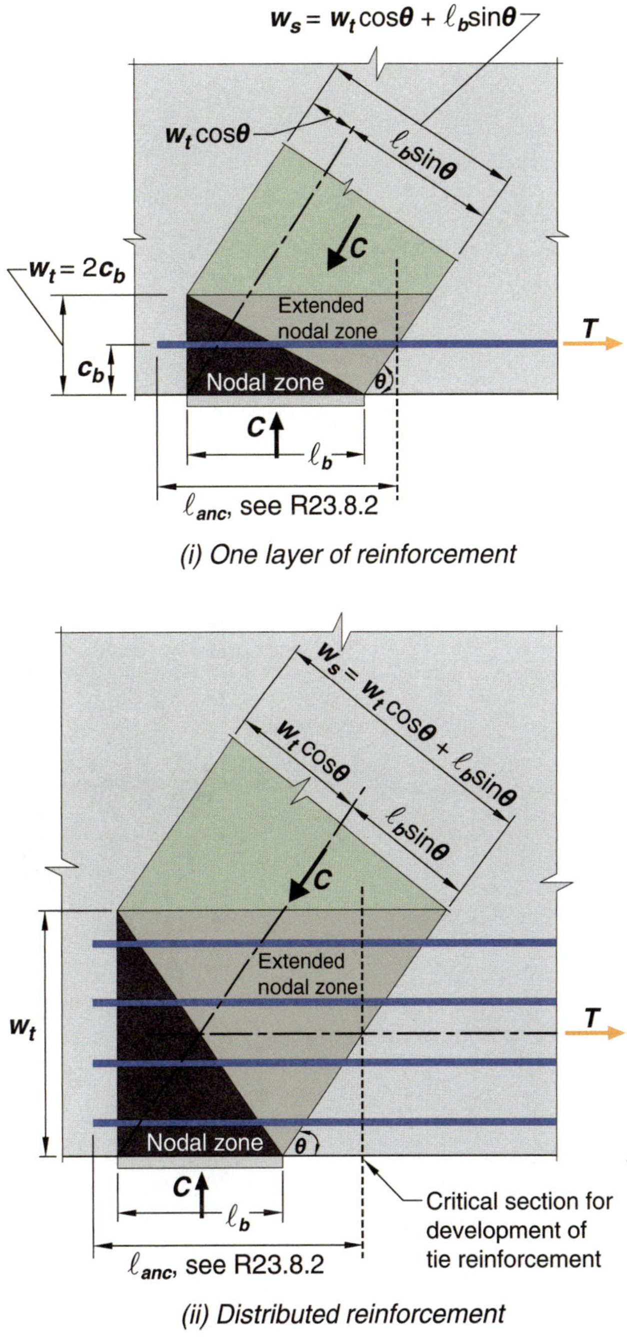

***Fig. R23.2.6b**—Extended nodal zone showing the effect of the distribution of the force.*

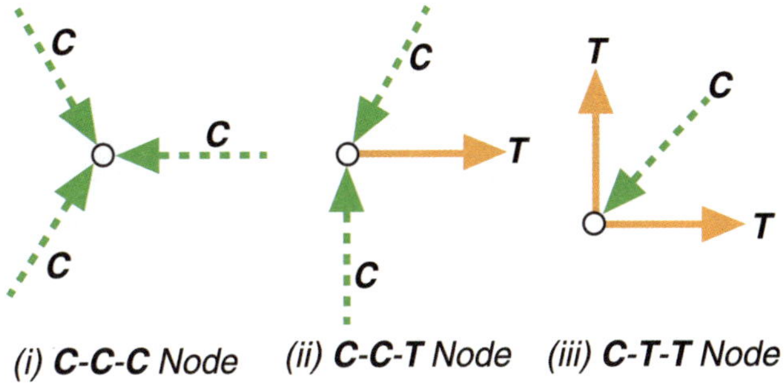

***Fig. R23.2.6c**—Classification of nodes.*

CODE

23.2.7 The angle between the axes of any strut and any tie entering a single node shall be at least 25 degrees.

COMMENTARY

R23.2.7 The angle between the axes of a strut and a tie acting on a node should be large enough to mitigate cracking and to avoid incompatibilities due to shortening of the strut and lengthening of the tie occurring in approximately the same direction. This limitation on the angle prevents modeling shear spans in slender beams using struts inclined at less than 25 degrees from the longitudinal reinforcement (Muttoni et al. 1997).

In some cases, strut-and-tie models can be adjusted to satisfy this requirement without excluding transverse reinforcement close to concentrated loads or reactions as illustrated in Fig. R23.2.7.

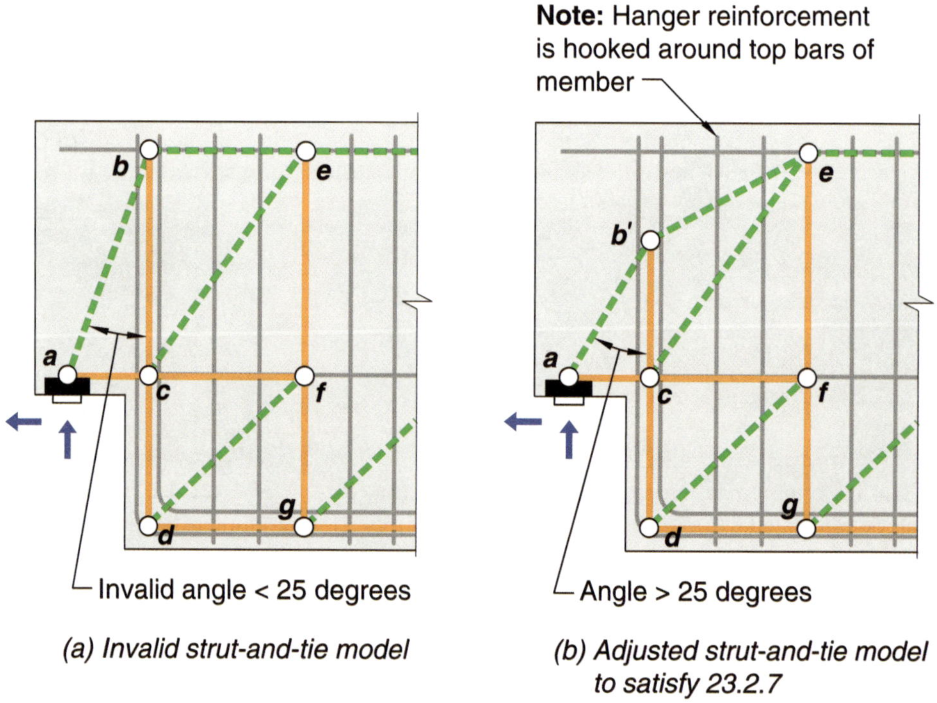

Fig. R23.2.7—Strut and-tie model of dapped connection illustrating adjustment to comply with 23.2.7.

23.2.8 The effects of prestressing shall be included in the strut-and-tie model as external loads with load factors in accordance with 5.3.11. For pretensioned members, it shall be permitted to assume that the prestress force is applied at the end of the strand transfer length.

R23.2.8 The flow of forces in the strut-and-tie model is unrealistic if prestressing effects are not considered as external loads. Including prestressing effects as external loads is required to identify regions where the effects of other external loads exceed the precompression force and vice versa. Prestressing effects are simulated by concentrated loads at the anchorages and transverse loads equivalent to the effects of tendon deviation or curvature. Provision 5.3.11 requires different load factors depending on the effects of prestressing on the strut-and-tie model. Applying the prestressing force at the end of the transfer length may require a deformed bar tie where the prestress force is being transferred.

23.2.9 Deep beams designed using the strut-and-tie method shall satisfy 9.9.2.1, 9.9.3.1, and 9.9.4.

CODE

23.2.10 Brackets and corbels with shear span-to-depth ratio $a_v/d < 2.0$ designed using the strut-and-tie method shall satisfy 16.5.2, 16.5.6, and Eq. (23.2.10).

$$A_{sc} \geq 0.04(f_c'/f_y)(b_w d) \qquad (23.2.10)$$

23.2.11 The shear friction requirements of 22.9 shall apply where it is appropriate to consider shear transfer across any given plane, such as an existing or potential crack, an interface between dissimilar materials, or an interface between two concretes cast at different times.

23.2.12 Members designed using strut-and-tie models that are part of seismic-force-resisting system shall meet the additional requirements of 23.11, if applicable.

23.3—Design strength

23.3.1 For each applicable factored load combination, design strength of each strut, tie, and nodal zone in a strut-and-tie model shall satisfy $\phi S_n \geq U$, including (a) through (c):

(a) Struts: $\phi F_{ns} \geq F_{us}$
(b) Ties: $\phi F_{nt} \geq F_{ut}$
(c) Nodal zones: $\phi F_{nn} \geq F_{us}$

23.3.2 ϕ shall be in accordance with 21.2.

23.4—Strength of struts

23.4.1 The nominal compressive strength of a strut, F_{ns}, shall be calculated by (a) or (b):

(a) Strut without longitudinal reinforcement

$$F_{ns} = f_{ce} A_{cs} \qquad (23.4.1a)$$

(b) Strut with longitudinal reinforcement

$$F_{ns} = f_{ce} A_{cs} + A_s' f_s' \qquad (23.4.1b)$$

where F_{ns} shall be evaluated at each end of the strut and taken as the lesser value; A_{cs} is the cross-sectional area at the end of the strut under consideration; f_{ce} is given in 23.4.3; A_s' is the area of compression reinforcement along the length of the strut; and f_s' is the stress in the compression reinforcement at the nominal axial strength of the strut. It shall be permitted to take f_s' equal to f_y for Grade 40 or 60 reinforcement.

23.4.2 Effective compressive strength of concrete in a strut, f_{ce}, shall be calculated in accordance with 23.4.3 or 23.4.4.

COMMENTARY

R23.2.11 A construction joint between a corbel and face of a column is an example of an interface where the shear friction requirements of 22.9 apply.

R23.3—Design strength

R23.3.1 Factored loads are applied to the strut-and-tie model, and the forces in all the struts, ties, and nodal zones are calculated. If several load combinations exist, each should be investigated separately. For a given strut, tie, or nodal zone, F_u is the largest force in that element for all load combinations considered.

R23.4—Strength of struts

R23.4.1 The width of strut, w_s, used to calculate A_{cs} is the dimension perpendicular to the axis of the strut at the ends of the strut. This strut width is illustrated in Fig. R23.2.6a(i) and Fig. R23.2.6b. If two-dimensional strut-and-tie models are appropriate, such as for deep beams, the thickness of the struts may be taken as the width of the member except at bearing supports where the thickness of the strut must equal the least thickness of the member or supporting element.

The contribution of reinforcement to the strength of the strut is given by the last term in Eq. (23.4.1b). The stress f_s' in the reinforcement in a strut at nominal strength can be obtained from the strains in the strut when the strut crushes. Detailing requirements in 23.6 must be met including confinement reinforcement to prevent buckling of the strut reinforcement.

R23.4.2 In design, struts are usually idealized as prismatic compression members. If the area of a strut differs at its two ends, due either to different nodal zone strengths at the two ends or to different bearing lengths, the strut is idealized as a uniformly tapered compression member.

Strut-and-Tie 23

CODE

23.4.3 Effective compressive strength of concrete in a strut, f_{ce}, shall be calculated by:

$$f_{ce} = 0.85\,\beta_c\beta_s f_c' \qquad (23.4.3)$$

where β_s is in accordance with Table 23.4.3(a) and β_c is in accordance with Table 23.4.3(b).

Table 23.4.3(a)—Strut coefficient β_s

Strut location	Strut type	Criteria	β_s	
Tension members or tension zones of members	Any	All cases	0.4	(a)
All other cases	Boundary struts	All cases	1.0	(b)
	Interior struts	Reinforcement satisfying (a) or (b) of Table 23.5.1	0.75	(c)
		Located in regions satisfying 23.4.4	0.75	(d)
		Beam-column joints	0.75	(e)
		All other cases	0.4	(f)

Table 23.4.3(b)—Strut and node confinement modification factor β_c

Location	β_c		
• End of a strut connected to a node that includes a bearing surface • Node that includes a bearing surface	Lesser of	$\sqrt{A_2/A_1}$, where A_1 is defined by the bearing surface	(a)
		2.0	(b)
Other cases	1.0		(c)

23.4.4 If use of β_s of 0.75 is based on line (d) of Table 23.4.3(a), member dimensions shall be selected to satisfy Eq. (23.4.4), where λ_s is defined in 23.4.4.1.

$$V_u \le \phi 5 \tan\theta \lambda\lambda_s \sqrt{f_c'} b_w d \qquad (23.4.4)$$

COMMENTARY

R23.4.3 The strength coefficient $\mathbf{0.85}f_c'$ in Eq. (23.4.3) represents the effective concrete strength under sustained compression, similar to that used in Eq. (22.4.2.2) and (22.4.2.3).

The value of β_s in (a) of Table 23.4.3(a) applies, for example, to a transverse model of a ledger beam used to proportion hanger and ledge reinforcement, where longitudinal tension in the flange reduces the strength of the transverse struts. The low value of β_s reflects that these struts need to transfer compression in a zone where tensile stresses act perpendicular to the plane of the strut-and-tie model.

The value of β_s in (b) of Table 23.4.3(a) applies to a boundary strut and results in a stress state that is comparable to the rectangular stress block in the compression zone of a beam or column. Boundary struts are not subject to transverse tension and therefore have a higher effective strength, f_{ce}, than interior struts (Fig. R23.2.1).

The value of β_s in (c) of Table 23.4.3(a) reflects the beneficial effect of distributed reinforcement.

The value of β_s in (d) of Table 23.4.3(a) applies to interior struts in regions with sufficient diagonal tension strength to satisfy Eq. (23.4.4).

The value of β_s in (e) of Table 23.4.3(a) reflects the requirements for reinforcement or confinement of beam-column joints in Chapter 15.

The value of β_s in (f) of Table 23.4.3(a) is reduced to preclude diagonal tension failure in regions without transverse reinforcement that do not meet or are not evaluated under 23.4.4. Evaluation of test results from the ACI shear database for members without transverse reinforcement indicates that diagonal tension failures are precluded if struts are proportioned based on β_s of 0.4 (Reineck and Todisco 2014). The ACI shear database includes test results for specimens with an average d of 15 in. and not exceeding 38 in.; therefore, size effect would not be expected to significantly reduce the strength of members of this size. Because size effect may be significant for deeper members without transverse reinforcement, evaluation in accordance with Eq. (23.4.4) is considered appropriate.

The influence of concrete confinement on the effective compressive strength of a strut or node is taken into account by β_c. The bearing surface can be a bearing plate or the area from a well-defined compressive load from another member, such as a column. It is the same confining effect as used for bearing areas in 22.8.3. The increase in compressive strength associated with the confinement provided by surrounding concrete for a strut-and-tie model is described by Tuchscherer et al. (2010) and Breen et al. (1994).

23.4.4 Equation (23.4.4) is intended to preclude diagonal tension failure. In discontinuity regions, diagonal tension strength increases as the strut angle increases. For very steeply inclined struts, V_u can exceed $\phi 10\lambda\lambda_s\sqrt{f_c'}\,b_w d$.

CODE

23.4.4.1 The size effect modification factor, λ_s, shall be determined by (a) or (b), as applicable:

(a) If distributed reinforcement is provided in accordance with 23.5, λ_s shall be taken as 1.0.
(b) If distributed reinforcement is not provided in accordance with 23.5, λ_s shall be taken in accordance with Eq. (23.4.4.1).

$$\lambda_s = \sqrt{\frac{2}{1+\frac{d}{10}}} \leq 1 \qquad (23.4.4.1)$$

23.5—Minimum distributed reinforcement

23.5.1 In D-regions designed using the strut-and-tie method, minimum distributed reinforcement shall be provided across the axes of interior struts in accordance with Table 23.5.1.

Table 23.5.1—Minimum distributed reinforcement

Lateral restraint of strut	Reinforcement configuration	Minimum distributed reinforcement ratio	
Not restrained	Orthogonal grid	0.0025 in each direction	(a)
	Reinforcement in one direction crossing strut at angle α_1	$\frac{0.0025}{\sin^2 \alpha_1}$	(b)
Restrained	Distributed reinforcement not required		(c)

23.5.2 Distributed reinforcement required by 23.5.1 shall satisfy (a) and (b):

(a) Spacing shall not exceed 12 in.
(b) Angle α_1 shall not be less than 40 degrees.

23.5.3 Struts are considered laterally restrained if they are restrained perpendicular to the plane of the strut-and-tie model in accordance with (a), (b), or (c):

(a) The discontinuity region is continuous perpendicular to the plane of the strut-and-tie model.
(b) The concrete restraining the strut extends beyond each side face of the strut a distance not less than half the width of the strut.
(c) The strut is in a joint that is restrained in accordance with 15.2.5 or 15.2.6.

23.5.4 Reinforcement required in 23.5.1 shall be developed beyond the extent of the strut in accordance with 25.4.

COMMENTARY

R23.5—Minimum distributed reinforcement

The strut-and-tie method is derived from the lower-bound theorem of plasticity; therefore, a member designed using this method requires sufficient reinforcement to promote redistribution of the internal forces in the cracked state (Marti 1985). In addition to allowing force redistribution, distributed reinforcement controls cracking at service loads and promotes ductile behavior (Smith and Vantsiotis 1982; Rogoswky and MacGregor 1986; Tan et al. 1977).

Interior struts are typically oriented parallel to compression fields and are therefore oriented perpendicular to diagonal tension fields. Tensile stresses across the strut may also develop where compressive stress at the node spreads out along the length of a strut. Minimum distributed reinforcement helps control cracking from these tensile stresses.

The distributed reinforcement ratio required by 23.5.1 is the total on both faces plus any interior layers placed in wide members. Figure R23.5.1 illustrates unidirectional distributed reinforcement crossing interior struts at angle α_1.

Although minimum distributed reinforcement is not required where interior struts are laterally restrained, distributed reinforcement may be beneficial in large discontinuity regions. A continuous corbel supporting a slab is an example of a discontinuity region that includes struts that are laterally restrained in accordance with 23.5.3(a). Pile caps and beam ledges supporting concentrated loads are examples of discontinuity regions that include struts that are laterally restrained in accordance with 23.5.3(b). The side faces of the strut in 23.5.3(b) are the faces parallel to the plane of the model. For pile caps evaluated using three-dimensional strut-and-tie models, the plane of the model in 23.5.3 is defined by the strut in question and the pile to which it connects.

Strut-and-Tie 23

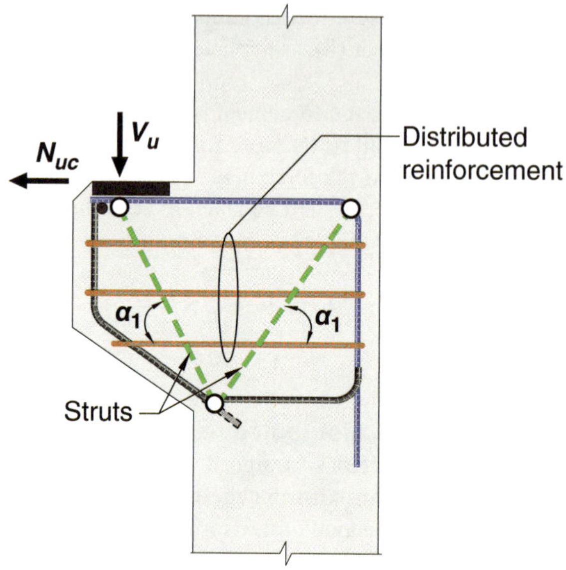

Distributed reinforcement crossing interior struts. Note that α_1 *is different for the two struts above; the minimum distributed reinforcement ratio is controlled by the smaller angle* α_1.

Fig. R23.5.1*—Distributed reinforcement crossing interior struts.*

23.6—Strut reinforcement detailing

23.6.1 Compression reinforcement in struts shall be parallel to the axis of the strut and enclosed along the length of the strut by closed ties in accordance with 23.6.3 or by spirals in accordance with 23.6.4.

23.6.2 Compression reinforcement in struts shall be anchored to develop f_s' at the face of the nodal zone, where f_s' is calculated in accordance with 23.4.1.

23.6.3 Closed ties enclosing compression reinforcement in struts shall satisfy 25.7.2 and this section.

23.6.3.1 Spacing of closed ties, s, along the length of the strut shall not exceed the smallest of (a) through (c):

(a) Smallest dimension of cross section of strut
(b) $48d_b$ of bar or wire used for closed tie reinforcement
(c) $16d_b$ of compression reinforcement

23.6.3.2 The first closed tie shall be located not more than $0.5s$ from the face of the nodal zone at each end of a strut.

23.6.3.3 Closed ties shall be arranged such that every corner and alternate longitudinal bar shall have lateral support provided by crossties or the corner of a tie with an included angle of not more than 135 degrees and no longitudinal bar shall be farther than 6 in. clear on each side along the tie from such a laterally supported bar.

R23.6—Strut reinforcement detailing

R23.6.1 Refer to R23.4.1.

R23.6.3.3 Refer to R25.7.2.3.

CODE

23.6.4 Spirals enclosing compression reinforcement in struts shall satisfy 25.7.3.

23.7—Strength of ties

23.7.1 Tie reinforcement shall be nonprestressed or prestressed.

23.7.2 The nominal tensile strength of a tie, F_{nt}, shall be calculated by:

$$F_{nt} = A_{ts} f_y + A_{tp} \Delta f_p \quad (23.7.2)$$

where A_{tp} is zero for nonprestressed members.

23.7.2.1 In Eq. (23.7.2), it shall be permitted to take Δf_p equal to 60,000 psi for bonded prestressed reinforcement and 10,000 psi for unbonded prestressed reinforcement. Higher values of Δf_p shall be permitted if justified by analysis, but Δf_p shall not be taken greater than $(f_{py} - f_{se})$.

23.8—Tie reinforcement detailing

23.8.1 The centroidal axis of the tie reinforcement shall coincide with the axis of the tie assumed in the strut-and-tie model.

23.8.2 Tie reinforcement shall be anchored by mechanical devices, post-tensioning anchorage devices, standard hooks, or straight bar development in accordance with 23.8.3, except for ties extending from curved-bar nodes designed in accordance with 23.10.

COMMENTARY

R23.7—Strength of ties

R23.7.2 The tie strength in 23.7.2 is based on including any effects of prestressing as external loads in accordance with 23.2.8. The total strength of a prestressed tie is $A_{tp}(f_{se} + \Delta f_p)$.

R23.8—Tie reinforcement detailing

R23.8.1 The effective tie width assumed in design, w_t, can vary between the following limits, depending on the distribution of the tie reinforcement:

(a) If the bars in the tie are in one layer, the effective tie width can be taken as the diameter of the bars in the tie plus twice the cover to the surface of the bars, as shown in Fig. R23.2.6b(i).

(b) A practical upper limit of the tie width can be taken as the width corresponding to the width in a hydrostatic nodal zone, calculated as $w_{t,max} = F_{nt}/(f_{ce} b_s)$, where f_{ce} is calculated for the nodal zone in accordance with 23.9.2.

If the tie width exceeds the value from (a), the tie reinforcement should be distributed approximately uniformly over the width and thickness of the tie, as shown in Fig. R23.2.6b(ii).

R23.8.2 Anchorage of ties often requires special attention in nodal zones of corbels or in nodal zones adjacent to exterior supports of deep beams. The reinforcement in a tie should be anchored before it exits the extended nodal zone at the point defined by the intersection of the centroid of the bars in the tie and the extensions of the outlines of either the strut or the bearing area. This length is ℓ_{anc}. In Fig. R23.2.6b, this occurs where the outline of the extended nodal zone is crossed by the centroid of the reinforcement in the tie. Some of the anchorage may be achieved by extending the reinforcement through the nodal zone, as shown in Fig. R23.2.6a(iii) and R23.2.6b, and developing it beyond the nodal zone. If the tie is anchored using 90-degree hooks, the hooks should be confined within reinforcement to avoid cracking along the outside of the hooks in the support region.

COMMENTARY

In deep beams, hairpin bars spliced with the tie reinforcement can be used to anchor the tie forces at exterior supports, provided the beam width is large enough to accommodate such bars.

Figure R23.8.2 shows two ties anchored at a nodal zone. Development is required where the centroid of the tie crosses the outline of the extended nodal zone.

The development length of the tie reinforcement can be reduced through hooks, headed bars, mechanical devices, additional confinement, or by splicing it with layers of smaller bars.

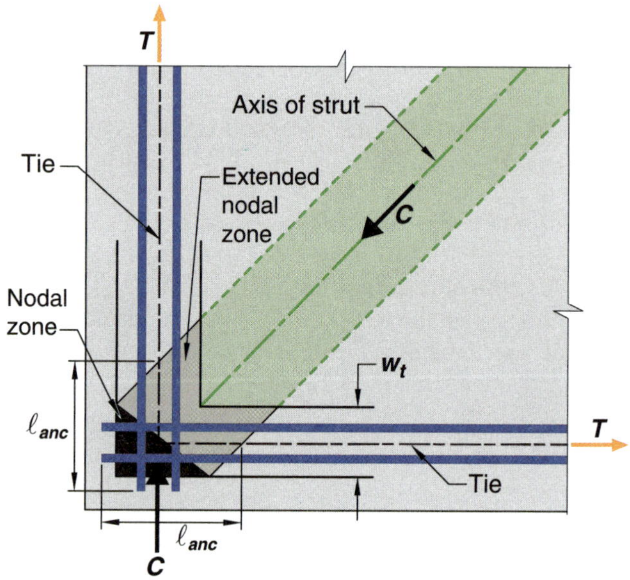

Fig. R23.8.2—Extended nodal zone anchoring two ties.

23.8.3 Tie force shall be developed in each direction at the point where the centroid of the reinforcement in the tie leaves the extended nodal zone.

23.9—Strength of nodal zones

23.9.1 The nominal compressive strength of a nodal zone, F_{nn}, shall be calculated by:

$$F_{nn} = f_{ce}A_{nz} \qquad (23.9.1)$$

where f_{ce} is defined in 23.9.2 or 23.9.3 and A_{nz} is given in 23.9.4 or 23.9.5.

23.9.2 The effective compressive strength of concrete at a face of a nodal zone, f_{ce}, shall be calculated by:

$$f_{ce} = 0.85\beta_c\beta_n f_c' \qquad (23.9.2)$$

where β_n shall be in accordance with Table 23.9.2 and β_c is in accordance with Table 23.4.3(b).

R23.9—Strength of nodal zones

R23.9.2 The nodes in two-dimensional models can be classified as shown in Fig. R23.2.6c. The effective compressive strength of the nodal zone is given by Eq. (23.9.2) where the value for β_n is given in Table 23.9.2.

Lower β_n values reflect the increasing degree of disruption of the nodal zones due to the incompatibility of tensile strains in the ties and compressive strains in the struts. The stress on any face of the nodal zone or on any section through the nodal zone should not exceed the value given by Eq. (23.9.2).

CODE

Table 23.9.2—Nodal zone coefficient β_n

Configuration of nodal zone	β_n	
Nodal zone bounded by struts, bearing areas, or both	1.0	(a)
Nodal zone anchoring one tie	0.80	(b)
Nodal zone anchoring two or more ties	0.60	(c)

COMMENTARY

As described in R23.4.3, $\boldsymbol{\beta_c}$ accounts for the effect of concrete confinement on the effective compressive strength of a node containing a bearing surface. $\boldsymbol{\beta_c}$ is the same for the node as for the node-strut interface.

23.9.3 If confining reinforcement is provided within the nodal zone and its effect is documented by tests and analyses, it shall be permitted to use an increased value of f_{ce} when calculating F_{nn}.

23.9.4 The area of each face of a nodal zone, A_{nz}, shall be taken as the smaller of (a) and (b):

(a) Area of the face of the nodal zone perpendicular to the line of action of F_{us}

(b) Area of a section through the nodal zone perpendicular to the line of action of the resultant force on the section

R23.9.4 If the stresses in all the struts meeting at a node are equal, a hydrostatic nodal zone can be used. The faces of such a nodal zone are perpendicular to the axes of the struts, and the widths of the faces of the nodal zone are proportional to the forces in the struts.

Stresses on nodal faces that are perpendicular to the axes of struts and ties are principal stresses, and 23.9.4(a) is used. If, as shown in Fig. R23.2.6b(ii), the face of a nodal zone is not perpendicular to the axis of the strut, there will be both shear stresses and normal stresses on the face of the nodal zone. Typically, these stresses are replaced by the normal (principal compressive) stress acting on the cross-sectional area, A_{nz}, of the strut, taken perpendicular to the axis of the strut as given in 23.9.4(a).

23.9.5 In a three-dimensional strut-and-tie model, the area of each face of a nodal zone shall be at least that given in 23.9.4, and the shape of each face of the nodal zone shall be similar to the shape of the projection of the end of the strut onto the corresponding face of the nodal zone.

23.10—Curved-bar nodes

23.10.1 Curved-bar nodes shall be designed and detailed in accordance with this section.

R23.10—Curved-bar nodes

R23.10.1 A curved-bar node is formed by the bend region of a continuous reinforcing bar (or bars) where two ties extending from the bend region are intersected by a strut or the resultant of two or more struts (Fig. R23.10.5), or where a single tie is anchored by a 180-degree bend (Fig. R23.10.2).

23.10.2 If specified clear cover normal to plane of bend is $2d_b$ or greater, the bend radius r_b shall be in accordance with (a) or (b), but shall not be less than half the minimum bend diameter specified in 25.3.

(a) Curved bar nodes with bends less than 180 degrees:

$$r_b \geq \frac{2A_{ts}f_y}{b_s f_c'} \quad (23.10.2a)$$

(b) Ties anchored by 180-degree bends:

$$r_b \geq \frac{1.5A_{ts}f_y}{w_t f_c'} \quad (23.10.2b)$$

R23.10.2 Equation (23.10.2a) is intended to avoid f_{ce} exceeding the limit for C-T-T nodes given by 23.9.2 (Klein 2008). b_s is the width of the strut transverse to the plane of the strut-and-tie model. Equation (23.10.2a) applies whether the tie forces at the node are equal or different; where the tie forces are different, ℓ_{cb} required by 23.10.6 must also be satisfied.

Ties anchored by 180-degree bends can be used at C-C-T nodes, as illustrated in Fig. R23.10.2. The parallel straight legs of the bar(s) that extend into the member form a single tie. Equation (23.10.2b) is intended to ensure that f_{ce} does not exceed the limit for C-C-T nodes given by 23.9.2. Width w_t is the effective tie width as illustrated in Fig. R23.10.2.

Strut-and-Tie 23

CODE

COMMENTARY

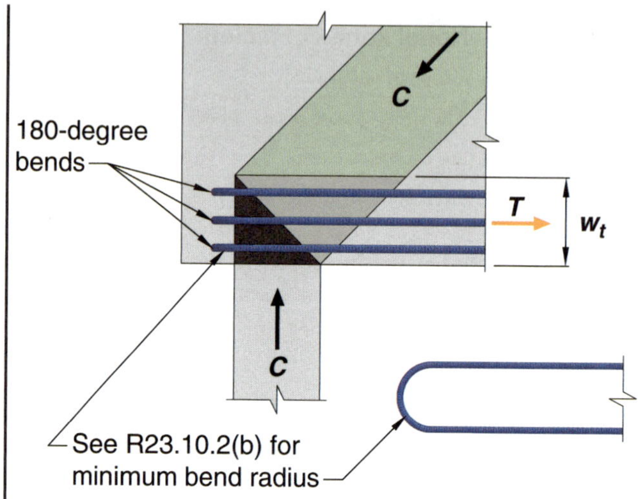

Fig. R23.10.2—C-C-T node using ties anchored by 180-degree bends.

23.10.3 If specified clear cover normal to plane of bend is less than $2d_b$, r_b required by 23.10.2 shall be multiplied by the ratio $2d_b/c_c$, where c_c is the specified clear cover to the side face.

R23.10.3 Larger bar bend radii at curved-bar nodes are required to reduce the likelihood of side splitting where concrete cover perpendicular to the plane of the bend is limited.

23.10.4 If curved-bar nodes are formed by more than one layer of reinforcement, A_{ts} shall be taken as the total area of tie reinforcement, and r_b shall be taken as the bend radius of the innermost layer.

R23.10.4 Figure R23.10.4 illustrates the use of a curved-bar node with two layers of reinforcing bars. In such cases, the total area of tie reinforcement contributes to the compressive stress on the face of the nodal zone (Face ab in Fig. R23.10.4).

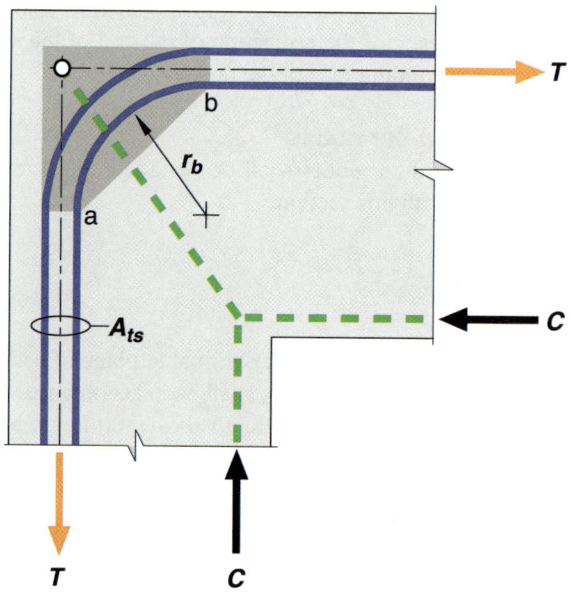

Fig. R23.10.4—Curved-bar node with two layers of reinforcement (nodal zone is shaded).

23.10.5 At frame corners, the joint and reinforcement shall be proportioned such that the center of bar curvature is located within the joint.

R23.10.5 The radius of the bend should be consistent with the geometry of the truss used for the strut-and-tie model. Figure R23.10.5 illustrates the region in which the center of curvature must be located for a typical frame corner.

CODE

23.10.6 ℓ_{cb} shall be sufficient to develop any difference in force between the straight legs of the bars extending from the bend region.

COMMENTARY

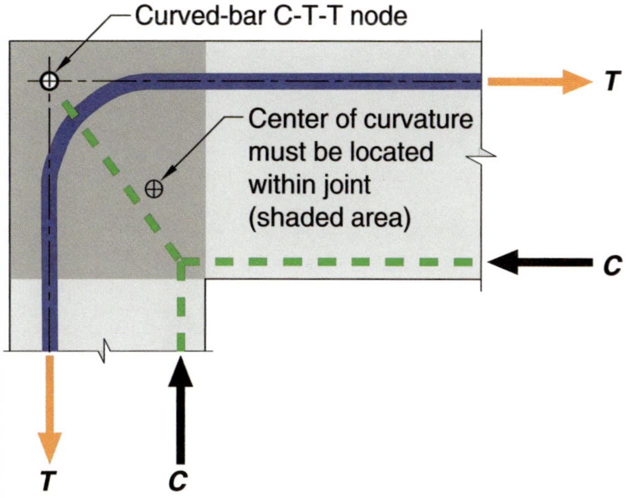

Fig. R23.10.5—Permissible zone for the center of curvature of a curved-bar node at a frame corner.

R23.10.6 Tie forces are unequal where the strut (or the resultant of two or more struts) does not bisect the angle formed by the ties at each end of the bend. Figure R23.10.6 shows a curved-bar node where the difference in tie force is developed in the bend region/nodal zone. Radial compressive stress acting on the node varies, and circumferential bond stress develops along the bar.

The difference in force between the two ties extending from the bend is developed over the length of bend ℓ_{cb} (the arc length of bar between c and b in Fig. R23.10.6). The following equation for ℓ_{cb} may be used at 90-degree corners:

$$\ell_{cb} > \ell_d(1 - \tan\theta_c)$$

where θ_c is the smaller of the two angles between the axis of the strut (or the resultant of two or more struts) and the ties extending from a curved-bar node, and ℓ_d is the development length calculated in accordance with 25.4.2.2 or 25.4.2.3 using the applicable modification factors of 25.4.2.4.

Strut-and-Tie 23

CODE | COMMENTARY

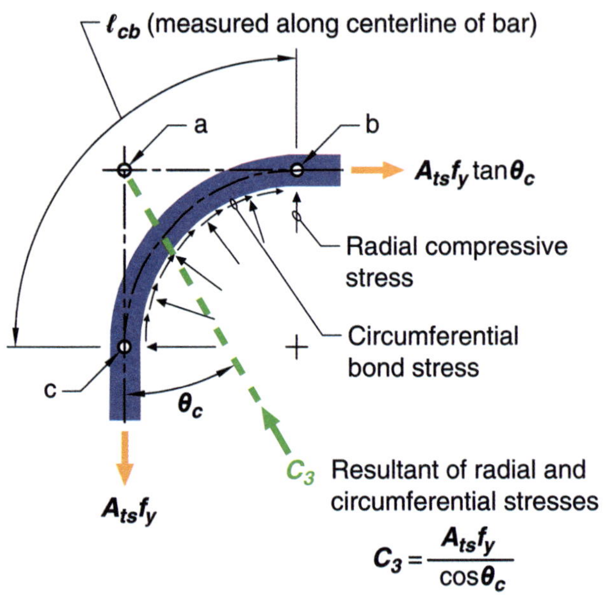

Fig. R23.10.6—Forces acting on a curved-bar node where there is a difference in tie forces.

23.11—Earthquake-resistant design using the strut-and-tie method

23.11.1 Regions of a seismic-force-resisting system assigned to Seismic Design Category (SDC) D, E, or F and designed with the strut-and-tie method shall be in accordance with (a) and (b):

(a) Chapter 18
(b) 23.11.2 through 23.11.5 unless design earthquake-induced force, E, in the strut-and-tie element is multiplied by an overstrength factor, Ω_o, not less than 2.5 unless a smaller value of Ω_o can be justified by a detailed analysis.

R23.11—Earthquake-resistant design using the strut-and-tie method

R23.11.1 Strut-and-tie elements of a seismic-force-resisting system may experience strength degradation due to force and displacement reversals. Strut-and-tie elements do not require seismic detailing when the design force is amplified by Ω_o. It is preferable that the strength of the seismic-force-resisting system not be limited by the strength of the discontinuity region designed by the strut-and-tie method. For diaphragm design that includes amplified seismic forces, an additional amplification factor is not required.

Load combinations are provided in 5.3.1, and Eq. (5.3.1e) and (5.3.1g) are used for seismic design. The effects of E may cause reversal of forces in strut and tie elements. In such cases, different strut-and-tie models are developed for each loading direction.

23.11.2 *Strut strength*

23.11.2.1 Effective compressive strength determined in accordance with 23.4 shall be multiplied by 0.8.

R23.11.2 *Strut strength*

R23.11.2.1 A reduction factor is applied to account for cracking that is likely to develop in struts located in a region subjected to force reversals.

23.11.3 *Strut detailing*

23.11.3.1 Struts shall have reinforcement satisfying the detailing requirements of 23.11.3.2 or 23.11.3.3.

R23.11.3 *Strut detailing*

R23.11.3.1 Two confinement options for struts are permitted. For 23.11.3.2, each strut contains longitudinal and transverse reinforcement as required for columns of special moment frames. For 23.11.3.3, the entire cross section of the region or member containing the struts are confined instead of the individual struts.

CODE

23.11.3.2 Struts shall be reinforced with a minimum of four longitudinal bars with a bar in each corner of the transverse reinforcement. Transverse reinforcement shall be placed perpendicular to the direction of the strut and satisfy (a) through (d):

(a) Detailed in accordance with 18.7.5.2(a) through (e)
(b) A_{sh}/sb_c determined in accordance with Table 23.11.3.2(a)
(c) Spacing satisfying 18.7.5.3(d) and not exceeding the values specified in Table 23.11.3.2(b)
(d) Continued through the nodal zone

Table 23.11.3.2(a)—Transverse reinforcement for struts[1][2]

Transverse reinforcement	Applicable expressions		
A_{sh}/sb_c for rectilinear hoops	Greater of	$0.3\left(\frac{A_{cs}}{A_{ch}}-1\right)\frac{f'_c}{f_{yt}}$	(a)
		$0.09\frac{f'_c}{f_{yt}}$	(b)

[1]A_{ch} is measured to the outside edges of the transverse reinforcement for the strut.
[2]It shall be permitted to configure hoops using two pieces of reinforcement as specified in 18.6.4.3.

Table 23.11.3.2(b)—Transverse reinforcement spacing limitation

Reinforcement	Maximum transverse bar spacing	
Grade 60	Lesser of	$6d_b$
		6 in.
Grade 80	Lesser of	$5d_b$
		6 in.
Grade 100	Lesser of	$4d_b$
		6 in.

Table 23.11.3.3—Transverse reinforcement for the entire member cross section

Transverse reinforcement	Applicable expressions		
A_{sh}/sb_c for rectilinear hoops	Greater of	$0.3\left(\frac{A_g}{A_{ch}}-1\right)\frac{f'_c}{f_{yt}}$	(a)
		$0.09\frac{f'_c}{f_{yt}}$	(b)

23.11.3.3 Transverse reinforcement shall be provided in each orthogonal direction and through the thickness of the

COMMENTARY

Expressions (a) and (b) in Table 23.11.3.2(a) are the same as those in Table 18.7.5.4 for columns of special moment frames with the exception of A_{cs} substituted for A_g.

23 Strut-and-Tie

CODE

entire member cross section or for the region of the member containing struts and shall satisfy (a) through (d).

(a) Detailed in accordance with 18.7.5.2(a) through (e)
(b) A_{sh}/sb_c determined in accordance with Table 23.11.3.3.
(c) Spacing measured along the longitudinal axis of the member not exceeding the values specified in Table 23.11.3.2(b).
(d) Spacing of crossties or legs of hoops both vertically and horizontally in the plane of the member cross section shall not exceed 8 in. Each crosstie and each hoop leg shall engage a longitudinal bar of equal or greater diameter.

23.11.4 *Strength of ties*

23.11.4.1 For tie reinforcement, development length shall be 1.25 times the length determined in accordance with 25.4.

23.11.5 *Strength of nodes*

23.11.5.1 The nominal compressive strength of a nodal zone calculated in accordance with 23.9 shall be multiplied by 0.8.

COMMENTARY

23.11.4 *Strength of ties*

R23.11.4.1 Because the actual yield strength of tie reinforcement may exceed the specified yield strength and strain hardening of the reinforcement is likely to occur, development lengths for tie reinforcement are determined considering a stress of $\mathbf{1.25}f_y$.

23.11.5 *Strength of nodes*

R23.11.5.1 A reduction of the nominal compressive strength at nodes is provided to account for tie yielding and the effect of reversed cyclic loading (Mansour and Hsu 2005; To et al. 2009; Ruggiero et al. 2016).

CHAPTER 24—SERVICEABILITY

CODE

24.1—Scope

24.1.1 This chapter shall apply to member design for minimum serviceability, including (a) through (d):

(a) Deflections due to service-level gravity loads (24.2)
(b) Distribution of flexural reinforcement in one-way slabs and beams to control cracking (24.3)
(c) Shrinkage and temperature reinforcement (24.4)
(d) Permissible stresses in prestressed flexural members (24.5)

COMMENTARY

R24.1—Scope

This chapter prescribes serviceability requirements that are referenced by other chapters of the Code, or are otherwise applicable to provide adequate performance of structural members. This chapter does not stand on its own as a complete and cohesive compilation of serviceability requirements for the design of structural members. This chapter has no specific requirements for vibrations.

Cast-in-place floor systems designed in accordance with the minimum thickness and deflection requirements of 7.3, 8.3, 9.3, and 24.2 have generally been found, through experience, to provide vibration performance suitable for human comfort under typical service conditions. However, there may be situations where serviceability conditions are not satisfied, for example:

(a) Long spans and open floor plans
(b) Floors with strict vibration performance requirements such as precision manufacturing and laboratory spaces
(c) Facilities subject to rhythmic loadings or vibrating mechanical equipment

Prestressed floor systems are not subject to the minimum thickness requirements of 7.3, 8.3, and 9.3, and if precast they are frequently simple span systems. Consequently, these floor systems can be more susceptible to vibration.

Guidance on the consideration of vibrations in the design of floor systems and the evaluation of vibration frequency and amplitude for concrete floor systems is contained in the *PCI Design Handbook* (PCI MNL 120), *ATC Design Guide 1* (Applied Technology Council 1999), Mast (2001), Fanella and Mota (2014), and Wilford and Young (2006). An example application is described by West et al. (2008).

CODE

24.2—Deflections due to service-level gravity loads

24.2.1 Members subjected to flexure shall be designed with adequate stiffness to limit deflections or deformations that adversely affect strength or serviceability of a structure.

COMMENTARY

R24.2—Deflections due to service-level gravity loads

This section is concerned only with deflections or deformations that may occur at service load levels. When time-dependent deflections are calculated, only the dead load and those portions of other loads that are sustained need be considered.

Two methods are given in the Code for controlling deflections (Sabnis et al. 1974). For nonprestressed one-way slabs and beams, including composite members, the minimum overall thickness required by 7.3.1 and 9.3.1 is considered to satisfy the requirements of the Code for members not supporting or attached to nonstructural elements likely to be damaged by large deflections. For nonprestressed two-way construction, the minimum thickness required by 8.3.1 is considered to satisfy the requirements of the Code.

For nonprestressed members that do not meet these minimum thickness requirements, for nonprestressed one-way members that support or are attached to nonstructural elements likely to be damaged by large deflections, and for prestressed flexural members, deflections are required to be calculated by 24.2.3 through 24.2.5. Calculated deflections are limited to the values in Table 24.2.2.

CODE

24.2.2 Deflections calculated in accordance with 24.2.3 through 24.2.5 shall not exceed the limits in Table 24.2.2.

COMMENTARY

R24.2.2 It should be noted that the limitations given in Table 24.2.2 relate only to supported or attached nonstructural elements. For those structures in which structural members are likely to be affected by deflection or deformation of members to which they are attached in such a manner as to affect adversely the strength of the structure, these deflections and the resulting forces should be considered explicitly in the analysis and design of the structures as required by 24.2.1 (ACI 209R).

When time-dependent deflections are calculated, the portion of the deflection before attachment of the nonstructural elements may be deducted. In making this correction, use may be made of the curve in Fig. R24.2.4.1 for members of usual sizes and shapes.

Table 24.2.2—Maximum permissible calculated deflections

<table>
<tr><th>Member</th><th colspan="2">Condition</th><th>Deflection to be considered</th><th>Deflection limitation</th></tr>
<tr><td>Flat roofs</td><td colspan="2" rowspan="2">Not supporting or attached to nonstructural elements likely to be damaged by large deflections
Immediate deflection due to L</td><td>Immediate deflection due to maximum of L_r, S, and R</td><td>ℓ/180[1]</td></tr>
<tr><td>Floors</td><td>ℓ/360</td><td></td></tr>
<tr><td rowspan="2">Roof or floors</td><td rowspan="2">Supporting or attached to nonstructural elements</td><td>Likely to be damaged by large deflections</td><td rowspan="2">That part of the total deflection occurring after attachment of nonstructural elements, which is the sum of the time-dependent deflection due to all sustained loads and the immediate deflection due to any additional live load[2]</td><td>ℓ/480[3]</td></tr>
<tr><td>Not likely to be damaged by large deflections</td><td>ℓ/240[4]</td></tr>
</table>

[1]Limit not intended to safeguard against ponding. Ponding shall be checked by calculations of deflection, including added deflections due to ponded water, and considering time-dependent effects of sustained loads, camber, construction tolerances, and reliability of provisions for drainage.

[2]Time-dependent deflection shall be calculated in accordance with 24.2.4, but shall be permitted to be reduced by amount of deflection calculated to occur before attachment of nonstructural elements. This amount shall be calculated on basis of accepted engineering data relating to time-deflection characteristics of members similar to those being considered.

[3]Limit shall be permitted to be exceeded if measures are taken to prevent damage to supported or attached elements.

[4]Limit shall not exceed tolerance provided for nonstructural elements.

24.2.3 *Calculation of immediate deflections*

24.2.3.1 Immediate deflections shall be calculated using methods or formulas for elastic deflections, considering effects of cracking and reinforcement on member stiffness.

24.2.3.2 Effect of variation of cross-sectional properties, such as haunches, shall be considered when calculating deflections.

24.2.3.3 Deflections in two-way slab systems shall be calculated taking into account size and shape of the panel, conditions of support, and nature of restraints at the panel edges.

R24.2.3 *Calculation of immediate deflections*

R24.2.3.1 For calculation of immediate deflections of uncracked prismatic members, the usual methods or formulas for elastic deflections may be used with a constant value of E_cI_g along the length of the member. However, if the member is expected to crack at one or more sections, or if its depth varies along the span, a more rigorous calculation becomes necessary.

R24.2.3.3 The calculation of deflections for two-way slabs is challenging even if linear elastic behavior can be assumed. For immediate deflections, the values of E_c and I_e specified in 24.2.3.4 and 24.2.3.5, respectively, may be used (ACI 209R). However, other procedures and other values of the stiffness E_cI_e may be used if they result in predictions of deflection in reasonable agreement with the results of comprehensive tests.

CODE

24.2.3.4 Modulus of elasticity, E_c, shall be permitted to be calculated in accordance with 19.2.2.

24.2.3.5 For nonprestressed members, unless obtained by a more comprehensive analysis, effective moment of inertia, I_e, shall be calculated in accordance with Table 24.2.3.5 using:

$$M_{cr} = \frac{f_r I_g}{y_t} \quad (24.2.3.5)$$

Table 24.2.3.5—Effective moment of inertia, I_e

Service moment	Effective moment of inertia, I_e, in.[4]	
$M_a \le (2/3)M_{cr}$	I_g	(a)
$M_a > (2/3)M_{cr}$	$\dfrac{I_{cr}}{1-\left(\dfrac{(2/3)M_{cr}}{M_a}\right)^2\left(1-\dfrac{I_{cr}}{I_g}\right)}$	(b)

24.2.3.6 For continuous one-way slabs and beams, I_e shall be permitted to be taken as the average of values obtained from Table 24.2.3.5 for the critical positive and negative moment sections.

24.2.3.7 For prismatic one-way slabs and beams, I_e shall be permitted to be taken as the value obtained from Table 24.2.3.5 at midspan for simple and continuous spans, and at the support for cantilevers.

24.2.3.8 For prestressed Class U slabs and beams as defined in 24.5.2, it shall be permitted to calculate deflections based on I_g.

24.2.3.9 For prestressed Class T and Class C slabs and beams as defined in 24.5.2, deflection calculations shall be based on a cracked transformed section analysis. It shall be permitted to base deflection calculations on a bilinear moment-deflection relationship or I_e in accordance with Eq. (24.2.3.9a)

$$I_e = \left(\frac{M_{cr}}{M_a}\right)^3 I_g + \left[1-\left(\frac{M_{cr}}{M_a}\right)^3\right] I_{cr} \quad (24.2.3.9a)$$

where M_{cr} is calculated as

$$M_{cr} = \frac{(f_r + f_{pe})I_g}{y_t} \quad (24.2.3.9b)$$

COMMENTARY

R24.2.3.5 The effective moment of inertia approximation, developed by Bischoff (2005), has been shown to result in calculated deflections that have sufficient accuracy for a wide range of reinforcement ratios (Bischoff and Scanlon 2007). M_{cr} is multiplied by two-thirds to consider restraint that can reduce the effective cracking moment as well as to account for reduced tensile strength of concrete during construction that can lead to cracking that later affects service deflections (Scanlon and Bischoff 2008).

Before 2019, ACI 318 used a different equation (Branson 1965) to calculate I_e. The previous equation has subsequently been shown to underestimate deflections for members with low reinforcement ratios, which often occur in slabs, and does not consider the effects of restraint. For members with greater than 1 percent reinforcement and a service moment at least twice the cracking moment, there is little difference between deflections calculated using the former and the current Code provisions.

R24.2.3.7 The use of the midspan section properties for continuous prismatic members is considered satisfactory in approximate calculations primarily because the midspan stiffness (including the effect of cracking) has the dominant effect on deflections, as shown by ACI 435.5R, ACI Committee 435 (1978), and Sabnis et al. (1974).

R24.2.3.8 Immediate deflections of Class U prestressed concrete members may be calculated by the usual methods or formulas for elastic deflections using the moment of inertia of the gross (uncracked) concrete section and the modulus of elasticity for concrete specified in 19.2.2.1.

R24.2.3.9 The effective moment of inertia equation in 24.2.3.5 was revised in the 2019 Code. The revision is not applicable to prestressed members. Equation (24.2.3.9a) maintains the provisions of previous editions of the Code for these types of members. The *PCI Design Handbook* (PCI MNL 120) gives information on deflection calculations using a bilinear moment-deflection relationship and using an effective moment of inertia. Mast (1998) gives additional information on deflection of cracked prestressed concrete members.

Shaikh and Branson (1970) shows that the I_e method can be used to calculate deflections of Class C and Class T prestressed members loaded above the cracking load. For this case, the cracking moment should take into account the effect of prestress as provided in Eq. (24.2.3.9).

CODE

24.2.4 *Calculation of time-dependent deflections*

24.2.4.1 *Nonprestressed members*

24.2.4.1.1 Unless obtained from a more comprehensive analysis, additional time-dependent deflection resulting from creep and shrinkage of flexural members shall be calculated as the product of the immediate deflection caused by sustained load and the factor λ_Δ

$$\lambda_\Delta = \frac{\xi}{1+50\rho'} \quad (24.2.4.1.1)$$

24.2.4.1.2 In Eq. (24.2.4.1.1), ρ' shall be calculated at midspan for simple and continuous spans, and at the support for cantilevers.

24.2.4.1.3 In Eq. (24.2.4.1.1), values of the time-dependent factor for sustained loads, ξ, shall be in accordance with Table 24.2.4.1.3.

Table 24.2.4.1.3—Time-dependent factor for sustained loads

Sustained load duration, months	Time-dependent factor ξ
3	1.0
6	1.2
12	1.4
60 or more	2.0

COMMENTARY

A method for predicting the effect of nonprestressed tension reinforcement in reducing creep camber is also given in Shaikh and Branson (1970), with approximate forms given in ACI 209R and Branson (1970).

R24.2.4 *Calculation of time-dependent deflections*

R24.2.4.1 *Nonprestressed members*

Shrinkage and creep cause time-dependent deflections in addition to the elastic deflections that occur when loads are first placed on the structure. Such deflections are influenced by temperature, humidity, curing conditions, age at time of loading, amount of compression reinforcement, and magnitude of the sustained load. The expression given in this section is considered satisfactory for use with the Code procedures for the calculation of immediate deflections, and with the limits given in Table 24.2.2. The deflection calculated in accordance with this section is the additional time-dependent deflection due to the dead load and those portions of other loads that will be sustained for a sufficient period to cause significant time-dependent deflections.

Equation (24.2.4.1.1) was developed in Branson (1971). In Eq. (24.2.4.1.1), the term $(1 + 50\rho')$ accounts for the effect of compression reinforcement in reducing time-dependent deflections. ξ = **2.0** represents a nominal time-dependent factor for a 5-year duration of loading. The curve in Fig. R24.2.4.1 may be used to estimate values of ξ for loading periods less than 5 years.

If it is desired to consider creep and shrinkage separately, approximate equations provided in Branson (1965, 1971, 1977) and ACI Committee 435 (1966) may be used.

Because available data on time-dependent deflections of two-way slabs are too limited to justify more elaborate procedures, calculation of the additional time-dependent deflection for two-way construction in accordance with Eq. (24.2.4.1.1) is required to use the multipliers given in 24.2.4.1.3.

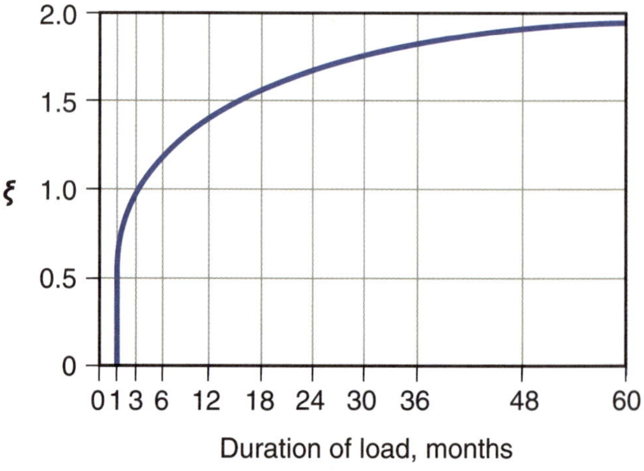

***Fig. R24.2.4.1**—Multipliers for time-dependent deflections.*

CODE

24.2.4.2 *Prestressed members*

24.2.4.2.1 Additional time-dependent deflection of prestressed concrete members shall be calculated considering stresses in concrete and reinforcement under sustained load, and the effects of creep and shrinkage of concrete and relaxation of prestressed reinforcement.

COMMENTARY

R24.2.4.2 *Prestressed members*

R24.2.4.2.1 Calculation of time-dependent deflections of prestressed concrete flexural members is challenging. The calculations should consider not only the increased deflections due to flexural stresses, but also the additional time-dependent deflections resulting from time-dependent shortening of the flexural member.

Prestressed concrete members shorten more with time than similar nonprestressed members due to the precompression in the slab or beam, which causes creep. This creep, together with concrete shrinkage, results in significant shortening of the flexural members that continues for several years after construction and should be considered in design. The shortening tends to reduce the tension in the prestressed reinforcement, reducing the precompression in the member and thereby causing increased time-dependent deflections.

Another factor that can influence time-dependent deflections of prestressed flexural members is adjacent concrete or masonry that is nonprestressed in the direction of the prestressed member. This can be a slab nonprestressed in the beam direction adjacent to a prestressed beam or a nonprestressed slab system. As the prestressed member tends to shrink and creep more than the adjacent nonprestressed concrete, the structure will tend to reach a compatibility of the shortening effects. This results in a reduction of the precompression in the prestressed member as the adjacent concrete absorbs the compression. This reduction in precompression of the prestressed member can occur over a period of years and will result in additional time-dependent deflections and an increase in tensile stresses in the prestressed member.

Any suitable method for calculating time-dependent deflections of prestressed members may be used, provided all effects are considered. Guidance may be found in ACI 209R, ACI Committee 435 (1963), Branson et al. (1970), and Ghali and Favre (1986).

CODE

24.2.5 *Calculation of deflections of composite concrete construction*

24.2.5.1 If composite concrete flexural members are shored during construction so that, after removal of temporary supports, the dead load is resisted by the full composite section, it shall be permitted to consider the composite member equivalent to a monolithically cast member for calculation of deflections.

24.2.5.2 If composite concrete flexural members are not shored during construction, the magnitude and duration of load before and after composite action becomes effective shall be considered in calculating time-dependent deflections.

24.2.5.3 Deflections resulting from differential shrinkage of precast and cast-in-place components, and of axial creep effects in prestressed members, shall be considered.

COMMENTARY

R24.2.5 *Calculation of deflections of composite concrete construction*

Composite concrete members are designed to meet the horizontal shear strength requirements of 16.4. Because few tests have been made to study the immediate and time-dependent deflections of composite members, the requirements given in this section are based on the judgment of ACI Committee 318 and on experience.

In 22.3.3.3, it is stated that distinction need not be made between shored and unshored members. This refers to strength calculations, not to deflections. Construction documents should indicate whether composite concrete design is based on shored or unshored construction, as required by 26.11.1.1.

CODE

24.3—Distribution of flexural reinforcement in one-way slabs and beams

24.3.1 Bonded reinforcement shall be distributed to control flexural cracking in tension zones of nonprestressed and Class C prestressed slabs and beams reinforced for flexure in one direction only.

24.3.2 Spacing of bonded reinforcement closest to the tension face shall not exceed the limits in Table 24.3.2, where c_c is the least distance from surface of deformed or prestressed reinforcement to the tension face. Calculated stress in deformed reinforcement, f_s, and calculated change in stress in bonded prestressed reinforcement, Δf_{ps}, shall be in accordance with 24.3.2.1 and 24.3.2.2, respectively.

Table 24.3.2—Maximum spacing of bonded reinforcement in nonprestressed and Class C prestressed one-way slabs and beams

Reinforcement type	Maximum spacing s	
Deformed bars or wires	Lesser of:	$15\left(\frac{40,000}{f_s}\right)-2.5c_c$
		$12\left(\frac{40,000}{f_s}\right)$
Bonded prestressed reinforcement	Lesser of:	$\left(\frac{2}{3}\right)\left[15\left(\frac{40,000}{\Delta f_{ps}}\right)-2.5c_c\right]$
		$\left(\frac{2}{3}\right)\left[12\left(\frac{40,000}{\Delta f_{ps}}\right)\right]$
Combined deformed bars or wires and bonded prestressed reinforcement	Lesser of:	$\left(\frac{5}{6}\right)\left[15\left(\frac{40,000}{\Delta f_{ps}}\right)-2.5c_c\right]$
		$\left(\frac{5}{6}\right)\left[12\left(\frac{40,000}{\Delta f_{ps}}\right)\right]$

COMMENTARY

R24.3—Distribution of flexural reinforcement in one-way slabs and beams

R24.3.1 Where service loads result in high stresses in the reinforcement, visible cracks should be expected, and steps should be taken in detailing of the reinforcement to control cracking. For reasons of durability and appearance, many fine cracks are preferable to a few wide cracks. Detailing practices limiting bar spacing will usually lead to adequate crack control where Grade 60 reinforcement is used.

Extensive laboratory work (Gergely and Lutz 1968; Kaar 1966; Base et al. 1966) involving deformed bars demonstrated that crack width at service loads is proportional to reinforcement stress. The significant variables reflecting reinforcement detailing were found to be thickness of concrete cover and the spacing of reinforcement.

Crack width is inherently subject to wide scatter even in careful laboratory work and is influenced by shrinkage and other time-dependent effects. Improved crack control is obtained where the reinforcement is well distributed over the zone of maximum concrete tension. Several bars at moderate spacing are much more effective in controlling cracking than one or two larger bars of equivalent area.

R24.3.2 The spacing of reinforcement is limited to control cracking (Beeby 1979; Frosch 1999; ACI Committee 318 1999). For the case of beams with Grade 60 reinforcement and 2 in. clear cover to the primary reinforcement, with f_s = **40,000 psi**, the maximum bar spacing is 10 in.

Crack widths in structures are highly variable. The Code provisions for spacing are intended to limit surface cracks to a width that is generally acceptable in practice but may vary widely in a given structure.

The role of cracks in the corrosion of reinforcement is controversial. Research (Darwin et al. 1985; Oesterle 1997) shows that corrosion is not clearly correlated with surface crack widths in the range normally found with reinforcement stresses at service load levels. For this reason, the Code does not differentiate between interior and exterior exposures.

Only tension reinforcement nearest the tension face need be considered in selecting the value of c_c used in calculating spacing requirements. To account for prestressed reinforcement, such as strand, having bond characteristics less effective than deformed reinforcement, a two-thirds effectiveness factor is used in Table 24.3.2.

For post-tensioned members designed as cracked members, it will usually be advantageous to provide crack control by the use of deformed reinforcement, for which the provisions in Table 24.3.2 for deformed bars or wires may be used. Bonded reinforcement required by other provisions of the Code may also be used as crack control reinforcement.

CODE

24.3.2.1 Stress f_s in deformed reinforcement closest to the tension face at service loads shall be calculated based on the unfactored moment, or it shall be permitted to take f_s as **(2/3)**f_y.

24.3.2.2 Change in stress, Δf_{ps}, in bonded prestressed reinforcement at service loads shall be equal to the calculated stress based on a cracked section analysis minus the decompression stress f_{dc}. It shall be permitted to take f_{dc} equal to the effective stress in the prestressed reinforcement f_{se}. The value of Δf_{ps} shall not exceed 36,000 psi. If Δf_{ps} does not exceed 20,000 psi, the spacing limits in Table 24.3.2 need not be satisfied.

24.3.3 If there is only one bonded bar, pretensioned strand, or bonded tendon nearest to the extreme tension face, the width of the extreme tension face shall not exceed s determined in accordance with Table 24.3.2.

24.3.4 If the flange of a T-beam is in tension, the portion of the bonded flexural tension reinforcement not located over the beam web shall be distributed within the lesser of the effective flange width, as defined in accordance with 6.3.2 and ℓ_n**/10**. If ℓ_n**/10** controls, additional bonded longitudinal reinforcement satisfying 24.4.3.1 shall be provided in the outer portions of the flange.

24.3.5 The spacing of bonded flexural reinforcement in nonprestressed and Class C prestressed one-way slabs and beams subject to fatigue, designed to be watertight, or exposed to corrosive environments, shall be selected based on investigations and precautions specific to those conditions and shall not exceed the limits of 24.3.2.

COMMENTARY

R24.3.2.1 For applications in which crack control is critical, the designer should consider reducing the value of f_s to help control cracking. Research by Frosch et al. (2014) and Puranam (2018) supports the use of these design provisions for Grade 100 reinforcement.

R24.3.2.2 It is conservative to take the decompression stress f_{dc} equal to f_{se}, the effective stress in the prestressed reinforcement. The maximum limitation of 36,000 psi for Δf_{ps} is intended to be similar to the maximum allowable stress in Grade 60 reinforcement **(**f_s **= 40,000 psi)**. The exemption for members with Δf_{ps} less than 20,000 psi reflects that many structures designed by working stress methods and with low reinforcement stress served their intended functions with very limited flexural cracking.

R24.3.4 In a T-beam, distribution of the negative moment reinforcement for control of cracking should take into account two considerations: 1) wide spacing of the reinforcement across the full effective width of flange may cause some wide cracks to form in the slab near the web; and 2) close spacing near the web leaves the outer regions of the flange unprotected. The one-tenth limitation is to guard against a spacing that is too wide, with some additional reinforcement required to protect the outer portions of the flange.

For T-beams designed to resist negative moments due to gravity and wind loads, all tensile reinforcement required for strength is located within the lesser of the effective flange width and ℓ_n**/10**. Common practice is to place more than half of the reinforcement over the beam web. For T-beams resisting load combinations including earthquake effects, all reinforcement placed within the effective flange width may contribute to the beam flexural strength for the anticipated drift (refer to 18.7.3).

R24.3.5 Although a number of studies have been conducted, clear experimental evidence is not available regarding the crack width beyond which a corrosion danger exists (ACI 222R). Exposure tests indicate that concrete quality, adequate consolidation, and ample concrete cover may be of greater importance for corrosion protection than crack width at the concrete surface (Schießl and Raupach 1997).

Provisions related to increased concrete cover and durability of reinforcement is covered in 20.5, while durability of concrete is covered in 19.3.

24.4—Shrinkage and temperature reinforcement

24.4.1 Reinforcement to resist shrinkage and temperature stresses shall be provided in one-way slabs in the direction

R24.4—Shrinkage and temperature reinforcement

R24.4.1 Shrinkage and temperature reinforcement is required at right angles to the principal reinforcement to minimize cracking and to tie the structure together to ensure

CODE

perpendicular to the flexural reinforcement in accordance with 24.4.3 or 24.4.4.

24.4.2 If shrinkage and temperature movements are restrained, the effects of $\boldsymbol{T}$ shall be considered in accordance with 5.3.6.

24.4.3 *Nonprestressed reinforcement*

24.4.3.1 Deformed reinforcement to resist shrinkage and temperature stresses shall conform to Table 20.2.2.4(a) and shall be in accordance with 24.4.3.2 through 24.4.3.5.

24.4.3.2 The ratio of deformed shrinkage and temperature reinforcement area to gross concrete area shall be greater than or equal to 0.0018.

24.4.3.3 The spacing of deformed shrinkage and temperature reinforcement shall not exceed the lesser of $5h$ and 18 in.

24.4.3.4 At all sections where required, deformed reinforcement used to resist shrinkage and temperature stresses shall develop f_y in tension.

24.4.3.5 For one-way precast slabs and one-way precast, prestressed wall panels, shrinkage and temperature rein-

COMMENTARY

it is acting as assumed in the design. The provisions of this section are intended for structural slabs only; they are not intended for slabs-on-ground.

R24.4.2 The area of shrinkage and temperature reinforcement required by 24.4.3.2 has been satisfactory where shrinkage and temperature movements are permitted to occur. Where structural walls or columns provide significant restraint to shrinkage and temperature movements, the restraint of volume changes causes tension in slabs, as well as displacements, shear forces, and flexural moments in columns or walls. In these cases, it may be necessary to increase the amount of slab reinforcement required by 24.4.3.2 due to the shrinkage and thermal effects in both principal directions (PCI MNL 120; Gilbert 1992). Top and bottom reinforcement are both effective in controlling cracks. Control strips during the construction period, which permit initial shrinkage to occur without causing an increase in stress, are also effective in reducing cracks caused by restraint.

Topping slabs also experience tension due to restraint of differential shrinkage between the topping and the precast elements or metal deck (which has zero shrinkage) that should be considered in reinforcing the slab. Consideration should be given to strain demands on reinforcement crossing joints of precast elements where most of the restraint is likely to be relieved.

R24.4.3 *Nonprestressed reinforcement*

R24.4.3.2 The minimum ratios of deformed bar or welded wire reinforcement area to gross concrete area of 0.0018 is empirical but has been used satisfactorily for many years. The resulting area of reinforcement may be distributed near the top or bottom of the slab, or may be distributed between the two faces of the slab as deemed appropriate for specific conditions. Previous editions of the Code permitted a reduction in shrinkage and temperature reinforcement for reinforcement with yield strength greater than 60,000 psi. However, the mechanics of cracking suggest that increased yield strength provides no benefit for the control of cracking. If crack width or leakage prevention is a design limit state, refer to ACI 224R or ACI 350 for recommended reinforcement ratios.

R24.4.3.4 Splices and end anchorages of shrinkage and temperature reinforcement are to be designed to develop the specified yield strength of the reinforcement in accordance with Chapter 25.

R24.4.3.5 For precast, prestressed concrete members not wider than 12 ft, such as hollow-core slabs, solid slabs, or

CODE

forcement is not required in the direction perpendicular to the flexural reinforcement if (a) through (c) are satisfied.

(a) Precast members are not wider than 12 ft
(b) Precast members are not mechanically connected to cause restraint in the transverse direction
(c) Reinforcement is not required to resist transverse flexural stresses

COMMENTARY

slabs with closely spaced ribs, there is usually no need to provide reinforcement to withstand shrinkage and temperature stresses in the short direction. This is generally also true for precast, nonprestressed floor and roof slabs. The 12 ft width is less than that in which shrinkage and temperature stresses can build up to a magnitude requiring reinforcement. In addition, much of the shrinkage occurs before the members are tied into the structure. Once in the final structure, the members are usually not as rigidly connected transversely as monolithic concrete, thus, the transverse restraint stresses due to both shrinkage and temperature change are significantly reduced.

The waiver does not apply where reinforcement is required to resist flexural stresses, such as in thin flanges of precast single and double tees.

24.4.4 *Prestressed reinforcement*

24.4.4.1 Prestressed reinforcement to resist shrinkage and temperature stresses shall conform to Table 20.3.2.2, and the effective prestress after losses shall provide an average compressive stress of at least 100 psi on gross concrete area.

R24.4.4 *Prestressed reinforcement*

R24.4.4.1 Prestressed reinforcement requirements have been selected to provide an effective force on the slab approximately equal to the force required to yield nonprestressed shrinkage and temperature reinforcement. This amount of prestressing—100 psi on the gross concrete area—has been used successfully on a large number of projects.

The effects of slab shortening should be evaluated to ensure serviceable behavior of the structure. In most cases, the low level of prestressing recommended should not cause difficulties in a properly detailed structure. Additional attention may be required where thermal effects or restraint become significant.

24.5—Permissible stresses in prestressed concrete flexural members

24.5.1 *General*

24.5.1.1 Concrete stresses in prestressed flexural members shall be limited in accordance with 24.5.2 through 24.5.4 unless it is shown by test or analysis that performance will not be impaired.

R24.5—Permissible stresses in prestressed concrete flexural members

R24.5.1 *General*

R24.5.1.1 Permissible stresses in concrete address serviceability but do not ensure adequate design strength, which should be checked in accordance with other Code requirements.

A mechanism is provided such that Code limits on stress need not inhibit the development of new products, materials, and techniques in prestressed concrete construction. Approvals for the design should be in accordance with 1.10 of the Code.

24.5.1.2 For calculation of stresses at transfer of prestress, at service loads, and at cracking loads, elastic theory shall be used with assumptions (a) and (b):

(a) Strains vary linearly with distance from neutral axis in accordance with 22.2.1.
(b) At cracked sections, concrete resists no tension.

24.5.2 *Classification of prestressed flexural members*

R24.5.2 *Classification of prestressed flexural members*

CODE

24.5.2.1 Prestressed flexural members shall be classified as Class U, T, or C in accordance with Table 24.5.2.1, based on the extreme fiber stress in tension f_t in the precompressed tension zone calculated at service loads assuming an uncracked section.

Table 24.5.2.1—Classification of prestressed flexural members based on f_t

Assumed behavior	Class	Limits of f_t
Uncracked	U[1]	$f_t \leq 7.5\sqrt{f_c'}$
Transition between uncracked and cracked	T	$7.5\sqrt{f_c'} < f_t \leq 12\sqrt{f_c'}$
Cracked	C	$f_t > 12\sqrt{f_c'}$

[1]Prestressed two-way slabs shall be designed as Class U with $f_t \leq 6\sqrt{f_c'}$.

COMMENTARY

R24.5.2.1 Three classes of behavior of prestressed flexural members are defined. Class U members are assumed to behave as uncracked members. Class C members are assumed to behave as cracked members. The behavior of Class T members is assumed to be in transition between uncracked and cracked. These classes apply to both bonded and unbonded prestressed flexural members, but prestressed two-way slab systems are required to be designed as Class U with $f_t \leq 6\sqrt{f_c'}$.

The serviceability requirements for each class are summarized in Table R24.5.2.1. For comparison, Table R24.5.2.1 also shows corresponding requirements for nonprestressed members. Due to lack of strain compatibility, it is inappropriate to include the area of unbonded prestressed reinforcement in the calculation of gross or cracked section properties, although the effective prestress force should be considered when determining the location of the neutral axis. Conversely, the calculation of section properties should account for the area of the voids created by the sheathing or duct for unbonded prestressed reinforcement. A method for evaluating stresses, deflections, and crack control in cracked prestressed members is given in Mast (1998).

The precompressed tension zone is that portion of a prestressed member where flexural tension, calculated using gross section properties, would occur under unfactored dead and live loads if the prestress force was not present. Prestressed concrete is usually designed so that the prestress force introduces compression into this zone, thus effectively reducing the magnitude of the tensile stress.

For corrosive environments, defined as an environment in which chemical attack (such as seawater, corrosive industrial atmosphere, or sewer gas) is encountered, cracking at service loads becomes more critical to long-term performance. For these conditions, cover should be increased in accordance with 20.5.1.4, and tensile stresses in the concrete reduced to minimize possible cracking at service loads.

Table R24.5.2.1—Serviceability design requirements

	Prestressed			Nonprestressed
	Class U	Class T	Class C	
Assumed behavior	Uncracked	Transition between uncracked and cracked	Cracked	Cracked
Section properties for stress calculation at service loads	Gross section 24.5.2.2	Gross section 24.5.2.2	Cracked section 24.5.2.3	No requirement
Allowable stress at transfer	24.5.3	24.5.3	24.5.3	No requirement
Allowable compressive stress based on uncracked section properties	24.5.4	24.5.4	No requirement	No requirement
Tensile stress at service loads 24.5.2.1	$\leq 7.5\sqrt{f_c'}$	$7.5\sqrt{f_c'} < f_t \leq 12\sqrt{f_c'}$	No requirement	No requirement
Deflection calculation basis	24.2.3.8, 24.2.4.2 Gross section	24.2.3.9, 24.2.4.2 Cracked section, bilinear	24.2.3.9, 24.2.4.2 Cracked section, bilinear	24.2.3, 24.2.4.1 Effective moment of inertia
Crack control	No requirement	No requirement	24.3	24.3
Computation of Δf_{ps} or f_s for crack control	—	—	Cracked section analysis	$M/(A_s \times$ lever arm), or $2/3f_y$
Side skin reinforcement	No requirement	No requirement	9.7.2.3	9.7.2.3

CODE

24.5.2.2 For Class U and T members, stresses at service loads shall be permitted to be calculated using the uncracked section.

24.5.2.3 For Class C members, stresses at service loads shall be calculated using the cracked transformed section.

24.5.3 *Permissible concrete stresses at transfer of prestress*

24.5.3.1 Calculated extreme concrete fiber stress in compression immediately after transfer of prestress, but before time-dependent prestress losses, shall not exceed the limits in Table 24.5.3.1.

Table 24.5.3.1—Concrete compressive stress limits immediately after transfer of prestress

Location	Concrete compressive stress limits
End of simply-supported members	$0.70f_{ci}'$
All other locations	$0.60f_{ci}'$

24.5.3.2 Calculated extreme concrete fiber stress in tension immediately after transfer of prestress, but before time-dependent prestress losses, shall not exceed the limits in Table 24.5.3.2, unless permitted by 24.5.3.2.1.

Table 24.5.3.2—Concrete tensile stress limits immediately after transfer of prestress, without additional bonded reinforcement in tension zone

Location	Concrete tensile stress limits
Ends of simply-supported members	$6\sqrt{f_{ci}'}$
All other locations	$3\sqrt{f_{ci}'}$

24.5.3.2.1 The limits in Table 24.5.3.2 shall be permitted to be exceeded where additional bonded reinforcement in the tension zone resists the total tensile force in the concrete calculated with the assumption of an uncracked section.

COMMENTARY

R24.5.2.3 Prestressed members are classified based on the magnitude of the stress in the precompressed tension zone, calculated assuming the section remains uncracked. Once it is determined that a member is Class C, with $f_t > 12\sqrt{f_c'}$, subsequent calculations of service load stresses are based on the cracked transformed section.

R24.5.3 *Permissible concrete stresses at transfer of prestress*

The concrete stresses at this stage are caused by the weight of the member and the force in the prestressed reinforcement after jacking reduced by the losses due to seating of the prestressed reinforcement and elastic shortening of the concrete. Shrinkage, creep, and relaxation effects are generally not included at this stage. These stresses apply to both pretensioned and post-tensioned concrete with proper modifications of the losses at transfer.

R24.5.3.1 The permissible concrete compressive stresses at transfer of prestress are higher at ends of simply supported members than at other locations based on research in the precast, prestressed concrete industry (Castro et al. 2004; Dolan and Krohn 2007; Hale and Russell 2006).

R24.5.3.2 The tensile stress limits of $3\sqrt{f_{ci}'}$ and $6\sqrt{f_{ci}'}$ refer to tensile stresses at transfer of prestress at locations other than the precompressed tension zone. Where tensile stresses exceed the permissible values, the total force in the tensile stress zone may be calculated and reinforcement proportioned on the basis of this force at a stress of $0.6f_y$, but not more than 30,000 psi. The effects of creep and shrinkage begin to reduce the tensile stress almost immediately; however, some tension remains in these locations after allowance is made for all prestress losses.

CODE

24.5.4 *Permissible concrete compressive stresses at service loads*

24.5.4.1 For Class U and T members, the calculated extreme concrete fiber stress in compression at service loads, after allowance for all prestress losses, shall not exceed the limits in Table 24.5.4.1.

Table 24.5.4.1—Concrete compressive stress limits at service loads

Load condition	Concrete compressive stress limits
Prestress plus sustained load	$0.45f_c'$
Prestress plus total load	$0.60f_c'$

COMMENTARY

R24.5.4 *Permissible concrete compressive stresses at service loads*

R24.5.4.1 The compressive stress limit of $0.45f_c'$ was originally established to decrease the probability of failure of prestressed concrete members due to repeated loads. This limit also seemed reasonable to preclude excessive creep deformation. At higher values of stress, creep strains tend to increase more rapidly as applied stress increases.

Fatigue tests of prestressed concrete beams have shown that concrete compressive failures are not the controlling criterion. Therefore, the stress limit of $0.60f_c'$ permits a one-third increase in allowable compressive stress for members subject to transient loads.

Sustained live load is any portion of the service live load that will be sustained for a sufficient period to cause significant time-dependent deflections. Thus, when the sustained live and dead loads are a large percentage of the total service load, the $0.45f_c'$ limit of Table 24.5.4.1 typically controls. On the other hand, when a large portion of the total service load consists of a transient or temporary service live load, the increased stress limit of $0.60f_c'$ typically controls.

The compression limit of $0.45f_c'$ for prestress plus sustained loads will continue to control the time-dependent behavior of prestressed members.

CHAPTER 25—REINFORCEMENT DETAILS

25.1—Scope

25.1.1 This chapter shall apply to reinforcement details, including:

(a) Minimum spacing
(b) Standard hooks, seismic hooks, and crossties
(c) Development of reinforcement
(d) Splices
(e) Bundled reinforcement
(f) Transverse reinforcement
(g) Post-tensioning anchorages and couplers

25.1.2 Provisions of 25.9 shall apply to anchorage zones for post-tensioned tendons.

R25.1—Scope

Recommended methods and standards for preparing design drawings, typical details, and drawings for the fabrication and placing of steel reinforcement in reinforced concrete structures are given in the *ACI Detailing Manual* (SP-66).

All provisions in the Code relating to bar, wire, or strand diameter (and area) are based on the nominal dimensions of the reinforcement as given in the appropriate ASTM specification. Nominal dimensions are equivalent to those of a circular area having the same weight per foot as the ASTM designated bar, wire, or strand sizes. Cross-sectional area of reinforcement is based on nominal dimensions.

R25.1.1 In addition to the requirements in this chapter that affect detailing of reinforcement, detailing specific to particular members is given in the corresponding member chapters. Additional detailing associated with structural integrity requirements is covered in 4.10.

25.2—Minimum spacing of reinforcement

25.2.1 For parallel nonprestressed reinforcement in a horizontal layer, clear spacing shall be at least the greatest of 1 in., $\boldsymbol{d_b}$, and $\boldsymbol{(4/3)d_{agg}}$.

25.2.2 For parallel nonprestressed reinforcement placed in two or more horizontal layers, reinforcement in the upper layers shall be placed directly above reinforcement in the bottom layer with a clear spacing between layers of at least 1 in.

25.2.3 For longitudinal reinforcement in columns, pedestals, struts, and boundary elements in walls, clear spacing between bars shall be at least the greatest of 1.5 in., $\boldsymbol{1.5d_b}$, and $\boldsymbol{(4/3)d_{agg}}$.

25.2.4 For pretensioned strands at ends of a member, minimum center-to-center spacing s shall be the greater of the value in Table 25.2.4, and $\boldsymbol{[(4/3)d_{agg} + d_b]}$.

R25.2—Minimum spacing of reinforcement

The minimum limits are set to permit concrete to flow readily into spaces between bars and between bars and forms without honeycombs, and to ensure against concentration of bars on a line that may cause shear or shrinkage cracking. Use of nominal bar diameter to define minimum spacing permits a uniform criterion for all bar sizes. In 2014, the size limitations on aggregates were translated to minimum spacing requirements, and are provided to promote proper encasement of reinforcement and to minimize honeycombing. The limitations associated with aggregate size need not be satisfied if, in the judgment of the licensed design professional, the workability and methods of consolidation of the concrete are such that the concrete can be placed without creating honeycombs or voids.

The development lengths given in 25.4 are a function of the bar spacing and cover. As a result, it may be desirable to use larger than minimum bar spacing or cover in some cases.

R25.2.4 The decreased spacing for transfer strengths of 4000 psi or greater is based on Deatherage et al. (1994) and Russell and Burns (1996).

Detailing 25

CODE

Table 25.2.4—Minimum center-to-center spacing of pretensioned strands at ends of members

f_{ci}', psi	Nominal strand diameter, in.	Minimum s
< 4000	All	$4d_b$
≥ 4000	< 0.5 in.	$4d_b$
	0.5 in.	1-3/4 in.
	0.6 in.	2 in.

25.2.5 For pretensioned wire at ends of a member, minimum center-to-center spacing, s, shall be the greater of $5d_b$ and $[(4/3)d_{agg} + d_b]$.

25.2.6 Reduced vertical spacing including bundling of prestressed reinforcement shall be permitted in the middle portion of a span.

25.2.7 For parallel nonprestressed reinforcement in shotcrete members, the clear spacing shall be in accordance with (a) or (b):

(a) The clear spacing between bars shall be at least the greater of $6d_b$ and 2-1/2 in.
(b) If two curtains of reinforcement are provided, the clear spacing between bars in the curtain nearer the nozzle shall be at least $12d_b$. The clear spacing between bars in the remaining curtain shall conform to (a).

25.2.7.1 It shall be permitted to use a clear spacing that does not meet 25.2.7(a) or 25.2.7(b) provided shotcrete mockup panels are used to demonstrate proper reinforcement encasement in accordance with (a) and (b):

(a) The shotcrete mockup panels shall be representative of the most complex reinforcement configurations to be encountered.
(b) The licensed design professional shall specify the shotcrete mock-up panel quantity, frequency of shooting per nozzleman and member type, and panel thickness to verify reinforcement encasement.

25.2.8 For prestressed strands in shotcrete members, minimum center-to-center spacing, s, shall satisfy 25.2.4, except as permitted in 25.2.6.

25.2.9 For prestressed wire in shotcrete members, minimum center-to-center spacing, s, shall satisfy the requirements for wire in 25.2.5, except as permitted in and 25.2.6

25.2.10 For ties, hoops, and spiral reinforcement in columns to be placed with shotcrete, minimum clear spacing shall be 3 in.

COMMENTARY

R25.2.7.1 Information on shotcrete mockup panels is provided in ACI 506R, and information on evaluating shotcrete is provided in ACI 506.4R.

R25.2.10 Shotcrete is usually not used in new construction for columns because the close spacing between ties, hoops, or spiral reinforcement makes it difficult to achieve adequate encasement of the column longitudinal reinforcement. Spacing closer than required in 25.2.10 requires approval by

CODE

25.2.10.1 It shall be permitted to use a clear spacing other than 3 in. provided shotcrete mockup panels are used to demonstrate proper encasement of the reinforcement in accordance with 25.2.7.1

25.3—Standard hooks, seismic hooks, crossties, and minimum inside bend diameters

25.3.1 Standard hooks for the development of deformed bars in tension shall conform to Table 25.3.1.

COMMENTARY

the licensed design professional based on shotcrete mockup panels demonstrating that the reinforcement can be encased without voids.

R25.3—Standard hooks, seismic hooks, crossties, and minimum inside bend diameters

R25.3.1 Standard bends in reinforcing bars are described in terms of the inside diameter of bend because the inside bend diameter is easier to measure than the radius of bend. The primary factors affecting the minimum bend diameter are feasibility of bending without breakage and avoidance of crushing the concrete inside the bend.

Table 25.3.1—Standard hook geometry for development of deformed bars in tension

Type of standard hook	Bar size	Minimum inside bend diameter, in.	Straight extension[1] ℓ_{ext}, in.	Type of standard hook
90-degree hook	No. 3 through No. 8	$6d_b$	$12d_b$	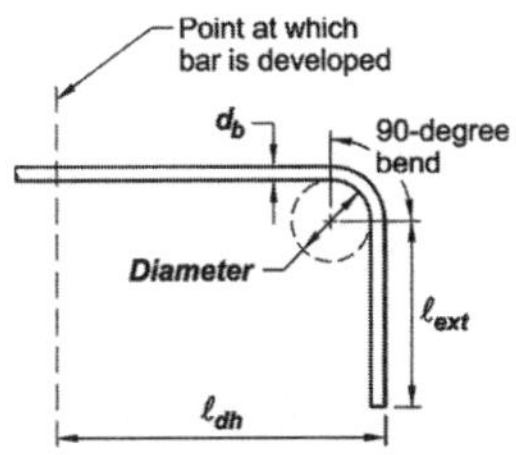
	No. 9 through No. 11	$8d_b$		
	No. 14 and No. 18	$10d_b$		
180-degree hook	No. 3 through No. 8	$6d_b$	Greater of $4d_b$ and 2.5 in.	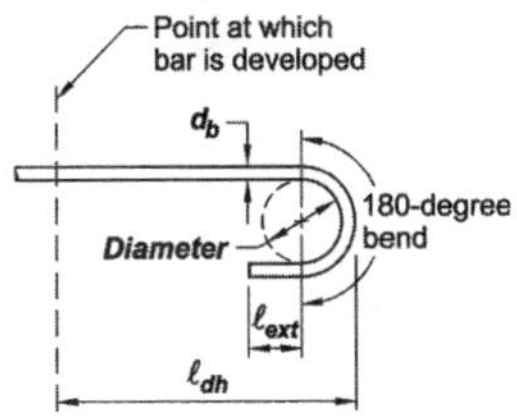
	No. 9 through No. 11	$8d_b$		
	No. 14 and No. 18	$10d_b$		

[1]A standard hook for deformed bars in tension includes the specific inside bend diameter and straight extension length. It shall be permitted to use a longer straight extension at the end of a hook. A longer extension shall not be considered to increase the anchorage capacity of the hook.

25.3.2 Minimum inside bend diameters for bars used as transverse reinforcement and standard hooks for bars used to anchor stirrups, ties, hoops, and spirals shall conform to Table 25.3.2. Standard hooks shall enclose longitudinal reinforcement.

R25.3.2 Standard stirrup, tie, and hoop hooks are limited to No. 8 bars and smaller, and the 90-degree hook with $6d_b$ extension is further limited to No. 5 bars and smaller, as the result of research showing that larger bar sizes with 90-degree hooks and $6d_b$ extensions tend to spall off the cover concrete when the reinforcement is stressed and the hook straightens.

The minimum $4d_b$ bend for the bar sizes commonly used for stirrups, ties, and hoops is based on accepted industry practice in the United States. Use of a stirrup bar size No. 5 or smaller for the 90-, 135-, or 180-degree standard stirrup hook will permit multiple bending on standard stirrup bending equipment.

Constructibility issues should be considered in selecting anchorage details. In particular, the use of 180-degree hooks

Detailing 25

CODE

COMMENTARY

should be avoided in closed stirrups, ties, and hoops made of continuous reinforcement.

Table 25.3.2—Minimum inside bend diameters and standard hook geometry for stirrups, ties, and hoops

Type of standard hook	Bar size	Minimum inside bend diameter, in.	Straight extension[1] ℓ_{ext}, in.	Type of standard hook
90-degree hook	No. 3 through No. 5	$4d_b$	Greater of $6d_b$ and 3 in.	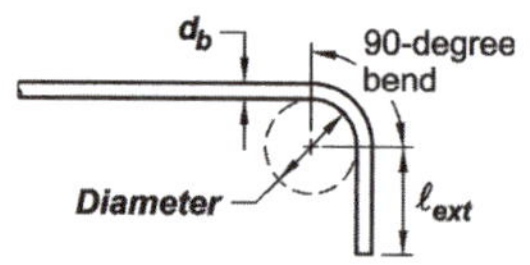
	No. 6 through No. 8	$6d_b$	$12d_b$	
135-degree hook	No. 3 through No. 5	$4d_b$	Greater of $6d_b$ and 3 in.	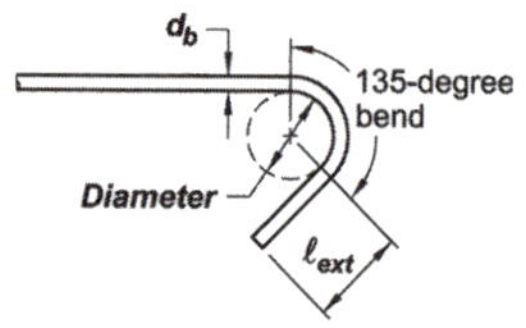
	No. 6 through No. 8	$6d_b$		
180-degree hook	No. 3 through No. 5	$4d_b$	Greater of $4d_b$ and 2.5 in.	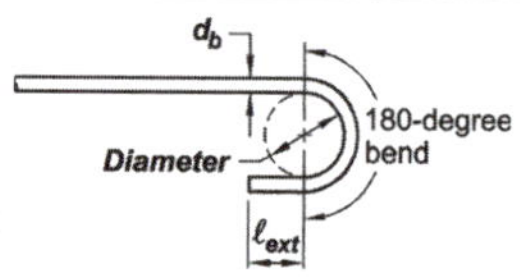
	No. 6 through No. 8	$6d_b$		

[1]A standard hook for stirrups, ties, and hoops includes the specific inside bend diameter and straight extension length. It shall be permitted to use a longer straight extension at the end of a hook. A longer extension shall not be considered to increase the anchorage capacity of the hook.

25.3.3 Minimum inside bend diameters for welded wire reinforcement used as stirrups or ties shall not be less than $4d_b$ for deformed wire larger than D6 and $2d_b$ for all other wires. Bends with inside diameter of less than $8d_b$ shall not be less than $4d_b$ from nearest welded intersection.

R25.3.3 Welded wire reinforcement can be used for stirrups and ties. The wire at welded intersections does not have the same uniform ductility and bendability as in areas that were not heated by welding in the manufacture of the welded wire reinforcement. These effects of the welding temperature are usually dissipated in a distance of approximately four wire diameters. Minimum bend diameters permitted are in most cases the same as those required in the ASTM bend tests for wire (ASTM A1064 and A1022).

25.3.4 Seismic hooks used to anchor stirrups, ties, hoops, and crossties shall be in accordance with (a) and (b):

(a) Minimum bend of 90 degrees for circular hoops and 135 degrees for all other hoops
(b) Hook shall engage longitudinal reinforcement and the extension shall project into the interior of the stirrup or hoop

25.3.5 Crossties shall be in accordance with (a) through (e):

(a) Crosstie shall be continuous between ends
(b) There shall be a seismic hook at one end
(c) There shall be a standard hook at other end with minimum bend of 90 degrees
(d) Hooks shall engage peripheral longitudinal bars
(e) 90-degree hooks of two successive crossties engaging the same longitudinal bars shall be alternated end for end, unless crossties satisfy 18.6.4.3 or 25.7.1.6.1

R25.3.5 Crossties are illustrated in Fig. R25.3.5.

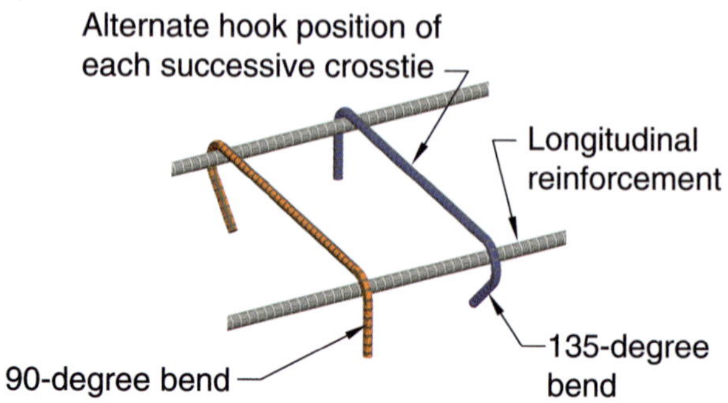

***Fig. R25.3.5**—Crosstie.*

CODE

25.4—Development of reinforcement

25.4.1 *General*

25.4.1.1 Calculated tension or compression in reinforcement at each section of a member shall be developed on each side of that section by embedment length; hook, headed deformed bar, mechanical device, or a combination thereof.

25.4.1.2 Hooks and heads shall not be used to develop bars in compression.

25.4.1.3 Development lengths do not require a strength reduction factor φ.

25.4.1.4 The values of $\sqrt{f_c'}$ used to calculate development length shall not exceed 100 psi.

25.4.2 *Development of deformed bars and deformed wires in tension*

25.4.2.1 Development length ℓ_d for deformed bars and deformed wires in tension shall be the greater of (a) and (b):

COMMENTARY

R25.4—Development of reinforcement

R25.4.1 *General*

R25.4.1.1 The development length concept is based on the attainable average bond stress over the length of embedment of the reinforcement (ACI Committee 408 1966). Development lengths are required because of the tendency of highly stressed bars to split relatively thin sections of restraining concrete. A single bar embedded in a mass of concrete should not require as great a development length, although a row of bars, even in mass concrete, can create a weakened plane with longitudinal splitting along the plane of the bars.

In application, the development length concept requires minimum lengths or extensions of reinforcement beyond all points of peak stress in the reinforcement. Such peak stresses generally occur at the points of maximum stress and points where reinforcement is bent or terminated. From a point of peak stress in reinforcement, some length of reinforcement or anchorage is necessary to develop the stress. This development length or anchorage is necessary on both sides of such peak stress points. Often, the reinforcement continues for a considerable distance on one side of a critical stress point so that calculations need involve only the other side, for example, the negative moment reinforcement continuing through a support to the middle of the next span. The requirement for a minimum value of K_{tr} along development and splice lengths in 9.7.1.4, 10.7.1.3, 25.4.2.2, and 25.5.1.5 improves ductility.

R25.4.1.2 Hooks and heads are ineffective in compression. No data are available to demonstrate that hooks and heads can reduce development length in compression.

R25.4.1.3 The strength reduction factor φ is not used in the development length and lap splice length equations. An allowance for strength reduction is already included in the expressions for determining development and splice lengths.

R25.4.1.4 Darwin et al. (1996) shows that the force developed in a bar in development and lap splice tests increases at a lesser rate than $\sqrt{f_c'}$ with increasing compressive strength. Using $\sqrt{f_c'}$, however, is sufficiently accurate for values of $\sqrt{f_c'}$ up to 100 psi, and because of the long-standing use of the $\sqrt{f_c'}$ in design, ACI Committee 318 has chosen not to change the exponent applied to the compressive strength used to calculate development and lap splice lengths, but rather to set an upper limit of 100 psi on $\sqrt{f_c'}$.

R25.4.2 *Development of deformed bars and deformed wires in tension*

R25.4.2.1 This provision gives a two-tier approach for the calculation of tension development length. The user can either use the simplified provisions of 25.4.2.3 or the general development length equation (Eq. (25.4.2.4a)), which is

25 Detailing

CODE

(a) Length calculated in accordance with 25.4.2.3 or 25.4.2.4 using the applicable modification factors of 25.4.2.5

(b) 12 in.

COMMENTARY

based on the expression previously endorsed by ACI 408.1R. In Table 25.4.2.3, ℓ_d is based on two preselected values of $(c_b + K_{tr})/d_b$, whereas ℓ_d from Eq. (25.4.2.4a) is based on the actual $(c_b + K_{tr})/d_b$.

Although there is no requirement for transverse reinforcement along the tension development or lap splice length, research (Azizinamini et al. 1999a,b) indicates that in concrete with very high compressive strength, brittle anchorage failure may occur for bars with inadequate transverse reinforcement. In lap splice tests of No. 8 and No. 11 bars in concrete with an f_c' of approximately 15,000 psi, transverse reinforcement improved ductile anchorage behavior.

25.4.2.2 For bars with $f_y \geq 80{,}000$ **psi** spaced closer than 6 in. on center, transverse reinforcement shall be provided such that K_{tr} shall not be smaller than $0.5d_b$.

25.4.2.3 For deformed bars or deformed wires, ℓ_d shall be calculated in accordance with Table 25.4.2.3.

Table 25.4.2.3—Development length for deformed bars and deformed wires in tension

Spacing and cover	No. 6 and smaller bars and deformed wires	No. 7 and larger bars
Clear spacing of bars or wires being developed or lap spliced not less than d_b, clear cover at least d_b, and stirrups or ties throughout ℓ_d not less than the Code minimum or Clear spacing of bars or wires being developed or lap spliced at least $2d_b$ and clear cover at least d_b	$\left(\dfrac{f_y\psi_t\psi_e\psi_g}{25\lambda\sqrt{f_c'}}\right)d_b$	$\left(\dfrac{f_y\psi_t\psi_e\psi_g}{20\lambda\sqrt{f_c'}}\right)d_b$
Other cases	$\left(\dfrac{3f_y\psi_t\psi_e\psi_g}{50\lambda\sqrt{f_c'}}\right)d_b$	$\left(\dfrac{3f_y\psi_t\psi_e\psi_g}{40\lambda\sqrt{f_c'}}\right)d_b$

R25.4.2.3 This provision recognizes that many current practical construction cases use spacing and cover values along with confining reinforcement, such as stirrups or ties, that result in a value of $(c_b + K_{tr})/d_b$ of at least 1.5. Examples include a minimum clear cover of d_b along with either minimum clear spacing of $2d_b$, or a combination of minimum clear spacing of d_b and minimum ties or stirrups. For these frequently occurring cases, the development length for larger bars can be taken as $\ell_d = [f_y\psi_t\psi_e\psi_g/(20\lambda\sqrt{f_c'})]d_b$. In the formulation of the provisions in ACI 318-95, a comparison with past provisions and a check of a database of experimental results maintained by ACI 408.1R indicated that for No. 6 deformed bars and smaller, as well as for deformed wire, the development lengths could be reduced 20 percent using $\psi_s = 0.8$. This is the basis for the *No. 6 and smaller bars and deformed wires* column of Table 25.4.2.3. With less cover and in the absence of minimum ties or stirrups, the minimum clear spacing limits of 25.2.1 and the minimum concrete cover requirements of 20.5.1.3 result in minimum values of c_b equal to d_b. Thus, for "other cases," the values are based on using $(c_b + K_{tr})/d_b = 1.0$ in Eq. (25.4.2.4a).

The user may easily construct simple, useful expressions. For example, in all members with normalweight concrete ($\lambda = 1.0$), uncoated reinforcement ($\psi_e = 1.0$), No. 7 and larger bottom bars ($\psi_t = 1.0$) with $f_c' = 4000$ **psi**, and Grade 60 reinforcement ($\psi_g = 1.0$), the expressions reduce to

$$\ell_d = \frac{(60{,}000)(1.0)(1.0)(1.0)}{20(1.0)\sqrt{4000}}d_b = 47d_b$$

or

$$\ell_d = \frac{3(60{,}000)(1.0)(1.0)(1.0)}{40(1.0)\sqrt{4000}}d_b = 71d_b$$

Thus, as long as minimum cover of d_b is provided along with a minimum clear spacing of $2d_b$, or a minimum clear

CODE

25.4.2.4 For deformed bars or deformed wires, ℓ_d shall be calculated by:

$$\ell_d = \left(\frac{3}{40} \frac{f_y}{\lambda\sqrt{f_c'}} \frac{\psi_t \psi_e \psi_s \psi_g}{\left(\dfrac{c_b + K_{tr}}{d_b}\right)} \right) d_b \quad (25.4.2.4a)$$

in which the confinement term $(c_b + K_{tr})/d_b$ shall not exceed 2.5, and

$$K_{tr} = \frac{40A_{tr}}{sn} \quad (25.4.2.4b)$$

where n is the number of bars or wires being developed or lap spliced along the plane of splitting. It shall be permitted to use $K_{tr} = 0$ as a design simplification even if transverse reinforcement is present or required.

25.4.2.5 For the calculation of ℓ_d, modification factors shall be in accordance with Table 25.4.2.5.

COMMENTARY

cover of d_b and a minimum clear spacing of d_b are provided along with minimum ties or stirrups, then $\ell_d = 47d_b$. The penalty for spacing bars closer or providing less cover is the requirement that $\ell_d = 71d_b$.

R25.4.2.4 Equation (25.4.2.4a) includes the effects of all variables controlling the development length. In Eq. (25.4.2.4a), c_b is a factor that represents the least of the side cover, the concrete cover to the bar or wire (in both cases measured to the center of the bar or wire), or one-half the center-to-center spacing of the bars or wires. K_{tr} is a factor that represents the contribution of confining reinforcement across potential splitting planes. ψ_t is the reinforcement location factor to reflect the effect of the casting position (that is, formerly denoted as "top bar effect"). ψ_e is a coating factor reflecting the effects of epoxy coating. There is a limit on the product $\psi_t\psi_e$. The reinforcement size factor ψ_s reflects the more favorable performance of smaller-diameter reinforcement. ψ_g is the reinforcement grade factor accounting for the yield strength of the reinforcement. A limit of 2.5 is placed on the term $(c_b + K_{tr})/d_b$. When $(c_b + K_{tr})/d_b$ is less than 2.5, splitting failures are likely to occur. For values above 2.5, a pullout failure is expected, and an increase in cover or transverse reinforcement is unlikely to increase the anchorage capacity.

Many practical combinations of side cover, clear cover, and confining reinforcement can be used with 25.4.2.4 to produce significantly shorter development lengths than allowed by 25.4.2.3. For example, bars or wires with minimum clear cover not less than $2d_b$ and minimum clear spacing not less than $4d_b$ and without any confining reinforcement would have a $(c_b + K_{tr})/d_b$ value of 2.5 and would require a development length of only $28d_b$ for the example in R25.4.2.3.

Before ACI 318-08, Eq. (25.4.2.4b) for K_{tr} included the yield strength of transverse reinforcement. The current expression includes only the area and spacing of the transverse reinforcement and the number of wires or bars being developed or lap spliced because tests demonstrate that transverse reinforcement rarely yields during a bond failure (Azizinamini et al. 1995).

Terms in Eq. (25.4.2.4a) may be disregarded if such omission results in longer and, hence, more conservative, development lengths.

R25.4.2.5 The lightweight factor λ for calculating development length of deformed bars and deformed wire in tension is the same for all types of lightweight concrete. Research does not support the variations of this factor in Codes prior to 1989 for all-lightweight and sand-lightweight concrete (ACI 408R).

The reinforcement grade factor ψ_g accounts for the effect of reinforcement yield strength on required development length. Research has shown that required development length increases disproportionately with increases in yield strength (Orangun et al. 1977; Canbay and Frosch 2005).

The epoxy factor ψ_e is based on studies (Treece and Jirsa 1989; Johnston and Zia 1982; Mathey and Clifton 1976) of

CODE

Table 25.4.2.5—Modification factors for development of deformed bars and deformed wires in tension

Modification factor	Condition	Value of factor
Lightweight λ	Lightweight concrete	0.75
	Normalweight concrete	1.0
Reinforcement grade ψ_g	Grade 40 or Grade 60	1.0
	Grade 80	1.15
	Grade 100	1.3
Epoxy[1] ψ_e	Epoxy-coated or zinc and epoxy dual-coated reinforcement with clear cover less than $3d_b$ or clear spacing less than $6d_b$	1.5
	Epoxy-coated or zinc and epoxy dual-coated reinforcement for all other conditions	1.2
	Uncoated or zinc-coated (galvanized) reinforcement	1.0
Size ψ_s	No. 7 and larger bars	1.0
	No. 6 and smaller bars and deformed wires	0.8
Casting position[1] ψ_t	More than 12 in. of fresh concrete placed below horizontal reinforcement	1.3
	Other	1.0

[1]The product $\psi_t\psi_e$ need not exceed 1.7.

COMMENTARY

the anchorage of epoxy-coated bars that show bond strength is reduced because the coating prevents adhesion and lowers the coefficient of friction between the bar and the concrete. The factors reflect the type of anchorage failure likely to occur. If the cover or spacing is small, a splitting failure can occur and the anchorage or bond strength is substantially reduced. If the cover and spacing between bars is large, a splitting failure is precluded and the effect of the epoxy coating on anchorage strength is not as large. Studies (Orangun et al. 1977) have shown that although the cover or spacing may be small, the anchorage strength may be increased by adding transverse reinforcement crossing the plane of splitting, and restraining the splitting crack.

Because the bond of epoxy-coated bars or zinc and epoxy dual-coated bars is already reduced due to the loss of adhesion and lower coefficient of friction between the bar and the concrete, an upper limit of 1.7 is established for the product of the factors for top reinforcement casting position and epoxy-coated reinforcement or zinc and epoxy dual-coated reinforcement.

The reinforcement size factor ψ_s reflects the more favorable performance of smaller-diameter reinforcement.

The reinforcement location or casting position factor ψ_t accounts for the position of the reinforcement in freshly placed concrete. The factor 1.3 is based on research (Jirsa and Breen 1981; Jeanty et al. 1988). The application of the casting position factor should be considered in determination of development lengths for inclined reinforcement.

25.4.3 *Development of standard hooks in tension*

25.4.3.1 Development length ℓ_{dh} for deformed bars in tension terminating in a standard hook shall be the greater of (a) through (c):

(a) $\left(\dfrac{f_y\psi_e\psi_r\psi_o\psi_c}{55\lambda\sqrt{f_c'}}\right)d_b^{1.5}$ with ψ_e, ψ_r, ψ_o, ψ_c, and λ given in 25.4.3.2

(b) $8d_b$

(c) 6 in.

25.4.3.2 For the calculation of ℓ_{dh}, modification factors ψ_e, ψ_r, ψ_o, ψ_c, and λ shall be in accordance with Table 25.4.3.2. At discontinuous ends of members, 25.4.3.4 shall apply.

R25.4.3 *Development of standard hooks in tension*

R25.4.3.1 The provisions for hooked bars are only applicable to standard hooks (refer to 25.3.1). The development length ℓ_{dh} is measured from the critical section to the outside end (or edge) of the hook.

In research by Sperry et al. (2017a), concrete breakout failure was the predominant failure mode of hooked bars. Closely-spaced hooks provide a lower strength per hooked bar than more widely-spaced hooked bars because the area of the breakout surface is reduced for the more closely-spaced bars (Ajaam et al. 2018). For bars located adjacent to the side of a member, the percentage of hooked bars exhibiting splitting failure increased with increasing bar size.

The effects of bar yield strength, spacing, and confinement by ties or stirrups have been updated to reflect test results (Sperry et al. 2018). The minimum values of ℓ_{dh} are specified to prevent failure by direct pullout in cases where a hook may be located near the critical section. Hooks in beam-column joints and corbels should be placed as close as practical to the back face of the joint.

R25.4.3.2 Unlike straight bar development, no distinction is made for casting position.

CODE

Table 25.4.3.2—Modification factors for development of hooked bars in tension

Modification factor	Condition	Value of factor
Lightweight λ	Lightweight concrete	0.75
	Normalweight concrete	1.0
Epoxy ψ_e	Epoxy-coated or zinc and epoxy dual-coated reinforcement	1.2
	Uncoated or zinc-coated (galvanized) reinforcement	1.0
Confining reinforcement ψ_r	For No. 11 and smaller bars with $A_{th} \geq 0.4A_{hs}$ or s[1] $\geq 6d_b$[2]	1.0
	Other	1.6
Location ψ_o	For No. 11 and smaller diameter hooked bars: (1) Terminating inside column core with side cover normal to plane of hook $\geq$ 2.5 in., or (2) With side cover normal to plane of hook $\geq 6d_b$	1.0
	Other	1.25
Concrete strength ψ_c	For $f_c' < 6000$ psi	$f_c'/15{,}000 + 0.6$
	For $f_c' \geq 6000$ psi	1.0

[1] s is minimum center-to-center spacing of hooked bars.
[2] d_b is nominal diameter of hooked bar.

COMMENTARY

The epoxy factor ψ_e is based on tests (Hamad et al. 1993) that indicate the development length for hooked bars should be increased by 20 percent to account for reduced bond when reinforcement is epoxy coated. The location factor ψ_o is based on tests (Johnson and Jirsa 1981; Sperry et al. 2017a,b) demonstrating that the development length of hooked bars anchored within a column core with side cover less than 2.5 in. or in other members with side cover less than $6d_b$ needs to be 25 percent longer than in similar members with larger cover.

The confining reinforcement factor ψ_r is based on test results reported by Ajaam et al. (2018). A value of 1.0 is used for ψ_r for widely-spaced hooked bars, $s \geq 6d_b$, and for hooked bars with $A_{th}/A_{hs} \geq 0.4$. Where bars are closely spaced or $A_{th}/A_{hs} < 0.4$, the confinement factor is 1.6. Because no test results are available for No. 14 and No. 18 bars, the values of ψ_r for hooked bars larger than No. 11 are the same as those for No. 11 and smaller diameter hooked bars without confining reinforcement. No tests were performed to verify extrapolation to large bars in concrete with strengths greater than 10,000 psi. When calculated using 25.4.3.1(a) and the factors in 25.4.3.2, development lengths are, however, as much as 50 percent longer than required by Codes prior to ACI 318-19.

Detailing 25

25.4.3.3 The total cross-sectional area of ties or stirrups confining hooked bars A_{th} shall consist of (a) or (b):

(a) Ties or stirrups that enclose the hook and satisfy 25.3.2.
(b) Other reinforcement enclosing the hook, that extends at least $0.75\ell_{dh}$ from the enclosed hook in the direction of the bar in tension, and is in accordance with (1) or (2). For members with confining reinforcement that is both parallel and perpendicular to ℓ_{dh}, it shall be permitted to use the value of A_{th} based on (1) or (2) that results in the lower value of ℓ_{dh}.

(1) Two or more ties or stirrups shall be provided parallel to ℓ_{dh} enclosing the hooks, evenly distributed with a center-to-center spacing not exceeding $8d_b$, and within $15d_b$ of the centerline of the straight portion of the hooked bars, where d_b is the nominal diameter of the hooked bar.

(2) Two or more ties or stirrups shall be provided perpendicular to ℓ_{dh}, enclosing the hooked bars, and evenly distributed along ℓ_{dh} with a center-to-center spacing not exceeding $8d_b$, where d_b is the nominal diameter of the hooked bar.

R25.4.3.3 Distribution of confining reinforcement is shown in Fig. R25.4.3.3a and 25.4.3.3b. Figure R25.4.3.3a shows placement of ties or stirrups parallel to the bar being developed along the length of the tail extension of the hook plus bend. This configuration would be typical in a beam-column joint. Tests show that confining reinforcement oriented parallel or perpendicular to the development length of the hooked bar, and located within the regions defined in 25.4.3.3 (a) or (b), contributes to anchorage strength in proportion to the area of the confining reinforcement for both 90- and 180-degree hooks (Sperry et al. 2017b). Figure R25.4.3.3b shows placement of ties or stirrups perpendicular to the bar being developed, spaced along $0.75\ell_{dh}$ of the hook. Tests used to establish these criteria were based on beam-column joints with perimeter ties and stirrups only (Sperry et al. 2017a; Ajaam et al. 2018). Both legs of individual stirrups and individual ties contribute to A_{th}.

$$\ell_{dh} = \frac{0.2\psi_e f_y}{\lambda\sqrt{f'_c}} d_b$$

d_b = bar diameter

CODE

COMMENTARY

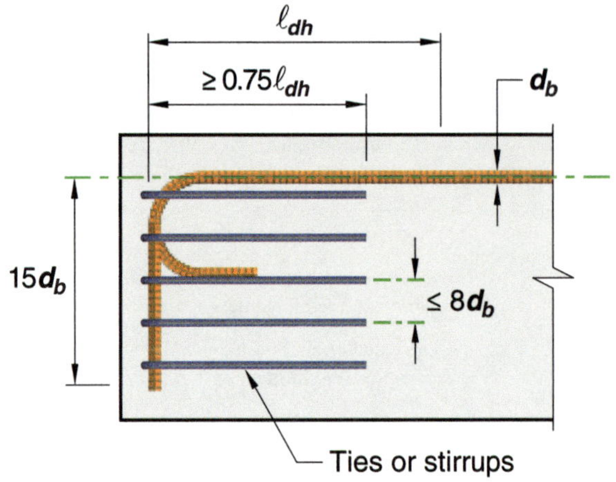

Fig. R25.4.3.3a—Confining reinforcement placed parallel to the bar being developed that contributes to anchorage strength of both 90- and 180-degree hooked bars.

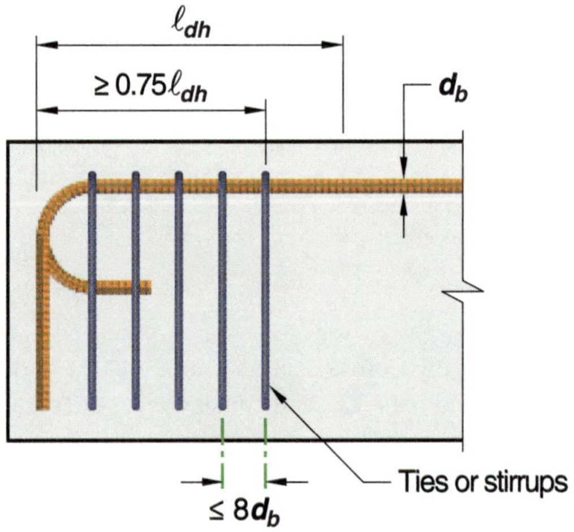

Fig. R25.4.3.3b—Confining reinforcement placed perpendicular to the bar being developed, spaced along the development length ℓ_{dh}, *that contributes to anchorage strength of both 90- and 180-degree hooked bars.*

25.4.3.4 For bars being developed by a standard hook at discontinuous ends of members with both side cover and top (or bottom) cover to hook less than 2-1/2 in., (a) and (b) shall be satisfied:

(a) The hook shall be enclosed along ℓ_{dh} within ties or stirrups perpendicular to ℓ_{dh} at $s \leq 3d_b$

(b) The first tie or stirrup shall enclose the bent portion of the hook within $2d_b$ of the outside of the bend

where d_b is the nominal diameter of the hooked bar.

R25.4.3.4 Hoooked bars are especially susceptible to a concrete splitting failure if both side cover (perpendicular to plane of hook) and top or bottom cover (in plane of hook) are small (refer to Fig. R25.4.3.4). Transverse reinforcement is required to provide additional splitting resistance. This provision applies at ends of simply-supported beams, at the free end of cantilevers, and at exterior joints for members framing into a joint where members do not extend beyond the joint. This provision does not apply for hooked bars at discontinuous ends of slabs where confinement is provided by the slab on both sides, perpendicular to the plane of the hook.

CODE

COMMENTARY

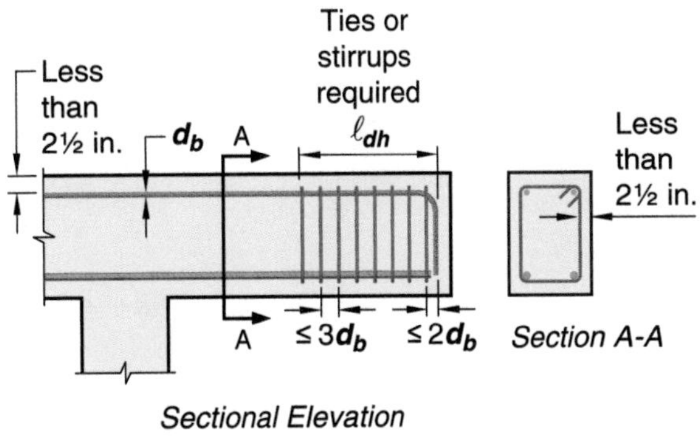

Fig. R25.4.3.4—Concrete cover according to 25.4.3.4.

25.4.4 *Development of headed deformed bars in tension*

25.4.4.1 Use of a head to develop a deformed bar in tension shall be permitted if conditions (a) through (f) are satisfied:

(a) Bar shall conform to 20.2.1.6
(b) Bar size shall not exceed No. 11
(c) Net bearing area of head A_{brg} shall be at least $4A_b$
(d) Concrete shall be normalweight
(e) Clear cover for bar shall be at least $2d_b$
(f) Center-to-center spacing between bars shall be at least $3d_b$

25.4.4.2 Development length ℓ_{dt} for headed deformed bars in tension shall be the longest of (a) through (c):

(a) $\left(\dfrac{f_y\psi_e\psi_p\psi_o\psi_c}{75\sqrt{f_c'}}\right)d_b^{1.5}$ with ψ_e, ψ_p, ψ_o, and ψ_c, given in 25.4.4.3
(b) $8d_b$
(c) 6 in.

R25.4.4 *Development of headed deformed bars in tension*

R25.4.4.1 As used in this section, development describes cases in which the force in the bar is transferred to the concrete through a combination of a bearing force at the head and bond forces along the bar. In contrast, Chapter 17 anchorage provisions describe cases in which the force in the bar is transferred through bearing to the concrete at the head alone. Headed bars are limited to those types that meet the criteria in 20.2.1.6 for Class HA heads.

The provisions for headed deformed bars were formulated with due consideration of the provisions for anchorage in Chapter 17 (Shao et al. 2016). Chapter 17 contains provisions for headed anchors related to the individual failure modes of concrete breakout, side-face blowout, and pullout. These failure modes were considered in the formulation of 25.4.4.2. The restrictions to maximum bar size of No. 11 and normalweight concrete are based on a lack of data for larger bars or lightweight concrete (Thompson et al. 2005, 2006a,b; Shao et al. 2016). The upper limit of 60,000 psi on f_y that appeared prior to the 2019 Code has been removed.

For bars in tension, heads allow the bars to be developed in a shorter length than required for standard hooks, but otherwise perform in a similar manner (Thompson et al. 2005, 2006a,b; Shao et al. 2016). The head is considered to be part of the bar for the purposes of satisfying the specified cover requirements in 20.5.1.3 and aggregate size requirements of 26.4.2.1(a)(5).

Headed bars with $A_{brg} < 4A_b$ have been used in practice, but their performance is not accurately represented by the provisions in 25.4.4.2, and they should be used only with designs that are supported by test results under 25.4.5. These provisions do not address the design of studs or headed stud assemblies used for shear reinforcement.

R25.4.4.2 The provisions for developing headed deformed bars give the length of bar, ℓ_{dt}, measured from the critical section to the bearing face of the head, as shown in Fig. R25.4.4.2a. The provisions are primarily based on tests of simulated beam-column joints (Shao et al. 2016).

If longitudinal headed deformed bars from a beam, slab, or corbel terminate in a supporting member, such as the column shown in Fig. R25.4.4.2b, the bars should extend through the joint to the far face of the confined core of the

Detailing 25

CODE

COMMENTARY

supporting member, allowing for cover and avoidance of interference with column reinforcement, even though the resulting anchorage length may exceed ℓ_{dt}. Extending the bar to the far side of the column core helps engage the entire joint in resisting the anchorage forces and thereby improves the performance of the joint.

If closely spaced headed bars are used, the potential for concrete breakout failure exists. For joints as shown in Fig. R25.4.4.2c and R25.4.4.2d, anchorage strengths will be generally higher if the anchorage length is equal to or greater than $d/1.5$ (Eligehausen 2006b), as shown in Fig. R25.4.4.2c, or by providing reinforcement in the form of hoops and ties to establish a load path in accordance with strut-and-tie modeling principles, as shown in Fig. R25.4.4.2d. Strut-and-tie models should be verified in accordance with Chapter 23. Note that the strut-and-tie models illustrated in Fig. R25.4.4.2c and R25.4.4.2d rely on a vertical strut from a column extending above the joint. Beam-column joints at roof-level and portal frames are vulnerable to joint failure and should be properly detailed to restrain diagonal cracking through the joint and breakout of the bars through the top surface.

For cases where development length cannot be designed in accordance with 25.4.4.2, use of the provisions of Chapter 17 should be considered.

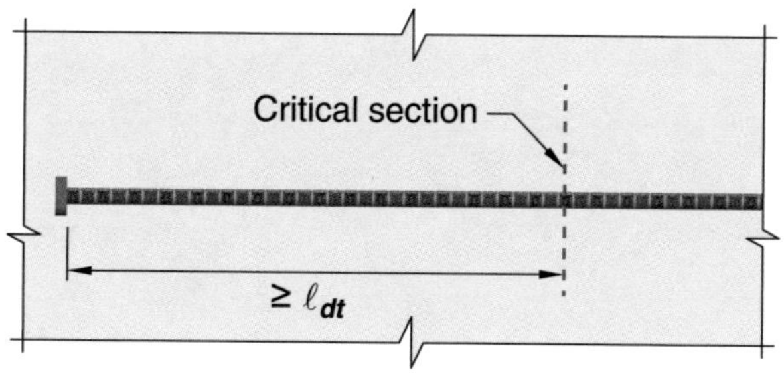

Fig. R25.4.4.2a*—Development of headed deformed bars.*

CODE

COMMENTARY

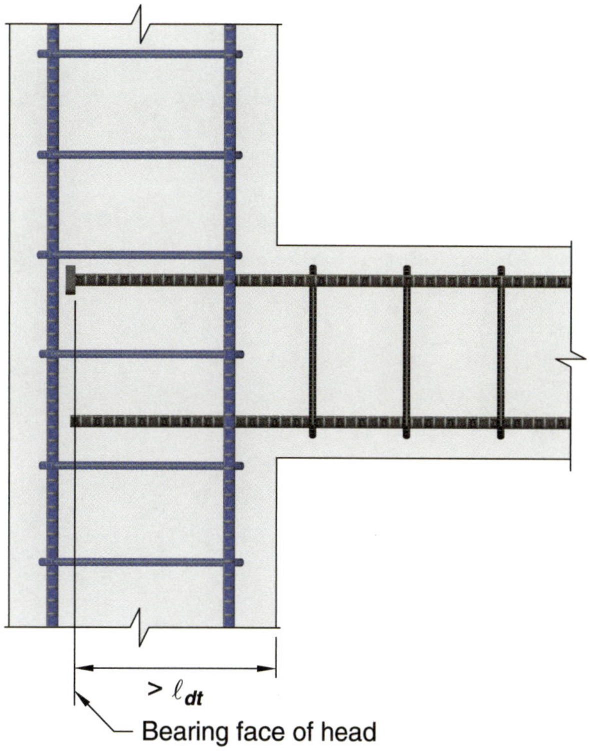

Fig. R25.4.4.2b—Headed deformed bar extended to far side of column core with anchorage length that exceeds ℓ_{dt}.

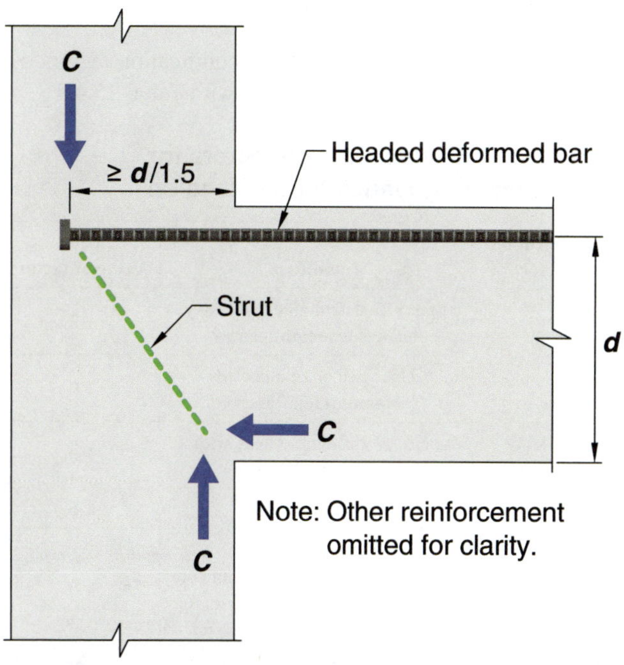

Fig. R25.4.4.2c—Breakout failure precluded in joint by keeping anchorage length greater than or equal to **d/1.5**.

CODE

COMMENTARY

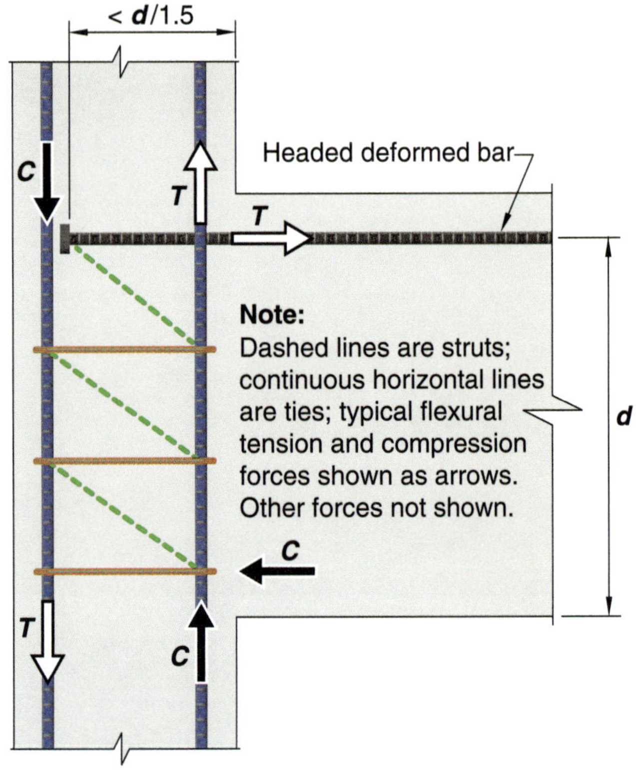

Fig. R25.4.4.2d—Breakout failure precluded in joint by providing transverse reinforcement to enable a strut-and-tie mechanism.

25.4.4.3 For the calculation of ℓ_{dt}, modification factors ψ_e, ψ_p, ψ_o, and ψ_c shall be in accordance with Table 25.4.4.3.

Table 25.4.4.3—Modification factors for development of headed bars in tension

Modification factor	Condition	Value of factor
Epoxy ψ_e	Epoxy-coated or zinc and epoxy dual-coated reinforcement	1.2
	Uncoated or zinc-coated (galvanized) reinforcement	1.0
Parallel tie reinforcement ψ_p	For No. 11 and smaller bars with $A_{tt} \geq 0.3A_{hs}$ or $s^{[1]} \geq 6d_b^{[2,3]}$	1.0
	Other	1.6
Location ψ_o	For headed bars: (1) Terminating inside column core with side cover to bar ≥ 2.5 in.; or (2) With side cover to bar $\geq 6d_b$	1.0
	Other	1.25
Concrete strength ψ_c	For $f_c' < 6000$ psi	$f_c'/15{,}000 + 0.6$
	For $f_c' \geq 6000$ psi	1.0

[1]s is minimum center-to-center spacing of headed bars.

[2]d_b is nominal diameter of headed bar.

[3]Refer to 25.4.4.5.

R25.4.4.3 The epoxy factor 1.2 is based conservatively on the value used for epoxy-coated standard hooks. The location factor ψ_o accounts for the confinement provided by the reinforcement within columns and large side cover for other members.

The factor ψ_p for headed reinforcement is similar to the confining reinforcement factor for hooked bars (Shao et al. 2016). Unlike hooked bars, however, test results indicate that only tie or hoop reinforcement parallel to headed bars contributes to anchorage strength and reduces development length (Thompson et al. 2005, 2006a,b).

CODE

25.4.4.4 For beam column joints, the total cross-sectional area of parallel tie reinforcement A_{tt} shall consist of ties or stirrups oriented parallel to ℓ_{dt} and located within $8d_b$ of the centerline of the headed bar toward the middle of the joint, where d_b is the nominal diameter of the headed bar.

COMMENTARY

R25.4.4.4 Reinforcement oriented parallel to the development length of the headed bars, located within the region defined in 25.4.4.4 (Fig. R25.4.4.4) contributes to anchorage strength in proportion to its area (Shao et al. 2016). This reinforcement serves to tie concrete near the head to concrete on the other side of the failure surface, thus mobilizing additional anchorage strength. With the exception of vertical joint reinforcement in the form of stirrups that are well anchored to the far side of the joint, reinforcement oriented perpendicular to the development length has been shown in a number of cases to be ineffective in improving the anchorage of headed deformed bars (Thompson et al. 2005, 2006a,b). Both legs of individual stirrups and ties parallel to the headed bars contribute to A_{tt}.

CODE

COMMENTARY

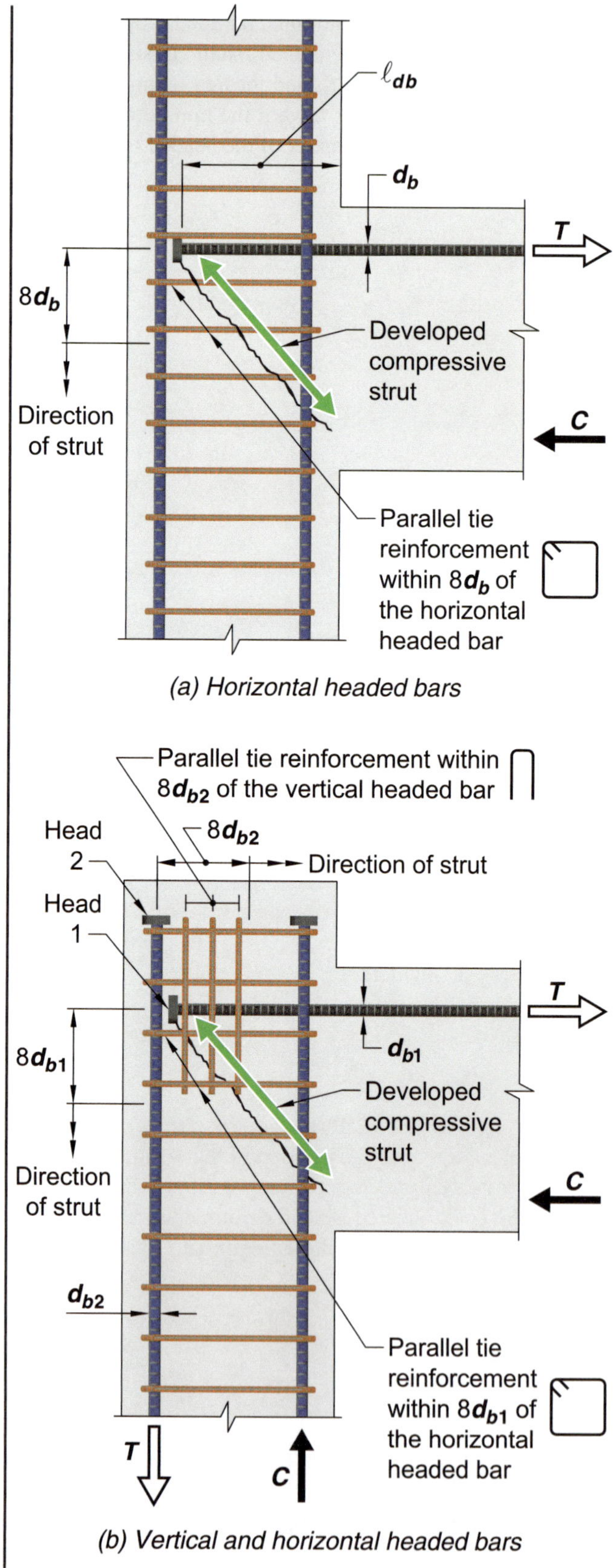

***Fig. R25.4.4.4**—Ties or stirrups placed parallel to the headed beam bars being developed in a beam-column joint that contribute to anchorage strength.*

CODE

25.4.4.5 For anchorages other than in beam-column joints, tie reinforcement, A_{tt}, shall not be considered, and ψ_p shall be taken as 1.0 provided the spacing is at least $6d_b$.

25.4.4.6 If beam negative moment reinforcement is provided by headed deformed bars that terminate in a joint, the column shall extend above the top of the joint a distance at least the depth h of the joint, where h is the horizontal dimension of the joint in the direction of the forces being considered. Alternatively, the beam reinforcement shall be enclosed by additional vertical joint reinforcement providing equivalent confinement to the top face of the joint.

25.4.5 *Development of mechanically anchored deformed bars in tension*

25.4.5.1 Any mechanical attachment or device capable of developing f_y of deformed bars shall be permitted, provided it is approved by the building official in accordance with 1.10. Development of deformed bars shall be permitted to consist of a combination of mechanical anchorage plus additional embedment length of the deformed bars between the critical section and the mechanical attachment or device.

25.4.6 *Development of welded deformed wire reinforcement in tension*

25.4.6.1 Development length ℓ_d for welded deformed wire reinforcement in tension measured from the critical section to the end of wire shall be the greater of (a) and (b), where wires in the direction of the development length shall all be deformed D31 or smaller.

(a) Length calculated in accordance with 25.4.6.2
(b) 8 in.

25.4.6.2 For welded deformed wire reinforcement, ℓ_d shall be calculated from 25.4.2.3 or 25.4.2.4, times welded deformed wire reinforcement factor ψ_w from 25.4.6.3 or 25.4.6.4. For epoxy-coated welded deformed wire reinforcement meeting 25.4.6.3, it shall be permitted to use $\psi_e = \mathbf{1.0}$ in 25.4.2.3 or 25.4.2.4.

25.4.6.3 For welded deformed wire reinforcement with at least one cross wire within ℓ_d that is at least 2 in. from the critical section, ψ_w shall be the greater of (a) and (b), and need not exceed 1.0:

(a) $\left(\dfrac{f_y - 35{,}000}{f_y}\right)$

COMMENTARY

R25.4.4.5 No evidence is available regarding the effect of parallel reinforcement on the development length of headed bars except in beam-column joints.

R25.4.4.6 Refer to R18.4.4.5.

R25.4.5 *Development of mechanically anchored deformed bars in tension*

R25.4.5.1 Anchorage of deformed bars through the use of mechanical devices within concrete that do not meet the requirements in 20.2.1.6, or are not developed in accordance with 25.4.4, may be used if tests demonstrate the ability of the head and bar system to develop or anchor the desired force in the bar, as described in this provision.

R25.4.6 *Development of welded deformed wire reinforcement in tension*

R25.4.6.1 ASTM A1064 for welded deformed wire reinforcement requires the same strength of the weld as required for welded plain wire reinforcement. Some of the development is assigned to welds and some assigned to the length of deformed wire.

R25.4.6.2 The welded deformed wire reinforcement factor ψ_w is applied to the deformed wire development length calculated from 25.4.2.3 or 25.4.2.4.

Tests (Bartoletti and Jirsa 1995) have indicated that epoxy-coated welded deformed wire reinforcement has essentially the same development and splice strengths as uncoated welded deformed wire reinforcement because the cross wires provide the primary anchorage for the wire. Therefore, ψ_e of 1.0 is used for development and splice lengths of epoxy-coated welded deformed wire reinforcement with cross wires within the splice or development length.

R25.4.6.3 Figure R25.4.6.3 shows the development requirements for welded deformed wire reinforcement with one cross wire within the development length.

Detailing 25

CODE

(b) $\left(\frac{5d_b}{s}\right)$

where s is the spacing between the wires to be developed.

25.4.6.4 For welded deformed wire reinforcement with no cross wires within ℓ_d or with a single cross wire less than 2 in. from the critical section, ψ_w shall be taken as 1.0.

25.4.6.5 Where any plain wires, or deformed wires larger than D31, are present in the welded deformed wire reinforcement in the direction of the development length, the reinforcement shall be developed in accordance with 25.4.7.

25.4.6.6 Zinc-coated (galvanized) welded deformed wire reinforcement shall be developed in accordance with 25.4.7.

25.4.7 *Development of welded plain wire reinforcement in tension*

25.4.7.1 Development length ℓ_d for welded plain wire reinforcement in tension measured from the critical section to the outermost cross wire shall be the greater of (a) and (b) and shall require a minimum of two cross wires within ℓ_d.

(a) Length calculated in accordance with 25.4.7.2
(b) 6 in.

25.4.7.2 ℓ_d shall be the greater of (a) and (b):
(a) spacing of cross wires + 2 in.

(b) $0.27\left(\frac{f_y}{\lambda\sqrt{f_c'}}\right)\left(\frac{A_b}{s}\right)$, where s is the spacing between the wires to be developed, and λ is given in Table 25.4.2.5.

COMMENTARY

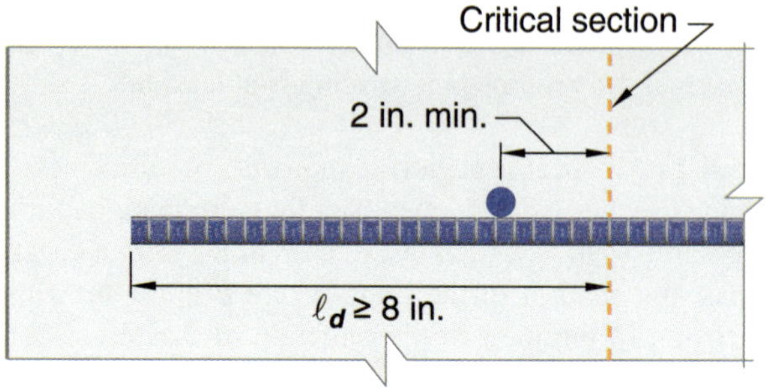

Fig. R25.4.6.3—Development of welded deformed wire reinforcement.

R25.4.6.5 Deformed wire larger than D31 is treated as plain wire because tests show that D45 wire will achieve only approximately 60 percent of the bond strength in tension given by Eq. (25.4.2.4a) (Rutledge and DeVries 2002).

R25.4.7 *Development of welded plain wire reinforcement in tension*

R25.4.7.1 ASTM A1064 for welded plain wire reinforcement requires the same strength of the weld as required for welded deformed wire reinforcement. All of the development is assigned to the welded cross wires; consequently, welded plain wire reinforcement requires at least two cross wires.

R25.4.7.2 Figure R25.4.7.2 shows the development requirements for welded plain wire reinforcement with development primarily dependent on the location of cross wires.

For welded plain wire reinforcement made with small wires, an embedment of at least two cross wires 2 in. or more beyond the point of critical section is adequate to develop the full yield strength of the anchored wires. However, for welded plain wire reinforcement made with larger closely spaced wires, a longer embedment is required with the development length controlled by 25.4.7.2(b).

CODE | COMMENTARY

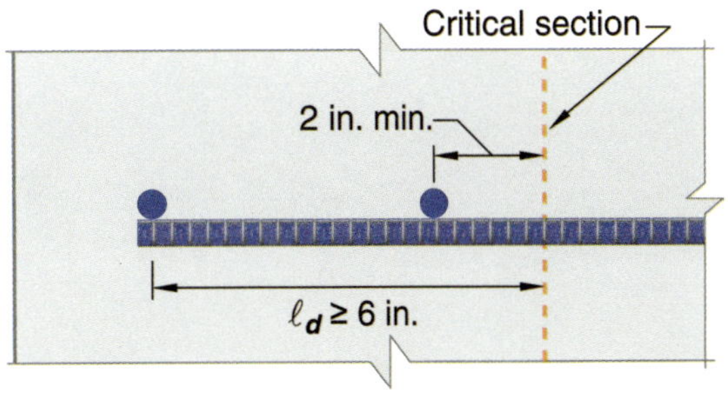

***Fig. R25.4.7.2**—Development of welded plain wire reinforcement.*

25.4.8 *Development of pretensioned seven-wire strands in tension*

R25.4.8 *Development of pretensioned seven-wire strands in tension*

Development requirements for pretensioned strand are intended to provide bond integrity for the strength of the member. Provisions are based on tests performed on normal-weight concrete members with a minimum cover of 2 in. These tests may not represent the behavior of strand in no-slump concrete. Concrete placement operations should ensure consolidation of concrete around the strand with complete contact between the steel and concrete.

The bond of strand is a function of a number of factors, including the configuration and surface condition of the steel, the stress in the steel, the depth of concrete beneath the strand, and the method used to transfer the force in the strand to the concrete. For bonded applications, quality assurance procedures should be used to confirm that the strand is capable of adequate bond (Rose and Russell 1997; Logan 1997). The precast concrete manufacturer may rely on certification from the strand manufacturer that the strand has bond characteristics that comply with this section.

This section does not apply to plain wires, to end-anchored tendons, or to unstressed strand. The development length for plain wire could be considerably greater due to the absence of mechanical interlock. Flexural bond failure would occur with plain wire when first slip occurred. Nontensioned prestressing steel is sometimes used as integrity reinforcement in precast concrete structures; however, there are limited data available regarding the bond length required to ensure development of the yield strength of the reinforcement (Salmons and McCrate 1977; PCA 1980).

25.4.8.1 Development length ℓ_d of pretensioned seven-wire strands in tension shall be in accordance with (a) and (b):

(a) $$\ell_d = \left(\frac{f_{se}}{3000}\right)d_b + \left(\frac{f_{ps} - f_{se}}{1000}\right)d_b \quad (25.4.8.1)$$

(b) If bonding of a strand does not extend to end of member, and design includes tension at service loads in the precompressed tension zone, ℓ_d calculated by Eq. (25.4.8.1) shall be doubled.

R25.4.8.1 The first term in Eq. (25.4.8.1) represents the transfer length of the strand, that is, the distance over which the strand should be bonded to the concrete to develop the effective prestress in the prestressed reinforcement, f_{se}. The second term represents the additional length over which the strand should be bonded so that the stress in the prestressed reinforcement at nominal strength of the member, f_{ps}, may develop.

Exploratory tests (Kaar and Magura 1965) that studied the effect of debonded strand (bond not permitted to extend to the ends of members) on performance of pretensioned girders indicated that the performance of these girders with

25 Detailing

CODE

COMMENTARY

embedment lengths twice those required by Eq. (25.4.8.1) closely matched the flexural performance of similar pretensioned girders with strand fully bonded to ends of girders. Accordingly, twice the development length is required for strand not bonded through to the end of a member. Subsequent tests (Rabbat et al. 1979) indicated that in pretensioned members designed for zero tension in the concrete under service load conditions (refer to 24.5.2), the development length for debonded strands need not be increased by a factor of 2. For analysis of sections with debonded strands at locations where strand is not fully developed, the procedure outlined in 21.2.3 is provided.

25.4.8.2 Seven-wire strand shall be bonded at least ℓ_d beyond the critical section except as provided in 25.4.8.3.

25.4.8.3 Embedment less than ℓ_d shall be permitted at a section of a member, provided the design strand stress at that section does not exceed values obtained from the bilinear relationship defined by Eq. (25.4.8.1).

R25.4.8.3 Figure R25.4.8.3 shows the relationship between steel stress and the distance over which the strand is bonded to the concrete represented by Eq. (25.4.8.1). This idealized variation of strand stress may be used for analyzing sections within the development region (Martin and Korkosz 1995; PCI MNL 120). The expressions for transfer length and for the additional bonded length necessary to develop an increase in stress of $(f_{ps} - f_{se})$ are based on tests of members prestressed with clean, 1/4, 3/8, and 1/2 in. diameter strands for which the maximum value of f_{ps} was 275,000 psi (Kaar and Magura 1965; Hanson and Kaar 1959; Kaar et al. 1963).

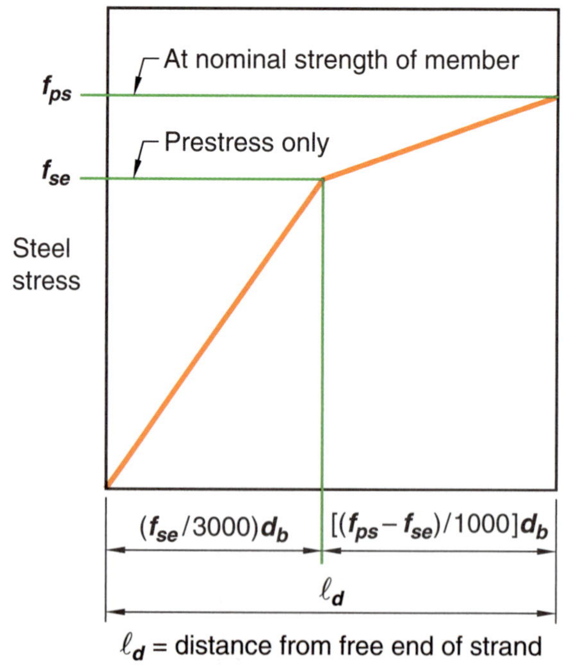

Fig. R25.4.8.3—Idealized bilinear relationship between steel stress and distance from the free end of strand.

dℓc

CODE

25.4.9 *Development of deformed bars and deformed wires in compression*

25.4.9.1 Development length ℓ_{dc} for deformed bars and deformed wires in compression shall be the greater of (a) and (b)

(a) Length calculated in accordance with 25.4.9.2
(b) 8 in.

25.4.9.2 ℓ_{dc} shall be the greater of (a) and (b), using the modification factors of 25.4.9.3:

(a) $\left(\dfrac{f_y\psi_r}{50\lambda\sqrt{f_c'}}\right)d_b$

(b) $0.0003 f_y \psi_r d_b$

25.4.9.3 For the calculation of ℓ_{dc}, modification factors shall be in accordance with Table 25.4.9.3, except ψ_r shall be permitted to be taken as 1.0.

Table 25.4.9.3—Modification factors for deformed bars and wires in compression

Modification factor	Condition	Value of factor
Lightweight λ	Lightweight concrete	0.75
	Normalweight concrete	1.0
Confining reinforcement ψ_r	Reinforcement enclosed within (1), (2), (3), or (4): (1) a spiral (2) a circular continuously wound tie with $d_b \geq 1/4$ in. and pitch 4 in. (3) No. 4 bar or D20 wire ties in accordance with 25.7.2 spaced ≤ 4 in. on center (4) hoops in accordance with 25.7.4 spaced ≤ 4 in. on center	0.75
	Other	1.0

25.4.10 *Reduction of development length for excess reinforcement*

25.4.10.1 Reduction of development lengths defined in 25.4.2.1(a), 25.4.6.1(a), 25.4.7.1(a), and 25.4.9.1(a) shall be permitted by use of the ratio $(A_{s,required})/(A_{s,provided})$, except where prohibited by 25.4.10.2. The modified development lengths shall not be less than the respective minimums specified in 25.4.2.1(b), 25.4.6.1(b), 25.4.7.1(b), and 25.4.9.1(b).

25.4.10.2 A reduction of development length in accordance with 25.4.10.1 is not permitted for (a) through (f)

(a) At noncontinuous supports

COMMENTARY

R25.4.9 *Development of deformed bars and deformed wires in compression*

R25.4.9.1 The weakening effect of flexural tension cracks is not present for bars and wires in compression, and usually end bearing of the bars on the concrete is beneficial. Therefore, shorter development lengths are specified for compression than for tension.

R25.4.9.2 The constant 0.0003 has units of in.^{2}/lb.

The term λ is provided in the expression for development in 25.4.9.2 recognizing that there are no known test data on compression development in lightweight concrete but that splitting is more likely in lightweight concrete.

R25.4.9.3 The development length may be reduced 25 percent when the reinforcement is enclosed within closely spaced spirals, ties, or hoops.

R25.4.10 *Reduction of development length for excess reinforcement*

R25.4.10.1 A reduction in development length is permitted in limited circumstances if excess reinforcement is provided.

R25.4.10.2 The excess reinforcement factor $(A_{s,required}/A_{s,provided})$, applicable to straight reinforcement is not applicable for hooked or headed bars where force is transferred through a combination of bearing at the hook or head and bond along the bar. Concrete breakout due to bearing at a

Detailing 25

CODE

(b) At locations where anchorage or development for f_y is required
(c) Where bars are required to be continuous
(d) For hooked, headed, and mechanically anchored deformed reinforcement
(e) In seismic-force-resisting systems in structures assigned to Seismic Design Categories C, D, E, or F
(f) Anchorage of concrete piles and concrete filled pipe piles to pile caps in structures assigned to Seismic Design Categories C, D, E, or F

COMMENTARY

hook or head was considered in developing the provisions of 25.4.3 and 25.4.4. Because the anchorage strength, and in particular the concrete breakout strength of a hooked or headed bar is a function of the embedment depth to a power slightly more than 1.0 (Shao et al. 2016; Sperry et al. 2017b), a reduction in development length with the application of the excess reinforcement factor could result in a potential concrete breakout failure.

Where a flexural member is part of the seismic-force-resisting-system, loads greater than those anticipated in design may cause reversal of moment at supports; some positive reinforcement should be fully developed into the support. This anchorage is required to ensure ductile response in the event of serious overstress, such as from earthquake or blast. It is not sufficient to use more reinforcement at lower stresses.

The reduction factor based on area is not to be used in those cases where anchorage development for full f_y is required. For example, the excess reinforcement factor does not apply for development of shrinkage and temperature reinforcement according to 24.4.3.4 or for development of reinforcement provided according to 7.7.7, 8.7.4.2, 8.8.1.6, 9.7.7, and 9.8.1.6.

25.5—Splices

25.5.1 *General*

25.5.1.1 Lap splices shall not be permitted for bars larger than No. 11, except as provided in 25.5.5.3.

25.5.1.2 For contact lap splices, minimum clear spacing between the contact lap splice and adjacent splices or bars shall be in accordance with the requirements for individual bars in 25.2.1.

25.5.1.3 For noncontact splices in flexural members, the transverse center-to-center spacing of spliced bars shall not exceed the lesser of one-fifth the required lap splice length and 6 in.

25.5.1.4 Reduction of development length in accordance with 25.4.10.1 is not permitted in calculating lap splice lengths.

R25.5—Splices

R25.5.1 *General*

Lap splice lengths of longitudinal reinforcement in columns should be calculated in accordance with 10.7.5, 18.7.4.4, and this section.

R25.5.1.1 Because of lack of adequate experimental data on lap splices of No. 14 and No. 18 bars in compression and in tension, lap splicing of these bar sizes is prohibited except as permitted in 25.5.5.3 for compression lap splices of No. 14 and No. 18 bars with smaller bars.

R25.5.1.3 If individual bars in noncontact lap splices are too widely spaced, an unreinforced section is created. Forcing a potential crack to follow a zigzag line (5-to-1 slope) is considered a minimum precaution. The 6 in. maximum spacing is added because most research available on the lap splicing of deformed bars was conducted with reinforcement within this spacing.

R25.5.1.4 The development length ℓ_d used to obtain lap length should be based on f_y because the splice classifications already reflect any excess reinforcement at the splice location; therefore, the factor from 25.4.10.1 for excess A_s should not be used.

CODE

25.5.1.5 For bars with $f_y \geq$ **80,000 psi** spaced closer than 6 in. on center, transverse reinforcement shall be provided such that K_{tr} shall not be smaller than $0.5d_b$.

25.5.1.6 Non-contact lap splices for reinforcement in shotcrete shall have clear spacing in accordance with (a) or (b):

(a) For. No. 6 and smaller bars, the clear spacing between bars shall be at least greater of $6d_b$ and 2-1/2 in.
(b) For. No. 7 and larger bars, the clear spacing shall be established using a shotcrete mockup panel to demonstrate that the reinforcement is properly encased.

25.5.1.7 Contact lap splices for reinforcement in shotcrete shall be oriented with the plane of the spliced bars perpendicular to the surface of the shotcrete and approved by the licensed design professional based on a shotcrete mockup panel to demonstrate that the reinforcement is properly encased.

25.5.1.8 Lap splices of bundled bars shall be in accordance with 25.6.1.7.

25.5.2 *Lap splice lengths of deformed bars and deformed wires in tension*

25.5.2.1 Tension lap splice length ℓ_{st} for deformed bars and deformed wires in tension shall be in accordance with Table 25.5.2.1, where ℓ_d shall be in accordance with 25.4.2.1(a).

Table 25.5.2.1—Lap splice lengths of deformed bars and deformed wires in tension

$A_{s,provided}/A_{s,required}$[1] over length of splice	Maximum percent of A_s spliced within required lap length	Splice type	ℓ_{st}	
≥ 2.0	50	Class A	Greater of:	$1.0\ell_d$ and 12 in.
≥ 2.0	100	Class B	Greater of:	$1.3\ell_d$ and 12 in.
< 2.0	All cases	Class B	Greater of:	$1.3\ell_d$ and 12 in.

[1]Ratio of area of reinforcement provided to area of reinforcement required by analysis at splice location.

25.5.2.2 If bars of different size are lap spliced in tension, ℓ_{st} shall be the greater of ℓ_d of the larger bar and ℓ_{st} of the smaller bar.

COMMENTARY

R25.5.1.6 and **R25.5.1.7** Information on shotcrete mockup panels is provided in ACI 506R, and information on evaluating shotcrete is provided in ACI 506.4R.

R25.5.2 *Lap splice lengths of deformed bars and deformed wires in tension*

R25.5.2.1 Lap splices in tension are classified as Class A or B, with length of lap a multiple of the tensile development length ℓ_d calculated in accordance with 25.4.2.3 or 25.4.2.4. The two-level lap splice requirements encourage splicing bars at points of minimum stress and staggering splices to improve behavior of critical details. For the purpose of calculating ℓ_d for staggered splices, the clear spacing is taken as the minimum distance between adjacent splices, as illustrated in Fig. R25.5.2.1.

The tension lap splice requirements encourage the location of splices away from regions of high tensile stress to locations where the area of steel provided is at least twice that required by analysis.

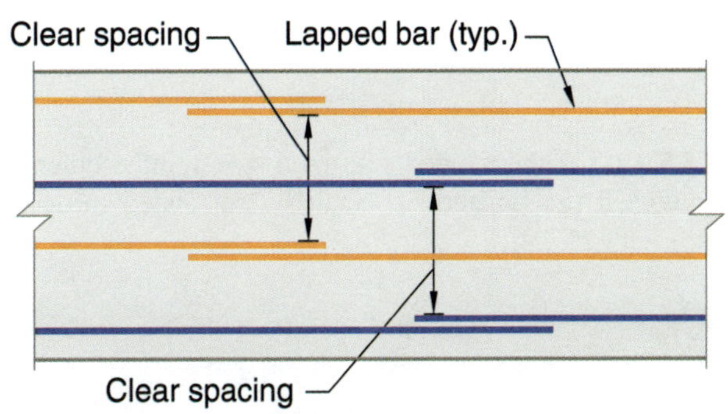

Fig. R25.5.2.1—Clear spacing of lap-spliced bars for determination of ℓ_d for staggered splices.

CODE

25.5.3 *Lap splice lengths of welded deformed wire reinforcement in tension*

25.5.3.1 Tension lap splice length ℓ_{st} of welded deformed wire reinforcement in tension with cross wires within the lap splice length shall be the greater of $1.3\ell_d$ and 8 in., where ℓ_d is calculated in accordance with 25.4.6.1(a), provided (a) and (b) are satisfied:

(a) Overlap between outermost cross wires of each reinforcement sheet shall be at least 2 in.
(b) Wires in the direction of the development length shall all be deformed D31 or smaller

25.5.3.1.1 If 25.5.3.1(a) is not satisfied, ℓ_{st} shall be calculated in accordance with 25.5.2.

25.5.3.1.2 If 25.5.3.1(b) is not satisfied, ℓ_{st} shall be calculated in accordance with 25.5.4.

25.5.3.1.3 If the welded deformed wire reinforcement is zinc-coated (galvanized), ℓ_{st} shall be calculated in accordance with 25.5.4.

25.5.4 *Lap splice lengths of welded plain wire reinforcement in tension*

25.5.4.1 Tension lap splice length ℓ_{st} of welded plain wire reinforcement in tension between outermost cross wires of each reinforcement sheet shall be at least the greatest of (a) through (c):

COMMENTARY

R25.5.3 *Lap splice lengths of welded deformed wire reinforcement in tension*

R25.5.3.1 Splice provisions for welded deformed wire reinforcement are based on available tests (Lloyd and Kesler 1969). Lap splices for welded deformed wire reinforcement meeting the requirements of this provision and 25.5.3.1.1 are illustrated in Fig. R25.5.3.1. If no cross wires are within the lap length, the provisions for deformed wire apply.

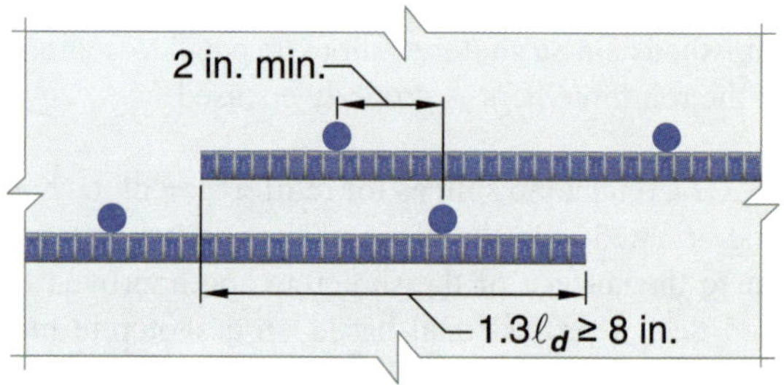

Lap splice satisfies R25.5.3.1a

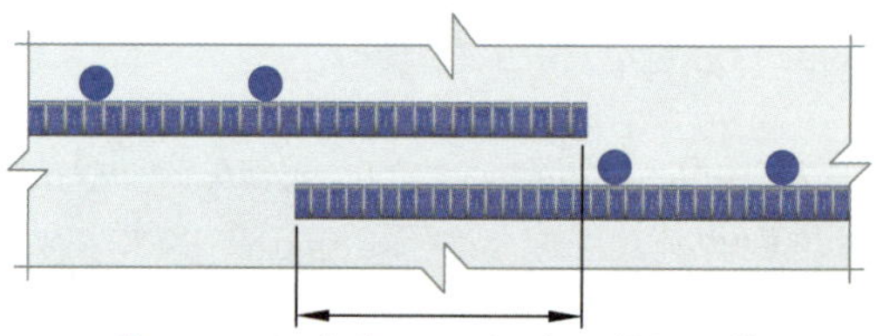

Lap splice satisfies R25.5.3.1.1

***Fig. R25.5.3.1**—Lap splices of welded deformed wire reinforcement.*

R25.5.3.1.2 Where any plain wires, or deformed wires larger than D31, are present in the welded deformed wire reinforcement in the direction of the lap splice or where welded deformed wire reinforcement is lap spliced to welded plain wire reinforcement, the reinforcement should be lap spliced in accordance with the plain wire reinforcement lap splice requirements. Deformed wire larger than D31 is treated as plain wire because tests show that D45 wire will achieve only approximately 60 percent of the bond strength in tension given by Eq. (25.4.2.4a) (Rutledge and DeVries 2002).

R25.5.4 *Lap splice lengths of welded plain wire reinforcement in tension*

R25.5.4.1 The strength of lap splices of welded plain wire reinforcement is dependent primarily on the anchorage obtained from the cross wires rather than on the length of wire in the splice. For this reason, the lap is specified in terms of overlap of cross wires (in inches) rather than in wire

CODE

(a) s + **2 in.**

(b) **1.5**ℓ_d

(c) 6 in.

where s is the spacing of cross wires and ℓ_d is calculated in accordance with 25.4.7.2(b).

COMMENTARY

diameters or length. The 2 in. additional lap required is to provide adequate overlap of the cross wires and to provide space for satisfactory consolidation of the concrete between cross wires. Research (Lloyd 1971) has shown an increased splice length is required when welded wire reinforcement of large, closely spaced wires is lapped and, as a consequence, additional splice length requirements are provided for this reinforcement in addition to an absolute minimum of 6 in. Splice requirements are illustrated in Fig. R25.5.4.1. If $A_{s,provided}/A_{s,required} \geq 2$ over the length of the splice, ℓ_{st} can be determined from 25.5.4.2.

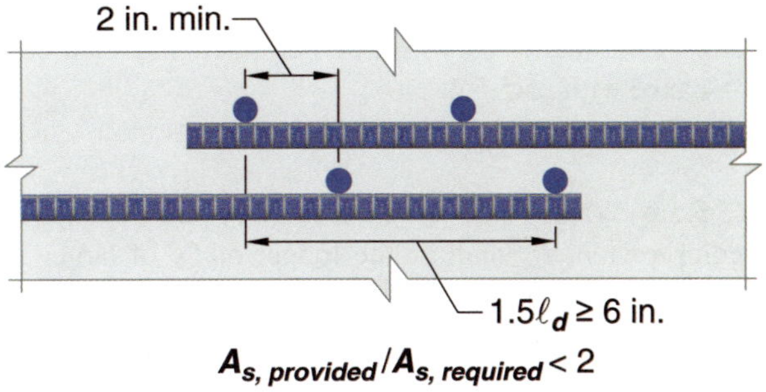

Fig. R25.5.4.1—Lap splices of plain welded wire reinforcement where $A_{s, provided}/A_{s, required} < 2$.

25.5.4.2 If $A_{s,provided}/A_{s,required} \geq 2.0$ over the length of the splice, ℓ_{st} measured between outermost cross wires of each reinforcement sheet shall be permitted to be the greater of (a) and (b).

(a) **1.5**ℓ_d

(b) 2 in.

where ℓ_d is calculated by 25.4.7.2(b).

R25.5.4.2 Where $A_{s,provided}/A_{s,required} \geq 2$, the lap splice for plain welded wire reinforcement is illustrated in Fig. R25.5.4.2.

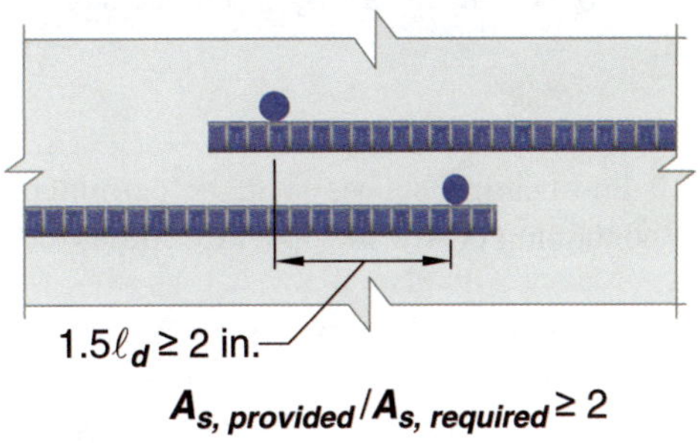

Fig. R25.5.4.2—Lap splices of plain welded wire reinforcement where $A_{s, provided}/A_{s, required} \geq 2$.

25.5.5 *Lap splice lengths of deformed bars in compression*

R25.5.5 *Lap splice lengths of deformed bars in compression*

Bond research has been primarily related to bars in tension. Bond behavior of compression bars is not complicated by the problem of transverse tension cracking and thus compression splices do not require provisions as strict as those specified for tension splices.

Lap splice requirements particular to columns are provided in Chapter 10.

25.5.5.1 Compression lap splice length ℓ_{sc} of No. 11 or smaller deformed bars in compression shall be calculated in accordance with (a), (b), or (c):

R25.5.5.1 Tests (ACI Committee 408 1966; Pfister and Mattock 1963) have shown that splice strengths in compression depend considerably on end bearing and do not increase proportionally in strength when the splice length is doubled.

Detailing 25

CODE

(a) For $f_y \leq$ **60,000 psi**: ℓ_{sc} is the longer of $0.0005 f_y d_b$ and 12 in.

(b) For 60,000 psi < $f_y \leq$ **80,000 psi**: ℓ_{sc} is the longer of $(0.0009 f_y - 24) d_b$ and 12 in.

(c) For f_y > **80,000 psi**, ℓ_{sc} is the longer of $(0.0009 f_y - 24) d_b$ and ℓ_{st} calculated in accordance with 25.5.2.1.

For f_c' < **3000 psi**, the length of lap shall be increased by one-third.

25.5.5.2 Compression lap splices shall not be used for bars larger than No. 11, except as permitted in 25.5.5.3.

25.5.5.3 Compression lap splices of No. 14 or No. 18 bars to No. 11 or smaller bars shall be permitted and shall be in accordance with 25.5.5.4.

25.5.5.4 Where bars of different size are lap spliced in compression, ℓ_{sc} shall be the longer of ℓ_{dc} of larger bar calculated in accordance with 25.4.9.1 and ℓ_{sc} of smaller bar calculated in accordance with 25.5.5.1 as appropriate.

25.5.6 *End-bearing splices of deformed bars in compression*

25.5.6.1 For bars required for compression only, transmission of compressive stress by end bearing of square-cut ends held in concentric contact by a suitable device shall be permitted.

25.5.6.2 End-bearing splices shall be permitted only in members containing closed stirrups, ties, spirals, or hoops.

25.5.6.3 Bar ends shall terminate in flat surfaces within 1.5 degrees of a right angle to the axis of the bars and shall be fitted within 3 degrees of full bearing after assembly.

25.5.7 *Mechanical and welded splices of deformed bars in tension or compression*

25.5.7.1 A mechanical or welded splice shall develop in tension or compression, as required, at least $1.25 f_y$ of the bar.

COMMENTARY

Accordingly, for specified yield strengths above 60,000 psi, compression lap lengths are significantly increased.

R25.5.5.3 Lap splices are generally prohibited for No. 14 or No. 18 bars. For compression only, however, lap splices are permitted between No. 14 or No. 18 bars and No. 11 or smaller bars.

R25.5.6 *End-bearing splices of deformed bars in compression*

R25.5.6.1 Experience with end-bearing splices has been almost exclusively with vertical bars in columns. If bars are significantly inclined from the vertical, attention is required to ensure that adequate end-bearing contact can be achieved and maintained.

R25.5.6.2 This limitation ensures a minimum shear resistance in sections containing end-bearing splices.

R25.5.6.3 These tolerances represent practice based on tests of full-size members containing No. 18 bars.

R25.5.7 *Mechanical and welded splices of deformed bars in tension or compression*

The 2014 Code eliminated mechanical and welded splices with strengths less than $1.25 f_y$. With the elimination of these mechanical and welded splices, the term "full" was deleted in reference to mechanical and welded splices that develop at least $1.25 f_y$.

R25.5.7.1 To ensure sufficient strength in splices so that yielding can be achieved in a member and thus brittle failure avoided, the 25 percent increase above the specified yield strength was selected as both an adequate minimum for safety and a practicable maximum for economy.

A welded splice is primarily intended for large bars (No. 6 and larger) in main members. The tensile strength require-

CODE

25.5.7.2 Welding of reinforcing bars shall conform to 26.6.4.

25.5.7.3 Mechanical or welded splices need not be staggered except as required by 25.5.7.4.

25.5.7.4 Splices in tension tie members shall be made with a mechanical or welded splice in accordance with 25.5.7.1. Splices in adjacent bars shall be staggered at least 30 in.

25.6—Bundled reinforcement

25.6.1 *Nonprestressed reinforcement*

25.6.1.1 Groups of parallel reinforcing bars bundled in contact to act as a unit shall be limited to four in any one bundle.

25.6.1.2 Bundled bars shall be enclosed within transverse reinforcement. Bundled bars in compression members shall be enclosed by transverse reinforcement at least No. 4 in size.

25.6.1.3 Bars larger than a No. 11 shall not be bundled in beams.

COMMENTARY

ment of 125 percent of specified yield strength is intended to provide sound welding that is also adequate for compression.

While direct butt welds are not required, AWS D1.4 states that wherever practical, direct butt welds are preferable for No. 7 bars and larger.

R25.5.7.3 Although mechanical and welded splices need not be staggered, staggering is encouraged and may be necessary for constructibility to provide enough space around the splice for installation or to meet the clear spacing requirements.

R25.5.7.4 A tension tie member has the following characteristics: member having an axial tensile force sufficient to create tension over the cross section; a level of stress in the reinforcement such that every bar should be fully effective; and limited concrete cover on all sides. Examples of members that may be classified as tension ties are arch ties, hangers carrying load to an overhead supporting structure, and main tension elements in a truss.

In determining if a member should be classified as a tension tie, consideration should be given to the importance, function, proportions, and stress conditions of the member related to the above characteristics. For example, a usual large circular tank, with many bars and with splices well staggered and widely spaced, should not be classified as a tension tie member, and Class B splices may be used.

R25.6—Bundled reinforcement

R25.6.1 *Nonprestressed reinforcement*

R25.6.1.1 The Code phrase "bundled in contact to act as a unit" is intended to preclude bundling more than two bars in the same plane. Typical bundle shapes in cross section are triangular, L-shaped, or square-shaped patterns for three- or four-bar bundles. As a practical caution, bundles more than one bar deep in the plane of bending should not be hooked or bent as a unit. Where end hooks are required, it is preferable to stagger the individual bar hooks within a bundle.

R25.6.1.3 A limitation that bars larger than No. 11 not be bundled in beams is a practical limit for application to building size members. (AASHTO LRFDUS Article 5.9.4 permits two-bar bundles for No. 14 and No. 18 bars in bridge girders.) Conformance to the crack control requirements of 24.3 will effectively preclude bundling of bars larger than No. 11 as tension reinforcement.

CODE

25.6.1.4 Individual bars within a bundle terminated within the span of flexural members shall terminate at different points with at least $40d_b$ stagger.

25.6.1.5 Development length for individual bars within a bundle, in tension or compression, shall be that of the individual bar, increased 20 percent for a three-bar bundle, and 33 percent for a four-bar bundle.

25.6.1.6 A unit of bundled bars shall be treated as a single bar with an area equivalent to that of the bundle and a centroid coinciding with that of the bundle. The diameter of the equivalent bar shall be used for d_b in (a) through (e):

(a) Spacing limitations based on d_b
(b) Cover requirements based on d_b
(c) Spacing and cover values in 25.4.2.3
(d) Confinement term in 25.4.2.4
(e) ψ_e factor in 25.4.2.5

25.6.1.7 Lap splices of bars in a bundle shall be based on the lap splice length required for individual bars within the bundle, increased in accordance with 25.6.1.5. Individual bar splices within a bundle shall not overlap. Entire bundles shall not be lap spliced.

25.6.2 *Post-tensioning ducts*

25.6.2.1 Bundling of post-tensioning ducts shall be permitted if shown that concrete can be satisfactorily placed and if provision is made to prevent the prestressed reinforcement from breaking through the duct.

25.7—Transverse reinforcement

25.7.1 *Stirrups*

25.7.1.1 Stirrups shall extend as close to the compression and tension surfaces of the member as cover requirements and proximity of other reinforcement permits and shall be anchored at both ends. Where used as shear reinforcement, stirrups shall extend a distance d from extreme compression fiber.

COMMENTARY

R25.6.1.4 Bond research (ACI Committee 408 1966) has shown that cutoff points within bundles should be staggered.

R25.6.1.5 An increased development length for individual bars is required when three or four bars are bundled together. The extra extension is needed because the grouping makes it more difficult to mobilize bond resistance from the core between the bars.

The development of bundled bars by a standard hook of the bundle is not covered by the provisions of 25.4.3.

R25.6.1.6 Although splice and development lengths of bundled bars are a multiple of the diameter of the individual bars being spliced increased by 20 or 33 percent, as appropriate, it is necessary to use an equivalent diameter of the entire bundle derived from the equivalent total area of bars for determining the spacing and cover values in 25.4.2.3, the confinement term, $[(c_b + K_{tr})/d_b]$, in 25.4.2.4, and the ψ_e factor in 25.4.2.5. For bundled bars, bar diameter d_b outside the brackets in the expressions of 25.4.2.3 and of Eq. (25.4.2.4a) is that of a single bar.

R25.6.1.7 The increased length of lap required for bars in bundles is based on the reduction in the exposed perimeter of the bars. Only individual bars are lap spliced along the bundle.

R25.6.2 *Post-tensioning ducts*

R25.6.2.1 Where ducts for prestressing reinforcement in a beam are arranged closely together vertically, provisions should be made to prevent the prestressed reinforcement from breaking through the duct when tensioned. Horizontal arrangement of ducts should allow proper placement of concrete. A clear spacing of one and one-third times the nominal maximum size of the coarse aggregate, but not less than 1 in., has proven satisfactory.

Where concentration of tendons or ducts tends to create a weakened plane in the concrete cover, reinforcement should be provided to control cracking.

R25.7—Transverse reinforcement

R25.7.1 *Stirrups*

R25.7.1.1 Stirrup legs should be extended as close as practicable to the compression face of the member because, near ultimate load, the flexural tension cracks penetrate deeply toward the compression zone.

It is essential that shear and torsional reinforcement be adequately anchored at both ends to be fully effective on either side of any potential inclined crack. This generally requires a hook or bend at the end of the reinforcement as provided by this section.

CODE

25.7.1.2 Between anchored ends, each bend in the continuous portion of a single or multiple U-stirrup and each bend in a closed stirrup shall enclose a longitudinal bar or strand.

25.7.1.3 Anchorage of deformed bar and wire shall be in accordance with (a), (b), or (c):

(a) For No. 5 bar and D31 wire, and smaller, and for No. 6 through No. 8 bars with $f_{yt} \leq$ **40,000 psi**, a standard hook around longitudinal reinforcement
(b) For No. 6 through No. 8 bars with $f_{yt} >$ **40,000 psi**, a standard hook around a longitudinal bar plus an embedment between midheight of the member and the outside end of the hook equal to or greater than $\mathbf{0.014 d_b f_{yt}/(\lambda \sqrt{f_c'})}$, with λ as given in Table 25.4.3.2
(c) In joist construction, for No. 4 bar and D20 wire and smaller, a standard hook

25.7.1.4 Anchorage of each leg of welded wire reinforcement forming a single U-stirrup shall be in accordance with (a) or (b):

(a) Two longitudinal wires spaced at a 2 in. spacing along the member at the top of the U
(b) One longitudinal wire located not more than $\boldsymbol{d/4}$ from the compression face and a second wire closer to the compression face and spaced not less than 2 in. from the first wire. The second wire shall be permitted to be located on the stirrup leg beyond a bend, or on a bend with an inside diameter of bend of at least $\mathbf{8d_b}$.

COMMENTARY

R25.7.1.3 Straight deformed bar and wire anchorage is not permitted because it is difficult to hold such a stirrup in position during concrete placement. Moreover, the lack of a standard stirrup hook may make the stirrup ineffective as it crosses shear cracks near the end of the stirrup.

For a No. 5 or D31 or smaller stirrup, anchorage is provided by a standard hook, as defined in 25.3.2, hooked around a longitudinal bar.

For a No. 6, No. 7, or No. 8 stirrup with f_{yt} of only 40,000 psi, a standard stirrup hook around a longitudinal bar provides sufficient anchorage. For a No. 6, No. 7, or No. 8 stirrup with higher strength, the embedment should be checked. A 135-degree or 180-degree hook is preferred, but a 90-degree hook may be used provided the free end of the 90-degree hook is extended the full 12 bar diameters as required in 25.3.2. Because it is not possible to bend a No. 6, No. 7, or No. 8 stirrup tightly around a longitudinal bar and due to the force in a bar with a design stress greater than 40,000 psi, stirrup anchorage depends on both the type of hook and whatever development length is provided. A longitudinal bar within a stirrup hook limits the width of any flexural cracks, even in a tension zone. Because such a stirrup hook cannot fail by splitting parallel to the plane of the hooked bar, the hook strength as used in 25.4.3.1(a) has been adjusted to reflect cover and confinement around the stirrup hook.

In joists, a small bar or wire can be anchored by a standard hook not engaging longitudinal reinforcement, allowing a continuously bent bar to form a series of single-leg stirrups along the length of the joist.

R25.7.1.4 The requirements for anchorage of welded wire reinforcement stirrups are illustrated in Fig. R25.7.1.4.

CODE

COMMENTARY

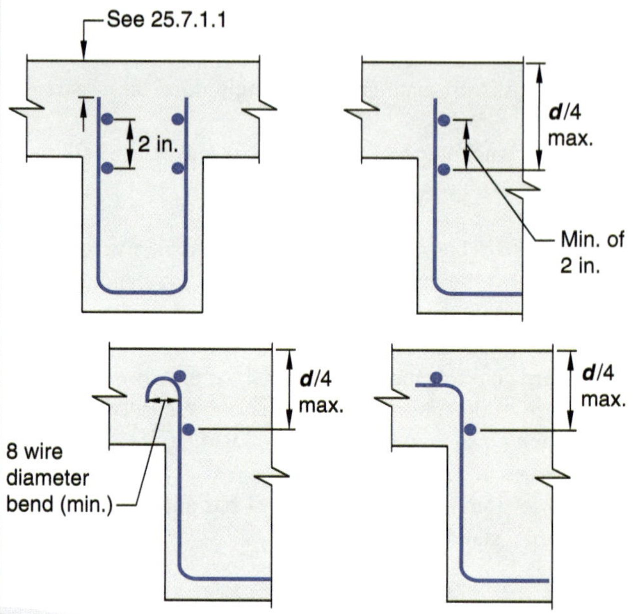

***Fig. R25.7.1.4**—Anchorage in compression zone of welded wire reinforcement U-stirrups.*

25.7.1.5 Anchorage of each end of a single leg stirrup of welded wire reinforcement shall be with two longitudinal wires at a minimum spacing of 2 in. in accordance with (a) and (b):

(a) Inner longitudinal wire at least the greater of ***d*/4** or 2 in. from ***d*/2**

(b) Outer longitudinal wire at tension face shall not be farther from the face than the portion of primary flexural reinforcement closest to the face

R25.7.1.5 Welded wire reinforcement for shear reinforcement is commonly used in the precast, prestressed concrete industry. The rationale for acceptance of straight sheets of welded wire reinforcement as shear reinforcement is presented in a report by the Joint PCI/WRI Ad Hoc Committee on Welded Wire Fabric for Shear Reinforcement (1980).

The provisions for anchorage of single-leg welded wire reinforcement in the tension face emphasize the location of the longitudinal wire at the same depth as the primary flexural reinforcement to avoid a splitting problem at the level of the tension reinforcement. Figure R25.7.1.5 illustrates the anchorage requirements for single-leg welded wire reinforcement. For anchorage of single-leg welded wire reinforcement, the Code permits hooks and embedment length in the compression and tension faces of members (refer to 25.7.1.3(a) and 25.7.1.4), and embedment only in the compression face (refer to 25.7.1.3(b)). This section provides for anchorage of straight, single-leg, welded wire reinforcement using longitudinal wire anchorage with adequate embedment length in compression and tension faces of members.

CODE

COMMENTARY

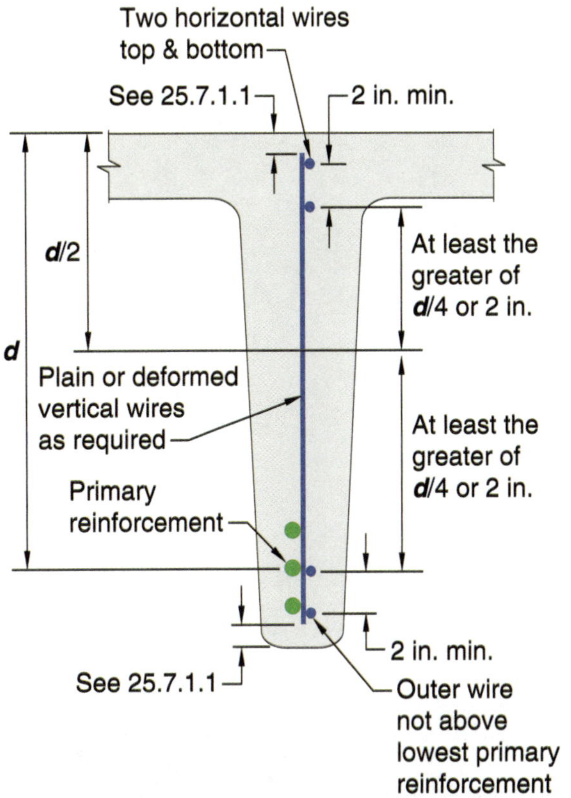

***Fig. R25.7.1.5**—Anchorage of single-leg welded wire reinforcement for shear.*

25.7.1.6 Stirrups used for torsion or integrity reinforcement shall be closed stirrups perpendicular to the axis of the member. Where welded wire reinforcement is used, transverse wires shall be perpendicular to the axis of the member. Such stirrups shall be anchored by (a) or (b):

(a) Ends shall terminate with 135-degree standard hooks around a longitudinal bar

(b) In accordance with 25.7.1.3(a) or (b) or 25.7.1.4, where the concrete surrounding the anchorage is restrained against spalling by a flange or slab or similar member

25.7.1.6.1 Stirrups used for torsion or integrity reinforcement shall be permitted to be made up of two pieces of reinforcement: a single U-stirrup anchored according to 25.7.1.6(a) closed by a crosstie where the 90-degree hook of the crosstie shall be restrained against spalling by a flange or slab or similar member.

R25.7.1.6 Both longitudinal and closed transverse reinforcement are required to resist the diagonal tension stresses due to torsion. The stirrups should be closed because inclined cracking due to torsion may occur on all faces of a member.

In the case of sections subjected primarily to torsion, the concrete side cover to the stirrups spalls off at high torsional moments (Mitchell and Collins 1976). This renders lap-spliced stirrups ineffective, leading to a premature torsional failure (Behera and Rajagopalan 1969). In such cases, closed stirrups should not be made up of pairs of U-stirrups lapping one another.

When a rectangular beam fails in torsion, the corners of the beam tend to spall off due to the inclined compressive stresses in the concrete diagonals of the space truss changing direction at the corner as shown in Fig. R25.7.1.6(a). In tests (Mitchell and Collins 1976), closed stirrups anchored by 90-degree hooks failed when this occurred. For this reason, 135-degree standard hooks or seismic hooks are preferable for torsional stirrups in all cases. In regions where this spalling is prevented by an adjacent slab or flange, 25.7.1.6(b) relaxes this requirement and allows 90-degree hooks because of the added confinement from the slab (refer to Fig. R25.7.1.6(b)).

CODE

COMMENTARY

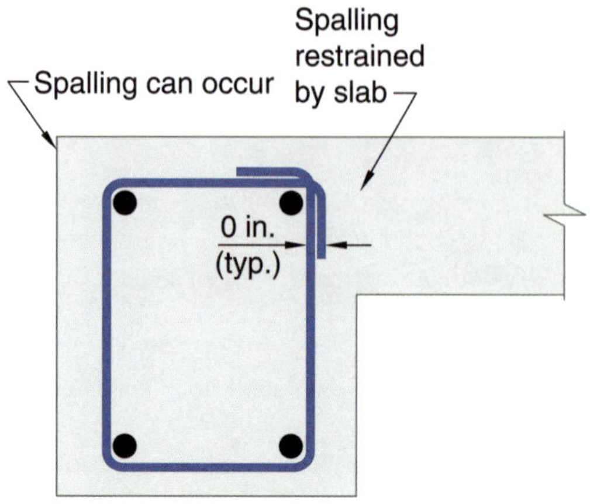

(a) Sectional elevation

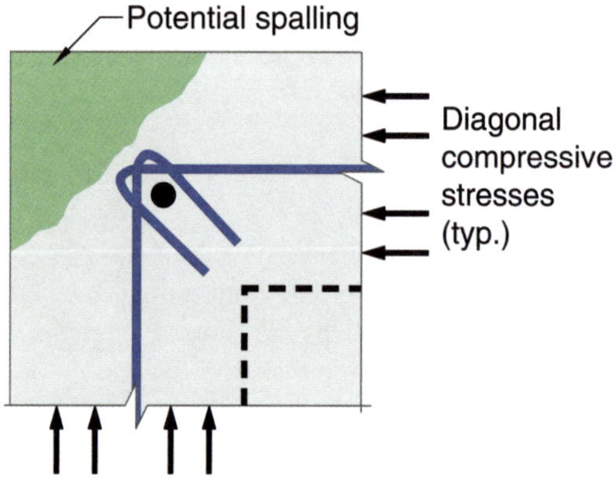

(b) Detail at corner

Fig. R25.7.1.6—Spalling of corners of beams subjected to torsion.

25.7.1.7 Except where used for torsion or integrity reinforcement, closed stirrups are permitted to be made using pairs of U-stirrups spliced to form a closed unit where lap lengths are at least $\mathbf{1.3\ell_d}$. In members with a total depth of at least 18 in., such splices with $\mathbf{A_b f_{yt} \leq 9000}$ **lb** per leg shall be considered adequate if stirrup legs extend the full available depth of member.

R25.7.1.7 Requirements for lapping of double U-stirrups to form closed stirrups control over the lap splice provisions of 25.5.2. Figure R25.7.1.7 illustrates closed stirrup configurations created with lap splices.

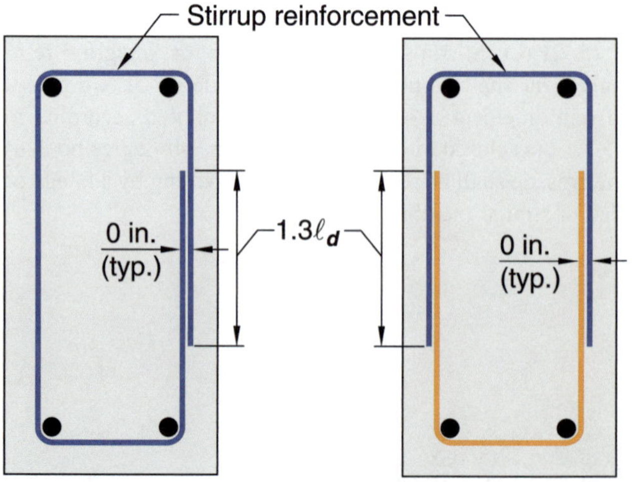

Fig. R25.7.1.7—Closed stirrup configurations.

CODE

25.7.2 *Ties*

25.7.2.1 Ties shall consist of a closed loop of deformed bar with spacing in accordance with (a) and (b):

(a) Clear spacing of at least $(4/3)d_{agg}$
(b) Center-to-center spacing shall not exceed the least of $16d_b$ of longitudinal bar, $48d_b$ of tie bar, and smallest dimension of member

25.7.2.2 Diameter of tie bar shall be at least (a) or (b):

(a) No. 3 enclosing No. 10 or smaller longitudinal bars
(b) No. 4 enclosing No. 11 or larger longitudinal bars or bundled longitudinal bars

25.7.2.2.1 As an alternative to deformed bars, deformed wire or welded wire reinforcement of equivalent area to that required in 25.7.2.1 shall be permitted subject to the requirements of Table 20.2.2.4(a).

25.7.2.3 Rectilinear ties shall be arranged to satisfy (a) and (b):

(a) Every corner and alternate longitudinal bar shall have lateral support provided by the corner of a tie with an included angle of not more than 135 degrees
(b) No unsupported bar shall be farther than 6 in. clear on each side along the tie from a laterally supported bar

COMMENTARY

R25.7.2 *Ties*

R25.7.2.2 These provisions apply to crossties as well as ties.

R25.7.2.3 The maximum permissible included angle of 135 degrees and the exemption of bars located within 6 in. clear on each side along the tie from adequately tied bars are illustrated in Fig. R25.7.2.3a. Limited tests (Pfister 1964) on full-size, axially-loaded, tied columns containing full-length bars (without splices) showed that ties on alternate longitudinal bars within 6 in. clear of a laterally supported longitudinal bar are adequate in columns subjected to axial force.

Continuously wound bars or wires can be considered as ties, provided their pitch and area are at least equivalent to the area and spacing of separate ties. Anchorage at the end of a continuously wound bar or wire should be by a standard hook as for separate bars or by one additional turn of the tie pattern (refer to Fig. R25.7.2.3b). A circular, continuously wound bar or wire is considered a spiral if it conforms to 25.7.3; otherwise, it is considered a tie.

CODE

COMMENTARY

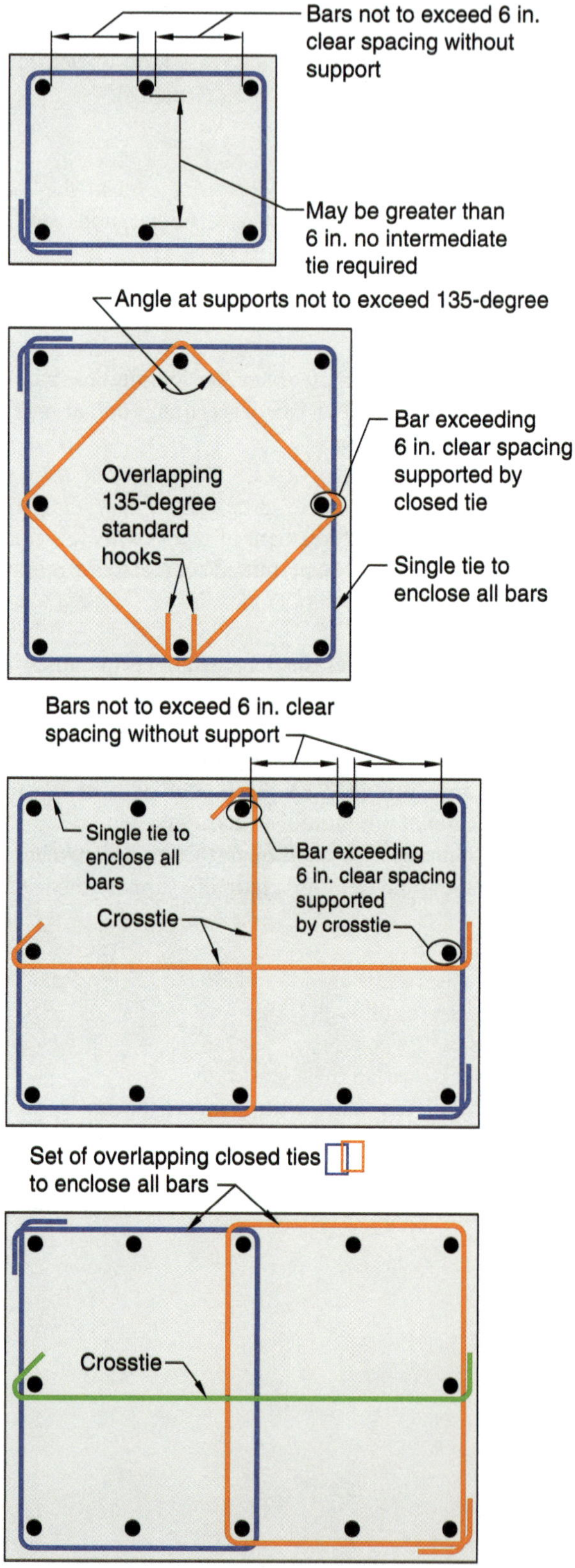

Fig. R25.7.2.3a—Illustrations to clarify measurements between laterally supported column bars and rectilinear tie anchorage.

CODE | COMMENTARY

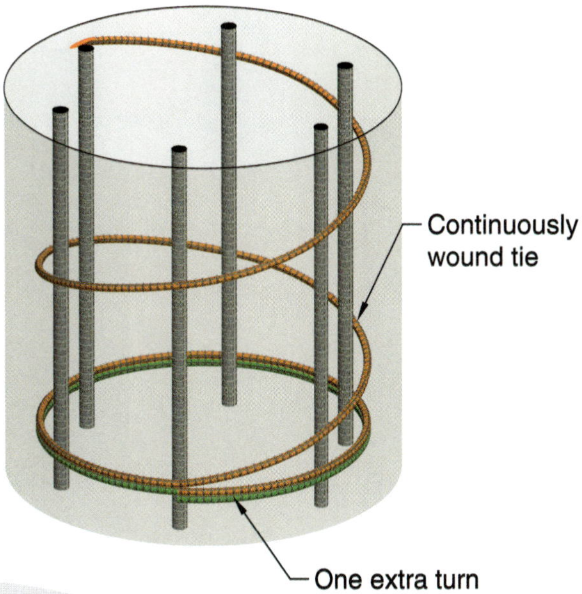

***Fig. R25.7.2.3b**—Continuous tie anchorage.*

25.7.2.3.1 Anchorage of rectilinear ties shall be provided by standard hooks that conform to 25.3.2 and engage a longitudinal bar. A tie shall not be made up of interlocking headed deformed bars.

R25.7.2.3.1 Standard tie hooks are intended for use with deformed bars only and should be staggered where possible.

25.7.2.4 Circular ties shall be permitted where longitudinal bars are located around the perimeter of a circle.

R25.7.2.4 While the transverse reinforcement in members with longitudinal bars located around the periphery of a circle can be either spirals or circular ties, spirals are usually more effective.

25.7.2.4.1 Anchorage of individual circular ties shall be in accordance with (a) through (c):

(a) Ends shall overlap by at least 6 in.
(b) Ends shall terminate with standard hooks in accordance with 25.3.2 that engage a longitudinal bar
(c) Overlaps at ends of adjacent circular ties shall be staggered around the perimeter enclosing the longitudinal bars

R25.7.2.4.1 Vertical splitting and loss of tie restraint are possible where the overlapped ends of adjacent circular ties are anchored at a single longitudinal bar. Adjacent circular ties should not engage the same longitudinal bar with end hook anchorages (refer to Fig. R25.7.2.4.1).

CODE

COMMENTARY

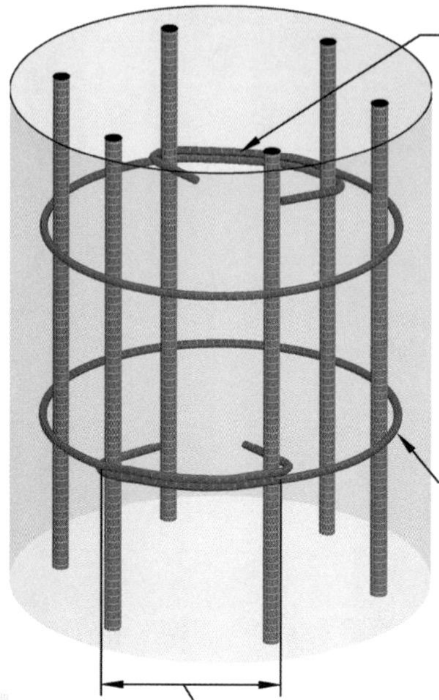

Fig. R25.7.2.4.1—Circular tie anchorage.

25.7.2.5 Ties to resist torsion shall be perpendicular to the axis of the member anchored by either (a) or (b):

(a) Ends shall terminate with 135-degree standard hooks or seismic hooks around a longitudinal bar
(b) In accordance with 25.7.1.3(a) or (b) or 25.7.1.4, where the concrete surrounding the anchorage is restrained against spalling

R25.7.2.5 Refer to R25.7.1.6.

25.7.3 *Spirals*

R25.7.3 *Spirals*

25.7.3.1 Spirals shall consist of evenly spaced continuous bar or wire with clear spacing conforming to (a) and (b):

(a) At least the greater of 1 in. and $(4/3)d_{agg}$
(b) Not greater than 3 in.

R25.7.3.1 Spirals should be held firmly in place, at proper pitch and alignment, to prevent displacement during concrete placement.

25.7.3.2 For cast-in-place construction, spiral bar or wire diameter shall be at least 3/8 in.

R25.7.3.2 For practical considerations in cast-in-place construction, the minimum diameter of spiral reinforcement is 3/8 in. (No. 3 deformed or plain bar, or D11 deformed or W11 plain wire).

Standard spiral sizes are 3/8, 1/2, and 5/8 in. diameter for hot-rolled or cold-drawn material, plain or deformed.

25.7.3.3 Except for transverse reinforcement in deep foundations, the volumetric spiral reinforcement ratio ρ_s shall satisfy Eq. (25.7.3.3).

$$\rho_s \geq 0.45\left(\frac{A_g}{A_{ch}}-1\right)\frac{f_c'}{f_{yt}} \quad (25.7.3.3)$$

R25.7.3.3 The effect of spiral reinforcement in increasing the strength of the concrete within the core is not fully realized until the column has been subjected to a load and deformation sufficient to cause the concrete shell outside the core to spall off. The amount of spiral reinforcement required by Eq. (25.7.3.3) is intended to provide additional strength for concentrically loaded columns equal to or slightly greater than the strength lost when the shell spalls off. The deriva-

CODE

where the value of f_{yt} shall not be taken greater than 100,000 psi.

25.7.3.4 Spirals shall be anchored by 1-1/2 extra turns of spiral bar or wire at each end.

25.7.3.5 Spirals are permitted to be spliced by (a) or (b):

(a) Mechanical or welded splices in accordance with 25.5.7
(b) Lap splices in accordance with 25.7.3.6 for f_{yt} not exceeding 60,000 psi

25.7.3.6 Spiral lap splices shall be at least the greater of 12 in. and the lap length in Table 25.7.3.6.

COMMENTARY

tion of Eq. (25.7.3.3) is given by Richart (1933). Tests and experience show that columns containing the amount of spiral reinforcement required by this section exhibit considerable toughness and ductility. Research (Richart et al. 1929; Richart 1933; Pessiki et al. 2001; Saatcioglu and Razvi 2002) has also indicated that up to 100,000 psi yield strength reinforcement can be effectively used for confinement.

R25.7.3.4 Spiral anchorage is illustrated in Fig. R25.7.3.4.

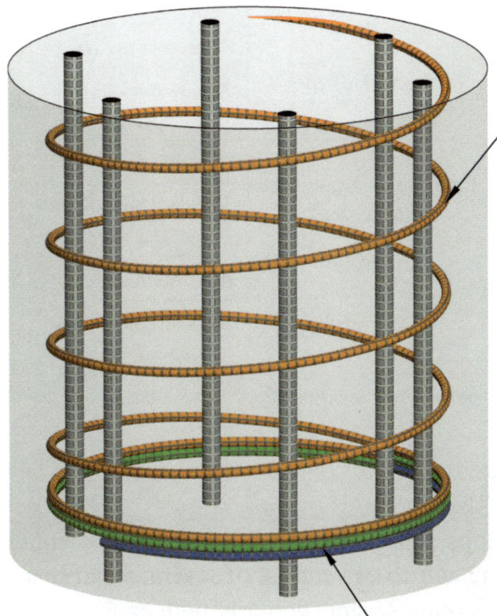

Fig. R25.7.3.4—Spiral anchorage.

CODE

COMMENTARY

Table 25.7.3.6—Lap length for spiral reinforcement

Reinforcement	Coating	Ends of lapped spiral bar or wire	Lap length in.
Deformed bar	Uncoated or zinc-coated (galvanized)	Hook not required	$48d_b$
	Epoxy-coated or zinc and epoxy dual-coated	Hook not required	$72d_b$
		Standard hook of 25.3.2[1]	$48d_b$
Deformed wire	Uncoated	Hook not required	$48d_b$
	Epoxy-coated	Hook not required	$72d_b$
		Standard hook of 25.3.2[1]	$48d_b$
Plain bar	Uncoated or zinc-coated (galvanized)	Hook not required	$72d_b$
		Standard hook of 25.3.2[1]	$48d_b$
Plain wire	Uncoated	Hook not required	$72d_b$
		Standard hook of 25.3.2[1]	$48d_b$

[1]Hooks shall be embedded within the core confined by the spiral.

25.7.4 *Hoops*

25.7.4.1 Hoops shall consist of a closed tie or continuously wound tie, which can consist of several reinforcement elements each having seismic hooks at both ends.

25.7.4.2 The ends of the reinforcement elements in hoops shall be anchored using seismic hooks that conform to 25.3.4 and engage a longitudinal bar. A hoop shall not be made up of interlocking headed deformed bars.

R25.7.4 *Hoops*

R25.7.4.1 Refer to R25.7.2.4.

25.8—Post-tensioning anchorages and couplers

25.8.1 Anchorages and couplers for tendons shall develop at least 95 percent of f_{pu} when tested in an unbonded condition, without exceeding anticipated set.

R25.8—Post-tensioning anchorages and couplers

R25.8.1 The required strength of the tendon-anchorage or tendon-coupler assemblies for both unbonded and bonded tendons, when tested in an unbonded state, is based on 95 percent of the specified tensile strength of the prestressing reinforcement in the test. The prestressing reinforcement is required to comply with the minimum provisions of the applicable ASTM standards as prescribed in 20.3.1. The specified strength of anchorages and couplers exceeds the maximum design strength of the prestressing reinforcement by a substantial margin and, at the same time, recognizes the stress-riser effects associated with most available post-tensioning anchorages and couplers. Anchorage and coupler strength should be attained with a minimum amount of permanent deformation and successive set, recognizing that some deformation and set will occur when testing to failure. Tendon assemblies should conform to the 2 percent elongation requirements in ACI 423.7.

Static and fatigue test methods for anchorage and couplers are provided in ICC-ES Acceptance Criteria AC303 (2011).

CODE

25.8.2 Anchorages and couplers for bonded tendons shall be located so that 100 percent of f_{pu} shall be developed at critical sections after the post-tensioned reinforcement is bonded in the member.

25.8.3 In unbonded construction subject to repetitive loads, the possibility of fatigue of prestressed reinforcement in anchorages and couplers shall be considered.

25.8.4 Couplers shall be placed at locations approved by the licensed design professional and enclosed in housings long enough to permit necessary movements.

25.9—Anchorage zones for post-tensioned tendons

25.9.1 *General*

25.9.1.1 Anchorage regions of post-tensioned tendons shall consist of two zones, (a) and (b):

(a) The local zone shall be assumed to be a rectangular prism (or equivalent rectangular prism for circular or oval anchorages) of concrete immediately surrounding the anchorage device and any confining reinforcement

(b) The general zone includes the local zone and shall be assumed to be the portion of the member through which the concentrated prestressing force is transferred to the concrete and distributed more uniformly across the section

COMMENTARY

R25.8.2 Anchorages and couplers for bonded tendons that develop less than 100 percent of the specified tensile strength of the prestressing reinforcement should be used only where the bond transfer length between the anchorage or coupler and critical sections equals or exceeds that required to develop the prestressed reinforcement strength. This bond length may be calculated based on the results of tests of bond characteristics of non-tensioned prestressing strand (Salmons and McCrate 1977; PCA 1980), or bond tests on other prestressing reinforcement, as appropriate.

R25.8.3 A discussion on fatigue loading is provided in ACI 215R.

Detailed recommendations on tests for static and cyclic loading conditions for tendons and anchorage fittings of unbonded tendons are provided in ACI 423.3R (Section 4.1.3) and ACI 301 (Section 15.2.2).

R25.9—Anchorage zones for post-tensioned tendons

R25.9.1 *General*

The detailed provisions in the AASHTO LRFD Bridge Design Specifications (AASHTO LRFDUS) for analysis and reinforcement detailing of post-tensioned anchorage zones are considered to satisfy the more general requirements of this Code. In the specific areas of anchorage device evaluation and acceptance testing, this Code references the detailed AASHTO provisions.

R25.9.1.1 Based on St. Venant's principle, the extent of the anchorage zone may be estimated as approximately equal to the largest dimension of the cross section. Local zones and general zones are shown in Fig. R25.9.1.1a.

When anchorage devices located away from the end of the member are tensioned, large local tensile stresses are generated ahead of and behind the device. These tensile stresses are induced by incompatibility of deformations. The entire shaded region shown in Fig. R25.9.1.1b should be considered in the design of the general zone.

CODE

COMMENTARY

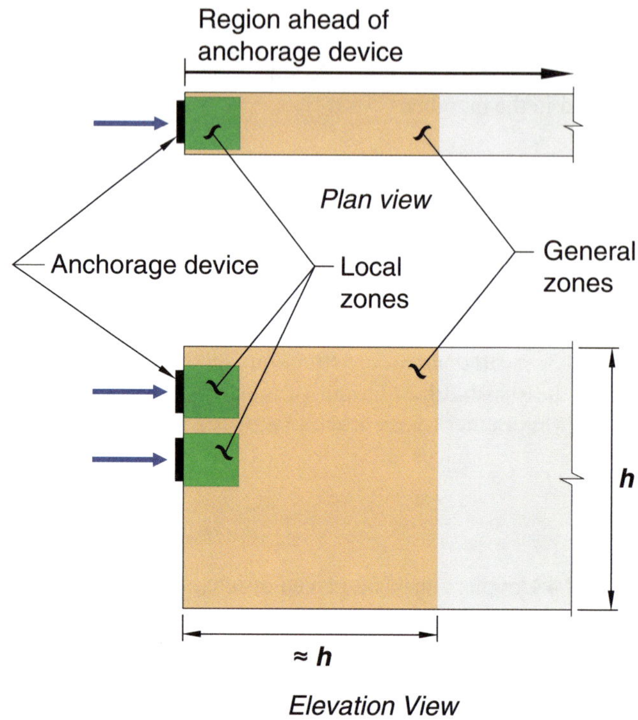

Fig. R25.9.1.1a—Local and general zones.

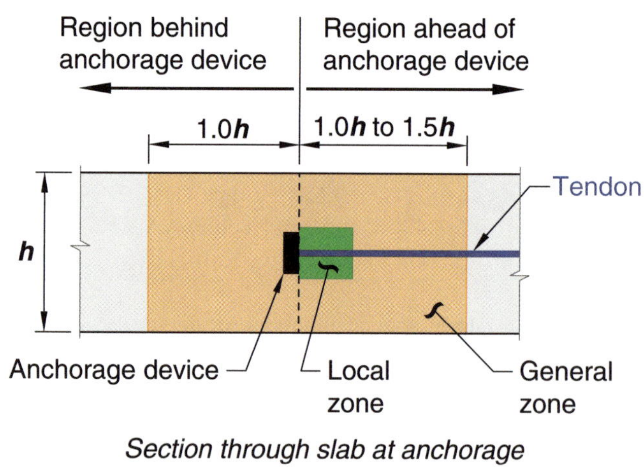

Fig. R25.9.1.1b—Local and general zones for anchorage device located away from the end of a member.

25.9.1.2 The local zone shall be designed in accordance with 25.9.3.

25.9.1.3 The general zone shall be designed in accordance with 25.9.4.

25.9.1.4 Compressive strength of concrete required at time of post-tensioning shall be specified as required by 26.10.

25.9.1.5 Stressing sequence shall be considered in the design process and specified as required by 26.10.

R25.9.1.5 The sequence of anchorage device stressing can have a significant effect on the general zone stresses. Therefore, it is important to consider not only the final stage of a stressing sequence with all tendons stressed, but also

CODE | COMMENTARY

intermediate stages during construction. The most critical bursting forces caused by each of the sequentially post-tensioned tendon combinations, as well as that of the entire group of tendons, should be taken into account.

25.9.2 *Required strength*

25.9.2.1 Factored prestressing force at the anchorage device, P_{pu}, shall exceed the least of (a) through (c), where 1.2 is the load factor from 5.3.12:

(a) $\mathbf{1.2(0.94}f_{py}\mathbf{)}A_{ps}$
(b) $\mathbf{1.2(0.80}f_{pu}\mathbf{)}A_{ps}$
(c) Maximum jacking force designated by the supplier of anchorage devices multiplied by 1.2

R25.9.2 *Required strength*

R25.9.2.1 The factored prestressing force is the product of the load factor and the maximum prestressing force permitted. The maximum permissible tensile stresses during jacking are defined in 20.3.2.5.1.

25.9.3 *Local zone*

25.9.3.1 The design of local zone in post-tensioned anchorages shall meet the requirements of (a), (b), or (c):

(a) Monostrand or single 5/8 in. or smaller diameter bar anchorage devices shall meet the bearing resistance and local zone requirements of ACI 423.7
(b) Basic multistrand anchorage devices shall meet the bearing resistance requirements of AASHTO LRFD Bridge Design Specifications, Article 5.8.4.4.2, except that the load factors shall be in accordance with 5.3.12 and ϕ shall be in accordance with 21.2.1
(c) Special anchorage devices shall satisfy the tests required in AASHTO LRFD Bridge Design Specifications, Article 5.8.4.4.3, and described in AASHTO LRFD Bridge Construction Specifications, Article 10.3.2.3

25.9.3.2 Where special anchorage devices are used, supplementary skin reinforcement shall be provided in addition to the confining reinforcement specified for the anchorage device.

25.9.3.2.1 Supplementary skin reinforcement shall be similar in configuration and at least equivalent in volumetric ratio to any supplementary skin reinforcement used in the qualifying acceptance tests of the anchorage device.

R25.9.3 *Local zone*

The local zone resists very high local stresses introduced by the anchorage device and transfers them to the remainder of the anchorage zone. The behavior of the local zone is strongly influenced by the specific characteristics of the anchorage device and its confining reinforcement, and is less influenced by the geometry and loading of the overall structure. Local-zone design sometimes cannot be completed until specific anchorage devices are selected. If special anchorage devices are used, the anchorage device supplier should furnish test information to demonstrate that the device is satisfactory under Article 10.3.2.3 of the AASHTO LRFD Bridge Construction Specifications (LRFDCONS) and provide information regarding necessary conditions for use of the device. The main considerations in local-zone design are the effects of high bearing pressure and the adequacy of any confining reinforcement provided to increase concrete bearing resistance.

R25.9.3.2.1 Skin reinforcement is placed near the outer faces in the anchorage zone to limit local crack width and spacing. Reinforcement in the general zone for other actions (such as shrinkage and temperature) may be used in satisfying the supplementary skin reinforcement requirement. Determination of the supplementary skin reinforcement depends on the anchorage device hardware used and frequently cannot be determined until the specific anchorage devices are selected.

25.9.4 *General zone*

R25.9.4 *General zone*

Within the general zone, the assumption that plane sections remain plane is not valid. Tensile stresses that can be caused by the tendon anchorage device, including bursting, spalling, and edge tension, as shown in Fig. R25.9.4, should be considered

CODE

COMMENTARY

in design. In addition, the compressive stresses immediately ahead of the local zone should be checked (Fig. R25.9.1.1b).

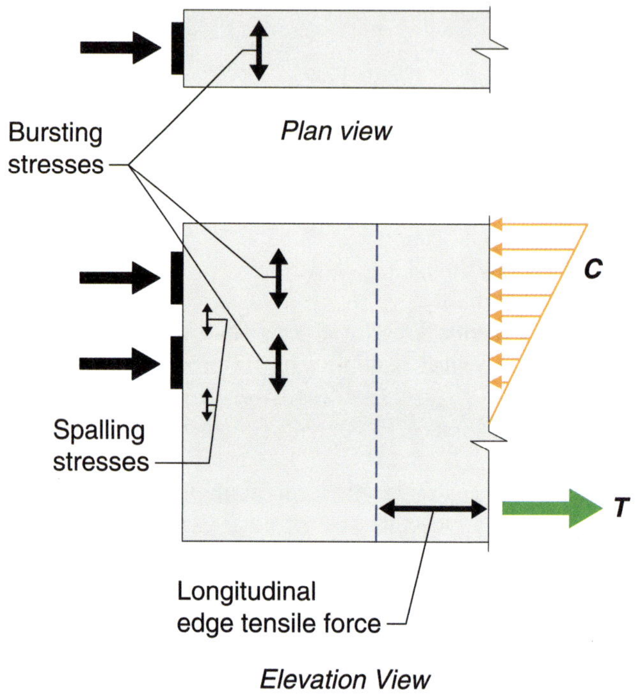

Fig. R25.9.4—Tensile stress zones within the general zone.

25.9.4.1 The extent of the general zone is equal to the largest dimension of the cross section. In the case of slabs with anchorages or groups of anchorages spaced along the slab edge, the depth of the general zone shall be taken as the spacing of the tendons.

R25.9.4.1 The depth of the general zone in slabs is defined in AASHTO LRFD Bridge Design Specifications (LRFDUS), Article 5.9.5.6 as the spacing of the tendons (Fig. R25.9.4.1). Refer to 25.9.4.4.6 for monostrand anchorages.

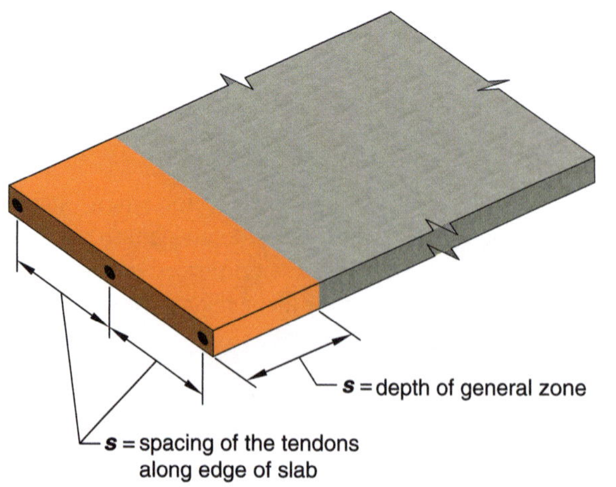

Fig. R25.9.4.1—Dimensions of general zone in post-tensioned slab.

25.9.4.2 For anchorage devices located away from the end of a member, the general zone shall include the disturbed regions ahead of and behind the anchorage devices.

R25.9.4.2 The dimensions of the general zone for anchorage devices located away from the end of the member are defined in Fig. R25.9.1.1b.

CODE

25.9.4.3 *Analysis of general zones*

25.9.4.3.1 Methods (a) through (c) shall be permitted for design of general zones:

(a) The strut-and-tie method in accordance with Chapter 23
(b) Linear stress analysis, including finite element analysis or equivalent
(c) Simplified equations in AASHTO LRFD Bridge Design Specifications, Article 5.8.4.5, except where restricted by 25.9.4.3.2

The design of general zones by other methods shall be permitted, provided that the specific procedures used for design result in prediction of strength in substantial agreement with results of comprehensive tests.

25.9.4.3.2 Simplified equations as permitted by 25.9.4.3.1(c) shall not be used for the design of a general zone if any of the situations listed in (a) through (g) occur:

COMMENTARY

R25.9.4.3 *Analysis of general zones*

R25.9.4.3.1 The design methods include those procedures for which guidelines have been given in AASHTO LRFDUS and Breen et al. (1994). These procedures have been shown to be conservative predictors of strength compared to test results (Breen et al. 1994). The use of the strut-and-tie method is especially helpful for general zone design (Breen et al. 1994). In many anchorage applications, where substantial or massive concrete regions surround the anchorages, simplified equations based on AASHTO LRFDUS and Breen et al. (1994) can be used except in the cases noted in 25.9.4.3.2.

Values for the magnitude of the bursting force, T_{burst}, and for its centroidal distance from the major bearing surface of the anchorage, d_{burst}, may be estimated from Eq. (R25.9.4.3.1a) and (R25.9.4.3.1b), respectively. The terms used in these equations are shown in Fig. R25.9.4.3.1 for a prestressing force with a small eccentricity. In the application of these equations, the specified stressing sequence should be considered if more than one tendon is present.

$$T_{burst} = 0.25\sum P_{pu}\left(1-\frac{h_{anc}}{h}\right) \quad \text{(R25.9.4.3.1a)}$$

$$d_{burst} = 0.5(h - 2e_{anc}) \quad \text{(R25.9.4.3.1b)}$$

where $\sum P_{pu}$ is the sum of the P_{pu} forces from the individual tendons; h_{anc} is the depth of the anchorage device or single group of closely spaced devices in the direction considered; and e_{anc} is the eccentricity (always taken as positive) of the anchorage device or group of closely spaced devices with respect to the centroid of the cross section (Fig. R25.9.4.3.1).

Anchorage devices should be treated as closely spaced if their center-to-center spacing does not exceed 1.5 times the width of the anchorage device in the direction considered.

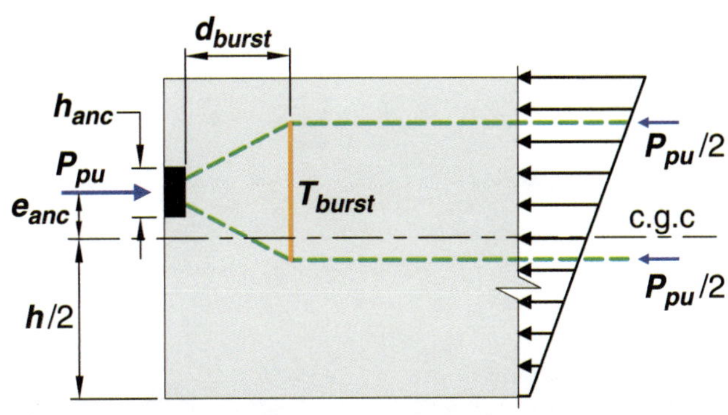

Fig. R25.9.4.3.1—Definition of terms used to define the general zone.

R25.9.4.3.2 The simplified equations in the AASHTO LRFDUS are not applicable in several common situations listed in 25.9.4.3.2. In these cases, a detailed analysis is required. In addition, in the post-tensioning of thin sections,

Detailing 25

CODE

(a) Member cross sections are nonrectangular
(b) Discontinuities in or near the general zone cause deviations in the force flow path
(c) Minimum edge distance is less than 1.5 times the anchorage device lateral dimension in that direction
(d) Multiple anchorage devices are used in other than one closely spaced group
(e) Centroid of the tendons is located outside the kern
(f) Angle of inclination of the tendon in the general zone is less than –5 degrees from the centerline of axis of the member, where the angle is negative if the anchor force points away from the centroid of the section
(g) Angle of inclination of the tendon in the general zone is greater than +20 degrees from the centerline of axis of the member, where the angle is positive if the anchor force points towards the centroid of the section

25.9.4.3.3 Three-dimensional effects shall be considered in design and analyzed by (a) or (b):

(a) Three-dimensional analysis procedures
(b) Approximated by considering the summation of effects for two orthogonal planes

25.9.4.4 *Reinforcement limits*

25.9.4.4.1 Tensile strength of concrete shall be neglected in calculations of reinforcement requirements.

25.9.4.4.2 Reinforcement shall be provided in the general zone to resist bursting, spalling, and longitudinal edge tension forces induced by anchorage devices, as applicable. Effects of abrupt changes in section and stressing sequence shall be considered.

COMMENTARY

flanged sections, or irregular sections, or where the tendons have appreciable curvature within the general zone, more general procedures such as those of AASHTO LRFDUS Articles 5.8.2.7 and 5.8.3 are required. Detailed recommendations for design principles that apply to all design methods are given in Article 5.9.5.6.5b of the AASHTO LRFDUS.

Groups of monostrand tendons with individual monostrand anchorage devices are often used in beams. If a beam has a single anchorage device or a single group of closely spaced anchorage devices, the use of simplified equations such as those given in R25.9.4.3.1 is permitted, unless 25.9.4.3.2 governs. More complex conditions can be designed using the strut-and-tie method. Detailed recommendations for use of such models are given in AASHTO LRFDUS and Breen et al. (1994).

R25.9.4.3.3 The provision for three-dimensional effects is to ensure that the effects perpendicular to the main plane of the member, such as bursting forces in the thin direction of webs or slabs are considered. In many cases, these effects can be determined independently for each direction, but some applications require a full three-dimensional analysis (for example, diaphragms for the anchorage of external tendons).

R25.9.4.4 *Reinforcement limits*

R25.9.4.4.2 In some cases, reinforcement requirements cannot be determined until specific tendon and anchorage device layouts are selected. Design and approval responsibilities should be clearly assigned in the construction documents.

Abrupt changes in section can cause substantial deviation in force paths. These deviations can greatly increase tensile forces, as shown in Fig. R25.9.4.4.2.

CODE

COMMENTARY

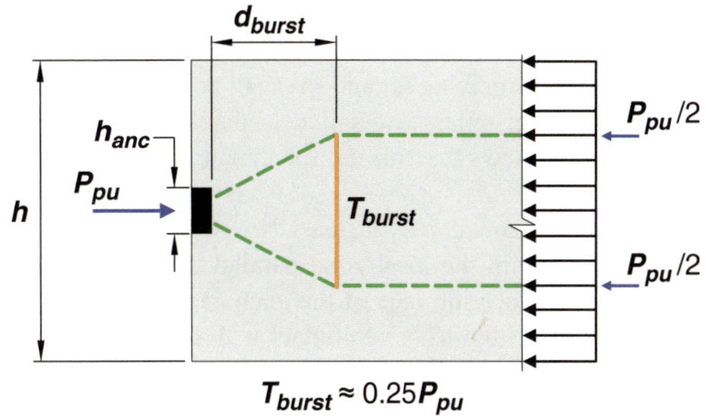

(a) Rectangular section

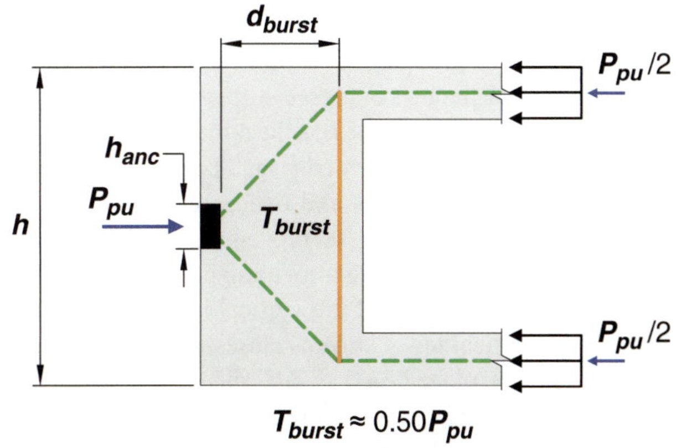

(b) Flanged section with end diaphragm

Fig. R25.9.4.4.2—*Effect of cross section change.*

25.9.4.4.3 For anchorage devices located away from the end of the member, bonded reinforcement shall be provided to transfer at least $\mathbf{0.35}P_{pu}$ into the concrete section behind the anchor. Such reinforcement shall be placed symmetrically around the anchorage device and shall be fully developed both behind and ahead of the anchorage device.

R25.9.4.4.3 Where anchorages are located away from the end of a member, local tensile stresses are generated behind these anchorages (Fig. R25.9.1.1b) due to compatibility of deformations ahead of and behind the anchorages. Bonded tie-back reinforcement parallel to the tendon is required in the immediate vicinity of the anchorage to limit the extent of cracking behind the anchorage. The requirement of $\mathbf{0.35}P_{pu}$ was derived using 25 percent of the unfactored prestressing force being resisted by reinforcement at $\mathbf{0.6}f_y$ considering a load factor of 1.2. Therefore, the full yield strength of the reinforcement, f_y, should be used in calculating the provided capacity.

25.9.4.4.4 If tendons are curved in the general zone, bonded reinforcement shall be provided to resist radial and splitting forces, except for monostrand tendons in slabs or where analysis shows reinforcement is not required.

25.9.4.4.5 Reinforcement with a nominal tensile strength equal to 2 percent of the factored prestressing force shall be provided in orthogonal directions parallel to the loaded face of the anchorage zone to limit spalling, except for monostrand tendons in slabs or where analysis shows reinforcement is not required.

R25.9.4.4.5 The spalling force for tendons for which the centroid lies within the kern of the section may be estimated as 2 percent of the total factored prestressing force, except for multiple anchorage devices with center-to-center spacing greater than 0.4 times the depth of the section.

Detailing 25

CODE

25.9.4.4.6 For monostrand anchorage devices for 1/2 in. or smaller diameter strands in normalweight concrete slabs, reinforcement satisfying (a) and (b) shall be provided in the anchorage zone, unless a detailed analysis in accordance with 25.9.4.3 shows that this reinforcement is not required:

(a) Two horizontal bars at least No. 4 in size shall be provided within the local zone parallel to the slab edge ahead of the bearing face of the anchorage device. They shall be permitted to be in contact with the bearing face of the anchorage device, the center of the bars shall be no farther than 4 in. ahead of the bearing face of the device, and the bars shall extend at least 6 in. either side of the outer edges of the device.

(b) If the center-to-center spacing of anchorage devices is 12 in. or less, the anchorage devices shall be considered as a group. For each group of six or more anchorage devices, at least $n + 1$ hairpin bars or closed stirrups at least No. 3 in size shall be provided, where n is the number of anchorage devices. One hairpin bar or stirrup shall be placed between adjacent anchorage devices and one on each side of the group. The hairpin bars or stirrups shall be placed with the horizontal legs extending into the slab perpendicular to the edge. The center line of the vertical leg of the hairpin bars, or the vertical leg of stirrups closest to the anchorage device, shall be placed $3h/8$ to $h/2$ ahead of the bearing face of the anchorage device. Hairpin bars or stirrups shall be detailed in accordance with 25.7.1.1 and 25.7.1.2.

COMMENTARY

R25.9.4.4.6 For monostrand slab tendons, the anchorage-zone minimum reinforcement requirements are based on the recommendations of Breen et al. (1994) and confirmed based on analysis of other test results by Roberts-Wollmann and Wollmann (2008). Typical details are shown in Fig. R25.9.4.4.6. For slabs not thicker than 8 in., with groups of anchors requiring hairpins, the bars parallel to the loaded face can satisfy 25.9.4.4.6(a) and also provide anchorage for the hairpin bars. Thicker slabs require two bars for 25.9.4.4.6 (a) and two additional bars to provide anchorage for the hairpins in accordance with 25.7.1.2. The horizontal bars parallel to the edge required by 25.9.4.4.6(a) should be continuous where possible.

The tests on which the recommendations of Breen et al. (1994) were based were limited to anchorage devices for 1/2 in. diameter, Grade 270 strand, and unbonded tendons in normalweight concrete. For larger strand anchorage devices or for use in lightweight concrete slabs, ACI Committee 423 recommends that the amount and spacing of reinforcement should be conservatively adjusted to provide for the larger anchorage force and smaller splitting tensile strength of lightweight concrete (ACI 423.3R).

ACI 423.3R and Breen et al. (1994) both recommend that hairpin bars also be furnished for anchorages located within 12 in. of slab corners to resist edge tension forces. The meaning of "ahead of" in 25.9.4.4.6 is illustrated in Fig. R25.9.1.1b.

In those cases where multistrand anchorage devices are used for slab tendons, all provisions of 25.9.4 are to be satisfied.

CODE | COMMENTARY

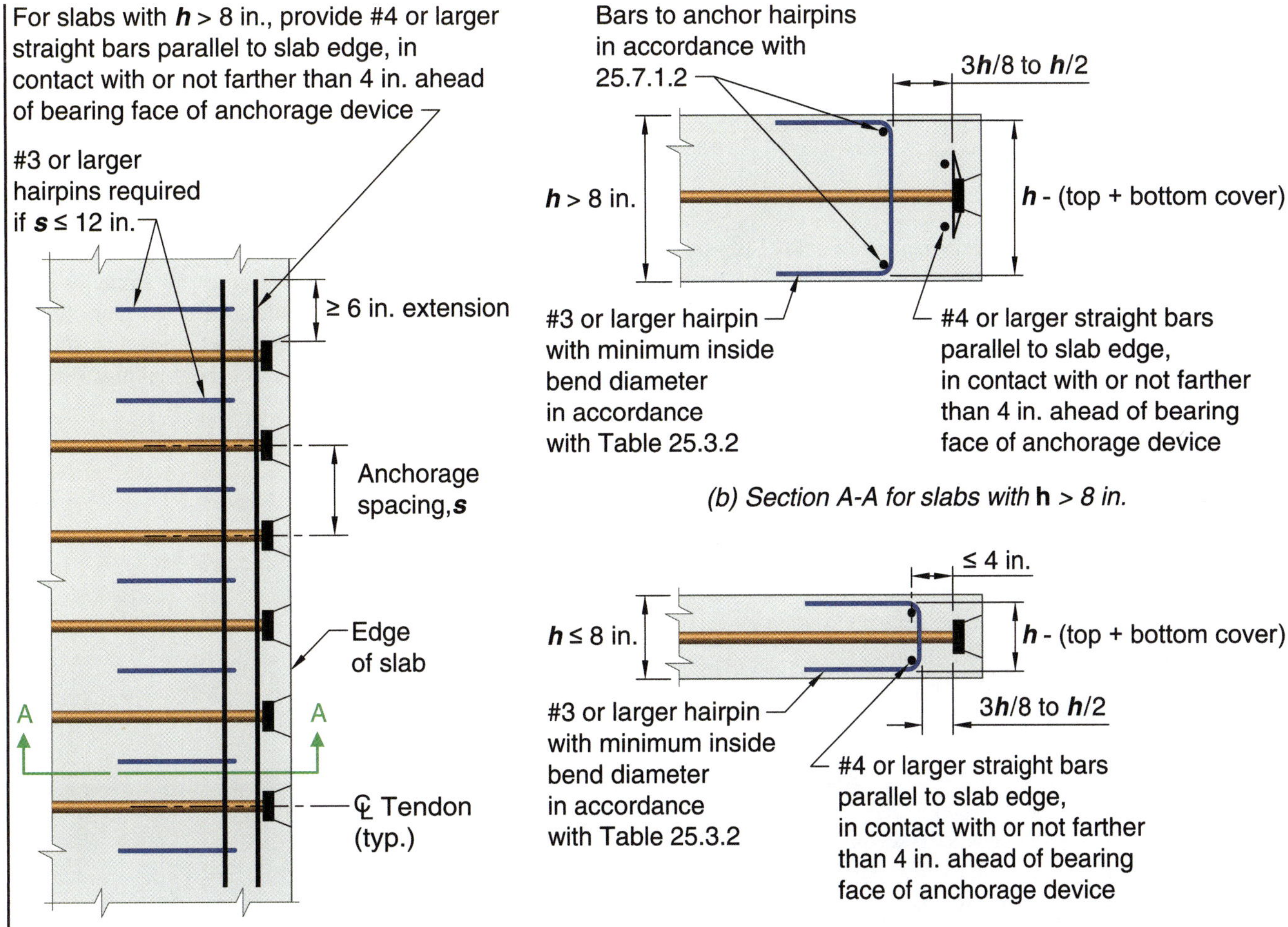

(a) Plan view

(b) Section A-A for slabs with **h** *> 8 in.*

(c) Section A-A for slabs with **h** *≤ 8 in.*

Fig. R25.9.4.4.6*—Anchorage zone reinforcement for groups of 1/2 in. or smaller diameter tendons in slabs (other reinforcement not shown).*

25.9.4.5 *Limiting stresses in general zones*

25.9.4.5.1 Maximum design tensile stress in reinforcement at nominal strength shall not exceed the limits in Table 25.9.4.5.1.

Table 25.9.4.5.1—Maximum design tensile stress in reinforcement

Type of reinforcement	Maximum design tensile stress
Nonprestressed reinforcement	f_y
Bonded, prestressed reinforcement	f_{py}
Unbonded, prestressed reinforcement	f_{se} + 10,000

25.9.4.5.2 Compressive stress in concrete at nominal strength shall not exceed $\mathbf{0.7\lambda f_{ci}'}$, where λ is defined in 19.2.4.

R25.9.4.5 *Limiting stresses in general zones*

R25.9.4.5.1 The value for maximum design tensile stress of bonded prestressed reinforcement is limited to the yield strength of the prestressing reinforcement because Eq. (20.3.2.3.1) may not apply to these nonflexural applications. The value for unbonded prestressed reinforcement is based on 20.3.2.4.1, but limited for these short-length, nonflexural applications.

R25.9.4.5.2 Some inelastic deformation of concrete within general zones is expected because anchorage zone design is based on a strength approach. Unless shown by tests, the λ factor for lightweight concrete should be applied to reflect a lower tensile strength, which is an indirect factor in limiting compressive stresses, as well as the wide scatter and brittleness exhibited in some lightweight concrete anchorage zone tests.

CODE

25.9.4.5.3 If concrete is confined by spirals or hoops and the effect of confining reinforcement is documented by tests and analysis, it shall be permitted to use an increased value of compressive stress in concrete when calculating the nominal strength of the general zone.

25.9.4.5.4 Prestressing reinforcement shall not be stressed until compressive strength of concrete, as indicated by tests of cylinders cured in a manner consistent with curing of the member, is at least 2500 psi for single-strand or bar tendons or at least 4000 psi for multistrand tendons unless 25.9.4.5.5 is satisfied.

25.9.4.5.5 Provisions of 25.9.4.5.4 need not be satisfied if (a) or (b) is satisfied:

(a) Oversized anchorage devices are used to compensate for a lower concrete compressive strength
(b) Prestressing reinforcement is stressed to no more than 50 percent of the final prestressing force

25.9.5 *Reinforcement detailing*

25.9.5.1 Selection of reinforcement size, spacing, cover, and other details for anchorage zones shall make allowances for tolerances on fabrication and placement of reinforcement; for the size of aggregate; and for adequate placement and consolidation of the concrete.

COMMENTARY

R25.9.4.5.3 For well-confined concrete, the effective compressive strength may be increased (Breen et al. 1994). Test results given in Breen et al. (1994) indicate that the compressive stress introduced by auxiliary prestressing applied perpendicular to the axis of the main tendons can be effective in increasing anchorage zone strength.

R25.9.4.5.4 To limit early shrinkage cracking, monostrand tendons are sometimes stressed at concrete strengths less than 2500 psi. In such cases, either oversized monostrand anchorages are used, or the strands are stressed in stages, often to levels one-third to one-half of the final prestressing force as permitted by 25.9.4.5.5.

CHAPTER 26—CONSTRUCTION DOCUMENTS AND INSPECTION

CODE

COMMENTARY

26.1—Scope

R26.1—Scope

This chapter establishes the minimum requirements for information that must be included in the construction documents as applicable to the project. The requirements include information developed in the structural design that must be conveyed to the contractor, provisions directing the contractor on required quality, and inspection requirements to verify compliance with the construction documents. Construction and inspection provisions for anchors were located in Chapter 17 of the 2014 Code. These provisions were moved to Chapter 26 of the 2019 Code.

This chapter is directed to the licensed design professional responsible for incorporating project requirements into the construction documents. The construction documents should contain all of the necessary design and construction requirements for the contractor to achieve compliance with the Code. It is not intended that the Contractor will need to read and interpret the Code.

A general reference in the construction documents requiring compliance with this Code is to be avoided because the contractor is rarely in a position to accept responsibility for design details or construction requirements that depend on detailed knowledge of the design. References to specific Code provisions should be avoided as well because it is the intention of the Code that all necessary provisions be included in the construction documents. For example, references to specific provisions within Chapter 26 are expected to be replaced with the appropriate references within the project construction documents. Reference to ACI and ASTM standards as well as to other documents is expected.

This chapter includes provisions for some of the information that is to be in the construction documents. This chapter is not intended as an all-inclusive list; additional items may be applicable to the Work or required by the building official. ACI 301 is a reference construction specification that is written to be consistent with the requirements of this Code.

Chapter 26 is organized as shown below:

Section	Coverage
26.1	Scope
26.2	Design criteria
26.3	Member information
26.4	Concrete materials and mixture requirements
26.5	Concrete production and construction
26.6	Reinforcement materials and construction requirements
26.7	Anchoring to concrete
26.8	Embedments
26.9	Additional requirements for precast concrete
26.10	Additional requirements for prestressed concrete
26.11	Formwork
26.12	Evaluation and acceptance of hardened concrete
26.13	Inspection

CODE

26.1.1 This chapter addresses (a) through (c):

(a) Design information that the licensed design professional shall specify in the construction documents, if applicable.
(b) Compliance requirements that the licensed design professional shall specify in the construction documents, if applicable.

(c) Inspection requirements that the licensed design professional shall specify in the construction documents, if applicable.

COMMENTARY

R26.1.1(a) and (b) Except for the inspection requirements of 26.13, the provisions of this chapter are organized by design information and compliance requirements.

Design information is project specific and developed during the structural design. It describes the basis of the design or provides information regarding the construction of the Work. Only design information that is applicable to the Work need be provided.

Compliance requirements are general provisions that provide a minimum acceptable level of quality for construction of the Work. It is not the intent of the Code to require the licensed design professional to incorporate verbatim the compliance requirements into the construction documents. Some of these requirements may not be applicable to a specific project.

Construction documents that incorporate the minimum applicable compliance requirements of this chapter are considered to comply with the Code, even if the requirements are stated differently, exceed these minimum requirements, or provide more detail.

R26.1.1(c) Section 26.13 provides inspection provisions to be used in the absence of general building code inspection provisions. These inspection requirements are intended to provide verification that the Work complies with the construction documents.

The inspection requirements of the governing jurisdiction or the general building code take precedence over those included in this chapter. Refer to 26.13.1. ACI 311.4R provides guidance for inspection of concrete construction, and ACI 311.6 is a reference specification for testing services for ready mixed concrete.

26.2—Design criteria

26.2.1 Design information:

(a) Name and year of issue of the Code, general building code, and any supplements governing design.
(b) Loads used in design.

(c) Design work delegated to the contractor including applicable design criteria.

26.2.2 Compliance requirements:

(a) Design work delegated to the contractor shall be performed by a specialty engineer.
(b) The contractor's specialty engineer, relying on the documents identifying the portion of design work assigned, shall produce design work that is compatible with the construction documents and the design criteria provided by the licensed design professional in charge of the design work.

R26.2—Design criteria

R26.2.1(a) and (b) Reference to the applicable version of the documents that govern the design including essential loading information, such as gravity and lateral loading, is to be included in the construction documents.

R26.2.1(c) Examples of design criteria include dimensions, loads, and other assumptions used during design that may affect the delegated portion of the Work.

CODE

(c) The contractor shall submit necessary information to the licensed design professional to confirm that the specialty engineer complied with the documents identifying the portion of the design work assigned.

26.3—Member information

26.3.1 Design information:

(a) Member size, location, and related tolerances.

(b) Members to be constructed using shotcrete.

(c) Identify structural members for which modulus of elasticity testing of concrete mixtures is required.

26.3.2 Compliance requirements:

(a) Use of shotcrete for structural members not identified in the construction documents as required to be placed by shotcrete shall be permitted in accordance with the project contract documents.

26.4—Concrete materials and mixture requirements

26.4.1 *Concrete materials*

26.4.1.1 *Cementitious materials*

26.4.1.1.1 Compliance requirements:

(a) Cementitious materials shall conform to the specifications in Table 26.4.1.1.1(a), except as permitted in 26.4.1.1.1(b).

Table 26.4.1.1.1(a)—Specifications for cementitious materials

Cementitious material	Specification
Portland cement	ASTM C150
Blended hydraulic cements	ASTM C595, excluding Type IS (≥70) and Type IT ($S \geq 70$)
Expansive hydraulic cement	ASTM C845
Hydraulic cement	ASTM C1157
Fly ash and natural pozzolan	ASTM C618
Slag cement	ASTM C989
Silica fume	ASTM C1240

(b) Alternative cements shall be permitted if approved by the licensed design professional and the building official.

COMMENTARY

R26.3—Member information

R26.3.1(a) Construction tolerances for member size and location can be incorporated in construction documents by reference to ACI 117 for cast-in-place construction or to ACI ITG-7 for precast construction. Specific project tolerances that are more restrictive or that are not covered in these references should also be included in the construction documents.

R26.3.2(a) If the contractor submits a request to use shotcrete for portions of the structure, the licensed design professional should make the contractor aware that the proposal must take into consideration provisions in governing shotcrete listed in R4.2.1.1.

R26.4—Concrete materials and mixture requirements

R26.4.1 *Concrete materials*

R26.4.1.1 *Cementitious materials*

R26.4.1.1.1(b) Provisions for strength and durability in Chapter 19 and many requirements in Chapter 26 are based

CODE

Approval shall be based upon test data documenting that the proposed concrete mixture made with the alternative cement meets the performance requirements for the application including structural, fire, and durability.

COMMENTARY

on test data and experience using concretes made with cementitious materials meeting the specifications in Table 26.4.1.1.1(a).

Some alternative cements may not be suitable for use in structural concrete covered by this Code. Therefore, requirements are included for evaluating the suitability of alternative cements. Recommendations for concrete properties to be evaluated are discussed in Becker et al. (2019), ITG-10R, and ITG-10.1R.

In addition to test data, documentation of prior successful use of the proposed alternative cement in structural concrete for conditions with essentially equivalent performance requirements as those of the project can be helpful to the licensed design professional determining whether to allow use of the material. As with all new technologies, a project owner should be informed of the risks and rewards.

26.4.1.2 *Aggregates*

26.4.1.2.1 Compliance requirements:

(a) Aggregates shall conform to (1) or (2):

(1) Normalweight aggregate: ASTM C33.

(2) Lightweight aggregate: ASTM C330.

(b) Aggregates not conforming to ASTM C33 or ASTM C330 are permitted if they have been shown by test or actual service to produce concrete of adequate strength and durability and are approved by the building official.

(c) Crushed hydraulic-cement concrete or recycled aggregate shall be permitted if approved by the licensed design professional and the building official based on documentation that demonstrates compliance with (1) and (2).

(1) Concrete incorporating the specific aggregate proposed for the Work has been demonstrated to provide the mechanical properties and durability required in structural design.

(2) A testing program to verify aggregate consistency and a quality control program to achieve consistency of properties of the concrete are conducted throughout the duration of the project.

R26.4.1.2 *Aggregates*

R26.4.1.2.1(b) Aggregates conforming to ASTM specifications are not always economically available and, in some instances, materials that do not conform to ASTM C33 or C330 may have a documented history of satisfactory performance under similar exposure. Such nonconforming materials are permitted if acceptable evidence of satisfactory performance is provided. Generally, aggregates conforming to the designated specifications should be used.

R26.4.1.2.1(c) This Code requires that concrete made with crushed hydraulic-cement concrete or recycled aggregate be specifically approved for use in a particular project. Properties of fresh and hardened concrete made with these aggregates are influenced by the nature, quality, and variability of the source concrete that is crushed to produce aggregate; nature and variability of the waste-stream from which recycled aggregate is extracted; and the grading, proportions, and uniformity of the resulting aggregate.

ASTM C33 notes that use of such aggregates "may require some additional precautions." These precautions include that any such aggregates meet the durability requirements of ASTM C33 and that the proposed concrete mixture meets the durability requirements of the Exposure Classes assigned for the Work. Areas of special concern include evidence of alkali-silica reactivity, chloride content, and sulfate content of concrete. Additionally, properties of concrete made with crushed hydraulic-cement concrete or recycled aggregate can be significantly more variable than those of comparable concretes made with conventional normalweight aggregates. (Bezaerra Cabral et al. 2010).

CODE

(d) For shotcrete, the aggregate gradation shall conform to ASTM C1436.

26.4.1.3 *Mineral fillers*

26.4.1.3.1 Compliance requirements:

(a) Mineral fillers shall confirm to ASTM C1797.

26.4.1.4 *Water*

26.4.1.4.1 Compliance requirements:

(a) Mixing water shall confirm to ASTM C1602.

26.4.1.5 *Admixtures*

26.4.1.5.1 Compliance requirements:

(a) Admixtures shall conform to (1) through (4):

COMMENTARY

This Code requires explicit documentation to verify that concrete made with crushed hydraulic-cement concrete or recycled aggregate can consistently provide the mechanical properties and durability required in design. Such properties may have been calculated or assumed in the design process, but may not have been specified in contract documents. Specific criteria for approval of concrete made with recycled aggregates including crushed hydraulic-cement concrete are expected to be unique to each project and set of exposure conditions. The project-specific test program and acceptance criteria should be established by the licensed design professional.

ACI 555R provides information on issues that should be considered in verifying required performance.

R26.4.1.3 *Mineral fillers*

R26.4.1.3.1(a) Mineral fillers are finely ground products derived from aggregate that can be used in self-consolidating concrete or in any concrete mixture to improve the properties of fresh and hardened concrete by optimizing particle packing. ASTM C1797 defines Types A and B mineral fillers derived from carbonate aggregate and Type C mineral fillers derived from quarried stone of any mineralogy. Refer to 26.4.2 for restrictions to use of carbonate-based mineral filler in concrete exposed to sulfates.

R26.4.1.4 *Water*

R26.4.1.4.1 Almost any natural water that is potable and has no pronounced taste or odor is satisfactory as mixing water for making concrete. Excessive impurities in mixing water may affect setting time, concrete strength, and volume stability, and may also cause efflorescence or corrosion of reinforcement.

ASTM C1602 allows the use of potable water without testing and includes methods for qualifying nonpotable sources of water, such as from concrete production operations, with consideration of effects on setting time and strength. Testing frequencies are established to ensure continued monitoring of water quality.

ASTM C1602 includes optional limits for chlorides, sulfates, alkalis, and solids in mixing water that can be invoked if appropriate.

R26.4.1.5 *Admixtures*

R26.4.1.5.1(a) ASTM C494 includes Type S—specific performance admixtures—that can be specified if perfor-

CODE

(1) Water reduction and setting time modification: ASTM C494.
(2) Producing flowing concrete: ASTM C1017.
(3) Air entrainment: ASTM C260.
(4) Inhibiting chloride-induced corrosion: ASTM C1582.

(b) Admixtures that do not conform to the specifications in 26.4.1.5.1(a) shall be subject to prior review by the licensed design professional.

(c) Admixtures used in concrete containing expansive cements conforming to ASTM C845 shall be compatible with the cement and produce no deleterious effects.

(d) Admixtures used in shotcrete shall conform to ASTM C1141.

26.4.1.6 *Steel fiber reinforcement*

26.4.1.6.1 Compliance requirements:

(a) Steel fiber reinforcement used for shear resistance shall satisfy (1) and (2):
(1) Be deformed and conform to ASTM A820.
(2) Have a length-to-diameter ratio of at least 50 and not exceeding 100.

26.4.1.7 *Packaged, preblended, dry, combined materials for shotcrete*

26.4.1.7.1 Compliance requirements:

(a) Packaged, preblended, dry, combined materials for shotcrete shall conform to ASTM C1480.

26.4.2 *Concrete mixture requirements*

26.4.2.1 Design information:

(a) Requirements (1) through (17) for each concrete mixture, based on assigned exposure classes or design of members:
(1) Minimum specified compressive strength of concrete, f_c'.
(2) Minimum modulus of elasticity of concrete, E_c, if specified in accordance with 19.2.2.2.
(3) Test age, if different from 28 days, for demonstrating compliance with f_c' and E_c if specified.

COMMENTARY

mance characteristics not listed in 26.4.1.5.1(a) are desired, such as viscosity-modifying admixtures. The basic requirement for a Type S admixture is that it will not have adverse effects on the properties of concrete when tested in accordance with ASTM C494. Meeting the requirements of Type S does not ensure that the admixture will perform its described function. The manufacturer of an admixture presented as conforming to Type S should also be required to provide data that the product will meet the performance claimed.

R26.4.1.5.1(c) In some cases, the use of admixtures in concrete containing ASTM C845 expansive cements has resulted in reduced levels of expansion or increased shrinkage values. Refer to ACI 223R.

R26.4.1.6 *Steel fiber reinforcement*

R26.4.1.6.1(a) Deformations in steel fibers enhance mechanical anchorage with the concrete. The limits for the fiber length-to-diameter ratio are based on available test data (Parra-Montesinos 2006). Because data are not available on the potential for corrosion problems due to galvanic action, the use of deformed steel fibers in members reinforced with stainless-steel bars or galvanized steel bars is not recommended.

R26.4.2 *Concrete mixture requirements*

R26.4.2.1(a) The requirements for each concrete mixture used for the Work are to be stated in the construction documents. These are determined from applicable concrete design requirements in 19.2 and durability requirements in 19.3. The most restrictive requirements that apply are to be stated.

CODE

(4) Maximum ***w/cm*** applicable to most restrictive assigned durability exposure class from 19.3.2.1.

(5) Nominal maximum size of coarse aggregate not to exceed the least of (i), (ii), and (iii):

(i) one-fifth the narrowest dimension between sides of forms

(ii) one-third the depth of slabs

(iii) three-fourths the minimum specified clear spacing between individual reinforcing bars or wires, bundles of bars, prestressed reinforcement, individual tendons, bundled tendons, or ducts

These limitations shall not apply if, in the judgment of the licensed design professional, workability and methods of consolidation are such that concrete can be placed without honeycombs or voids.

(6) Applicable air content for Exposure Category F from 19.3.3.1 or 19.3.3.3.

(7) For members assigned to Exposure Class F3, indicate that concrete mixtures shall meet the limits on supplementary cementitious materials in Table 26.4.2.2(b).

(8) For members assigned to Exposure Class S1, S2, or S3, indicate that mineral fillers derived from carbonate aggregate are prohibited unless approved by the licensed design professional.

(9) Applicable cementitious materials for Exposure Category S from 19.3.2.1.

(10) For members assigned to Exposure Category S, indicate if alternative combinations of cementitious materials qualified in accordance with 26.4.2.2(c) are permitted.

(11) Members in which calcium chloride is prohibited because of assignment to Exposure Class S2 or S3.

COMMENTARY

R26.4.2.1(a)(4) In accordance with Table 19.3.2.1, the ***w/cm*** is based on all cementitious and supplementary cementitious materials in the concrete mixture. The ***w/cm*** of concrete made with alternative cements may not reflect the strength and durability characteristics of the concrete made with portland cement and supplementary cementitious materials permitted in Table 26.4.1.1.1(a). As noted in R26.4.1.1.1(b), it is imperative that testing be conducted to determine the performance of concrete made with alternative cements and to develop appropriate project specification.

R26.4.2.1(a)(5) The size limitations on aggregates are provided to facilitate placement of concrete around the reinforcement without honeycombing due to blockage by closely-spaced reinforcement. It is the intent of the Code that the licensed design professional select the appropriate nominal maximum size aggregate and include this value in the construction documents for each concrete mixture. Because maximum aggregate size can impact concrete properties such as shrinkage, and also the cost of concrete, the largest aggregate size consistent with the requirements of 26.4.2.1 should be permitted. Increasing aggregate size will only decrease shrinkage if there is a concurrent reduction in paste volume.

R26.4.2.1(a)(6) ASTM C94 and ASTM C685 include a tolerance for air content as delivered of ±1.5 percentage points. This same tolerance is acceptable for shotcrete.

R26.4.2.1(a)(8) If concrete members are assigned to Exposure Class S1, S2, or S3, the use of mineral fillers derived from carbonate aggregate in concrete mixtures can result in a form of sulfate attack. Information is provided in ACI 201.2R. ASTM C1797 Type C mineral fillers that are derived from noncarbonate quarried stone can be used in concrete exposed to sulfates. If the quantity of Type A, B, or C mineral filler derived from carbonate aggregate proposed for use is such that the total calcium carbonate content from cement and mineral filler is equal to or less than 15 percent by mass of the cementitious materials, then sulfate resistance can be evaluated by ASTM C1012 to comply with the expansion criteria in Table 26.4.2.2(c).

CODE

(12) For members assigned to exposure class W1 or W2, requirements for the evaluation of the potential for alkali-aggregate reactivity.

(13) Applicable water-soluble chloride ion limits for Exposure Category C from 19.3.2.1.

(14) Equilibrium density of lightweight concrete.

(15) Requirement for submittal of the volumetric fractions of aggregate in lightweight concrete mixtures if Table 19.2.4.1(b) is used as the basis for λ in design.

(16) Requirements for steel fiber-reinforced concrete if used for shear resistance in accordance with 9.6.3.1.

(17) For shotcrete, nominal maximum size of coarse aggregate shall not exceed 1/2 in.

COMMENTARY

R26.4.2.1(a)(12) Members assigned to exposure class W1 or W2 are potentially susceptible to alkali-aggregate reaction. As noted in ASTM C1778, alkali-aggregate reaction (AAR) can occur between the alkali hydroxides in the pore solution of concrete and certain components found in some aggregates. Two types of AAR are recognized depending on the nature of the reactive component: alkali-silica reaction (ASR), which involves various types of reactive siliceous minerals; and alkali-carbonate reaction (ACR), which involves certain types of aggregates that contain dolomite. Both types of reaction can result in expansion and cracking of concrete elements under prolonged exposure to moisture, leading to a reduction in the structural strength and service life of a concrete structure. Options for mitigating ASR, including use of supplementary cementitious materials or limiting alkali content of the concrete, are provided in ASTM C1778. ACR can only be prevented by not using the reactive aggregate.

R26.4.2.1(a)(14 and 15) Equilibrium density is an estimate of the density of lightweight concrete assuming some degree of drying after initial construction. The equilibrium density of lightweight concrete is determined in accordance with ASTM C567. Acceptance of lightweight concrete at the time of delivery is based on a fresh density determined by the concrete supplier that has been correlated with the equilibrium density. The range of fresh densities can vary based on variations in moisture and air content, mixture proportion, and type of lightweight aggregate, and should be considered when establishing the fresh density that will result in the required equilibrium density. Acceptance of lightweight concrete based on density as well as strength is necessary because the value of λ and self-weight used for design is a function of equilibrium density.

R26.4.2.1(a)(16) If steel fibers are used for shear resistance, there are specific requirements for the steel fiber-reinforced concrete: 26.4.1.5.1(a) provides fiber requirements; 26.4.2.2(d) provides minimum dosage requirements; and 26.12.7.1(a) provides acceptance criteria. Fibers are typically specified by fiber type, fiber length, aspect ratio (ℓ/d), and dosage rate (ACI 544.3R).

For structural applications, the Code only addresses the use of discontinuous deformed steel fibers in resisting shear. For other structural applications where it is desired to use discontinuous deformed steel fibers, Section 1.10 provides a procedure for approval. Also, there are nonstructural applications or functional purposes where discontinuous steel fibers are used in concrete. The provisions of the Code that address use of steel fibers for shear strength are not intended for such nonstructural applications.

CODE

(b) At the option of the licensed design professional, exposure classes based on the severity of the anticipated exposure of members.

(c) The required compressive strength at designated stages of construction for each part of the structure designed by the licensed design professional.

26.4.2.2 Compliance requirements:

(a) The required compressive strength at designated stages of construction for each part of the structure not designed by the licensed design professional shall be submitted for review.

(b) For members identified in construction documents as subject to cycles of freezing and thawing and application of deicing chemicals, supplementary cementitious materials, including fly ash and natural pozzolans, silica fume, and slag cement, shall not exceed the maximum percentage allowed in Table 26.4.2.2(b) and shall satisfy (1) and (2).

(1) Supplementary cementitious materials, including fly ash and natural pozzolans, silica fume, and slag cement, used in the manufacture of ASTM C595 and C1157 blended cements shall be included in assessing compliance with the limits in Table 26.4.2.2(b).

(2) The individual limits in Table 26.4.2.2(b) shall apply regardless of the number of cementitious materials in a concrete mixture.

COMMENTARY

R26.4.2.1(b) Durability requirements for concrete are based on exposure classification of members as given in 19.3. Therefore, the exposure classes applicable to the members establish the basis for the requirements for concrete mixtures. Section 19.3.1 requires the licensed design professional to assign exposure classes for different members in the structure. Concrete mixtures should be specified accordingly, but the Code does not require the assigned exposure classes to be explicitly stated in the construction documents. If the licensed design professional is requiring the contractor to determine concrete properties by specifying ACI 301, the assigned exposure classes for all members will need to be stated explicitly in the construction documents.

R26.4.2.1(c) If design or construction requirements dictate that in-place strength of concrete be achieved at specific ages or stages of construction, these requirements should be stated explicitly in the construction documents. Typical stages of construction when the required compressive strength of concrete needs to be specified include at removal of formwork and shores. Additionally, required compressive strength of concrete should be specified for: 1) cast-in-place post-tensioned concrete at the application of post-tensioning; 2) precast concrete at stripping from the forms and during handling, shipping, and erection; and 3) precast, prestressed concrete at transfer of prestress, at stripping from the forms, and during handling, shipping, and erection.

For portions of the structure that are not designed by the licensed design professional, refer to 26.4.2.2(a).

R26.4.2.2(b) These limits on supplementary cementitious materials are applicable to concrete mixtures for members assigned to Exposure Class F3.

CODE

Table 26.4.2.2(b)—Limits on cementitious materials for concrete assigned to Exposure Class F3

Supplementary cementitious materials	Maximum percent of total cementitious materials by mass
Fly ash or natural pozzolans conforming to ASTM C618	25
Slag cement conforming to ASTM C989	50
Silica fume conforming to ASTM C1240	10
Total of fly ash or natural pozzolans and silica fume	35
Total of fly ash or natural pozzolans, slag cement, and silica fume	50

(c) For concrete mixtures for members identified in construction documents to be exposed to sulfate, alternative combinations of cementitious materials to those specified in 26.4.2.1(a)(9) are permitted if tests for sulfate resistance satisfy the criteria in Table 26.4.2.2(c).

Table 26.4.2.2(c)—Requirements for establishing suitability of combinations of cementitious materials for Exposure Class S

Exposure class		Maximum length change for tests in accordance with ASTM C1012, percent		
		At 6 months	At 12 months	At 18 months
S1		0.10	No requirement	No requirement
S2		0.05	0.10[1]	No requirement
S3	Option 1	No requirement	No requirement	0.10
	Option 2	0.05	0.10[1]	No requirement

[1]The 12-month expansion limit applies only if the measured expansion exceeds the 6-month maximum expansion limit.

(d) For concrete identified as being exposed to water in service, evidence shall be submitted that the concrete mixture complies with (1) and (2).

(1) Aggregates are not alkali-silica reactive or measures to mitigate alkali-silica reactivity have been established.

(2) Aggregates are not alkali-carbonate reactive.

(e) Compliance with the specified chloride ion content limits shall be demonstrated by (1) or (2).

(1) Calculating total chloride ion content of the concrete mixture on the basis of measured total chloride ion content from concrete materials and concrete mixture proportions.

COMMENTARY

R26.4.2.2(c) Mixture requirements for Exposure Category S are given in 19.3.2.1. ASTM C1012 may be used to evaluate the sulfate resistance of concrete mixtures using alternative combinations of cementitious materials to those listed in Table 19.3.2.1 for all classes of sulfate exposure. More detailed guidance on qualification of such mixtures using ASTM C1012 is given in ACI 201.2R. The expansion criteria in Table 26.4.2.2(c) for testing in accordance with ASTM C1012 are the same as those in ASTM C595 and C1157 for moderate sulfate resistance (Optional Designation MS) in Exposure Class S1 and for high sulfate resistance (Optional Designation HS) in Exposure Class S2 and Exposure Class S3 Option 2. The 18-month expansion limit only applies for Exposure Class S3, Option 1.

R26.4.2.2(d) Documentation that the potential for AAR has been evaluated can be provided by the concrete supplier. ASTM C1778 provides methods and criteria for determining the reactivity of aggregates and guidance for reducing the risk of deleterious alkali-aggregate reactions in concrete.

R26.4.2.2(e)(1) This procedure was discussed in the Commentary of Code editions before ACI 318-19 and moved into the Code to remove ambiguity over whether it is permitted. It is common practice for total chloride ion content of a proposed concrete mixture to be evaluated by combining total chloride ion content of the concrete materials based on the mixture proportions. Total chloride ion content of cementitious materials and mixing water can be determined in accordance with ASTM C114. Total chloride ion content of aggregates can be determined on an aggregate sample prepared as specified for concrete samples and

CODE / COMMENTARY

tested in accordance with ASTM C1152. Total chloride ion content of admixtures is reported by the supplier. Calculated total chloride ion content determined in this manner is conservative. If calculated total chloride ion content exceeds the limits in Table 19.3.2.1, the concrete materials can be adjusted until compliance is achieved, or water-soluble chloride ion content can be determined using 26.4.2.2(e)(2).

(2) Determining water-soluble chloride ion content of hardened concrete in accordance with ASTM C1218 at age between 28 and 42 days.

R26.4.2.2(e)(2) This option is to determine the water-soluble chloride ion content in hardened concrete by ASTM C1218 and is an alternative to 26.4.2.2(e)(1) if the total chloride ion content calculated in accordance with 26.4.2.2(e)(1) exceeds the limits of Table 19.3.2.1. The chloride ions present in the pore water solution impact the corrosion of reinforcement or embedded metal. To estimate the water-soluble chloride ion content in the concrete that can impact corrosion, ASTM C1218 is used after a period of hydration. The chlorides in some materials, like aggregates, are not available as water-soluble chlorides. Furthermore, some chlorides initially in solution will be bound by hydration of cementitious materials. Chlorides insoluble in water are not considered to accelerate corrosion of embedded metals.

(f) For prestressed concrete, admixtures containing calcium chloride are prohibited.

(g) For concrete placed on or against stay-in-place galvanized steel forms, maximum water soluble chloride ion content shall be 0.30 percent by mass of cementitious materials unless a more stringent limit for the member is specified.

R26.4.2.2(g) The contractor might select a construction option not shown in the construction documents. Because of the critical nature of placements against stay-in-place galvanized steel forms, the Code requires a more stringent chloride ion limit than what may be shown in the construction documents. For example, if a member was originally specified in the construction documents with a chloride limit of 1.00 percent, use of stay-in-place galvanized steel forms results in a change to the more stringent limit of 0.30 percent.

(h) For lightweight concrete, fresh density shall be determined in accordance with ASTM C138 that corresponds with the specified equilibrium density determined in accordance with ASTM C567. The fresh density corresponding to the specified equilibrium density shall be used as the basis of acceptance.

R26.4.2.2(h) ASTM C567 provides two methods for determining equilibrium density. To measure equilibrium density, specimens are maintained at 73°F and 50 percent relative humidity until they achieve constant mass. This measurement can take in excess of 2 months. Alternatively, the calculated equilibrium density can be more rapidly estimated from the oven-dry density. The licensed design professional can require the measurement of equilibrium density in accordance with ASTM C567.

(i) Steel fiber-reinforced concrete used for shear resistance shall satisfy (1) and (2):

(1) Conform to ASTM C1116.

(2) Contain at least 100 lb of deformed steel fibers per cubic yard of concrete.

26.4.3 *Proportioning of concrete mixtures*

26.4.3.1 Compliance requirements:

R26.4.3 *Proportioning of concrete mixtures*

The 2014 edition of the Code deleted the statistical requirements for proportioning concrete that were contained in previous editions. This information was removed from the Code because it is not the responsibility of the licensed

CODE

(a) Concrete mixture proportions shall be established so that the concrete satisfies (1) through (4):

(1) Can be placed without segregation and fully encase reinforcement.

(2) Meets durability requirements given in the construction documents.

(3) Conforms to strength test requirements for standard-cured specimens.

(4) Conforms to modulus of elasticity requirements (i) through (iii) for mixtures requiring testing in accordance with construction documents.

(i) The modulus of elasticity shall be determined as the average modulus obtained from at least three cylinders made from the same sample of concrete and tested at 28 days or at test age designated for E_c.

(ii) Cylinders used to determine modulus of elasticity shall be made and cured in the laboratory in accordance with ASTM C192 and tested in accordance with ASTM C469.

(iii) Modulus of elasticity of a concrete mixture shall be acceptable if the measured value equals or exceeds the specified value.

(b) Concrete mixture proportions shall be established in accordance with Article 4.2.3 of ACI 301 or by an alternative method acceptable to the licensed design professional. Alternative methods shall have a probability of satisfying the strength requirements for acceptance tests of standard-cured specimens that meets or exceeds the probability associated with the method in Article 4.2.3 of ACI 301. If Article 4.2.3 of ACI 301 is used, the strength test records used for establishing and documenting concrete mixture proportions shall not be more than 24 months old.

COMMENTARY

design professional to proportion concrete mixtures. Further, this information is available in other ACI documents, such as ACI 301 and ACI 214R. Finally, the quality control procedures of some concrete producers allow meeting the acceptance criteria of the Code without following the process included in previous editions of the Code.

R26.4.3.1(a) This section provides requirements for developing mixture proportions. The concrete is required to be workable and to meet the durability and strength requirements of the Code. The term "without segregation" is intended to provide for a cohesive mixture in which aggregates remain well distributed while the concrete is in its fresh state. It is recognized that some segregation in the form of bleeding will occur. The required workability will depend on reinforcement congestion, member geometry, and the placement and consolidation methods to be used. Construction requirements of the contractor should be considered in establishing required workability of the concrete.

The Code does not include provisions for especially severe exposures, such as chemical contact, high temperatures, temporary freezing-and-thawing conditions during construction, abrasive conditions, or other unique durability considerations pertinent to the structure. The Code also does not address aesthetic considerations such as surface finishes. If applicable, these items should be covered specifically in the construction documents.

Strength test requirements for standard-cured specimens are given in 26.12.3.

R26.4.3.1(a)(4) Modulus of elasticity testing may be required for the development of concrete mixtures to verify that specified modulus of elasticity can be obtained. It is necessary to specify both $\boldsymbol{E_c}$ and test age. Testing to verify that the specified modulus of elasticity is being attained during construction is at the discretion of the licensed design professional, including specification of acceptance criteria. Field testing may also be required by the local building official.

R26.4.3.1(b) Article 4.2.3 of ACI 301 contains the statistical procedures for selecting the required average strength that were included previously in the Code. Alternatively, the concrete producer may provide evidence acceptable to the licensed design professional that the concrete can be proportioned by another method to meet the project requirements and the acceptance criteria of 26.12.3. The Code presumes that the probability of not meeting the acceptance criteria in 26.12.3 is not more than 1 in 100. Following the method of proportioning in ACI 301 will maintain this level of risk. A key factor in evaluating any proposed alternative propor-

CODE

(c) The concrete materials used to develop the concrete mixture proportions shall correspond to those to be used in the proposed Work.

(d) If different concrete mixtures are to be used for different portions of proposed Work, each mixture shall comply with the concrete mixture requirements stated in the construction documents.

(e) Shotcrete mixture proportions shall be established so that shotcrete satisfies (1) through (3):

(1) Can be placed without segregation and fully encase reinforcement.

(2) Meets durability requirements given in the construction documents.

(3) Conforms to strength test requirements for shotcrete.

COMMENTARY

tioning method should be its ability to preserve this presumed level of risk. Refer to ACI 214R for additional information.

R26.4.3.1(d) If more than one concrete mixture is used for the project, each mixture is required to satisfy Code requirements. A change in concrete constituents, such as sources or types of cementitious materials, aggregates, or admixtures, is considered a different mixture. A minor change in mixture proportions made in response to field conditions is not considered a new mixture.

Concrete mixture requirements to be placed in the construction documents are given in 26.4.2.1(a).

26.4.4 *Documentation of concrete mixture characteristics*

26.4.4.1 Compliance requirements:

(a) Documentation of concrete mixture characteristics shall be submitted for review by the licensed design professional before the mixture is used and before making changes to mixtures already in use. Evidence of the ability of the proposed mixture to comply with the fresh and hardened concrete mixture requirements in the construction documents shall be included in the documentation. The evidence shall include records of consecutive strength tests, as defined in 26.12.1.1, of the same concrete mixture used in previous projects or the results of laboratory trial batches of the proposed mixture.

(b) If field or laboratory test data are not available, and f_c' **≤ 5000 psi**, concrete proportions shall be based on other experience or information, if approved by the licensed design professional. If f_c' **> 5000 psi**, test data documenting the characteristics of the proposed mixtures are required.

(c) It shall be permitted to modify mixtures during the course of the Work. Before using the modified mixture, evidence acceptable to the licensed design professional

R26.4.4 *Documentation of concrete mixture characteristics*

R26.4.4.1(a) Review of the proposed concrete mixture is necessary to ensure that it is appropriate for the project and meets all of the requirements for strength and durability as established by the licensed design professional. The licensed design professional typically reviews the documentation on a proposed concrete mixture to evaluate the likelihood that the concrete will meet the strength-test acceptance requirements of 26.12.3 and that acceptable materials are used. The statistical principles discussed in ACI 214R can be useful in evaluating the likelihood that a proposed mixture will meet the strength-test requirements of 26.12.3.

Concrete mixture requirements to be placed in the construction documents are given in 26.4.2.1(a).

R26.4.4.1(b) If f_c' **≤ 5000 psi** and test data are not available, concrete mixture proportions should be established to produce a sufficiently high average strength such that the likelihood that the concrete would not meet the strength acceptance criteria would be acceptably low. Guidance on an appropriate average strength is provided in ACI 214R. The purpose of this provision is to allow construction to continue when there is an unexpected interruption in concrete supply and there is not sufficient time for testing and evaluation. It also applies for a small project where the cost of trial mixture data is not justified.

R26.4.4.1(c) It is sometimes necessary or beneficial to adjust concrete mixtures during the course of a project. Conditions that could result in mixture adjustments include

CODE

shall be submitted to demonstrate that the modified mixture complies with the concrete mixture requirements in the construction documents.

(d) Documentation of shotcrete mixture characteristics shall be submitted for review by the licensed design professional before the mixture is used and before making changes to mixtures already approved. Evidence of the ability of the proposed shotcrete mixture to comply with the shotcrete mixture requirements in the construction documents shall be included in the documentation.

26.5—Concrete production and construction

26.5.1 *Concrete production*

26.5.1.1 Compliance requirements:

(a) Cementitious materials and aggregates shall be stored to prevent deterioration or contamination.
(b) Material that has deteriorated or has been contaminated shall not be used in concrete.
(c) Equipment for mixing and transporting concrete shall conform to ASTM C94 or ASTM C685.

(d) Ready-mixed and site-mixed concrete shall be batched, mixed, and delivered in accordance with ASTM C94 or ASTM C685.

26.5.2 *Concrete placement and consolidation*

26.5.2.1 Compliance requirements:

(a) Debris and ice shall be removed from spaces to be occupied by concrete before placement.

(b) Standing water shall be removed from place of deposit before concrete is placed unless a tremie is to be used or unless otherwise permitted by both the licensed design professional and the building official.

COMMENTARY

changes in concrete materials, seasonal temperature fluctuations, or changes in conveying and placing methods. Additionally, an adjustment to a concrete mixture may be required or appropriate if strength tests are lower or higher than required.

R26.5—Concrete production and construction

Detailed recommendations for mixing, handling, transporting, and placing concrete are given in ACI 304R.

R26.5.1 *Concrete production*

R26.5.1.1(c) ASTM C94 and ASTM C685 address operational requirements for equipment used to produce concrete.

R26.5.1.1(d) ASTM C94 is a specification for ready mixed concrete whereby materials are primarily measured by mass (weight) and production is by batches. This is the more common method of concrete production, and it is also used in precast concrete plants. ASTM C685 is a specification for concrete where materials are measured by volume and the production is by continuous mixing. These specifications include provisions for capacity of mixers, accuracy of measuring devices, batching accuracy, mixing and delivery, and tests for evaluating the uniformity of mixed concrete.

R26.5.2 *Concrete placement and consolidation*

R26.5.2.1(a) Forms need to be cleaned before beginning to place concrete. In particular, sawdust, nails, wood pieces, and other debris that may collect inside forms need to be removed.

R26.5.2.1(b) The tremie referred to in this provision is not a short tube or "elephant trunk." It is a full-depth pipe used in accordance with accepted procedures for placing concrete under water. Information regarding placing concrete using a tremie is given in ACI 304R.

CODE

(c) Equipment used to convey concrete from the mixer to the location of final placement shall have capabilities to achieve the placement requirements.

(d) Concrete shall not be pumped through pipe made of aluminum or aluminum alloys.

(e) Concrete shall be placed in accordance with (1) through (5):

(1) At a rate to provide an adequate supply of concrete at the location of placement.

(2) At a rate so concrete at all times has sufficient workability such that it can be consolidated by the intended methods.

(3) Without segregation or loss of materials.

(4) Without interruptions sufficient to permit loss of workability between successive placements that would result in cold joints.

(5) Deposited as near to its final location as practicable to avoid segregation due to rehandling or flowing.

(f) Concrete that has been contaminated or has lost its initial workability to the extent that it can no longer be consolidated by the intended methods shall not be used.

(g) Retempering concrete in accordance with the limits of ASTM C94 shall be permitted unless otherwise restricted by the licensed design professional.

(h) After starting, concreting shall be carried on as a continuous operation until the completion of a panel or section, as defined by its boundaries or predetermined joints.

(i) Concrete shall be consolidated by suitable means during placement and shall be worked around reinforcement and embedments and into corners of forms.

COMMENTARY

R26.5.2.1(c) The Code requires the equipment for handling and transporting concrete to be capable of supplying concrete to the place of deposit continuously and reliably under all conditions and for all methods of placement. This applies to all placement methods, including pumps, belt conveyors, pneumatic systems, wheelbarrows, buggies, crane buckets, and tremies.

R26.5.2.1(d) Loss of strength can result if concrete is pumped through pipe made of aluminum or aluminum alloy. This loss is caused by the formation of hydrogen gas generated by the reaction between the cement alkalies and the aluminum eroded from the interior of the pipe surface. The strength reduction has been shown to be as much as 50 percent (Newlon and Ozol 1969). Hence, equipment made of aluminum or aluminum alloys should not be used for pump lines, tremies, or chutes other than short chutes such as those used to convey concrete from a truck mixer.

R26.5.2.1(e) Concrete should be available at a supply rate consistent with the capacity of the placement equipment and the placement crew. Concrete supplied at a faster rate than can be accommodated by placement equipment or crew can result in loss of workability of concrete in equipment waiting to discharge. Excessive delays in the supply of concrete can cause previous placements to stiffen and result in the formation of cold joints.

Each step in the handling and transporting of concrete needs to be controlled to maintain uniformity within a batch and from batch to batch. It is important to minimize segregation of the coarse aggregate from the mortar or of water from the other ingredients.

Rehandling and transferring concrete over large distances from delivery vehicles to the point of placement in the structure can cause segregation of materials. The Code therefore requires that concrete be deposited as close to its final location as possible. However, self-consolidating concrete mixtures can be developed to flow longer distances and maintain their stability with minimal segregation. Guidance on self-consolidating concrete is provided in ACI 237R.

R26.5.2.1(g) ASTM C94 permits water addition to mixed concrete before concrete is discharged to bring it up to the specified slump range as long as prescribed limits on the maximum mixing time and ***w/cm*** are not violated.

R26.5.2.1(i) Detailed recommendations for consolidation of concrete are given in ACI 309R. This guide presents information on the mechanism of consolidation and provides recommendations on equipment characteristics and procedures for various types of concrete mixtures.

CODE

(j) Prior to placement of a new layer of shotcrete, rebound and overspray from adjacent placements shall be removed.
(k) Cuttings and rebound shall not be incorporated into the Work.

(l) Shotcrete surfaces intended to receive subsequent shotcrete placement shall be roughened to a full amplitude of approximately 1/4 in. before the shotcrete has reached final set.
(m) Before placing additional material onto hardened shotcrete, laitance shall be removed, joints shall be cleaned, and the surface shall be dampened.
(n) In-place fresh shotcrete that exhibits sags, sloughs, segregation, honeycombing, or sand pockets shall be removed and replaced.

(o) A certified shotcrete nozzle operator shall place all shotcrete.

(p) If a project-specific shotcrete mockup panel is required, each nozzle operator shall have demonstrated the ability to shoot an approved shotcrete mockup panel.

26.5.3 *Curing*

26.5.3.1 Design information:

(a) If supplementary tests of field-cured specimens are required to verify adequacy of curing and protection, the number and size of test specimens and the frequency of these supplementary tests.

26.5.3.2 Compliance requirements:

(a) Concrete, other than high-early-strength, shall be maintained at a temperature of at least 50°F and in a moist condition for at least the first 7 days after placement, except if accelerated curing is used.
(b) High-early-strength concrete shall be maintained at a temperature of at least 50°F and in a moist condition for at

COMMENTARY

R26.5.2.1(j and k) Rebound material is loose aggregate and cement paste that bounces off after colliding with formwork, reinforcement, or a hardened shotcrete surface.

Overspray is the paste-rich material that separates from the stream during shotcreting and adheres to nearby reinforcement and formwork. Adjacent surfaces are typically protected from overspray.

Cuttings refers to shotcrete that has been applied beyond the finish face and is cut off during trimming or rod finishing.

Rod finishing refers to the use of a hard-edged tool or rod to cut excess material by trimming, slicing, or scraping the exposed shotcrete to a true line and grade.

R26.5.2.1(n) If the shotcrete sags because of improper consistency, adjacent vibration, or improper finishing, those sections should also be removed and replaced. ACI 506.4R provides additional recommendations for repairing shotcrete.

R26.5.2.1(o) Nozzle operators become certified through testing and training programs that include written and performance examinations. Each shotcrete nozzle operator should be certified in accordance with the applicable ACI certification program for dry-mix or wet-mix shotcrete (both are covered by CPP 660.1-15).

R26.5.3 *Curing*

Detailed recommendations for curing concrete are given in ACI 308R. This guide presents basic principles of proper curing and describes the various methods, procedures, and materials for curing of concrete.

CODE

least the first 3 days after placement, except if accelerated curing is used.

(c) Accelerated curing to accelerate strength gain and reduce time of curing is permitted using high-pressure steam, steam at atmospheric pressure, heat and moisture, or other process acceptable to the licensed design professional. If accelerated curing is used, (1) and (2) shall apply:

(1) Compressive strength at the load stage considered shall be at least the strength required at that load stage.

(2) Accelerated curing shall not impair the durability of the concrete.

(d) If required by the building official or licensed design professional, test results for cylinders made and cured in accordance with (1) and (2) shall be provided in addition to test results for standard-cured cylinders.

(1) At least two 6 x 12 in. or at least three 4 x 8 in. cylinders to be field cured shall be molded at the same time and from the same samples as standard-cured cylinders.

(2) Field-cured cylinders shall be cured in accordance with the field curing procedure of ASTM C31 and tested in accordance with ASTM C39.

(e) Procedures for protecting and curing concrete shall be considered adequate if (1) or (2) are satisfied:

(1) Average strength of field-cured cylinders at test age designated for determination of f_c' is equal to or at least 85 percent of that of companion standard-cured cylinders.

(2) Average strength of field-cured cylinders at test age exceeds f_c' by more than 500 psi.

COMMENTARY

R26.5.3.2(c) This section applies whenever an accelerated curing method is used, whether for precast or cast-in-place elements. EB-001.15, and PCI MNL 116, and PCI MNL 117 provide general information on accelerated curing. Accelerated curing procedures require careful attention to obtain uniform and satisfactory results. Preventing moisture loss during the curing is essential.

The compressive strength of accelerated-cured concrete is not as high at later ages as that of nominally identical concrete continuously cured under moist conditions at moderate temperatures. Also, the modulus of elasticity, E_c, of accelerated-cured specimens may vary from that of specimens moist-cured at normal temperatures.

R26.5.3.2(d) Strengths of cylinders cured under field conditions may be required to evaluate the adequacy of curing and protection of concrete in the structure.

The Code provides a specific criterion in 26.5.3.2(e) for determining the adequacy of curing and protection afforded to the structure. For a valid comparison, field-cured cylinders and companion standard-cured cylinders need to be made from the same sample. Field-cured cylinders are to be cured, as nearly as possible, under the same conditions as the structure. The field-cured cylinders should not be treated more favorably than the structural members they represent.

In evaluating test results of field-cured cylinders, it should be recognized that even if cylinders are protected in the same manner as the structure, they may not experience the same temperature history as the concrete in the structure. This different temperature history occurs because heat of hydration may be dissipated differently in a cylinder compared with the structural member.

R26.5.3.2(e) Research (Bloem 1968) has shown that the strength of cylinders protected and cured to simulate good field practice should be at least about 85 percent of standard-cured cylinders if both are tested at the age designated for f_c'. Thus, a value of 85 percent has been set as a rational basis for determining the adequacy of field curing. The comparison is made between the measured strengths of companion field-cured and standard-cured cylinders, not between the strength of field-cured cylinders and the value of f_c'. Test results for the field-cured cylinders are considered satisfactory, however, if the strength of field-cured cylinders exceeds f_c' by more than 500 psi, even though they fail to reach 85 percent of the strength of companion standard-cured cylinders.

The 85 percent criterion is based on the assumption that concrete is maintained above 50°F and in a moist condition for at least the first 7 days after placement, or high-early-strength concrete is maintained above 50°F and in a moist condition for at least the first 3 days after placement.

CODE

COMMENTARY

If the field-cured cylinders do not provide satisfactory strength by this comparison, steps need to be taken to improve the curing. If the tests indicate a possible serious deficiency in strength of concrete in the structure, core tests may be required, with or without supplemental wet curing, to evaluate the structural adequacy, as provided in 26.12.6.

(f) Shotcrete shall be cured in accordance with (1) through (3).

(1) For 24 hours from completion of placement, initial curing shall be provided by one of the following methods:

(i) Ponding, fogging, or continuous sprinkling;

(ii) Absorptive mat, fabric, or other protective covering kept continuously moist;

(iii) Application of a membrane-forming curing compound.

(2) After 24 hours from completion of placement, final curing shall be provided by one of the following methods:

(i) Same method used in the initial curing process;

(ii) Sheet materials;

(iii) Other moisture-retaining covers kept continuously moist.

(3) Final curing shall be maintained for a minimum duration of:

(i) 7 days,

(ii) 3 days if either a high-early-strength cement or an accelerating admixture is used.

R26.5.3.2(f) If using a curing compound, it will usually be necessary to apply the compound at a higher rate than the manufacturer's recommendation because of the rougher surface of many shotcrete applications.

26.5.4 *Concreting in cold weather*

R26.5.4 *Concreting in cold weather*

Detailed recommendations for cold weather concreting are given in ACI 306R. Specification requirements for concreting in cold weather are provided in ACI 301 and ACI 306.1. If both ACI 301 and ACI 306.1 are referenced in construction documents, the governing requirements should be identified.

26.5.4.1 Design information:

(a) Temperature limits for concrete as delivered in cold weather.

R26.5.4.1(a) ASTM C94, ACI 306R, and ACI 301 contain requirements and recommendations for concrete temperature based on section size.

26.5.4.2 Compliance requirements:

(a) Adequate equipment shall be provided for heating concrete materials and protecting concrete during freezing or near-freezing weather.

(b) Frozen materials or materials containing ice shall not be used.

(c) Forms, fillers, and ground with which concrete is to come in contact shall be free from frost and ice.

CODE

(d) Concrete materials and production methods shall be selected so that the concrete temperature at delivery complies with the specified temperature limits.

26.5.5 *Concreting in hot weather*

26.5.5.1 Design information:

(a) Temperature limits for concrete as delivered in hot weather.

26.5.5.2 Compliance requirements:

(a) Concrete materials and production methods shall be selected so that the concrete temperature at delivery complies with the specified temperature limits.
(b) Handling, placing, protection, and curing procedures shall limit concrete temperatures or water evaporation that could reduce strength, serviceability, and durability of the member or structure.

26.5.6 *Construction, contraction, and isolation joints*

26.5.6.1 Design information:

(a) If required by the design, locations and details of construction, isolation, and contraction joints.

(b) Details required for transfer of shear and other forces through construction joints.

(c) Surface preparation, including intentional roughening of hardened concrete surfaces where concrete is to be placed against previously hardened concrete.

(d) Locations where shear is transferred between as-rolled steel and concrete using headed studs or welded reinforcing bars requiring steel to be clean and free of paint.
(e) Surface preparation including intentional roughening if composite topping slabs are to be cast in place on a precast floor or roof intended to act structurally with the precast members.

COMMENTARY

R26.5.5 *Concreting in hot weather*

Detailed recommendations for hot weather concreting are given in ACI 305R. This guide identifies the hot weather factors that affect concrete properties and construction practices and recommends measures to eliminate or minimize undesirable effects. Specification requirements for concreting in hot weather are provided in ACI 301 and ACI 305.1.

R26.5.5.1(a) ACI 301 and ACI 305.1 limit the maximum concrete temperature to 95°F at the time of placement.

R26.5.6 *Construction, contraction, and isolation joints*

For the integrity of the structure, it is important that joints in the structure be located and constructed as required by the design. Any deviations from locations indicated in construction documents should be approved by the licensed design professional.

Construction or other joints should be located where they will cause the least weakness in the structure. Lateral force design may require additional consideration of joints during design.

R26.5.6.1(b) Shear keys, intermittent shear keys, diagonal dowels, or shear friction may be used where force transfer is required. If shear friction at a joint interface in accordance with 22.9 is invoked in the design, include applicable construction requirements in the construction documents.

R26.5.6.1(c) The preparations referenced are applicable if design for shear friction is in accordance with 22.9 and for contact surfaces at construction joints for structural walls.

R26.5.6.1(d) The locations referenced are those for which design for shear friction is in accordance with 22.9.

CODE

(f) For shotcrete, location of construction joints for which square joints are permitted.

26.5.6.2 Compliance requirements:

(a) Joint locations or joint details not shown or that differ from those indicated in construction documents shall be submitted for review by the licensed design professional.

(b) Except for prestressed concrete, construction joints in floor and roof systems shall be located within the middle third of spans of slabs, beams, and girders unless otherwise approved by the licensed design professional.

(c) Construction joints in girders shall be offset a distance of at least two times the width of intersecting beams, measured from the face of the intersecting beam, unless otherwise approved by the licensed design professional.
(d) Construction joints shall be cleaned and laitance removed before new concrete is placed.
(e) Surface of concrete construction joints shall be intentionally roughened if specified.
(f) Immediately before new concrete is placed, construction joints shall be prewetted and standing water removed.
(g) For shotcrete, construction joint surfaces shall be cut at a 45-degree angle to the finished surface, unless a square joint is designated in the construction documents.
(h) For shotcrete, construction joints proposed at locations not shown on the construction documents shall be submitted to the licensed design professional for approval prior to shotcrete placement.

26.5.7 *Construction of concrete members*

26.5.7.1 Design information:

(a) Details required to accommodate dimensional changes resulting from prestressing, creep, shrinkage, and temperature.
(b) Identify if a slab-on-ground is designed as a structural diaphragm or part of the seismic-force-resisting system.

COMMENTARY

R26.5.6.2(a) If the licensed design professional does not designate specific joint locations, the contractor should submit joint locations for construction to the licensed design professional for review to determine that the proposed locations do not impact the performance of the structure.

R26.5.6.2(b) Tendons of continuous post-tensioned slabs and beams are usually stressed at a point along the span where the tendon profile is at or near the centroid of the concrete cross section. Therefore, interior construction joints are usually located within the end thirds of the span rather than the middle third of the span. Construction joints located within the end thirds of continuous post-tensioned slab and beam spans have a long history of satisfactory performance; therefore, 26.5.6.2(b) is not applicable to prestressed concrete.

R26.5.7 *Construction of concrete members*

R26.5.7.1(b) A slab-on-ground may be designed to act as a structural diaphragm or to provide required ties between foundations. The construction documents should clearly identify any slab on ground that is a structural diaphragm, and state that saw cutting or joints are prohibited unless approved by the licensed design professional. Joints can affect the integrity of the slab and its ability to act as a structural diaphragm, unless structural repairs are made. Refer also to 26.5.7.2(d).

CODE

(c) Details for construction of sloped or stepped footings designed to act as a unit.
(d) Locations where floor system and column concrete placements are required to be integrated during placement in accordance with 15.5.
(e) Locations where steel fiber-reinforced concrete is required for shear resistance in accordance with 9.6.3.1.

26.5.7.2 Compliance requirements:

(a) Beams, girders, or slabs supported by columns or walls shall not be cast until concrete in the vertical support members is no longer plastic.

(b) Beams, girders, haunches, drop panels, shear caps, and capitals shall be placed monolithically as part of a slab system, unless otherwise shown in construction documents.

(c) At locations where floor system and column concrete placements are required to be integrated during placement, column concrete shall extend full depth of the floor system at least 2 ft into the floor system from face of column and be integrated with floor system concrete.

(d) Saw cutting or construction of joints that can affect the integrity of a slab-on-ground identified in the construction documents as structural diaphragms or part of the seismic-force-resisting system shall not be permitted unless specifically indicated or approved by the licensed design professional.

26.6—Reinforcement materials and construction requirements

26.6.1 *General*

26.6.1.1 Design information:

(a) ASTM designation and grade of reinforcement, including applicable requirements for special seismic systems in accordance with 20.2.2.5.

(b) Type, size, location requirements, detailing, and embedment length of reinforcement.
(c) Concrete cover to reinforcement.

(d) Location and length of lap splices.

(e) Type and location of mechanical splices.

COMMENTARY

R26.5.7.2(a) Delay in placing concrete in members supported by columns and walls is necessary to minimize potential cracking at the interface of the slab and supporting member caused by bleeding and settlement of plastic concrete in the supporting member.

R26.5.7.2(b) Separate placement of slabs and beams, haunches, or similar elements is permitted if shown in the construction documents and if provision has been made to transfer forces as required in 22.9.

R26.5.7.2(c) Application of the concrete placement procedure described in 15.5 may require the placing of two different concrete mixtures in the floor system. It is the responsibility of the licensed design professional to indicate in the construction documents where the higher- and lower-strength concretes are to be placed.

R26.5.7.2(d) This restriction applies to slabs identified as structural diaphragms in 26.5.7.1(b).

R26.6—Reinforcement materials and construction requirements

R26.6.1 *General*

R26.6.1.1(a) If ASTM A615 reinforcement is used in place of ASTM A706 reinforcement in special seismic systems, the strength and minimum elongation requirements of 20.2.1.3 and 20.2.2.5(b) should be included.

R26.6.1.1(d) Splices should, if possible, be located away from points of maximum tensile stress. The lap splice requirements of 25.5.2 encourage this practice.

26 Construction

CODE

(f) Type and location of end-bearing splices.
(g) Type and location of welded splices and other required welding of reinforcing bars.
(h) ASTM designation for protective coatings of nonprestressed reinforcement.
(i) Corrosion protection for exposed reinforcement intended to be bonded with extensions on future Work.

26.6.1.2 Compliance requirements:

(a) Mill test reports for reinforcement shall be submitted.
(b) Nonprestressed reinforcement with rust, mill scale, or a combination of both shall be considered satisfactory, provided a hand-wire-brushed representative test specimen of the reinforcement complies with the applicable ASTM specification for the minimum dimensions (including height of deformations) and weight per unit length.

(c) Prestressing reinforcement shall be free of mill scale, pitting, and excessive rust. A light coating of rust shall be permitted.
(d) At the time concrete is placed, reinforcement to be bonded shall be clean of ice, mud, oil, or other deleterious coatings that decrease bond.

26.6.2 *Placement*

26.6.2.1 Design information:

(a) Tolerances on location of reinforcement taking into consideration tolerances on d and specified concrete cover in accordance with Table 26.6.2.1(a).

Table 26.6.2.1(a)—Tolerances on d and specified cover

d, in.	Tolerance on d, in.	Tolerance on specified concrete cover, in.[1]	
≤ 8	±3/8	Smaller of:	–3/8
			– (1/3) · specified cover
> 8	±1/2	Smaller of:	–1/2
			– (1/3) · specified cover

[1]Tolerance for cover to formed soffits is –1/4 in.

(b) Tolerance for longitudinal location of bends and ends of reinforcement in accordance with Table 26.6.2.1(b). The tolerance for specified concrete cover in Table 26.6.2.1(a) shall also apply at discontinuous ends of members.
(c) Tolerance for spacing of hoops in members of intermediate and special seismic systems:

COMMENTARY

R26.6.1.1(g) Refer to R26.6.4.

R26.6.1.2(b) Specific limits on rust are based on tests (Kemp et al. 1968) plus a review of earlier tests and recommendations. Kemp et al. (1968) provides guidance with regard to the effects of rust and mill scale on bond characteristics of deformed reinforcing bars. Research has shown that a normal amount of rust increases bond. Normal rough handling generally removes rust that is loose enough to impair the bond between the concrete and reinforcement.

R26.6.1.2(c) Guidance for evaluating the degree of rusting on strand is given in Sason (1992).

R26.6.1.2(d) The use of epoxy coating in accordance with 20.5.2 is permitted. Materials used for the protection of prestressed reinforcement against corrosion in unbonded tendons are not considered to be contaminants as described in this provision.

R26.6.2 *Placement*

R26.6.2.1 Generally accepted practice, as reflected in ACI 117, has established tolerances on total depth (formwork or finish) and fabrication of closed ties, stirrups, spirals, and truss bent reinforcing bars. The licensed design professional should specify more restrictive tolerances than those permitted by the Code when necessary to minimize the accumulation of tolerances resulting in excessive reduction in effective depth or cover.

More restrictive tolerances have been placed on minimum clear distance to formed soffits because of their importance for durability and fire protection and because reinforcement is usually supported in such a manner that the specified tolerance is practical.

More restrictive tolerances than those required by the Code may be desirable for prestressed concrete. In such cases, the construction documents should specify the necessary tolerances. Recommendations are provided in ACI ITG-7.

The Code permits a reinforcement placement tolerance on effective depth d that is directly related to the flexural and shear strength of the member. Because reinforcement is placed with respect to edges of members and formwork surfaces, d is not always conveniently measured in the field. This provision is included in the design information section

CODE

(1) Lesser of +1-1/2 in. and +1.5d_b of the smallest longitudinal bar.

(2) Lesser of –1 in. per ft of least side dimension of member and –3 in.

(3) Spacing adjustments shall result in no more than two hoops being in contact with each other.

Table 26.6.2.1(b)—Tolerances for longitudinal location of bends and ends of reinforcement

Location of bends or reinforcement ends	Tolerances, in.
Discontinuous ends of brackets and corbels	±1/2
Discontinuous ends of other members	±1
Other locations	±2

26.6.2.2 Compliance requirements:

(a) Reinforcement, including bundled bars, shall be placed within required tolerances and supported to prevent displacement beyond required tolerances during concrete placement.

(b) Spiral units shall be continuous bar or wire placed with even spacing and without distortion beyond the tolerances for the specified dimensions.

(c) Splices of reinforcement shall be made only as permitted in the construction documents, or as authorized by the licensed design professional.

(d) For longitudinal column bars forming an end-bearing splice, the bearing of square cut ends shall be held in concentric contact.

COMMENTARY

because tolerances on ***d*** should be considered in member design. Placement tolerances for cover are also provided.

Tolerances for placement of reinforcement should be specified in accordance with ACI 117 unless stricter tolerances are required. The more restrictive tolerance for spacing of hoops in members of intermediate and special seismic systems is to provide better control against premature buckling of longitudinal bars.

R26.6.2.2(a) Reinforcement, including bundled bars, should be adequately supported in the forms to prevent displacement by concrete placement or workers. Bundled bars should be tied or otherwise fastened together to maintain their position, whether vertical or horizontal. Beam stirrups should be supported on the bottom form of the beam by supports such as continuous longitudinal beam bolsters. If only the longitudinal beam bottom reinforcement is supported, construction traffic can dislodge the stirrups as well as any top beam reinforcement tied to the stirrups.

R26.6.2.2(b) Spirals should be held firmly in place, at proper pitch and alignment, to prevent displacement during concrete placement. The Code has traditionally required spacers to hold the fabricated spiral cage in place, but alternate methods of installation are also permitted. If spacers are used, the following may be used for guidance: for spiral bar or wire smaller than 5/8 in. diameter, a minimum of two spacers should be used for spirals less than 20 in. in diameter, three spacers for spirals 20 to 30 in. in diameter, and four spacers for spirals greater than 30 in. in diameter. For spiral bar or wire 5/8 in. diameter or larger, a minimum of three spacers should be used for spirals 24 in. or less in diameter, and four spacers for spirals greater than 24 in. in diameter.

R26.6.2.2(d) Experience with end-bearing splices has been almost exclusively with vertical bars in columns. If bars are significantly inclined from the vertical, attention is required to ensure that adequate end-bearing contact can be achieved and maintained.

CODE

(e) Bar ends shall terminate in flat surfaces within 1.5 degrees of a right angle to the axis of the bars and shall be fitted within 3 degrees of full bearing after assembly.

26.6.3 *Bending*

26.6.3.1 Design information:

(a) Nonstandard bend geometry.

26.6.3.2 Compliance requirements:

(a) Reinforcement shall be bent cold prior to placement, unless otherwise permitted by the licensed design professional.

(b) Field bending of reinforcement partially embedded in concrete shall not be permitted, except as shown in the construction documents or permitted by the licensed design professional.

(c) Offset bars shall be bent before placement in the forms.

26.6.4 *Welding*

26.6.4.1 Design information:

(a) Details for welding of anchor bars at the front face of brackets or corbels designed by the licensed design professional in accordance with 16.5.6.3(a).

COMMENTARY

R26.6.2.2(e) These tolerances represent practice based on tests of full-size members containing No. 18 bars.

R26.6.3 *Bending*

R26.6.3.1 Bend radii larger than the minimums of Tables 25.3.1 and 25.3.2 may be required by geometric constraints or by 23.10 for discontinuity regions designed using the strut-and-tie method with curved-bar nodes. Nonstandard bends should be indicated on the drawings.

R26.6.3.2(b) Construction conditions may make it necessary to bend bars that have been embedded in concrete. Such field bending should not be done without authorization of the licensed design professional. Construction documents should specify whether the bars will be permitted to be bent cold or if heating should be used. Bends should be gradual and should be straightened as required.

Tests (Black 1973; Stecich et al. 1984) have shown that ASTM A615 Grade 40 and Grade 60 reinforcing bars can be cold bent and straightened up to 90 degrees at or near the minimum diameter specified in 25.3. If cracking or breakage is encountered, heating to a maximum temperature of 1500°F may avoid this condition for the remainder of the bars. Bars that fracture during bending or straightening can be spliced outside the bend region.

Heating should be performed in a manner that will avoid damage to the concrete. If the bend area is within approximately 6 in. of the concrete, some protective insulation may need to be applied. Heating of the bar should be controlled by temperature-indicating crayons or other suitable means. The heated bars should not be artificially cooled (with water or forced air) until after cooling to at least 600°F.

R26.6.4 *Welding*

If welding of reinforcing bars is required, the weldability of the steel and compatible welding procedures need to be considered. The provisions in AWS D1.4 cover aspects of welding reinforcing bars, including criteria to qualify welding procedures.

Weldability of the steel is based on its carbon equivalent (CE), calculated from the chemical composition of the steel. AWS D1.4 establishes preheat and interpass temperatures for a range of carbon equivalents and reinforcing bar sizes. AWS D1.4 has two expressions for calculating CE. The expression considering only the elements carbon and

CODE

COMMENTARY

manganese is to be used for bars other than ASTM A706. A more comprehensive CE expression is given for ASTM A706 bars, which is identical to the CE formula presented in ASTM A706.

ASTM A706 covers low-alloy steel reinforcing bars intended for applications that require controlled tensile properties, welding, or both. Weldability is accomplished in ASTM A706 by requiring the CE not to exceed 0.55 percent and controlling the chemical composition. The manufacturer is required by ASTM A706 to report the chemical analysis and carbon equivalent (Gustafson and Felder 1991). When welding reinforcing bars other than ASTM A706, the construction documents should specifically require that the mill test report include chemical analysis results to permit calculation of the carbon equivalent.

It is often necessary to weld to existing reinforcing bars in a structure when no mill test report of the existing reinforcement is available. This condition is particularly common in alterations or building expansions. AWS D1.4 states for such bars that a chemical analysis may be performed on representative bars. If the chemical composition is not known or obtained, AWS D1.4 requires a minimum preheat. For bars other than ASTM A706, the minimum preheat required is 300°F for No. 6 bars or smaller, and 500°F for No. 7 bars or larger. The required preheat for all sizes of ASTM A706 bars is to be the temperature given in the Welding Code's table for minimum preheat corresponding to the range of CE "over 0.45 percent to 0.55 percent." Welding of the particular bars should be performed in accordance with AWS D1.4. It should also be determined if additional precautions are necessary, based on other considerations such as stress level in the bars, consequences of failure, and heat damage to existing concrete due to welding operations.

AWS D1.4 requires the contractor to prepare welding procedure specifications (WPSs) conforming to the requirements of the Welding Code. Appendix A in AWS D1.4 contains a suggested form that shows the information required for a WPS.

Welding of wire to wire, and of wire or welded wire reinforcement to reinforcing bars or structural steel elements is not covered by AWS D1.4. If welding of this type is required on a project, the construction documents should specify requirements or performance criteria for this welding. If cold-drawn wires are to be welded, the welding procedures should address the potential loss of yield strength and ductility achieved by the cold-working process (during manufacture) when such wires are heated by welding. These potential concerns are not an issue for machine and resistance welding as used in the manufacture of welded plain and deformed wire reinforcement covered by ASTM A1064.

26.6.4.2 Compliance requirements:

(a) Welding of all nonprestressed bars shall conform to the requirements of AWS D1.4. ASTM specifications for bar

CODE

reinforcement, except for ASTM A706, shall be supplemented to require a mill test report of material properties that demonstrate conformance to the requirements in AWS D1.4.

(b) Welding of crossing bars shall not be used for assembly of reinforcement except at the front face of brackets or corbels or as otherwise permitted by the licensed design professional.

26.7—Anchoring to concrete

26.7.1 Design information:

(a) Requirements for assessment and qualification of anchors for the applicable conditions of use shall be in accordance with 17.1.2.

(b) Type, size, location requirements, effective embedment depth, and installation requirements for anchors.

(c) Type, size, and location or location requirements for anchor reinforcement designed to develop the anchor strength in accordance with 17.5.2.1, as well as transverse confinement reinforcement for anchors installed in the tops of columns or pedestals in accordance with 10.7.6.1.5.

(d) Type, size, and location for shear lugs designed to develop shear strength in accordance with 17.11.

(e) Size and location of base plate holes to permit inspection and vent air when placing concrete or grout per 17.11.1.2.

(f) Minimum edge distance of anchors in accordance with 17.9.

(g) Corrosion protection for exposed anchors intended for attachment with future Work.

(h) For post-installed anchors, parameters associated with the design strength in accordance with 17.5, including anchor category, concrete strength, aggregate type, type of lightweight concrete, required installation torque, and requirements for hole drilling and preparation.

(i) For adhesive anchors in tension, parameters associated with the characteristic bond stress used for design in accordance with 17.6.5, including concrete temperature range, moisture condition of concrete at time of installation, type

COMMENTARY

R26.6.4.2(b) "Tack" welding (welding crossing bars) can seriously weaken a bar at the point welded by creating a metallurgical notch effect. This operation can be performed safely only when the material welded and welding operations are under continuous competent control, as in the manufacture of welded wire reinforcement. Welding of anchor bars at the front face of brackets or corbels is addressed in R16.5.6.3.

R26.7—Anchoring to concrete

R26.7.1 Minimum requirements for specification of anchors in the construction documents for conformance with this Code are listed.

R26.7.1(a) Post-installed anchor strength and deformation capacity are assessed by acceptance testing under ACI 355.2 or ACI 355.4. These tests are carried out assuming installation in accordance with the manufacturer's recommended procedures (in the case of adhesive anchors, the Manufacturer's Printed Installation Instructions [MPII]).

R26.7.1(h) Certain types of anchors can be sensitive to variations in hole diameter, cleaning conditions, orientation of the axis, magnitude of the installation torque, crack width, and other variables. Some of this sensitivity is indirectly accounted for in the assigned ϕ values for the different anchor categories, which depend in part on the results of the installation safety tests. If anchor components are altered or if anchor installation procedures deviate from those specified, the anchor may fail to comply with the acceptance criteria of ACI 355.2 or 355.4.

R26.7.1(i) Due to the sensitivity of bond strength to installation, on-site quality control is important for adhesive anchors. The construction documents must provide all parameters relevant to the characteristic bond stress used in design. These parameters may include, but are not limited to:

CODE

of lightweight concrete, if applicable, and requirements for hole drilling and preparation.

(j) Identification of adhesive anchors installed in a horizontal or upwardly inclined orientation to resist sustained tensile loads.

(k) Identification of adhesive anchors requiring proof loading in accordance with ACI 355.4 or the inspection program established by the licensed design professional.

(l) Specify certification required for installers of adhesive anchors including adhesive anchors that are installed in a horizontal or upwardly inclined orientation to resist sustained tensile loads.

COMMENTARY

(a) Acceptable anchor installation environment (dry or saturated concrete; concrete temperature range)

(b) Acceptable drilling methods

(c) Required hole cleaning procedures

(d) Anchor type and size range (threaded rod or reinforcing bar)

Hole cleaning is intended to ensure that drilling debris and dust do not impair bond. Depending on the Manufacturer's Printed Installation Instructions (MPII), type of qualified anchor, and on-site conditions, hole cleaning may involve operations to remove drilling debris from the hole with vacuum or compressed air mechanical brushing of the hole wall to remove surface dust, and a final step to evacuate any remaining dust or debris, usually with compressed air. If wet core drilling is used, holes may be flushed with water and then dried with compressed air. Compressed air must be free of oil and moisture. For anchors installed in locations where the concrete is saturated (for example, outdoor locations exposed to rainfall), the resulting drilling mud must be removed by other means. In all cases, the procedures used should be clearly described by the MPII accompanying the product. If the installation procedures are not clearly described, contact the manufacturer. These printed installation instructions, which also describe the limits on concrete temperature and the presence of water during installation as well as the procedures necessary for void-free adhesive injection and adhesive cure requirements, constitute an integral part of the adhesive anchor system and are part of the assessment performed in accordance with ACI 355.4.

R26.7.1(l) Adhesive anchors are sensitive to installation orientation. This sensitivity, combined with variability in strength of adhesive anchors subjected to sustained tensile loading, requires installation by certified installers. Certification may also be appropriate for other safety-related applications. Installers can become certified through testing and training programs that include written and performance examinations as defined by the ACI Adhesive Anchor Installer Certification program (ACI CPP 680.1-17) or similar programs with equivalent requirements. The acceptability of certification other than the ACI Adhesive Anchor Installer Certification should be determined by the Licensed Design Professional. In addition, installers should obtain instruction through product-specific training offered by manufacturers of qualified adhesive anchor systems.

An equivalent certified installer program should test the adhesive anchor installer's knowledge and skill by an objectively fair and unbiased administration and grading of a written and performance exam. Programs should reflect the knowledge and skill required to install available commercial anchor systems. The effectiveness of a written exam should be verified through statistical analysis of the questions and answers. An equivalent program should provide a responsive and accurate mechanism to verify credentials, which are renewed on a periodic basis.

26.7.2 Compliance requirements:

(a) Cast-in anchors, their attachments, and anchor reinforcement, shall be securely positioned in the formwork and oriented in accordance with the construction documents. Concrete shall be consolidated around anchors and anchor reinforcement using suitable means during placement.
(b) Proper consolidation of concrete or grout around shear lugs shall be verified by use of base plate inspection holes.
(c) Post-installed anchors shall be installed in accordance with the manufacturer's instructions. Post-installed adhesive anchors shall be installed in accordance with the Manufacturer's Printed Installation Instructions (MPII).

R26.7.2(c) The Manufacturer's Printed Installation Instructions (MPII) contain information required for the proper installation of post-installed adhesive anchors. Additional requirements may apply for specific cases in accordance with 26.7.1(f) and 26.7.1(g). For adhesive anchors, application-dependent requirements for qualification of installers and inspection requirements may apply.

(d) Post-installed anchors shall be installed by qualified installers.
(e) Adhesive anchors identified in the construction documents as installed in a horizontal or upwardly inclined orientation to resist sustained tensile loads shall be installed by certified installers.

R26.7.2(e) Many anchor performance characteristics depend on proper installation of the anchor. Horizontally or upwardly inclined adhesive anchors resisting sustained tension load are required to be installed by personnel certified for the adhesive anchor system and installation procedures being used. Construction personnel can establish qualifications by becoming certified through certification programs.

(f) Adhesive anchors shall be installed in concrete having a minimum age of 21 days at time of anchor installation.

R26.7.2(f) Adhesive anchors qualified in accordance with ACI 355.4 are tested in concrete with compressive strengths within two ranges: 2500 to 4000 psi and 6500 to 8500 psi. Bond strength is, in general, not highly sensitive to concrete compressive strength. The design performance of adhesive anchors cannot be ensured by establishing a minimum concrete compressive strength at the time of installation in early-age concrete. Therefore, a minimum concrete age of 21 days at the time of adhesive anchor installation was adopted.

26.8—Embedments

26.8.1 Design information:

(a) Type, size, details, and location of embedments designed by the licensed design professional.
(b) Reinforcement required to be placed perpendicular to pipe embedments.
(c) Specified concrete cover for pipe embedments with their fittings.

CODE

(d) Corrosion protection for exposed embedments intended to be connected with future Work.

26.8.2 Compliance requirements:

(a) Type, size, details, and location of embedments not shown in the construction documents shall be submitted for review by the licensed design professional.
(b) Aluminum embedments shall be coated or covered to prevent aluminum-concrete reaction and electrolytic action between aluminum and steel.
(c) Pipes and fittings not shown in the construction documents shall be designed to resist effects of the material, pressure, and temperature to which they will be subjected.
(d) No liquid, gas, or vapor, except water not exceeding 90°F or 50 psi pressure, shall be placed in the pipes until the concrete has attained its specified strength.
(e) In solid slabs, piping, except for radiant heating or snow melting, shall be placed between top and bottom reinforcement.
(f) Conduit and piping shall be fabricated and installed so that cutting, bending, or displacement of reinforcement from its specified location is not required.

26.9—Additional requirements for precast concrete

26.9.1 Design information:

(a) Dimensional tolerances for precast members and interfacing members.

(b) Details of lifting devices, embedments, and related reinforcement required to resist temporary loads from handling, storage, transportation, and erection, if designed by the licensed design professional.

26.9.2 Compliance requirements:

(a) Members shall be marked to indicate location and orientation in the structure and date of manufacture.
(b) Identification marks on members shall correspond to erection drawings.

COMMENTARY

R26.9—Additional requirements for precast concrete

R26.9.1(a) Design of precast members and connections is particularly sensitive to tolerances on the dimensions of individual members and on their location in the structure. To prevent misunderstanding, the tolerances used in design should be specified in the construction documents. Instead of specifying individual tolerances, the standard industry tolerances assumed in design may be specified. It is important to specify any deviations from standard industry tolerances.

The tolerances required by 26.6.2 are considered to be a minimum acceptable standard for reinforcement in precast concrete. Industry-standard product and erection tolerances are provided in ACI ITG-7. Interfacing tolerances for precast concrete with cast-in-place concrete are provided in ACI 117.

R26.9.1(b) If the devices, embedments, or related reinforcement are not designed by the licensed design professional, these details should be provided in shop drawings in accordance with 26.9.2(c).

CODE

(c) Design and details of lifting devices, embedments, and related reinforcement required to resist temporary loads from handling, storage, transportation, and erection shall be provided if not designed by the licensed design professional.

(d) During erection, precast members and structures shall be supported and braced to ensure proper alignment, strength, and stability until permanent connections are completed.

(e) If approved by the licensed design professional, items embedded while the concrete is in a plastic state shall satisfy (1) through (4):

(1) Embedded items shall protrude from the precast concrete members or remain exposed for inspection.
(2) Embedded items are not required to be hooked or tied to reinforcement within the concrete.
(3) Embedded items shall be maintained in the correct position while the concrete remains plastic.
(4) The concrete shall be consolidated around embedded items.

26.10—Additional requirements for prestressed concrete

26.10.1 Design information:

(a) Magnitude and location of prestressing forces.
(b) Stressing sequence of tendons.
(c) Type, size, details, and location of post-tensioning anchorages for systems selected by the licensed design professional.
(d) Tolerances for placement of tendons and post-tensioning ducts in accordance with Table 26.6.2.1(a).
(e) Materials and details of corrosion protection for tendons, couplers, end fittings, post-tensioning anchorages, and anchorage regions.

COMMENTARY

R26.9.2(c) Refer to R26.9.1(b). At the option of the licensed design professional, specifications can require that shop drawings, calculations, or both be submitted for the items included in this provision when their design is delegated to the contractor.

R26.9.2(d) All temporary erection connections, bracing, and shoring as well as the sequencing of removal of these items should be shown in construction documents or erection drawings, depending on the assignment of responsibility for the means and methods of construction.

R26.9.2(e) Many precast products are manufactured in such a way that it is difficult, if not impossible, to position reinforcement that protrudes from the concrete before the concrete is placed. Such items as ties for horizontal shear and inserts can be placed while the concrete is plastic, if proper precautions are taken. This provision is not applicable to reinforcement that is completely embedded, or to embedded items that will be hooked or tied to embedded reinforcement.

R26.10—Additional requirements for prestressed concrete

R26.10.1(b) The sequence of anchorage device stressing can have a significant effect on general zone stresses. Therefore, it is important to consider not only the final stage of a stressing sequence with all tendons stressed, but also intermediate stages during construction. The most critical bursting forces caused by each of the sequentially post-tensioned tendon combinations, as well as that of the entire group of tendons, should be taken into account.

R26.10.1(e) For recommendations regarding protection, refer to Sections 4.2 and 4.3 of ACI 423.3R, and Sections 3.4, 3.6, 5, 6, and 8.3 of ACI 423.7. Also refer to 20.5.1.4.2 for corrosion protection requirements.

Corrosion protection can be achieved by a variety of methods. The corrosion protection provided should be suitable for the environment in which the tendons are located. Some conditions will require that the prestressed reinforcement be protected by concrete cover or by cement grout in metal or plastic duct; other conditions will permit the protection provided by coatings such as paint or grease. Corrosion protection methods should meet the fire protection require-

CODE

(f) Requirements for ducts for bonded tendons.

(g) Requirements for grouting of bonded tendons, including maximum water-soluble chloride ion (Cl^-) content requirements in 19.4.1.

26.10.2 Compliance requirements:

(a) Type, size, details, and location of post-tensioning anchorage systems not shown in the construction documents shall be submitted to the licensed design professional for review.
(b) Tendons and post-tensioning ducts shall be placed within required tolerances and supported to prevent displacement beyond required tolerances during concrete placement.
(c) Couplers shall be placed in areas approved by the licensed design professional and enclosed in housings long enough to permit necessary movements.
(d) Burning or welding operations in the vicinity of prestressing reinforcement shall be performed in such a manner that prestressing reinforcement is not subject to welding sparks, ground currents, or temperatures that degrade the properties of the reinforcement.
(e) Prestressing force and friction losses shall be verified by (1) and (2).

(1) Measured elongation of prestressed reinforcement compared with elongation calculated using the modulus of elasticity determined from tests or as reported by the manufacturer.

(2) Jacking force measured using calibrated equipment such as a hydraulic pressure gauge, load cell, or dynamometer.

(f) The cause of any difference in force determination between (1) and (2) of 26.10.2(e) that exceeds 5 percent for pretensioned construction or 7 percent for post-tensioned construction shall be ascertained and corrected, unless approved by the licensed design professional.

(g) Loss of prestress force due to unreplaced broken prestressed reinforcement shall not exceed 2 percent of

COMMENTARY

ments of the general building code unless the installation of external post-tensioning is to only improve serviceability.

R26.10.1(f) Guidance for specifying duct requirements for bonded tendons is provided in PTI M50.3 and PTI M55.1.

R26.10.1(g) Guidance for specifying grouting requirements for bonded tendons is provided in PTI M55.1.

R26.10.2(e) Elongation measurements for prestressing should be in accordance with the procedures outlined in the *Manual for Quality Control for Plants and Production of Structural Precast Concrete Products* (MNL 117), published by the Precast/Prestressed Concrete Institute.

R26.10.2(f) The 5 percent tolerance for pretensioned construction reflects experience with production of those members. Because prestressing reinforcement for pretensioned construction is usually stressed in air with minimal friction effects, a 5 percent tolerance is considered reasonable. For post-tensioned construction, a slightly higher tolerance is permitted. Elongation measurements for post-tensioned construction are affected by several factors that are less significant or that do not exist for pretensioned construction. The friction along prestressing reinforcement in post-tensioning applications may be affected to varying degrees by placing tolerances and small irregularities in tendon profile due to tendon and concrete placement. The friction coefficients between the prestressing reinforcement and the duct are also subject to variation.

R26.10.2(g) This provision applies to all prestressed concrete members. For cast-in-place post-tensioned slab

CODE

the total prestress force in prestressed concrete members, unless approved by the licensed design professional.

(h) If the transfer of force from the anchorages of the pretensioning bed to the concrete is accomplished by flame cutting prestressed reinforcement, the cutting locations and cutting sequence shall be selected to avoid undesired temporary stresses in pretensioned members.

(i) Long lengths of exposed pretensioned strand shall be cut near the member to minimize shock to the concrete.

(j) Prestressing reinforcement in post-tensioned construction shall not be stressed until the concrete compressive strength is at least 2500 psi for single-strand or bar tendons, 4000 psi for multistrand tendons, or a higher strength, if required. An exception to these strength requirements is provided in 26.10.2(k).

(k) Lower concrete compressive strength than required by 26.10.2(j) shall be permitted if (1) or (2) is satisfied:

(1) Oversized anchorage devices are used to compensate for a lower concrete compressive strength.

(2) Prestressing reinforcement is stressed to no more than 50 percent of the final prestressing force.

26.11—Formwork

26.11.1 *Design of formwork*

26.11.1.1 Design information:

(a) Requirement for the contractor to design, fabricate, install, and remove formwork.

(b) Location of composite members requiring shoring.

(c) Requirements for removal of shoring of composite members.

COMMENTARY

systems, a member should be that portion considered as an element in the design, such as the joist and effective slab width in one-way joist systems, or the column strip or middle strip in two-way flat plate systems. Some members can be shown to accommodate more than 2 percent loss of prestress due to unreplaced broken prestressed reinforcement.

R26.10.2(k) To limit early shrinkage cracking, monostrand tendons are sometimes stressed at concrete strengths less than 2500 psi. In such cases, either oversized monostrand anchorages are used, or the strands are stressed in stages, often to levels one-third to one-half the final prestressing force.

R26.11—Formwork

R26.11.1 *Design of formwork*

Typically, the contractor is responsible for formwork design, and the Code provides the minimum formwork performance requirements necessary for public health and safety. Concrete formwork design, construction, and removal demands sound judgment and planning to achieve adequate safety. Detailed information on formwork for concrete is given in "Guide to Formwork for Concrete" (ACI 347). This guide is directed primarily to contractors for design, construction, materials for formwork, and forms for unusual structures, but it should aid the licensed design professional in preparing the construction documents.

Formwork for Concrete, ACI SP-4, is a practical handbook for contractors, engineers, and architects. It follows the guidelines established in ACI 347 and includes information on planning, building, and using formwork. It also includes tables, diagrams, and formulas for formwork design loads.

ACI 301 Section 2 provides specification requirements for design and construction of formwork.

R26.11.1.1 Section 24.2.5 covers the requirements pertaining to deflections of shored and unshored members.

CODE

26.11.1.2 Compliance requirements:

(a) Design of formwork shall consider (1) through (6):

(1) Method of concrete placement.

(2) Rate of concrete placement.

(3) Construction loads, including vertical, horizontal, and impact.

(4) Avoidance of damage to previously constructed members.

(5) For post-tensioned members, allowance for movement of the member during tensioning of the prestressing reinforcement without damage to the member.

(6) For post-tensioned members, allowance for load redistribution on formwork resulting from tensioning of the prestressing reinforcement.

(b) Formwork fabrication and installation shall result in a final structure that conforms to shapes, lines, and dimensions of the members as required by the construction documents.

(c) Formwork shall be sufficiently tight to inhibit leakage of paste or mortar.

(d) Formwork shall be braced or tied together to maintain position and shape.

26.11.2 *Removal of formwork*

26.11.2.1 Compliance requirements:

(a) Before starting construction, the contractor shall develop a procedure and schedule for removal of formwork and installation of reshores, and shall calculate the loads transferred to the structure during this process.

(b) Structural analysis and concrete strength requirements used in planning and implementing the formwork removal and reshore installation shall be furnished by the contractor to the licensed design professional and to the building official, when requested.

(c) No construction loads shall be placed on, nor any formwork removed from, any part of the structure under construction except when that portion of the structure in combination with remaining formwork has sufficient strength to support safely its weight and loads placed thereon and without impairing serviceability.

(d) Sufficient strength shall be demonstrated by structural analysis considering anticipated loads, strength of formwork, and an estimate of in-place concrete strength.

COMMENTARY

R26.11.2 *Removal of formwork*

R26.11.2.1 In determining the time for removal of formwork, consideration should be given to the construction loads, in-place strength of concrete, and possibility of deflections greater than acceptable to the licensed design professional (ACI 347 and ACI 347.2R). Construction loads may be greater than the specified live loads. Even though a structure may have adequate strength to support the applied loads at early ages, deflections can cause serviceability problems.

The removal of formwork for multistory construction should be a part of a planned procedure developed by the contractor that considers the temporary support of the entire structure as well as each individual member. Such a procedure should be planned before construction and should be based on a structural analysis taking into account at least (a) through (e):

(a) The structural system that exists at the various stages of construction, and the construction loads corresponding to those stages;

(b) The in-place strength of the concrete at the various stages during construction;

(c) The influence of deformations of the structure and shoring system on the distribution of dead loads and construction loads during the various stages of construction;

(d) The strength and spacing of shores or shoring systems used, as well as the method of shoring, bracing, shore removal, and reshoring including the minimum time interval between the various operations;

CODE

COMMENTARY

(e) Any other loading or condition that affects the safety or serviceability of the structure during construction.

ACI 347.2R provides information for shoring and reshoring multistory buildings.

(e) The estimate of in-place concrete strength shall be based on tests of field-cured cylinders or on other procedures to evaluate concrete strength approved by the licensed design professional and, when requested, approved by the building official.

R26.11.2.1(e) Evaluation of concrete strength during construction may be demonstrated by field-cured test cylinders or other procedures approved by the licensed design professional and, when requested, approved by the building official, such as (a) though (d):

(a) Tests of cast-in-place cylinders in accordance with ASTM C873. This method is limited to use for slabs where the depth of concrete is between 5 to 12 in.

(b) Penetration resistance in accordance with ASTM C803

(c) Pullout strength in accordance with ASTM C900

(d) Maturity index measurements and correlation in accordance with ASTM C1074

Procedures (b), (c), and (d) require sufficient data for the materials used in the Work to demonstrate correlation of measurements on the structure with the compressive strength of molded cylinders or drilled cores. ACI 228.1R discusses the use of these methods to evaluate the in-place strength of concrete

(f) Formwork shall be removed in such a manner not to impair safety and serviceability of the structure.

(g) Concrete exposed by formwork removal shall have sufficient strength not to be damaged by the removal.

(h) Formwork supports for post-tensioned members shall not be removed until sufficient post-tensioning has been applied to enable post-tensioned members to support their dead load and anticipated construction loads.

(i) No construction loads exceeding the combination of superimposed dead load plus live load including reduction shall be placed on any unshored portion of the structure under construction, unless analysis indicates adequate strength to support such additional loads and without impairing serviceability.

R26.11.2.1(i) The nominal live load specified on the drawings is frequently reduced for members supporting large floor areas, and the limit on construction loads needs to account for such reductions.

26.12—Evaluation and acceptance of hardened concrete

26.12.1 *General*

26.12.1.1 Compliance requirements:

(a) Evaluation of hardened concrete shall be based on strength tests. A strength test is the average of the compressive strengths of at least two 6 x 12 in. cylinders or at least three 4 x 8 in. cylinders made from the same sample of concrete taken in accordance with ASTM C172 at the point of delivery, handled and standard-cured in accordance with ASTM C31, and tested in accordance with ASTM C39 at 28 days or at test age designated for f_c'.

R26.12—Evaluation and acceptance of hardened concrete

R26.12.1 *General*

R26.12.1.1(a) Casting and testing more than the minimum number of specimens may be desirable in case it becomes necessary to discard an outlying individual cylinder strength in accordance with ACI 214R. If individual cylinder strengths are discarded in accordance with ACI 214R, a strength test is valid provided at least two individual 6 x 12 in. cylinder strengths or at least three 4 x 8 in. cylinder strengths are averaged. All individual cylinder strengths that are not discarded in accordance with ACI 214R are to be used to calculate the average strength. The size and number

CODE | COMMENTARY

of specimens representing a strength test should be the same for each concrete mixture. The cylinder size should be agreed upon by the owner, licensed design professional, and testing agency before construction.

Testing three instead of two 4 x 8 in. cylinders preserves the confidence level of the average strength because 4 x 8 in. cylinders tend to have approximately 20 percent higher within-test variability than 6 x 12 in. cylinders (Carino et al. 1994).

Representative concrete samples for making strength-test specimens are obtained from concrete as delivered to the project site. For example, samples of concrete delivered in a truck mixer would be obtained from the truck chute at discharge. ASTM C172 provides requirements for sampling concrete from different equipment used in the production or transportation of concrete.

Note that the term "strength test" does not apply to results of tests on cylinders field cured in or on the structure as described in ASTM C31, nor does it apply to results of tests on cylinders from laboratory trial batches.

(b) For shotcrete, a strength test shall be the average strength of at least three 3 in. nominal diameter cores taken from a test panel prepared in accordance with ASTM C1140 and tested at 28 days from time of placement or at test age designated for f_c'.

(c) The testing agency performing acceptance testing shall comply with ASTM C1077.

R26.12.1.1(c) ASTM C1077 defines the duties, responsibilities, and minimum technical requirements of testing agency personnel and defines the technical requirements for equipment used in testing concrete and concrete aggregates. Agencies that test cylinders or cores to determine compliance with Code requirements should be accredited or inspected for conformance to the requirements of ASTM C1077 by a recognized evaluation authority.

(d) Certified field testing technicians shall perform tests on fresh concrete at the job site, prepare specimens for standard curing, prepare specimens for field curing, if required, and record the temperature of the fresh concrete when preparing specimens for strength tests.

R26.12.1.1(d) Technicians can become certified through testing and training programs that include written and performance examinations. Field technicians in charge of sampling concrete; testing for slump, density (unit weight), yield, air content, and temperature; and making and curing test specimens should be certified in accordance with the ACI Concrete Field Testing Technician—Grade 1 Certification Program (ACI CPP 610.1-18) or an equivalent program meeting the requirements of ASTM C1077.

(e) Certified laboratory technicians shall perform required laboratory tests.

R26.12.1.1(e) Concrete laboratory testing technicians performing strength testing should be certified in accordance with the ACI Concrete Strength Testing Technician Certification Program (ACI CPP 620.2-12) or an equivalent program meeting the requirements of ASTM C1077.

(f) All reports of acceptance tests shall be provided to the licensed design professional, contractor, concrete producer, and, if requested, to the owner and the building official.

R26.12.1.1(f) The Code requires testing reports to be distributed to the parties responsible for the design, construction, and approval of the Work. Such distribution of test reports should be indicated in contracts for inspection and testing services. Prompt distribution of testing reports allows

CODE | COMMENTARY

for timely identification of either compliance or the need for corrective action. A complete record of testing allows the concrete producer to reliably establish appropriate mixture proportions for future work.

26.12.2 *Frequency of testing*

26.12.2.1 Compliance requirements:

(a) Samples for preparing strength-test specimens of each concrete mixture placed each day shall be taken in accordance with (1) through (3):

(1) At least once a day.

(2) At least once for each 150 yd^3 of concrete.

(3) At least once for each 5000 ft^2 of surface area for slabs or walls.

(b) On a given project, if total volume of concrete is such that frequency of testing would provide fewer than five strength tests for a given concrete mixture, strength test specimens shall be made from at least five randomly selected batches or from each batch if fewer than five batches are used.

(c) If the total quantity of a given concrete mixture is less than 50 yd^3, strength tests are not required if evidence of satisfactory strength is submitted to and approved by the building official.

(d) For shotcrete, prepare a shotcrete test panel for each mixture and each nozzle operator at least once per day or for every 50 yd^3 placed, whichever results in the greater number of panels.

R26.12.2 *Frequency of testing*

R26.12.2.1(a) Concrete samples for preparing strength-test specimens are to be taken on a strictly random basis if they are to measure properly the acceptability of the concrete. To be representative within the period of placement, the choice of sampling times, or the concrete batches to be sampled, is to be made on the basis of chance alone. Batches are not sampled on the basis of appearance, convenience, or another possibly biased criterion, because the statistical analyses will lose their validity. ASTM D3665 describes procedures for random selection of the batches to be tested. Specimens for one strength test (as defined in 26.12.1.1(a)) are to be made from a single batch, and ASTM C172 requires that the sample be taken only after all adjustments to the batch are made.

In calculating surface area, only one side of the slab or wall is considered. Criterion (3) will require more frequent sampling than once for each 150 yd^3 placed if average wall or slab thickness is less than 9-3/4 in.

26.12.3 *Acceptance criteria for standard-cured specimens*

26.12.3.1 Compliance requirements:

(a) Strength level of a concrete mixture shall be acceptable if (1) and (2) are satisfied:

(1) Every average of any three consecutive strength tests equals or exceeds f_c'.

(2) No strength test falls below f_c' by more than 500 psi if f_c' is 5000 psi or less; or by more than $0.10f_c'$ if f_c' exceeds 5000 psi.

(b) If either of the requirements of 26.12.3.1(a) is not satisfied, steps shall be taken to increase subsequent strength tests.

R26.12.3 *Acceptance criteria for standard-cured specimens*

R26.12.3.1 Evaluation and acceptance of the concrete can be determined as test results are received during the course of the Work. Strength tests failing to meet these criteria will occur occasionally, with a probability of approximately once in 100 tests (ACI 214R) even though concrete strength and uniformity are satisfactory. Allowance should be made for such statistically expected variations in deciding whether the strength being produced is adequate. The strength acceptance criteria of 26.12.3.1(a) apply to test results from either 4 x 8 in. or 6 x 12 in. test cylinders permitted in 26.12.1.1(a). The average difference (Carino et al. 1994) between test results obtained by the two specimen sizes is not considered to be significant in design.

R26.12.3.1(b) The steps taken to increase the values of subsequent strength tests will depend on the particular circumstances but could include one or more of (a) through (g):

(a) Increase in cementitious materials content;

(b) Reduction in or better control of water content;

(c) Use of a water-reducing admixture to improve the dispersion of cementitious materials;

CODE

COMMENTARY

(d) Other changes in mixture proportions;
(e) Reduction in delivery time;
(f) Closer control of air content;
(g) Improvement in the quality of the testing, including strict compliance with ASTM C172, ASTM C31, and ASTM C39.

Such changes in operating procedures or small changes in cementitious materials content or water content should not require a formal resubmission of mixture proportions; however, changes in sources of cement, aggregates, or admixtures should be accompanied by evidence submitted to the licensed design professional that the concrete strength will be improved.

(c) Requirements of 26.12.6 for investigating strength tests shall apply if the requirements of 26.12.3.1(a)(2) are not met.

26.12.4 *Acceptance criteria for shotcrete*

26.12.4.1 Compliance requirements:

(a) Specimens for acceptance tests shall be in accordance with (1) and (2):
(1) Test panels shall be prepared in the same orientation and by the same nozzle operator placing shotcrete.
(2) Cores shall be obtained, conditioned, and tested in accordance with ASTM C1604.
(b) Strength of a shotcrete mixture shall be acceptable if (1) and (2) are satisfied:
(1) Every arithmetic average of the strengths from three consecutive test panels equals or exceeds f_c'.
(2) The average compressive strength of three cores from a single test panel is not less than $\mathbf{0.85}f_c'$ with no core having a strength less than $\mathbf{0.75}f_c'$.
(c) If either of the requirements of 26.12.4.1(b) are not satisfied, steps shall be taken to increase the average of subsequent strength results.
(d) Requirements for investigating low strength-test results shall apply if the requirements of 26.12.6.1(b)(2) are not met.

R26.12.4 *Acceptance criteria for shotcrete*

R26.12.4.1(a) Cores taken from shotcrete test panels, made in accordance with ASTM C1140, typically have length-to-diameter ratios less than 1.75. Therefore the core strengths used for comparison with the acceptance criteria are the values after correction for the length to diameter ratio in accordance with ASTM C1604.

26.12.5 *Acceptance criteria for density of lightweight concrete*

R26.12.5 *Acceptance criteria for density of lightweight concrete*

26.12.5.1 Compliance requirements:

(a) Frequency of sampling for determining fresh density shall be according to 26.12.2.
(b) Sampling of lightweight concrete for determining fresh density shall be at the point of delivery in accordance with ASTM C172.
(c) Fresh density of lightweight concrete shall be determined in accordance with ASTM C138.

CODE

(d) Unless otherwise permitted by the licensed design professional, fresh density of lightweight concrete shall be acceptable if within ±4.0 lb/ft^3 of the fresh density corresponding to the specified equilibrium density.

COMMENTARY

R26.12.5(d) The permitted tolerance for fresh concrete density for a mixture designed for the specified equilibrium density, w_c, is intended to account for variations in aggregate moisture, air content, and batch quantities. The impact of the tolerance in density on the value of λ assumed in design is minimal and deemed to be acceptable. The Licensed Design Professional can consider permitting a larger tolerance on fresh density to accommodate these expected variations when appropriate.

26.12.6 *Investigation of strength tests*

26.12.6.1 Compliance requirements:

R26.12.6 *Investigation of strength tests*

R26.12.6.1 Requirements are provided if strength tests have failed to meet the acceptance criterion of 26.12.3.1(a)(2) or if the average strengths of field-cured cylinders do not comply with 26.5.3.2(e). These requirements are applicable only for evaluation of in-place strength at the time of construction. Strength evaluation of existing structures is covered by Chapter 27. The building official should apply judgment as to the significance of low test results and whether they indicate need for concern. If further investigation is deemed necessary, such investigation may include in-place tests as described in ACI 228.1R or, in extreme cases, measuring the compressive strength of cores taken from the structure.

In-place tests of concrete, such as probe penetration (ASTM C803), rebound hammer (ASTM C805), or pullout test (ASTM C900), may be useful in determining whether a portion of the structure actually contains low-strength concrete. Unless these in-place tests have been correlated with compressive strength using accepted procedures, such as described in ACI 228.1R, they are of value primarily for comparisons within the same structure rather than as quantitative estimates of strength.

For cores, if required, conservative acceptance criteria are provided that should ensure structural adequacy for virtually any type of construction (Bloem 1965, 1968; Malhotra 1976, 1977). Lower strength may be tolerated under many circumstances, but this is a matter of judgment on the part of the licensed design professional and building official. If the strengths of cores obtained in accordance with 26.12.6.1(d) fail to comply with 26.12.6.1(e), it may be practicable, particularly in the case of floor or roof systems, for the building official to require a strength evaluation as described in Chapter 27. Short of a strength evaluation, if time and conditions permit, an effort may be made to improve the strength of the concrete in place by supplemental wet curing. Effectiveness of supplemental curing should be verified by further strength evaluation using procedures previously discussed.

The Code, as stated, concerns itself with achieving structural safety, and the requirements for investigation of low strength-test results (26.12.6) are aimed at that objective. It is not the function of the Code to assign responsibility for strength deficiencies.

CODE

(a) If any strength test of standard-cured cylinders falls below f_c' by more than the limit allowed for acceptance, or if tests of field-cured cylinders indicate deficiencies in protection and curing, steps shall be taken to ensure that structural adequacy of the structure is not jeopardized.

(b) If the likelihood of low-strength concrete is confirmed and calculations indicate that structural adequacy is significantly reduced, tests of cores drilled from the area in question in accordance with ASTM C42 shall be permitted. In such cases, three cores shall be taken for each strength test that falls below f_c' by more than the limit allowed for acceptance.

(c) The licensed design professional or the building official shall be permitted to modify details of core tests as stated in ASTM C42.

(d) Cores shall be obtained, moisture-conditioned by storage in watertight bags or containers, transported to the testing agency, and tested in accordance with ASTM C42. Cores shall be tested between 5 days after last being wetted and 7 days after coring unless otherwise approved by the licensed design professional or building official.

(e) Concrete in an area represented by core tests shall be considered structurally adequate if (1) and (2) are satisfied:

(1) The average of three cores is equal to at least 85 percent of f_c'.

(2) No single core is less than 75 percent of f_c'.

COMMENTARY

R26.12.6.1(a) If the strength of field-cured cylinders does not conform to 26.5.3.2(e), steps need to be taken to improve the curing. If supplemental in-place tests confirm a possible deficiency in strength of concrete in the structure, core tests may be required to evaluate structural adequacy.

R26.12.6.1(c) Some default requirements in ASTM C42 are permitted to be altered by the "specifier of the tests," who is defined in ASTM C42 as "the individual responsible for analysis or review and acceptance of core test results." For the purposes of ACI 318, the "specifier of the tests" is the licensed design professional or the building official.

R26.12.6.1(d) The use of a water-cooled core barrel or a water-cooled saw for end trimming results in a core with a moisture gradient between the exterior surface and the interior. This gradient lowers the apparent compressive strength of the core (Bartlett and MacGregor 1994). The requirement of at least 5 days between the time of last being wetted and time of testing provides time for the moisture gradient to be reduced. If a water-cooled saw is used for end trimming, the conditioning period begins when sawing is completed. The maximum time of 7 days between coring and testing is intended to ensure timely testing of cores if strength of concrete is in question. If end trimming with a water-cooled saw is necessary, it should be done within 2 days of drilling the core to meet the time limits established by the testing criterion.

Research (Bartlett and MacGregor 1994) has also shown that other moisture conditioning procedures, such as soaking or air drying, affect measured core strengths and result in conditions that are not representative of the in-place concrete. Therefore, to provide reproducible moisture conditions that are representative of in-place conditions, a standard moisture conditioning procedure that permits dissipation of moisture gradients is prescribed for cores. ASTM C42 permits the specifier of the tests to modify the default duration of moisture conditioning before testing. The specifier of the tests, however, must be aware of the potential reduction in strength if cores are tested before moisture gradients are allowed to dissipate.

R26.12.6.1(e) An average core strength of 85 percent of the specified strength is realistic (Bloem 1968). It is not realistic, however, to expect the average core strength to be equal to f_c', because of differences in the size of specimens, conditions of obtaining specimens, degree of consolidation, and curing conditions. The acceptance criteria for core strengths have been established with consideration that cores for investigating low strength-test results will typically be extracted at an age later than specified for f_c'. For the purpose of satisfying 26.12.4.1(e), this Code does not intend that core strengths be adjusted for the age of the cores.

(f) Additional testing of cores extracted from locations represented by erratic core strength results shall be permitted.

(g) If criteria for evaluating structural adequacy based on core strength results are not met, and if the structural adequacy remains in doubt, the responsible authority shall be permitted to order a strength evaluation in accordance with Chapter 27 for the questionable portion of the structure or take other appropriate action.

26.12.7 *Acceptance of steel fiber-reinforced concrete*

26.12.7.1 Compliance requirements:

(a) Steel fiber-reinforced concrete used for shear resistance shall satisfy (1) through (3):

(1) The compressive strength acceptance criteria for standard-cured specimens

(2) The residual strength obtained from flexural testing in accordance with ASTM C1609 at a midspan deflection of 1/300 of the span length is at least the greater of (i) and (ii):

(i) 90 percent of the measured first-peak strength obtained from a flexural test and

(ii) 90 percent of the strength corresponding to $7.5\sqrt{f_c'}$

(3) The residual strength obtained from flexural testing in accordance with ASTM C1609 at a midspan deflection of 1/150 of the span length is at least the greater of (i) and (ii):

(i) 75 percent of the measured first-peak strength obtained from a flexural test and

(ii) 75 percent of the strength corresponding to $7.5\sqrt{f_c'}$

R26.12.7 *Acceptance of steel fiber-reinforced concrete*

R26.12.7.1 The performance criteria for the ASTM C1609 tests are based on results from flexural tests (Chen et al. 1995) conducted on steel fiber-reinforced concretes with fiber types and contents similar to those used in the tests of beams that served as the basis for 9.6.3.1.

The term "residual strength" is defined in ASTM C1609 and is related to the ability of cracked fiber-reinforced concrete to resist tension. The strength of $7.5\sqrt{f_c'}$ is consistent with the design modulus of rupture of the concrete provided by Eq. (19.2.3.1).

26.13—Inspection

26.13.1 *General*

R26.13—Inspection

R26.13.1 *General*

The quality of concrete structures depends largely on workmanship in construction. The best materials and design practices will not be effective unless construction is performed well. Inspection is necessary to verify that construction is in accordance with construction documents. Proper performance of the structure depends on construction that accurately represents the design and meets the requirements of this Code.

Some general building codes have incorporated inspection requirements based upon established procedures such as PCI Plant Certification.

CODE

26.13.1.1 Concrete construction shall be inspected as required by the general building code, and as a minimum, the inspection shall comply with the requirements provided in 26.13. In the absence of a general building code, concrete construction shall be inspected in accordance with the provisions of this Code.

COMMENTARY

R26.13.1.1 By inspection, this Code does not intend that the inspector should supervise the construction. Rather, it means the inspector should visit the project as necessary to observe the various stages of Work and determine that it is being performed in conformance with the construction documents. The frequency of inspections should follow 26.13.3 for items requiring continuous or periodic inspection.

Inspection does not relieve the contractor from the obligation to follow the construction documents and to provide the designated quality and quantity of materials and workmanship for all stages of the Work.

This Code prescribes minimum requirements for inspection of all structures within its scope. This Code is not a construction specification and any user of this Code may require higher standards of inspection than cited in the general building code or this Code if additional requirements are necessary. ACI 311.4R describes the recommended procedure for organizing and conducting concrete inspection and serves as a guide to owners, architects, and engineers. ACI SP-2 describes methods of inspecting concrete construction that are generally accepted as good practice and serves as a guide in matters not covered by construction documents.

26.13.1.2 Inspection of concrete construction shall be conducted by the licensed design professional responsible for the design, a person under the supervision of the licensed design professional, or a qualified inspector. The inspection shall verify conformance with construction documents throughout the various Work stages. If an inspector conducts inspection of formwork, concrete placement, reinforcement, and embedments, the inspector shall be certified.

R26.13.1.2 The licensed design professional responsible for the design is in the best position to determine if construction is in conformance with the construction documents. However, if the licensed design professional responsible for the design is not retained, inspection of construction through other licensed design professionals or through separate inspection organizations with demonstrated capability for performing the inspection may be used.

Inspectors should establish their capability of performing inspection requirements by becoming certified to inspect and record the results of concrete construction, including pre-placement, placement, and post-placement through the ACI Concrete Construction Special Inspector Certification Program (ACI CPP 630.1-15) or equivalent.

In some jurisdictions, legislation has established registration or licensing procedures for persons performing certain inspection functions. The general building code should be reviewed, or the building official should be consulted to determine if any such requirements exist within a specific jurisdiction. The building official may be contacted for clarification of the inspection requirements if not clearly identified in the general building code.

If inspection is conducted independently of the licensed design professional responsible for the design, it is recommended that the licensed design professional responsible for the design review inspection reports and observe portions of the Work to verify that the design requirements are properly executed.

Inspection reports should be distributed promptly to the owner, licensed design professional responsible for the design, contractor, appropriate subcontractors, appropriate

CODE

26.13.1.3 Inspection of concrete placement and reinforcement for special moment frames, boundary elements of special structural walls, coupling beams, and precast concrete diaphragms assigned to SDC C, D, E, or F using moderate or high-deformability connections, shall be performed under the supervision of the licensed design professional responsible for the design, by a person under the supervision of a licensed design professional with demonstrated capability to supervise inspection of these elements or by a certified inspector. Installation tolerances of precast concrete diaphragm connections shall be inspected for compliance with ACI 550.5.

26.13.1.4 Inspection of reinforcement welding shall be performed by a qualified welding inspector in accordance with AWS D1.4. The weldability of reinforcement other than ASTM A706 shall be confirmed by documentation in accordance with 26.6.4.

26.13.1.5 Inspection of the installation of post-installed expansion (torque-controlled and displacement-controlled), screw, and undercut anchors shall be performed by a certified inspector or a qualified inspector specifically approved for that purpose by the Licensed Design Professional and the building official.

26.13.1.6 The installation inspection of all adhesive anchors shall be performed by a certified inspector.

26.13.2 *Inspection reports*

COMMENTARY

suppliers, and the building official to allow timely identification of compliance or the need for corrective action.

Inspection responsibility and the degree of inspection required should be set forth in the contracts between the owner, architect, engineer, contractor, and inspector. Adequate resources should be provided to properly perform and oversee the inspection.

R26.13.1.3 The purpose of this requirement is to verify that the detailing required in special moment frames and special structural walls is properly executed through inspection by personnel who are certified to inspect these elements. Certification of inspectors should be acceptable to the jurisdiction enforcing the general building code and as described in R26.13.1.2.

Continuous construction inspection is needed for completion of connections for precast concrete diaphragms designed in accordance with 18.12.1.1 to verify the tolerances specified in ACI 550.5 are met.

R26.13.1.5 The International Building Code (IBC 2018) requires inspection of all post-installed anchors. For post-installed expansion (torque-controlled and displacement-controlled), screw, and undercut anchors, monitoring of installation by a certified inspector is recommended to ensure required installation procedures are followed. Certification is established through an independent assessment such as the ACI Post-Installed Concrete Anchor Installation Inspector program (ACI CPP 681.2-19), or similar program with equivalent requirements.

R26.13.1.6 The installation of all adhesive anchors requires inspection by a certified inspector. Certification is established through an independent assessment such as the ACI Adhesive Anchor Installation Inspector program (ACI CPP 681.1-17), or similar program with equivalent requirements.

The installation of adhesive anchors identified in the construction documents as resisting sustained tensile loads in horizontal or upwardly inclined orientations (clockwise from 9 o'clock to 3 o'clock) poses challenges to the installer and requires particular attention to execution quality as well as an enhanced level of oversight. It is required that these anchor installations be inspected by a certified inspector who is continuously present when and where the installations are being performed.

R26.13.2 *Inspection reports*

CODE

26.13.2.1 Inspection reports shall document inspected items and be developed throughout each construction Work stage. Records of the inspection shall be preserved by the party performing the inspection for at least 2 years after completion of the project.

COMMENTARY

R26.13.2.1 A record of inspection is required in case questions subsequently arise concerning the performance or safety of the structure or members. Photographs documenting construction progress are also desirable.

The general building code or other legal documents may require these records be preserved longer than two years after completion of the project.

26.13.2.2 Inspection reports shall document (a) through (e):

(a) General progress of the Work.
(b) Any significant construction loadings on completed floors, members, or walls.
(c) The date and time of mixing, quantity of concrete placed, identification of mixtures used, approximate placement location in the structure, and results of tests for fresh and hardened concrete properties for all concrete mixtures used in the Work.
(d) Concrete temperatures and protection given to concrete during placement and curing if the ambient temperature falls below 40°F or rises above 95°F.

R26.13.2.2(d) The term "ambient temperature" means the temperature of the environment to which the concrete is directly exposed. Concrete temperature as used in this section may be taken as the surface temperature of the concrete. Surface temperatures may be determined by placing temperature sensors in contact with concrete surfaces or between concrete surfaces and covers used for curing, such as insulation blankets or plastic sheeting.

(e) Placement of reinforcement and tensioning of prestressing reinforcement including measurement and recording of tendon elongation and force from a calibrated gauge.

26.13.2.3 For ASTM A615 deformed reinforcement used in special seismic systems, verify mill test reports for compliance with the construction documents.

R26.13.2.3 If ASTM A615 reinforcement is used for special seismic applications, it is important that the inspector review the mill certificates for compliance with the applicable requirements provided in the construction documents.

26.13.2.4 Test reports shall be verified to confirm weldability of reinforcement other than ASTM A706, if weldability is required.

26.13.2.5 For post-installed expansion (torque-controlled and displacement-controlled), screw, and undercut anchors and adhesive anchors, materials, and installation procedures shall be verified for conformance with the approved construction documents and the manufacturer's recommended procedures, which are the Manufacturer's Printed Installation Instructions (MPII) in the case of adhesive anchors. Confirm procedures and results of proof loading where required in accordance with 26.7.1(k).

26.13.3 *Items requiring inspection*

R26.13.3 *Items requiring inspection*

26.13.3.1 Unless otherwise specified in the general building code, items shall be continuously or periodically inspected in accordance with 26.13.3.2 and 26.13.3.3, respectively.

R26.13.3.1 Table 1705 in Chapter 17 of the 2012 IBC was used to determine which items of Work require continuous or periodic inspection.

CODE

26.13.3.2 Items requiring verification and continuous inspection shall include (a) through (e):

(a) Prior to placement, concrete mixture for intended location.

(b) Tensioning of prestressing reinforcement and grouting of bonded tendons.

(c) Placement of reinforcement for special moment frames, boundary elements of special structural walls, and coupling beams.

(d) Welding of reinforcement for special moment frames, boundary elements of special structural walls, and coupling beams.

(e) Post-installed anchor installation, if required as a condition of the anchor assessment or if adhesive anchors are installed in horizontal or upwardly inclined orientations to resist sustained tensile loads.

26.13.3.3 Items requiring verification and periodic inspection shall include (a) through (j):

(a) Placement of reinforcement, embedments, and post-tensioning tendons.

(b) Welding of reinforcement except as required in 26.13.3.2(d).

(c) Curing method and duration of curing for each member.

(d) Construction and removal of forms and reshoring.

(e) Sequence of erection and connection of precast members.

(f) Verification of in-place strength of concrete before stressing post-tensioned tendons and before removal of shores and formwork from beams and structural slabs.

(g) Placement of cast-in-anchors and anchor reinforcement, including tolerances required for location of anchor reinforcement.

(h) Installation of post-installed expansion (torque-controlled and displacement-controlled) screw, and undercut anchors.

(i) Installation of adhesive anchors, except as required in 26.13.3.2(e).

(j) Proof loading of anchors if required in accordance with 26.13.2.5.

COMMENTARY

R26.13.3.3(e) Some jurisdictions may require continuous inspection of sequence of erection and connection of precast members, and also may require inspection of the shoring, bracing, or other temporary measures.

R26.13.3.3(i) Inspection requirements for adhesive anchors are different from other post-installed anchors and are derived from four sources: a) the general building code, which requires periodic inspection for anchors in concrete; b) the assessment and qualification of the anchor under the provisions of ACI 355.4, which may require either periodic inspection or continuous inspection with proof loading depending on the strength reduction factors assigned to the anchor; c) the requirements of 26.13.3.2(e), which mandate continuous inspection for anchors in a horizontal or upwardly inclined orientation to resist sustained tension loads; and d) the proof loading requirement of 26.13.2.5.

CHAPTER 27—STRENGTH EVALUATION OF EXISTING STRUCTURES

CODE

27.1—Scope

27.1.1 Provisions of this chapter shall apply to strength evaluation of existing structures by analytical means or by load testing.

27.2—General

27.2.1 If there is doubt that a part or all of a structure meets the safety requirements of this Code and the structure is to remain in service, a strength evaluation shall be carried out as required by the licensed design professional or building official.

27.2.2 If the effect of a strength deficiency is well understood and it is practical to measure the dimensions and determine the material properties of the members required for analysis, an analytical evaluation of strength based on this information is permitted. Required data shall be determined in accordance with 27.3.

27.2.3 If the effect of a strength deficiency is not well understood or it is not practical to measure the dimensions and determine the material properties of the members required for analysis, a load test is required in accordance with 27.4.

27.2.4 If uncertainty about the strength of part or all of a structure involves deterioration, and if the observed response during the load test satisfies the acceptance criteria in 27.5 or 27.6 for the selected load test procedure, the structure or part of the structure is permitted to remain in service for a time period specified by the licensed design professional. If deemed necessary by the licensed design professional, periodic reevaluations shall be conducted.

COMMENTARY

R27.1—Scope

R27.1.1 Provisions of this chapter may be used to evaluate whether a structure or a portion of a structure satisfies the safety requirements of the Code. A strength evaluation may be required if the materials are considered to be deficient in quality, if there is evidence indicating faulty construction, if a building will be used for a new function, or if, for any reason, a structure or a portion of it does not appear to satisfy the requirements of the Code. In such cases, this chapter provides guidance for investigating the safety of the structure. This chapter does not cover load testing for the approval of new design or construction methods. Acceptance of alternative materials or systems is covered in 1.10.

R27.2—General

R27.2.1 If a load test is described as part of the strength evaluation process, it is desirable for all parties to agree on the region to be loaded, the magnitude of the load, the load test procedure, and acceptance criteria before any load tests are conducted. If the safety concerns are related to an assemblage of members or an entire structure, it is not feasible to load test every member and section. In such cases, it is appropriate that an investigation plan be developed to address the specific safety concerns.

R27.2.2 Strength considerations related to axial load, flexure, and combined axial load and flexure are well understood. There are reliable theories relating strength and short-term displacement to load in terms of member dimensional and material data. To determine the strength of the structure by analysis, calculations should be based on data gathered on the actual dimensions of the structure, properties of the materials in place, and all pertinent details.

R27.2.3 If the shear or bond strength of a member is critical in relation to the doubt expressed about safety, a test may be the most efficient solution to eliminate or confirm the doubt. A test may also be appropriate if it is not feasible to determine the material and dimensional properties required for analysis, even if the cause of the concern relates to flexure or axial load. Wherever possible and appropriate, the results of the load test should be supported by analysis.

R27.2.4 For a deteriorating structure, acceptance provided by the load test is, by necessity, limited in terms of future service life. In such cases, a periodic inspection program is useful. A program that involves physical tests and periodic inspection can justify a longer period in service. Another option for maintaining the structure in service, while the periodic inspection program continues, is to limit the live load to a level determined to be appropriate in accordance with 27.2.5. The length of the specified time period between inspections should be based on consideration of: a) the nature of the deterioration; b) environmental and load effects; c) service history of the structure; and d) scope of the

CODE

COMMENTARY

periodic inspection program. At the end of a specified time period, further strength evaluation is required if the structure is to remain in service. With the agreement of all concerned parties, procedures may be devised for periodic testing that do not necessarily conform to the loading and acceptance criteria specified within this chapter.

27.2.5 If the structure under investigation does not satisfy conditions or criteria of 27.3, 27.5, or 27.6, the structure shall be permitted for use at a lower load rating, based on the results of the load test or analysis, and if approved by the building official.

R27.2.5 Except for load tested members that have failed under a test (refer to 27.4.5), the building official may permit the use of a structure or member at a lower load rating that is judged to be safe and appropriate on the basis of the strength evaluation.

27.3—Analytical strength evaluation

27.3.1 *Verification of as-built condition*

R27.3—Analytical strength evaluation

R27.3.1 *Verification of as-built condition*

27.3.1.1 As-built dimensions of members shall be field-verified at critical sections.

R27.3.1.1 As-built dimensions at critical locations requiring field verification are those dimensions necessary to quantify the performance at those sections. Critical sections for different load effects, such as moment, shear force, and axial force, are locations where stresses caused by such effects reach their maximum value and as further defined for various member types in the Code. Additionally, critical sections may be defined by specific conditions in the structure being evaluated, such as localized member deterioration.

27.3.1.2 Locations and sizes of reinforcement shall be determined by measurement. It shall be permitted to base reinforcement locations on available drawings if field-verified at representative locations to confirm the information on the drawings.

R27.3.1.2 If investigating individual members, the amount, size, arrangement, and location of reinforcement designed to resist applied load should be determined at the critical sections. Nondestructive investigation methods are generally acceptable. In structures with many critical sections, the frequency of measurements may be reduced if the field measurements are consistent.

27.3.1.3 If required, an estimated equivalent f_c' shall be based on analysis of results of cylinder tests from the original construction, tests of cores removed from the structure, or both sets of data. Original cylinder data and core test data shall be representative of the area of concern.

R27.3.1.3 Guidance on estimating equivalent f_c' from original cylinder data can be found in Bartlett (2012).

ACI Committee 214 has developed two methods for determining an equivalent f_c' from cores taken from an existing structure. These methods are described in ACI 214.4R and rely on statistical analysis techniques. The procedures described are only appropriate where the determination of an equivalent f_c' is necessary for the strength evaluation of an existing structure and should not be used to investigate low cylinder strength test results in new construction, which is considered in 26.12.4. The number of core tests may depend on the size of the structure and the sensitivity of structural safety to concrete strength.

27.3.1.4 The method for obtaining and testing cores shall be in accordance with ASTM C42.

27.3.1.5 The properties of reinforcement are permitted to be based on tensile tests of representative samples of the material in the structure.

R27.3.1.5 The number of tests required depends on the uniformity of the material within the structure and should be

CODE

COMMENTARY

determined by the licensed design professional responsible for the evaluation.

27.3.2 *Strength reduction factors*

27.3.2.1 If dimensions, size, and location of reinforcement, and material properties are determined in accordance with 27.3.1, it is permitted to increase ϕ from the design values elsewhere in this Code; however, ϕ shall not exceed the limits in Table 27.3.2.1.

Table 27.3.2.1—Maximum permissible strength reduction factors

Strength	Classification	Transverse reinforcement	Maximum permissible ϕ
Flexure, axial, or both	Tension controlled	All cases	1.0
	Compression controlled	Spirals[1]	0.9
		Other	0.8
Shear, torsion, or both			0.8
Bearing			0.8

[1]Spirals shall satisfy 10.7.6.3, 20.2.2, and 25.7.3.

R27.3.2 *Strength reduction factors*

R27.3.2.1 The strength reduction factors are larger than those defined in Chapter 21. These increased values are justified by the use of field-obtained material properties and actual in-place dimensions.

27.4—Strength evaluation by load test

27.4.1 Load tests shall be conducted either monotonically in accordance with 27.5 or cyclically in accordance with 27.6.

27.4.2 Load tests shall be conducted in a manner that provides for safety of life and the structure during the test.

27.4.3 Safety measures shall not interfere with the load test or affect the results.

27.4.4 The portion of the structure subject to the test load shall be at least 56 days old. If the owner of the structure, the contractor, the licensed design professional, and all other involved parties agree, it shall be permitted to perform the load test at an earlier age.

27.4.5 A precast member to be made composite with cast-in-place concrete shall be permitted to be tested in flexure as a precast member alone in accordance with (a) and (b):

(a) Test loads shall be applied only when calculations indicate the isolated precast member will not fail by compression or buckling.

R27.4—Strength evaluation by load test

R27.4.1 If the strength of the structure being evaluated may be limited by the strength of concrete or the expected failure of the structure is controlled by shear or development of the reinforcement, the monotonic load test procedure is recommended. The monotonic procedure is recommended because the sustained load applied during the monotonic test allows greater time for widening and propagation of cracks, creep, and slip of reinforcement compared with the cyclic procedure.

R27.4.4 Other involved parties may include building officials, concrete subcontractors, and persons with a future interest in the structure.

CODE

(b) The test load, when applied to the precast member alone, shall induce the same total force in the tensile reinforcement as would be produced by loading the composite member with the test load in accordance with 27.4.6.

27.4.6 *Test load arrangement and load factors*

27.4.6.1 Test load arrangements shall be selected to maximize the load effects in the critical regions of the members being evaluated.

27.4.6.2 The total test load $\boldsymbol{T_t}$, including dead load already in place, shall be at least the greatest of (a), (b), and (c):

(a) $T_t = 1.0D_w + 1.1D_s + 1.6L + 0.5(L_r \text{ or } S \text{ or } R)$ (27.4.6.2a)

(b) $T_t = 1.0D_w + 1.1D_s + 1.0L + 1.6(L_r \text{ or } S \text{ or } R)$ (27.4.6.2b)

(c) $T_t = 1.3(D_w + D_s)$ (27.4.6.2c)

27.4.6.3 It is permitted to reduce $\boldsymbol{L}$ in 27.4.6.2 in accordance with the general building code.

27.4.6.4 The load factor on the live load $\boldsymbol{L}$ in 27.4.6.2(b) shall be permitted to be reduced to 0.5 except for parking structures, areas occupied as places of public assembly, or areas where $\boldsymbol{L}$ is greater than 100 lb/ft^2.

27.4.6.5 Unless documentation or tests are available to confirm the density of normalweight concrete used in the structure, the density shall be taken as 150 lb/ft^3. For other types of concrete materials, the density shall be determined based upon test results or from other documentation.

27.5—Monotonic load test procedure

27.5.1 *Test load application*

27.5.1.1 Total test load $\boldsymbol{T_t}$ shall be applied in at least four approximately equal increments.

COMMENTARY

R27.4.6 *Test load arrangement and load factors*

R27.4.6.1 It is important to apply the load at locations so the effects on the suspected deficiency are a maximum and sharing of the applied load with unloaded members is minimized. In cases where it is shown by analysis that adjoining unloaded members will help resist some of the load, the test load should be adjusted to produce appropriate load effects in the critical region of the members being evaluated.

R27.4.6.2 Test loads were changed in ACI 318-19 to be consistent with the requirements in ACI 437.2 for tests on a portion of a structure and for statically indeterminate structures. The test load separates the dead load into self-weight dead load and the superimposed dead load on the structure during the load test. ACI 437.1R provides additional discussion of test loads for concrete structures.

R27.4.6.3 The live load $\boldsymbol{L}$ may be reduced as permitted by the general building code governing safety considerations for the structure. The test load should be increased to compensate for resistance provided by unloaded portions of the structure in question. The increase in test load is determined from analysis of the loading conditions in relation to the selected pass/fail criterion for the test.

R27.4.6.5 Documentation to support a different unit weight may include test results showing concrete unit weight during placement or measured unit weight of concrete core samples. For other types of concrete materials (such as lightweight concrete), the unit weight should be determined based upon concrete core test results or other documentation. The calculation of D_w may include determination of the weight of bonded concrete materials, such as a topping slab to be placed on precast members, not present during a load test. D_s may also include the weight from structural framing members.

R27.5—Monotonic load test procedure

R27.5.1 *Test load application*

R27.5.1.1 Inspecting the area of the structure subject to test loading for signs of distress after each load increment is advisable (refer to R27.5.3.1).

CODE

27.5.1.2 Uniform T_t shall be applied in a manner that ensures uniform distribution of the load transmitted to the structure or portion of the structure being tested. Arching action in the test load apparatus shall be avoided.

27.5.1.3 After the final load increment is applied, T_t shall remain on the structure for at least 24 hours unless signs of distress, as noted in 27.5.3, are observed.

27.5.1.4 After all response measurements are recorded, the test load shall be removed as soon as practical.

27.5.2 *Response measurements*

27.5.2.1 Response measurements, such as deflection, strain, slip, and crack width, shall be made at locations where maximum response is expected. Additional measurements shall be made if required.

27.5.2.2 The initial value for all applicable response measurements shall be obtained not more than 1 hour before applying the first load increment.

27.5.2.3 A set of response measurements shall be recorded after each load increment is applied and after T_t has been applied on the structure for at least 24 hours.

27.5.2.4 A set of final response measurements shall be made 24 hours after T_t is removed.

27.5.3 *Acceptance criteria*

27.5.3.1 The portion of the structure tested shall show no spalling or crushing of concrete, or other evidence of failure.

COMMENTARY

R27.5.1.2 Arching refers to the tendency for the load to be transmitted nonuniformly to the flexural member being tested. For example, if a slab is loaded by a uniform arrangement of bricks, arching of bricks in contact would result in reduction of the load on the slab near the midspan of the slab.

R27.5.3 *Acceptance criteria*

R27.5.3.1 Evidence of failure includes distress (cracking, spalling, or deflection) of such magnitude and extent that the observed result is obviously excessive and incompatible with the safety requirements of the structure. No simple rules have been developed for application to all types of structures and conditions. If sufficient damage has occurred so that the structure is considered to have failed that test, retesting is not permitted because it is considered that damaged members should not be put into service even at a lower load rating.

Local spalling or flaking of the compressed concrete in flexural members related to casting imperfections need not indicate overall structural distress. Crack widths are good indicators of the state of the structure and should be observed to help determine whether the structural strength and behavior are satisfactory. However, accurate prediction or measurement of crack widths in structural concrete members is not likely to be achieved under field conditions. It is advisable to establish criteria before the test relative to the types of cracks anticipated; where the cracks will be measured; how they will be measured; and approximate limits or criteria to evaluate new cracks or limits for the changes in crack width.

CODE

27.5.3.2 Members tested shall not exhibit cracks indicating imminent shear failure.

27.5.3.3 In regions of members without transverse reinforcement, structural cracks inclined to the longitudinal axis and having a horizontal projection greater than the depth of the member shall be evaluated. For variable-depth members, the depth shall be measured at the midlength of the crack.

27.5.3.4 In regions of anchorage and lap splices of reinforcement, short inclined cracks or horizontal cracks along the line of reinforcement shall be evaluated.

27.5.3.5 Measured deflections shall satisfy:

$$\Delta_r \leq \frac{\Delta_1}{4} \qquad (27.5.3.5)$$

27.5.3.6 If the maximum deflection measured during the test, Δ_1, does not exceed the larger of 0.05 in. or ℓ_t**/2000**, the residual deflection requirements in 27.5.3.5 shall be permitted to be waived.

27.5.3.7 If 27.5.3.5 or 27.5.3.6 is not satisfied, it shall be permitted to repeat the load test, provided that the second load test begins no earlier than 72 hours after removal of externally applied loads from the first load test.

27.5.3.8 Portions of the structure tested in the second load test shall be considered acceptable if:

$$\Delta_r \leq \frac{\Delta_2}{5} \qquad (27.5.3.8)$$

27.6—Cyclic load test procedure

27.6.1 A cyclic load test in accordance with ACI 437.2 shall be permitted to be used to evaluate the strength of an existing structure.

COMMENTARY

R27.5.3.2 Forces are transmitted across a shear crack plane by aggregate interlock at the interface of the crack that is enhanced by clamping action of transverse reinforcement and by dowel action of stirrups crossing the crack. The member is assumed to be approaching imminent shear failure when crack lengths increase to approach a horizontal projected length equal to the depth of the member and concurrently widen to the extent that aggregate interlock cannot occur, and as transverse stirrups, if present, begin to yield or display loss of anchorage so as to threaten their integrity.

R27.5.3.3 Inclined cracks may lead to brittle failure of members without transverse reinforcement. Assessment of all inclined cracks is advisable where transverse reinforcement is not present.

R27.5.3.4 Cracking along the axis of the reinforcement in anchorage zones may be related to high stresses associated with the transfer of forces between the reinforcement and the concrete. These cracks may be indicators of impending brittle failure of the member if they are associated with the development of main reinforcement. It is important that their causes and consequences be evaluated.

R27.5.3.5 If the structure shows no evidence of failure, recovery of deflection after removal of the test load is used to determine whether the strength of the structure is satisfactory.

R27.5.3.6 In the case of a very stiff structure, errors in measurements under field conditions may be of the same order as the actual deflections and recovery. To avoid penalizing a satisfactory structure in such a case, recovery measurements are waived if the maximum deflection does not exceed the larger of 0.05 in. or ℓ_t**/2000**.

R27.6—Cyclic load test procedure

R27.6.1 Cyclic load testing involves the cyclic application and removal of load to a structure or structural element. The cyclic load test protocol described in ACI 437.2 involves the application of increasing levels of load to a structure in repeated load cycles. The measured load-deformation

CODE

COMMENTARY

response of the structure is used to evaluate the performance of the tested element. The acceptance criteria for the cyclic test are based upon deviation of the load deformation response from linear elastic behavior, permanency of deflections during each cycle of the load test, and recovery of deflection after completion of the load test.

27.6.2 Acceptance criteria for cyclic load test results shall be in accordance with ACI 437.2 .

27.6.3 If a member fails a cyclic load test, it shall be permitted to retest the member or structure in accordance with ACI 437.2. It shall be permitted to waive the maximum deflection limit (ℓ_t**/180**) in ACI 437.2 that precludes a retest.

R27.6.3 ACI 437.2 precludes a retest if the member exceeds a maximum deflection limit of ℓ_t**/180** (Section 6.4.4.2 in ACI 437.2-13). For consistency with the monotonic testing protocol, this limit is waived.

Notes

APPENDIX A—DESIGN VERIFICATION USING NONLINEAR RESPONSE HISTORY ANALYSIS

CODE

A.1—Notation and terminology

A.1.1 *Notation*

B = bias factor to adjust nominal strength to seismic target reliabilities

D_u = ultimate deformation capacity; the largest deformation at which the hysteresis model is deemed valid given available laboratory data or other substantiating evidence

f_{ce}' = expected compressive strength of concrete, psi

f_{ue} = expected tensile strength for nonprestressed reinforcement, psi

f_{ye} = expected yield strength for nonprestressed reinforcement, psi

ℓ_p = plastic-hinge length for analysis purposes, in.

R_{ne} = expected yield strength

V_{ne} = expected shear strength, lb

θ_y = yield rotation, radians

ϕ_s = seismic resistance factor for force-controlled actions

A.1.2 *Terminology*

distributed plasticity (fiber) model—component model consisting of discrete fibers explicitly representing nonlinear stress-strain or force-deformation responses.

structural wall panel zone—portion of a structural wall common to intersecting wall segments where forces from adjacent wall segments are resolved.

The following actions shall be as defined by ASCE/SEI 7 Chapter 16:

action, deformation-controlled
action, force-controlled
action, force-controlled critical
action, force-controlled ordinary
action, force-controlled noncritical

A.2—Scope

A.2.1 This appendix shall supplement the requirements of Chapter 16 of ASCE/SEI 7 when performing nonlinear response history analysis to determine the design of earthquake-resistant concrete structures.

A.2.2 The provisions of Appendix A shall be in addition to the provisions of Chapters 1 through 26.

A.2.3 This appendix shall be used in conjunction with Chapter 16 of ASCE/SEI 7 for additional general requirements, ground motions, load combinations, modeling, and analysis for design of new reinforced concrete structures, including:

(a) Structural systems designated as part of the seismic force-resisting system, including diaphragms, moment-resisting frames, structural walls, and foundations.

(b) Members not designated as part of the seismic force-resisting system but required to support other loads while

COMMENTARY

RA.1—Notation and terminology

RA.1.2 *Terminology*

Force-controlled and deformation-controlled actions are classified in A.7 for design using nonlinear analysis of concrete structures.

RA.2—Scope

RA.2.3 This appendix is intended to complement documents such as Chapter 16 of ASCE/SEI 7, TBI (2017), and LATBSDC (2017). This appendix provides requirements specific to nonlinear response history analysis and design of concrete members. For additional analysis and modeling requirements that are not specific to concrete members, refer to Chapter 16 of ASCE/SEI 7, TBI (2017), and LATBSDC (2017).

CODE

sustaining deformations and forces associated with earthquake effects.

A.2.4 All concrete structures designed or verified by this Appendix shall be proportioned and detailed as required by Chapter 18 and the requirements of A.12 when applicable.

A.2.5 It shall be permitted to use the provisions of Appendix A to demonstrate the adequacy of a structural system as required by 18.2.1.7.

A.2.6 Independent structural design review consistent with A.13 shall be required for use of Appendix A.

A.2.7 The licensed design professional shall provide justification for any interpretation required for the application of Appendix A, and if accepted by the independent structural design reviewers, justification shall be provided to the building official for acceptance.

A.3—General

A.3.1 Action Classification and Criticality in A.7, and Acceptance Criteria in A.10 and A.11 provide a comprehensive design approach following the intent of Chapter 16 of ASCE/SEI 7 and the general building code, and shall take precedence over those of Chapter 16 of ASCE/SEI 7.

A.4—Earthquake ground motions

A.4.1 Nonlinear response history analysis shall include the effects of horizontal earthquake ground motions.

A.4.2 Vertical earthquake ground motion shall be considered simultaneously with horizontal earthquake ground motions where inclusion of vertical ground motion will substantially affect the structural design requirements.

A.4.3 Earthquake ground motion acceleration histories shall be selected and modified in accordance with procedures established by the general building code.

COMMENTARY

RA.2.7 It is anticipated that the initial design of a earthquake-resistant structure will be performed using elastic analysis combined with engineering judgment. A nonlinear response history analysis following the requirements of this Appendix can then be performed to demonstrate the design, which may not fully comply with all provisions of ASCE/SEI 7 or the general building code.

RA.3—General

RA.3.1 Due to inconsistencies between ACI 318 and Chapter 16 of ASCE/SEI 7-16 in the approach to Action Classification and Acceptance Criteria for concrete members, the requirements in this Appendix take precedence over those of ASCE. The requirements of this Appendix are closely aligned with those in TBI (2017) and LATBSDC (2017).

RA.4—Earthquake ground motions

RA.4.1 Nonlinear response history analysis commonly is performed using two horizontal components of earthquake ground motion applied to a three-dimensional model of the building.

RA.4.2 Structures with vertical discontinuities in the gravity-load-resisting systems can experience vertical earthquake response that can affect building performance. Examples include columns or walls that terminate on beams or slabs. Some structures with long spans or long cantilevers can be sensitive to vertical ground motion. Engineering judgment should be exercised when considering the sensitivity of structures to vertical ground motions.

RA.4.3 The analysis procedures in Appendix A are based on ground motion selection and scaling consistent with Chapter 16 of ASCE/SEI 7, which includes scaling to a risk-targeted maximum considered earthquake ground acceleration. ASCE/SEI 7 describes appropriate procedures for selection and modification of earthquake ground motions in terms of acceptable hazard and risk levels.

CODE

A.5—Load factors and combinations

A.5.1 Load combinations for nonlinear response history analysis shall conform to the requirements of the general building code.

COMMENTARY

RA.5—Load factors and combinations

RA.5.1 Load combinations for response history analysis used in conjunction with this Appendix are intended to be similar to those of Chapter 16 of ASCE/SEI 7-16, TBI (2017), or LATBSDC (2017).

For nonlinear response history analysis, the principles of linear superposition do not apply. Therefore, it would be incorrect to conduct separate analyses considering various loads and then combine the load effects. Instead, it is necessary to conduct an analysis for each factored load combination and take the design value as the envelope of the analysis results. For any nonlinear analysis including earthquake effects, gravity loads are to be applied to the model first and then the ground shaking simulations are applied in the presence of the gravity loads.

There is a low probability that maximum considered earthquake shaking and factored design gravity load combinations of the general building code will occur simultaneously. A more representative load combination is the occurrence of expected, realistic gravity loading combined with maximum considered earthquake shaking.

One load combination is typically considered for analysis, which includes expected dead load concurrent with expected live load and Maximum Considered Earthquake shaking. Chapter 16 of ASCE/SEI 7-16 requires consideration of a second load combination without live load. It should be noted that this case will seldom govern the design of a tall building.

Accidental torsion is not commonly considered in cases where linear analysis indicates that torsional irregularities are negligible.

Load combinations used in the nonlinear analysis may differ from load combinations used to evaluate force-controlled actions (refer to A.11).

A.6—Modeling and analysis

A.6.1 Models for analysis shall be three-dimensional and shall conform to the requirements of the general building code.

RA.6—Modeling and analysis

A.6.2 Modeling of member nonlinear behavior, including effective stiffness, expected strength, expected deformation capacity, and hysteresis under force or deformation reversals, shall be substantiated by applicable physical test data and shall not be extrapolated beyond the limits of testing.

RA.6.2 Multiple element formulations and material models are appropriate for use in inelastic dynamic analysis of concrete structures. ASCE/SEI 41, ACI 374.3R, ACI 369.1, and NIST GCR 17-917-46 provide guidance on modeling and defining model parameters. Selecting model parameters at the mean value of experimental data, as is recommended by the aforementioned documents, avoids skewing analysis results and produces a more reliable evaluation of concrete building response.

A.6.3 Degradation in member strength or stiffness shall be included in the numerical models unless it can be demonstrated that the demand is not sufficiently large to produce these effects. If degradation in component strength is included in the numerical model, the model formulation shall be such

RA.6.3 The model mesh size selected should allow determination of the structural responses in sufficient detail and with sufficient accuracy. Some systems will exhibit mesh-dependent response, with a reduction in mesh size resulting in reduced deformation capacity and more rapid strength

A Nonlinear

CODE

that structural deformation at onset of strength loss is not affected by mesh configuration in the finite element model.

A.6.4 For structural walls with aspect ratio $h_w/\ell_w \geq 2$, the numerical model of the wall and its connection to surrounding elements shall represent kinematic effects associated with wall rotation and uplift, including the effect of migration of the neutral axis as a function of applied axial force and lateral deformation, unless it can be demonstrated that such effects do not affect the structural design requirements.

A.7—Action classification and criticality

A.7.1 All actions shall be classified as deformation-controlled or force-controlled in accordance with A.7.2 and A.7.3.

A.7.2 *Deformation-controlled actions*

A.7.2.1 Deformation-controlled actions shall satisfy the requirements of A.10.

A.7.2.2 The following shall be designated as deformation-controlled actions:

(a) Moment in beams, structural walls, coupling beams, and slab-column connections
(b) Shear in diagonally reinforced coupling beams that meet the requirements of 18.10.7.4
(c) Moment in columns when combined with axial force for columns meeting the requirements of 18.7.4, 18.7.5, and 18.7.6

A.7.3 *Force-controlled actions*

A.7.3.1 Force-controlled actions shall satisfy the requirements of A.11.

A.7.3.2 The following shall be designated as ordinary force-controlled actions:

(a) Shear and moment in perimeter basement walls
(b) In-plane shear in non-transfer diaphragms
(c) In-plane normal forces in diaphragms other than collectors
(d) Moment in shallow foundation members, including spread footings and mat foundations
(e) Moment in deep foundation members

A.7.3.3 Noncritical force-controlled actions shall be designated as actions in any component where failure will not result in: (a) collapse of the structure; (b) loss of the earthquake resistance of the structure; and (c) falling hazard.

COMMENTARY

loss. For these systems, material softening should be defined using a measure of mesh size, or the chosen material model parameters and mesh size should be shown, using an appropriate experimental data set, to provide accurate simulation of onset of strength loss.

RA.7—Action classification and criticality

RA.7.2 *Deformation-controlled actions*

RA.7.2.2 Similar to the requirements of 18.14.3.3, if columns are detailed with sufficient confinement and reinforcement detailing, column moment can be evaluated as a deformation-controlled action rather than as a force-controlled action.

RA.7.3 *Force-controlled actions*

RA.7.3.2 For diaphragm shear to be considered an ordinary force-controlled action, the shear should not be related to a transfer of force between lateral-force-resisting system components.

CODE

A.7.3.4 All actions not designated as deformation controlled, ordinary force-controlled, or noncritical force-controlled shall be classified as critical-force controlled.

A.8—Effective stiffness

A.8.1 Member stiffness shall include effects of deformations due to flexure, shear, axial elongation or shortening, and reinforcement slip along its development length.

A.8.2 If cracking is anticipated as a result of combined effects of applied forces, displacements, and volume change effects associated with shrinkage, temperature, or creep, effects of concrete cracking on effective member stiffness shall be modeled.

A.8.3 If yielding of reinforcement or nonlinear response of concrete is anticipated as a result of combined effects of applied forces, displacements, and volume change effects associated with shrinkage, temperature, or creep, the structural model shall be capable of representing member stiffness for loading near the onset of inelastic response, as well as behavior past the onset of inelastic response.

COMMENTARY

RA.8—Effective stiffness

RA.8.1 Software for nonlinear analysis generally is capable of directly modeling deformations due to flexure, shear, and axial elongation or shortening. Additional deformation may occur due to slip of longitudinal reinforcement from adjacent anchorages. Such effects commonly occur where beams frame into beam-column joints or walls, where columns frame into beam-column joints or foundations, and where walls frame into foundations. If such effects are considered important to the performance of the structure, appropriate assumptions should be included in the analytical model, either directly or by adjustment of flexural stiffness.

RA.8.2 Effects of cracking on stiffness reduction can be considered directly by using models that represent stiffness reduction as calculated stress reaches the cracking stress or indirectly by reducing the effective stiffness relative to the gross-section stiffness. Where the latter approach is used, the degree of stiffness reduction should be consistent with the degree of cracking anticipated under earthquake loading. Structural walls that are lightly cracked, including basement walls, have traditionally been modeled using effective flexural stiffness in the range 0.5 to 1.0 times gross-section stiffness. Diaphragms at major force transfer levels are commonly modeled using effective axial stiffness in the range 0.10 to 0.5 times gross-section stiffness. TBI (2017) and LATBSDC (2017) provide additional effective stiffness recommendations while NIST GCR 17-917-46v1 (NIST 2017a) and NIST GCR 17-917-46v3 (NIST 2017b) provide more detailed guidance on modeling of diaphragms and frame elements.

For stiffness of beams, columns, and structural walls other than basement walls, refer to RA.8.3.

RA.8.3 If calculations indicate nonlinear response under load combinations including earthquake effects, the nonlinear model should be capable of representing an effective secant stiffness from zero loading to a point corresponding to yield-level forces (slope from A to B in Fig. RA.8.3). The model should also be capable of representing stiffness reduction past the yield point. Degradation in element strength or stiffness should be included in the analytical model unless it can be demonstrated that the demand is not sufficiently large as to produce these effects.

CODE

A.8.4 It shall be permitted to represent member stiffness near the onset of inelastic response using an effective stiffness based on analysis substantiated by physical test data. Alternatively, it shall be permitted to represent member stiffness near the onset of inelastic response using the effective stiffness values in Table A.8.4.

COMMENTARY

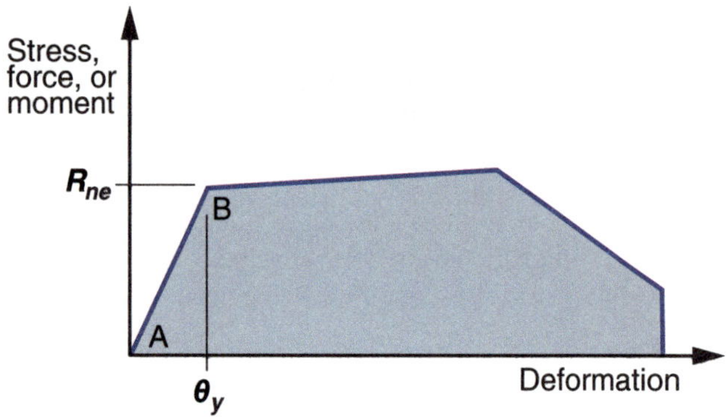

Fig. RA.8.3—Generalized force-deformation relations.

RA.8.4 The effective stiffness values are intended to represent the slope from A to B in Fig. RA.8.3, where B corresponds to expected yield strength. Effective stiffness values for beams and columns are based on Elwood et al. (2007), and incorporate the effects of reinforcement slip along the development length. Tabulated values for structural walls are appropriate to use where the wall is represented by a line element. In some building models, structural walls will be represented by distributed fiber models, in which case the fiber model should directly represent effects of concrete cracking and reinforcement yielding, such that the stiffness values in Table A.8.4 do not apply. Basement walls are unlikely to respond at yield-level forces; therefore, larger stiffness values may be more applicable than those in Table A.8.4 for walls. Diaphragm stiffnesses provided in Table A.8.4 represent typical values. Prestressed and nonprestressed diaphragms mainly resisting single-floor in-plane earthquake forces are commonly modeled as rigid, as allowed by ASCE/SEI 7. Diaphragms transferring relatively large in-plane earthquake forces from multiple floor levels can have effective stiffnesses somewhat lower than those represented in Table A.8.4. In cases where analysis results are sensitive to diaphragm stiffness assumptions, it may be prudent to "bound" the solution by analyzing the structure using a range of diaphragm stiffnesses and selecting the design values as the larger forces from the two analyses. Coupling beam effective stiffnesses are intended to represent values for beams cast monolithically with floor slabs. Values are based on equations presented by Vu et al. (2014), but are adjusted to account for the presence of a slab, differences in modeling approach, and typical shear levels (TBI 2017). Engineering judgment should be used to evaluate effective shear stiffness values, noting that due to typical software implementation limitations, gross area is used in lieu of effective area.

CODE | COMMENTARY

Table A.8.4—Effective stiffness values[1]

Component		Axial	Flexural	Shear
Beams	nonprestressed	$1.0E_cA_g$	$0.3E_cI_g$	$0.4E_cA_g$
	prestressed	$1.0E_cA_g$	$1.0E_cI_g$	$0.4E_cA_g$
Columns with compression caused by design gravity loads[2]	$\geq 0.5A_gf_c'$	$1.0E_cA_g$	$0.7E_cI_g$	$0.4E_cA_g$
	$\leq 0.1A_gf_c'$ or with tension	$1.0E_cA_g$ (compression) $1.0E_sA_{st}$ (tension)	$0.3E_cI_g$	$0.4E_cA_g$
Structural walls[3]	in-plane	$1.0E_cA_g$	$0.35E_cI_g$	$0.2E_cA_g$
	out-of-plane	$1.0E_cA_g$	$0.25E_cI_g$	$0.4E_cA_g$
Diaphragms (in-plane only)[4]	nonprestressed	$0.25E_cA_g$	$0.25E_cI_g$	$0.25E_cA_g$
	prestressed	$0.5E_cA_g$	$0.5E_cI_g$	$0.4E_cA_g$
Coupling beams	with or without diagonal reinforcement	$1.0E_cA_g$	$0.07\left(\frac{\ell_n}{h}\right)E_cI_g$ $\leq 0.3E_cI_g$	$0.4E_cA_g$
Mat foundations	in-plane	$0.5E_cA_g$	$0.5E_cI_g$	$0.4E_cA_g$
	out-of-plane[5]		$0.5E_cI_g$	

[1]Tabulated values for axial, flexural, and shear shall be applied jointly in defining effective stiffness of an element, unless alternative combinations are justified.

[2]For columns with axial compression falling between the limits provided, flexural stiffness shall be determined by linear interpolation.

[3]Tabulated values are appropriate where members are modeled using line elements to represent their properties.

[4]Diaphragms shall be permitted to be modeled as rigid in-plane if this does not result in differences in analysis outcomes.

[5]Specified stiffness values for mat foundations pertain for the general condition of the mat. Where the wall or other vertical members imposed sufficiently large forces, including local force reversals across stacked wall openings, the stiffness values may need to be reduced.

A.8.5 In beam-column joints if joint flexibility is not modeled explicitly, it shall be permitted to model joint flexibility implicitly by defining the effective stiffness of beams and columns framing into the joint to include joint flexibility and by introducing beam and column rigid end offsets that extend to the center of the joint.

RA.8.5 In reinforced concrete frames detailed to resist earthquake forces, joints are not expected to experience significant degradation. In lieu of a more rigorous representation of joint shear stiffness, rigid offsets of beam and column members extending the length of the joint dimensions are permitted (Birely et al. 2012). A sensitivity study on stiffness assumptions indicates that overall building stiffness may be more sensitive to the choice of effective stiffness for frame and wall members than for joints (Kwon and Ghannoum 2016). The rigid joint offset approach is compatible with the effective stiffness values presented in Table A.8.4, which account for the softening effects of longitudinal bar slip within the joints.

A.8.6 If beams other than coupling beams are cast monolithically with slabs, the effective slab width defined in 6.3.2 shall be included in the evaluation of beam flexural and axial stiffnesses.

A.9—Expected material strength

A.9.1 Expected material strength shall be defined based on applicable project-specific data or data from projects using similar materials and construction. If applicable data are not available, the expected material strengths in Table A.9.1 shall be permitted.

RA.9—Expected material strength

RA.9.1 The multiplier on f_c' may be smaller for high-strength concrete when higher quality control measures are in place or when fly ash and other additives are. Refer to ACI 232.2R for discussion of impacts of fly ash. Default values for other steel grades have not been provided in Table A.9.1 due to insufficient data.

CODE

COMMENTARY

Table A.9.1—Expected material strengths

Material		Expected strength	
Concrete		$f_{ce}' = 1.3f_c'$[1]	
Reinforcing steel		Expected yield strength, f_{ye}, psi	Expected tensile strength, f_{ue}, psi
A615	Grade 60	70,000	106,000
A706	Grade 60	69,000	95,000
	Grade 80	85,000	112,000

[1]Expected strength f_{ce}' is strength expected at approximately 1 year or longer.

A.10—Acceptance criteria for deformation-controlled actions

A.10.1 Deformations in any of the response history analyses shall not exceed the ultimate deformation capacity $\boldsymbol{D_u}$ unless (a) or (b) is satisfied.

(a) The analysis assumes the strength associated with this mode of deformation is negligible for the remainder of that analysis, and the structure is evaluated for stability and strength.
(b) The analysis is considered to have an unacceptable response as defined by ASCE/SEI 7.

A.10.2 $\boldsymbol{D_u}$ shall be determined by (a), (b), or (c):

(a) $\boldsymbol{D_u}$ of the component shall be taken as the valid range of modeling as demonstrated by comparison of the hysteresis model with suitable laboratory test data including the appropriate gravity load effect.
(b) If special structural walls are modeled using distributed plasticity (fiber) models, $\boldsymbol{D_u}$ shall be evaluated using the average vertical strain. The strain shall be evaluated over a height of the plastic hinge length, $\boldsymbol{\ell_p}$, where $\boldsymbol{\ell_p}$ is the longer of (i) and (ii):

(i) $\ell_p = 0.2\ell_w + 0.03h_w$ (A.10.2.a)

(ii) $\ell_p = 0.08h_w + 0.00015f_yd_b$ (A.10.2.b)

but not exceeding the story height, where $\boldsymbol{d_b}$ and $\boldsymbol{f_y}$ are determined based on the wall longitudinal reinforcement.

(c) If structural components are modeled using lumped plasticity (concentrated hinge) or distributed plasticity (fiber) models, $\boldsymbol{D_u}$ shall be permitted to be in accordance with ACI 369.1 or as substantiated by laboratory testing.

RA.10—Acceptance criteria for deformation-controlled actions

RA.10.1 These acceptance criteria are consistent with the component acceptance criteria in TBI (2017), which are different from those in ASCE/SEI 7 and LATBSDC (2017). More detailed discussion regarding the differences of evaluation approaches of deformation-controlled actions in ASCE/SEI 7 and TBI (2017) are provided in TBI (2017).

RA.10.2 Ultimate deformation capacity, $\boldsymbol{D_u}$, is typically obtained from statistical analysis of the available test data and can be closely related to Collapse Prevention Acceptance Criteria in ACI 369.1 and ASCE/SEI 41. $\boldsymbol{D_u}$ is based on the deformation where substantial loss of gravity load-carrying capacity occurs or, if tests do not progress to this deformation, the limiting deformation for which testing was performed. An example of $\boldsymbol{D_u}$ in the hysteresis curve of an analysis model is shown in Fig. RA.10.2. The Collapse Prevention Acceptance Criteria in ACI 369.1 and ASCE/SEI 41 are typically less than mean experimental values due to scatter in data sets used to develop these criteria. The ASCE/SEI 41 approach also evaluates deformation as the mean of the maximum absolute response from each response history analysis. Appendix A, however, evaluates deformation as the maximum of any of the response history analyses.

Hysteresis behavior of the structural component simulated using fiber material models should be evaluated and adjusted using experimental data for the range of deformation demands and behaviors simulated in the analyses. ACI 374.3R and ACI 369.1 provide nonlinear modeling parameters that can be the basis for hysteresis shape based on experimental data. Figure RA.10.2 shows a hysteresis curve generated using adjusted fiber material models based on such nonlinear modeling parameters to simulate the component behavior observed in laboratory tests.

If $\boldsymbol{D_u}$ is defined by average strain, the length over which strain is defined in the analysis should be consistent with the length over which strain limits are established from experi-

CODE | COMMENTARY

mental data or are specified in documents such as ASCE/SEI 41, ACI 369.1, TBI, or LATBSDC.

Sufficient number of fibers along the cross section should be used to allow the strain values at fiber centerlines to be extrapolated to locations where strain values are calculated to compare with strain limits, such as, at the extreme edge of the wall compression zone.

For structural walls or coupling beams modeled using fiber elements, deformation acceptance criteria can be represented in either a strain or member deformation basis. The strain results can be obtained directly from the fiber model. The member deformation results, such as plastic hinge rotation, story drift, or chord rotation, can be obtained by aggregated deformation over a group of fiber elements representing the member. Plastic hinge length Eq. (A.10.2a) and (A.10.2b) for walls are from Paulay and Priestley (1992).

An example of acceptance criteria for strain limits is provided in TBI (2017). The unconfined concrete model includes a peak stress at a compressive strain of 0.002, with a descending backbone to 50 percent of the peak stress value at a compressive strain of 0.003 (the ultimate deformation capacity, D_u). The confined concrete model, used where confinement meeting the requirements of 18.10.6.4(e) and (f) are provided, includes a peak stress at a compressive strain 0.008, with a descending backbone to 80 percent of the peak stress value at a compressive strain of 0.015 (the ultimate deformation capacity, D_u). The longitudinal reinforcement tensile strain limit of 0.05 (the ultimate deformation capacity, D_u) is based on tensile rupture with consideration of low-cycle fatigue effects, which is corroborated by Segura and Wallace (2018).

Additional references for ultimate deformation capacity, such as ACI 369.1, TBI (2017), and LATBSDC (2017), may be used subject to approval of the independent structural design review.

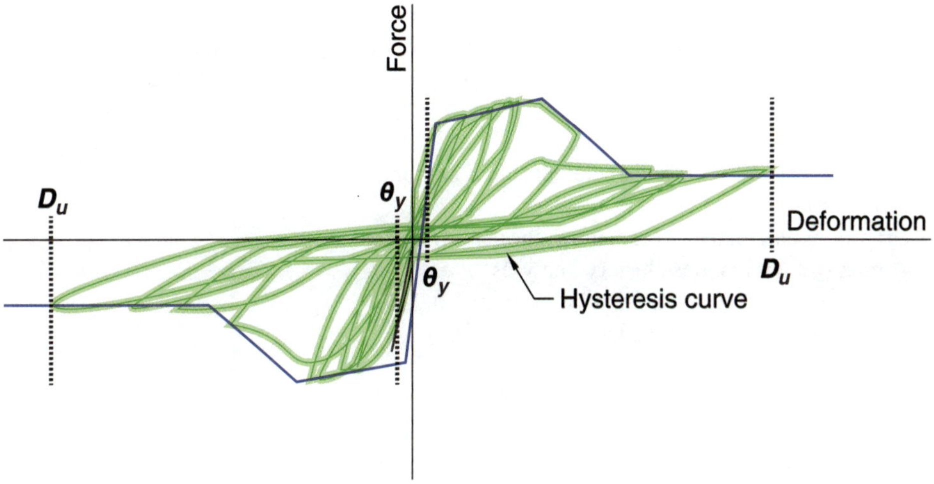

Fig. RA.10.2—D_u *in response hysteresis from an analysis model.*

CODE

A.11—Expected strength for force-controlled actions

A.11.1 Force-controlled actions shall be evaluated in accordance with the general building code, with expected strength taken as $\phi_s B R_n$.

A.11.2 ϕ_s shall be in accordance with Table A.11.2, with ϕ determined in accordance with Chapter 21, except that 21.2.4.1 shall not apply.

Table A.11.2—Seismic resistance factor

Force-controlled action	ϕ_s
Critical	ϕ
Ordinary	$\phi/0.9 \leq 1.0$
Noncritical	$\phi/0.85 \leq 1.0$

A.11.3 Bias factor, B, shall be taken as 1.0. Alternatively, it shall be permitted to calculate B using Eq. (A.11.3):

$$B = 0.9R_{ne}/R_n \geq 1.0 \tag{A.11.3}$$

A.11.3.1 Nominal strength, R_n, shall be in accordance with Chapter 18, 22, or 23.

A.11.3.2 The expected strength, R_{ne}, is permitted to be defined in accordance with the nominal strength provisions of Chapters 18, 22, or 23, with f_{ce}' substituted for f_c' and f_{ye} substituted for f_y or f_{yt}, except as provided in A.11.3.2.1 and A.11.3.2.2.

A.11.3.2.1 For structural walls where $h_w/\ell_w \geq 2$ meeting (a) through (d), the requirements of A.11.3.2.1.1 and A.11.3.2.1.2 shall apply.

(a) Wall is modeled with fiber elements in accordance with A.10.2(b)
(b) Strains calculated as the mean of the maximum demand from a suite of response history analyses
(c) Calculated concrete compressive strain < 0.005
(d) Calculated longitudinal tensile strain < 0.01

A.11.3.2.1.1 $V_{ne} = 1.5A_{cv}(2\lambda\sqrt{f_{ce}'} + \rho_t f_{ye})$

A.11.3.2.1.2 For all vertical wall segments sharing a common lateral force, V_{ne} shall not be taken greater than

COMMENTARY

RA.11—Expected strength for force-controlled actions

RA.11.1 Currently, strength reduction factors, ϕ, are not specifically calibrated to the seismic reliability targets specified in ASCE/SEI 7. Rather, these strength reduction factors are calibrated to the target reliabilities for other loads (ASCE/SEI 7-16 Table 1.3-1). The bias factor, B, is provided to adjust the resistance factors specified by the materials standards to the seismic target reliabilities, considering the inherent bias in the nominal strength equations contained in the materials standards. This bias is a function of both the ratio of expected material strength to minimum specified strength and also inherent conservatism in the predictive equations specified by the materials standards.

RA.11.2 For ordinary and noncritical actions, the resistance factors are relaxed in order to accept a higher probability of failure.

More detailed discussion regarding the differences of evaluation approaches of force-controlled actions in ASCE/SEI 7, TBI (2017), and LATBSDC (2017) are provided in TBI (2017) and LATBSDC (2017). Additional background on this approach is provided in Wallace et al. (2013) and Kim and Wallace (2017).

RA.11.3.2.1 The shear strength determined from these provisions is applicable only to walls with relatively low flexural ductility demands (Wallace 2013; LATBSDC 2017).

CODE

$12A_{cv}\sqrt{f'_{ce}}$. For any individual vertical wall segments, V_{ne} shall not be taken greater than $15A_{cv}\sqrt{f'_{ce}}$.

A.11.3.2.2 For structural wall panel zones, V_{ne} shall be calculated in accordance with A.11.3. 2.1(a). V_{ne} shall not be taken greater than $25A_{cv}\sqrt{f'_{ce}}$.

A.12—Enhanced detailing requirements

A.12.1 If the mean maximum deformation from the set of response history analyses exceeds $0.5D_u$ of confined concrete, members shall be subject to the added detailing requirements of this section.

A.12.2 *Special moment frames*

A.12.2.1 For beams of special moment frames, the spacing of transversely supported flexural reinforcing bars as required by 18.6.4.2 shall not exceed 8 in.

A.12.2.2 The sum of the column strengths at any joint as required by 18.7.3.2 shall be at least 1.4 times the sum of the beam strengths at the joint.

A.12.2.3 For tied columns of special moment frames, every longitudinal bar shall have lateral support by a corner of a hoop or a seismic hook as required in 18.7.5.2(f) regardless of axial load or concrete strength.

A.12.2.4 When deformations of beams of special moment frames exceed $0.5D_u$, the column dimension parallel to the beam longitudinal reinforcement required in 18.8.2.3 shall be increased by 20 percent.

A.12.3 *Special structural walls*

A.12.3.1 Boundary elements shall be provided in accordance with 18.10.6 with transverse reinforcement conforming with A.12.2.3.

COMMENTARY

RA.12—Enhanced detailing requirements

RA.12.1 The requirements for earthquake-resisting systems and detailing have been developed over many years using actual earthquake damage observations, research, and engineering judgment. These requirements are codified in ASCE/SEI 7, IBC, and ACI 318. In recent years, enhanced computational abilities allow engineers to model and calculate seismic response in great detail.

Designs that exceed the prescriptive limits of the general building code are sometimes prepared, verified, and justified. In some instances, these new designs have not been tested in strong ground shaking, and there is some concern that these designs may be extrapolating beyond the collective knowledge. Therefore, these enhanced details are provided to improve inelastic response ductility and are appropriate when using Appendix A for designs beyond prescriptive code limits.

RA.12.2 *Special moment frames*

RA.12.2.3 This code has allowed crossties in compression members with a seismic hook at only one end and with crossties alternated recognizing their ease in construction. However, recent earthquakes and research tests have shown that 90-degree hooks do not always provide adequate support (Moehle and Cavanagh 1985).

A.12.3 *Special structural walls*

CODE

A.12.3.2 If boundary elements are required, splices of shear reinforcement shall be made with mechanical or welded splices, or lap splices enclosed in transverse reinforcement spaced at the smaller of $6d_b$ of the spliced bars or 6 in.

A.12.3.3 If the floor or roof slab is shown by analysis to undergo inelastic response at a slab-wall connection, the slab flexural reinforcement shall be extended through the slab-wall joint and anchored for structural integrity.

A.12.3.4 If shear force exceeds $4A_{cv}\lambda\sqrt{f_c'}$, enhanced construction joint detailing shall be provided with thorough roughening of concrete, intermittent shear keys in the concrete, or both, to reduce the possibility of slip along the construction joint.

A.13—Independent structural design review

A.13.1 The analysis and design shall be reviewed by an independent structural design reviewer. The independent structural design reviewer shall act under the direction of the building official.

A.13.2 The independent structural design review shall be performed by one or more individuals acceptable to the building official and possessing knowledge of (a) through (d):

(a) Selection and scaling of ground motions for use in nonlinear response history analysis.
(b) Behavior of structural systems of the type under consideration when subjected to earthquake loading.
(c) Analytical structural modeling for use in nonlinear response history analysis, including use of physical tests in the creation and calibration of the structural analysis models, and knowledge of soil-structure interaction if used in the analysis or in the development of ground motions.
(d) The requirements of Appendix A as they pertain to design of the type of structure under consideration.

A.13.3 The scope of the independent structural design review shall be approved by the building official and shall include a minimum of (a) through (h):

COMMENTARY

RA.12.3.3 Analysis of tall buildings with structural core wall systems have shown inelastic response in slabs at their connection to core walls. Integrity of this connection is critical to the overall performance of the structure. Enhanced details, which include properly anchored or continuous reinforcement and post-tensioning tendons, providing additional integrity are required.

RA.12.3.4 Sliding at horizontal construction joints of walls has been observed in earthquakes and in laboratory testing of structural walls. Enhanced detailing is required in regions of high shear to minimize slip or sliding at construction joints.

RA.13—Independent structural design review

RA.13.1 The independent structural design reviewer provides an independent, objective, technical review of those aspects of the structural design of the building that relate to earthquake-performance and advises the building official whether the design meets the acceptance criteria and the expected building performance.

Review by the independent structural design reviewer is not intended to replace quality assurance measures ordinarily exercised by the licensed design professional. Responsibility for the structural design remains solely with the licensed design professional in responsible charge of the structural design.

RA.13.2 On many projects, independent structural design review may be provided by a review team approved by the building official. Each member of the review team may possess specialized knowledge and expertise, and jointly meet the requirements of A.13.2.

An independent structural design reviewer should not have conflicts of interest with respect to the project and should not be part of the design team for the project.

RA.13.3 The scope of the independent structural design review should be clearly defined and acceptable to the building official.

CODE

(a) Basis of design document, including the earthquake-performance objectives, the overall earthquake-resistant design methodology, and acceptance criteria
(b) Proposed structural system
(c) Earthquake hazard determination, and selection and modification of earthquake ground motions
(d) Modeling approaches for components
(e) Structural analysis model, including soil-structure interaction as applicable, and verification that the structural analysis model adequately represents the properties of the structural system
(f) Review of structural analysis results and determination of whether calculated response meets approved acceptance criteria
(g) Design and detailing of structural components
(h) Drawings, specifications, and quality control/quality assurance and inspection provisions in the design documents

A.13.4 The independent structural design review shall be documented as follows:

(a) The independent structural design reviewer shall issue comments and questions to the licensed design professional.
(b) The licensed design professional shall provide written responses to the independent structural design reviewer.
(c) The independent structural design reviewer shall summarize the review in a letter addressed to the building official that shall include a log of all questions or comments and responses. Any items that lack resolution or consensus shall be clearly explained with reasons for lack of agreement.

COMMENTARY

RA.13.4 A statement of agreement with the design should be provided. However, there may be occasions where complete agreement between the independent structural design reviewer and the licensed design professional cannot be reached. These items should be documented in the summary review letter.

Notes

APPENDIX B—STEEL REINFORCEMENT INFORMATION

As an aid to users of the ACI Building Code, information on sizes, areas, and weights of various steel reinforcement is presented.

ASTM STANDARD REINFORCING BARS

Bar size, no.	Nominal diameter, in.	Nominal area, in.^{2}	Nominal weight, lb/ft
3	0.375	0.11	0.376
4	0.500	0.20	0.668
5	0.625	0.31	1.043
6	0.750	0.44	1.502
7	0.875	0.60	2.044
8	1.000	0.79	2.670
9	1.128	1.00	3.400
10	1.270	1.27	4.303
11	1.410	1.56	5.313
14	1.693	2.25	7.65
18	2.257	4.00	13.60

ASTM STANDARD PRESTRESSING STRANDS, WIRES, AND BARS

Type*	Nominal diameter, in.	Nominal area, in.^{2}	Nominal weight, lb/ft
Seven-wire strand (Grade 250)	1/4 (0.250)	0.036	0.122
	5/16 (0.313)	0.058	0.197
	3/8 (0.375)	0.080	0.272
	7/16 (0.438)	0.108	0.367
	1/2 (0.500)	0.144	0.490
	(0.600)	0.216	0.737
Seven-wire strand (Grade 270)	3/8 (0.375)	0.085	0.290
	7/16 (0.438)	0.115	0.390
	1/2 (0.500) (0.520) (0.563)	0.153 0.167 0.192	0.520 0.570 0.650
	(0.600) (0.620) (0.700)	0.217 0.231 0.294	0.740 0.780 1.000
Prestressing wire	0.192	0.029	0.098
	0.196	0.030	0.102
	0.250	0.049	0.170
	0.276	0.060	0.204
Prestressing bars (Type I, plain)	3/4	0.44	1.50
	7/8	0.60	2.04
	1	0.78	2.67
	1-1/8	0.99	3.38
	1-1/4	1.23	4.17
	1-3/8	1.48	5.05
Prestressing bars (Type II, deformed)	5/8	0.28	0.98
	3/4	0.42	1.49
	1	0.85	3.01
	1-1/4	1.25	4.39
	1-3/8 1-3/4 2-1/2 3	1.58 2.58 5.16 6.85	5.56 9.10 18.20 24.09

*Availability of some strand, wire, and bar sizes should be investigated in advance.

WRI STANDARD WIRE REINFORCEMENT*

W & D size		Nominal diameter, in.	Nominal area, in.^{2}	Nominal weight, lb/ft	Area, in.^{2}/ft of width for various spacings						
			Center-to-center spacing, in.								
Plain	Deformed				2	3	4	6	8	10	12
W31	D31	0.628	0.310	1.054	1.86	1.24	0.93	0.62	0.46	0.37	0.31
W30	D30	0.618	0.300	1.020	1.80	1.20	0.90	0.60	0.45	0.36	0.30
W28	D28	0.597	0.280	0.952	1.68	1.12	0.84	0.56	0.42	0.33	0.28
W26	D26	0.575	0.260	0.884	1.56	1.04	0.78	0.52	0.39	0.31	0.26
W24	D24	0.553	0.240	0.816	1.44	0.96	0.72	0.48	0.36	0.28	0.24
W22	D22	0.529	0.220	0.748	1.32	0.88	0.66	0.44	0.33	0.26	0.22
W20	D20	0.505	0.200	0.680	1.20	0.80	0.60	0.40	0.30	0.24	0.20
W18	D18	0.479	0.180	0.612	1.08	0.72	0.54	0.36	0.27	0.21	0.18
W16	D16	0.451	0.160	0.544	0.96	0.64	0.48	0.32	0.24	0.19	0.16
W14	D14	0.422	0.140	0.476	0.84	0.56	0.42	0.28	0.21	0.16	0.14
W12	D12	0.391	0.120	0.408	0.72	0.48	0.36	0.24	0.18	0.14	0.12
W11	D11	0.374	0.110	0.374	0.66	0.44	0.33	0.22	0.16	0.13	0.11
W10.5		0.366	0.105	0.357	0.63	0.42	0.315	0.21	0.15	0.12	0.105
W10	D10	0.357	0.100	0.340	0.60	0.40	0.30	0.20	0.15	0.12	0.10
W9.5		0.348	0.095	0.323	0.57	0.38	0.285	0.19	0.14	0.11	0.095
W9	D9	0.338	0.090	0.306	0.54	0.36	0.27	0.18	0.13	0.10	0.09
W8.5		0.329	0.085	0.289	0.51	0.34	0.255	0.17	0.12	0.10	0.085
W8	D8	0.319	0.080	0.272	0.48	0.32	0.24	0.16	0.12	0.09	0.08
W7.5		0.309	0.075	0.255	0.45	0.30	0.225	0.15	0.11	0.09	0.075
W7	D7	0.299	0.070	0.238	0.42	0.28	0.21	0.14	0.10	0.08	0.07
W6.5		0.288	0.065	0.221	0.39	0.26	0.195	0.13	0.09	0.07	0.065
W6	D6	0.276	0.060	0.204	0.36	0.24	0.18	0.12	0.09	0.07	0.06
W5.5		0.265	0.055	0.187	0.33	0.22	0.165	0.11	0.08	0.06	0.055
W5	D5	0.252	0.050	0.170	0.30	0.20	0.15	0.10	0.07	0.06	0.05
W4.5		0.239	0.045	0.153	0.27	0.18	0.135	0.09	0.06	0.05	0.045
W4	D4	0.226	0.040	0.136	0.24	0.16	0.12	0.08	0.06	0.04	0.04
W3.5		0.211	0.035	0.119	0.21	0.14	0.105	0.07	0.05	0.04	0.035
W3		0.195	0.030	0.102	0.18	0.12	0.09	0.06	0.04	0.03	0.03
W2.9		0.192	0.029	0.098	0.174	0.116	0.087	0.058	0.04	0.03	0.029
W2.5		0.178	0.025	0.085	0.15	0.10	0.075	0.05	0.03	0.03	0.025
W2		0.160	0.020	0.068	0.12	0.08	0.06	0.04	0.03	0.02	0.02
W1.4		0.134	0.014	0.049	0.084	0.056	0.042	0.028	0.02	0.01	0.014

*Reference "Structural Welded Wire Reinforcement Manual of Standard Practice," Wire Reinforcement Institute, Hartford, CT, sixth edition, Apr. 2001, 38 pp.

Notes

APPENDIX C—EQUIVALENCE BETWEEN SI-METRIC, MKS-METRIC, AND U.S. CUSTOMARY UNITS OF NONHOMOGENOUS EQUATIONS IN THE CODE

Provision number	SI-metric stress in MPa	mks-metric stress in kgf/cm²	U.S. Customary units stress in pounds per square inch (psi)
	1 MPa	10 kgf/cm²	145 psi
	$f_c' = 21$ MPa	$f_c' = 210$ kgf/cm²	$f_c' = 3000$ psi
	$f_c' = 28$ MPa	$f_c' = 280$ kgf/cm²	$f_c' = 4000$ psi
	$f_c' = 35$ MPa	$f_c' = 350$ kgf/cm²	$f_c' = 5000$ psi
	$f_c' = 40$ MPa	$f_c' = 420$ kgf/cm²	$f_c' = 6000$ psi
	$f_y = 280$ MPa	$f_y = 2800$ kgf/cm²	$f_y = 40{,}000$ psi
	$f_y = 420$ MPa	$f_y = 4200$ kgf/cm²	$f_y = 60{,}000$ psi
	$f_y = 550$ MPa	$f_y = 5600$ kgf/cm²	$f_y = 80{,}000$ psi
	$f_y = 690$ MPa	$f_y = 7000$ kgf/cm²	$f_y = 100{,}000$ psi
	$f_{pu} = 1725$ MPa	$f_{pu} = 17{,}600$ kgf/cm²	$f_{pu} = 250{,}000$ psi
	$f_{pu} = 1860$ MPa	$f_{pu} = 19{,}000$ kgf/cm²	$f_{pu} = 270{,}000$ psi
	$\sqrt{f_c'}$ in MPa	$3.18\sqrt{f_c'}$ in kgf/cm²	$12\sqrt{f_c'}$ in psi
	$0.313\sqrt{f_c'}$ in MPa	$\sqrt{f_c'}$ in kgf/cm²	$3.77\sqrt{f_c'}$ in psi
	$0.083\sqrt{f_c'}$ in MPa	$0.27\sqrt{f_c'}$ in kgf/cm²	$\sqrt{f_c'}$ in psi
	$0.17\sqrt{f_c'}$ in MPa	$0.53\sqrt{f_c'}$ in kgf/cm²	$2\sqrt{f_c'}$ in psi
6.6.4.5.4	$M_{2,min} = P_u(15 + 0.03h)$	$M_{2,min} = P_u(1.5 + 0.03h)$	$M_{2,min} = P_u(0.6 + 0.03h)$
7.3.1.1.1	$\left(0.4+\frac{f_y}{700}\right)$	$\left(0.4+\frac{f_y}{7000}\right)$	$\left(0.4+\frac{f_y}{100{,}000}\right)$
7.3.1.1.2	$(1.65 - 0.0003w_c)$	$(1.65 - 0.0003w_c)$	$(1.65 - 0.005w_c)$
7.7.3.5(c)	$0.41\frac{b_w s}{f_{yt}}$	$4.2\frac{b_w s}{f_{yt}}$	$60\frac{b_w s}{f_{yt}}$
8.3.1.1	$f_r = 0.4\sqrt{f_c'}$	$f_r = 1.33\sqrt{f_c'}$	$f_r = 5\sqrt{f_c'}$
8.3.1.2(b)	$h = \frac{\ell_n\left(0.8+\frac{f_y}{1400}\right)}{36+5\beta(\alpha_{fm}-0.2)} \geq 125$ mm	$h = \frac{\ell_n\left(0.8+\frac{f_y}{14{,}000}\right)}{36+5\beta(\alpha_{fm}-0.2)} \geq 12.5$ cm	$h = \frac{\ell_n\left(0.8+\frac{f_y}{200{,}000}\right)}{36+5\beta(\alpha_{fm}-0.2)} \geq 5$ in.
8.3.1.2(d)	$h = \frac{\ell_n\left(0.8+\frac{f_y}{1400}\right)}{36+9\beta} \geq 90$ mm	$h = \frac{\ell_n\left(0.8+\frac{f_y}{14{,}000}\right)}{36+9\beta} \geq 9$ cm	$h = \frac{\ell_n\left(0.8+\frac{f_y}{200{,}000}\right)}{36+9\beta} \geq 3.5$ in.
8.3.4.1	$f_t \leq 0.50\sqrt{f_c'}$	$f_t \leq 1.6\sqrt{f_c'}$	$f_t \leq 6\sqrt{f_c'}$
8.6.2.3	$0.17\sqrt{f_c'}$ $0.50\sqrt{f_c'}$	$0.53\sqrt{f_c'}$ $1.6\sqrt{f_c'}$	$2\sqrt{f_c'}$ $6\sqrt{f_c'}$

C Conv. Tables

Provision number	SI-metric stress in MPa	mks-metric stress in kgf/cm²	U.S. Customary units stress in pounds per square inch (psi)
8.7.5.6.3.1(a) and (b)	$A_s = \frac{0.37\sqrt{f'_c}c_2d}{f_y}$ $A_s = \frac{2.1c_2d}{f_y}$	$A_s = \frac{1.2\sqrt{f'_c}c_2d}{f_y}$ $A_s = \frac{21c_2d}{f_y}$	$A_s = \frac{4.5\sqrt{f'_c}c_2d}{f_y}$ $A_s = \frac{300c_2d}{f_y}$
8.7.7.1.2	$\phi 0.5\sqrt{f'_c}$	$\phi 1.6\sqrt{f'_c}$	$\phi 6\sqrt{f'_c}$
9.3.1.1.1	$\left(0.4+\frac{f_y}{700}\right)$	$\left(0.4+\frac{f_y}{7000}\right)$	$\left(0.4+\frac{f_y}{100,000}\right)$
9.3.1.1.2	$(1.65 - 0.0003w_c)$	$(1.65 - 0.0003w_c)$	$(1.65 - 0.005w_c)$
9.6.1.2(a) and (b)	$\frac{0.25\sqrt{f'_c}}{f_y}b_wd$ $\frac{1.4}{f_y}b_wd$	$\frac{0.80\sqrt{f'_c}}{f_y}b_wd$ $\frac{14}{f_y}b_wd$	$\frac{3\sqrt{f'_c}}{f_y}b_wd$ $\frac{200}{f_y}b_wd$
9.6.3.1	$V_u > 0.083\phi\lambda\sqrt{f'_c}\,b_wd$	$V_u > 0.27\phi\lambda\sqrt{f'_c}\,b_wd$	$V_u > \phi\lambda\sqrt{f'_c}\,b_wd$
9.6.3.4	$A_{v,min} \geq 0.062\sqrt{f'_c}\frac{b_ws}{f_{yt}}$ $A_{v,min} \geq 0.35\frac{b_ws}{f_{yt}}$	$A_{v,min} \geq 0.2\sqrt{f'_c}\frac{b_ws}{f_{yt}}$ $A_{v,min} \geq 3.5\frac{b_ws}{f_{yt}}$	$A_{v,min} \geq 0.75\sqrt{f'_c}\frac{b_ws}{f_{yt}}$ $A_{v,min} \geq 50\frac{b_ws}{f_{yt}}$
9.6.4.2(a) and (b)	$(A_v + 2A_t)/s \geq 0.062\sqrt{f'_c}\frac{b_w}{f_{yt}}$ $(A_v + 2A_t)/s \geq 0.35\frac{b_w}{f_{yt}}$	$(A_v + 2A_t)/s \geq 0.2\sqrt{f'_c}\frac{b_w}{f_{yt}}$ $(A_v + 2A_t)/s \geq 3.5\frac{b_w}{f_{yt}}$	$(A_v + 2A_t)/s \geq 0.75\sqrt{f'_c}\frac{b_w}{f_{yt}}$ $(A_v + 2A_t)/s \geq 50\frac{b_w}{f_{yt}}$
9.6.4.3(a) and (b)	$A_{\ell,min} \leq \frac{0.42\sqrt{f'_c}A_{cp}}{f_y} - \left(\frac{A_t}{s}\right)p_h\frac{f_{yt}}{f_y}$ $A_{\ell,min} \leq \frac{0.42\sqrt{f'_c}A_{cp}}{f_y} - \left(\frac{0.175b_w}{f_{yt}}\right)p_h\frac{f_{yt}}{f_y}$	$A_{\ell,min} \leq \frac{1.33\sqrt{f'_c}A_{cp}}{f_y} - \left(\frac{A_t}{s}\right)p_h\frac{f_{yt}}{f_y}$ $A_{\ell,min} \leq \frac{1.33\sqrt{f'_c}A_{cp}}{f_y} - \left(\frac{1.75b_w}{f_{yt}}\right)p_h\frac{f_{yt}}{f_y}$	$A_{\ell,min} \leq \frac{5\sqrt{f'_c}A_{cp}}{f_y} - \left(\frac{A_t}{s}\right)p_h\frac{f_{yt}}{f_y}$ $A_{\ell,min} \leq \frac{5\sqrt{f'_c}A_{cp}}{f_y} - \left(\frac{25b_w}{f_{yt}}\right)p_h\frac{f_{yt}}{f_y}$
9.7.3.5(c)	$0.41\frac{b_ws}{f_{yt}}$	$4.2\frac{b_ws}{f_{yt}}$	$60\frac{b_ws}{f_{yt}}$
9.7.6.2.2	$0.33\sqrt{f'_c}\,b_wd$	$1.1\sqrt{f'_c}\,b_wd$	$4\sqrt{f'_c}\,b_wd$
9.9.2.1	$V_u \leq \phi 0.83\sqrt{f'_c}\,b_wd$	$V_u \leq \phi 2.65\sqrt{f'_c}\,b_wd$	$V_u \leq \phi 10\sqrt{f'_c}\,b_wd$
10.6.2.2	$A_{v,min} \geq 0.062\sqrt{f'_c}\frac{b_ws}{f_{yt}}$ $A_{v,min} \geq 0.35\frac{b_ws}{f_{yt}}$	$A_{v,min} \geq 0.2\sqrt{f'_c}\frac{b_ws}{f_{yt}}$ $A_{v,min} \geq 3.5\frac{b_ws}{f_{yt}}$	$A_{v,min} \geq 0.75\sqrt{f'_c}\frac{b_ws}{f_{yt}}$ $A_{v,min} \geq 50\frac{b_ws}{f_{yt}}$

Provision number	SI-metric stress in MPa	mks-metric stress in kgf/cm²	U.S. Customary units stress in pounds per square inch (psi)
10.7.6.5.2	$0.33\sqrt{f_c'}\,b_w d$	$1.1\sqrt{f_c'}\,b_w d$	$4\sqrt{f_c'}\,b_w d$
11.5.4.2	$0.67\sqrt{f_c'}\,A_{cv}$	$2.12\sqrt{f_c'}\,A_{cv}$	$8\sqrt{f_c'}\,A_{cv}$
11.5.4.3	$V_n = (\alpha_c\lambda\sqrt{f_c'} + \rho_t f_y)A_{cv}$ $\alpha_c = 0.25$ for $\frac{h_w}{\ell_w} \le 1.5$ $\alpha_c = 0.17$ for $\frac{h_w}{\ell_w} \ge 2.0$	$V_n = (\alpha_c\lambda\sqrt{f_c'} + \rho_t f_y)A_{cv}$ $\alpha_c = 0.80$ for $\frac{h_w}{\ell_w} \le 1.5$ $\alpha_c = 0.53$ for $\frac{h_w}{\ell_w} \ge 2.0$	$V_n = (\alpha_c\lambda\sqrt{f_c'} + \rho_t f_y)A_{cv}$ $\alpha_c = 3.0$ for $\frac{h_w}{\ell_w} \le 1.5$ $\alpha_c = 2.0$ for $\frac{h_w}{\ell_w} \ge 2.0$
11.5.4.4	$\alpha_c = 0.17\left(1+\frac{N_u}{3.5A_g}\right) \ge 0.0$	$\alpha_c = 0.53\left(1+\frac{N_u}{35A_g}\right) \ge 0.0$	$\alpha_c = 2\left(1+\frac{N_u}{500A_g}\right) \ge 0.0$
11.6.1	$V_u \le 0.04\phi\alpha_c\lambda\sqrt{f_c'}\,A_{cv}$	$V_u \le 0.13\phi\alpha_c\lambda\sqrt{f_c'}\,A_{cv}$	$V_u \le 0.5\phi\alpha_c\lambda\sqrt{f_c'}\,A_{cv}$
11.6.2	$V_u \ge 0.04\phi\alpha_c\lambda\sqrt{f_c'}\,A_{cv}$	$V_u \ge 0.13\phi\alpha_c\lambda\sqrt{f_c'}\,A_{cv}$	$V_u \ge 0.5\phi\alpha_c\lambda\sqrt{f_c'}\,A_{cv}$
12.5.3.3	$V_n = A_{cv}(0.17\lambda\sqrt{f_c'} + \rho_t f_y)$ $\sqrt{f_c'} \le 8.3$ MPa	$V_n = A_{cv}(0.53\lambda\sqrt{f_c'} + \rho_t f_y)$ $\sqrt{f_c'} \le 27$ kgf/cm²	$V_n = A_{cv}(2\lambda\sqrt{f_c'} + \rho_t f_y)$ $\sqrt{f_c'} \le 100$ psi
12.5.3.4	$V_u \le \phi 0.66 A_{cv}\sqrt{f_c'}$ $\sqrt{f_c'} \le 8.3$ MPa	$V_u \le \phi 2.1 A_{cv}\sqrt{f_c'}$ $\sqrt{f_c'} \le 27$ kgf/cm²	$V_u \le \phi 8 A_{cv}\sqrt{f_c'}$ $\sqrt{f_c'} \le 100$ psi
14.5.2.1(a)	$M_n = 0.42\lambda\sqrt{f_c'}\,S_m$	$M_n = 1.33\lambda\sqrt{f_c'}\,S_m$	$M_n = 5\lambda\sqrt{f_c'}\,S_m$
14.5.4.1(a)	$\frac{M_u}{S_m} - \frac{P_u}{A_g} \le \phi 0.42\lambda\sqrt{f_c'}$	$\frac{M_u}{S_m} - \frac{P_u}{A_g} \le \phi 1.33\lambda\sqrt{f_c'}$	$\frac{M_u}{S_m} - \frac{P_u}{A_g} \le \phi 5\lambda\sqrt{f_c'}$
14.5.5.1(a)	$V_n = 0.11\lambda\sqrt{f_c'}\,b_w h$	$V_n = 0.35\lambda\sqrt{f_c'}\,b_w h$	$V_n = \frac{4}{3}\lambda\sqrt{f_c'}\,b_w h$
14.5.5.1(b) and (c)	$V_n = 0.11\left[1+\frac{2}{\beta}\right]\lambda\sqrt{f_c'}b_o h$ $V_n = 0.22\lambda\sqrt{f_c'}b_o h$	$V_n = 0.35\left[1+\frac{2}{\beta}\right]\lambda\sqrt{f_c'}b_o h$ $V_n = 0.71\lambda\sqrt{f_c'}b_o h$	$V_n = \left[1+\frac{2}{\beta}\right]\frac{4}{3}\lambda\sqrt{f_c'}b_o h$ $V_n = 2\left(\frac{4}{3}\lambda\sqrt{f_c'}b_o h\right)$
15.4.2.3	$2.0\lambda\sqrt{f_c'}\,A_j$ $1.7\lambda\sqrt{f_c'}\,A_j$ $1.3\lambda\sqrt{f_c'}\,A_j$ $1.0\lambda\sqrt{f_c'}\,A_j$	$6.4\lambda\sqrt{f_c'}\,A_j$ $5.3\lambda\sqrt{f_c'}\,A_j$ $4.0\lambda\sqrt{f_c'}\,A_j$ $3.2\lambda\sqrt{f_c'}\,A_j$	$24\lambda\sqrt{f_c'}\,A_j$ $20\lambda\sqrt{f_c'}\,A_j$ $15\lambda\sqrt{f_c'}\,A_j$ $12\lambda\sqrt{f_c'}\,A_j$
16.4.4.1	$\phi(3.5b_v d)$	$\phi(35b_v d)$	$\phi(500b_v d)$
16.4.4.2	$\lambda\left(1.8+0.6\frac{A_v f_{yt}}{b_v s}\right)b_v d$ $3.5b_v d$ $0.55b_v d$	$\lambda\left(18+0.6\frac{A_v f_{yt}}{b_v s}\right)b_v d$ $35b_v d$ $5.6b_v d$	$\lambda\left(260+0.6\frac{A_v f_{yt}}{b_v s}\right)b_v d$ $500b_v d$ $80b_v d$

C Conv. Tables

Provision number	SI-metric stress in MPa	mks-metric stress in kgf/cm²	U.S. Customary units stress in pounds per square inch (psi)
16.4.6.1	$A_{v,min} \geq 0.062\sqrt{f_c'}\frac{b_w s}{f_{yt}}$ $A_{v,min} \geq 0.35\frac{b_w s}{f_{yt}}$	$A_{v,min} \geq 0.2\sqrt{f_c'}\frac{b_w s}{f_{yt}}$ $A_{v,min} \geq 3.5\frac{b_w s}{f_{yt}}$	$A_{v,min} \geq 0.75\sqrt{f_c'}\frac{b_w s}{f_{yt}}$ $A_{v,min} \geq 50\frac{b_w s}{f_{yt}}$
16.5.2.4(b) and (c)	$(3.3 + 0.08f_c')b_w d$ $11b_w d$	$(34 + 0.08f_c')b_w d$ $110b_w d$	$(480 + 0.08f_c')b_w d$ $1600b_w d$
16.5.2.5(b)	$\left(5.5-1.9\frac{a_v}{d}\right)b_w d$	$\left(55-20\frac{a_v}{d}\right)b_w d$	$\left(800-280\frac{a_v}{d}\right)b_w d$
17.6.2.2.3	$N_b = 3.9\lambda_a\sqrt{f_c'}\,h_{ef}^{5/3}$	$N_b = 5.8\lambda_a\sqrt{f_c'}\,h_{ef}^{5/3}$	$N_b = 16\lambda_a\sqrt{f_c'}\,h_{ef}^{5/3}$
17.6.2.2.1	$N_b = k_c\lambda_a\sqrt{f_c'}\,h_{ef}^{1.5}$ $k_c = 10$ or 7	$N_b = k_c\lambda_a\sqrt{f_c'}\,h_{ef}^{1.5}$ $k_c = 10$ or 7	$N_b = k_c\lambda_a\sqrt{f_c'}\,h_{ef}^{1.5}$ $k_c = 24$ or 17
17.6.4.1	$N_{sb} = 13c_{a1}\sqrt{A_{brg}}\,\lambda_a\sqrt{f_c'}$	$N_{sb} = 42.5c_{a1}\sqrt{A_{brg}}\,\lambda_a\sqrt{f_c'}$	$N_{sb} = 160c_{a1}\sqrt{A_{brg}}\,\lambda_a\sqrt{f_c'}$
17.6.5.1.2b	$10d_a\sqrt{\frac{\tau_{uncr}}{7.6}}$	$10d_a\sqrt{\frac{\tau_{uncr}}{76}}$	$10d_a\sqrt{\frac{\tau_{uncr}}{1100}}$
17.7.2.2.1a	$V_b = 0.6\left(\frac{\ell_e}{d_a}\right)^{0.2}\sqrt{d_a}\lambda_a\sqrt{f_c'}(c_{a1})^{1.5}$	$V_b = 1.9\left(\frac{\ell_e}{d_a}\right)^{0.2}\sqrt{d_a}\lambda_a\sqrt{f_c'}(c_{a1})^{1.5}$	$V_b = 7\left(\frac{\ell_e}{d_a}\right)^{0.2}\sqrt{d_a}\lambda_a\sqrt{f_c'}(c_{a1})^{1.5}$
17.7.2.2.1b	$V_b = 3.7\lambda_a\sqrt{f_c'}(c_{a1})^{1.5}$	$V_b = 3.8\lambda_a\sqrt{f_c'}(c_{a1})^{1.5}$	$V_b = 9\lambda_a\sqrt{f_c'}(c_{a1})^{1.5}$
17.7.2.2.2	$V_b = 0.66\left(\frac{\ell_e}{d_a}\right)^{0.2}\sqrt{d_a}\lambda_a\sqrt{f_c'}(c_{a1})^{1.5}$	$V_b = 2.1\left(\frac{\ell_e}{d_a}\right)^{0.2}\sqrt{d_a}\lambda_a\sqrt{f_c'}(c_{a1})^{1.5}$	$V_b = 8\left(\frac{\ell_e}{d_a}\right)^{0.2}\sqrt{d_a}\lambda_a\sqrt{f_c'}(c_{a1})^{1.5}$
18.7.5.3	$s_o = 100+\left(\frac{350-h_x}{3}\right)$	$s_o = 10+\left(\frac{35-h_x}{3}\right)$	$s_o = 4+\left(\frac{14-h_x}{3}\right)$
18.7.5.4(a)	$k_f = \frac{f_c'}{175}+0.6 \geq 1.0$	$k_f = \frac{f_c'}{1750}+0.6 \geq 1.0$	$k_f = \frac{f_c'}{25,000}+0.6 \geq 1.0$
18.8.4.3	$1.7\lambda\sqrt{f_c'}\,A_j$ $1.3\lambda\sqrt{f_c'}\,A_j$ $1.0\lambda\sqrt{f_c'}\,A_j$	$5.3\lambda\sqrt{f_c'}\,A_j$ $4.0\lambda\sqrt{f_c'}\,A_j$ $3.2\lambda\sqrt{f_c'}\,A_j$	$20\lambda\sqrt{f_c'}\,A_j$ $15\lambda\sqrt{f_c'}\,A_j$ $12\lambda\sqrt{f_c'}\,A_j$
18.8.5.1	$\ell_{dh} = f_y d_b/(5.4\lambda\sqrt{f_c'})$	$\ell_{dh} = f_y d_b/(17\lambda\sqrt{f_c'})$	$\ell_{dh} = f_y d_b/(65\lambda\sqrt{f_c'})$
18.10.2.1	$0.083\lambda\sqrt{f_c'}\,A_{cv}$	$0.27\lambda\sqrt{f_c'}\,A_{cv}$	$\lambda\sqrt{f_c'}\,A_{cv}$
18.10.2.2	$0.17\lambda\sqrt{f_c'}\,A_{cv}$	$0.53\lambda\sqrt{f_c'}\,A_{cv}$	$2\lambda\sqrt{f_c'}\,A_{cv}$
18.10.2.4	$0.50\frac{\sqrt{f_c'}}{f_y}$	$1.6\frac{\sqrt{f_c'}}{f_y}$	$6\frac{\sqrt{f_c'}}{f_y}$

Provision number	SI-metric stress in MPa	mks-metric stress in kgf/cm^2	U.S. Customary units stress in pounds per square inch (psi)
18.10.4.1	$V_n = A_{cv}(\alpha_c\lambda\sqrt{f_c'} + \rho_t f_y)$ $\alpha_c = 0.25$ for $\frac{h_w}{\ell_w} \leq 1.5$ $\alpha_c = 0.17$ for $\frac{h_w}{\ell_w} \geq 2.0$	$V_n = A_{cv}(\alpha_c\lambda\sqrt{f_c'} + \rho_t f_y)$ $\alpha_c = 0.80$ for $\frac{h_w}{\ell_w} \leq 1.5$ $\alpha_c = 0.53$ for $\frac{h_w}{\ell_w} \geq 2.0$	$V_n = A_{cv}(\alpha_c\lambda\sqrt{f_c'} + \rho_t f_y)$ $\alpha_c = 3.0$ for $\frac{h_w}{\ell_w} \leq 1.5$ $\alpha_c = 2.0$ for $\frac{h_w}{\ell_w} \geq 2.0$
18.10.4.4	$0.66\sqrt{f_c'}A_{cv}$ $0.83\sqrt{f_c'}A_{cw}$	$2.12\sqrt{f_c'}A_{cv}$ $2.65\sqrt{f_c'}A_{cw}$	$8\sqrt{f_c'}A_{cv}$ $10\sqrt{f_c'}A_{cw}$
18.10.4.5	$0.83\sqrt{f_c'}A_{cw}$	$2.65\sqrt{f_c'}A_{cw}$	$10\sqrt{f_c'}A_{cw}$
18.10.6.2b	$\frac{\delta_c}{h_{wcs}} = \frac{1}{100}\left(4 - \frac{1}{50}\left(\frac{l_w}{b}\right)\left(\frac{c}{b}\right) - \frac{V_u}{0.66\sqrt{f_c'}A_{cv}}\right)$	$\frac{\delta_c}{h_{wcs}} = \frac{1}{100}\left(4 - \frac{1}{50}\left(\frac{l_w}{b}\right)\left(\frac{c}{b}\right) - \frac{V_u}{2.1\sqrt{f_c'}A_{cv}}\right)$	$\frac{\delta_c}{h_{wcs}} = \frac{1}{100}\left(4 - \frac{1}{50}\left(\frac{l_w}{b}\right)\left(\frac{c}{b}\right) - \frac{V_u}{8\sqrt{f_c'}A_{cv}}\right)$
18.10.6.5(a)	$0.083\lambda\sqrt{f_c'}A_{cv}$	$0.27\lambda\sqrt{f_c'}A_{cv}$	$\lambda\sqrt{f_c'}A_{cv}$
18.10.6.5(b)	$2.8/f_y$	$28/f_y$	$400/f_y$
18.10.7.2	$0.33\lambda\sqrt{f_c'}A_{cw}$	$1.1\lambda\sqrt{f_c'}A_{cw}$	$4\lambda\sqrt{f_c'}A_{cw}$
18.10.7.4	$V_n = 2A_{vd}f_y\sin\alpha \leq 0.83\sqrt{f_c'}A_{cw}$	$V_n = 2A_{vd}f_y\sin\alpha \leq 2.65\sqrt{f_c'}A_{cw}$	$V_n = 2A_{vd}f_y\sin\alpha \leq 10\sqrt{f_c'}A_{cw}$
18.12.7.7	$A_{v,min} \geq 0.062\sqrt{f_c'}\frac{b_w s}{f_{yt}}$ $A_{v,min} \geq 0.35\frac{b_w s}{f_{yt}}$	$A_{v,min} \geq 0.2\sqrt{f_c'}\frac{b_w s}{f_{yt}}$ $A_{v,min} \geq 3.5\frac{b_w s}{f_{yt}}$	$A_{v,min} \geq 0.75\sqrt{f_c'}\frac{b_w s}{f_{yt}}$ $A_{v,min} \geq 50\frac{b_w s}{f_{yt}}$
18.12.9.1	$V_n = A_{cv}(0.17\lambda\sqrt{f_c'} + \rho_t f_y)$	$V_n = A_{cv}(0.53\lambda\sqrt{f_c'} + \rho_t f_y)$	$V_n = A_{cv}(2\lambda\sqrt{f_c'} + \rho_t f_y)$
18.12.9.2	$0.66\sqrt{f_c'}A_{cv}$	$2.12\sqrt{f_c'}A_{cv}$	$8\sqrt{f_c'}A_{cv}$
18.14.5.3	$0.29\sqrt{f_c'}$	$0.93\sqrt{f_c'}$	$3.5\sqrt{f_c'}$
19.2.2.1(a)	$E_c = w_c^{1.5}0.043\sqrt{f_c'}$	$E_c = w_c^{1.5}0.14\sqrt{f_c'}$	$E_c = w_c^{1.5}33\sqrt{f_c'}$
19.2.2.1(b)	$E_c = 4700\sqrt{f_c'}$	$E_c = 15,100\sqrt{f_c'}$	$E_c = 57,000\sqrt{f_c'}$
19.2.3.1	$f_r = 0.62\lambda\sqrt{f_c'}$	$f_r = 2\lambda\sqrt{f_c'}$	$f_r = 7.5\lambda\sqrt{f_c'}$
20.3.2.4.1	$f_{se} + 70 + \frac{f_c'}{100\rho_p}$ $f_{se} + 420$ $f_{se} + 70 + \frac{f_c'}{300\rho_p}$ $f_{se} + 210$	$f_{se} + 700 + \frac{f_c'}{100\rho_p}$ $f_{se} + 4200$ $f_{se} + 700 + \frac{f_c'}{300\rho_p}$ $f_{se} + 2100$	$f_{se} + 10,000 + \frac{f_c'}{100\rho_p}$ $f_{se} + 60,000$ $f_{se} + 10,000 + \frac{f_c'}{300\rho_p}$ $f_{se} + 30,000$
21.2.3	$\ell_{tr} = \left(\frac{f_{se}}{21}\right)d_b$	$\ell_{tr} = \left(\frac{f_{se}}{210}\right)d_b$	$\ell_{tr} = \left(\frac{f_{se}}{3000}\right)d_b$

Provision number	SI-metric stress in MPa	mks-metric stress in kgf/cm²	U.S. Customary units stress in pounds per square inch (psi)
22.2.2.4.3(b)	$0.85-\dfrac{0.05(f_c'-28)}{7}$	$0.85-\dfrac{0.05(f_c'-280)}{70}$	$0.85-\dfrac{0.05(f_c'-4000)}{1000}$
22.5.1.2	$V_u \le \phi(V_c + 0.66\sqrt{f_c'}\, b_w d)$	$V_u \le \phi(V_c + 2.2\sqrt{f_c'}\, b_w d)$	$V_u \le \phi(V_c + 8\sqrt{f_c'}\, b_w d)$
22.5.5.1(a)	$\left(0.17\lambda\sqrt{f_c'}+\dfrac{N_u}{6A_g}\right)b_w d$	$\left(0.53\lambda\sqrt{f_c'}+\dfrac{N_u}{6A_g}\right)b_w d$	$\left(2\lambda\sqrt{f_c'}+\dfrac{N_u}{6A_g}\right)b_w d$
22.5.5.1(b)	$\left(0.66\lambda(\rho_w)^{\frac{1}{3}}\sqrt{f_c'}+\dfrac{N_u}{6A_g}\right)b_w d$	$\left(2.1\lambda(\rho_w)^{\frac{1}{3}}\sqrt{f_c'}+\dfrac{N_u}{6A_g}\right)b_w d$	$\left(8\lambda(\rho_w)^{\frac{1}{3}}\sqrt{f_c'}+\dfrac{N_u}{6A_g}\right)b_w d$
22.5.5.1(c)	$\left(0.66\lambda_s\lambda(\rho_w)^{\frac{1}{3}}\sqrt{f_c'}+\dfrac{N_u}{6A_g}\right)b_w d$	$\left(2.1\lambda_s\lambda(\rho_w)^{\frac{1}{3}}\sqrt{f_c'}+\dfrac{N_u}{6A_g}\right)b_w d$	$\left(8\lambda_s\lambda(\rho_w)^{\frac{1}{3}}\sqrt{f_c'}+\dfrac{N_u}{6A_g}\right)b_w d$
22.5.5.1.1	$0.42\lambda\sqrt{f_c'}b_w d$	$1.33\lambda\sqrt{f_c'}b_w d$	$5\lambda\sqrt{f_c'}b_w d$
22.5.6.2	$0.17\lambda\sqrt{f_c'}b_w d$	$0.53\lambda\sqrt{f_c'}b_w d$	$2\lambda\sqrt{f_c'}b_w d$
22.5.6.2(a), (b), and (c)	$V_c=\left(0.05\lambda\sqrt{f_c'}+4.8\dfrac{V_u d_p}{M_u}\right)b_w d$ $V_c=(0.05\lambda\sqrt{f_c'}+4.8)b_w d$ $V_c=0.42\lambda\sqrt{f_c'}\, b_w d$	$V_c=\left(0.16\lambda\sqrt{f_c'}+49\dfrac{V_u d_p}{M_u}\right)b_w d$ $V_c=(0.16\lambda\sqrt{f_c'}+49)b_w d$ $V_c=1.33\lambda\sqrt{f_c'}\, b_w d$	$V_c=\left(0.6\lambda\sqrt{f_c'}+700\dfrac{V_u d_p}{M_u}\right)b_w d$ $V_c=(0.6\lambda\sqrt{f_c'}+700)b_w d$ $V_c=5\lambda\sqrt{f_c'}\, b_w d$
22.5.6.3.1a	$V_{ci}=0.05\lambda\sqrt{f_c'}\, b_w d_p+V_d+\dfrac{V_i M_{cre}}{M_{max}}$	$V_{ci}=0.16\lambda\sqrt{f_c'}\, b_w d_p+V_d+\dfrac{V_i M_{cre}}{M_{max}}$	$V_{ci}=0.6\lambda\sqrt{f_c'}\, b_w d_p+V_d+\dfrac{V_i M_{cre}}{M_{max}}$
22.5.6.3.1b	$V_{ci}=0.14\lambda\sqrt{f_c'}\, b_w d$	$V_{ci}=0.45\lambda\sqrt{f_c'}\, b_w d$	$V_{ci}=1.7\lambda\sqrt{f_c'}\, b_w d$
22.5.6.3.1c	$V_{ci}=0.17\lambda\sqrt{f_c'}\, b_w d$	$V_{ci}=0.53\lambda\sqrt{f_c'}\, b_w d$	$V_{ci}=2\lambda\sqrt{f_c'}\, b_w d$
22.5.6.3.1d	$M_{cre}=\left(\dfrac{I}{y_t}\right)(0.5\lambda\sqrt{f_c'}+f_{pe}-f_d)$	$M_{cre}=\left(\dfrac{I}{y_t}\right)(1.6\lambda\sqrt{f_c'}+f_{pe}-f_d)$	$M_{cre}=\left(\dfrac{I}{y_t}\right)(6\lambda\sqrt{f_c'}+f_{pe}-f_d)$
22.5.6.3.2	$V_{cw}=(0.29\lambda\sqrt{f_c'}+0.3f_{pc})b_w d_p+V_p$	$V_{cw}=(0.93\lambda\sqrt{f_c'}+0.3f_{pc})b_w d_p+V_p$	$V_{cw}=(3.5\lambda\sqrt{f_c'}+0.3f_{pc})b_w d_p+V_p$
22.5.6.3.3	$0.33\lambda\sqrt{f_c'}$	$1.1\lambda\sqrt{f_c'}$	$4\lambda\sqrt{f_c'}$
22.5.8.6.2(b)	$V_s=0.25\sqrt{f_c'}\, b_w d$	$V_s=0.8\sqrt{f_c'}\, b_w d$	$V_s=3\sqrt{f_c'}\, b_w d$
22.6.3.1	$\sqrt{f_c'} \le 8.3$ MPa	$\sqrt{f_c'} \le 27$ kgf/cm²	$\sqrt{f_c'} \le 100$ psi
22.6.5.2(a)	$v_c=0.33\lambda_s\lambda\sqrt{f_c'}$	$v_c=1.1\lambda_s\lambda\sqrt{f_c'}$	$v_c=4\lambda_s\lambda\sqrt{f_c'}$
22.6.5.2(b)	$v_c=0.17\left(1+\dfrac{2}{\beta}\right)\lambda_s\lambda\sqrt{f_c'}$	$v_c=0.53\left(1+\dfrac{2}{\beta}\right)\lambda_s\lambda\sqrt{f_c'}$	$v_c=\left(2+\dfrac{4}{\beta}\right)\lambda_s\lambda\sqrt{f_c'}$
22.6.5.2(c)	$v_c=0.083\left(2+\dfrac{\alpha_s d}{b_o}\right)\lambda_s\lambda\sqrt{f_c'}$	$v_c=0.27\left(2+\dfrac{\alpha_s d}{b_o}\right)\lambda_s\lambda\sqrt{f_c'}$	$v_c=\left(2+\dfrac{\alpha_s d}{b_o}\right)\lambda_s\lambda\sqrt{f_c'}$

Provision number	SI-metric stress in MPa	mks-metric stress in kgf/cm^2	U.S. Customary units stress in pounds per square inch (psi)
22.6.5.5	$\sqrt{f'_c} \leq 5.8$ MPa $f_{pc} \leq 3.5$ MPa	$\sqrt{f'_c} \leq 19$ kgf/cm^2 $f_{pc} \leq 35$ kgf/cm^2	$\sqrt{f'_c} \leq 70$ psi $f_{pc} \leq 500$ psi
22.6.5.5a	$v_c = (0.29\lambda_s\sqrt{f'_c} + 0.3f_{pc}) + V_p/(b_o d)$	$v_c = (0.93\lambda_s\sqrt{f'_c} + 0.3f_{pc}) + V_p/(b_o d)$	$v_c = (3.5\lambda_s\sqrt{f'_c} + 0.3f_{pc}) + V_p/(b_o d)$
22.6.5.5b	$v_c = 0.083\left(1.5 + \frac{\alpha_s d}{b_o}\right)\lambda\sqrt{f'_c}$ $+ 0.3f_{pc} + V_p/(b_o d)$	$v_c = 0.27\left(1.5 + \frac{\alpha_s d}{b_o}\right)\lambda\sqrt{f'_c}$ $+ 0.3f_{pc} + V_p/(b_o d)$	$v_c = \left(1.5 + \frac{\alpha_s d}{b_o}\right)\lambda\sqrt{f'_c}$ $+ 0.3f_{pc} + V_p/(b_o d)$
22.6.6.1(a) and (e)	$0.17\lambda_s\lambda\sqrt{f'_c}$	$0.53\lambda_s\lambda\sqrt{f'_c}$	$2\lambda_s\lambda\sqrt{f'_c}$
22.6.6.1(b)	$0.25\lambda_s\lambda\sqrt{f'_c}$	$0.80\lambda_s\lambda\sqrt{f'_c}$	$3\lambda_s\lambda\sqrt{f'_c}$
22.6.6.1(c)	$\left(0.17 + \frac{0.33}{\beta}\right)\lambda_s\lambda\sqrt{f'_c}$	$\left(0.53 + \frac{1.06}{\beta}\right)\lambda_s\lambda\sqrt{f'_c}$	$\left(2 + \frac{4}{\beta}\right)\lambda_s\lambda\sqrt{f'_c}$
22.6.6.1(d)	$\left(0.17 + \frac{0.083\alpha_s d}{b_o}\right)\lambda_s\lambda\sqrt{f'_c}$	$\left(0.53 + \frac{0.27\alpha_s d}{b_o}\right)\lambda_s\lambda\sqrt{f'_c}$	$\left(2 + \frac{\alpha_s d}{b_o}\right)\lambda_s\lambda\sqrt{f'_c}$
22.6.6.2(a)	$\frac{A_v}{s} \geq 0.17\sqrt{f'_c}\frac{b_o}{f_{yt}}$	$\frac{A_v}{s} \geq 0.53\sqrt{f'_c}\frac{b_o}{f_{yt}}$	$\frac{A_v}{s} \geq 2\sqrt{f'_c}\frac{b_o}{f_{yt}}$
22.6.6.2(b)	$\frac{A_v}{s} \geq 0.17\sqrt{f'_c}\frac{b_o}{f_{yt}}$	$\frac{A_v}{s} \geq 0.53\sqrt{f'_c}\frac{b_o}{f_{yt}}$	$\frac{A_v}{s} \geq 2\sqrt{f'_c}\frac{b_o}{f_{yt}}$
22.6.6.3(a)	$\phi 0.5\sqrt{f'_c}$	$\phi 1.6\sqrt{f'_c}$	$\phi 6\sqrt{f'_c}$
22.6.6.3(b)	$\phi 0.66\sqrt{f'_c}$	$\phi 2.1\sqrt{f'_c}$	$\phi 8\sqrt{f'_c}$
22.6.8.3	$\left(\frac{A_v}{s}\right) \geq 0.17\sqrt{f'_c}\left(\frac{b_o}{f_{yt}}\right)$	$\left(\frac{A_v}{s}\right) \geq 0.53\sqrt{f'_c}\left(\frac{b_o}{f_{yt}}\right)$	$\left(\frac{A_v}{s}\right) \geq 2\sqrt{f'_c}\left(\frac{b_o}{f_{yt}}\right)$
22.7.2.1	$\sqrt{f'_c} \leq 8.3$ MPa	$\sqrt{f'_c} \leq 27$ kgf/cm^2	$\sqrt{f'_c} \leq 100$ psi
22.7.4.1(a)(a)	$T_{th} = 0.083\lambda\sqrt{f'_c}\left(\frac{A_{cp}^2}{p_{cp}}\right)$	$T_{th} = 0.27\lambda\sqrt{f'_c}\left(\frac{A_{cp}^2}{p_{cp}}\right)$	$T_{th} = \lambda\sqrt{f'_c}\left(\frac{A_{cp}^2}{p_{cp}}\right)$
22.7.4.1(a)(b)	$T_{th} = 0.083\lambda\sqrt{f'_c}\left(\frac{A_{cp}^2}{p_{cp}}\right)\sqrt{1 + \frac{f_{pc}}{0.33\lambda\sqrt{f'_c}}}$	$T_{th} = 0.27\lambda\sqrt{f'_c}\left(\frac{A_{cp}^2}{p_{cp}}\right)\sqrt{1 + \frac{f_{pc}}{1.1\lambda\sqrt{f'_c}}}$	$T_{th} = \lambda\sqrt{f'_c}\left(\frac{A_{cp}^2}{p_{cp}}\right)\sqrt{1 + \frac{f_{pc}}{4\lambda\sqrt{f'_c}}}$
22.7.4.1(a)(c)	$T_{th} = 0.083\lambda\sqrt{f'_c}\left(\frac{A_{cp}^2}{p_{cp}}\right)\sqrt{1 + \frac{N_u}{0.33A_g\lambda\sqrt{f'_c}}}$	$T_{th} = 0.27\lambda\sqrt{f'_c}\left(\frac{A_{cp}^2}{p_{cp}}\right)\sqrt{1 + \frac{N_u}{1.1A_g\lambda\sqrt{f'_c}}}$	$T_{th} = \lambda\sqrt{f'_c}\left(\frac{A_{cp}^2}{p_{cp}}\right)\sqrt{1 + \frac{N_u}{4A_g\lambda\sqrt{f'_c}}}$
22.7.4.1(b)(a)	$T_{th} = 0.083\lambda\sqrt{f'_c}\left(\frac{A_g^2}{P_{cp}}\right)$	$T_{th} = 0.27\lambda\sqrt{f'_c}\left(\frac{A_g^2}{P_{cp}}\right)$	$T_{th} = \lambda\sqrt{f'_c}\left(\frac{A_g^2}{P_{cp}}\right)$

Provision number	SI-metric stress in MPa	mks-metric stress in kgf/cm²	U.S. Customary units stress in pounds per square inch (psi)
22.7.4.1(b)(b)	$T_{th} = 0.083\lambda\sqrt{f_c'}\left(\frac{A_g^2}{P_{cp}}\right)\sqrt{1+\frac{f_{pc}}{0.33\lambda\sqrt{f_c'}}}$	$T_{th} = 0.27\lambda\sqrt{f_c'}\left(\frac{A_g^2}{P_{cp}}\right)\sqrt{1+\frac{f_{pc}}{1.1\lambda\sqrt{f_c'}}}$	$T_{th} = \lambda\sqrt{f_c'}\left(\frac{A_g^2}{P_{cp}}\right)\sqrt{1+\frac{f_{pc}}{4\lambda\sqrt{f_c'}}}$
22.7.4.1(b)(c)	$T_{th} = 0.083\lambda\sqrt{f_c'}\left(\frac{A_g^2}{P_{cp}}\right)\sqrt{1+\frac{N_u}{0.33A_g\lambda\sqrt{f_c'}}}$	$T_{th} = 0.27\lambda\sqrt{f_c'}\left(\frac{A_g^2}{P_{cp}}\right)\sqrt{1+\frac{N_u}{1.1A_g\lambda\sqrt{f_c'}}}$	$T_{th} = \lambda\sqrt{f_c'}\left(\frac{A_g^2}{P_{cp}}\right)\sqrt{1+\frac{N_u}{4A_g\lambda\sqrt{f_c'}}}$
22.7.5.1(a)	$T_{cr} = 0.33\lambda\sqrt{f_c'}\left(\frac{A_{cp}^2}{p_{cp}}\right)$	$T_{cr} = \lambda\sqrt{f_c'}\left(\frac{A_{cp}^2}{p_{cp}}\right)$	$T_{cr} = 4\lambda\sqrt{f_c'}\left(\frac{A_{cp}^2}{p_{cp}}\right)$
22.7.5.1(b)	$T_{cr} = 0.33\lambda\sqrt{f_c'}\left(\frac{A_{cp}^2}{p_{cp}}\right)\sqrt{1+\frac{f_{pc}}{0.33\lambda\sqrt{f_c'}}}$	$T_{cr} = \lambda\sqrt{f_c'}\left(\frac{A_{cp}^2}{p_{cp}}\right)\sqrt{1+\frac{f_{pc}}{1.1\lambda\sqrt{f_c'}}}$	$T_{cr} = 4\lambda\sqrt{f_c'}\left(\frac{A_{cp}^2}{p_{cp}}\right)\sqrt{1+\frac{f_{pc}}{4\lambda\sqrt{f_c'}}}$
22.7.5.1(c)	$T_{cr} = 0.33\lambda\sqrt{f_c'}\left(\frac{A_{cp}^2}{p_{cp}}\right)\sqrt{1+\frac{N_u}{0.33A_g\lambda\sqrt{f_c'}}}$	$T_{cr} = \lambda\sqrt{f_c'}\left(\frac{A_{cp}^2}{p_{cp}}\right)\sqrt{1+\frac{N_u}{1.1A_g\lambda\sqrt{f_c'}}}$	$T_{cr} = 4\lambda\sqrt{f_c'}\left(\frac{A_{cp}^2}{p_{cp}}\right)\sqrt{1+\frac{N_u}{4A_g\lambda\sqrt{f_c'}}}$
22.7.7.1a	$\sqrt{\left(\frac{V_u}{b_wd}\right)^2+\left(\frac{T_up_h}{1.7A_{oh}^2}\right)^2} \le \phi\left(\frac{V_c}{b_wd}+0.66\sqrt{f_c'}\right)$	$\sqrt{\left(\frac{V_u}{b_wd}\right)^2+\left(\frac{T_up_h}{1.7A_{oh}^2}\right)^2} \le \phi\left(\frac{V_c}{b_wd}+2\sqrt{f_c'}\right)$	$\sqrt{\left(\frac{V_u}{b_wd}\right)^2+\left(\frac{T_up_h}{1.7A_{oh}^2}\right)^2} \le \phi\left(\frac{V_c}{b_wd}+8\sqrt{f_c'}\right)$
22.7.7.1b	$\left(\frac{V_u}{b_wd}\right)+\left(\frac{T_up_h}{1.7A_{oh}^2}\right) \le \phi\left(\frac{V_c}{b_wd}+0.66\sqrt{f_c'}\right)$	$\left(\frac{V_u}{b_wd}\right)+\left(\frac{T_up_h}{1.7A_{oh}^2}\right) \le \phi\left(\frac{V_c}{b_wd}+2\sqrt{f_c'}\right)$	$\left(\frac{V_u}{b_wd}\right)+\left(\frac{T_up_h}{1.7A_{oh}^2}\right) \le \phi\left(\frac{V_c}{b_wd}+8\sqrt{f_c'}\right)$
22.9.4.4(b), (c), and (e)	$(3.3 + 0.08f_c')A_c$ $11A_c$ $5.5A_c$	$(34 + 0.08f_c')A_c$ $110A_c$ $55A_c$	$(480 + 0.08f_c')A_c$ $1600A_c$ $800A_c$
23.4.4	$V_u \le \phi 0.42\tan(\theta)\lambda\lambda_s\sqrt{f_c'}b_wd$	$V_u \le \phi 1.33\tan(\theta)\lambda\lambda_s\sqrt{f_c'}b_wd$	$V_u \le \phi 5\tan(\theta)\lambda\lambda_s\sqrt{f_c'}b_wd$
23.4.4.1	$\lambda_s = \sqrt{\frac{2}{1+0.004d}} \le 1.0$	$\lambda_s = \sqrt{\frac{2}{1+0.04d}} \le 1.0$	$\lambda_s = \sqrt{\frac{2}{1+\frac{d}{10}}} \le 1.0$
24.3.2	$s = 380\left(\frac{280}{f_s}\right) - 2.5c_c$ $s = 300\left(\frac{280}{f_s}\right)$	$s = 38\left(\frac{2800}{f_s}\right) - 2.5c_c$ $s = 30\left(\frac{2800}{f_s}\right)$	$s = 15\left(\frac{40,000}{f_s}\right) - 2.5c_c$ $s = 12\left(\frac{40,000}{f_s}\right)$
24.3.2.2	$\Delta f_{ps} \le 250$ MPa $\Delta f_{ps} < 140$ MPa	$\Delta f_{ps} \le 2500$ kgf/cm² $\Delta f_{ps} < 1400$ kgf/cm²	$\Delta f_{ps} \le 36,000$ lb/in.² $\Delta f_{ps} < 20,000$ lb/in.²
24.5.2.1	$f_t \le 0.62\sqrt{f_c'}$ $0.62\sqrt{f_c'} < f_t \le 1.0\sqrt{f_c'}$ $f_t > 1.0\sqrt{f_c'}$ $f_t \le 0.50\sqrt{f_c'}$	$f_t \le 2\sqrt{f_c'}$ $2\sqrt{f_c'} < f_t \le 3.2\sqrt{f_c'}$ $f_t > 3.2\sqrt{f_c'}$ $f_t \le 1.6\sqrt{f_c'}$	$f_t \le 7.5\sqrt{f_c'}$ $7.5\sqrt{f_c'} < f_t \le 12\sqrt{f_c'}$ $f_t > 12\sqrt{f_c'}$ $f_t \le 6\sqrt{f_c'}$
24.5.3.2	$0.50\sqrt{f_{ci}'}$ $0.25\sqrt{f_{ci}'}$	$1.6\sqrt{f_{ci}'}$ $0.8\sqrt{f_{ci}'}$	$6\sqrt{f_{ci}'}$ $3\sqrt{f_{ci}'}$

Provision number	SI-metric stress in MPa	mks-metric stress in kgf/cm²	U.S. Customary units stress in pounds per square inch (psi)
25.4.1.4	$\sqrt{f_c'} \leq 8.3$ MPa	$\sqrt{f_c'} \leq 27$ kgf/cm²	$\sqrt{f_c'} \leq 100$ psi
25.4.2.3a	$\ell_d = \dfrac{f_y}{1.1\lambda\sqrt{f_c'}}\dfrac{\Psi_t\Psi_e\Psi_s}{\left(\dfrac{c_b+K_{tr}}{d_b}\right)}d_b$	$\ell_d = \dfrac{f_y}{3.5\lambda\sqrt{f_c'}}\dfrac{\Psi_t\Psi_e\Psi_s}{\left(\dfrac{c_b+K_{tr}}{d_b}\right)}d_b$	$\ell_d = \dfrac{3f_y}{40\lambda\sqrt{f_c'}}\dfrac{\Psi_t\Psi_e\Psi_s}{\left(\dfrac{c_b+K_{tr}}{d_b}\right)}d_b$
25.4.2.3 (top left)	$\left(\dfrac{f_y\Psi_t\Psi_e\Psi_g}{2.1\lambda\sqrt{f_c'}}\right)d_b$	$\left(\dfrac{f_y\Psi_t\Psi_e\Psi_g}{6.6\lambda\sqrt{f_c'}}\right)d_b$	$\left(\dfrac{f_y\Psi_t\Psi_e\Psi_g}{25\lambda\sqrt{f_c'}}\right)d_b$
25.4.2.3 (top right)	$\left(\dfrac{f_y\Psi_t\Psi_e\Psi_g}{1.7\lambda\sqrt{f_c'}}\right)d_b$	$\left(\dfrac{f_y\Psi_t\Psi_e\Psi_g}{5.3\lambda\sqrt{f_c'}}\right)d_b$	$\left(\dfrac{f_y\Psi_t\Psi_e\Psi_g}{20\lambda\sqrt{f_c'}}\right)d_b$
25.4.2.3 (lower left)	$\left(\dfrac{f_y\Psi_t\Psi_e\Psi_g}{1.4\lambda\sqrt{f_c'}}\right)d_b$	$\left(\dfrac{f_y\Psi_t\Psi_e\Psi_g}{4.4\lambda\sqrt{f_c'}}\right)d_b$	$\left(\dfrac{3f_y\Psi_t\Psi_e\Psi_g}{50\lambda\sqrt{f_c'}}\right)d_b$
25.4.2.3 (lower right)	$\left(\dfrac{f_y\Psi_t\Psi_e\Psi_g}{1.1\lambda\sqrt{f_c'}}\right)d_b$	$\left(\dfrac{f_y\Psi_t\Psi_e\Psi_g}{3.5\lambda\sqrt{f_c'}}\right)d_b$	$\left(\dfrac{3f_y\Psi_t\Psi_e\Psi_g}{40\lambda\sqrt{f_c'}}\right)d_b$
25.4.2.4	$\ell_d = \dfrac{f_y}{1.1\lambda\sqrt{f_c'}}\dfrac{\Psi_t\Psi_e\Psi_s\Psi_g}{\left(\dfrac{c_b+K_{tr}}{d_b}\right)}d_b$	$\ell_d = \dfrac{f_y}{3.5\lambda\sqrt{f_c'}}\dfrac{\Psi_t\Psi_e\Psi_s\Psi_g}{\left(\dfrac{c_b+K_{tr}}{d_b}\right)}d_b$	$\ell_d = \dfrac{3f_y}{40\lambda\sqrt{f_c'}}\dfrac{\Psi_t\Psi_e\Psi_s\Psi_g}{\left(\dfrac{c_b+K_{tr}}{d_b}\right)}d_b$
25.4.3.1(a)	$\left(\dfrac{f_y\Psi_e\Psi_r\Psi_o\Psi_c}{23\sqrt{f_c'}}\right)d_b^{1.5}$	$\left(\dfrac{f_y\Psi_e\Psi_r\Psi_o\Psi_c}{23\sqrt{f_c'}}\right)d_b^{1.5}$	$\left(\dfrac{f_y\Psi_e\Psi_r\Psi_o\Psi_c}{55\sqrt{f_c'}}\right)d_b^{1.5}$
25.4.4.2(a)	$\left(\dfrac{f_y\Psi_e\Psi_p\Psi_o\Psi_c}{31\sqrt{f_c'}}\right)d_b^{1.5}$	$\left(\dfrac{f_y\Psi_e\Psi_p\Psi_o\Psi_c}{32\sqrt{f_c'}}\right)d_b^{1.5}$	$\left(\dfrac{f_y\Psi_e\Psi_p\Psi_o\Psi_c}{75\sqrt{f_c'}}\right)d_b^{1.5}$
25.4.4.3	$0.01f_c' + 0.6$	$\dfrac{f_c'}{1055}+0.6$	$\dfrac{f_c'}{15{,}000}+0.6$
25.4.6.3(a)	$\left(\dfrac{f_y-240}{f_y}\right)$	$\left(\dfrac{f_y-2460}{f_y}\right)$	$\left(\dfrac{f_y-35{,}000}{f_y}\right)$
25.4.7.2(b)	$3.3\left(\dfrac{f_y}{\lambda\sqrt{f_c'}}\right)\left(\dfrac{A_b}{s}\right)$	$\left(\dfrac{f_y}{\lambda\sqrt{f_c'}}\right)\left(\dfrac{A_b}{s}\right)$	$0.27\left(\dfrac{f_y}{\lambda\sqrt{f_c'}}\right)\left(\dfrac{A_b}{s}\right)$
25.4.8.1(a)	$\left(\dfrac{f_{se}}{21}\right)d_b + \left(\dfrac{f_{ps}-f_{se}}{7}\right)d_b$	$\left(\dfrac{f_{se}}{210}\right)d_b + \left(\dfrac{f_{ps}-f_{se}}{70}\right)d_b$	$\left(\dfrac{f_{se}}{3000}\right)d_b + \left(\dfrac{f_{ps}-f_{se}}{1000}\right)d_b$
25.4.9.2(a)	$\left(\dfrac{0.24f_y}{\lambda\sqrt{f_c'}}\right)d_b$	$\left(\dfrac{0.075f_y}{\lambda\sqrt{f_c'}}\right)d_b$	$\left(\dfrac{f_y}{50\lambda\sqrt{f_c'}}\right)d_b$
25.4.9.2(b)	$(0.043f_y)d_b$	$(0.0044f_y)d_b$	$(0.0003f_y)d_b$
25.5.5.1(a), (b) and (c)	$0.071f_yd_b$ $(0.13f_y - 24)d_b$	$0.0073f_yd_b$ $(0.013f_y - 24)d_b$	$0.0005f_yd_b$ $(0.0009f_y - 24)d_b$

C Conv. Tables

Provision number	SI-metric stress in MPa	mks-metric stress in kgf/cm²	U.S. Customary units stress in pounds per square inch (psi)
25.7.1.3(b)	$0.17\frac{d_b f_{yt}}{\lambda\sqrt{f_c'}}$	$0.053\frac{d_b f_{yt}}{\lambda\sqrt{f_c'}}$	$0.014\frac{d_b f_{yt}}{\lambda\sqrt{f_c'}}$
25.7.1.7	$A_b f_{yt} \leq 40{,}000$ N	$A_b f_{yt} \leq 4000$ kgf	$A_b f_{yt} \leq 9000$ lb
25.9.4.5.1	$f_{ps} = f_{se} + 70$	$f_{ps} = f_{se} + 700$	$f_{ps} = f_{se} + 10{,}000$
26.12.7.1	$0.62\sqrt{f_c'}$	$2\sqrt{f_c'}$	$7.5\sqrt{f_c'}$
A.10.2b(ii)	$\ell_p = 0.08h_w + 0.022f_y d_b$	$\ell_p = 0.08h_w + 0.0021f_y d_b$	$\ell_p = 0.08h_w + 0.00015f_y d_b$
A.11.3.2.1.1	$V_{ne} = 1.5A_{cv}(0.17\lambda\sqrt{f_{ce}'} + \rho_t f_{ye})$	$V_{ne} = 1.5A_{cv}(0.53\lambda\sqrt{f_{ce}'} + \rho_t f_{ye})$	$V_{ne} = 1.5A_{cv}(2\lambda\sqrt{f_{ce}'} + \rho_t f_{ye})$
A.11.3.2.1.2	$1.0A_{cv}\sqrt{f_{ce}'}$ $1.25A_{cv}\sqrt{f_{ce}'}$	$3.2A_{cv}\sqrt{f_{ce}'}$ $4.0A_{cv}\sqrt{f_{ce}'}$	$12A_{cv}\sqrt{f_{ce}'}$ $15A_{cv}\sqrt{f_{ce}'}$
A.11.3.2.2	$2.1A_{cv}\sqrt{f_{ce}'}$	$6.6A_{cv}\sqrt{f_{ce}'}$	$25A_{cv}\sqrt{f_{ce}'}$
A.12.3.4	$0.33A_{cv}\lambda\sqrt{f_c'}$	$1.1A_{cv}\lambda\sqrt{f_c'}$	$4A_{cv}\lambda\sqrt{f_c'}$

COMMENTARY REFERENCES

ACI Committee documents and documents published by other organizations that are cited in the commentary are listed first by document number, year of publication, and full title, followed by authored documents listed alphabetically.

American Association of State Highway and Transportation Officials (AASHTO)

LRFDCONS-4-2017—LRFD Bridge Construction Specifications, Fourth Edition

LRFDUS-8-2017—LRFD Bridge Design Specifications, Eighth Editioin

American Concrete Institute (ACI)

117-10—Specification for Tolerances for Concrete Construction and Materials

201.2R-08—Guide to Durable Concrete

209R-92(08)—Prediction of Creep, Shrinkage, and Temperature Effects in Concrete Structures

211.1-91(09)—Standard Practice for Selecting Proportions for Normal, Heavyweight, and Mass Concrete

213R-03—Guide for Structural Lightweight-Aggregate Concrete

213R-14—Guide for Structural Lightweight-Aggregate Concrete

214R-11—Guide to Evaluation of Strength Test Results of Concrete

214.4R-10—Guide for Obtaining Cores and Interpreting Compressive Strength Results

215R-92(97)—Considerations for Design of Concrete Structures Subjected to Fatigue Loading

216.1-07—Code Requirements for Determining Fire Resistance of Concrete and Masonry Construction Assemblies

222R-01—Protection of Metals in Concrete against Corrosion

223R-10—Guide for the Use of Shrinkage-Compensating Concrete

224R-01(08)—Control of Cracking in Concrete Structures

228.1R-03—In-Place Methods to Estimate Concrete Strength

232.2R-18—Report on the Use of Fly Ash in Concrete

233R-03—Slag Cement in Concrete and Mortar

234R-06—Guide for the Use of Silica Fume in Concrete

237R-07—Self-Consolidating Concrete

301-16—Specifications for Structural Concrete

304R-00(09)—Guide for Measuring, Mixing, Transporting, and Placing Concrete

305R-10—Guide to Hot Weather Concreting

305.1-06—Specification for Hot Weather Concreting

306R-10—Guide to Cold Weather Concreting

306.1-90(02)—Standard Specification for Cold Weather Concreting

307-08—Code Requirements for Reinforced Concrete Chimneys (ACI 307-08) and Commentary

308R-01(08)—Guide to Curing Concrete

309R-05—Guide for Consolidation of Concrete

311.4R-05—Guide for Concrete Inspection

311.6-09—Specification for Ready Mixed Concrete Testing Services

313-97—Standard Practice for Design and Construction of Concrete Silos and Stacking Tubes for Storing Granular Materials

318-63—Commentary on Building Code Requirements for Reinforced Concrete

318-11—Building Code Requirements for Structural Concrete (ACI 318-11) and Commentary

318.2-14—Building Code Requirements for Concrete Thin Shells (ACI 318.2-14) and Commentary

332-14—Requirements for Residential Concrete Construction (ACI 332-14) and Commentary

334.1R-92(02)—Concrete Shell Structures – Practice and Commentary

334.2R-91—Reinforced Concrete Cooling Tower Shells – Practice and Commentary

336.2R-88—Suggested Analysis and Design Procedures for Combined Footings and Mats

336.3R-93(06)—Design and Construction of Drilled Piers

347-04—Guide to Formwork for Concrete

347.2R-05—Guide for Shoring/Reshoring of Concrete Multistory Buildings

349-13—Code Requirements for Nuclear Safety-Related Concrete Structures (ACI 349-13) and Commentary

350-06—Code Requirements for Environmental Engineering Concrete Structures (ACI 350-06) and Commentary

352R-02—Recommendations for Design of Beam-Column Connections in Monolithic Reinforced Concrete Structures

352.1R-11—Guide for Design of Slab-Column Connections in Monolithic Concrete Structures

355.2-19—Qualifications of Post Installed Mechanical Anchors in Concrete and Commentary

355.4-11—Qualification of Post-Installed Adhesive Anchors in Concrete (ACI 355.4-11) and Commentary

359-13—Code for Concrete Containments

360R-10—Guide to Design of Slabs-on-Ground

362.1R-97(02)—Guide for the Design of Durable Parking Structures

363R-10—Report on High-Strength Concrete

369.1-17—Standard Requirements for Seismic Evaluation and Retrofit of Existing Concrete Buildings (ACI 369.1-17) and Commentary

372R-13—Guide to Design and Construction of Circular Wire- and Strand-Wrapped Prestressed Concrete Structures

374.1-05—Acceptance Criteria for Moment Frames Based on Structural Testing and Commentary

374.3R-16—Guide to Nonlinear Modeling Parameters for Earthquake-Resistant Structures

408R-03(12)—Bond and Development of Straight Reinforcing Bars in Tension

R Comm. Ref.

408.1R-90—Suggested Development, Splice, and Standard Hook Provisions for Deformed Bars in Tension
408.2R-12—Report on Bond of Steel Reinforcing Bars Under Cyclic Loads
421.1R-08—Guide to Shear Reinforcement for Slabs
423.3R-05—Recommendations for Concrete Members Prestressed with Unbonded Tendons
423.7-14—Specification for Unbonded Single-Strand Tendon Materials
423.10R-16—Guide to Estimating Prestress Losses
435R-95(00)—Control of Deflection in Concrete Structures
435.5R-73(89)—Deflections of Continuous Concrete Beams
437.1R-07—Load Tests of Concrete Structures: Methods, Magnitude, Protocols, and Acceptance Criteria
437.2-13—Code Requirements for Load Testing of Existing Concrete Structures and Commentary
440.1R-06—Guide for the Design and Construction of Structural Concrete Reinforced with FRP Bars
440.2R-08—Guide for the Design and Construction of Externally Bonded FRP Systems for Strengthening Concrete Structures
445R-99(09)—Recent Approaches to Shear Design of Structural Concrete
506R-16—Guide to Shotcrete
506.2-13—Specification for Shotcrete
506.4R-94(04)—Guide for the Evaluation of Shotcrete
543R-00—Guide to Design, Manufacture, and Installation of Concrete Piles
544.3R-08—Guide for Specifying, Proportioning, and Production of Fiber-Reinforced Concrete
550.3-13—Design Specification for Unbonded Post-Tensioned Precast Concrete Special Moment Frames Satisfying ACI 374.1 (ACI 550.3-13) and Commentary
550.4-18—Qualification of Precast Concrete Diaphragm Connections and Reinforcement at Joints for Earthquake Loading (ACI 550.4-18) and Commentary (ACI 550.4R-18)
550.5-18—Code Requirements for the Design of Precast Concrete Diaphragms for Earthquake Motions (ACI 550.5-18) and Commentary (ACI 550.5R-18)
551.2R-10—Design Guide for Tilt-Up Concrete Panels
555R-01—Removal and Reuse of Hardened Concrete
560R-16—Report on Design and Construction with Insulating Concrete Forms (ICFs)
562-19—Code Requirements for Assessment, Repair, and Rehabilitation of Existing Concrete Structures and Commentary (ACI 562-19).
CPP 610.1-18—American Concrete Institute Certification Policies for Concrete Field Testing Technician - Grade I. doi: 10.14359/51716912
CPP 620.2-12—American Concrete Institute Certification Policies for Concrete Strength Testing Technician. doi: 10.14359/51716913
CPP 630.1-15—American Concrete Institute Certification Policies for Concrete Construction Special Inspector. doi: 10.14359/51716916
CPP 660.1-17—American Concrete Institute Certification Policies for Shotcrete Nozzleman and Shotcrete Nozzleman-in-Training. doi: 10.14359/51716915
CPP 680.1-17—American Concrete Institute Certification Policies for Adhesive Anchor Installer. doi: 10.14359/51716917
CPP 681.1-17—American Concrete Institute Certification Policies for Adhesive Anchor Installation Inspector. doi: 10.14359/51716918
CPP 681.2-19—American Concrete Institute Certification Policies for Post-Installed Concrete Anchor Installation Inspector. doi: 10.14359/51716936
CT-18—Concrete Terminology
ITG-5.1-07—Acceptance Criteria for Special Unbonded Post-Tensioned Precast Structural Walls Based on Validation Testing and Commentary
ITG-5.2-09—Requirements for Design of a Special Unbonded Post-Tensioned Precast Shear Wall Satisfying ACI ITG-5.1 (ACI 5.2-09) and Commentary
ITG-7-09—Specification for Tolerances for Precast Concrete
ITG-10R-18—Practitioner's Guide for Alternative Cements
ITG-10.1R-18—Report on Alternative Cements
SP-2(07)—Manual of Concrete Inspection, Tenth Edition
SP-4(05)—Formwork for Concrete, Seventh Edition
SP-17(09)—ACI Design Handbook
SP-66(04)—ACI Detailing Manual

American Institute of Steel Construction (AISC)
341-10—Seismic Provisions for Structural Steel Buildings
360-10—Specification for Structural Steel Buildings

American Iron and Steel Institute (AISI)
D100-08—Cold-Formed Steel Design Manual
S100-07—North American Specification for the Design of Cold-Formed Steel Structural Members

American Society of Civil Engineers (ASCE)
7-05—Minimum Design Loads for Buildings and Other Structures
7-16—Minimum Design Loads for Buildings and Other Structures
41-17—Seismic Evaluation and Retrofit of Existing Buildings
61-14—ASCE/COPRI Standard for the Seismic Design of Piers and Wharves

American Society of Mechanical Engineers (ASME)
B1.1-03—Unified Inch Screw Threads (UN and UNR Thread Form)
B18.2.1-96—Square and Hex Bolts and Screws, Inch Series
B18.2.6-96—Fasteners for Use in Structural Applications
B31.1-92—Power Piping
B31.3-90—Chemical Plant and Petroleum Refinery Piping

American Welding Society (AWS)
D1.1/D1.1M:2015—Structural Welding Code – Steel

D1.4/D1.4M:2005—Structural Welding Code – Reinforcing Steel

ASTM International

A36/A36M-14—Standard Specification for Carbon Structural Steel

A307-14$^{\varepsilon1}$—Standard Specification for Carbon Steel Bolts, Studs, and Threaded Rod 60000 PSI Tensile Strength

A370-18—Standard Test Methods and Definitions for Mechanical Testing of Steel Products

A416/A416M-18—Standard Specification for Steel Strand, Uncoated Seven-Wire for Prestressed Concrete

A421/A421M-15—Standard Specification for Uncoated Stress-Relieved Steel Wire for Prestressed Concrete, including Supplementary Requirement SI, Low-Relaxation Wire and Relaxation Test

A615/A615M-18$^{\varepsilon1}$—Standard Specification for Deformed and Plain Carbon-Steel Bars for Concrete Reinforcement

A706/A706M-16—Standard Specification for Deformed and Plain Low-Alloy Steel Bars for Concrete Reinforcement

A767/A767M-16—Standard Specification for Zinc-Coated (Galvanized) Steel Bars for Concrete Reinforcement

A775/A775M-17—Standard Specification for Epoxy-Coated Steel Reinforcing Bars

A934/A934M-16—Standard Specification for Epoxy-Coated Prefabricated Steel Reinforcing Bars

A955/A955M-18b—Standard Specification for Deformed and Plain Stainless-Steel Bars for Concrete Reinforcement

A970/A970M-18—Standard Specification for Headed Steel Bars for Concrete Reinforcement, including Annex A1 Requirements for Class HA Head Dimensions

A996/A996M-16—Standard Specification for Rail-Steel and Axle-Steel Deformed Bars for Concrete Reinforcement

A1022/A1022M-16b—Standard Specification for Deformed and Plain Stainless-Steel Wire and Welded Wire for Concrete Reinforcement

A1035/A1035M-15—Standard Specification for Deformed and Plain, Low-Carbon, Chromium, Steel Bars for Concrete Reinforcement

A1044/A1044M-16a—Standard Specification for Steel Stud Assemblies for Shear Reinforcement of Concrete

A1055/A1055M-16—Standard Specification for Zinc and Epoxy Dual-Coated Steel Reinforcing Bars

A1064/A1064M-18a—Standard Specification for Carbon-Steel Wire and Welded Wire Reinforcement, Plain and Deformed, for Concrete

A1077/A1077M-14—Standard Specification for Structural Steel with Improved Yield Strength at High Temperature for Use in Buildings

C31/C31M-19—Standard Practice for Making and Curing Concrete Test Specimens in the Field

C33/C33M-18—Standard Specification for Concrete Aggregates

C39/C39M-18—Standard Test Method for Compressive Strength of Cylindrical Concrete Specimens

C42/C42M-18a—Standard Test Method for Obtaining and Testing Drilled Cores and Sawed Beams of Concrete

C94/C94M-18—Standard Specification for Ready-Mixed Concrete

C114-18—Standard Test Methods for Chemical Analysis of Hydraulic Cement

C150/C150M-19a—Standard Specification for Portland Cement

C172/C172M-17—Standard Practice for Sampling Freshly Mixed Concrete

C173/C173M-16—Standard Test Method for Air Content of Freshly Mixed Concrete by the Volumetric Method

C231/C231M-17a—Standard Test Method for Air Content of Freshly Mixed Concrete by the Pressure Method

C330/C330M-17a—Standard Specification for Lightweight Aggregates for Structural Concrete

C457/C457M-16—Standard Test Method for Microscopial Determination of Parameters of the Air-Void System in Hardened Concrete

C469/C469M-14—Standard Test Method for Static Modulus of Elasticity and Poisson's Ratio of Concrete in Compression

C494/C494M-17—Standard Specification for Chemical Admixtures for Concrete

C567/567M-14—Standard Test Method for Determining Density of Structural Lightweight Concrete

C595/C595M-19—Standard Specification for Blended Hydraulic Cements

C618-19—Standard Specification for Coal Fly Ash and Raw or Calcined Natural Pozzolan for Use in Concrete

C685/C685M-17a—Standard Specification for Concrete Made by Volumetric Batching and Continuous Mixing

C803/803M-18—Standard Test Method for Penetration Resistance of Hardened Concrete

C805/C805M-18—Standard Test Method for Rebound Number of Hardened Concrete

C845/C845M-18—Standard Specification for Expansive Hydraulic Cement

C873/873CM-15—Standard Test Method for Compressive Strength of Concrete Cylinders Cast in Place in Cylindrical Molds

C900-15—Standard Test Method for Pullout Strength of Hardened Concrete

C989/C989M-18a—Standard Specification for Slag Cement for Use in Concrete and Mortars

C1012/C1012M-18b—Standard Test Method for Length Change of Hydraulic-Cement Mortars Exposed to a Sulfate Solution

C1017/C1017M-13$^{\varepsilon1}$—Standard Specification for Chemical Admixtures for Use in Producing Flowing Concrete

C1074-17—Standard Practice for Estimating Concrete Strength by the Maturity Method

C1077-17—Standard Practice for Laboratories Testing Concrete and Concrete Aggregates for Use in Construction and Criteria for Testing Agency Evaluation

C1140/C1140M-11—Standard Practice for Preparing and Testing Specimens from Shotcrete Test Panels

C1152/C1152M-04(2012)$^{\varepsilon1}$—Standard Test Method for Acid-Soluble Chloride in Mortar and Concrete

C1157/C1157M-17—Standard Performance Specification for Hydraulic Cement

C1202-19—Standard Test Method for Electrical Indication of Concrete's Ability to Resist Chloride Ion Penetration

C1218/C1218M-17—Standard Test Method for Water-Soluble Chloride in Mortar and Concrete

C1240-15—Standard Specification for Silica Fume Used in Cementitious Mixtures

C1602/C1602M-18—Standard Specification for Mixing Water Used in Production of Hydraulic Cement Concrete

C1604/C1604M-05(2012)—Standard Test Method for Obtaining and Testing Drilled Cores of Shotcrete

C1609/C1609M-12—Standard Test Method for Flexural Performance of Fiber-Reinforced Concrete (Using Beam With Third-Point Loading)

C1778-16—Standard Guide for Reducing the Risk of Deleterious Alkali-Aggregate Reaction in Concrete

C1797-17—Standard Specification for Ground Calcium Carbonate and Aggregate Mineral Fillers for use in Hydraulic Cement Concrete

D3665-12(2017)—Standard Practice for Random Sampling of Construction Materials

E8/E8M-16a—Standard Test Methods for Tension Testing of Metallic Materials

F1554-18—Standard Specification for Anchor Bolts, Steel, 36, 55, and 105-ksi Yield Strength

Federal Emergency Management Agency (FEMA)

P749-10—Earthquake-Resistant Design Concepts: An Introduction to the NEHRP Recommended Provisions Seismic Provisions

P750-10—NEHRP Recommended Seismic Provisions for New Buildings and Other Structures (2009 edition)

P751-12—NEHRP Recommended Seismic Provisions: Design Examples (2009 edition)

International Code Council (ICC)

2018 IBC—International Building Code

ES AC193-15—Mechanical Anchors in Concrete Elements

National Fire Protection Association (NFPA)

5000-2012—Building Construction Safety Code

National Institute of Standards and Technology (NIST)

CGR 17-917-46—Guidelines for Nonlinear Structural Analysis for Design of Buildings

Portland Cement Association (PCA)

EB001.15-11—Design and Control of Concrete Mixtures, 16th edition

PCA 100-2017—Prescriptive Design of Exterior Concrete Walls

Precast/Prestressed Concrete Institute (PCI)

MNL 116-99—Manual for Quality Control for Plants and Production of Structural Precast Concrete Products

MNL 117-13—Manual for Quality Control for Plants and Production of Architectural Precast Concrete Products

MNL 120-10—Design Handbook: Precast and Prestressed Concrete, Seventh Edition

MNL 120-17—Design Handbook: Precast and Prestressed Concrete, Eighth Edition

MNL 123-88—Design and Typical Details of Connections for Precast and Prestressed Concrete

MNL 126-15—PCI Manual for the Design of Hollow Core Slabs and Walls

MNL 133-04—Bridge Design Manual

Post-Tensioning Institute (PTI)

DC10.5-12—Standard Requirements for Design and Analysis of Shallow Post-Tensioned Concrete Foundations of Expansive Soils

DC20.8-04—Design of Post-Tensioned Slabs Using Unbonded Tendons

M50.3-12—Guide Specification for Grouted Post-Tensioning

M55.1-12—Specification for Grouting of Post-Tensioned Structures

TAB.1-06—Post-Tensioning Manual, Sixth Edition

Standards New Zealand

NZS 3101-2006—Concrete Structure Standard, Part 1: The Design of Concrete Structures: Part 2: Commentary on the Design of Concrete Structures

Steel Deck Institute (SDI)

C-2011—Standard for Composite Steel Floor Deck – Slabs

NC-2010—Standard for Non-Composite Steel Floor Deck

Authored documents

Aaleti, S.; Brueggen, B. L.; Johnson, B.; French, C. E.; and Sritharan, S., 2013, "Cyclic Response of Reinforced Concrete Walls with Different Anchorage Details: Experimental Investigation," *Journal of Structural Engineering*, V. 139, No. 7, July, pp. 1181-1191. doi: 10.1061/(ASCE)ST.1943-541X.0000732

Abdullah, S. A., and Wallace, J. W., 2019, "Drift Capacity of Reinforced Concrete Structural Walls with Special Boundary Elements," *ACI Structural Journal*, V. 116, No. 1, Jan., pp. 183-194. doi: 10.14359/51710864

AC303, 2011, "Acceptance Criteria for Post-Tensioning Anchorages and Couplers of Prestressed Concrete," Apr. 2011, editorially revised Aug. 2012, ICC-ES Evaluation Service, LLC, Whittier, CA.

ACI Committee 318, 1965, "Commentary on Building Code Requirements for Reinforced Concrete (ACI 318-63)," SP-10, American Concrete Institute, Farmington Hills, MI, pp. 78-84.

ACI Committee 318, 1999, "Closure to Public Comments on ACI 318-99," *Concrete International*, V. 21, No. 5, May, pp. 318-1 to 318-50. doi: 10.14359/19333

ACI Committee 408, 1966, "Bond Stress—The State of the Art," *ACI Journal Proceedings*, V. 63, No. 11, Nov., pp. 1161-1188. doi: 10.14359/7665

ACI Committee 435, 1963, "Deflections of Prestressed Concrete Members (ACI 435.1R-63)," *ACI Journal*

Proceedings, V. 60, No. 12, Dec., pp. 1697-1728. doi: 10.14359/ 19413

ACI Committee 435, 1966, "Deflections of Reinforced Concrete Flexural Members (ACI 435.2R-66)," *ACI Journal Proceedings*, V. 63, No. 6, June, pp. 637-674. doi: 10.14359/7643

ACI Committee 435, 1968, "Allowable Deflections (ACI 435.3R-68)," *ACI Journal Proceedings*, V. 65, No. 6, June, pp. 433-444. doi: 10.14359/7482

ACI Committee 435, 1978, "Proposed Revisions by Committee 435 to ACI Building Code and Commentary Provisions on Deflections," *ACI Journal Proceedings*, V. 75, No. 6, June, pp. 229-238. doi: 10.14359/10935

Adebar, P.; Kuchma, D.; and Collins, M. P., 1990, "Strut-and-Tie Models for the Design of Pile Caps: An Experimental Study," *ACI Structural Journal*, V. 87, No. 1, Jan.-Feb., pp. 81-92. doi: 10.14359/2945

Ajaam, A.; Yasso, S.; Darwin, D.; O'Reilly, M.; and Sperry, J., 2018, "Anchorage Strength of Closely Spaced Hooked Bars," *ACI Structural Journal*, V. 115, No. 4, pp. 1143-1152. doi: 10.14359/51702065

Anderson, A. R., 1978, "Shear Strength of Hollow Core Members," *Technical Bulletin* 78-81, Concrete Technology Associates, Tacoma, WA, Apr., 33 pp. doi: 10.14359/19270

Anderson, N. S., and Meinheit, D. F., 2005, "Pryout Capacity of Cast-In Headed Stud Anchors," *PCI Journal*, V. 50, No. 2, Mar.-Apr., pp. 90-112. doi: 10.15554/pcij.03012005.90.112

Anderson, N. S., and Meinheit, D. F., 2007, "A Review of Headed Stud-Design Criteria," *PCI Journal*, V. 52, No. 1, Jan.-Feb., pp. 82-100. doi: 10.15554/pcij.01012007.82.100

Anderson, N. S., and Ramirez, J. A., 1989, "Detailing of Stirrup Reinforcement," *ACI Structural Journal*, V. 86, No. 5, Sept.-Oct., pp. 507-515. doi: 10.14359/3005

Angelakos, D.; Bentz, E. C.; and Collins, M. D., 2001, "Effect of Concrete Strength and Minimum Stirrups on Shear Strength of Large Members," *ACI Structural Journal*, V. 98, No. 3, May-June, pp. 290-300. doi: 10.14359/10220

Aoyama, H., ed., 2001, *Design of Modern Highrise Reinforced Concrete Structures*, Imperial College Press, London, UK, 442 pp.

Applied Technology Council, 1999, *ATC Design Guide 1: Minimizing Floor Vibration*, Redwood City, CA, 64 pp.

Arteta, C. A., 2015, "Seismic Response Assessment of Thin Boundary Element Specimens of Special Concrete Shear Walls," PhD dissertation, University of California, Berkeley, Berkeley, CA, 240 pp.

ASCE Joint Committee, 1940, "Recommended Practice and Standard Specification for Concrete and Reinforced Concrete," *Proceedings*, ASCE, V. 66, No. 6, Part 2, June, 81 pp.

Asmus, J., 1999, "Verhalten von Befestigungen bei der Versagensart Spalten des Betons (Behavior of Fastenings with the Failure Mode Splitting of Concrete)," dissertation, Universität Stuttgart, Stuttgart, Germany.

Aswad, A., and Jacques, F. J., 1992, "Behavior of Hollow-Core Slabs Subject to Edge Loads," *PCI Journal*, V. 37, No. 2, Mar.-Apr., pp. 72-86. doi: 10.15554/pcij.03011992.72.84

Athey, J. W., ed., 1982, "Test Report on Slender Walls," Southern California Chapter of the American Concrete Institute and Structural Engineers Association of Southern California, Los Angeles, CA, 129 pp.

Azizinamini, A.; Chisala, M.; and Ghosh, S. K., 1995, "Tension Development Length of Reinforcing Bars Embedded in High-Strength Concrete," *Engineering Structures*, V. 17, No. 7, pp. 512-522. doi: 10.1016/0141-0296(95)00096-P

Azizinamini, A.; Darwin, D.; Eligehausen, R.; Pavel, R.; and Ghosh, S. K., 1999b, "Proposed Modifications to ACI 318-95 Development and Splice Provisions for High-Strength Concrete," *ACI Structural Journal*, V. 96, No. 6, Nov.-Dec., pp. 922-926. doi: 10.14359/766

Azizinamini, A.; Pavel, R.; Hatfield, E.; and Ghosh, S. K., 1999a, "Behavior of Spliced Reinforcing Bars Embedded in High-Strength Concrete," *ACI Structural Journal*, V. 96, No. 5, Sept.-Oct., pp. 826-835. doi: 10.14359/737

Barda, F.; Hanson, J. M.; and Corley, W. G., 1977, "Shear Strength of Low-Rise Walls with Boundary Elements," *Reinforced Concrete Structures in Seismic Zones*, SP-53, American Concrete Institute, Farmington Hills, MI, pp. 149-202. doi: 10.14359/17697

Barney, G. B.; Corley, W. G.; Hanson, J. M.; and Parmelee, R. A., 1977, "Behavior and Design of Prestressed Concrete Beams with Large Web Openings," *PCI Journal*, V. 22, No. 6, Nov.-Dec., pp. 32-61. doi: 10.15554/pcij.11011977.32.61

Barney, G. B.; Shiu, K. N.; Rabbat, B. G.; Fiorato, A. E.; Russell, H. G.; and Corley, W. G., 1980, "Behavior of Coupling Beams under Load Reversals (RD068.01B)," Portland Cement Association, Skokie, IL. doi: 10.14359/51685458

Bartlett, F. M., 2012, "Using Historical Cylinder Data for Structural Evaluation," *Andy Scanlon Symposium on Serviceability and Safety of Concrete Structures: From Research to Practice*, SP-284, P. H. Bischoff, E. Musselman, S. Gross, and H. Nassif, eds., American Concrete Institute, Farmington Hills, MI, 12 pp. (CD-ROM) doi: 10.14359/51683800

Bartlett, M. F., and MacGregor, J. G., 1994, "Effect of Moisture Condition on Concrete Core Strengths," *ACI Materials Journal*, V. 91, No. 3, May-June, pp. 227-236. doi: 10.14359/4328

Bartoletti, S. J., and Jirsa, J. O., 1995, "Effects of Epoxy Coating on Anchorage and Development of Welded Wire Fabric," *ACI Structural Journal*, V. 92, No. 6, Nov.-Dec., pp. 757-764. doi: 10.14359/9669

Base, G. D.; Reed, J. B.; Beeby, A. W.; and Taylor, H. P. J., 1966, "An Investigation of the Crack Control Characteristics of Various Types of Bar in Reinforced Concrete Beams," *Research Report* No. 18, Cement and Concrete Association, London, UK, Dec., 44 pp. doi: 10.14359/51685142

Bažant, Z. P.; Yu, Q.; Gerstle, W.; Hanson, J.; and Ju, J., 2007, "Justification of ACI 446 Code Provisions for Shear Design of Reinforced Concrete Beams," *ACI Structural Journal*, V. 104, No. 5, Sept.-Oct., pp. 601-610. doi: 10.14359/18862

Becker, R. J., and Buettner, D. R., 1985, "Shear Tests of Extruded Hollow Core Slabs," *PCI Journal*, V. 30, No. 2, Mar.-Apr., pp. 40-54. doi: 10.15554/pcij.03011985.40.54

Becker, R. J.; Holland, T. C.; and Malits, F. S., 2019, "Structural Concrete Using Alternative Cements," *Concrete International*, V. 41, No. 6, June, pp. 39-44. doi: 10.14359/51716910

Beeby, A. W., 1979, "The Prediction of Crack Widths in Hardened Concrete," *The Structural Engineer*, V. 57A, No. 1, Jan., pp. 9-17. doi: 10.14359/51685143

Behera, U., and Rajagopalan, K. S., 1969, "Two-Piece U-Stirrups in Reinforced Concrete Beams," *ACI Journal Proceedings*, V. 66, No. 7, July, pp. 522-524. doi: 10.14359/51685214

Bertrand, J., and Vezina, D., 1994, "The Development of Air Entrained Durable Shotcrete for Structural Repairs," Proceedings of Shotcrete for Underground Support VII, pp. 58-65.

Bezerra Cabral, A. E.; Schalch, V.; Dal Molin, D. C. C.; and Ribeiro, J. L. D., 2010, "Mechanical Properties Modeling of Recycled Aggregate Concrete," *Construction and Building Materials*, V. 24, No. 4, Apr., pp. 421-430. doi: 10.1016/j.conbuildmat.2009.10.011

Bianchini, A. C.; Woods, R. E.; and Kesler, C. E., 1960, "Effect of Floor Concrete Strength on Column Strength," *ACI Journal Proceedings*, V. 56, No. 11, May, pp. 1149-1169. doi: 10.14359/8135

Birely, A. C.; Lowes, L. N.; and Lehman, D. E., 2012, "Linear Analysis of Concrete Frames Considering Joint Flexibility," *ACI Structural Journal*, V. 109, No. 3, May-June, pp. 381-391. doi: 10.14359/51683752

Birkeland, P. W., and Birkeland, H. W., 1966, "Connections in Precast Concrete Construction," *ACI Journal Proceedings*, V. 63, No. 3, Mar., pp. 345-368. doi: 10.14359/7627

Bischoff, P. H., 2005, "Revaluation of Deflection Prediction for Concrete Beams Reinforced with Steel and Fiber Reinforced Polymer Bars," *Journal of Structural Engineering*, V. 131, No. 5, May, pp. 752-767. doi: 10.1061/(ASCE)0733-9445(2005)131:5(752)

Bischoff, P. H., and Scanlon, A., 2007, "Effective Moment of Inertia for Calculating Deflections of Concrete Members Containing Steel Reinforcement and Fiber-Reinforced Polymer Reinforcement," *ACI Structural Journal*, V. 104, No. 1, Jan.-Feb., pp. 68-75. doi: 10.14359/18434

Black, W. C., 1973, "Field Corrections to Partially Embedded Reinforcing Bars," *ACI Journal Proceedings*, V. 70, No. 10, Oct., pp. 690-691. doi: 10.14359/51684023

Bloem, D. L., 1965, "Concrete Strength Measurement—Cores vs. Cylinders," *Proceedings, ASTM*, V. 65, pp. 668-696.

Bloem, D. L., 1968, "Concrete Strength in Structures," *ACI Journal Proceedings*, V. 65, No. 3, Mar., pp. 176-187. doi: 10.14359/7465

Blume, J. A.; Newmark, N. M.; and Corning, L. H., 1961, *Design of Multistory Reinforced Concrete Buildings for Earthquake Motions*, Portland Cement Association, Skokie, IL, 318 pp. doi: 10.14359/16796

BOCA, 1999, "BOCA National Building Code," 13th edition, Building Officials and Code Administration International, Inc., Country Club Hills, IL.

Bondy, K. B., 2003, "Moment Redistribution: Principles and Practice Using ACI 318-02," *PTI Journal*, V. 1, No. 1, Jan., pp. 3-21. doi: 10.14359/19230

Branson, D. E., 1965, "Instantaneous and Time-Dependent Deflections on Simple and Continuous Reinforced Concrete Beams," *HPR Report* No. 7, Part 1, Alabama Highway Department, Bureau of Public Roads, Aug., pp. 1-78.

Branson, D. E., 1970, discussion of "Proposed Revision of ACI 318-63: Building Code Requirements for Reinforced Concrete," *ACI Journal Proceedings*, V. 67, No. 9, Sept., pp. 692-695.

Branson, D. E., 1971, "Compression Steel Effect on Long-Time Deflections," *ACI Journal Proceedings*, V. 68, No. 8, Aug., pp. 555-559. doi: 10.14359/16519

Branson, D. E., 1977, *Deformation of Concrete Structures,* McGraw-Hill Book Co., New York, 546 pp.

Branson, D. E.; Meyers, B. L.; and Kripanarayanan, K. M., 1970, "Time-Dependent Deformation of Noncomposite and Composite Prestressed Concrete Structures," *Symposium on Concrete Deformation*, Highway Research Record 324, Highway Research Board, pp. 15-43. doi: 10.14359/19393

Breen, J. E.; Burdet, O.; Roberts, C.; Sanders, D.; Wollmann, G.; and Falconer, B., 1994, "Anchorage Zone Requirements for Post-Tensioned Concrete Girders," *NCHRP Report* 356, Transportation Research Board, National Academy Press, Washington, DC. doi: 10.14359/19236

Briss, G. R.; Paulay, T.; and Park, R., 1978, "Elastic Behavior of Earthquake Resistant R. C. Interior Beam-Column Joints," *Report* 78-13, University of Canterbury, Department of Civil Engineering, Christchurch, New Zealand, Feb. doi: 10.14359/19340

Broms, C. E., 1990, "Shear Reinforcement for Deflection Ductility of Flat Plates," *ACI Structural Journal*, V. 87, No. 6, Nov.-Dec., pp. 696-705. doi: 10.14359/2988

Brown, M. D.; Bayrak, O.; and Jirsa, J. O., 2006, "Design for Shear Based on Loading Conditions," *ACI Structural Journal*, V. 103, No. 4, July-Aug., pp. 541-550. doi: 10.14359/16430

Budek, A.; Priestley, M.; and Lee, C., 2002, "Seismic Design of Columns with High-Strength Wire and Strand as Spiral Reinforcement," *ACI Structural Journal*, V. 99, No. 5, Sept.-Oct., pp. 660-670. doi: 10.14359/12306

Burns, N. H., and Hemakom, R., 1977, "Test of Scale Model Post-Tensioned Flat Plate," *Journal of the Structural Division*, V. 103, June, pp. 1237-1255. doi: 10.14359/16870

Canbay, E., and Frosch, R. J., 2005, "Bond Strength of Lap-Spliced Bars," *ACI Structural Journal*, V. 102, No. 4, Jul.-Aug., pp. 605-614. doi: 10.14359/14565

Carino, N. J.; Guthrie, W. F.; Lagergren, E. S.; and Mullings, G. M., 1994, "Effects of Testing Variables on the Strength of High-Strength (90 MPa) Concrete Cylinders," *High-Performance Concrete*, SP-149, V. M. Malhotra, ed., American Concrete Institute, Farmington Hills, MI, pp. 589-632. doi: 10.14359/4176

Carter III, J. W.; Hawkins, N. M.; and Wood, S. L., 1993, "Seismic Response of Tilt-Up Construction," SRS No. 581, Civil Engineering Series, University of Illinois, Urbana, IL, Dec., 224 pp. doi: 10.14359/19661

Castro, A.; Kreger, M.; Bayrak, O.; Breen, J. E.; and Wood, S. L., 2004, "Allowable Design Release Stresses for Pretensioned Concrete Beams," *Report* No. FHWA/TX-04/0-4086-2, Center for Transportation Research, University of Texas at Austin, Austin, TX, Aug., 127 pp.

CEB, 1994, "Fastenings to Concrete and Masonry Structures, State of the Art Report," Comite Euro-International du Beton (CEB), *Bulletin* No. 216, Thomas Telford Services Ltd., London, UK, 249 pp.

CEB, 1997, *Design of Fastenings in Concrete*, Comite Euro-International du Beton (CEB), Thomas Telford Services Ltd., London, UK, Jan.

Chen, L.; Mindess, S.; Morgan, D. R.; Shah, S. P.; Johnston, C. D.; and Pigeon, M., 1995, "Comparative Toughness Testing of Fiber Reinforced Concrete," *Testing of Fiber Reinforced Concrete*, SP-155, American Concrete Institute, Farmington Hills, MI, pp. 41-69. doi: 10.14359/928

Cheng, M. Y.; Hung, S. C.; Lequesne, R. D.; and Lepage, A., 2016, "Earthquake-Resistant Squat Walls Reinforced with High-Strength Steel," *ACI Structural Journal*, V. 113, No. 5, Sep.-Oct., pp. 1065-1076. doi: 10.14359/51688825

Chow, L.; Conway, H.; and Winter, G., 1953, "Stresses in Deep Beams," *Transactions of the American Society of Civil Engineers*, V. 118, pp. 686-708. doi: 10.14359/19672

Clough, R. W., 1960, "Dynamic Effects of Earthquakes," *Journal of the Structural Division*, V. 86, Apr., pp. 49-65. doi: 10.14359/19323

Cohn, M. A., 1965, "Rotational Compatibility in the Limit Design of Reinforced Concrete Continuous Beams," *Flexural Mechanics of Reinforced Concrete*, SP-12, American Concrete Institute/American Society of Civil Engineers, Farmington Hills, MI, pp. 35-46. doi: 10.14359/16724

Collins, M. P., and Lampert, P., 1973, "Redistribution of Moments at Cracking—The Key to Simpler Torsion Design?" *Analysis of Structural Systems for Torsion*, SP-35, American Concrete Institute, Farmington Hills, MI, pp. 343-383. doi: 10.14359/17486

Collins, M. P., and Mitchell, D., 1991, *Prestressed Concrete Structures*, Prentice Hall Inc., Englewood Cliffs, NJ, 766 pp.

Collins, M. P., and Mitchell, D., 1997, *Prestressed Concrete Structures*, Response Publications, Canada, pp. 517-518. doi: 10.14359/19232

Column Research Council, 1966, "Guide to Design Criteria for Metal Compression Members," second edition, Fritz Engineering Laboratory, Lehigh University, Bethlehem, PA, 117 pp. doi: 10.14359/19273

Cook, R. A., and Klingner, R. E., 1992a, "Behavior of Ductile Multiple-Anchor Steel-to-Concrete Connections with Surface-Mounted Baseplates," *Anchors in Concrete: Design and Behavior*, SP-130, G. A. Senkiw and H. B. Lancelot III, eds., American Concrete Institute, Farmington Hills, MI, pp. 61-122. doi: 10.14359/1268

Cook, R. A., and Klingner, R. E., 1992b, "Ductile Multiple-Anchor Steel-to-Concrete Connections," *Journal of Structural Engineering*, V. 118, No. 6, June, pp. 1645-1665. doi: 10.1061/(ASCE)0733-9445(1992)118:6(1645)

Cook, R. A.; Kunz, J.; Fuchs, W.; and Konz, R. C., 1998, "Behavior and Design of Single Adhesive Anchors under Tensile Load in Uncracked Concrete," *ACI Structural Journal*, V. 95, No. 1, Jan.-Feb., pp. 9-26. doi: 10.14359/522

Cook, R. A., and Michler, H., 2017, "Behavior and Design of Anchorages with Shear Lugs," *3rd International Symposium on Connections between Steel and Concrete*, Stuttgart, Germany, pp. 560-570.

Corley, W. G., and Jirsa, J. O., 1970, "Equivalent Frame Analysis for Slab Design," *ACI Journal Proceedings*, V. 67, No. 11, Nov., pp. 875-884. doi: 10.14359/7317

Corley, W. G.; Sozen, M. A.; and Siess, C. P., 1961, "Equivalent-Frame Analysis for Reinforced Concrete Slabs," *Structural Research Series* No. 218, Civil Engineering Studies, University of Illinois, June, 166 pp. doi: 10.14359/16677

Crist, R. A., 1966, "Shear Behavior of Deep Reinforced Concrete Beams," *Proceedings*, Symposium on the Effects of Repeated Loading of Materials and Structural Elements (Mexico City), V. 4, RILEM, Paris, France, 31 pp.

CRSI Handbook, 1984, sixth edition, Concrete Reinforcing Steel Institute, Schaumburg, IL.

Dam, T. X.; Wight, J. K.; and Parra-Montesinos, G. J., 2017, "Behavior of Monotonically Loaded Slab-Column Connections Reinforced with Shear Studs," *ACI Structural Journal*, V. 114, No. 1, Jan.-Feb., pp. 221-232. doi: 10.14359/51689165

Darwin, D.; Manning, D. G.; and Hognestad, E., 1985, "Debate: Crack Width, Cover, and Corrosion," *Concrete International*, V. 7, No. 5, May, pp. 20-35. doi: 10.14359/16539

Darwin, D.; Zuo, J.; Tholen, M. L.; and Idun, E. K., 1996, "Development Length Criteria for Conventional and High Relative Rib Area Reinforcing Bars," *ACI Structural Journal*, V. 93, No. 3, May-June, pp. 347-359. doi: 10.14359/9694

Deatherage, J. H.; Burdette, E. G.; and Chew, C. K., 1994, "Development Length and Lateral Spacing Requirements of Prestressing Strand for Prestressed Concrete Bridge Girders," *PCI Journal*, V. 39, No. 1, Jan.-Feb., pp. 70-83. doi: 10.15554/pcij.01011994.70.83

Dolan, C. W., and Krohn, J. J., 2007, "A Case for Increasing the Allowable Compressive Release Stress for Prestressed Concrete," *PCI Journal*, V. 52, No. 1, Jan.-Feb., pp. 102-105. doi: 10.15554/pcij.01012007.102.105

Dönmez, A., and Bažant, Z. P., 2017, "Size Effect on Punching Strength of Reinforced Concrete Slabs Without and With Shear Reinforcement," *ACI Structural Journal*, V. 114, No. 4, July-Aug., pp. 876-886. doi: 10.14359/51689719

Dovich, L. M., and Wight, J. K., 2005, "Effective Slab Width Model for Seismic Analysis of Flat Slab Frames," *ACI Structural Journal*, V. 102, No. 6, Nov.-Dec., pp. 868-875. doi: 10.14359/14795

Durrani, A. J., and Wight, J. K., 1982, "Experimental and Analytical Study of Internal Beam to Column Connections Subjected to Reversed Cyclic Loading," *Report* No. UMEE 82R3, Department of Civil Engineering, University of Michigan, Ann Arbor, MI, July, 275 pp. doi: 10.14359/16809

Ehsani, M. R., 1982, "Behavior of Exterior Reinforced Concrete Beam to Column Connections Subjected to Earthquake Type Loading," *Report* No. UMEE 82R5, Department of Civil Engineering, University of Michigan, Ann Arbor, MI, July, 275 pp.

Elgabry, A. A., and Ghali, A., 1987, "Tests on Concrete Slab-Column Connections with Stud Shear Reinforcement Subjected to Shear-Moment Transfer," *ACI Structural Journal*, V. 84, No. 5, Sept-Oct., pp. 433-442. doi: 10.14359/1656

Eligehausen, R., and Balogh, T., 1995, "Behavior of Fasteners Loaded in Tension in Cracked Reinforced Concrete," *ACI Structural Journal*, V. 92, No. 3, May-June, pp. 365-379. doi: 10.14359/1137

Eligehausen, R., and Fuchs, W., 1988, "Load Bearing Behavior of Anchor Fastenings under Shear, Combined Tension and Shear or Flexural Loadings," *Betonwerk + Fertigteil-Technik*, V. 54, pp. 48-56. doi: 10.14359/16854

Eligehausen, R.; Cook, R. A.; and Appl, J., 2006a, "Behavior and Design of Adhesive Bonded Anchors," *ACI Structural Journal*, V. 103, No. 6, Nov.-Dec., pp. 822-831. doi: 10.14359/18234

Eligehausen, R.; Fuchs, W.; and Mayer, B., 1987, "Load Bearing Behavior of Anchor Fastenings in Tension," *Betonwerk + Fertigteiltechnik*, V. 12, pp. 826-832, and 1988, V. 1, pp. 29-35. doi: 10.14359/19643

Eligehausen, R.; Mallée, R.; and Silva, J., 2006b, *Anchorage in Concrete Construction*, Ernst & Sohn (J. T. Wiley), Berlin, Germany, May, 380 pp.

Elwood, K. J.; Maffei, J. M.; Riederer, K. A.; and Telleen, K., 2009, "Improving Column Confinement—Part 2: Proposed New Provisions for the ACI 318 Building Code," *Concrete International*, V. 31, No. 12, Dec., pp. 41-48.

Elwood, K. J.; Matamoros, A. B.; Wallace, J. W. ; Lehman, D. E.; Heintz, J. A.; Mitchell, A. D.; Moore, M. A.; Valley, M. T.; Lowes, L. N.; Comartin, C. D. ; and Moehle, J. P., 2007, "Update to ASCE/SEI 41 Concrete Provisions," *Earthquake Spectra*, V. 23, No. 3, pp. 493-523, doi: 10.1193/1.2757714.

Elzanaty, A. H.; Nilson, A. H.; and Slate, F. O., 1986, "Shear Capacity of Reinforced Concrete Beams Using High Strength Concrete," *ACI Journal Proceedings*, V. 83, No. 2, Mar.-Apr., pp. 290-296. doi: 10.14359/10433

Fanella, D. A., and Mota, M., 2014, *Design Guide for Vibrations of Reinforced Concrete Floor Systems*, 10-DG-Vibration, Concrete Reinforcing Steel Institute, Schaumburg, IL, 60 pp.

Faradji, M. J., and Diaz de Cossio, R., 1965, "Diagonal Tension in Concrete Members of Circular Section," (in Spanish) Institut de Ingenieria, Mexico (translation by Portland Cement Association, Foreign Literature Study No. 466), 61 pp. doi: 10.14359/19496

Farrow, C. B., and Klingner, R. E., 1995, "Tensile Capacity of Anchors with Partial or Overlapping Failure Surfaces: Evaluation of Existing Formulas on an LRFD Basis," *ACI Structural Journal*, V. 92, No. 6, Nov.-Dec., pp. 698-710. doi: 10.14359/9664

Fennel, A. W.; Line, P.; Mochizuki, G. L.; Moore, K. S.; Van Dorpe, T. D.; and Voss, T. A., 2009, "Report on Laboratory Testing of Anchor Bolts Connecting Wood Sill Plates to Concrete with Minimum Edge Distances," SEAONC, San Francisco, CA, Mar., 50 pp.

Fintel, M.; Ghosh, S. K.; and Iyengar, H., 1986, *Column Shortening in Tall Buildings—Prediction and Compensation*, EB108D, Portland Cement Association, Skokie, IL, 34 pp. doi: 10.14359/51685083

fib, 2011, "Design of Anchorages in Concrete. Guide to Good Practice," Bulletin No.58, International Federation for Structural Concrete, Lausanne, Switzerland, 280 pp.

FIP, 1999, *FIP Recommendations, Practical Design of Structural Concrete*, FIP-Commission 3, "Practical Design," Pub.: SETO, London, UK, Sept., 112 pp.

Fling, R. S., 1987, *Practical Design of Reinforced Concrete*, John Wiley & Sons, Inc., New York, 536 pp.

Ford, J. S.; Chang, D. C.; and Breen, J. E., 1981, "Design Indications from Tests of Unbraced Multipanel Concrete Frames," *Concrete International*, V. 3, No. 3, Mar., pp. 37-47. doi: 10.14359/16551

Foutch, D. A.; Gamble, W. L.; and Sunidja, H., 1990, "Tests of Post-Tensioned Concrete Slab-Edge Column Connections," *ACI Structural Journal*, V. 87, No. 2, Mar.-Apr., pp. 167-179. doi: 10.14359/2689

Frantz, G. C., and Breen, J. E., 1980, "Cracking on the Side Faces of Large Reinforced Concrete Beams," *ACI Journal Proceedings*, V. 77, No. 5, Sept.-Oct., pp. 307-313. doi: 10.14359/7007

French, C. W., and Moehle, J. P., 1991, "Effect of Floor Slab on Behavior of Slab-Beam-Column Connections," *Design of Beam-Column Joints for Seismic Resistance*, SP-123, J. O. Jirsa, ed., American Concrete Institute, Farmington Hills, MI, pp. 225-258. doi: 10.14359/19328

Frosch, R. J., 1999, "Another Look at Cracking and Crack Control in Reinforced Concrete," *ACI Structural Journal*, V. 96, No. 3, May-June, pp. 437-442. doi: 10.14359/679

Frosch, R. J., 2002, "Modeling and Control of Side Face Beam Cracking," *ACI Structural Journal*, V. 99, No. 3, May-June, pp. 376-385. doi: 10.14359/11922

Frosch, R. J.; Labi, S.; and Sim, C., 2014, "Increasing Bridge Deck Service Life: Volume 1 – Technical Evaluation," *Publication* No. FHWA/IN/JTRP-2014/16. Joint Transportation Research Program, Indiana Department of Transportation and Purdue University, West Lafayette, IN.

Frosch, R. J.; Yu, Q.; Cusatis, G.; and Bažant, Z. P., 2017, "A Unified Approach to Shear Design," *Concrete International*, V. 39, No. 9, pp. 47-52. doi: 10.14359/51701014

Fuchs, W.; Eligehausen, R.; and Breen, J., 1995, "Concrete Capacity Design (CCD) Approach for Fastening to Concrete," *ACI Structural Journal*, V. 92, No. 1, Jan.-Feb. 1995, pp. 73-93. Also discussion, *ACI Structural Journal*, V. 92, No. 6, Nov.-Dec., pp. 787-802. doi: 10.14359/1533

Furche, J., and Eligehausen, R., 1991, "Lateral Blow-out Failure of Headed Studs Near a Free Edge," *Anchors in Concrete—Design and Behavior*, SP-130, G. A. Senkiw and H. B. Lancelot III, eds., American Concrete Institute, Farmington Hills, MI, pp. 235-252. doi: 10.14359/1276

Furlong, R. W.; Fenves, G. L.; and Kasl, E. P., 1991, "Welded Structural Wire Reinforcement for Columns," *ACI*

Structural Journal, V. 88, No. 5, Sept.-Oct., pp. 585-591. doi: 10.14359/9452

Furlong, R. W.; Hsu, C.-T. T.; and Mirza, S. A., 2004, "Analysis and Design of Concrete Columns for Biaxial Bending—Overview," *ACI Structural Journal*, V. 101, No. 3, May-June, pp. 413-423. doi: 10.14359/13101

Gamble, W. L., 1972, "Moments in Beam Supported Slabs," *ACI Journal Proceedings*, V. 69, No. 3, Mar., pp. 149-157. doi: 10.14359/11258

Gamble, W. L.; Sozen, M. A.; and Siess, C. P., 1969, "Tests of a Two-Way Reinforced Concrete Floor Slab," *Journal of the Structural Division*, V. 95, June, pp. 1073-1096. doi: 10.14359/19673

Genikomsou, A. S., and Polak, M. A., 2017, "Effect of Openings on Punching Shear Strength of Reinforced Concrete Slabs—Finite Element Investigation," *ACI Structural Journal*, V. 114, No. 5, Sept.-Oct., pp. 1249-1261. doi: 10.14359/51689871

Gerber, L. L., and Burns, N. H., 1971, "Ultimate Strength Tests of Post-Tensioned Flat Plates," *PCI Journal*, V. 16, No. 6, Nov.-Dec., pp. 40-58. doi: 10.15554/pcij.11011971.40.58

Gergely, P., and Lutz, L. A., 1968, "Maximum Crack Width in Reinforced Concrete Flexural Members," *Causes, Mechanism, and Control of Cracking in Concrete*, SP-20, American Concrete Institute, Farmington Hills, MI, pp. 87-117. doi: 10.14359/17348

Ghali, A., and Favre, R., 1986, *Concrete Structures: Stresses and Deformations*, Chapman and Hall, New York, 348 pp.

Ghimire, K.; Darwin, D.; and O'Reilly, M., 2018, "Anchorage of Headed Reinforcing Bars," *SM Report* No. 127, University of Kansas Center for Research, Lawrence, KS, Jan., 278 pp.

Gilbert, R. I., 1992, "Shrinkage Cracking in Fully Restrained Concrete Members," *ACI Structural Journal*, V. 89, No. 2, Mar.-Apr., pp. 141-149. doi: 10.14359/2917

Gomez, I. R.; Kanvinde, A. M.; Smith, C.; and Deierlein, G. G., 2009, "Shear Transfer in Exposed Column Base Plates," Report Presented to American Institute of Steel Construction, Chicago, IL, Mar., 159 pp.

Goto, Y., 1971, "Cracks Formed in Concrete around Deformed Tension Bars in Concrete," *ACI Journal Proceedings*, V. 68, No. 4, Apr., pp. 244-251. doi: 10.14359/11325

Graybeal, B., 2014, "Lightweight Concrete: Development of Mild Steel in Tension," *Technical Brief* No. FHWA-HRT-14-030, Federal Highway Administration, Washington, DC.

Greene, G., and Graybeal, B., 2013, "Lightweight Concrete: Mechanical Properties," *Report* No. FHWA-HRT-13-062, Federal Highway Administration, Washington, DC, 12 pp.

Greene, G., and Graybeal, B., 2015, "Lightweight Concrete: Shear Performance," *Report* No. FHWA-HRT-15-022, Federal Highway Administration, Washington, DC, 20 pp.

Griezic, A.; Cook, W. D.; and Mitchell, D., 1994, "Tests to Determine Performance of Deformed Welded-Wire Fabric Stirrups," *ACI Structural Journal*, V. 91, No. 2, Mar.-Apr., pp. 211-220. doi: 10.14359/4597

Grossfield, B., and Birnstiel, C., 1962, "Tests of T-Beams with Precast Webs and Cast-in-Place Flanges," *ACI Journal Proceedings*, V. 59, No. 6, June, pp. 843-851. doi: 10.14359/16709

Grossman, J. S., 1987, "Reinforced Concrete Design," *Building Structural Design Handbook*, R. N. White and C. G. Salmon, eds., John Wiley and Sons, Inc., New York.

Grossman, J. S., 1990, "Slender Concrete Structures—The New Edge," *ACI Structural Journal*, V. 87, No. 1, Jan.-Feb., pp. 39-52. doi: 10.14359/3212

Guimares, G. N.; Kreger, M. E.; and Jirsa, J. O., 1992, "Evaluation of Joint-Shear Provisions for Interior Beam-Column-Slab Connections Using High Strength Materials," *ACI Structural Journal*, V. 89, No. 1, Jan.-Feb., pp. 89-98. doi: 10.14359/1299

Gulkan, P., and Sozen, M. A., 1974, "Inelastic Response of Reinforced Concrete Structures to Earthquake Motions," *ACI Journal Proceedings*, V. 71, No. 12, Dec., pp. 604-610. doi: 10.14359/7110

Guralnick, S. A., and LaFraugh, R. W., 1963, "Laboratory Study of a 45-Foot Square Flat Plate Structure," *ACI Journal Proceedings*, V. 60, No. 9, Sept., pp. 1107-1185. doi: 10.14359/7893

Gustafson, D. P., and Felder, A. L., 1991, "Questions and Answers on ASTM A706 Reinforcing Bars," *Concrete International*, V. 13, No. 7, July, pp. 54-57. doi: 10/14359/51685022

Hale, W. M., and Russell, B. W., 2006, "Effect of Allowable Compressive Stress at Release on Prestress Losses and on the Performance of Precast, Prestressed Concrete Bridge Girders," *PCI Journal*, V. 51, No. 2, Mar.-Apr., pp. 14-25. doi: 10.15554/pcij.03012006.14.25

Hamad, B. S.; Jirsa, J. O.; and D'Abreu, N. I., 1993, "Anchorage Strength of Epoxy-Coated Hooked Bars," *ACI Structural Journal*, V. 90, No. 2, Mar.-Apr., pp. 210-217. doi: 10.14359/4127

Hansell, W., and Winter, G., 1959, "Lateral Stability of Reinforced Concrete Beams," *ACI Journal Proceedings*, V. 56, No. 3, Sept., pp. 193-214. doi: 10.14359/8091

Hanson, J. A., 1961, "Tensile Strength and Diagonal Tension Resistance of Structural Lightweight Concrete," *ACI Journal Proceedings*, V. 58, No. 1, July, pp. 1-40. doi: 10.14359/7972

Hanson, N. W., 1960, "Precast-Prestressed Concrete Bridges: Horizontal Shear Connections," Bulletin, PCA Research and Development Laboratories, V. 2, No. 2, May, pp. 38-58. doi: 10.14359/16708

Hanson, N. W., and Connor, H. W., 1967, "Seismic Resistance of Reinforced Concrete Beam-Column Joints," *Journal of the Structural Division*, V. 93, No. ST5 Oct., pp. 533-560. doi: 10.14359/19667

Hanson, N. W., and Hanson, J. M., 1968, "Shear and Moment Transfer between Concrete Slabs and Columns," *Bulletin,* PCA Research and Development Laboratories, V. 10, No. 1, Jan., pp. 2-16. doi: 10.14359/19463

Hanson, N. W., and Kaar, P. H., 1959, "Flexural Bond Tests Pretensioned Beams," *ACI Journal Proceedings*, V. 55, No. 7, Jan., pp. 783-802. doi: 10.14359/11389

Hardisty, J.; Villalobos, E.; Richter, B.; and Pujol, S., 2015, "Lap Splices in Unconfined Boundary Elements," *Concrete International*, V. 37, No. 1, Jan., pp. 51-58.

Hatcher, D. S.; Sozen, M. A.; and Siess, C. P., 1965, "Test of a Reinforced Concrete Flat Plate," *Journal of the Structural Division*, V. 91, May, pp. 205-232.

Hatcher, D. S.; Sozen, M. A.; and Siess, C. P., 1969, "Test of a Reinforced Concrete Flat Slab," *Journal of the Structural Division*, V. 95, June, pp. 1051-1072. doi: 10.14359/51685284

Hawkins, N. M., 1968, "Bearing Strength of Concrete Loaded through Rigid Plates," *Magazine of Concrete Research*, V. 20, No. 62, Mar., pp. 31-40. doi: 10.1680/macr.1968.20.62.31

Hawkins, N. M., 1974, "Shear Strength of Slabs with Shear Reinforcement," *Shear in Reinforced Concrete*, SP-42, American Concrete Institute, Farmington Hills, MI, pp. 785-815. doi: 10.14359/17306

Hawkins, N. M., 1981, "Lateral Load Resistance of Unbonded Post-Tensioned Flat Plate Construction," *PCI Journal*, V. 26, No. 1, Jan.-Feb., pp. 94-116. doi: 10.15554/pcij.01011981.94.117

Hawkins, N. M.; Mitchell, D.; and Hanna, S. N., 1975, "The Effects of Shear Reinforcement on Reversed Cyclic Loading Behavior of Flat Plate Structures," *Canadian Journal of Civil Engineering*, V. 2, No. 4, pp. 572-582. doi: 10.1139/l75-052

Hawkins, N. M., and Ospina, C. E., 2017, "Effect of Slab Flexural Reinforcement and Depth on Punching Strength," *Joint ACI-fib International Symposium on Punching Shear of Structural Concrete Slabs*, SP-315, American Concrete Institute, Farmington Hills, MI, pp. 117-140. doi: 10.14359/51700935

Henry, R. E.; Dizhur, D.; Elwood, K. J.; Hare, J.; and Brunsdon, D., 2017, "Damage to Concrete Buildings with Precast Floors During the 2016 Kaikoura Earthquake," *Bulletin of New Zealand Society for Earthquake Engineering*, V. 50, No. 2, pp. 174-187. http://www.nzsee.org.nz/db/Bulletin/Archive/50(2)0174.pdf

Hirosawa, M., 1977, "Strength and Ductility of Reinforced Concrete Members," *Report* No. 76, Building Research Institute, Ministry of Construction, Tokyo, Mar. (in Japanese). Also, data in Civil Engineering Studies, 1978, *Structural Research Series* No. 452, University of Illinois, Champaign, IL, 141 pp. doi: 10.14359/19325

Hoehler, M., and Eligehausen, R., 2008, "Behavior and Testing of Anchors in Simulated Seismic Cracks," *ACI Structural Journal*, V. 105, No. 3, May-June, pp. 348-357. doi: 10.14359/19794

Hsu, T. T. C., 1968, "Torsion of Structural Concrete—Behavior of Reinforced Concrete Rectangular Members," *Torsion of Structural Concrete*, SP-18, American Concrete Institute, Farmington Hills, MI, pp. 291-306. doi: 10.14359/17572

Hsu, T. T. C., 1990, "Shear Flow Zone in Torsion of Reinforced Concrete," *Journal of Structural Engineering*, V. 116, No. 11, Nov., pp. 3206-3226. doi: 10.1061/(ASCE)0733-9445(1990)116:11(3206)

Hsu, T. T. C., 1997, "ACI Shear and Torsion Provisions for Prestressed Hollow Girders," *ACI Structural Journal*, V. 94, No. 6, Nov.-Dec., pp. 787-799. doi: 10.14359/9738

Hsu, T. T. C., and Burton, K. T., 1974, "Design of Reinforced Concrete Spandrel Beams," *Journal of the Structural Division*, V. 100, Jan., pp. 209-229. doi: 10.14359/16866

Huq, M. S.; Burgos, E. A.; Lequesne, R. D.; and Lepage, A., 2018, "High-Strength Steel Bars in T-Shaped Concrete Walls," Eleventh U.S. National Conference on Earthquake Engineering, Los Angeles, CA.

Hwang, S., and Moehle, J. P., 2000, "Models for Laterally Loaded Slab-Column Frames," *ACI Structural Journal*, V. 97, No. 2, Mar.-Apr., pp. 345-353. doi: 10.14359/866

ICBO, 1997, "Uniform Building Code," V. 2, Structural Engineering Design Provisions, International Conference of Building Officials, Whittier, CA, 492 pp.

Ichinose, T., 1995, "Splitting Bond Failure of Columns under Seismic Action," *ACI Structural Journal*, V. 92, No. 5, pp. 535-541. doi: 10.14359/904

Iguro, M.; Shioya, T.; Nojiri, Y.; and Akiyama, H.; 1985, "Experimental Studies on Shear Strength of Large Reinforced Concrete Beams under Uniformly Distributed Load," Concrete Library International, Japan Society of Civil Engineers, Tokyo, No. 5, pp. 137-154. doi: 10.2208/jscej.1984.348_175

Ishizuka, T., and Hawkins, N. M., 1987, "Effect of Bond Deterioration on the Seismic Response of Reinforced and Partially Prestressed Concrete Ductile Moment Resistant Frames," *Report* SM 87-2, Department of Civil Engineering, University of Washington, Seattle, WA.

Ivey, D. L., and Buth, E., 1967, "Shear Capacity of Lightweight Concrete Beams," *ACI Journal Proceedings*, V. 64, No. 10, Oct., pp. 634-643. doi: 10.14359/7591

Jeanty, P. R.; Mitchell, D.; and Mirza, M. S., 1988, "Investigation of 'Top Bar' Effects in Beams," *ACI Structural Journal*, V. 85, No. 3, May-June, pp. 251-257. doi: 10.14359/2613

Jirsa, J. O., and Breen, J. E., 1981, "Influence of Casting Position and Shear on Development and Splice Length—Design Recommendations," *Research Report* 242-3F, Center for Transportation Research, Bureau of Engineering Research, University of Texas at Austin, Austin, TX, Nov., 50 pp. doi: 10.14359/19469

Jirsa, J. O.; Lutz, L. A.; and Gergely, P., 1979, "Rationale for Suggested Development, Splice, and Standard Hook Provisions for Deformed Bars in Tension," *Concrete International*, V. 1, No. 7, July, pp. 47-61. doi: 10.14359/15038

Jirsa, J. O.; Sozen, M. A.; and Siess, C. P., 1963, "Effects of Pattern Loadings on Reinforced Concrete Floor Slabs," *Structural Research Series* No. 269, Civil Engineering Studies, University of Illinois, Urbana, IL, July. doi: 10.14359/51685300

Jirsa, J. O.; Sozen, M. A.; and Siess, C. P., 1966, "Test of a Flat Slab Reinforced with Welded Wire Fabric," *Journal of the Structural Division*, V. 92, June, pp. 199-224. doi: 10.14359/16861

Jirsa, J. O.; Sozen, M. A.; and Siess, C. P., 1969, "Pattern Loadings on Reinforced Concrete Floor Slabs," *Journal of the Structural Division*, V. 95, No. ST6, June, pp. 1117-1137.

Johnson, L. A., and Jirsa, J. O., 1981, "The Influence of Short Embedment and Close Spacing on the Strength of Hooked Bar Anchorages," *PMFSEL Report* No. 81-2, Department of Civil Engineering-Structures Research Laboratory, University of Texas, Austin, TX, 93 pp.

Johnson, M. K., and Ramirez, J. A., 1989, "Minimum Amount of Shear Reinforcement in High Strength Concrete Members," *ACI Structural Journal*, V. 86, No. 4, July-Aug., pp. 376-382. doi: 10.14359/2896

Johnson, T., and Ghadiali, Z., 1972, "Load Distribution Test on Precast Hollow Core Slabs with Openings," *PCI Journal*, V. 17, No. 5, Sept.-Oct., pp. 9-19. doi: 10.15554/pcij.09011972.9.19

Johnston, D. W., and Zia, P., 1982, "Bond Characteristics of Epoxy-Coated Reinforcing Bars," *Report* No. FHWA/NC/82-002, Department of Civil Engineering, North Carolina State University, Raleigh, NC, Aug., 176 pp. doi: 10.14359/16632

Joint ACI-ASCE Committee 326, 1962, "Shear and Diagonal Tension," *ACI Journal Proceedings*, V. 59, No. 1, Jan. 1962, pp. 1-30; No. 2, Feb. 1962, pp. 277-334; and No. 3, Mar., pp. 352-396. doi: 10.14359/18461

Joint ACI-ASCE Committee 423, 1958, "Tentative Recommendations for Prestressed Concrete," *ACI Journal Proceedings*, V. 54, No. 7, Jan., pp. 545-578. doi: 10.14359/11455

Joint ACI-ASCE Committee 423, 1974, "Tentative Recommendations for Prestressed Concrete Flat Plates," *ACI Journal Proceedings*, V. 71, No. 2, Feb., pp. 67-71. doi: 10.14359/7344

Joint ACI-ASCE Committee 426, 1973, "Shear Strength of Reinforced Concrete Members (ACI 426R-74)," *Proceedings*, ASCE, V. 99, No. ST6, June, pp. 1148-1157.

Joint ACI-ASCE Committee 426, 1974, "The Shear Strength of Reinforced Concrete Members—Slabs," *Journal of the Structural Division*, V. 100, Aug., pp. 1543-1591.

Joint PCI/WRI Ad Hoc Committee on Welded Wire Fabric for Shear Reinforcement, 1980, "Welded Wire Fabric for Shear Reinforcement," *PCI Journal*, V. 25, No. 4, July-Aug., pp. 32-36. doi: 10.14359/16661

Kaar, P. H., 1966, "High Strength Bars as Concrete Reinforcement, Part 8: Similitude in Flexural Cracking of T-Beam Flanges," *PCA Research and Development Laboratories Bulletin*, V. 8, No. 2, May, pp. 2-12. doi: 10.14359/51685141

Kaar, P. H.; Hanson, N. W.; and Capell, H. T., 1978, "Stress-Strain Characteristics of High Strength Concrete," *Douglas McHenry International Symposium on Concrete and Concrete Structures*, SP-55, American Concrete Institute, Farmington Hills, MI, pp. 161-185. doi: 10.14359/6613

Kaar, P. H.; Kriz, L. B.; and Hognestad, E., 1960, "Precast-Prestressed Concrete Bridges: (1) Pilot Tests of Continuous Girders," *PCA Research and Development Laboratories Bulletin*, V. 2, No. 2, May, pp. 21-37. doi: 10.14359/51685340

Kaar, P. H.; La Fraugh, R. W.; and Mass, M. A., 1963, "Influence of Concrete Strength on Strand Transfer Length," *PCI Journal*, V. 8, No. 5, Oct., pp. 47-67. doi: 10.15554/pcij.10011963.47.67

Kaar, P., and Magura, D., 1965, "Effect of Strand Blanketing on Performance of Pretensioned Girders," *PCI Journal*, V. 10, No. 6, Dec., pp. 20-34. doi: 10.15554/pcij.12011965.20.34

Kahn, L. F., and Mitchell, A. D., 2002, "Shear Friction Tests with High-Strength Concrete," *ACI Structural Journal*, V. 99, No. 1, Jan.-Feb., pp. 98-103. doi: 10.14359/11040

Kang, T. H.-K.; LaFave, J. M.; Robertson, I. N.; and Hawkins, N. M., 2007, "Post-Tensioned Slab-Column Connections—Drift Capacity at Punching of Connections Subjected to Lateral Loading," *Concrete International*, V. 29, No. 4, Apr., pp. 61-68. doi: 10.14359/18569

Kang, T. H.-K.; Shin, M.; Mitra, N.; and Bonacci, J. F., 2009, "Seismic Design of Reinforced Concrete Beam-Column Joints with Headed Bars," *ACI Structural Journal*, V. 106, No. 6, Nov.-Dec., pp. 868-877. doi: 10.14359/51663188

Kang, T. H.-K., and Wallace, J. W., 2006, "Punching of Reinforced and Post-Tensioned Concrete Slab-Column Connections," *ACI Structural Journal*, V. 103, No. 4, July-Aug., pp. 531-540. doi: 10.14359/16429

Kemp, E. L.; Brezny, F. S.; and Unterspan, J. A., 1968, "Effect of Rust and Scale on the Bond Characteristics of Deformed Reinforcing Bars," *ACI Journal Proceedings*, V. 65, No. 9, Sept., pp. 743-756. doi: 10.14359/7509

Khalifa, J. U., and Collins, M. P., 1981, "Circular Reinforced Concrete Members Subjected to Shear," *Publications* No. 81-08, Department of Civil Engineering, University of Toronto, Toronto, ON, Canada, Dec., 117 pp. doi: 10.14359/16572

Khuntia, M., and Ghosh, S. K., 2004a, "Flexural Stiffness of Reinforced Concrete Columns and Beams: Analytical Approach," *ACI Structural Journal*, V. 101, No. 3, May-June, pp. 351-363. doi: 10.14359/13095

Khuntia, M., and Ghosh, S. K., 2004b, "Flexural Stiffness of Reinforced Concrete Columns and Beams: Experimental Verification," *ACI Structural Journal*, V. 101, No. 3, May-June, pp. 364-374. doi: 10.14359/13096

Kim, S., and Wallace, J., 2017, "Shear Design of Structural Walls for Tall Reinforced Concrete Core Wall Buildings," 16th World Conference on Earthquake Engineering, Santiago, Chile, 10 pp.

Klein, G. J., 1986, "Design of Spandrel Beams," *PCI Specially Funded Research Project* No. 5, Precast/Prestressed Concrete Institute, Chicago, IL, 104 pp. doi: 10.14359/51685218

Klein, G. J., 2008, "Curved-Bar Nodes: A Detailing Tool for Strut-and-Tie Models," *Concrete International*, V. 30, No. 9, Sept., pp. 42-47. doi: 10.14359/19949

Klein, G. J., and Lindenberg, R. J., 2009, "Volume Change Response of Precast Concrete Buildings," *PCI Journal*, V. 54, No. 4, Fall, pp. 112-131. doi: 10.15554/pcij.09012009.112.131

Klemencic, R.; McFarlane, I. S.; Hawkins, N. M.; and Nikolau, S., 2012, "Seismic Design of Reinforcing Concrete Mat Foundations: A Guide for Practicing Engineers," *NEHRP Seismic Design Technical Brief No. 7*, National

Institute of Standards and Technology, Gaithersburg, MD, pp. 11-12.

Klingner, R.; Mendonca, J.; and Malik, J., 1982, "Effect of Reinforcing Details on the Shear Resistance of Anchor Bolts under Reversed Cyclic Loading," *ACI Journal Proceedings*, V. 79, No. 1, Jan.-Feb., pp. 3-12. doi: 10.14359/10455

Kosmatka, S. H., and Wilson, M. L., 2016, *Design and Control of Concrete Mixtures*, EB001, 16th edition, Portland Cement Association, Skokie, IL, 632 pp.

Kramrisch, F., and Rogers, P., 1961, "Simplified Design of Combined Footings," *Journal of the Soil Mechanics and Foundations Division*, V. 87, Oct., pp. 19-44.

Kriz, L. B., and Raths, C. H., 1965, "Connections in Precast Concrete Structures—Strength of Corbels," *PCI Journal*, V. 10, No. 1, Feb., pp. 47-61. doi: 10.15554/pcij.02011965.16.61

Kuchma, D.; Wei, S.; Sanders, D.; Belarbi, A.; and Novak, L., 2019, "The Development of the One-Way Shear Design Provisions of ACI 318-19," *ACI Structural Journal*, V. 116 No. 4, July, doi: 10.14359/51716739

Kuhn, D., and Shaikh, F., 1996, "Slip-Pullout Strength of Hooked Anchors," *Research Report*, University of Wisconsin-Milwaukee, submitted to the National Codes and Standards Council, 55 pp. doi: 10.14359/51685520

Kurose, Y.; Nagami, K.; and Saito, Y., 1991, "Beam-Column Joints in Precast Concrete Construction in Japan," *Design of Beam-Column Joints for Seismic Resistance*, SP-123, J. O. Jirsa, ed., American Concrete Institute, Farmington Hills, MI, pp. 493-514. doi: 10.14359/2907

Kwon, K., and Ghannoum, W. M., 2016, "Assessment of International Standard Provisions on Stiffness of Reinforced Concrete Moment Frame and Shear Wall Buildings," *Engineering Structures*, V. 128, pp. 149-160. doi: 10.1016/j.engstruct.2016.09.025

LaGue, D. J., 1971, "Load Distribution Tests on Precast Prestressed Hollow-Core Slab Construction," *PCI Journal*, V. 16, No. 6, Nov.-Dec., pp. 10-18. doi: 10.15554/pcij.11011971.10.18

Lai, S. M. A., and MacGregor, J. G., 1983, "Geometric Nonlinearities in Unbraced Multistory Frames," *Journal of Structural Engineering*, V. 109, No. 11, Nov., pp. 2528-2545. doi: 10.1061/(ASCE)0733-9445(1983)109:11(2528)

LATBSDC, 2017, "An Alternative Procedure For Seismic Analysis and Design of Tall Buildings Located in the Los Angeles Region," Los Angeles Tall Buildings Structural Design Council, 72 pp.

Laughery, L., and Pujol, S., 2015, "Compressive Strength of Unreinforced Struts," *ACI Structural Journal*, V. 112, No. 5, pp. 617-624. doi: 10.14359/51687711

Lee, N. H.; Kim, K. S.; Bang, C. J.; and Park, K. R., 2007, "Tensile-Headed Anchors with Large Diameter and Deep Embedment in Concrete," *ACI Structural Journal*, V. 104, No. 4, July-Aug., pp. 479-486. doi: 10.14359/18778

Lee, N. H.; Park, K. R.; and Suh, Y. P., 2010, "Shear Behavior of Headed Anchors with Large Diameters and Deep Embedments," *ACI Structural Journal*, V. 107, No. 2, Mar.-Apr., pp. 146-156. doi: 10.14359/51663530

Lenz, K. A., 1992, "Concrete Materials Investigation for Gardiner Dam: Final Report," Canada Department of Agriculture, Prairie Farm Rehabilitation Administration Development Service, Geotechnical Division, Saskatoon, SK, Canada, Mar., 56 pp.

Leon, R. T., 1989, "Interior Joints with Variable Anchorage Lengths," *Journal of Structural Engineering*, V. 115, No. 9, Sept., pp. 2261-2275. doi: 10.1061/(ASCE)0733-9445(1989)115:9(2261)

Leonhardt, F., and Walther, R., 1964, "The Stuttgart Shear Tests," *C&CA Translation*, No. 111, Cement and Concrete Association, London, UK, 134 pp. doi: 10.14359/19344

Lepage, A., 1998, "Nonlinear Drift of Multistory RC Structures during Earthquakes," Sixth National Conference on Earthquake Engineering, Seattle, WA.

Leslie, K. E.; Rajagopalan, K. S.; and Everard, N. J., 1976, "Flexural Behavior of High-Strength Concrete Beams," *ACI Journal Proceedings*, V. 73, No. 9, Sept., pp. 517-521. doi: 10.14359/11093

Li, S., and Roy, D. M., 1986, "Investigation of Relations between Porosity, Pore Structure and CL Diffusion of Fly Ash and Blended Cement Pastes," *Cement and Concrete Research*, V. 16, No. 5, Sept., pp. 749-759. doi: 10.1016/0008-8846(86)90049-9

Lin, C. M.; Restrepo, J. I.; and Park, R., 2000, "Seismic Behaviour and Design of Reinforced Concrete Interior Beam Column Joints," *Research Report 2000-1*, Department of Civil Engineering, University of Canterbury, Christchurch, New Zealand, 471 pp. http://hdl.handle.net/10092/15092.

Lin, T. Y., and Thornton, K., 1972, "Secondary Moment and Moment Redistribution in Continuous Prestressed Beams," *PCI Journal*, V. 17, No. 1, Jan.-Feb., pp. 8-20 and comments by A. H. Mattock and author's closure, *PCI Journal*, V. 17, No. 4, July-Aug., pp. 86-88. doi: 10.14359/19321

Lloyd, J. P., 1971, "Splice Requirements for One-Way Slabs Reinforced with Smooth Welded Wire Fabric," *Publication* No. R(S)4, Civil Engineering, Oklahoma State University, Stillwater, OK, June, 37 pp. doi: 10.14359/19630

Lloyd, J. P., and Kesler, C. E., 1969, "Behavior of One-Way Slabs Reinforced with Deformed Wire and Deformed Wire Fabric," *T&AM Report* No. 323, University of Illinois, Urbana, IL, 129 pp. doi: 10.14359/16663

Logan, D. R., 1997, "Acceptance Criteria for Bond Quality of Strand for Pretensioned Prestressed Concrete Applications," *PCI Journal*, V. 42, No. 2, Mar.-Apr., pp. 52-90. doi: 10.15554/pcij.03011997.52.90

Lotze, D.; Klingner, R. E.; and Graves III, H. L., 2001, "Static Behavior of Anchors under Combinations of Tension and Shear Loading," *ACI Structural Journal*, V. 98, No. 4, July-Aug., pp. 525-536. doi: 10.14359/10295

Lu, Y.; Henry, R. S.; Gultom, R.; and Ma, Q. T., 2017, "Cyclic Testing of Reinforced Concrete Walls with Distributed Minimum Vertical Reinforcement," *Journal of Structural Engineering*, V. 143, No. 5, p. 04016225 doi: 10.1061/(ASCE)ST.1943-541X.0001723

Lubell, A. S.; Bentz, E. C.; and Collins, M. P., 2009, "Shear Reinforcement Spacing in Wide Members," *ACI*

Structural Journal, V. 106, No. 2, Mar.-Apr., pp. 205-214. doi: 10.14359/56359

Lubell, A. S.; Sherwood, E. G.; Bentz, E. C.; and Collins, M. P., 2004, "Safe Shear Design of Large Wide Beams," *Concrete International*, V. 26, No. 1, Jan., pp. 66-78. doi: 10.14359/19136

Lucier, G.; Walter, C.; Rizkalla, S.; Zia, P.; and Klein, G., 2011a, "Development of a Rational Design Methodology for Precast Concrete Slender Spandrel Beams, Part 1: Experimental Results," *PCI Journal*, V. 56, No. 2, Spring, pp. 88-112. doi: 10.15554/pcij.03012011.88.112

Lucier, G.; Walter, C.; Rizkalla, S.; Zia, P.; and Klein, G., 2011b, "Development of a Rational Design Methodology for Precast Concrete Slender Spandrel Beams, Part 2: Analysis and Design Guidelines," *PCI Journal*, V. 56, No. 4, Fall, pp. 106-133. doi: 10.15554/pcij.09012011.106.133

Lutz, L., 1995, discussion to "Concrete Capacity Design (CCD) Approach for Fastening to Concrete," *ACI Structural Journal*, Nov.-Dec., pp. 791-792. Also, authors' closure, pp. 798-799. doi: 10.14359/5165518

MacGregor, J. G., 1976, "Safety and Limit States Design for Reinforced Concrete," *Canadian Journal of Civil Engineering*, V. 3, No. 4, Dec., pp. 484-513. doi: 10.1139/l76-055

MacGregor, J. G., 1993, "Design of Slender Concrete Columns—Revisited," *ACI Structural Journal*, V. 90, No. 3, May-June, pp. 302-309. doi: 10.14359/4237

MacGregor, J. G., 1997, *Reinforced Concrete: Mechanics and Design*, third edition, Prentice Hall, Englewood Cliffs, NJ, 939 pp.

MacGregor, J. G., and Ghoneim, M. G., 1995, "Design for Torsion," *ACI Structural Journal*, V. 92, No. 2, Mar.-Apr., pp. 211-218. doi: 10.14359/1152

MacGregor, J. G., and Hage, S. E., 1977, "Stability Analysis and Design of Concrete Frames," *Journal of the Structural Division*, V. 103, Oct., pp. 1953-1970. doi: 10.14359/51685164

MacGregor, J. G., and Hanson, J. M., 1969, "Proposed Changes in Shear Provisions for Reinforced and Prestressed Concrete Beams," *ACI Journal Proceedings*, V. 66, No. 4, Apr., pp. 276-288. doi: 10.14359/7360

MacGregor, J. G.; Breen, J. E.; and Pfrang, E. O., 1970, "Design of Slender Concrete Columns," *ACI Journal Proceedings*, V. 67, No. 1, Jan., pp. 6-28. doi: 10.14359/7254

Malhotra, V. M., 1976, *Testing Hardened Concrete: Nondestructive Methods*, ACI Monograph No. 9, American Concrete Institute/Iowa State University Press, Farmington Hills, MI, 188 pp.

Malhotra, V. M., 1977, "Contract Strength Requirements—Cores Versus In Situ Evaluation," *ACI Journal Proceedings*, V. 74, No. 4, Apr., pp. 163-172. doi: 10.14359/10998

Mansour, M., and Hsu, T., 2005, "Behavior of Reinforced Concrete Elements under Cyclic Shear. II: Theoretical Model," *Journal of Structural Engineering*, V. 131, No. 11, pp. 54-65. doi: 10.1061/(ASCE)0733-9445(2005)131:1(54)

Marti, P., 1985, "Basic Tools of Reinforced Concrete Beam Design," *ACI Journal Proceedings*, V. 82, No. 1, Jan.-Feb., pp. 46-56. doi: 10.14359/10314

Martin, L., and Korkosz, W., 1995, "Strength of Prestressed Concrete Members at Sections Where Strands Are Not Fully Developed," *PCI Journal*, V. 40, No. 5, Sept.-Oct., pp. 58-66. doi: 10.15554/pcij.09011995.58.66

Mast, R. F., 1968, "Auxiliary Reinforcement in Concrete Connections," *Journal of the Structural Division*, V. 94, June, pp. 1485-1504. doi: 10.14359/16710

Mast, R. F., 1992, "Unified Design Provision for Reinforced and Prestressed Concrete Flexural and Compression Members," *ACI Structural Journal*, V. 89, No. 2, Mar.-Apr., pp. 185-199. doi: 10.14359/3209

Mast, R. F., 1998, "Analysis of Cracked Prestressed Concrete Sections: A Practical Approach," *PCI Journal*, V. 43, No. 4, July-Aug., pp. 80-91. doi: 10.15554/pcij.07011998.80.91

Mast, R. F., 2001, "Vibration of Precast Prestressed Concrete Floors," *PCI Journal*, V. 46, No. 6, pp. 76-86. doi: 10.15554/pcij.11012001.76.86

Mathey, R. G., and Clifton, J. R., 1976, "Bond of Coated Reinforcing Bars in Concrete," *Journal of the Structural Division*, V. 102, Jan., pp. 215-228. doi: 10.14359/16871

Mattock, A. H., 1959, "Redistribution of Design Bending Moments in Reinforced Concrete Continuous Beams," *Proceedings—Institution of Civil Engineers*, V. 13, No. 1, pp. 35-46. doi: 10.1680/iicep.1959.12087

Mattock, A. H., 1974, "Shear Transfer in Concrete Having Reinforcement at an Angle to the Shear Plane," *Shear in Reinforced Concrete*, SP-42, American Concrete Institute, Farmington Hills, MI, pp. 17-42. doi: 10.14359/18149

Mattock, A. H., 1977, discussion of "Considerations for the Design of Precast Concrete Bearing Wall Buildings to Withstand Abnormal Loads," by PCI Committee on Precast Concrete Bearing Wall Buildings, *PCI Journal*, V. 22, No. 3, May-June, pp. 105-106. doi: 10.14359/51685223

Mattock, A. H., 2001, "Shear Friction and High-Strength Concrete," *ACI Structural Journal*, V. 98, No. 1, Jan.-Feb., pp. 50-59. doi: 10.14359/10146

Mattock, A. H., and Hawkins, N. M., 1972, "Shear Transfer in Reinforced Concrete—Recent Research," *PCI Journal*, V. 17, No. 2, Mar.-Apr., pp. 55-75. doi: 10.15554/pcij.03011972.55.75

Mattock, A. H., and Shen, J. F., 1992, "Joints between Reinforced Concrete Members of Similar Depth," *ACI Structural Journal*, V. 89, No. 3, pp. 290-295. doi: 10.14359/3235

Mattock, A. H.; Chen, K. C.; and Soongswang, K., 1976a, "The Behavior of Reinforced Concrete Corbels," *PCI Journal*, V. 21, No. 2, Mar.-Apr., pp. 52-77. doi: 10.15554/pcij.03011976.52.77

Mattock, A. H.; Johal, L.; and Chow, H. C., 1975, "Shear Transfer in Reinforced Concrete with Moment or Tension Acting Across the Shear Plane," *PCI Journal*, V. 20, No. 4, July-Aug., pp. 76-93. doi: 10.15554/pcij.07011975.76.93

Mattock, A. H.; Kriz, L. B.; and Hognestad, E., 1961, "Rectangular Concrete Stress Distribution in Ultimate Strength Design," *ACI Journal Proceedings*, V. 57, No. 8, Feb., pp. 875-928. doi: 10.14359/8051

Mattock, A. H.; Li, W. K.; and Wang, T. C., 1976b, "Shear Transfer in Lightweight Reinforced Concrete," *PCI*

Journal, V. 21, No. 1, Jan.-Feb., pp. 20-39. doi: 10.15554/pcij.01011976.20.39

Mattock, A. H.; Yamazaki, J.; and Kattula, B. T., 1971, "Comparative Study of Prestressed Concrete Beams, with and without Bond," *ACI Journal Proceedings*, V. 68, No. 2, Feb., pp. 116-125. doi: 10.14359/11298

Megally, S., and Ghali, A., 2002, "Punching Shear Design of Earthquake-Resistant Slab-Column Connections," *ACI Structural Journal*, V. 97, No. 5, Sept.-Oct., pp. 720-730. doi: 10.14359/8807

Meinheit, D. F., and Jirsa, J. O., 1977, "Shear Strength of Reinforced Concrete Beam-Column Joints," *Report* No. 77-1, Department of Civil Engineering, Structures Research Laboratory, University of Texas at Austin, Austin, TX, Jan., 291 pp. doi: 10.14359/51685433

Meinheit, D. F., and Jirsa, J. O., 1981, "Shear Strength of R/C Beam-Column Connections," *Journal of the Structural Division*, V. 107, Nov., pp. 2227-2244.

Menn, C., 1986, *Prestressed Concrete Bridges*, B. Birkhäuser, ed., 535 pp.

Mirza, S. A., 1990, "Flexural Stiffness of Rectangular Reinforced Concrete Columns," *ACI Structural Journal*, V. 87, No. 4, July-Aug., pp. 425-435. doi: 10.14359/3056

Mirza, S. A.; Lee, P. M.; and Morgan, D. L., 1987, "ACI Stability Resistance Factor for RC Columns," *Journal of Structural Engineering*, V. 113, No. 9, Sept., pp. 1963-1976. doi: 10.1061/(ASCE)0733-9445(1987)113:9(1963)

Moehle, J. P., 1996, "Seismic Design Considerations for Flat Plate Construction," *Mete A. Sozen Symposium: A Tribute from his Students*, SP-162, J. K. Wight and M. E. Kreger, eds., American Concrete Institute, Farmington Hills, MI, pp. 1-35. doi: 10.14359/1509

Moehle, J. P., 1988, "Strength of Slab-Column Edge Connections," *ACI Structural Journal*, V. 85, No. 1, Jan.-Feb., pp. 89-98. doi: 10.14359/2995

Moehle, J. P., 1992, "Displacement-Based Design of RC Structures Subjected to Earthquakes," *Earthquake Spectra*, V. 8, No. 3, Aug., pp. 403-428. doi: 10.1193/1.1585688

Moehle, J. P., and Cavanagh, T., 1985, "Confinement Effectiveness of Crossties in RC," *Journal of Structural Engineering*, V. 111, No. 10, pp. 2105-2120. doi: 10.1061/(ASCE)0733-9445(1985)111:10(2105)

Moehle, J. P.; Ghodsi, T.; Hooper, J. D.; Fields, D. C.; and Gedhada, R., 2011, "Seismic Design of Cast-in-Place Concrete Special Structural Walls and Coupling Beams: A Guide for Practicing Engineers," *NEHRP Seismic Design Technical Brief No. 6*, National Institute of Standards and Technology, Gaithersburg, MD, 41 pp.

Moehle, J. P.; Hooper, J. D.; Kelly, D. J.; and Meyer, T. R., 2010, "Seismic Design of Cast-in-Place Concrete Diaphragms, Chords, and Collectors: A Guide for Practicing Engineers," *NEHRP Seismic Design Technical Brief No. 3*, National Institute of Standards and Technology, Gaithersburg, MD, 33 pp.

Mitchell, D., and Collins, M. P., 1976, "Detailing for Torsion," *ACI Journal Proceedings*, V. 73, No. 9, Sept., pp. 506-511. doi: 10.14359/11091

Mitchell, D., and Cook, W. D., 1984, "Preventing Progressive Collapse of Slab Structures," *Journal of Structural Engineering*, V. 110, No. 7, July, pp. 1513-1532. doi: 10.1061/(ASCE)0733-9445(1984)110:7(1513)

Mojtahedi, S., and Gamble, W. L., 1978, "Ultimate Steel Stresses in Unbonded Prestressed Concrete," *Journal of the Structural Division*, V. 104, No. ST7, July, pp. 1159-1165. doi: 10.14359/19419

Mphonde, A. G., and Frantz, G. C., 1984, "Shear Tests of High- and Low-Strength Concrete Beams without Stirrups," *ACI Journal Proceedings*, V. 81, No. 4, July-Aug., pp. 350-357. doi: 10.14359/10690

Muguruma, H., and Watanabe, F., 1990, "Ductility Improvement of High-Strength Concrete Columns with Lateral Confinement," *Proceedings, Second International Symposium on High-Strength Concrete*, SP-121, American Concrete Institute, Farmington Hills, MI, pp. 47-60. doi: 10.14359/2783

Muttoni, A., 2008, "Punching Shear Strength of Reinforced Concrete Slabs without Shear Reinforcement," *ACI Structural Journal*, V. 105, No. 4, July-Aug., pp. 440-450. doi: 10.14359/19858

Muttoni, A.; Schwartz, J.; and Thürlimann, B., 1997, *Design of Concrete Structures with Stress Fields*, Birkhauser, Boston, MA, 143 pp.

Nakaki, S. D.; Stanton, J. F.; and Sritharan, S., 1999, "An Overview of the PRESSS Five-Story Precast Test Building," *PCI Journal*, V. 44, No. 2, Apr., pp. 26-39. doi: 10.15554/pcij.03011999.26.39

National Transportation Safety Board (NTSB), 2007, "Ceiling Collapse in the Interstate 90 Connector Tunnel, Boston, Massachusetts, July 10, 2006," *Highway Accident Report* NTSB/HAR-07/02, Washington, DC, 120 pp.

NEHRP, 1994, "The NEHRP Recommended Provisions for Seismic Regulations for New Buildings," Building Seismic Safety Council, Washington, DC.

Newlon, H. J., and Ozol, A., 1969, "Delayed Expansion of Concrete Delivered by Pumping through Aluminum Pipe Line," Concrete Case Study No. 20; Virginia Highway Research Council, Charlottesville, VA, Oct., 39 pp. doi: 10.14359/16493

Nichols, J. R., 1914, "Statical Limitations upon the Steel Requirement in Reinforced Concrete Flat Slab Floors," *Transactions of the American Society of Civil Engineers*, V. 77, pp. 1670-1736.

Nilsson, I. H. E., and Losberg, A., 1976, "Reinforced Concrete Corners and Joints Subjected to Bending Moment," *Journal of the Structural Division*, V. 102, June, pp. 1229-1254. doi: 10.14359/51685465

NIST, 2014, "Use of High-Strength Reinforcement in Earthquake-Resistant Concrete Structures," NIST GCR 14-917-30, National Institute of Standards and Technology, Gaithersburg, MD.

NIST, 2017a, "Guidelines for Nonlinear Structural Analysis for Design of Buildings, Part I—General," NIST GCR 17-917-46v1, prepared by the Applied Technology Council for the National Institute of Standards and Technology, Gaithersburg, MD.

NIST, 2017b, "Guidelines for Nonlinear Structural Analysis for Design of Buildings, Part IIb—Concrete Moment Frames," NIST GCR 17-917-46v3, prepared by the Applied Technology Council for the National Institute of Standards and Technology, Gaithersburg, MD.

Odello, R. J., and Mehta, B. M., 1967, "Behavior of a Continuous Prestressed Concrete Slab with Drop Panels," *Graduate Student Research Report*, Division of Structural Engineering and Structural Mechanics, University of California, Berkeley, Berkeley, CA, 63 pp. doi: 10.14359/19581

Oesterle, R. G., 1997, "The Role of Concrete Cover in Crack Control Criteria and Corrosion Protection," PCA R&D Serial No. 2054, Portland Cement Association, Skokie, IL, 87 pp. doi: 10.14359/16540

Olesen, S. E.; Sozen, M. A.; and Siess, C. P., 1967, "Investigation of Prestressed Reinforced Concrete for Highway Bridges, Part IV: Strength in Shear of Beams with Web Reinforcement," *Bulletin* No. 493, Engineering Experiment Station, University of Illinois, Urbana, IL., 115pp. doi: 10.14359/16580

Olsen, J.; Pregartner, T.; and Lamanna, A. J., 2012, "Basis for Design of Screw Anchors in Concrete," *ACI Structural Journal*, V. 109, No. 4, July-Aug., pp. 559-568. doi: 10.14359/51683875

Orakcal, K.; Massone, L. M.; and Wallace, J. W., 2009, "Shear Strength of Lightly Reinforced Wall Piers and Spandrels," *ACI Structural Journal*, V. 106, No. 4, July-Aug., pp. 455-465. doi: 10.14359/56611

Orangun, C. O.; Jirsa, J. O.; and Breen, J. E., 1977, "A Reevaluation of Test Data on Development Length and Splices," *ACI Journal Proceedings*, V. 74, No. 3, Mar., pp. 114-122. doi: 10.14359/10993

Ospina, C. E., and Alexander, S. D. B., 1998, "Transmission of Interior Concrete Column Loads through Floors," *Journal of Structural Engineering*, V. 124, No. 6, June, pp. 602-610. doi: 10.1061/(ASCE)0733-9445(1998)124:6(602)

Ožbolt, J.; Eligehausen, R.; Periškić, G.; and Mayer, U., 2007, "3D FE Analysis of Anchor Bolts with Large Embedment Depths," *Engineering Fracture Mechanics*, V. 74, No. 1-2, Jan., pp. 168-178. doi: 10.1016/j.engfracmech.2006.01.019

Ozcebe, G.; Ersoy, U.; and Tankut, T., 1999, "Evaluation of Minimum Shear Reinforcement for Higher Strength Concrete," *ACI Structural Journal*, V. 96, No. 3, May-June, pp. 361-368. doi: 10.14359/51701132

Ozyildirim, C., and Halstead, W., 1988, "Resistance to Chloride Ion Penetration of Concretes Containing Fly Ash, Silica Fume, or Slag," *Permeability of Concrete*, SP-108, American Concrete Institute, Farmington Hills, MI, pp. 35-61. doi: 10.14359/2158

Pacific Earthquake Engineering Center, 2017, "Guidelines for Performance-Based Seismic Design of Tall Buildings," *Report* No. 2017/06, University of California, Berkeley, Berkeley, CA.

Palmieri, L.; Saqan, E.; French, C.; and Kreger, M., 1996, "Ductile Connections for Precast Concrete Frame Systems," *Mete A. Sozen Symposium: A Tribute from his Students*, SP-162, American Concrete Institute, Farmington Hills, MI, pp. 315-335. doi: 10.14359/1515

Pan, A., and Moehle, J. P., 1989, "Lateral Displacement Ductility of Reinforced Concrete Flat Plates," *ACI Structural Journal*, V. 86, No. 3, May-June, pp. 250-258. doi: 10.14359/2889

Park, R., and Paulay, T., 1975, *Reinforced Concrete Structures,* Wiley-Interscience, New York, 769 pp.

Park, R., and Thompson, K. J., 1977, "Cyclic Load Tests on Prestressed and Partially Prestressed Beam-Column Joints," *PCI Journal*, V. 22, No. 5, pp. 84-110. doi: 10.15554/pcij.09011977.84.110

Parra-Montesinos, G. J., 2006, "Shear Strength of Beams with Deformed Steel Fibers," *Concrete International*, V. 28, No. 11, Nov., pp. 57-66. doi: 10.14359/18323

Paulay, T., and Binney, J. R., 1974, "Diagonally Reinforced Coupling Beams of Shear Walls," *Shear in Reinforced Concrete*, SP-42, American Concrete Institute, Farmington Hills, MI, pp. 579-598. doi: 10.14359/17302

Paulay, T., and Priestley, M. J. N., 1992, *Seismic Design of Reinforced Concrete and Masonry Buildings*, John Wiley and Sons, New York, 768 pp.

Paulson, C.; Graham, S. K.; and Rautenberg, J. M., 2013, "Determination of Yield Strength for Nonprestressed Steel Reinforcement," Charles Pankow Foundation RGA #04-13, WJE No. 2013.4171, Wiss, Janney, Elstner Associates, Inc., Pasadena, CA, Dec. 31, 100 pp. doi: 10.14359/51702081

Paultre, P., and Légeron, F., 2008, "Confinement Reinforcement Design for Reinforced Concrete Columns," *Journal of Structural Engineering*, V. 134, No. 5, pp. 738-749. doi: 10.1061/(ASCE)0733-9445(2008)134:5(738)

Pauw, A., 1960, "Static Modulus of Elasticity of Concrete as Affected by Density," *ACI Journal Proceedings*, V. 57, No. 6, Dec., pp. 679-687. doi: 10.14359/17302

PCI, 1993, "Recommended Practice for Design, Manufacture, and Installation of Prestressed Concrete Piling," *PCI Journal*, V. 38, No. 2, Mar.-Apr., pp. 14-41. doi: 10.14359/19099

PCI Committee on Precast Concrete Bearing Wall Buildings, 1976, "Considerations for the Design of Precast Concrete Bearing Wall Buildings to Withstand Abnormal Loads," *PCI Journal*, V. 21, No. 2, Mar.-Apr., pp. 18-51. doi: 10.14359/16702

Peiris, C., and Ghali, A., 2012, "Flexural Reinforcement Essential for Punching Shear Resistance of Slabs," *Recent Developments in Reinforced Concrete Slab Analysis, Design and Serviceability*, SP-287, American Concrete Institute, Farmington Hills, MI, May, pp. 1-16. doi: 10.14359/51683860

Perez, F. J.; Pessiki, S.; Sause, R.; and Lu, L.-W., 2003, "Lateral Load Tests of Unbonded Post-Tensioned Precast Concrete Walls," *Large Scale Structural Testing*, SP-211, American Concrete Institute, Farmington Hills, MI, pp. 161-182. doi: 10.14359/12589

Pessiki, S.; Graybeal, B.; and Mudlock, M., 2001, "Proposed Design of High-Strength Spiral Reinforcement in Compression Members," *ACI Structural Journal*, V. 98, No. 6, Nov.-Dec., pp. 799-810. doi: 10.14359/10747

Pfeifer, D. W., and Nelson, T. A., 1983, "Tests to Determine the Lateral Distribution of Vertical Loads in a Long-Span Hollow-Core Floor Assembly," *PCI Journal*, V. 28, No. 6, Nov.-Dec., pp. 42-57. doi: 10.15554/pcij.11011983.42.57

Pfister, J. F., 1964, "Influence of Ties on the Behavior of Reinforced Concrete Columns," *ACI Journal Proceedings*, V. 61, No. 5, May, pp. 521-537. doi: 10.14359/7795

Pfister, J. F., and Mattock, A. H., 1963, "High Strength Bars as Concrete Reinforcement, Part 5: Lapped Splices in Concentrically Loaded Columns," *PCA Research and Development Laboratories Bulletin*, V. 5, No. 2, May, pp. 27-40. doi: 10.14359/51685279

Popov, E. P.; Bertero, V. V.; and Krawinkler, H., 1972, "Cyclic Behavior of Three R/C Flexural Members with High Shear," EERC *Report* No. 72-5, Earthquake Engineering Research Center, University of California, Berkeley, Berkeley, CA, Oct., 91 pp. doi: 10.14359/19326

Portland Cement Association (PCA), 1946, "Design of Deep Girders," IS079D, Portland Cement Association, Skokie, IL, 10 pp. doi: 10.14359/51685151

Portland Cement Association (PCA), 1972, *Handbook of Frame Constants*, Skokie, IL, 1972, 34 pp. doi: 10.14359/16511

Portland Cement Association (PCA), 1980, "Design and Construction of Large-Panel Concrete Structures," six reports, 762 pp., EB 100D; three studies, 300 pp., 1980, EB 102D, 1976-1980, Portland Cement Association, Skokie, IL.

Portland Cement Association (PCA), 2011, *Design and Control of Concrete Mixtures*, 15th edition, Portland Cement Association, Skokie, IL, 444 pp.

Primavera, E. J.; Pinelli, J.-P.; and Kalajian, E. H., 1997, "Tensile Behavior of Cast-in-Place and Undercut Anchors in High-Strength Concrete," *ACI Structural Journal*, V. 94, No. 5, Sept.-Oct., pp. 583-594. doi: 10.14359/507

Priestley, M. J. N.; Sritharan, S.; Conley, J.; and Pampanin, S., 1999, "Preliminary Results and Conclusions from the PRESSS Five-Story Precast Concrete Test Building," *PCI Journal*, V. 44, No. 6, Nov.-Dec., pp. 42-67. doi: 10.15554/pcij.11011999.42.67

Powers, T. C., 1975, "Freezing Effects in Concrete," *Durability of Concrete*, SP-47, American Concrete Institute, Farmington Hills, MI, pp. 1-11. doi: 10.14359/17603

Puranam, A., 2018, "Strength and Serviceability of Concrete Elements Reinforced with High-Strength Steel," PhD dissertation, Purdue University, West Lafayette, IN.

Rabbat, B. G.; Kaar, P. H.; Russell, H. G.; and Bruce Jr., R. N., 1979, "Fatigue Tests of Pretensioned Girders with Blanketed and Draped Strands," *PCI Journal*, V. 24, No. 4, July-Aug., pp. 88-114. doi: 10.15554/pcij.07011979.88.114

Reineck, K.-H., ed., 2002, *Strut-and-Tie Models*, SP-208, American Concrete Institute, Farmington Hills, MI, 250 pp.

Reineck, K.-H., and Novak, L., C., eds., 2010, *Further Examples for the Design of Structural Concrete with Strut-and-Tie Models*, SP-273, American Concrete Institute, Farmington Hills, MI, 288 pp.

Reineck, K., and Todisco, L., 2014, "Database of Shear Tests for Non-slender Reinforced Concrete Beams without Stirrups," *ACI Structural Journal*, V. 111, No. 6, pp. 1363-1372. doi: 10.14359/51686820

Restrepo, J. I., 2002, "New Generation of Earthquake Resisting Systems," *Proceedings*, First *fib* Congress, Session 6, Osaka, Japan, Oct., pp. 41-60.

Restrepo, J. I.; Park, R.; and Buchanan, A. H., 1995a, "Tests on Connections of Earthquake Resisting Precast Reinforced Concrete Perimeter Frames of Buildings," *PCI Journal*, V. 40, No. 4, July-Aug., pp. 44-61. doi: 10.15554/pcij.07011995.44.61

Restrepo, J.; Park, R.; and Buchanan, A. H., 1995b, "Design of Connections of Earthquake Resisting Precast Reinforced Concrete Perimeter Frames," *PCI Journal*, V. 40, No. 5, pp. 68-80. doi: 10.15554/pcij.09011995.68.80

Richart, F. E., 1933, "Reinforced Concrete Column Investigation—Tentative Final Report of Committee 105," *ACI Journal Proceedings*, V. 29, No. 2, Feb., pp. 275-282. doi: 10.14359/8273

Richart, F. E.; Brandzaeg, A.; and Brown, R. L., 1929, "The Failure of Plain and Spirally Reinforced Concrete in Compression," *Bulletin* No. 190, University of Illinois Engineering Experiment Station, Urbana, IL, Apr., 74 pp. doi: 10.14359/19264

Roberts-Wollmann, C. L., and Wollmann, G. P., 2008, "Bearing Strength of Unbonded Monostrand Post-Tensioning Anchors in Slab Edges," *PTI Journal*, V. 6, No. 2, Aug., pp. 29-39.

Rogowsky, D. M., and MacGregor, J. G., 1986, "Design of Reinforced Concrete Deep Beams," *Concrete International*, V. 8, No. 8, Aug., pp. 46-58. doi: 10.14359/51685228

Rogowsky, D., and Wight, K., 2010, "Load Factors are Load Factors," *Concrete International*, V. 32, No. 7, July, pp. 75-76.

Roller, J. J., and Russell, H. G., 1990, "Shear Strength of High-Strength Concrete Beams with Web Reinforcement," *ACI Structural Journal*, V. 87, No. 2, Mar.-Apr., pp. 191-198. doi: 10.14359/2682

Rose, D. R., and Russell, B. W., 1997, "Investigation of Standardized Tests to Measure the Bond Performance of Prestressing Strand," *PCI Journal*, V. 42, No. 4, July-Aug., pp. 56-80. doi: 10.15554/pcij.07011997.56.80

Rotz and Reifschneider, 1984, "Experimental Evaluation of Steel Embedments Subjected to Combined Axial and Shear," Bechtel Power Corp., Ann Arbor, MI, 100 pp.

Ruggiero, D. M.; Bentz, E. C.; Calvi, G. M.; and Collins, M. P., 2016, "Shear Response under Reversed Cyclic Loading," *ACI Structural Journal*, V. 113, No. 6, Nov.-Dec., pp. 1313-1324. doi: 10.14359/51689033

Russell, B. W., and Burns, N. H., 1996, "Measured Transfer Lengths of 0.5 and 0.6 in. Strands in Pretensioned Concrete," *PCI Journal*, V. 41, No. 5, Sept.-Oct., pp. 44-65. doi: 10.15554/pcij.09011996.44.65

Rutledge, S., and DeVries, R. A., 2002, "Development of D45 Wire in Concrete," *Report*, School of Civil and Environmental Engineering, Oklahoma State University, Stillwater, OK, Jan., 28 pp.

Saatcioglu, M., and Razvi, S. R., 2002, "Displacement-Based Design of Reinforced Concrete Columns for

Confinement," *ACI Structural Journal*, V. 99, No. 1, Jan.-Feb., pp. 3-11. doi: 10.14359/11030

Sabnis, G. M., Meyers; B. L.; and Roll, F., eds., 1974, *Deflections of Concrete Structures*, SP-43, American Concrete Institute, Farmington Hills, MI, 637 pp.

Saemann, J. C., and Washa, G. W., 1964, "Horizontal Shear Connections between Precast Beams and Cast-in-Place Slabs," *ACI Journal Proceedings*, V. 61, No. 11, Nov., pp. 1383-1409. doi: 10.14359/7832. Also see discussion," *ACI Journal Proceedings*, V. 62, June 1965, pp. 1807-1810.

Sakai, K., and Sheikh, S. A., 1989, "What Do We Know about Confinement in Reinforced Concrete Columns? (A Critical Review of Previous Work and Code Provisions)," *ACI Structural Journal*, V. 86, No. 2, Mar.-Apr., pp. 192-207. doi: 10.14359/2705

Salmons, J. R., and McCrate, T. E., 1977, "Bond Characteristics of Untensioned Prestressing Strand," *PCI Journal*, V. 22, No. 1, Jan.-Feb., pp. 52-65. doi: 10.15554/pcij.01011977.52.65

Sant, J. K., and Bletzacker, R. W., 1961, "Experimental Study of Lateral Stability of Reinforced Concrete Beams," *ACI Journal Proceedings*, V. 58, No. 6, Dec., pp. 713-736. doi: 10.14359/8004

Sason, A. S., 1992, "Evaluation of Degree of Rusting on Prestressed Concrete Strand," *PCI Journal*, V. 37, No. 3, May-June, pp. 25-30. doi: 10.15554/pcij.05011992.25.30

SBC, 1999, "Standard Building Code," Southern Building Code Congress International, Inc., Birmingham, AL.

Scanlon, A., and Bischoff, P. H., 2008, "Shrinkage Restraint and Loading History Effects on Deflections of Flexural Members," *ACI Structural Journal*, V. 105, No. 4, July-Aug., pp. 498-506. doi: 10.14359/19864

Schießl, P., and Raupach, M., 1997, "Laboratory Studies and Calculations on the Influence of Crack Widths on Chloride-Induced Corrosion of Steel in Concrete," *ACI Materials Journal*, V. 94, No. 1, Jan.-Feb., pp. 56-62. doi: 10.14359/285

Schlaich, J.; Schafer, K.; and Jennewein, M., 1987, "Toward a Consistent Design of Structural Concrete," *PCI Journal*, V. 32, No. 3, May-June, pp. 74-150. doi: 10.15554/pcij.05011987.74.150

Scordelis, A. C., 1990, "Non-Linear Material, Geometric, and Time Dependent Analysis of Reinforced and Prestressed Concrete Shells," *Bulletin*, International Association for Shells and Spatial Structures, Madrid, Spain, No. 102, Apr., pp. 57-90. doi: 10.14359/51685392

Scordelis, A. C.; Lin, T. Y.; and Itaya, R., 1959, "Behavior of a Continuous Slab Prestressed in Two Directions," *ACI Journal Proceedings*, V. 56, No. 6, Dec., pp. 441-459. doi: 10.14359/8105

Seegebrecht, G. W.; Litvan, A.; and And Gebler, S. H., 1989, "Durability of Dry-Mix Shotcrete," *Concrete International*, V. 11, No. 10, Oct., pp. 47-50. doi: 10.14359/2481

Segura, C. L., and Wallace, W. J., 2017, "Seismic Performance Limitations and Detailing of Slender RC Walls," *ACI Structural Journal*, V. 115, No. 3, May-June, pp. 849-859. doi: 10.14359/51701918

Segura, C. L., and Wallace, J. W., 2018, "Impact of Geometry and Detailing on Drift Capacity of Slender Walls," *ACI Structural Journal*, V. 115, No. 3, May, pp. 885-895. doi: 10.14359/51702046

Seismology Committee of the Structural Engineers Association of California (SEAOC), 1996, "Recommended Lateral Force Requirements and Commentary," sixth edition, Seismology Committee of the Structural Engineers Association of California, Sacramento, CA, 504 pp.

Shaikh, A. F., and Branson, D. E., 1970, "Non-Tensioned Steel in Prestressed Concrete Beams," *PCI Journal*, V. 15, No. 1, Feb., pp. 14-36. doi: 10.15554/pcij.02011970.14.36

Shaikh, A. F., and Yi, W., 1985, "In-Place Strength of Welded Headed Studs," *PCI Journal*, V. 30, No. 2, Mar.-Apr., pp. 56-81. doi: 10.15554/pcij.03011985.56.81

Shao, Y.; Darwin, D.; O'Reilly, M.; Lequesne, R. D.; Ghimire, K.; and Hano, M., 2016, "Anchorage of Conventional and High-Strength Headed Reinforcing Bars," *SM Report* No. 117, University of Kansas Center for Research, Lawrence, KS, Aug., 234 pp. https://kuscholarworks.ku.edu/handle/1808/21738

Sivakumar, B.; Gergely, P.; and White, R. N., 1983, "Suggestions for the Design of R/C Lapped Splices for Seismic Loading," *Concrete International*, V. 5, No. 2, Feb., pp. 46-50. doi: 10.14359/9145

Slavin, C. M., and Ghannoum, W. M., 2015, "Defining Structurally Acceptable Properties of High-Strength Steel Bars through Material and Column Testing, Part I: Material Testing Report," Charles Pankow Foundation RGA #05-14, Aug., 135 pp. doi: 10.14359/51702075

Smith K. N. and Vantsiotis, A. S., 1982, "Shear Strength of Deep Beams," *ACI Journal Proceedings*, V. 79, No. 3, May-June 1982, pp. 201-213. doi: 10.14359/10899

Smith, S. W., and Burns, N. H., 1974, "Post-Tensioned Flat Plate to Column Connection Behavior," *PCI Journal*, V. 19, No. 3, May-June, pp. 74-91. doi: 10.15554/pcij.05011974.74.91

Sneed, L. H., and Ramirez, J. A., 2010, "Influence of Effective Depth on Shear Strength of Concrete Beams—Experimental Study," *ACI Structural Journal*, V. 107, No. 5, Sept.-Oct., pp. 554-562. doi: 10.14359/51663906

Sokoli, D., and Ghannoum, W. M., 2016, "High-Strength Reinforcement in Columns under High Shear Stresses," *ACI Structural Journal*, V. 113, No. 3, May-June, pp. 605-614. doi: 10.14359/51688203

Sperry, J.; Darwin, D.; O'Reilly, M.; Lepage, A.; Lequesne, R. D.; Matamoros, A.; Feldman, L. R.; Yasso, S.; Searle, N.; DeRubeis, M.; and Ajaam, A., 2018, "Conventional and High-Strength Steel Hooked Bars: Detailing Effects," *ACI Structural Journal*, V. 115, No. 1, Jan.-Feb., pp. 247-257. doi: 10.14359/51700920

Sperry, J.; Darwin, D.; O'Reilly, M.; Lequesne, R. D.; Yasso, S.; Matamoros, A.; Feldman, L. R.; and Lepage, A., 2017b, "Conventional and High-Strength Hooked Bars—Part 2: Data Analysis," *ACI Structural Journal*, V. 114, No. 1, Jan.-Feb., pp. 267-276. doi: 10.14359/51689457

Sperry, J.; Yasso, S.; Searle, N.; DeRubeis, M.; Darwin, D.; O'Reilly, M.; Matamoros, A.; Feldman, L.; Lepage, A.; Lequesne, R.; and Ajaam, A., 2017a, "Conventional and High-Strength Hooked Bars—Part 1: Anchorage Tests," *ACI*

R Comm. Ref.

Structural Journal, V. 114, No. 1, Jan.-Feb., pp. 255-266. doi: 10.14359/51689456

Spragg, R. P.; Castro, J.; Li, W.; Pour-Ghaz, M.; Huang, P.-T.; and Weiss, J., 2011, "Wetting and Drying of Concrete in the Using Aqueous Solutions Containing Deicing Salt," *Cement and Concrete Composites*, V. 33, No. 5, May, pp. 535-542. doi: 10.1016/j.cemconcomp.2011.02.009

Sritharan, S.; Beyer, K.; Henry, R. S.; Chai, Y. H.; Kowalsky, M.; and Bull, D., 2014, "Understanding Poor Seismic Performance of Concrete Walls and Design Implications," *Earthquake Spectra*, V. 30, No. 1, pp. 307-334. doi: 10.1193/021713EQS036M

Sritharan, S.; Cox, A. M.; Huang, J.; Suleiman, M.; and Arulmoli, K., 2016, "Minimum Confinement Reinforcement for Prestressed Concrete Piles and a Rational Seismic Design Framework," *PCI Journal*, V. 61, No. 1, pp. 51-69. doi: 10.15554/pcij.01012016.51.69

Stanton, J. F., 1987, "Proposed Design Rules for Load Distribution in Precast Concrete Decks," *ACI Structural Journal*, V. 84, No. 5, Sept.-Oct., pp. 371-382. doi: 10.14359/1683

Stanton, J. F., 1992, "Response of Hollow-Core Floors to Concentrated Loads," *PCI Journal*, V. 37, No. 4, July-Aug., pp. 98-113. doi: 10.15554/pcij.07011992.98.113

Stecich, J.; Hanson, J. M.; and Rice, P. F., 1984, "Bending and Straightening of Grade 60 Reinforcing Bars," *Concrete International*, V. 6, No. 8, Aug., pp. 14-23. doi: 10.14359/16496

Stone, W.; Cheok, G.; and Stanton, J., 1995, "Performance of Hybrid Moment-Resisting Precast Beam-Column Concrete Connections Subjected to Cyclic Loading," *ACI Structural Journal*, V. 92, No. 2, Mar.-Apr., pp. 229-249. doi: 10.14359/1145

Sugano, S.; Nagashima, T.; Kimura, H.; Tamura, A.; and Ichikawa, A., 1990, "Experimental Studies on Seismic Behavior of Reinforced Concrete Members of High Strength Concrete," *Proceedings of the Second International Symposium on High-Strength Concrete*, SP-121, American Concrete Institute, Farmington Hills, MI, pp. 61-87. doi: 10.14359/2791

Tan, K. H.; Kong, F. K.; Teng, S.; and Wang L. W., 1997, "Effect of Web Reinforcement on High Strength Concrete Deep Beams," *ACI Structural Journal*, V. 94, No. 5, Sept.-Oct., pp 572-582. doi: 10.14359/506

Taylor, C. P.; Cote, P. A.; and Wallace, J. W., 1998, "Design of Slender Reinforced Concrete Walls with Openings," *ACI Structural Journal*, V. 95, No. 4, July-Aug., pp. 420-433. doi: 10.14359/558

TBI, 2017, "Guidelines for Performance-Based Seismic Design of Tall Buildings," *Report* No. 2017/06, Pacific Earthquake Engineering Center, University of California, Berkeley, Berkeley, CA, May, 147 pp.

Tepke, D.; Trejo, D.; and Isgor, B., eds., 2016, *Chloride Thresholds and Limits for New Construction*, SP-308, American Concrete Institute, Farmington Hills, MI.

Thompson, K. J., and Park, R., 1980, "Seismic Response of Partially Prestressed Concrete," *Journal of the Structural Division*, V. 106, pp. 1755-1775.

Thompson, M. K.; Jirsa, J. O.; and Breen, J. E., 2006a, "CCT Nodes Anchored by Headed Bars—Part 2: Capacity of Nodes," *ACI Structural Journal*, V. 103, No. 1, Jan.-Feb., pp. 65-73. doi: 10.14359/15087

Thompson, M. K.; Ledesma, A.; Jirsa, J. O.; and Breen, J. E., 2006b, "Lap Splices Anchored by Headed Bars," *ACI Structural Journal*, V. 103, No. 2, Mar.-Apr., pp. 271-279. doi: 10.14359/15185

Thompson, M. K.; Ziehl, M. J.; Jirsa, J. O.; and Breen, J. E., 2005, "CCT Nodes Anchored by Headed Bars—Part 1: Behavior of Nodes," *ACI Structural Journal*, V. 102, No. 6, Nov.-Dec., pp. 808-815. doi: 10.14359/14788

Thomsen IV, J. H., and Wallace, J. W., 2004, "Displacement-Based Design of Slender Reinforced Concrete Structural Walls—Experimental Verification," *Journal of Structural Engineering*, V. 130, No. 4, pp. 618-630. doi: 10.1061/(ASCE)0733-9445(2004)130:4(618)

To, N. H. T.; Sritharan, S.; and Ingham, J., 2009, "Strut-and-Tie Nonlinear Cyclic Analysis of Concrete Frames," *Journal of Structural Engineering*, V. 135, No. 10, pp. 1259-1268. doi: 10.1061/(ASCE)0733-9445(2009)135:10(1259)

Treece, R. A., and Jirsa, J. O., 1989, "Bond Strength of Epoxy-Coated Reinforcing Bars," *ACI Materials Journal*, V. 86, No. 2, Mar.-Apr., pp. 167-174. doi: 10.14359/2341

Tuchscherer, R. G.; Birrcher, D. B.; Huizinga, M. H.; and Bayrak, O., 2010, "Confinement of Deep Beam Nodal Regions," *ACI Structural Journal*, V. 107, No. 6, Nov.-Dec., pp. 709-717. doi: 10.14359/51664019

Umehara, H., and Jirsa, J. O., 1984, "Short Rectangular RC Columns Under Bidirectional Loadings," *Journal of Structural Engineering*, V. 110, No. 3, pp. 605-618. doi: 10.1061/(ASCE)0733-9445(1984)110:3(605)

Uzel, A.; Podgorniak, B.; Bentz, E.; and Collins, M., 2011, "Design of Large Footings for One-Way Shear," *ACI Structural Journal*, V. 108, No. 2, pp. 131-138. doi: 10.14359/51664247

Vanderbilt, M. D., 1972, "Shear Strength of Continuous Plates," *Journal of the Structural Division*, V. 98, May, pp. 961-973. doi: 10.14359/51685238

Vanderbilt, M. D., and Corley, W. G., 1983, "Frame Analysis of Concrete Buildings," *Concrete International*, V. 5, No. 12, Dec., pp. 33-43. doi: 10.14359/9202

Vanderbilt, M. D.; Sozen, M. A.; and Siess, C. P., 1969, "Test of a Modified Reinforced Concrete Two-Way Slab," *Journal of the Structural Division*, V. 95, No. ST6, June, pp. 1097-1116.

Vintzileou, E., and Eligehausen, R., 1992, "Behavior of Fasteners under Monotonic or Cyclic Shear Displacements," *Anchors in Concrete: Design and Behavior*, SP-130, American Concrete Institute, Farmington Hills, MI, pp. 181-203. doi: 10.14359/1272

Vu, S.; Li, N. B.; and Beyer, K., 2014, "Effective Stiffness of Reinforced Concrete Coupling Beams," *Engineering Structures*, V. 76, pp. 371-382. doi: 10.1016/j.engstruct.2014.07.014

Wallace, J. W., 1996, "Evaluation of UBC-94 Provisions for Seismic Design of RC Structural Walls," *Earth-*

quake Spectra, V. 12, No. 2, May, pp. 327-348. doi: 10.1193/1.1585883

Wallace, J. W., 1998, "Behavior and Design of High-Strength RC Walls," *High-Strength Concrete in Seismic Regions*, SP-176, C. W. French and M. E. Kreger, eds., American Concrete Institute, Farmington Hills, MI, pp. 259-279. doi: 10.14359/5903

Wallace, J. W., 2012, "Behavior, Design, and Modeling of Structural Walls and Coupling Beams—Lessons from Recent Laboratory Tests and Earthquakes," *International Journal of Concrete Structures and Materials*, V. 6, No. 1, pp. 3-18. doi: 10.1007/s40069-012-0001-4

Wallace, J. W., and Orakcal, K., 2002, "ACI 318-99 Provisions for Seismic Design of Structural Walls," *ACI Structural Journal*, V. 99, No. 4, July-Aug., pp. 499-508. doi: 10.14359/12119

Wallace, J. W.; Massone, L. M.; Bonelli, P.; Dragovich, J.; Lagos, R.; Lüders, C.; and Moehle, J. P., 2012, "Damage and Implications for Seismic Design of RC Structural Wall Buildings," *Earthquake Spectra*, V. 28, pp. 281-299. doi: 10.1193/1.4000047

Wallace, J.; Segura, C.; and Tran, T., 2013, "Shear Design of Structural Walls," Los Angeles Tall Buildings Structural Design Council Annual Meeting, 13 pp.

Watson, S.; Zahn, F. A.; and Park, R., 1994, "Confining Reinforcement for Concrete Columns," *Journal of Structural Engineering*, V. 120, No. 6, June, pp. 1798-1824. doi: 10.1061/(ASCE)0733-9445(1994)120:6(1798)

Weber-Kamin, A. S.; Lequesne, R. D.; and Lepage, A., 2019, "RC Coupling Beams with High-Strength Steel Bars: Summary of Test Results," *SL Report 19-1*, The University of Kansas Center for Research, Inc., Lawrence, KS, Jan., 133 pp. http://hdl.handle.net/1808/27676

Welt, T. S., 2015, "Detailing for Compression in Reinforced Concrete Wall Boundary Elements: Experiments, Simulations, and Design Recommendations," PhD dissertation, University of Illinois at Urbana-Champaign, Champaign, IL, 530 pp.

Welt, T. S.; Massone, L. M.; LaFave, J. M.; Lehman, D. E.; McCabe, S. L.; and Polanco, P., 2017, "Confinement Behavior of Rectangular Reinforced Concrete Prisms Simulating Wall Boundary Elements," *Journal of Structural Engineering*, V. 143, No. 4, p. 04016204 doi: 10.1061/(ASCE)ST.1943-541X.0001682

West, J. S.; Innocenzi, M. J.; Ulloa, F. V.; and Poston, R. W., 2008, "Assessing Vibrations," *Concrete International*, V. 30, No. 10, Oct., pp. 43-49. doi: 10.14359/19985

Widianto; Bayrak, O.; and Jirsa, J. O., 2009, "Two-Way Shear Strength of Slab-Column Connections: Reexamination of ACI 318 Provisions," *ACI Structural Journal*, V. 106, No. 2, Mar.-Apr., pp. 160-170. doi: 10.14359/56354

Wilford, M. R., and Young, P., 2006, *A Design Guide for Footfall Induced Vibration of Structures,* The Concrete Centre, Surrey, UK, 83 pp.

Wilson, E. L., 1997, "Three-Dimensional Dynamic Analysis of Structures—With Emphasis on Earthquake Engineering," Computers and Structures, Inc., Berkeley, CA.

Winter, G., 1979, "Safety and Serviceability Provisions in the ACI Building Code," *Concrete Design: U.S. and European Practices*, SP-59, American Concrete Institute, Farmington Hills, MI, pp. 35-49. doi: 10.14359/17764

Wire Reinforcement Institute, 2001, *Structural Welded Wire Reinforcement Manual of Standard Practice*, sixth edition, Hartford, CT, Apr., 38 pp.

Wood, S. L.; Stanton, J. F.; and Hawkins, N. M., 2000, "Development of New Seismic Design Provisions for Diaphragms Based on the Observed Behavior of Precast Concrete Parking Garages during the 1994 Northridge Earthquake," *PCI Journal*, V. 45, No. 1, Jan.-Feb., pp. 50-65. doi: 10.15554/pcij.01012000.50.65

Wyllie Jr., L. A., 1987, "Structural Walls and Diaphragms—How They Function," *Building Structural Design Handbook*, R. N. White and C. G. Salmon, eds., John Wiley & Sons, Inc., New York, pp. 188-215.

Xanthakis, M., and Sozen, M. A., 1963, "An Experimental Study of Limit Design in Reinforced Concrete Flat Slabs," *Structural Research Series* No. 277, Civil Engineering Studies, University of Illinois, Urbana, IL, Dec., 159 pp. doi: 10.14359/16666

Yamada, T.; Nanni, A.; and Endo, K., 1991, "Punching Shear Resistance of Flat Slabs: Influence of Reinforcement Type and Ratio," *ACI Structural Journal*, V. 88, No. 4, July-Aug., pp. 555-563. doi: 10.14359/2984

Yoshioka, K., and Sekine, M., 1991, "Experimental Study of Prefabricated Beam-Column Subassemblages," *Design of Beam-Column Joints for Seismic Resistance*, SP-123, J. O. Jirsa, ed., American Concrete Institute, Farmington Hills, MI, pp. 465-492. doi: 10.14359/2876

Zhang, L., 2015, "Testing Air Content of Dry-Mix Shotcrete," *Shotcrete*, V. 17, No. 2, pp. 22-24.

Zhang, Y.; Klingner, R. E.; and Graves III, H. L., 2001, "Seismic Response of Multiple-Anchor Connections to Concrete," *ACI Structural Journal*, V. 98, No. 6, Nov.-Dec., pp. 811-822. doi: 10.14359/10748

Zhu, S., and Jirsa, J. O., 1983, "Study of Bond Deterioration in Reinforced Concrete Beam-Column Joints," PMFSEL *Report* No. 83-1, Department of Civil Engineering, University of Texas at Austin, Austin, TX, July, 79 pp. doi: 10.14359/51685438

Zia, P., and Hsu, T. T. C., 2004, "Design for Torsion and Shear in Prestressed Concrete Flexural Members," *PCI Journal*, V. 49, No. 3, May-June, pp. 34-42. doi: 10.15554/pcij.05012004.34.42

Zsutty, T. C., 1971, "Shear Strength Prediction for Separate Categories of Simple Beams Tests," *ACI Journal Proceedings*, V. 68, No. 2, Feb., pp. 138-143. doi: 10.14359/11300

Notes

INDEX

Index I

American Concrete Institute
Always advancing

The American Concrete Institute envisions a future where everyone has the knowledge needed to use concrete effectively to meet the demands of a changing world.

Founded in 1904 with a headquarters in Farmington Hills, Michigan, USA, and a regional office in Dubai, UAE, the American Concrete Institute is always advancing by developing educational programs, publishing technical documents, managing various certification programs, and hosting industry-wide events. With 99 chapters, 65 student chapters, and nearly 30,000 members spanning over 120 countries, the American Concrete Institute has always retained the same basic mission — to develop, disseminate, and advance the adoption of its consensus-based knowledge on concrete and its uses.

In today's market, it is imperative to be knowledgeable and have an edge over the competition. ACI membership provides concrete industry professionals the chance to save money and time, while increasing productivity, competitiveness, and awareness of new technology and cutting-edge research.

ACI 318-19 Resources

ACI offers a comprehensive slate of resources for designing and constructing according to *ACI 318-19 Building Code Requirements for Structural Concrete*. These resources include:

318-19 Seminars that provide technical updates through comprehensive day-long seminars at your office or a location near you;

318-19 Webinars that provide technical updates from your desktop;

ACI's *Reinforced Concrete Design Manual* including explanations, analyses, examples, and design aids for reinforced concrete structures; and

ACI's *Detailing Manual* providing examples and considerations for conveying your design intent through structural details.

Learn more about these resources at http://www.concrete.org.

American Concrete Institute
38800 Country Club Drive
Farmington Hills, MI 48331
Phone: +1.248.848.3700
www.concrete.org